ORIGINS OF THE HIGHER GROUPS OF TETRAPODS

ORIGINS OF THE HIGHER GROUPS OF TETRAPODS

Controversy and Consensus

Edited by

Hans-Peter Schultze *and* Linda Trueb

Museum of Natural History and Department of Systematics and Ecology
The University of Kansas

Comstock Publishing Associates
a division of Cornell University Press
ITHACA AND LONDON

Library of Congress Cataloging-in-Publication Data
Origins of the higher groups of tetrapods : controversy and consensus
  / edited by Hans-Peter Schultze and Linda Trueb.
      p.   cm.
    Includes bibliographical references and indexes.
    ISBN 0-8014-2497-6 (cloth : alkaline paper)
      1. Vertebrates—Evolution—Congresses.   I. Schultze, Hans-Peter.
    II. Trueb, Linda.
    QL607.5.075 1991
    596'.015—dc20   90-55752

**To our students of Evolutionary Morphology at
The University of Kansas—past and present.
Thank you.**

First published 1991 by Cornell University Press.

Printed in the United States of America

♾ The paper in this book meets the minimum requirements of the
American National Standard for Information Sciences—
Permanence of Paper for Printed Library Materials, ANSI Z39.48-1984.

Contents

Preface

It is something of an academic anachronism that in this, a technological age characterized by sophisticated computer simulations, space travel, and genetic engineering, biologists remain at odds with one another with respect to their hypotheses of the phylogenetic relationships of major groups of fossil and living tetrapods—i.e., amphibians, reptiles, birds, and mammals. We seem to know more about the geological history of the earth, its past climates, and mobile tectonic plates than we know about the origins and histories of the animals inhabiting these land masses. Lest we be accused of being alarmists and overstating the problem, we hasten to point out that the disputes concerning the origins of amniotes and mammals seem to be relatively minor. The origins of anamniotes and birds, however, remain matters of substantial disagreement.

Presumably, contemporary biologists subscribe to the scientific method and are the intellectual products of the Darwinian revolution. Why then should there be unresolved conflicts? The answers seem to involve historical constraints that have determined the methodological approaches of the biologists analyzing these data and the kinds of data available for their analyses. An important constraint is the biologist's training as a paleontologist or neontologist. Because few are equally knowledgeable of both disciplines, their data sets tend to be limited to those organisms with which they are familiar. Thus, the majority of neontologists gather most of their data on living organisms, whereas paleontologists concentrate on fossils. Exclusion of either extant or extinct organisms obviously produces an incomplete and therefore biased data set. Moreover, paleontologists work at a distinct disadvantage owing to the fragmentary nature of the fossil record and, not infrequently, the incomplete nature of the individual specimens that they examine. Although nearly all biologists accept evolution as the operative phylogenetic process, they can differ markedly in their deductive approach to unraveling historical relationships and sequences—a procedure that usually is based on the analysis of anatomical features of the organisms they study. Some seek ancestor-descendant relationships based on analysis of morphological similarities observed among groups of organisms. In contrast, biologists subscribing to cladistic principles eschew primitive characters and seek to discover assemblages of organisms that on the bases of shared-derived characters seem to have a common ancestor.

Owing to the varying backgrounds of the investigators and the analytical approaches they have employed, there is little or no consensus as to the origin(s) and relationships of tetrapods. Thus, Westoll and Romer proposed a monophyletic origin of tetrapods from osteolepiform fish, whereas Jarvik posited a diphyletic origin of tetrapods from porolepiform and osteolepiform fishes. More recently, Rosen et al. concluded that lungfish are related more closely to tetrapods than osteolepiform crossopterygian fish—a view reminiscent of the proposition that tetrapods were derived from dipnoans advanced by Kesteven. The phylogenetic relationships of birds are similarly disputed. Thus, some authorities such as Ostrom and Gauthier hold that birds are derived from dinosaurs, whereas others such as Walker think that they are related to crocodilomorphs.

Given this diversity of opinion with respect to the origins and relationships of tetrapods, a review of the evidence seemed to be appropriate and timely. A seminar series at The University of Kansas in the spring of 1985 provided a convenient vehicle for our purpose. An international group of scholars, representing a variety of specialties and points of view, was invited to participate. The seminars formed the basis for the chapters included herein and briefly introduced below. We wish to emphasize that we encouraged interaction among the participants and requested each to focus on the presentation and evaluation of his or her data. We are not so presumptuous as to suggest that we have solved the problems of the origins of the higher groups of tetrapods. Rather, our aim has been to focus attention on some of the alternatives in comparable ways. This approach has been adopted in the hope that we will prompt the discriminating reader to independent decisions and perhaps stimulate future research efforts.

Each of the collected papers represents a chapter, and chapters are arranged in five major sections addressing the origin of tetrapods as a group, the monophyly and relationship of amphibians, the origin of amniotes and relationships of diapsid reptiles, avian origins, and the origins of mammal-like reptiles and mammals, respectively. Each section contains an overview chapter intended either to provide a summary of or a perspective on the development of various schools of thought regarding the origin of the tetrapod group (Trueb and Cloutier on amphibians, Benton on reptiles, Witmer on birds, and Miao on mammals) or to summarize critically the arguments presented in the section (Forey et al. on tetrapods).

In the first contribution, Chang proposes a close relationship between sarcopterygians as a whole and tetrapods. Her views contrast sharply with two currently popular hypotheses—one of which proposes dipnoans as the sister-group of tetrapods, and the other of which proposes osteolepiform fish to be the sister-group of tetrapods. Schultze (Chapter 2) reviews the various schemes that have been advanced to account for the origin of tetrapods. Vorobyeva and Schultze elaborate on their proposed close relationship between panderichthyid fish and tetrapods in the third chapter, drawing attention to the fact that in some respects the differences between the groups are so minimal as to require a redefinition of tetrapods. Taking a controversial position, Panchen (Chapter 4) defends his view that *Crassigyrinus* is allied closely with one specific group of tetrapods, the batrachosaurs; in so doing, he takes issue with cladistic methodology. A wrap-up commentary on the preceding four chapters is provided by Forey, Gardiner, and Patterson (Chapter 5).

The second section, which deals with amphibians, contains three chapters. By way of introduction, Trueb and Cloutier (Chapter 6) present an overview of the history of the development of amphibian systematics during the past two centuries. This is followed by Bolt's (Chapter 7) reinvestigation of the positions of dissorophids and the microsaur *Hapsidopareion* with respect to the Lissamphibia. Chapter 8 by Trueb and Cloutier is a major phylogenetic study based on osteological features as well as soft-anatomical characters. The authors attempt to document the monophyly of the Lissamphibia, and relationships within this group and between this group and fossil amphibians.

Benton (Chapter 9) prefaces the third section, on reptiles, with a review of the controversies involving the historical relationships of amniotes. Carroll follows with an essay (Chapter 10) on his thoughts on the origin of amniotes and a reassessment of the taxonomic status and the phylogenetic position of the proterothyrids in relation to the primitive amniotes. Carroll and Currie (Chapter 11) present a scheme of relationships among diapsid reptiles.

The spirited disputes concerning avian origins are cast in a historical perspective by Witmer (Chapter 12). The three chapters that follow represent diametrically opposed points of view. Ostrom adheres to his proposed close relationship between birds and dinosaurs (Chapter 13), whereas Martin (Chapter 14) defends the position that they are related to crocodilomorphs, and Tarsitano (Chapter 15) favors a relationship between birds and thecodonts.

The final section, on mammals, is introduced by Miao (Chapter 16), who discusses controversial views on the origin and early diversification of mammals. In Chapter 17, Hotton discusses the diversity of the mammal-like reptiles from an ecological point of view, whereas Hopson (Chapter 18) analyzes the phylogenetic patterns with the synapsids.

As in any undertaking of this magnitude, we are indebted to a great many individuals for their cooperation and support of our efforts. Travel funds for Dr. Emilia Vorobyeva were provided by the Advisory Committee on the USSR and Eastern Europe Interacademy Exchange Program of the National Research Council of Washington, D.C. The majority of the funding was provided by The University of Kansas. We express our gratitude particularly to the Museum of Natural History, the Department of Systematics and Ecology, the Department of Geology, the Office of Research, Graduate Studies and Public Service, the College of Liberal Arts and Sciences, and the Graduate Student Organization of the Department of Systematics and Ecology. Without the generous support of these agencies, the seminar series could not have been realized. Anne Musser, Scientific Illustrator of the Museum of Natural History, skillfully prepared many of the original illustrations, and amended, repaired, and relettered many others—an onerous task that we acknowledge gratefully. We owe a special debt of gratitude to the graduate students of the Department of Systematics and Ecology. It was they who encouraged us to implement this project as an outgrowth of an ongoing seminar series, Topics in Evolutionary Morphology. Without their unflagging support and spirited participation, the discussions with the participants would have been far less productive, and the resulting papers far less provocative.

HANS-PETER SCHULTZE AND LINDA TRUEB

Lawrence, Kansas

Contributors

Michael J. Benton, Department of Geology, University of Bristol, Bristol B58 1TR, UK

John R. Bolt, Department of Geology, Field Museum of Natural History, Roosevelt Road at Lake Shore Drive, Chicago, Illinois 60605, USA

Robert L. Carroll, Redpath Museum, McGill University, 859 Sherbrook Street West, Montreal, Quebec H3A 2K6, Canada

Chang Mee-Mann, Institute of Vertebrate Paleontology and Paleoanthropology, Academica Sinica, Beijing, PRC

Richard Cloutier, Museum of Natural History, and Department of Systematics and Ecology, The University of Kansas, Lawrence, Kansas 66045-2454, USA (Present address: Department of Palaeontology, Natural History Museum, Cromwell Road, London SW7 5BD, UK)

Philip J. Currie, Tyrrell Museum of Palaeontology, P.O. Box 7500, Drumheller, Alberta T0J 0Y0, Canada

P. L. Forey, Department of Palaeontology, Natural History Museum, Cromwell Road, London SW7 5BD, UK

B. G. Gardiner, Department of Biology, Kings College, Kensington Campus, Campden Hill Road, London W8 7AH, UK

James A. Hopson, Department of Organismal Biology and Anatomy, The University of Chicago, 1025 East 57th Street, Chicago, Illinois 60637, USA

NICHOLAS HOTTON III, Department of Paleobiology, Smithsonian Institution, Washington, D.C. 20560, USA

LARRY D. MARTIN, Museum of Natural History, and Department of Systematics and Ecology, The University of Kansas, Lawrence, Kansas 66045-2454, USA

MIAO DESUI, Museum of Natural History, The University of Kansas, Lawrence, Kansas 66045-2454, USA

JOHN H. OSTROM, Department of Geology and Geophysics, and Peabody Museum of Natural History, Yale University, New Haven, Connecticut 06511, USA

A. L. PANCHEN, Department of Biology, The University, Newcastle upon Tyne, NE1 7RU, UK

COLIN PATTERSON, Department of Palaeontology, Natural History Museum, Cromwell Road, London SW7 5BD, UK

HANS-PETER SCHULTZE, Division of Vertebrate Paleontology, Museum of Natural History, and Department of Systematics and Ecology, The University of Kansas, Lawrence, Kansas 66045-2454, USA

SAMUEL TARSITANO, Department of Biology, Southwest Texas State University, San Marcos, Texas 78666, USA

LINDA TRUEB, Division of Herpetology, Museum of Natural History, and Department of Systematics and Ecology, The University of Kansas, Lawrence, Kansas 66045-2454, USA

EMILIA VOROBYEVA, Institute of Evolutionary Morphology and Ecology of Animals, Academy of Science, Moscow, USSR

LAWRENCE M. WITMER, Department of Cell Biology and Anatomy, The Johns Hopkins University School of Medicine, Baltimore, Maryland 21205, USA

Section I

TETRAPODS

1 "Rhipidistians," Dipnoans, and Tetrapods

Chang Mee-Mann

Currently there are two competing views regarding the phylogenetic relationships among "rhipidistians," dipnoans, and tetrapods. Proponents of the first hypothesis argue for a close relationship between "rhipidistians" (or at least osteolepiforms) and tetrapods by deemphasizing putative similarities between dipnoans and tetrapods (Fig. 1). The second hypothesis is based on synapomorphies that unite dipnoans with tetrapods (Fig. 2). Proponents of this view contend that the characters used to support a "rhipidistian"-tetrapod group either have been misinterpreted anatomically or represent symplesiomorphies. The first hypothesis has been accepted nearly universally since the middle of this century. In a recent revival of the debate, however, Rosen et al. (1981) demonstrated that the argument for a "rhipidistian"-tetrapod group that excludes dipnoans is strongly influenced by a priori assumption of an ancestor-descendant relationship between the well-studied osteolepiform *Eusthenopteron* and primitive tetrapods. Anatomical details of *Eusthenopteron* have been interpreted in the light of primitive tetrapods, and primitive tetrapods, in turn, have been interpreted in the light of *Eusthenopteron*. However, it seems that Rosen et al. (1981) similarly were predisposed to favor a close dipnoan-tetrapod affinity in their consideration of the evidence.

The recent discovery of two new Lower Devonian sarcopterygians, *Youngolepis* (Chang and Yu, 1981; Chang, 1982) and *Diabolichthys* (subsequently renamed *Diabolepis* because of a senior homonym given to a

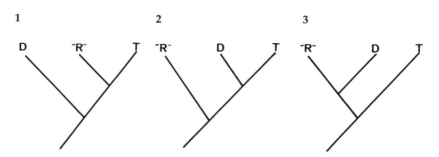

Figs. 1–3. Three schemes of interrelationships among "rhipidistians," dipnoans, and tetrapods, given "rhipidistians" as a monophyletic taxon. (1) Traditional arrangement. (2) Arrangement of Rosen et al. (1981). (3) Arrangement preferred herein. D = dipnoans; "R" = "rhipidistians"; T = tetrapods.

recent batoid; Chang and Yu, 1984, 1987), forces reconsideration of the problem. The new forms possess characters shared by both porolepiforms and dipnoans—a circumstance that suggests that we should reexamine Romer's (1966) proposal of a close affinity between "rhipidistians" and dipnoans. If we assume "rhipidistians" to be monophyletic, this phylogenetic scheme represents the only alternative for a three-taxon statement among "rhipidistians," dipnoans, and tetrapods (Fig. 3). The purpose of this chapter is to discuss the most important evidence cited in support of the competing hypotheses and to comment on the likelihood of a more parsimonious solution.

This chapter presents the main features that commonly are used to support the relationship of tetrapods to different groups of sarcopterygians. The state of each feature has been confirmed on actual specimens, including serial sections and wax-model reconstructions in the Paleozoologiska sectionen of the Naturhistoriska Riksmuseet, Stockholm, Sweden (SMNH), the Institut of Vertebrate Paleontology and Paleoanthropology, Beijing, P.R. China (IVPP), and the American Museum of Natural History, New York, United States (AMNH). The state of each character is evaluated, and a cladistic scheme of interrelationships of sarcopterygians and their relationships to tetrapods is hypothesized.

REVIEW OF "RHIPIDISTIAN"-TETRAPOD AND DIPNOAN-TETRAPOD CHARACTERS

Choana

"Rhipidistians"

The cranial anatomy of the "rhipidistian"-like *Youngolepis* was reconstructed in detail from serial ground sections (Chang, 1982). Examination of the wax reconstruction and whole specimens demonstrates conclusively the absence in *Youngolepis* of any kind of passageway in the snout that opened on the palate as the choana. Removal of the dermal palatal bones reveals the presence of a fairly large fenestra, the fenestra ventrolateralis (= fenestra ventralis of Chang, 1982) lateral to the posteroventral portion of the ethmoid region of *Youngolepis* (Fig. 4). Fenestrae in similar positions in *Glyptolepis* and *Eusthenopteron* are identified as the fenestrae ventrolateralis (Fig. 5) and endochoanalis (Fig. 6), respectively. Each of these fenestrae is located in the posterior floor (i.e., solum nasi) of

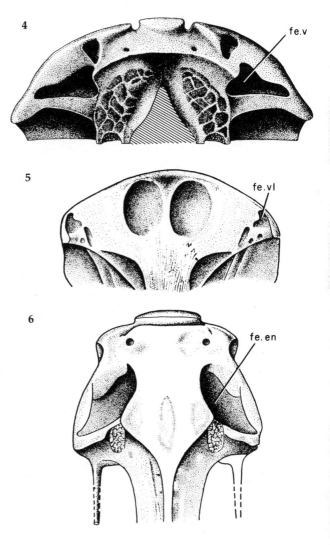

Figs. 4–6. Anterior part of the endocranium in ventral view of (4) *Youngolepis* (after Chang, 1982), (5) *Porolepis* (after Jarvik, 1972), and (6) *Eusthenopteron* (after Jarvik, 1942). See Appendix I for key to abbreviations.

the nasal cavity and is bounded posteriorly by the lower margin of the postnasal wall. In specimens of *Youngolepis* in which articulated premaxillae are preserved, the fenestra ventrolateralis usually is divided into medial and lateral portions by the posterolateral premaxilla (Fig. 7). The size of the medial part of the fenestra is more variable than the lateral part (Figs. 7–8), and the medial part is covered ventrally by the vomer, which

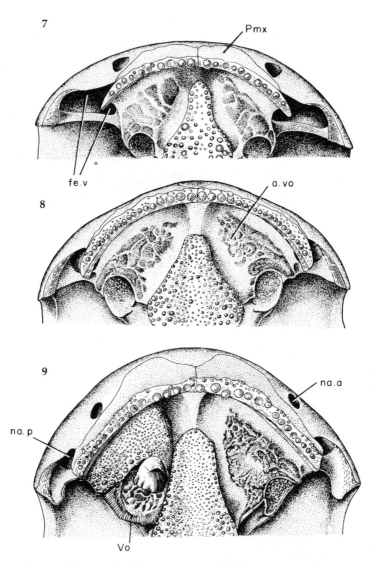

Figs. 7–9. Snout of *Youngolepis*. (7) In ventral view. (8) With both vomers removed. (9) With vomer on left side removed. See Appendix I for key to abbreviations. (After Chang, 1982.)

partially floors the nasal cavity. Only the lateral part of the fenestra is visible in a palatal view of the articulated skull; the fenestra lies lateral to the premaxilla and communicates with the posterior external nasal opening (Fig. 9). Thus, there is no evidence of a direct passage between the nasal and mouth cavities in *Youngolepis*. If a palatal opening bounded by the vomer, dermopalatine, premaxilla, and maxilla is present (as it is in some "rhipidistians"), the sizable vomer that extends posterior to the nasal cavity in *Youngolepis* obviates the existence of a passage between such an opening and the nasal cavity. Consequently, there is no evidence of a choana in *Youngolepis*.

The palatal morphology of *Powichthys* seems to resemble that of *Youngolepis*. In the holotype of *P. thorsteinssoni* Jessen, the vomers and posterior portions of the premaxillae are missing, thereby exposing the entire fenestra ventrolateralis (Jessen, 1980:Pl. 5, Fig. 1). If the posterolateral premaxilla is reconstructed along the anterolateral margin of the vomeral area, the portion of the fenestra ventrolateralis that would lie medial to the premaxilla would be rather small. Although the vomer of *Powichthys* is unknown, it surely would have covered the medial part of the fenestra, as it does in *Youngolepis*.

The morphology of the porolepiform snout is known from sections and a reconstruction of *Glyptolepis groenlandica* (Jarvik, 1972, 1980) and from well-preserved specimens of *Porolepis brevis, P. elongata,* and *P. spitsbergensis*. In these taxa, the fenestra ventrolateralis opens posterolaterally and clearly is situated lateral to the premaxillary-maxillary arch; it communicates only with the posterior external naris and does not open ventrally toward the palate (Fig. 10; Jarvik, 1972:Fig. 13A). However, as reconstructed by Jarvik, an opening is present medial to the premaxillary-maxillary arch between the maxilla, premaxilla, vomer, and dermopalatine (cf. Jarvik, 1972:Fig. 31; and fenestra exochoanalis in Fig. 11, herein). Had an opening on the dermal palate existed as shown by Jarvik, there still would be no evidence for a passage extending dorsolaterally from such an opening across the premaxillary-maxillary arch to the fenestra ventrolateralis.

The articular area for the ethmoidal process of the palatoquadrate in porolepiforms is located on the posteroventral margin of the postnasal wall just medial to the premaxilla (Fig. 10). In the sectioned specimen of *Glyptolepis,* the palatoquadrate is fused to the endocranium in this area. Obviously, the association of the palatoquadrate with the endocranium would block any possible nasal passage. Thus, despite Jarvik's assertions, there is no evidence for a passage that leads from the nasal cavity to a choana on the palate in porolepiforms.

Rosen et al. (1981:193–194) also questioned the presence of a choana in *Glyptolepis* as reconstructed by Jarvik. They pointed out the small size of

10

fe. vl

ar. et. pq

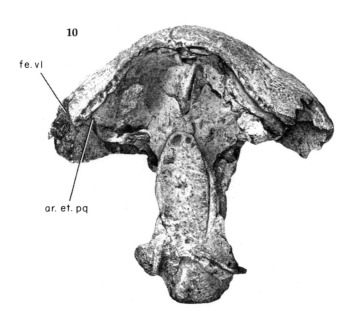

11

fe. ex

Figs. 10–11. Ventral views of anterior cranium of *Porolepis* showing (10) imperfect cranial division, and (11) restoration of snout. (After Jarvik, 1972.) See Appendix I for key to abbreviations.

the opening (a notch in the margin of the palate) and suggested that it might have resulted from postmortem displacement of the dermal bones of the palate. More important than the presence and size of the putative opening is the fact that, morphologically, a passage could not have existed between the opening and the nasal cavity. It follows that if a choana were absent in porolepiforms, there could not have been three narial openings—i.e., an incurrent naris, excurrent naris, and choana. Jarvik (1942, 1962, 1972, 1980) repeatedly described these three narial openings, using their presence as one of his primary arguments against the homology of the choana and the excurrent naris of fish, in general, and the homology of the choana and the internal naris of dipnoans, in particular (Panchen, 1967:379–388).

The presence of a choana still is regarded as one of the most important synapomorphies of osteolepiforms (or *Eusthenopteron*) and tetrapods (Holmes, 1985:387). Among currently recognized osteolepiforms, a choana has been restored in *Eusthenopteron* (Jarvik, 1942, 1980), *Megalichthys* (Jarvik, 1966), *Ectosteorhachis* (Romer, 1937; Thomson, 1964), and *Thursius* (Vorobyeva, 1977). As evidenced by Jarvik's illustrations (1966:Fig. 15A, Pls. 2–3) of *Megalichthys*, the lateral portions of the snout of this taxon are preserved poorly; the fenestra endochoanalis (shown here in Fig. 13) doubtless was reconstructed entirely (Fig. 12). The state of preservation of *Ectosteorhachis*, a North American taxon related to *Megalichthys*, is similar. According to Romer (1937) and Thomson (1964), the ventral wall of the nasal capsule of *Ectosteorhachis* is not preserved in any of the material available for examination. Both authors suggested that a cartilaginous solum nasi might have been present in living *Ectosteorhachis*, and the presence of a choana was assumed a priori. The positions of the posteroventral fenestra ("p.v.f." of Thomson, 1964:Fig. 4) and choana were inferred from the relative positions of the dermal bones of the palate.

Vorobyeva (1971) reported similarities between the Middle Devonian osteolepidid *Thursius estonicus* Vorobyeva, 1977 and porolepiforms. The fenestra endochoanalis is exposed on the right side of the cranium of *T. estonicus*, whereas that part of the fenestra medial to the premaxilla is covered by the vomer on the left side (Fig. 14). If the maxilla is restored as a continuation of the premaxilla, the fenestra would be covered com-

12   13

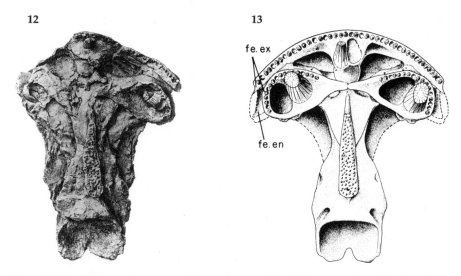

Figs. 12–13. Ventral views of anterior cranium of *Megalichthys* showing (12) imperfect cranial division, and (13) restoration of anterior cranium. (After Jarvik, 1966.) See Appendix I for key to abbreviations.

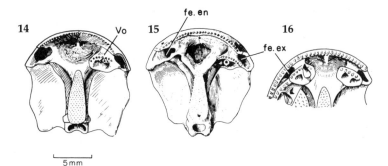

Figs. 14–16. Restoration of anterior cranial division of *Thursius* in ventral view. (14) After Vorobyeva (1971). (15–16) After Vorobyeva (1977). See Appendix I for key to abbreviations.

pletely medial to the premaxillary-maxillary arch, and a passage between the nasal and mouth cavities would be obstructed by the vomer. The configuration of this region of the skull varies in subsequent restorations by Vorobyeva (cf. Figs. 3, 25B in Vorobyeva, 1977). Thus, part (Fig. 15) or all (Fig. 16) of the fenestra lies medial to the premaxillary-maxillary arch, and it is uncertain which, if either, interpretation is correct.

In most specimens of *Eusthenopteron*, the cheek bones and palate are disarticulated from the remainder of the cranium. Complete skulls usually are compressed transversely; thus, separation of the lower jaw is difficult and the palate is obscured. The right side of the palate is preserved in SMNH P.341 (Jarvik, 1942:Pl. 8, Fig. 1). The anterior part of the dermopalatine is broken, and the space labeled "fe.exch" is much smaller than that in Jarvik's reconstruction (1942:Fig. 56). If the entire specimen were illustrated, it could be seen that the palate on the right side is dislodged to a position nearly perpendicular to the ventral surface of the endocranium. Thus, the "choanal" configuration in this figure is suspect. Although the specimen (ground series 2, SMNH P.222) on which Jarvik's reconstruction is based is otherwise well preserved, the tip of the snout is incomplete and the left nasal cavity is missing almost entirely (Jarvik, 1942:Pls. 11–12; 1980:Fig. 73). In Jarvik's reconstruction, a small depressed area of the dermopalatine overlaps, but does not articulate with, the vomer. A rather extensive drop-shaped opening in the palate is interpreted to be the fenestra exochoanalis (Fig. 17).

Rosen et al. (1981) discussed in detail the presence or absence of the choana in "rhipidistians" and provided a new reconstruction based on additional specimens of *Eusthenopteron*. The preservation of the snout in one specimen (BMNH P.60310) illustrated by Rosen et al. (1981:Fig. 13) is better than that of specimens from the Swedish Museum. Thus, one would suspect that their reconstruction (Fig. 18) is more accurate than Jarvik's with respect to (1) the shape, border, and size of the palatal

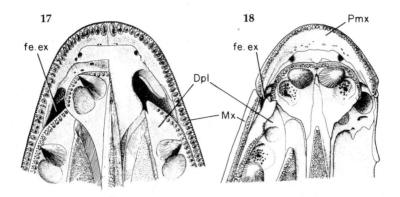

Figs. 17–18. Restoration of the snout of *Eusthenopteron* in ventral view with the dermopalatine and maxillae in articulation. (17) Left vomer removed (after Jarvik, 1942). (18) Restoration of the same part of *Eusthenopteron* with right dermopalatine and vomer in articulation (after Rosen et al., 1981). See Appendix I for key to abbreviations.

opening; (2) the structure of the head of the dermopalatine; and (3) the anterior extension of the marginal tooth row of the dermopalatine. Their interpretation of the mobile articulation between the palatoquadrate-cheek unit and the snout also is more reasonable. Another specimen of *Eusthenopteron* (AMNH 9827) has an articulation between the dermopalatine and the vomer that is similar to that of the specimen illustrated by Rosen et al. (1981).

It is clear that preconceptions of an ancestor-descendant relationship between "rhipidistians" and tetrapods have strongly influenced interpretations of the "rhipidistian" palate, particularly with respect to the presence or absence of a choana. Confirmation of the existence of a choana in fossil "rhipidistians" requires the presence of the following morphological conditions: (1) an opening in the dermal palate that is bordered by the bones of the premaxillary-maxillary arch, (2) a passage that extends from this opening to the nasal cavity, and (3) no blockage of the opening or the passage by other structures. Because the fenestra ventrolateralis communicates only with the posterior external nostril and lacks any opening into the mouth cavity, a choana could not have been present in *Youngolepis*, *Powichthys*, or any other known porolepiform. Similarly, some osteolepiforms such as *Thursius* seem to have lacked a choana; others, such as *Megalichthys* and *Ectosteorhachis*, are known too poorly to determine whether a choana is present. The only "rhipidistian" that might have possessed a passage between the nasal and mouth cavities is *Eusthenopteron foordi*, but as Rosen et al. (1981) pointed out, its presence is questionable. Thus, a choana has yet to be identified unequivocally in any "rhipidistian."

Dipnoans

The logic that led Rosen et al. (1981) to speculate that dipnoans possess a choana is similar to that which led Jarvik to the conclusion that "rhipidistians" possessed a choana. In order to interpret the dipnoan internal nostril as a choana, Rosen et al. attempted to establish homologies between the dermal bones of the palate in dipnoans and tetrapods. Other authors, however, have documented the homology of the dipnoan internal nostril to the posterior external nostril of other fishes on the basis of morphological and embryological evidence (Bertmar, 1965, 1966; Jarvik, 1942, 1980). Owing to the presence of the opening of the posterior excurrent nostril in the roof of the mouth cavity, dipnoans possess a pattern of dermal bones on the snout and palate different from that of other osteichthyans. Moreover, the pattern of dermal bones of the anterior part of the palate is variable among dipnoans; thus, it is extremely difficult to establish the homologies of individual palatal elements among different genera of lungfish or between dipnoans and other osteichthyans. (For a review of this problem, see Campbell and Barwick, 1984.) Although the Lower Devonian *Diabolepis* resembles dipnoans in its skull-roof pattern and palatal bite, it differs from them in the presence of a reduced, but unquestionably distinct, premaxilla (Fig. 19). The anterior part of the infraorbital sensory canal and the ethmoidal commissure (absent in currently known dipnoans) lie between the premaxillae and posteriorly adjacent elements of the skull roof. The relative proportions of the premaxilla of *Diabolepis* and the extent of the bone that lies within the mouth cavity vary with the degree of ventral and posteroventral recurvature of the snout—i.e., the greater the recurvature of the snout, the greater the proportion of the premaxilla that is included within the mouth cavity. On the basis of the attenuate posterior mandibular sector and edentate upper margin of the posterior part of the disarticulated lower jaw in *Diabolepis*, it seems that the maxilla was either toothless or absent. Alternatively, there might have been a gap between the premaxilla and maxilla.

If we accept *Diabolepis* as the sister-group of currently known dipnoans and take into account the absence of the infraorbital sensory canal, ethmoidal commissure, and the dermal bones on the ventral (i.e., posterior)

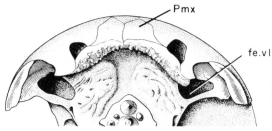

Fig. 19. Snout of *Diabolepis* in ventral view. See Appendix I for key to abbreviations. (After Chang and Yu, 1984.)

margin of the incurrent (i.e., anterior) nostril in dipnoans, it is reasonable to conclude that dipnoans lost the premaxilla and part, or all, of the maxilla in addition to several posterior dermal bones.

Rosen et al. (1981) justified their identification of a premaxilla in *Griphognathus whitei* on the basis of the presence of denticles (or tuberosities) on the element, and its relationship to adjacent bones. However, in osteichthyans, it is quite common for the rostral, which borders the mouth and lies between the premaxillae medially, to bear teeth similar to those of the premaxillae (Schaeffer, 1984; Gardiner, 1984:Figs. 45, 48). Perhaps Rosen et al. (1981) identified the dermal bones bordering and roofing the mouth cavity in *G. whitei* as the premaxilla, maxilla, vomer, and dermopalatine in order to establish a tetrapod pattern for the dermal elements of the palate in dipnoans. Their identifications of the premaxilla and maxilla are not convincing. Furthermore, it has long been known that the internal nares of living dipnoans are not used for breathing (Atz, 1952). The existence of a choana in *Eusthenopteron* and the supposed homology of the internal naris of dipnoans to the choana of tetrapods remain ambiguous. Despite the assertions of some optimistic authors (e.g., Holmes, 1985), we lack convincing evidence for the presence of choanae in any fossil osteichthyan.

Nasolacrimal Duct

In his description of *Eusthenopteron wenjukowi*, Jarvik (1937) reported the presence of a canal that extended from the orbit through the lower part of the postnasal wall to the nasal cavity; he assumed that the canal afforded passage for the ramus maxillaris trigemini and the ramus buccalis of the facial nerve and its accompanying vessels. This interpretation is consistent with the observations of Rosen et al. (1981:196–197) that the endocranial canal of *Eusthenopteron* "exactly parallels the course of the maxillary and buccal nerve and concomitant vessels as they pass through the postnasal wall in elasmobranchs, *Polypterus*, *Amia*, dipnoans and urodeles." Jarvik (1942) subsequently altered his interpretation, however, and argued that the canal is homologous to the nasolacrimal duct that is universally present in tetrapods.

The nasolacrimal duct of tetrapods lies within dermal bones or external to them (Fig. 20) and is not similar to the canal in *Eusthenopteron*, in which it penetrates the postnasal endocranial wall (Fig. 21). On the basis of reconstructions of ground serial sections, we know that the course of the canal in *Eusthenopteron* is similar to that of the canal in the ventromedial part of the postnasal wall of *Youngolepis* (Figs. 22, 25) and that of the posterior canal of the orbitorostral passage of *Glyptolepis* (Fig. 24). The position of the posterior opening of the canal in the postnasal wall of both

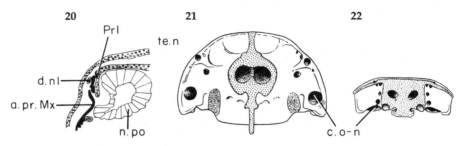

Figs. 20–22. (20) Transverse section through the nasal region of a larval *Salamandrella keyserlingii* (after Lebedkina, 1979:Fig. 59B). (21) Simplified drawing of snout of *Eusthenopteron* in posterior view (after Jarvik, 1980). (22) Simplified drawing of snout of *Youngolepis* in posterior view (after Chang, 1982). See Appendix I for key to abbreviations.

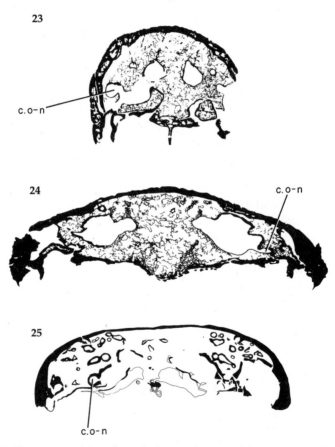

Figs. 23–25. Transverse sections through the anterior part of the cranium in (23) *Eusthenopteron* (after Jarvik, 1942), (24) *Glyptolepis* (after Jarvik, 1966), and (25) *Youngolepis* (after Chang, 1982). See Appendix I for key to abbreviations.

14

Youngolepis and *Glyptolepis* is more medial than that of *Eusthenopteron* (Figs. 21, 23). However, the position of the posterior opening is variable in *Eusthenopteron*. In some specimens of *Eusthenopteron* and other osteolepiforms (e.g., *Thursius;* Vorobyeva, 1977:Fig. 25B), the canal opening is often located at some distance from the orbital margin. Likewise, a shallow groove extending from the posterior nasal opening to the orbit in the lacrimal bone was observed only in one specimen of *Glyptolepis* (SMNH P.510; Jarvik, 1972:Pl. 17, Fig. 2). Thus, there is a lack of convincing evidence to support the homology of the canal in the postnasal wall of *Eusthenopteron* and the posterior nasal openings of other osteichthyans. It is still more unreasonable to argue that the canal in *Eusthenopteron* represents a nasolacrimal duct uniquely shared by "rhipidistians" and tetrapods. The homology of the dipnoan labial cavity to the tetrapod nasolacrimal duct also is problematic, but this issue is not considered herein.

Fenestra Ovalis and the Footplate of the Columella Auris

The ventral wall of the otic capsule of *Eusthenopteron* and some other osteolepiforms (e.g., *Gyroptychius;* Jarvik, 1980:Fig. 147A) has an extensive opening (Fig. 26) that has been termed the *vestibular fontanelle.* The posteroventral ossification of the endocranium of porolepiforms usually is preserved incompletely, or absent. Only one specimen of *Porolepis* (Jarvik, 1972:Pl. 3) and one specimen of *Holoptychius* (Jarvik, 1972:Pl. 24, Fig. 2) have a notch that presumably represents the anterior boundary of

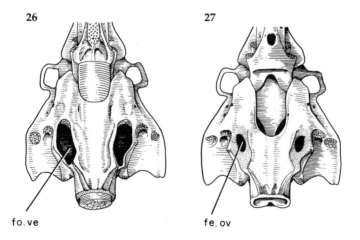

26   27

fo.ve   fe.ov

Figs. 26–27. Posterior endocranium of *Eusthenopteron* in ventral view showing extensive openings. (26) Vestibular fontanelle in the ventral wall of the otic capsule. (27) Fenestra ovalis in cartilage filling vestibular fontanelle. See Appendix I for key to abbreviations. (After Jarvik, 1980.)

the vestibular fontanelle. Because the fontanelle is not perichondrally lined in *Eusthenopteron*, it seems likely that it was filled with cartilage in life. However, Jarvik (1954, 1980) conjectured that an opening—the fenestra ovalis—was present in the center of the cartilage (Fig. 27). Jarvik (1972:210, 1981:160) also suggested that in *Eusthenopteron* a thin lamina projecting posteriorly from the anterodorsal part of the margin of the vestibular fontanelle represents the footplate of the columella auris.

In *Youngolepis*, the opening in the ventral wall of the otic capsule is proportionally smaller than that of *Eusthenopteron* (Chang, 1982:Fig. 15). Moreover, there is a well-developed ridge, the subjugular ridge, that lies dorsolateral to the opening under the jugular groove in *Youngolepis*. The posterior part of the subjugular ridge is configured into a prominent, projecting process; the distal end of the process lacks perichondral bone. In its position, the process resembles the structure in *Nesides* that Bjerring (1972) termed the *adotic eminence*. Chang (1982) suggested that in *Youngolepis* the adotic process served for attachment of the basicranial portions of the trunk muscles as in many osteichthyans. That process is comparable to the restored portion of the bulla acustica in *Eusthenopteron* for the attachment of the same muscle (Jarvik, 1980:126). A large foramen pierces the ventromedial wall of the posterior section of the jugular groove dorsal and anterior to the subjugular ridge in *Youngolepis*; a similar foramen occurs in the lateral wall of the otic capsule of *Nesides*. Such a foramen is absent in *Latimeria* and thus far has been reported only in *Youngolepis* and *Nesides*.

Eusthenopteron lacks a subjugular ridge. The only ossified structure in this region is the posteriorly directed lamina (discussed above) that lies between the anterior section of the jugular groove and the vestibular fontanelle. In *Eusthenopteron*, there is no ossification between the lamina and the fissura occipitalis lateralis. The portions corresponding to (1) the posterior part of the subjugular ridge with its posterior process and (2) the lateral wall of the otic capsule posterior to the large foramen in *Youngolepis* are unossified. Thus, *Eusthenopteron* possesses a huge vestibular fontanelle (Fig. 26) that is bounded posteriorly by the fissura occipitalis lateralis and posterolaterally by the jugular groove.

Although *Nesides* has a large foramen in the lateral wall of the otic capsule, the ventral wall is complete and lacks a vestibular fontanelle. *Latimeria* and *Megalichthys* lack both a foramen in the lateral wall of the otic capsule and a vestibular fontanelle, whereas *Youngolepis* possesses both perforations. In known specimens of *Eusthenopteron*, the vestibular fontanelle is a single large opening that incorporates the foramen in the lateral wall of the otic capsule. Because the presence and sizes of these two openings vary among closely related groups and among different forms of the same group, it seems inappropriate to accord much significance to the observed differences, which simply may reflect varying degrees of ossification.

The vestibular fontanelle is not perichondrally lined in *Youngolepis* and *Eusthenopteron* and probably was filled with cartilage in life. At present, there is no unequivocal evidence as to whether the fenestra ovalis existed in osteolepiforms and porolepiforms. The thin lamina projecting backward from the anterodorsal margin of the vestibular fontanelle in *Eusthenopteron* probably has no relationship to the footplate of the columella auris in amphibians. Instead, the structure probably was simply a part of the lateral wall of the otic capsule of which the portion posterior to the lamina remained unossified.

Intracranial Joint

Herein, the intracranial joint is considered to be confined to the walls of the cranial cavity (Fig. 28). Thus, it does not include the fissure in the base of the endocranium in *Youngolepis* (Chang, 1982:Fig. 10) or the anterior margin of the basicranial fenestra in osteolepiforms, porolepiforms, and coelacanthiforms (Jarvik, 1980:Figs. 88B, 189A, 212B). It has been suggested that the basicranial fissure is homologous to the ventral otic fis-

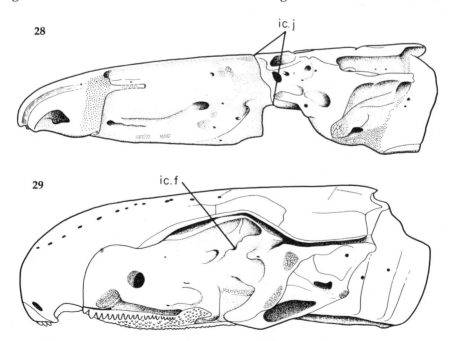

Figs. 28–29. Neurocranium in lateral view of (28) *Eusthenopteron* (after Jarvik, 1980), and (29) *Youngolepis* (after Chang, 1982). See Appendix I for key to abbreviations.

sure in the cranial floor of Devonian palaeoniscids and dipnoans and thus is a plesiomorphic character of osteichthyans (Gardiner, 1973, 1984; Patterson, 1975; Gardiner and Bartram, 1977; Miles, 1977; Chang, 1982). Various proposals as to the homology and distribution of the intracranial joint are discussed below.

Hypothesis 1. Bjerring (1973, 1978) and Jarvik (1980) argued that the intracranial joint is homologous to either intermetameric (in coelacanths) or intrametameric (in osteolepiforms and porolepiforms) vertebral articulations and therefore represents a primitive gnathostome character. This hypothesis seems implausible because it requires that disparate groups such as the placoderms, chondrichthyans, acanthodians, dipnoans, and actinopterygians form a monophyletic assemblage united by the loss of the intracranial joint (Miles, 1977). The alternative interpretation requires independent loss of the intracranial joint in each of the five aforementioned groups (Forey, 1980).

Hypothesis 2. Because the intracranial joint (sensu stricto) is a feature shared by coelacanths, osteolepiforms, and porolepiforms, its presence in these taxa has been regarded as a shared derived character that unites them as crossopterygians (e.g., Romer, 1966; Colbert, 1969). However, the validity of this hypothesis is questionable, as is discussed below.

Hypothesis 3. Miles (1975) identified the presence of an intracranial joint as a character uniting coelacanths and choanates (i.e., "rhipidistians" and tetrapods fide Miles). A similar view was espoused by Schultze (1987), who united coelacanths, onychodontiforms, *Youngolepis*, *Powichthys*, porolepiforms, osteolepiforms, panderichthyids, and tetrapods on the basis of their possession of an intracranial joint. However, in a later paper, Miles (1977) modified his position because he noted that the distribution of such characters as possession of a supraotic cavity and cosmine, as well as the presence or absence of submandibular bones in coelacanths, choanates, and dipnoans, was incongruous with that of the intracranial joint, which is absent in dipnoans. As a consequence, Miles (1977) opted to accept Westoll's (1949) hypothesis (number 4, below) in lieu of his own earlier hypothesis.

Hypothesis 4. Westoll (1949) proposed that the intracranial joint is a derived feature of sarcopterygians—one that is present in coelacanths and choanates but that has been lost in dipnoans (Miles, 1977). Thus, in both Hypotheses 3 and 4, tetrapods are believed to retain a vestige of the intracranial joint, which Jarvik (1980:234, Figs. 171B, 172) considered to be homologous to a transverse suture, the fissura preoticalis, in the cranial base of *Ichthyostega*. However, so far as can be determined, the preotic fissure separates the basisphenoid and basioccipital in *Ichthyostega*. As such, it can be compared only with the ventral otic fissure in osteichthyans and has no relation to the intracranial joint (sensu stricto) of coelacanths, porolepiforms, and osteolepiforms.

Hypothesis 5. Herein, I propose that the presence of an intracranial joint is a plesiomorphic character of sarcopterygians (i.e., coelacanths, osteolepiforms, porolepiforms, and dipnoans). Mobility of the joint is reduced in porolepiforms and absent in dipnoans.

The partial immobilization of the intracranial joint in porolepiforms was noted by Gardiner (1984) in *Glyptolepis* and *Youngolepis*. In the sectioned Stockholm specimen of *G. groenlandica*, the palatoquadrate is fused to the endocranium in several places. Thus, the ethmoidal process of the palatoquadrate is fused to the ethmosphenoid lateral to the orbitorostral passage (= orbitonasal canal), and the anteromedial articular lamina is fused to the dorsal face of the suborbital ledge in two places. The medial margin of the palatoquadrate is fused to the lateral side of the internasal wall dorsal to the most posterior part of the fossa autopalatina. More posteriorly, the basipterygoid process is laterally continuous with the palatoquadrate. Still farther posterior, the basal process of the palatoquadrate is thought to be synchondrotically united to the lateral part of the anterior face of the posterior vertical portion of the basipterygoid process (Jarvik, 1972).

Although *Youngolepis* lacks a straight transverse suture, it has a rather uneven suture between the frontoethmoidal and parietal shields (Figs. 30–32). In some instances, the suture is obliterated by fusion of the cosmine layer of the shields (Fig. 33). The endocranium is a single ossification that has only two fissures—a posterodorsally directed fissure in the lateral wall, and the basicranial fissure (= ventral otic fissure). The roof, floor, and the upper and lower parts of the lateral wall of the cranial cavity are continuous (Fig. 29; Chang, 1982:Fig. 15).

Because *Ichthyostega* possesses only a faint suture at the base of the endocranium that does not extend dorsally to divide the endocranial wall, tetrapods cannot be considered to possess an intracranial joint in the strict sense of the term. The absence of an intracranial joint in dipnoans is interpreted as a loss, owing to the number of other characters discussed below that dipnoans share with osteolepiforms and porolepiforms.

Paired Appendages

The structure of the paired appendages long has been cited as evidence supporting a close relationship between sarcopterygians (or crossop-

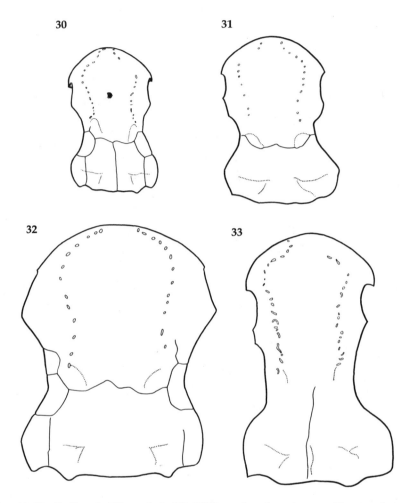

Figs. 30–33. Skull roof of *Youngolepis*. (30–32) Examples of specimens with suture between the frontoethmoidal and parietal shields. (33) Example of a specimen lacking such a suture. (After Chang, 1982.)

terygians, or "rhipidistians," or osteolepiforms) and tetrapods (Holmgren, 1933, 1949a,b; Andrews and Westoll, 1970a,b; Jarvik, 1965, 1981). Rosen et al. (1981:204, 256) pointed out that *Eusthenopteron*, coelacanths (*Latimeria*), porolepiforms, dipnoans, and tetrapods share exclusively metapterygial pectoral and pelvic appendages in which all radials anterior to the base of the metapterygium are lost and each appendage is supported by a single basal element. Depending on the phylogenetic view in favor, the similarities and differences between the paired appendages of particular sarcopterygian groups and tetrapods often have been either exaggerated or discounted.

In order to demonstrate the similarities between dipnoans and tetrapods, Rosen et al. (1981) emphasized the subequal lengths of the paired subbasal elements during early development in *Neoceratodus*, despite the fact that the subbasal elements later fuse in this taxon. Concomitantly, the authors cited the subequal lengths and distal divergence of the two subbasal elements in adult osteolepiforms as evidence of the difference between osteolepiforms and tetrapods. Rosen et al. also distinguished between the articular surface between the first and second metapterygial segments in *Eusthenopteron* and the ball-and-socket joint surface in *Neoceratodus*. However, the authors regarded the high number of axial elements and the presence of true postaxial radials in living and many fossil dipnoans as autapomorphies. In contrast, other authors who favored an alternate phylogenetic scheme (Fig. 34; Jarvik, 1980, 1981; Andrews and Westoll, 1970a,b) attempted to make every detail of osteolepiform paired appendages consistent with that of tetrapods.

Recently, Shubin and Alberch (1986) studied the paired appendages of tetrapods and sarcopterygians from a developmental perspective. During development, the specific branching events that occur within the dipnoan metapterygial axis and the tetrapod digital arch are similar; the similarity suggests that the axis and arch are homologous. The comparison was extended to osteolepiform fins as well. A characteristic developmental pattern is shared by all living sarcopterygians and tetrapods with limbs. The first event is the appearance of a single proximal element (i.e., humerus or femur). The second involves bifurcation into two additional distal elements (i.e., radius and ulna or tibia and fibula) that results in the fundamental asymmetry between preaxial and postaxial sides. Subsequent branching events produce preaxial radials and axial segments. There is no further developmental event that is shared by any one sarcopterygian group and tetrapods to the exclusion of other sarcopterygian groups. Thus, on the basis of the development and morphology of the paired appendages, neither sarcopterygian group (i.e., dipnoans or osteolepiforms) can be shown to be allied more closely with tetrapods than the other.

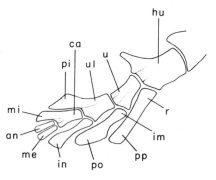

Fig. 34. Interpretation of the pectoral fin of *Eusthenopteron* in light of the tetrapod limb. See Appendix I for key to abbreviations. (After Jarvik, 1980.)

The anocleithrum is present in the dermal shoulder girdle of coelacanths, porolepiforms, and dipnoans, but the element is absent in tetrapods. It is simpler to consider the presence of an anocleithrum to be a synapomorphy of those groups possessing it than to argue for its loss in tetrapods.

PHYLOGENETIC RELATIONSHIPS

Generally, it has been assumed that features that are uniquely shared by "rhipidistians" (i.e., osteolepiforms and porolepiforms) and dipnoans and that are absent in tetrapods are (1) plesiomorphic and (2) have been lost in tetrapods. The basis for this reasoning is rooted in preconceived notions that restrict the closest phylogenetic relative of the tetrapods to either a "rhipidistian" or dipnoan.

On the basis of a reevaluation of the characters and their distribution in relevant groups, a tentative phylogenetic scheme describing the relationships among "rhipidistians" (i.e., osteolepiforms and porolepiforms), dipnoans, and tetrapods is proposed as a challenge to the two traditional and mutually exclusive arrangements. In this scheme (Fig. 35), "rhipidistians" are considered to be a paraphyletic group. Thus, porolepiforms (previously referred to "rhipidistians") are regarded as the sister-group of dipnoans, and osteolepiforms are the sister-group of porolepiforms and dipnoans. Collectively, these groups along with the coelacanths approxi-

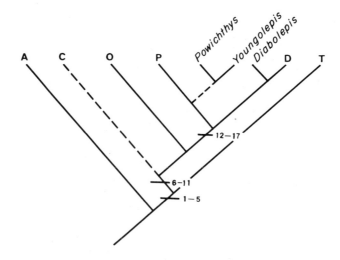

Fig. 35. Tentative phylogenetic scheme for the interrelationships among "rhipidistians," dipnoans, and tetrapods. A = actinopterygians; C = coelacanths; D = dipnoans; O = osteolepiforms; P = porolepiforms; T = tetrapods.

mate the taxon Sarcopterygii of Romer (1966). The Sarcopterygii, as defined herein, is regarded as the sister-group of tetrapods. The characters defining the various groups at different levels are listed below.

Sarcopterygii + Tetrapoda

The following five characters that unite sarcopterygians and tetrapods are not particularly controversial:

1. An extensive metapterygial pectoral and pelvic fin supported by a single basal element that, in turn, articulates with two subbasal elements to form a pre- and postaxial series with postaxial dominance. (Rosen et al., 1981—Character 24, shared by *Eusthenopteron*, actinistians, porolepiforms, dipnoans, and tetrapods; Gardiner, 1984—Character 35, shared by the same groups as mentioned by Rosen et al., 1981; Shubin and Alberch, 1986—as shared by "rhipidistians," dipnoans, and tetrapods.)
2. Enamel on teeth. (Rosen et al., 1981—Character 26; Gardiner, 1984—Character 36; Schultze, 1987—Character 27.)
3. Sclerotic ring composed of more than four plates. (Rosen et al., 1981—Character 27; Gardiner, 1984—Character 37; Schultze, 1987—Character 28.)
4. Four infradentaries; three coronoids.
5. Supraotic cavity with a paired anterior and median posterior division. (Miles, 1977:100–102, 311.)

Sarcopterygii

For reasons discussed above, many of the characters defining the Sarcopterygii (i.e., osteolepiforms, porolepiforms, and dipnoans) previously were used to define a group that included the tetrapods; the absence of features such as Characters 6–9 in tetrapods was explained by loss. The position of coelacanths in the cladogram (Fig. 35) is tentative, but they share Characters 7 and 9–11 with osteolepiforms, porolepiforms, and dipnoans.

6. Cosmine. (Miles, 1977:312—as a character shared by dipnoans and choanates [i.e., osteolepiforms, porolepiforms, and tetrapods fide Miles]; Schultze, 1987—Character 30, as shared by dipnoans and crossopterygians [i.e., Actinistia, Onychodontiformes, *Youngolepis* and *Powichthys*, Porolepiformes, Osteolepiformes, Panderichthyidae, and Tetrapoda].)
7. Intracranial joint. (Cf. discussion above; Miles, 1977:312–313—as a

sarcopterygian [i.e., Actinistia + Dipnoi + Choanata] character lost in dipnoans; Schultze, 1987—Character 33, as a crossopterygian character [see point 6, above].)

8. Submandibular series. (Gardiner, 1984—Character 39, as shared by *Eusthenopteron*, coelacanths, porolepiforms, dipnoans, and tetrapods; Schultze, 1987—Character 31, as a crossopterygian character [see points 6–7, above].)

9. Incorporation of anocleithrum in the dermal shoulder girdle between supracleithrum and cleithrum. (Rosen et al., 1981—Character 24, as a sarcopterygian synapomorphy subsequently lost in tetrapods; Schultze, 1987—Character 26, the same consideration as Rosen et al.)

10. Median extrascapular = A-bone.

11. Hyomandibular facet extensive, straddling jugular vein at a high angle. (Miles, 1977:90, 310).

Porolepiforms + Dipnoans

Until Maisey (1986:Figs. 12, 14), there had been no serious attempt to unite porolepiforms and dipnoans as a monophyletic assemblage. *Diabolepis* seems to be the primitive sister-taxon of known dipnoans. The following characters support this view—(1) the skull-roof pattern (i.e., the presence of a median element, the B-bone); (2) the palatal bite and related change of upper and lower tooth-bearing bones in the mouth cavity into toothplate-like structures of a dipnoan type; and (3) the ventral position of both the external nostrils and restriction of the dentary to the anterior part of the mouth margin.

12. Rostral tubuli (Figs. 36–37). (Maisey, 1986:233, T1.) Thus far, rostral tubuli have been described only in dipnoans (Thomson and Campbell, 1971; Miles, 1977), *Powichthys* (Jessen, 1975, 1980), and *Youngolepis* (Chang, 1982). Quite recently, rostral tubuli were discovered in *Porolepis* as well (D. Goujet, pers. comm.).

13. Details of the cosmine pore-canal system—mesh canals without horizontal partition, pore-canals enamel-lined. (Gross, 1935; Ørvig, 1969; Schultze, 1969; Gardiner, 1984—Character 52, as shared by porolepiforms, dipnoans, and tetrapods; Maisey, 1986:233, T3.)

14. Long, leaf-shaped pectoral fin; skeleton with central axis. (Jarvik, 1972:Fig. 59C; Rosen et al., 1981:257.)

15. Median extrascapular (A-bone) covering lateral extrascapular (or bones lateral to it). (Jarvik, 1980; Miles, 1977:Fig. 116.)

16. Immobilization of intracranial joint; partial or complete fusion of palatoquadrate to neurocranium. (Gardiner, 1984—Character 51,

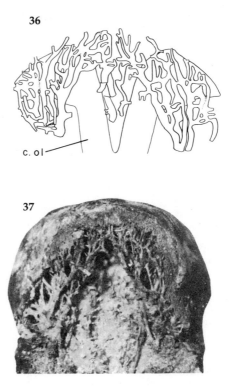

Figs. 36–37. Rostral tubuli in (36) *Youngolepis* (after Chang, 1982), and (37) *Chirodipterus* (after Miles, 1977). See Appendix I for key to abbreviations.

as shared by porolepiforms, dipnoans, and tetrapods; Maisey, 1986:233, T2; this chapter.)

17. Unornamented anocleithrum. (Rosen et al., 1981:218; Gardiner, 1984—Character 41, as shared by actinistians, porolepiforms, dipnoans, and tetrapods.)

CONCLUSIONS

Three major conclusions can be drawn from this study. First, *Diabolepis* is the most primitive dipnoan known. Second, dipnoans are related most closely to porolepiforms; therefore, "rhipidistians" do not constitute a monophyletic group. Finally, neither "rhipidistians" (or osteolepiforms) nor dipnoans individually constitute the sister-group of tetrapods. Instead, "rhipidistians," dipnoans, and probably coelacanths collectively constitute the monophyletic group Sarcopterygii, which is the sister-group of the Tetrapoda.

Acknowledgments Sincere gratitude is extended to Dr. M. J. Novacek for kindly providing me the opportunity to work at the Department of Vertebrate Paleontology of the American Museum of Natural History, without which this paper would not have been realized. I am particularly obliged to Drs. J. G. Maisey, Bobb Schaeffer, and M. K. Hecht for stimulating discussions, and to the latter and Drs. C. Patterson and P. L. Forey for commentaries on this chapter. Similarly, I thank Dr. H.-P. Schultze for his valuable commentary and cooperation, and Dr. D. E. Rosen for his encouragement. Thanks are extended to Dr. D. Goujet for providing information on *Porolepis*, and Paul Sereno for stylistic revision of the draft of this chapter.

APPENDIX I. Anatomical Abbreviations Used in the Figures

an = anularis; a.pr.Mx = ascending process of maxilla; ar.et.pq = articular area for ethmoidal process of palatoquadrate; a.vo = vomeral area

ca = carpales; c.ol = olfactory canal; c.o-n = orbitonasal canal

d.nl = nasolacrimal duct; Dpl = dermopalatine

fe.en = fenestra endochoanalis; fe.ex = fenestra exochoanalis; fe.ov = fenestra ovalis; fe.v = fenestra ventralis; fe.vl = fenestra ventrolateralis; fo.ve = vestibular fontanelle

hu = humerus

ic.f = intracranial fissure; ic.j = intracranial joint; im = intermedium; in = index

me = medius; mi = minimus; Mx = maxilla

na.a, p = anterior, posterior external nasal opening; n.po = nasal pouch

pi = pisiform; po = pollex; pp = prepollex; Pmx = premaxilla; Prl = prefrontolacrimal

r = radius

te.n = tectum nasi

u = ulna; ul = ulnare

Vo = vomer

Literature Cited

Andrews, S. M., and T. S. Westoll. 1970a. The postcranial skeleton of *Eusthenopteron foordi* Whiteaves. Trans. R. Soc. Edinburgh, 68(9):207–329.

Andrews, S. M., and T. S. Westoll. 1970b. The postcranial skeleton of rhipidistian fishes excluding *Eusthenopteron*. Trans. R. Soc. Edinburgh, 68(12):391–489.

Atz, J. W. 1952. Narial breathing in fishes and the evolution of internal nares. Q. Rev. Biol., 27:365–377.

Bertmar, G. 1965. The olfactory organ and upper lips in Dipnoi, an embryological study. Acta Zool. Stockholm, 46:1–40.

Bertmar, G. 1966. On the ontogeny and homology of the choanal tubes and choanae in Urodela. Acta Zool. Stockholm, 47:43–59.

Bjerring, H. C. 1972. The *nervus rarus* in coelacanthiform phylogeny. Zool. Scr., 1:57–68.

Bjerring, H. C. 1973. Relationships of coelacanthiforms. Pp. 179–205 *in* Greenwood, P. H., R. S. Miles, and C. Patterson (eds.), *Interrelationships of Fishes*. London: Academic Press.

Bjerring, H. C. 1978. The "intracranial joint" versus the "ventral otic fissure." Acta Zool. Stockholm, 59:203–214.

Campbell, K. S. W., and R. E. Barwick. 1984. The choana, maxillae, premaxillae and anterior palatal bones of early dipnoans. Proc. Linn. Soc. New South Wales, 107:147–170.

Chang M.-M. 1982. The Braincase of *Youngolepis*, a Lower Devonian crossopterygian from Yunnan, South-western China. Doctoral dissertation. Stockholm: Dept. of Geology, Univ. of Stockholm.

Chang M.-M., and X. B. Yu. 1981. A new crossopterygian, *Youngolepis praecursor*, gen. et sp. nov., from the Lower Devonian of E. Yunnan, China. Sci. Sin., 24:89–97.

Chang M.-M., and X. B. Yu. 1984. Structure and phylogenetic significance of *Diabolichthys speratus* gen. et sp. nov., a new dipnoan of eastern Yunnan, China. Proc. Linn. Soc. New South Wales, 107:171–184.

Chang M.-M., and X. B. Yu. 1987. A *nomen novum* for *Diabolichthys* Chang et Yu, 1984. Vertebr. PalAsiatica, 25:79.

Colbert, E. H. 1969. *Evolution of the Vertebrates. A History of the Backboned Animals Through Time.* 2nd ed. New York: Wiley and Sons.

Forey, P. L. 1980. *Latimeria:* a paradoxical fish. Proc. R. Soc. London, Ser. B, 208:369–384.

Gardiner, B. G. 1973. Interrelationships of teleostomes. Pp. 105–135 *in* Greenwood, P. H., R. S. Miles, and C. Patterson (eds.), *Interrelationships of Fishes.* London: Academic Press.

Gardiner, B. G. 1984. The relationship of the palaeoniscid fishes, a review based on new specimens of *Mimia* and *Moythomasia* from the Upper Devonian of Western Australia. Bull. Br. Mus. (Nat. Hist.) Geol., 37:173–428.

Gardiner, B. G., and A. W. H. Bartram. 1977. The homologies of ventral fissures in osteichthyans. Pp. 227–245 *in* Andrews, S. M., R. S. Miles, and A. D. Walker (eds.), *Problems in Vertebrate Evolution.* Linn. Soc. Symp. Ser. 4.

Gross, W. 1935. Histologische Studien am Aussenskelett fossiler Agnathen und Fische. Palaeontogr., Ser. A, 83:1–60.

Holmes, E. B. 1985. Are lungfishes the sister group of tetrapods? Biol. J. Linn. Soc. London, 25:379–397.

Holmgren, N. 1933. On the origin of the tetrapod limb. Acta Zool. Stockholm, 14:186–295.

Holmgren, N. 1949a. On the tetrapod limb problem—again. Acta Zool. Stockholm, 30:485–508.

Holmgren, N. 1949b. Contributions to the question of the origin of tetrapods. Acta Zool. Stockholm, 30:459–484.

Jarvik, E. 1937. On the species of *Eusthenopteron* found in Russia and the Baltic states. Bull. Geol. Inst. Univ. Uppsala, 27:63–127.

Jarvik, E. 1942. On the structure of the snout of crossopterygians and lower gnathostomes in general. Zool. Bijdr., 21:235–675.

Jarvik, E. 1954. On the visceral skeleton in *Eusthenopteron*, with a discussion of the parasphenoid and palatoquadrate in fishes. K. Sven. VetenskAkad. Handl. (4)5:1–104.

Jarvik, E. 1962. Les porolépiformes et l'origine des urodèles. Colloq. Int. C. N. R. S., 104:87–101.

Jarvik, E. 1965. On the origin of girdles and paired fins. Israel J. Zool., 14:141–172.

Jarvik, E. 1966. Remarks on the structure of the snout in *Megalichthys* and certain other rhipidistid crossopterygians. Ark. Zool., (2)19:41–98.

Jarvik, E. 1972. Middle and Upper Devonian Porolepiformes from East Greenland with special references to *Glyptolepis groenlandica* n. sp. Medd. Groenland, 187:1–295.

Jarvik, E. 1980. *Basic Structure and Evolution of Vertebrates.* Vol. 1. New York and London: Academic Press.

Jarvik, E. 1981. *Basic Structure and Evolution of Vertebrates.* Vol. 2. New York and London: Academic Press.

Jessen, H. 1975. A new choanate fish, *Powichthys thorsteinssoni* n. g., n. sp., from the early Lower Devonian of the Canadian Arctic Archipelago. Colloq. Int. C. N. R. S., 218:213–222.

Jessen, H. 1980. Lower Devonian Porolepiformes from the Canadian Arctic with special reference to *Powichthys thorsteinssoni* Jessen. Palaeontogr., Ser. A, 167:180–214.

Lebedkina, N. S. 1979. *Evolution of the skull of amphibians. On the Problem of Morphological Integration.* Moscow: NAUKA. [In Russian].

Maisey, J. 1986. Heads and tails: a chordate phylogeny. Cladistics, 2:201–256.

Miles, R. S. 1975. The relationships of the Dipnoi. Colloq. Int. C. N. R. S., 218:133–148.

Miles, R. S. 1977. Dipnoan (lungfish) skulls and the relationships of the group: a study based on new species from the Devonian of Australia. J. Linn. Soc., London Zool., 61:1–328.

Ørvig, T. 1969. Cosmine and cosmine growth. Lethaia, 2:219–239.

Panchen, A. L. 1967. The nostrils of choanate fish and early tetrapods. Biol. Rev. Cambridge Philos. Soc., 42:374–420.

Patterson, C. 1975. The braincase of pholidophorid and leptolepid fishes, with a review of the actinopterygian braincase. Philos. Trans. R. Soc. London, Ser. B, 269:275–579.

Romer, A. S. 1937. The braincase of the Carboniferous crossopterygian *Megalichthys nitidus.* Bull. Mus. Comp. Zool. Harvard Univ., 82:1–73.

Romer, A. S. 1966. *Vertebrate Paleontology.* 3rd ed. Chicago: Univ. Chicago Press.

Rosen, D. E., P. L. Forey, B. G. Gardiner, and C. Patterson. 1981. Lungfish, tetrapods, paleontology and plesiomorphy. Bull. Am. Mus. Nat. Hist., 167:159–276.

Schaeffer, B. 1984. On the relationships of the Triassic-Liassic redfieldiiform fishes. Am. Mus. Novit., 2795:1–18.

Schultze, H.-P. 1969. *Griphognathus* Gross, ein langschnauziger Dipnoer aus dem Oberdevon von Bergisch-Gladbach (Rheinisches Schiefergebirge) und von Lettland. Geologica Palaeontologica, 3:21–79.

Schultze, H.-P. 1987. Dipnoans as sarcopterygians. Pp. 39–74 in Bemis, W. E., W. W. Burggren, and N. E. Kemp (eds.), *The Biology and Evolution of Lungfishes.* J. Morphol., Suppl. 1.

Shubin, N. H., and P. Alberch. 1986. A morphogenetic approach to the origin and basic organization of the tetrapod limb. Evol. Biol., 20:319–387.

Thomson, K. S. 1964. The comparative anatomy of the snout in rhipidistian fishes. Bull. Mus. Comp. Zool. Harvard Univ., 131:315–357.

Thomson, K. S., and K. S. W. Campbell. 1971. The structure and relationships of the primitive Devonian lungfish—*Dipnorhynchus sussmilchi* (Etheridge). Bull. Peabody Mus. Nat. Hist., 38:1–109.

Vorobyeva, E. 1971. Evolution of Rhipidistia (Crossopterygii). Palaeontol. Zh., 1971(3):3–16.

Vorobyeva, E. 1977. Morphology and nature of evolution of crossopterygian fish. Tr. Paleontol. Inst. Akad. Nauk SSSR, 163:1–239. [In Russian].

Westoll, T. S. 1949. On the evolution of the Dipnoi. Pp. 121–184 in Jepsen, G. L., E. Mayr, and G. G. Simpson (eds.), *Genetics, Paleontology and Evolution.* Princeton: Princeton Univ. Press.

2 A Comparison of Controversial Hypotheses on the Origin of Tetrapods

Hans-Peter Schultze

The question of tetrapod origins is a recurring theme of evolutionary studies, as discussion is renewed periodically by new discoveries. (For an extensive historical review, see Rosen et al., 1981.) When the South American and African lungfishes were discovered, this group was considered to be the closest relatives of tetrapods (Bischoff, 1840). Although this view was attacked adamantly, it was only after the discovery of *Eusthenopteron* (Whiteaves, 1881) that Cope (1892) proposed another group, the crossopterygians, to be the closest relatives of tetrapods. This hypothesis was accepted widely, although authors such as Holmgren (1933, 1939, 1949a,b), Kesteven (1950), and Fox (1965) continued to assert that lungfishes were the closest relatives of either all tetrapods or a subset of them (assuming a diphyletic origin of the group).

On the basis of a detailed study of the nasal region of different fish groups and tetrapods, Jarvik (1942) opposed the diphyletic origin of tetrapods from lungfishes and crossopterygians (Holmgren, 1933), because unlike tetrapods, lungfishes lack a true choana. Jarvik (1942) proposed a different diphyletic arrangement in which urodeles were derived from porolepiform crossopterygians, and anurans and all other tetrapods from osteolepiform crossopterygians.

The description of exceptionally well preserved material of the Late Devonian lungfish *Griphognathus* (Miles, 1977) inspired Gardiner (1980) and Rosen (cf. Rosen et al., 1981), independently, to reconsider close relationships between lungfishes and tetrapods. This view has been opposed by Schultze (1981, 1987), Jarvik (1981b), Holmes (1985), Maisey (1986), and Panchen and Smithson (1987).

Here, I compare the three current hypotheses of tetrapod relationships and origins: (1) the diphyletic origin of tetrapods from lungfish and crossopterygians (Jarvik, 1942, 1972, 1981a, 1986); (2) a close relationship between tetrapods and lungfish (Rosen et al., 1981); and (3) a close relationship between tetrapods and osteolepiforms (Szarski, 1962, 1977; Schmalhausen, 1968; Worobjewa, 1975; Vorobyeva, 1977a, 1984, 1985, 1986; Schultze, 1977a, 1981, 1987; Holmes, 1985; Maisey, 1986; Panchen and Smithson, 1987). Further, I compare these hypotheses with the sarcopterygian relationship proposed by Chang (this volume). I focus on the two features that have played a central role in the transition from fishes to tetrapods—the nasal region and the limb.

CHARACTERIZATION OF TETRAPODS

In order to discuss the origin of a group, one first must define the group by unique features. This can be done by in-group comparison (i.e., searching for characters common to all members of the group), or by establishing the relationship of members of the group and using the features of the most plesiomorphic (i.e., basal) taxon, or by out-group comparison. Out-group comparison, the commonly used method, is based on an a priori assumption of the identity of the most closely related groups. In the case of tetrapods, two features (breathing through nasal openings and walking on feet) set the group apart from any piscine relative; thus, out-group comparison with crossopterygians (Gaffney, 1979) or lungfish (Rosen et al., 1981) leads to similar conclusions.

Gaffney (1979), using osteolepiform rhipidistians as an out-group, defined tetrapods as a monophyletic unit possessing the following characters: (1) anterior and posterior moieties of braincase intimately united in adults; (2) otico-occipital region of skull longitudinally compact; (3) otic notch; (4) single pair of nasals with or without internasal; (5) stapes; (6) fenestra ovalis; (7) carpus, tarsus, and dactyly; (8) dermal pectoral skeleton free from skull with absence of posttemporal, supracleithrum, and anocleithrum; (9) iliac blade of pelvis attached to vertebral column by an ossified sacral rib; (10) well-developed ischiac ramus of pelvis and pubic symphysis; and (11) ossified ribs well developed and directed ventrally.

Rosen et al. (1981), using dipnoans as an out-group, accepted Gaffney's synapomorphies of tetrapods with the exception of Characters 1, 8, 10, and 11. Panchen and Smithson (1987), using osteolepiform rhipidistians as out-group as did Gaffney (1979), accepted Gaffney's synapomorphies except for Characters 1–3 and 11. They divided Gaffney's Character 7 into six separate synapomorphies and added two new features—the presence of a distinct scapular blade and a distinct arrangement of bones in the cheek region. The braincase of tetrapods (Gaffney's Character 1) develops by fusion of an anterior and posterior moiety. This mode of development might indicate that the braincase was divided in the ancestor of tetrapods (Roček, 1986; Schultze, 1987), as Jarvik (1952) suggested that it was in the case of *Ichthyostega*.

There is no indication of a divided braincase in dipnoans either ontogenetically or in primitive forms. Thus, the union of the anterior and posterior parts of the braincase *in adults* is a reliable synapomorphy of tetrapods, given crossopterygians as out-group, because an intracranial joint is present in actinistians, onychodonts, porolepiforms, and osteolepiforms. The compact braincase in dipnoans, then, must be interpreted as a parallel evolutionary event.

The suturing of the anterior and posterior dermal shields of the skull roof is a feature uniting tetrapods with panderichthyids, and one must assume that the pattern of the skull roof bones (i.e., frontals, parietals, postparietals) is a synapomorphy of panderichthyids and tetrapods. The otico-occipital region is longitudinally shortened in panderichthyids but not compacted; panderichthyids are intermediate between osteolepiforms and tetrapods in this feature. Panchen and Smithson (1987) cited additional proportions shared by rhipidistians and early tetrapods. As defined by Gaffney (1979), the otic notch is homologous to the spiracular slit in panderichthyids; it has the same shape and position between the squamosal and tabular or supratemporal. The hyomandibular is visible in the spiracular slit, and it is in close contact with the palatoquadrate. Godfrey et al. (1987) proposed the use of the term *squamosal embayment* for the emargination between the squamosal and tabular or supratemporal, because it has not been demonstrated that the emargination contains a tympanum and that the stapes is directed toward it as it would be if the emargination were an otic notch. They compared this embayment with the region of the postspiracular plate in *Eusthenopteron*. The postspiracular plate is known in *Eusthenopteron* and *Eusthenodon* (Jarvik, 1980), but not in osteolepids and panderichthyids. In addition, the postspiracular plate belonging to the extrascapular-opercular series does not extend anteriorly to the squamosal embayment. The spiracular slit occupies the position between squamosal, tabular, and supratemporal (Jarvik, 1980:Fig. 151; Vorobyeva and Schultze, this volume:Fig. 4), and therefore, it is topographically homologous to the "otic notch" (= squamosal embayment) of primitive tetrapods. A squamosal embayment is absent in amniotes, aïstopods, nectrideans, microsaurs, and colosteid temnospondyls (Panchen and Smithson, 1987).

In tetrapods, the shoulder girdle has no connection to the skull roof (Gaffney's Character 8) such as that which exists between Bone Z (same position as the posttemporal in other osteichthyans) and the anocleithrum in dipnoans, a connection between the shoulder girdle and skull roof similar to that found in other osteichthyans. An anocleithrum and supracleithrum are described in only one primitive tetrapod, *Tulerpeton* (Lebedev, 1984). In Recent dipnoans, the right and left halves of the pelvic girdle are fused to form one cartilaginous element, whereas in fossil forms ossified elements form right and left halves of the pelvic girdle that do not meet in the midline. The tetrapod pubic symphysis is unknown in fishes, as is an ischiac ramus. The comparison of advanced representatives of both Recent lungfishes and urodeles by Rosen et al. (1981:Fig. 61) is misleading because, in each case, we deal with terminal members of the group, whereas the primitive members of each group are not comparable in these features (cf. above). *Crassigyrinus* is problematic in this character (Panchen, 1985).

Ossified ribs (Gaffney's Character 11) occur in some actinopterygians, actinistians, and dipnoans. It is unclear if ribs are a feature of primitive members of these three groups. The occurrence of ribs is widespread within actinopterygians, but they are missing in the most primitive forms—*Cheirolepis* (Pearson and Westoll, 1979), *Mimia* (Gardiner, 1984), and *Moythomasia* (Jessen, 1968). Ribs are restricted to a few genera of actinistians; therefore, they are thought to be autapomorphies for these taxa (Gaffney, 1979). The situation in dipnoans is similar to that of actinopterygians; ribs are widespread within the group, but *Uranolophus*, one of the most primitive dipnoans, lacks them. This suggests that ossified ribs have developed in parallel in these groups, and that their presence is not a synapomorphous character for all osteichthyans (contra Gardiner, 1984). As defined by Gaffney (1979), Character 11 is ambiguous and also occurs at least in panderichthyids.

Of Gaffney's (1979) 11 characters, 1–3 are gradational, having been achieved partially by panderichthyids; Characters 8 and 9 are not ubiquitous among tetrapods because they are missing in *Tulerpeton* and *Crassigyrinus*, respectively. Characters 4–7 and 10 define tetrapods, but they are not useful in identifying their sister-group. In order to do this, one must consider plesiomorphous characters of primitive tetrapods— i.e., characters that occur in all or most primitive tetrapods but are not restricted to them. These characters include the presence of (1) three pairs of median cranial bones (i.e., postparietal, parietal, frontal) between the lateral-line bones; (2) a pineal foramen between the parietals and posterior to the orbits; (3) one pair of large squamosals; (4) lacrimal, jugal, postorbital, postfrontal, and prefrontal bones around the orbit; (5) narrow, elongate parasphenoid with buccohypophysial foramen ending anterior to the otico-occipital region; (6) paired premaxillae, maxillae, vomers, dermopalatines, and ecto- and endopterygoids; (7) lower jaw with long dentary, four infradentaries, three coronoids, and one prearticular; (8) pointed, conical teeth with enamel and plicidentine; (9) one external nasal opening surrounded by premaxilla, maxilla, lacrimal, and tectals; (10) one palatal opening (choana) surrounded by premaxilla, maxilla, vomer, and dermopalatine; (11) appendages composed of one proximal, two distal, and many more distal elements; (12) a proximal limb element that articulates with the endoskeletal girdle via a ball-and-socket joint; and (13) centra formed by two paired splintlike elements (i.e., pleuro- and intercentrum).

Although panderichthyids and tetrapods share all of these characters except Character 13 (Vorobyeva and Schultze, this volume), most are synapomorphies of lower levels of sarcopterygians. In comparing the four hypotheses of the origin of tetrapods, I have concentrated on the nasal region (Characters 9–10) and the appendages (Characters 11–12).

Nasal Region

Tetrapods (Figs. 1–4) possess one external nasal opening for air intake. The air passes over the olfactory tissue within the nasal capsule and enters the mouth cavity via the internal nasal opening, the choana. Thus, the olfactory capsule of tetrapods functions in the passage of air (a secondary function), as well as olfaction (its primary function). The nasal capsule is an endoskeletal structure covered by exoskeletal bones such that the external and internal openings are formed by both exo- and endoskeletal structures. Consequently, Jarvik (1942) distinguished the fenestra exonarina from the endonarina, and the fenestra exochoanalis from the endochoanalis. In early tetrapods, the fenestra exochoanalis (Fig. 1) is surrounded by four bones; laterally, there are the maxilla and premaxilla (except in *Ichthyostega*, Fig. 2), and anteriorly, medially, and posteriorly, there are the vomer and dermopalatine. The fenestra exonarina (Fig. 3) lies outside the outer dental arcade above the premaxilla and maxilla, and it is bordered by the lacrimal, anterior tectal, premaxilla, and maxilla in ichthyostegids (Fig. 4). In the absence of the anterior tectal (Fig. 3), the nasal borders the fenestra exonarina. In tetrapods, the nasolacrimal duct connects the nasal capsule with the orbit. The comparisons that follow are based on these topographic definitions. Soft anatomical features such as fleshy tubes extending distally from the external nasal opening (Rosen et al., 1981) or the nasolabial groove (Panchen, 1967) are not considered herein because such structures are preserved neither in early tetrapods nor in related early osteichthyans.

Endoskeleton of Appendages

Tetrapods possess two pairs of appendages that support the body. The hind limb and forelimb have the same general Bauplan. A single, proximal element (humerus or femur) bears a convex surface that articulates

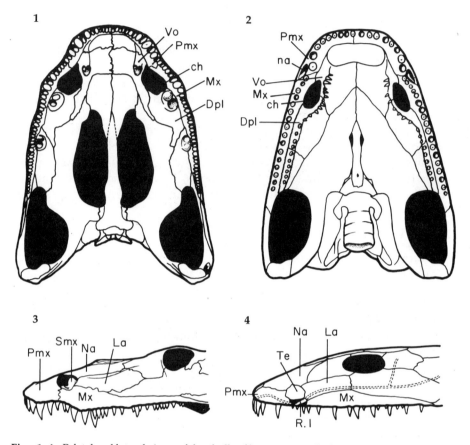

Figs. 1–4. Palatal and lateral views of the skulls of lower tetrapods showing the positions of the choana (internal nasal opening) and external nasal opening. Palatal views of (1) *Eryops megacephalus* (after Sawin, 1941:Pl. 2) and (2) *Ichthyostega* sp. (after Jarvik, 1980:Fig. 171B). Lateral views of the left anterior portions of the skulls of (3) *Eryops megacephalus* (after Sawin, 1941:Pl. 4, Fig. b) and (4) *Ichthyostega* sp. (after Jarvik, 1980:Fig. 171C). See Appendix I for key to abbreviations.

with a socket in the endoskeletal girdle (scapulocoracoid or pelvic girdle). Two elements follow distally (radius and ulna or tibia and fibula). The most distal part of the limb consists of many elements (carpalia, etc. or tarsalia, etc.) that constitute the manus or pes, respectively. A pentadactyl appendage (Fig. 5), with or without prepollex, generally is accepted as the basic Bauplan of tetrapods, although this may be true only of the hind limb. Temnospondyls and extant amphibians have four fingers in the hand, whereas batrachosaurian amphibians and amniotes have five. The hand of the Late Devonian *Ichthyostega* is unknown, but Lebedev (1984) described a Late Devonian tetrapod, *Tulerpeton*, with six digits on the

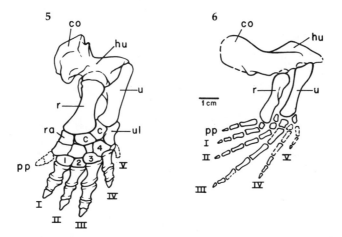

Figs. 5–6. Left forelimbs of lower tetrapods. (5) *Eryops megacephalus* (after Gregory and Raven, 1941:Fig. 24C). (6) *Tulerpeton curtum* (after Lebedev, 1984:Fig. 2). See Appendix I for key to abbreviations.

hand (Fig. 6)—i.e., a prepollex and five fingers. Additional elements distal to the radius and to the ulna (and prepollex and postminimus, respectively) are known, and apparently, the prepollex is fully developed in *Tulerpeton*. The hind limb of *Tulerpeton* has the basic pentadactyl pattern of early tetrapods.

Panchen and Smithson (1987:369) cited six synapomorphies involving tetrapod appendages, as follows: (1) radius and ulna (and tibia and fibula) as parallel (and primitively independent) ossifications, both bearing articular surfaces for the carpus (tarsus) distally; (2) carpus and tarsus consisting of several bones that articulate with one another medially and laterally, as well as proximally and distally; (3) manus and pes terminating in a series of separate load-bearing digits, each with a skeleton of articulated phalanges joined proximally to a metacarpal or metatarsal; (4) wrist joint incorporating the radius and forming a hinge; (5) knee joint between femur and tibiofibula a hinge; and (6) rotary ankle joint incorporating the tibia.

HYPOTHESES

Diphyletic Origin of Tetrapods

The first hypothesis I will examine asserts a diphyletic origin of tetrapods (Fig. 7). Holmgren (1933, 1939, 1949a,b) and Jarvik (1942, 1972, 1981a, 1986) argued that the tetrapods do not form a monophyletic group. Whereas Holmgren derived the urodeles from dipnoans, Jarvik

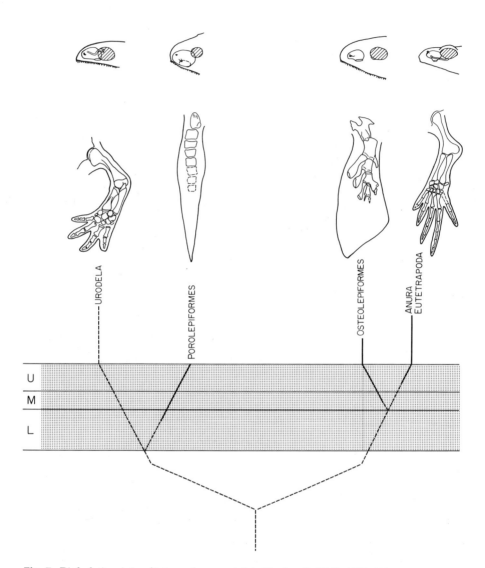

Fig. 7. Diphyletic origin of tetrapods as postulated by Jarvik (1942, 1972, 1981a, 1986). Top row of figures shows the left nasal capsule with openings of (*from left to right*) a urodele with one external and one internal nasal opening, and a nasolacrimal duct with bend; a porolepiform with two external nasal openings and ventrally closed ventrolateral fenestra; an osteolepiform with one external and one internal nasal opening; and an anuran with one external and one internal nasal opening, and a straight nasolacrimal duct. Lower row figures depict left anterior appendage of (*from left to right*) a urodele; a porolepiform pectoral fin with axially arranged elements and fishlike articulation between scapulocoracoid and first axial element; an osteolepiform pectoral fin with radials on preaxial side only and with tetrapod-like articulation between scapulocoracoid and first axial element; and an eutetrapodan forelimb with a pentadactyl hand. Stipples indicate Devonian time: U = Late, M = Middle, L = Early Devonian.

36

derived them from porolepiforms and considered osteolepiforms to be the ancestors of eutetrapods (i.e., Anura + all other Eutetrapoda).

Nasal Region

Jarvik (1942) based the diphyly of tetrapods on the interpretation of the snout of fishes and tetrapods. He distinguished a porolepiform type with which the snout of urodeles "agrees . . . almost completely" from an osteolepiform type found in anurans. In contrast to tetrapods and osteolepiforms, porolepiforms have two external nasal openings, the primitive condition for osteichthyans. The two nasal openings are separated from one another by a bone, the nariodal, known only from some porolepiforms. Thus, the nasal capsule of porolepiforms has two functional external openings—an incurrent and an excurrent opening. The nasal capsule of porolepiforms also possesses an endochoanal ventrolateral fenestra, but it is not certain that a tube could extend from this opening to an exochoanal opening in the palate, as postulated by Jarvik (1942, 1972, 1980). The endochoanal part of the ventrolateral fenestra is covered by bone on the palatal side (Chang, 1982, this volume), whereas it opens laterally to the fenestra exonarina posterior, the excurrent opening of the nasal capsule. The fenestrae exonarina anterior and posterior are widely separated in Early and Middle Devonian porolepiforms (e.g., *Youngolepis, Porolepis, Glyptolepis*) but located near one another in Late Devonian porolepiforms (e.g., *Holoptychius, Quebecius*). Jarvik (1964) postulated a reversed chronological sequence (Schultze, 1969) by which the nasal openings were separated gradually and the posterior nasal opening eventually reached the orbit to form the nasolacrimal duct. He considered a groove posterior to the posterior external nasal opening in the Middle Devonian *Glyptolepis* (Jarvik, 1972:Fig. 79D) to be homologous to the nasolacrimal duct. In Jarvik's opinion, the wide separation of the two external openings in the Early Devonian *Porolepis* indicates the closest similarity to urodeles, because the posterior nasal opening is close to the orbit in the process of its transformation into the nasolacrimal duct, which terminates in the orbit. Jarvik considered the nariodal, a bone separating anterior and posterior external nasal openings in the Late Devonian *Holoptychius* and *Quebecius*, to be a homologue of the septomaxilla, which is associated with the nasolacrimal duct in urodeles. The shape of the nasolacrimal duct may be unique to urodeles, but there is no justification for comparing it to the groove extending posteriorly from the posterior nasal opening in porolepiforms. Also, Jarvik's methodology is questionable because he uses more than one porolepiform taxon from which to derive the various nasal features of urodeles.

Jarvik (1942, 1972, 1981a) supported his derivation of the nasal characters of urodeles from porolepiforms by asserting that their nasal capsules were similar in structure. In both groups, the internal structure of the nasal capsule is simple (specifically, lacking ridges and divisions), and the course of canals for nerves and vessels in the wall and area around the nasal capsule supposedly is similar. However, according to Szarski (1962:231–232), the arrangement of nerve branches used by Jarvik (1942) for comparison with porolepiforms occurs only in some groups of urodeles (e.g., hynobiids). Jarvik (1964:58, note) discarded this argument by referring to other authorities.

Jurgens (1971), in an extensive study of the ontogeny of the nasal region of the Lissamphibia, recognized intermediate stages among the three groups and, more important, structures shared by the groups. All living amphibians possess a simple, sacklike olfactory organ in early ontogeny; the differences in the structure of the nasal capsule among adults are derived from variation in differential growth between the brain and the nasal sacs. Primitive genera of the three lissamphibian groups, especially urodeles and anurans, are more similar to each other than they are to the more advanced representatives to which Jarvik referred.

Jarvik (1981a) cited 11 character complexes of the snout which he believed distinguish his Osteolepipoda [Osteolepiformes + Eutetrapoda] from the Urodelomorpha [Porolepiformes + Urodela]. The formation of the septum nasi (Jarvik's Character 1) varies. A septum nasi occurs in the internasal region of nearly all anurans, caecilians, and some urodeles (Figs. 8–10); according to Jurgens, it is absent in the bufonoid anuran *Brachycephalus* (Brachycephalidae) (Jurgens, 1971:Fig. 41a) and many of the urodeles that he examined (Fig. 11; Jurgens, 1971:Fig. 38A). Jurgens (1971) argued that the formation of the septum nasi depends upon the degree of ontogenetic retraction of the brain, and that urodeles and anurans are not distinguished by the absence and presence of a septum nasi. The relationship between internasal cavities in rhipidistians and nasal capsules depends on the proportions of the snout (Kulczycki, 1960; Vorobyeva, 1977a,b, 1980). The internasal cavity lies between the nasal capsules in broad-snouted forms (i.e., porolepiforms and a few osteolepiforms) but anterior to the nasal capsules in narrow-snouted forms (i.e., most osteolepiforms).

Jarvik distinguished between the internasal gland in the deep internasal cavity of Urodelomorpha and an intermaxillary gland in Osteolepipoda. Intermaxillary glands occur in the buccal roof in urodeles and anurans (Jurgens, 1971:Fig. 41A,a); in urodeles, they lie between the medial walls of the olfactory capsules (Fig. 11), whereas in anurans and caecilians they lie ventromedial to the narrowly separated olfactory capsules. This slight difference in position seems to be correlated with the configuration of the septum nasi and has little to do with the presumed homology of the glands (Jurgens, 1971). It is doubtful that porolepiforms or osteolepiforms possessed an intermaxillary gland because the gland occurs in terrestrially adapted urodeles, in which it facilitates swallowing of prey; it is absent in aquatic urodeles (Jurgens, 1971).

The subdivision of the anterior part of the nasal cavity differs in the Osteolepipoda and Urodelomorpha (Jarvik's Character 2). In porolepiforms, the simple anterior part of the nasal cavity has two external nasal openings as in primitive gnathostomes. However, a derived condition characterizes urodeles, in which the anterior part of the nasal cavity has an internal nasal opening.

Jarvik (Character 3) contrasted the septomaxilla of anurans (Fig. 12; presumably derived from the processus dermintermedius of the lateral rostral of osteolepiforms) with his "so-called" septomaxilla of urodeles (Fig. 11; supposedly derived from the nariodal of porolepiforms). A processus dermintermedius exists in *Youngolepis* (Chang, 1982), which is a relative of porolepiforms, its out-group; thus, it is a primitive feature of porolepiforms and osteolepiforms, not a character that distinguishes the groups.

In contrast to Jarvik (1942), who argued that the "so-called septomaxilla" of urodeles is cartilaginous in origin, Schmalhausen (1958) and Medvedeva (1959, 1975) showed that the septomaxilla is a dermal bone in all lissamphibians, in which it lies "in approximately the corresponding place" (Jarvik, 1942:249) with respect to the nasolacrimal duct and nasal muscles (Figs. 11–13).

Jarvik's Characters 4, 7, and 8 do not support a relationship between urodeles and porolepiforms. Character 4 is the presence of an external posterior nasal opening in Urodelomorpha. Character 7 states that the fenestra ventrolateralis includes the fenestrae endochoanalis and endonarina posterior in Urodelomorpha; Character 8 maintains that the nasolacrimal duct in urodeles is homologous to the posterior external nasal opening and its posteriorly associated external groove in porolepiforms. Each of the latter characters is correlated with the presence of an external posterior nasal opening in porolepiforms (Jarvik, 1981a). Because urodeles lack this opening, I cannot see how it can be used to relate them to porolepiforms.

Contrary to Jarvik (1981a), Jurgens (1971) perceived similarities, rather than differences, between the structures of the nasal cavities of urodeles and anurans. Thus, Jarvik distinguished between the Urodelomorpha and Osteolepidoda on the basis of the presence of a crista rostrocaudalis in the Urodelomorpha (Character 5), the limits of the lateral recess and presence of Seydel's palatal process in Urodelomorpha (Character 6), and differences in the course of nerves between Urodelomorpha and Osteolepidoda (Characters 9–10). He chose advanced members (e.g., *Salamandra* and *Rana*) as representatives of both groups. Jurgens, in contrast,

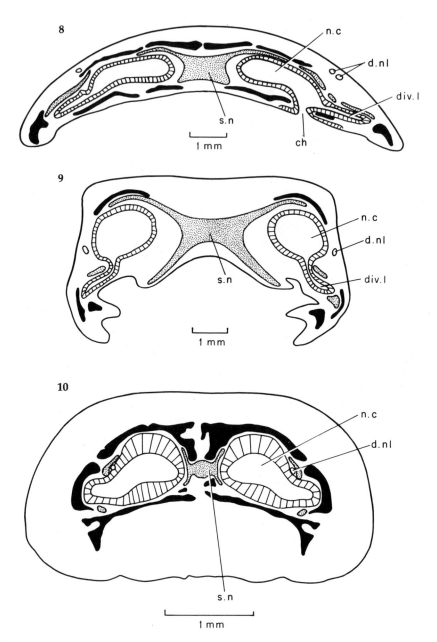

Figs. 8–10. Transverse sections through the nasal region of lissamphibians. Septum nasi in (8) ambystomatid urodele *Ambystoma maculatum* (after Jurgens, 1971:Fig. 41B), (9) ascaphid anuran *Ascaphus truei* (after Jurgens, 1971:Fig. 41b), and (10) ichthyophiid caecilian *Ichthyophis glutinosus* (after Visser, 1963:Fig. 6C). See Appendix I for key to abbreviations.

40

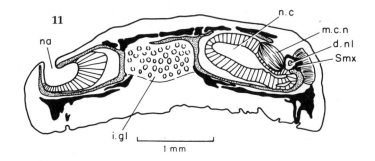

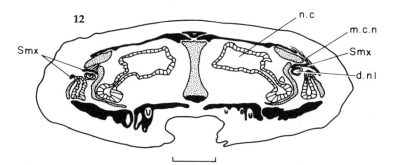

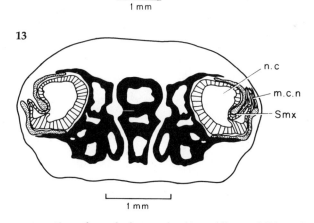

Figs. 11–13. Transverse sections through the nasal region of lissamphibians. Position of septomaxilla in (11) hynobiid urodele *Hynobius leechii* (after Lapage, 1928a:Fig. 18), (12) pipid anuran *Silurana tropicalis* ([= *Xenopus calcaratus* of Lapage] after Lapage, 1928b:Fig. 12), and (13) ichthyophiid caecilian *Ichthyophis glutinosus* (after Lapage, 1928b:Fig. 14). See Appendix I for key to abbreviations.

pointed out that the crista rostrocaudalis of urodeles is homologous to crista intermedia of anurans, and that the course of nerves is similar in both groups (Figs. 14–16), especially in primitive representatives of the groups. Jurgens showed, by comparisons within Recent amphibians, that many of Jarvik's characters occur in both or all three living groups; therefore, these characters cannot be used to separate urodeles from anurans.

Appendages

According to Holmgren (1933), the eutetrapod appendage evolved from a bifurcating fin skeleton, whereas a short, biserially arranged archipterygium gave rise to the urodele extremity (Figs. 17–19). Steiner (1935) and others dismissed Holmgren's arguments by demonstrating that the elements in the extremity of urodeles and other tetrapods are arranged in the same ontogenetic bifurcate branching plan. Jarvik (1964, 1965) reiterated the differences between both groups by comparing an early ontogenetic stage in the development of the hind limb of a urodele

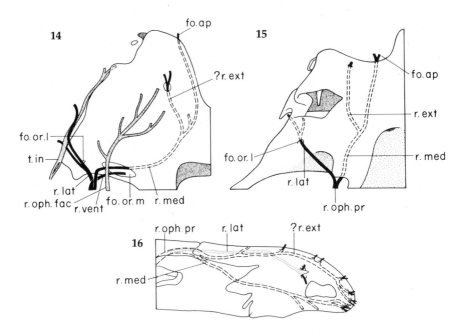

Figs. 14–16. Course of ramus ophthalmicus profundus branch of trigeminal nerve around the nasal capsule of lissamphibians. (14) Dorsal view of cryptobranchid urodele *Cryptobranchus alleganiensis* (after Jurgens, 1971:Fig. 22A). (15) Dorsal view of ascaphid anuran *Ascaphus truei* (after Jurgens, 1971:Fig. 1A). (16) Lateral view of typhlonectid caecilian *Typhlonectes compressicauda* (after Jurgens, 1971:Fig. 35). See Appendix I for key to abbreviations.

with that of the forelimb of an anuran or squamate reptile. In addition, he compared the forelimb of urodeles with that of the rhipidistian *Saurip-terus* (Figs. 20–21), which in Jarvik's (1964:69) opinion was "most probably a porolepiform." Andrews and Westoll (1970) placed *Sauripterus* within the Rhizodontida, which is related more closely to osteolepiform than to porolepiform rhipidistians (Vorobyeva and Schultze, this volume).

The discovery of the endoskeleton in the pectoral fin of *Glyptolepis groenlandica* (Jarvik, 1972:Fig. 59C), a porolepiform rhipidistian, indirectly supports the placement of *Sauripterus* within the Rhizodontida. The pectoral fin of porolepiform rhipidistians is elongated like that of dipnoans, and an axial skeleton comparable to the archipterygium of dipnoans was predicted (Gross, 1964). It is difficult, if not impossible, to derive the pentadactyl limb of urodeles from the axial elements in the pectoral fin of *G. groenlandica* (Fig. 22). Nonetheless, Jarvik (1981a) argued that the limbs of urodeles, at a very early stage with undifferentiated cartilage, could be compared to the fin shape of porolepiforms. However, in these early stages, no element is formed that can be compared to the axial elements in the pectoral fin of *Glyptolepis*. A branching pattern basic to the pentadactyl extremity of tetrapods is absent in the pectoral fin of *Glyptolepis*. In addition, Jarvik (1981a) accepted another difference between porolepiforms and urodeles; porolepiforms possess, as do other fishes, an articular condyle on the endoskeletal shoulder girdle which fits into an articular fossa in the proximal element of the pectoral endoskeleton (Figs. 17, 22). This differs from the articulation in osteolepiform and rhizodontiform rhipidistians and tetrapods (Figs. 19–21) in which the articular condyle of the most proximal element (humerus or femur) fits into an articular fossa of the endoskeletal girdle (Schultze, 1977a).

In her reinvestigation of the ontogeny of the amphibian limb, Saint-Aubain (1981) demonstrated, as did Steiner (1934, 1935), that the basic morphological (morphotype) plan of urodele and anuran limbs is the same (Figs. 23–24); in a proximal-to-distal sequence lie the radius (tibia), radiale (tibiale), centrale, and prepollex (prehallux). All authors draw the first line of their branching pattern through this sequence. All five digits are related to the ulna (fibula). Authors disagree only with regard to the branching pattern of the lines relating carpals to the five digits. The fusion of Carpale (Tarsale) 1 + 2 in urodeles accounts for the difference in Jarvik's (1964, 1965) arrangements of lines for urodeles as contrasted to anurans. Ontogenetically, Carpalia (Tarsalia) 1 and 2 are formed separately—as is shown in the salamander *Ambystoma* by Saint-Aubain (1981); they fuse later in ontogeny, and the fusion of Carpalia 1 + 2 could be used as a synapomorphy for all urodeles. Saint-Aubain (1981:182) considered the line pattern drawn by so many authors "much too ques-

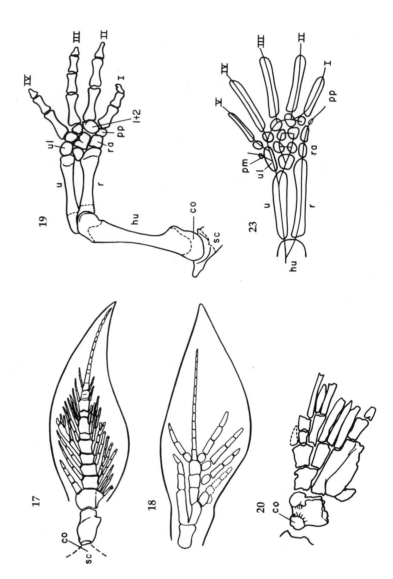

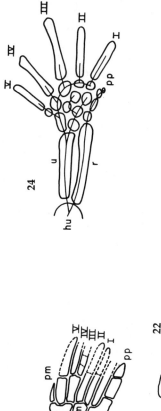
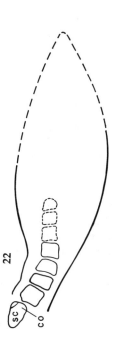

Figs. 17–24. Left anterior appendages of sarcopterygians. (17) Dipnoan *Neoceratodus forsteri* (after Howes, 1887:Pl. 1, Fig. 5). (18) Urodele hand as biserial archipterygium of *Neoceratodus* type (after Holmgren, 1933:Fig. 37). (19) Forelimb of the hynobiid urodele *Batrachuperus mustersi*, CAS 147041 (courtesy of R. Cloutier). (20) Rhizodont crossopterygian *Sauripterus taylori* (after Andrews and Westoll, 1970:Fig. 15). (21) Interpretation of pectoral fin of *S. taylori* by Jarvik (1964:Fig. 23A). (22) Porolepiform *Glyptolepis groenlandica* (after Jarvik, 1980:Fig. 200C). (23) Morphotype of urodele forelimb (after Saint-Aubain, 1981:Fig. 8) with metapterygial axis (after Shubin and Alberch, 1986). (24) Morphotype of anuran forelimb (after Saint-Aubain, 1981:Fig. 19) with metapterygial axis (after Shubin and Alberch, 1986). See Appendix I for key to abbreviations.

45

tionable to be homologized one by one with the radii of a hypothetical fin." She accepted lines only in the limbs of urodeles in which the elements differentiate from strings of mesenchyme; in anurans, each element appears separately. Because the basic pattern (morphotype) of the limb elements of urodeles and anurans is identical (Figs. 23–24), Saint-Aubain correctly concluded that the various limb elements of the two groups are homologous, and that it is improbable that the limbs of urodeles and anurans developed twice. Although not all limb components may be homologous on a one-to-one basis to the elements in the fins of "rhipidistian" fishes, the basic branching pattern (i.e., humerus [femur] to radius [tibia] and ulna [fibula], to ulnare [tibiale] and intermedium) is found only in osteolepiform and rhizodontiform "rhipidistian" fishes, and not in porolepiforms.

Remarks

The differences postulated by Jarvik (1942, 1972, 1981a) separating porolepiforms from osteolepiforms are to some extent differences in proportions, as shown by Kulczycki (1960), Worobjewa (1975), and Vorobyeva (1977a,b, 1980). The most devastating evidence against a wide separation of porolepiforms from osteolepiforms has been the discovery of a combination of characters of both groups in one fish, *Youngolepis* (Chang, 1982). Chang (1982, this volume) demonstrated that the fenestra ventrolateralis of *Youngolepis* and porolepiforms is covered ventrally by the vomer, premaxilla, and maxilla (Fig. 30). This agrees with the functional aspect of a nasal capsule with two external nasal openings, one incurrent and one excurrent. The lack of a choana and the pattern of bones in the pectoral fin make it unlikely that porolepiforms are closely related to any tetrapod group.

Dipnoan-Tetrapod Sister-Group Relationship

In reasserting a close relationship between dipnoans and tetrapods (Fig. 25), Rosen et al. (1981) emphasized the existence of homologous structures in the nasal region and the appendages. Their initial idea for such a relationship was based on observations of the palate of the Late Devonian lungfish *Griphognathus* in which more bones occur in the anterior part of the palate than in other dipnoans (Miles, 1977; Campbell and Barwick, 1984). Campbell and Barwick (1984) considered *Griphognathus* to be a highly derived dipnoan "with neomorphic bones formed in response to an unusual mode of feeding" in the palate, whereas Rosen et al. (1981) treated *Griphognathus* as a primitive dipnoan.

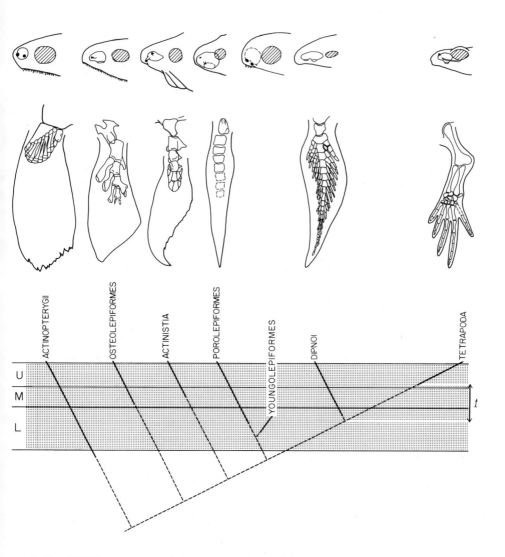

Fig. 25. Dipnoan-tetrapod sister-group relationship, as proposed by Rosen et al. (1981). Top row of figures illustrates the left nasal capsule with openings of (*from left to right*) an actinopterygian with two external nasal openings; an osteolepiform with one external and one internal nasal opening; an actinistian with two external nasal openings; a porolepiform and a youngolepiform with two external nasal openings and ventrally closed ventrolateral fenestra; a dipnoan with two palatal openings; and an anuran with one external and one internal nasal opening. Lower row of figures depicts left anterior appendage of (*from left to right*) an actinopterygian condensed endoskeleton of pectoral fin; an osteolepiform pectoral fin with radials on preaxial side only and with tetrapod-like articulation between scapulocoracoid and first axial element; an actinistian and porolepiform pectoral fin with axially arranged elements and fishlike articulation between scapulocoracoid and first axial element; a dipnoan pectoral fin with archipterygium and fishlike articulation between scapulocoracoid and first axial element; and a tetrapod forelimb with pentadactyl hand. Stipples indicate Devonian time: U = Late, M = Middle, L = Early Devonian; t = 30 million years between common ancestor with dipnoans and first fossil record of tetrapods.

47

Nasal Region

In dipnoans, both nasal openings are located inside the mouth in the anterior part of the palate. Even in the reinterpretation of the palatal and marginal bones by Rosen et al. (1981), in which they homologize these elements to those of tetrapods, the anterior nasal opening lies lingual to the "premaxilla" and "maxilla." In dipnoans, unlike tetrapods, there is no external nasal opening (fenestra exonarina) outside the outer dental arcade above the premaxilla and maxilla. Furthermore, dipnoans lack the elements that normally surround this opening—i.e., the lacrimal and nasal. Consequently, the nasal capsule cannot serve as a passage for water or air from the outside to the inside of the mouth cavity while the mouth is closed (Broman, 1939; Panchen and Smithson, 1987). Dipnoans have a unique adaptation—grooves on the ventral side of the snout that permit water to be drawn toward the anterior nasal openings inside the mouth when the mouth is closed. Furthermore, the posterior nasal opening is in a different position relative to the surrounding bones than is the choana of tetrapods. Attempts to rename the palatal bones of *Griphognathus* (Figs. 26–28) by Rosen et al. (1981) were unsuccessful; the authors did not demonstrate that the inferred position of the posterior nasal opening in *Griphognathus* was homologous to that of tetrapods. This opening is placed between the posterior end of "maxilla" and "dermopalatine," whereas the "vomer" and "premaxilla" are far anterior. The "maxilla" (= ectopterygoid of Miles) and "dermopalatine" (= dermopalatine 2 of Miles) are separate bones in Miles's 1977 reconstruction (Fig. 26; Miles, 1977:Fig. 6), but not in the drawing of the same specimen by Rosen et al. (1981:Fig. 7B) nor in the holotype (Fig. 28) or in specimens figured by Campbell and Barwick (1984:Fig. 2, 1987:Fig. 8 = Fig. 28, herein). These differences among specimens indicate great intraspecific variability in the numbers and shapes of these bones.

One can invoke functional reasons (Campbell and Barwick, 1984) to explain the variation in the number, shapes, sizes, and positions of palatal bones of *Griphognathus* (Figs. 26–28), which sometimes resembles the pattern around the choana of tetrapods. The relative position of the palatal bones in *Griphognathus* also demonstrates that they are not homologous to those of tetrapods (Schultze, 1987). Jarvik (1942) demonstrated (on the basis of a survey of older literature) that the cartilages surrounding the nasal capsule in Recent dipnoans are homologous to dorsal cartilages close to the external nasal openings in tetrapods and not to those surrounding the choana. He concluded that, on the basis of the available evidence (e.g., position of cartilages, embryological development of the nasal openings, course of nerves around the nasal capsule), dipnoans possess incurrent (anterior) and excurrent (posterior) palatal nasal openings that are homologous to the two external nasal openings in

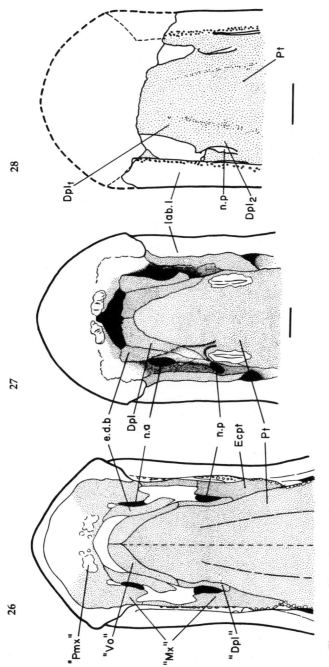

Figs. 26–28. Ventral views of anterior part of skull of *Griphognathus whitei* showing variation in bone pattern. (26) After Miles (1977:Fig. 6) with terminology of Rosen et al. (1981) at left. (27) After Campbell and Barwick (1987:Fig. 18). (28) Holotype with bones fused around the posterior nasal opening (after Miles, 1977:Fig. 80). See Appendix I for key to abbreviations.

other fishes. Expressing it in terms of Panchen and Smithson (1987), the posterior nasal opening of dipnoans does not fulfill the criterion of functional similarity and of topographic similarity.

With the description of the Early Devonian fish *Diabolepis* (Chang and Yu, 1984) and its placement as the sister-group to dipnoans (Chang and Yu, 1984; Maisey, 1986; Chang, this volume), the interpretation of the posterior nasal opening of dipnoans as the choana became obsolete (Maisey, 1986), or as Forey (1987) interpreted it, a parallelism. *Diabolepis* has two external nasal openings outside of the premaxilla and maxilla. The anterior nasal opening is bordered by the premaxilla and the ossified snout; the posterior nasal opening is bordered by the premaxilla, the ossified snout, and supposedly the lacrimal and maxilla, neither of which is preserved. The medial part of the fenestra ventrolateralis is covered by the premaxilla and vomer as in *Youngolepis* (Fig. 30). It follows that possession of a choana is not a synapomorphy uniting dipnoans and tetrapods. The anterior and posterior nasal openings of dipnoans do not lie in the same position relative to surrounding bones as in tetrapods, and the sister-group of dipnoans, *Diabolepis*, does not have a nasal opening in the palate.

Appendages

The interpretation of the archipterygium of dipnoans as the precursor of the endoskeleton of the tetrapod limb is fraught with problems. Two primary joints occur in lungfishes—one between the shoulder girdle and the proximal element, and another between the proximal element and the next distal subbasal element. These joints are not comparable to those in tetrapods. The proximal joint is typical of fish, which possess a ball on the shoulder girdle that articulates in a socket formed by the proximal element ("fish joint" of Wiedersheim, 1892). This condition characterizes porolepiforms, actinistians, and other fishes, whereas the reverse (socket on shoulder girdle, ball on proximal element) characterizes osteolepiforms, panderichthyids, and tetrapods (see above).

In order to support the close relationship of lungfish and tetrapods, Rosen et al. (1981) employed an early ontogenetic stage of *Neoceratodus* in which joints are undeveloped. In this stage, the second axial element (subbasal element) shows two chondrification centers that Rosen et al. (1981) homologized with the radius and ulna of tetrapods; they contrasted these centers to the situation in osteolepiforms. The sizes of the radius and ulna of tetrapods are approximately equal, whereas the radius (their "ulna") in early ontogenetic stages of *Neoceratodus* (Rosen et al., 1981:Fig. 31) is 1.7 times larger than the ulna (their "radius"); this is not significantly different from the ratio in *Eusthenopteron*, in which the radius is 2.0 to 2.3 times larger than the ulna (Andrews and Westoll, 1970:Figs. 6, 8).

The occurrence of paired chondrifications in the second axial element in the early ontogeny of *Neoceratodus* was interpreted as plesiomorphic for all sarcopterygians (Maisey, 1986; Shubin and Alberch, 1986). Nevertheless, such a structure is unknown in porolepiforms, actinistians, and primitive lungfishes; however, paired chondrifications in the second axial element may occur in the early ontogeny of all sarcopterygians (Shubin and Alberch, 1986). The metapterygial axis of the sarcopterygian fin endoskeleton is homologous to the digital arch in the developing tetrapod limb (Shubin and Alberch, 1986). "Gregory's Ulna Pyramid" (Westoll, 1943) of *Eusthenopteron* correlates well with the developmental sequence of branching and segmentation in the tetrapod limb. The archipterygium of dipnoans carries postaxial radials serially arranged in one-to-one correspondence with the axial segments; these postaxial radials may have been formed de novo (Shubin and Alberch, 1986) and thereby distinguish dipnoans from other sarcopterygians. Holmes (1985) and Maisey (1986) argued convincingly that the rotation of the pectoral appendages in dipnoans and *Latimeria* does not occur in tetrapods. The postaxial and preaxial sides of the tetrapod forelimb, along with the internal structure, are directly comparable to those of the osteolepiform pectoral fin. Holmes (1985), Maisey (1986), and Schultze (1987) declined to accept as synapomorphies certain other characters of the appendages used by Rosen et al. (1981), because those characters also appear in other groups or are distributed sporadically throughout the dipnoans.

Remarks

Forey (1987) attempted to reconfirm the phylogenetic arrangement of Rosen et al. (1981) using a slightly different set of characters. He added soft anatomical characters that Rosen et al. (1981) had united into a single character because soft structures could not be determined in fossil forms. Forey's approach employed features that could be demonstrated only in the derived, terminal members of various lineages. They could not be shown to exist in the primitive members of these lineages and might represent convergences. Burggren and Johansen (1987) and Northcutt (1987) showed that lepidosirenids and tetrapods are more similar to each other in some features of soft anatomy than either is to *Neoceratodus*, the primitive extant dipnoan. Forey (1987:89) was "unwilling to accept that at present," and he argued that features that place lepidosirenids closer to tetrapods than either to *Neoceratodus* would "deny a group Dipnoi." The simplest and most parsimonious conclusion would be to explain the situation as parallelism, as suggested by Burggren and Johansen (1987) and Northcutt (1987), produced by paedomorphosis. This clearly demonstrates the perils of using features of extant terminal members of groups with long geological histories. As succinctly expressed by Maisey (1986:242), "Any discussion of actinopterygian, sarcopterygian, chondrichthyan, or tetrapod interrelationships involves characters acquired at least 350–400 million years ago. Under these circumstances we should take particular care to ascertain whether the putative synapomorphies of such antique higher taxa are not merely artifacts whose significance has become exacerbated with the passage of geologic time." In the specific case of similarities in features of lepidosirenid lungfishes and tetrapods, the parallelism simply may reflect terrestrial adaptations of these lung-fishes.

Finally, the renewal of the old hypothesis of a dipnoan-tetrapod inter-relationship by Rosen et al. (1981) has triggered a reexamination of the characters relating tetrapods to sarcopterygian fishes. All other recent analyses dealing with the subject (Jarvik, 1981b; Schultze, 1981, 1987; Holmes, 1985; Maisey, 1986; Panchen and Smithson, 1987) agree that osteolepiforms are the sister-group of tetrapods. Even Forey (1987:85), one of the authors of Rosen et al. (1981), now concurs "that the extinct Panderichthyidae (*Elpistostege, Panderichthys*) may be more closely related to tetrapods than dipnoans." The primary source of disagreement now involves the position of the dipnoans within the sarcopterygians. They are placed variously as the sister-group of porolepiforms on the basis of features found in *Diabolepis* and *Youngolepis* (Maisey, 1986; Chang, this volume), or as the sister-group of crossopterygian fishes and tetrapods (Schultze, 1987), but this question is not addressed herein.

Sarcopterygian-Tetrapod Sister-Group Relationship

During her visit in the United States in 1985, Chang Mee-Mann hy-pothesized that dipnoans are the closest relatives of porolepiforms. Her views derive largely from her study of two Chinese forms, *Diabolepis*, the most primitive dipnoan, and *Youngolepis*, the most primitive porolepi-form (Chang, this volume; cf. Maisey, 1986). She considers tetrapods to be the sister-group of all sarcopterygian fishes (Fig. 29), and of these fishes, osteolepiforms to be the most plesiomorphic group having the most similarities to tetrapods. It follows from Chang's hypothesis that the common ancestor of sarcopterygian fishes and tetrapods would have possessed a single external nasal opening combined with a choana, and a tetrapod-like articulation of the humerus with the shoulder girdle, along with a radiating endoskeleton. Alternatively, these features could have been acquired in parallel in osteolepiforms and tetrapods.

Nasal Region

Chang (this volume) convincingly demonstrates that porolepiforms, *Youngolepis* (Fig. 30), and *Diabolepis* do not possess a choana and instead have two external nasal openings, as in actinistians and actinopteryg-

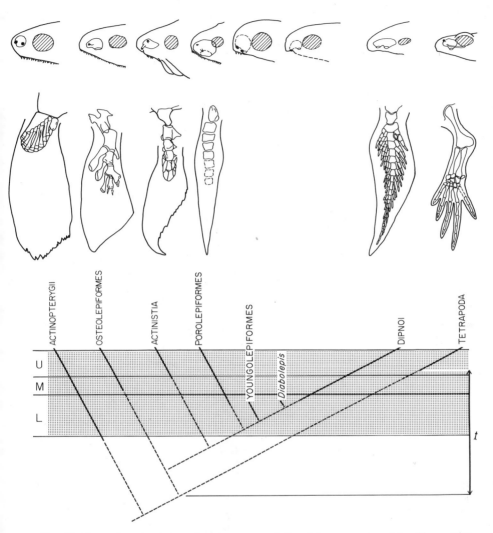

Fig. 29. Sarcopterygian-tetrapod sister-group relationship, as proposed by Chang (this volume). Top row of figures shows the left nasal capsule with openings of (*from left to right*) an actinopterygian with two external nasal openings; an osteolepiform with one external and one internal nasal opening; an actinistian with two external nasal openings; a porolepiform and a youngolepiform with two external nasal openings and ventrally closed ventrolateral fenestra; *Diabolepis* with two ventrally positioned external nasal openings and ventrally closed ventrolateral fenestra; a dipnoan with two palatal openings; and an anuran with one external and one internal nasal opening. Lower row of figures illustrates the left anterior appendage of (*from left to right*) an actinopterygian condensed endoskeleton of pectoral fin; an osteolepiform pectoral fin with radials on preaxial side only and with tetrapod-like articulation between scapulocoracoid and first axial element; an actinistian and a porolepiform pectoral fin with axially arranged elements and fishlike articulation between scapulocoracoid and first axial element; a dipnoan pectoral fin with archipterygium and fishlike articulation between scapulocoracoid and first axial element; and a tetrapod forelimb with pentadactyl hand. Stipples indicate Devonian time: U = Late, M = Middle, L = Early Devonian; t = 60 million years between common ancestor with sarcopterygian fishes and first fossil record of tetrapods.

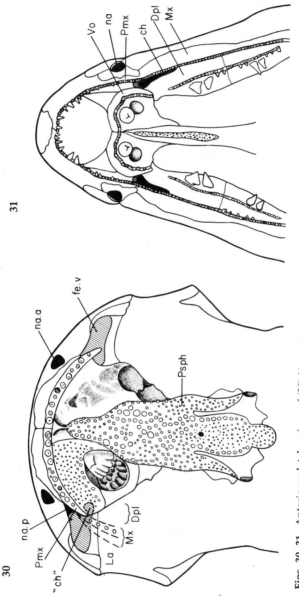

Figs. 30–31. Anterior palatal regions of (30) *Youngolepis praecursor* (after Chang 1982:Figs. 10, 7B), and (31) *Panderichthys rhombolepis*. See Appendix I for key to abbreviations.

54

ians. Two nasal openings (incurrent and excurrent) are required to permit water flow over the olfactory tissue. Porolepiforms and *Youngolepis* possess a small anterior fenestra endonarina and a large posterior fenestra ventrolateralis. Most of the latter is covered by the exoskeleton, specifically the ventral portion by the vomer; only the lateral portion of the opening for the posterior nostril is not covered by bone. Osteolepiforms, panderichthyids, and tetrapods have only one external nasal opening on a side. In addition, another nasal opening, the choana, appears in the palate. Chang (this volume) argues that the existence of a choana has not been proven in osteolepiforms. Nevertheless, the choana is present in *Eusthenopteron;* Jarvik's (1980) reconstruction of the palate of this taxon does not differ from that of the acid-prepared specimen of Rosen et al. (1981), save for the configuration of the anterior part of the dermopalatine. Presence of the choana is independent of the function of the movable anterior end of the dermopalatine.

The long slit-shaped choana in *Panderichthys* (Fig. 31) is even more convincing than the choana of *Eusthenopteron*. Specimens of *Panderichthys* were discovered in Late Devonian clays in Latvia. Preservation in these easily removed clays is excellent; hence, a connection between the external nasal opening, nasal capsule, and choana is evident, and there the snout is immovable relative to the palate because the palate is rigid (Vorobyeva and Schultze, this volume).

A second opening of the nasal capsule is necessary for functional reasons, and the existence of a choana can be documented at least in *Eusthenopteron* and *Panderichthys*. Therefore, osteolepiforms and panderichthyids share with tetrapods the existence of one external nasal opening combined with a choana.

Appendages

Chang (this volume) discards all previous attempts to relate the internal fin structure of any sarcopterygian with the limbs of tetrapods, basing her argument on the findings of the ontogenetic study by Shubin and Alberch (1986). The first appearance of a simple proximal element is shared by all sarcopterygian fishes and tetrapods. Subsequently, the limb skeleton bifurcates with the formation of two distal elements preserved in adult osteolepiforms, panderichthyids, and tetrapods. This bifurcation disappears in dipnoans and has not been shown to occur in actinistians and porolepiforms. On the basis of parsimony and contrary to Chang (this volume), I favor the view that only osteolepiforms, panderichthyids, and tetrapods share the bifurcation of two distal elements, and that the metapterygial axis is the plesiomorphous feature in porolepiforms, actinistians, and dipnoans. Chang's (this volume) hypothesis postulates a tetrapod-like common ancestor for sarcopterygian fishes and tetrapods. The metapterygial fin would have been more tetrapod-like with a branching pattern like that of *Eusthenopteron;* within the sarcopterygian fishes the fin skeleton becomes more elongated. This sequence represents a reversal from a fishlike shoulder articulation in actinopterygians to a tetrapod-like articulation in the common ancestor of sarcopterygians and tetrapods, and back to a fishlike articulation in actinistians, porolepiforms, and dipnoans.

Remarks

Reversals such as those described above also are necessary to explain many cranial as well as postcranial features. Among the paired skull roof bones, two reversals are needed to account for the presence of parietals and postparietals in actinopterygians, three for the frontals, parietals, and postparietals in the common ancestor of sarcopterygians and tetrapods, and two for the parietals and postparietals in actinistians, porolepiforms, and *Diabolepis*. Phylogenetic sequences from two to one to two squamosals and external nasal openings, and from absence to presence to absence of a choana and vertebrae are implied. Another feature, plicidentine, has an even more complicated history in Chang's cladogram. The common ancestor had folded plicidentine; within sarcopterygian fishes, plicidentine is lost then regained (and highly complicated in porolepiforms) and is never shown to exist in dipnoans. The four infradentaries of osteolepiforms, porolepiforms, and tetrapods are not in the same relative position or sequence as the external bones of the lower jaw of dipnoans, with which they have been homologized. The lower jaw of *Diabolepis* (assuming that the lower jaws are correctly associated with the skulls) is more similar to those of dipnoans than to those of any rhipidistian. This is one of the features that distinguishes dipnoans from the group composed of crossopterygians and tetrapods. Another such feature is the intracranial joint. Dipnoans lack an intracranial joint, but an ontogenetic division of the endocranium may occur in tetrapods (Roček, 1986). In *Ichthyostega,* the notochord extends forward below the occipital and otic regions of the braincase to the ventral fissure or to the region where the ventral fissure could be expected to lie. This arrangement is found in osteolepiforms and porolepiforms but not in dipnoans.

A sister-group relationship among all sarcopterygian fishes and tetrapods would imply that the common ancestor was more similar to a tetrapod than it was to the derived, fishlike descendants. Such an arrangement would entail numerous character reversals that seem most unlikely.

Osteolepiform-Tetrapod Sister-Group Relationship

The osteolepiform-tetrapod sister-group relationship (Fig. 32) was proposed initially by Cope (1892) and became the most widely accepted

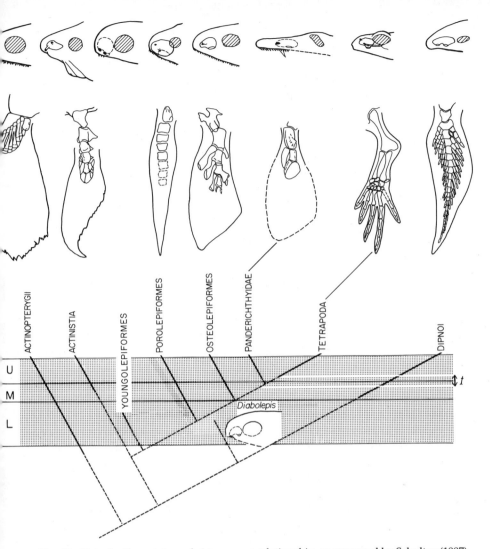

Fig. 32. Osteolepiform-tetrapod sister-group relationship, as proposed by Schultze (1987). Top row of figures illustrates the left nasal capsule with openings of (*from left to right*) an actinopterygian and actinistian with two external nasal openings; a youngolepiform and a porolepiform with two external nasal openings and ventrally closed ventrolateral fenestra; an osteolepiform, a panderichthyid and a tetrapod with one external and one internal nasal opening (choana); a dipnoan with two palatal openings; and *Diabolepis* (*in lower part of figure*) with two ventrally positioned external nasal openings and ventrally closed ventrolateral fenestra. Lower row of figures depicts the left anterior appendage of (*from left to right*) an actinopterygian condensed endoskeleton of pectoral fin; an actinistian and a porolepiform pectoral fin with axially arranged elements and fishlike articulation between scapulocoracoid and first axial element; an osteolepiform and panderichthyid pectoral fin with radials on preaxial side only and with tetrapod-like articulation between scapulocoracoid and first axial element; a tetrapod forelimb with pentadactyl hand; and a dipnoan pectoral fin with archipterygium and fishlike articulation between scapulocoracoid and first axial element. Stipples indicate Devonian time: U = Late, M = Middle, L = Early Devonian; t = 10 million years between common ancestor with panderichthyids and first fossil record of tetrapods. For characters supporting branches from Porolepiformes to Tetrapoda, see Vorobyeva and Schultze, this volume:Fig. 15.

hypothesis until reestablishment of the older hypothesis of dipnoan-tetrapod relationship by Rosen et al. (1981). Subsequently, the osteolepi-form-tetrapod relationship has been reexamined and reconfirmed by Schultze (1981), Holmes (1985), and Maisey (1986). Schultze (1969, 1970), Worobjewa (1973), Schultze and Arsenault (1985), and Vorobyeva and Schultze (this volume) have singled out the panderichthyids as the clos-est sister-group of tetrapods.

Nasal Region

Panderichthyids and at least some osteolepiforms possess a true cho-ana surrounded by a premaxilla, maxilla, vomer, and dermopalatine (Figs. 30–31), as in tetrapods. The choana of *Panderichthys* forms an anteroposteriorly elongated slit. A straw can be passed through the choana into the nasal capsule and out through the external nasal open-ing. The position of the external nasal opening is ventral near the margin of the upper jaw, as is typical for primitive tetrapods (Panchen, 1967). I discuss above the rationale for considering the palatal opening of *Eus-thenopteron* to be a choana. I would argue as well that all osteolepiforms have a choana (contra Chang, this volume) because functionally they require an excurrent opening of the nasal capsule to complement the small external incurrent nasal opening. Chang (this volume) may be correct that preservation in most cases is insufficient to determine the size and shape of the choana, and that the size of the choana is overempha-sized in some reconstructions. Still, the vomer does not align laterally with the maxilla, and the dermopalatine does not extend between vomer and maxilla in osteolepiforms. Therefore, space for an excurrent nasal opening (i.e., the choana) is available. In addition, the position of the fenestra endochoanalis (homologous to the fenestra ventrolateralis of porolepiforms) is ventral in osteolepiforms rather than ventrolateral to lateral, as the fenestra ventrolateralis is in porolepiforms. Contrary to Chang (this volume), I cannot imagine that reconstructing the maxilla and dermopalatine on the palate of *Thursius estonicus* or *Megalichthys hibberti* would close the fenestra endochoanalis. The choana would re-main as in *Megalichthys* (reconstructed by Schultze, 1974:Fig. 9), in which the fenestra endochoanalis reaches farther posteriorly than it does in Jarvik's (1966:Fig. 19) reconstruction. The fenestra endochoanalis has a posterior extension in *Gogonasus andrewsi* (Long, 1985) also.

The presence of a fenestra ventrolateralis unites porolepiforms, os-teolepiforms, panderichthyids, and tetrapods; in the latter three groups, the fenestra is more ventral and is termed the *fenestra endochoanalis*. The palatal part of the fenestra ventrolateralis is covered by the vomer and possibly the dermopalatine, so that there is no choana present in por-olepiforms. In osteolepiforms and panderichthyids, the internal opening is functionally required as an excurrent opening. It is present as space between the vomer and premaxilla and maxilla, and it continues posteriorly, between the dermopalatine and maxilla. The choana has the same position in early tetrapods. The ventromarginal position of the external nasal opening unites panderichthyids and tetrapods.

Appendages

The phylogenetic arrangement of the sarcopterygians as presented in Figure 32 postulates an archipterygium with a fishlike articulation between the shoulder girdle and humerus, and a common ontogenetic development by segmentation and branching of axial elements (metapterygial axis homologous to digital arch in tetrapods) as features for the common ancestor. The axial elements are more robust in sarcopterygians (i.e., actinistians, porolepiforms, osteolepiforms, panderichthyids, and tetrapods) than in actinopterygians. The length and the number of elements in the axis and pre- and postaxial radials also are reduced. A tetrapod-like articulation between the shoulder girdle (with fossa) and humerus (with ball) and the presence of two bones as second element are features common to osteolepiforms, panderichthyids, and tetrapods. Further division into carpals, metacarpals, and phalangeal bones and the articulation between them is unique to tetrapods (Panchen and Smithson, 1987).

Remarks

The panderichthyid-tetrapod relationship is supported by other characters such as the skull-roof pattern involving three pairs of bones along the midline, the position of frontals and parietals in relation to orbits, the lack of external intracranial joint, a dorsoventrally flattened skull with high, dorsally situated orbits (Schultze, 1987), labyrinthodont plicidentine (Schultze, 1969, 1970), and a squamosal that is separated from the maxilla, and other features (cf. Vorobyeva and Schultze, this volume).

The existence of close similarities between panderichthyids and tetrapods was accepted by Forey (1987), one of the coauthors of Rosen et al. (1981). Thus, in Forey's cladogram (1987:Fig. 2), panderichthyids "bridge the phylogenetic distance" occupied by actinistians, porolepiforms, and dipnoans between osteolepiforms and tetrapods. This suggests to me that his cladogram would collapse if panderichthyids were included as the sister-group of tetrapods, and a scheme of relationships similar to that in Figure 32 would result.

CONCLUSIONS

Four different phylogenetic arrangements derived from analyses of the same features have been discussed. Each author interpreted the characters such that they favored his or her preferred relationship of tetrapods. Jarvik emphasized a distinction between eutetrapods and urodeles and the similarities between urodeles and porolepiforms. Rosen et al. downplayed the significance of similarities between osteolepiforms and tetrapods, and then interpreted the features of dipnoans in tetrapod fashion. Chang held to the view that all sarcopterygian fishes, including dipnoans, lack a choana. Thus, in her opinion, the history of tetrapods is separate from that of all other sarcopterygians. Vorobyeva and Schultze argued that the internal opening on the palate of osteolepiforms and panderichthyids is a true choana, and that only in osteolepiforms is the internal structure of the fins comparable to tetrapods.

Except for Jarvik, all of these workers used similar methods of phylogenetic analysis. A more elaborate cladistic analysis ("parsimony analysis," cf. Forey, 1987) would not alter these basic differences, which also exist in the interpretation of other characters and their polarities. The differences in the postulated relationships are based on differences in the evaluation of the same characters—different codings of the same characters. For example, the dipnoans are closely related to tetrapods if their posterior internal nasal opening is interpreted as a choana (Rosen et al., 1981), whereas they are unrelated to tetrapods if the internal nasal openings of dipnoans are interpreted as unique (Jarvik, 1942; Schultze, 1987; Chang, this volume). The same is true of most other characters. The entire question of relationships turns on an evaluation of similarities and dissimilarities of features in order to assess their homologies accurately. The most frequently used criterion for homology is topography—i.e., the spatial relationship of an element to its surrounding structures (Remane, 1952). Accordingly, a palatal opening would be homologous to the choana of tetrapods only if it is surrounded by the same bones as found in primitive tetrapods. The opening also must be associated with a single external nasal opening. The sister-group relationship of *Diabolepis* with dipnoans precludes the possibility of using the choana as a synapomorphy of dipnoans and tetrapods, because *Diabolepis* possesses two external but no internal nasal opening. Acceptance of panderichthyids as the sister-group of tetrapods (Forey, 1987:89) implies one of two things. Either panderichthyids also possess the many characters that place dipnoans as the sister-group of tetrapods (Forey, 1987:85), or a rearrangement of all sarcopterygian groups is required to explain the similarities between osteolepiforms and panderichthyids. In his cladogram, Forey (1987:Fig. 2) clearly shows that panderichthyids bridge the phylogenetic distance between osteolepiforms and tetrapods; thus, a rearrangement of the groups obviously is necessary.

The inadequacy of the fossil record frequently is invoked as a reason not to use fossils in the reconstruction of phylogenetic relationships. Each of the four schemes of tetrapod relationships reviewed here requires a different temporal gap in the fossil record. Jarvik (1981a) thought that all important evolutionary events in vertebrate history occurred in the Precambrian—a view that requires a gap in the fossil record of tetrapods of more than 300 million years. On the other hand, the unrecorded fossil record of eutetrapods goes back only about 20 million years to their common ancestor with osteolepiforms in Jarvik's scheme (Fig. 7). The minimum duration of the unrecorded fossil record of tetrapods in Chang's (this volume) scheme would be 60 million years (Fig. 29). Warren et al. (1986) described impressions from Upper Silurian or Early Devonian rocks of Australia as probable tetrapod trackways. These impressions are irregular and do not show digit impressions; there is no difference between footprints of the hind and forelimbs. Thus, they are unconvincing as tetrapod trackways. Tetrapods were distributed worldwide at least by the Late Devonian as shown by their fossil record—*Ichthyostega* in Greenland (Säve-Söderbergh, 1932; Jarvik, 1952), *Tulerpeton* in Russia (Lebedev, 1984), and trackways in Australia (Warren and Wakefield, 1972) and Brazil (Leonardi, 1982, 1983). Thus, the age of the common ancestor of tetrapods and any sarcopterygian fish group must be at least Middle Devonian. The fossil record of panderichthyids extends into the earliest Late Devonian or late Middle Devonian so that the minimal time span for the gap in the fossil record of tetrapods is only 10 million years in the scheme of Schultze (Fig. 32). Rosen et al. (1981) have to reconcile a time span of at least 30 million years. In addition, each group has a long unrecorded fossil history in the scheme of Rosen et al. (1981) (Fig. 25). This is also the case in Chang's arrangement (Fig. 29) and especially in Jarvik's (Fig. 7). The traditionally accepted phylogenetic arrangement (Fig. 32) requires a relatively late dichotomy between tetrapods and their sister-group; thus, only a short time elapses between the origin of tetrapods and the beginning of their fossil record. The gap between the first record of actinistians in the Middle Devonian to the occurrence of the common ancestor of this group and other crossopterygians in the Late Silurian is problematic regardless of the phylogenetic pattern adopted. The earliest fossil record of sarcopterygians is in the Late Silurian. Only one known form could connect all osteichthyans—i.e., *Lophosteus* (Gross, 1969; Schultze, 1977b).

The comparison of phylogenetic schemes based on analysis of morphological characters of extant and fossil taxa indicates that, in fact, the fossil record is sufficiently complete to justify analysis of this material. The fragmentary nature of the fossil record often is used as a scapegoat. Thus, Jarvik (1964, 1981a, 1986), extrapolating on the basis of the slow speed of evolutionary changes between Devonian and Recent times, did not hesitate to place the origin of tetrapods in the Precambrian, despite the fact that there is no record of a gnathostome before the Silurian.

Rosen et al. (1981) argued that an analysis of relationship must be based first on extant forms, with fossils being introduced in the analysis secondarily. The incompleteness of the fossil record is a given. As a result, the fossils contribute nothing to the resolution of the phylogenetic scheme. In contrast, the phylogenetic arrangement preferred herein in which osteolepiforms are the sister-group of tetrapods (e.g., Maisey, 1986; Panchen and Smithson, 1987) coincides more closely with the fossil record and explains more satisfactorily the gap in the fossil record between the origin and discovery of tetrapods. As succinctly stated by Maisey (1984:49), "The fossil record seems to be sending a strong signal that the pattern of anatomical data accurately reflects ancestor-descendant relationships above the species level."

Acknowledgments I thank Colleen Z. Gregoire, Donna Stevens, and Anne Musser (Museum of Natural History, The University of Kansas) for preparation of the figures. I am grateful to Drs. J. Chorn and Linda Trueb (Museum of Natural History, The University of Kansas), who edited a draft of the manuscript, and Dr. P. Janvier (Muséum National d'Histoire Naturelle, Paris), who provided a thoughtful review of the essay. Thanks are extended to Jan Elder and Judy Wiglesworth (Word Processing Service, Division of Biology, The University of Kansas), who typed the manuscript.

APPENDIX I. Anatomical Abbreviations Used in the Figures and Text

c = centrale; CAS = California Academy of Sciences; ch = choana (= internal nasal opening); "ch" = covered ventral part of ventrolateral fenestra; co = condylus (ball articulation)

d.nl = nasolacrimal duct; div.l = diverticulum laterale; $Ddl_{1,2}$ = dermopalatine 1,2; "Dpl" = dermopalatine of Rosen et al. (1981)

Ecpt = ectopterygoid; e.d.b = extra dermal bone

fe.v = ventrolateral fenestra; fo.ap = foramen apicale; fo.or.l, m = foramen orbitonasale laterale, mediale

hu = humerus

im = intermedium; i.gl = intermaxillary gland

La = lacrimal; lab.l = labial lamina

m.c.n = nasal constrictor muscle; Mx = maxilla; "Mx" = maxilla of Rosen et al. (1981)

Na = nasal; na = external nasal opening; na.a,p = anterior, posterior external nasal opening; n.a,p = anterior, posterior nasal opening in palate of dipnoans; n.c = nasal cavity

pm = postminimus; Pmx = premaxilla; "Pmx" = premaxilla of Rosen et al. (1981); pp = prepollex; Psph = parasphenoid; Pt = pterygoid r = radius; ra = radiale; r.ext = ramus externus narium; R.l = lateral rostral; r.lat = ramus lateralis narium; r.med = ramus medialis narium; r.oph.fac = ramus ophthalmicus superficialis facialis; r.oph.pr = ramus ophthalmicus profundus; r.vent = ramus ventralis narium

sc = scapulocoracoid; Smx = septomaxilla; s.n = septum nasi

Te = tectal; t.in = truncus infraorbitalis

u = ulna; ul = ulnare

Vo = vomer; "Vo" = vomer of Rosen et al. (1981)

1 + 2 = fused Carpal 1 + 2; 1–4 = Carpals 1–4

I–V = Phalanges I–V

Literature Cited

Andrews, S. M., and T. S. Westoll. 1970. The postcranial skeleton of rhipidistian fishes excluding *Eusthenopteron*. Trans. R. Soc. Edinburgh, 68(12):391–489.

Badenhorst, A. 1978. The development and the phylogeny of the organ of Jacobson and the tentacular apparatus of Ichthyophis glutinosus (Linné). Ann. Univ. Stellenbosch, Ser. A2 (Sool.), 1(1):1–26.

Bischoff, T. L. W., von. 1840. Lepidosiren paradoxa. *Anatomisch untersucht und beschrieben*. Leipzig: Leopold Voss.

Broman, J. 1939. Passiert bei den Lungenfischen die Atemluft durch die Nasenhöhlen? Anat. Anz., 88:139–145.

Burggren, W. W., and K. Johansen. 1987. Circulation and Respiration in Lungfishes (Dipnoi). Pp. 217–236 *in* Bemis, W. E., W. W. Burggren, and N. E. Kemp (eds.), *The Biology and Evolution of Lungfishes*. J. Morphol., Suppl. 1.

Campbell, K. S. W., and R. E. Barwick. 1984. The choana, maxillae, premaxillae and anterior palatal bones of early dipnoans. Proc. Linn. Soc. New South Wales, 107:147–170.

Campbell, K. S. W., and R. E. Barwick. 1987. Paleozoic lungfishes—a review. Pp. 93–131 *in* Bemis, W. E., W. W. Burggren, and N. E. Kemp (eds.), *The Biology and Evolution of Lungfishes*. J. Morphol., Suppl. 1.

Chang M.-M. 1982. The braincase of *Youngolepis*, a Lower Devonian crossopterygian from Yunnan, south-western China. Doctoral dissertation. Stockholm: Dept. of Geology, Univ. of Stockholm.

Chang M.-M., and X. B. Yu. 1984. Structure and phylogenetic significance of *Diabolichthys separatus*. gen. et sp. nov., a new dipnoan-like form from the Lower Devonian of eastern Yunnan, China. Proc. Linn. Soc. New South Wales, 107:171–184.

Cope, E. D. 1892. On the phylogeny of the Vertebrata. Proc. Am. Philos. Soc., 30:278–281.

Forey, P. 1987. Relationships of lungfishes. Pp. 75–91 *in* Bemis, W. E., W. W. Burggren, and N. E. Kemp (eds.), *The Biology and Evolution of Lungfishes*. J. Morphol., Suppl. 1.

Fox, H. 1965. Early development of the head and pharynx of *Neoceratodus* with a consideration of its phylogeny. J. Zool. London, 146:470–554.

Gaffney, E. S. 1979. Tetrapod monophyly: a phylogenetic analysis. Bull. Carnegie Mus. Nat. Hist., 13:92–105.

Gardiner, B. G. 1980. Tetrapod ancestry: a reappraisal. Pp. 177–185 *in* A. L. Panchen (ed.), *The Terrestrial Environment and the Origin of Land Vertebrates*. Newcastle-upon-Tyne: Syst. Assoc. Spec. Vol. 15.

Gardiner, B. G. 1984. The relationships of the palaeoniscid fishes, a review based on new specimens of *Mimia* and *Moythomasia* from the Upper Devonian of Western Australia. Bull. Br. Mus. Nat. Hist. Geol., 37(4):173–428.

Godfrey, S. J., A. R. Fiorillo, and R. L. Carroll. 1987. A newly discovered skull of the temnospondyl amphibian *Dendrerpeton acadianum* Owen. Can. J. Earth Sci., 24:796–805.

Gregory, W. K., and H. C. Raven. 1941. Studies on the origin and early evolution of paired fins and limbs. Part III. On the transformation of pectoral and pelvic paddles of *Eusthenopteron* type into pentadactylate limbs. Ann. New York Acad. Sci., 42:313–353.

Gross, W. 1964. Polyphyletische Stämme im System der Wirbeltiere? Zool. Anz., 173:1–22.

Gross, W. 1969. *Lophosteus superbus* Pander, ein Teleostome aus dem Silur Oesels. Lethaia, 2:15–47.

Holmes, E. B. 1985. Are lungfishes the sister group of tetrapods? Biol. J. Linn. Soc., 25:379–397.

Holmgren, N. 1933. On the origin of the tetrapod limb. Acta Zool. Stockholm, 14:195–292.

Holmgren, N. 1939. Contribution to the question of the origin of the tetrapod limb. Acta Zool. Stockholm, 20:89–124.

Holmgren, N. 1949a. Contributions to the question of the origin of tetrapods. Acta Zool. Stockholm, 30:459–484.

Holmgren, N. 1949b. On the tetrapod limb problem again. Acta Zool. Stockholm, 30:485–508.

Howes, G. B. 1887. On the skeleton and affinities of the paired fins of *Ceratodus*, with observations upon those of the elasmobranchii. Proc. Zool. Soc. London, 1887:3–26.

Jarvik, E. 1942. On the structure of the snout of crossopterygians and lower gnathostomes in general. Zool. Bidr. Uppsala, 21:235–675.

Jarvik, E. 1952. On the fish-like tail in the ichthyostegid stegocephalians with descriptions of a new stegocephalian and a new crossopterygian from the Upper Devonian of East Greenland. Medd. Groenland, 114:1–90.

Jarvik, E. 1964. Specializations in early vertebrates. Ann. Soc. R. Zool. Belgique, 94(1):11–95.

Jarvik, E. 1965. Die Raspelzunge der Cyclostomen und die pentadactyle Extremität der Tetrapoden als Beweise für monophyletische Herkunft. Zool. Anz., 175:101–143.

Jarvik, E. 1966. Remarks on the structure of the snout in *Megalichthys* and certain other rhipidistid crossopterygians. Ark. Zool. Stockholm, 19(2):41–98.

Jarvik, E. 1972. Middle and Upper Devonian Porolepiformes from East Greenland with special reference to *Glyptolepis groenlandica* n. sp. and a discussion on the structure of the head in the Porolepiformes. Medd. Groenland, 187:1–307.

Jarvik, E. 1980. *Basic Structure and Evolution of Vertebrates*. Vol. 1. New York and London: Academic Press.

Jarvik, E. 1981a. *Basic Structure and Evolution of Vertebrates*. Vol. 2. New York and London: Academic Press.

Jarvik, E. 1981b. Lungfishes, Tetrapods, Paleontology, and Plesiomorphy.—Donn E. Rosen, Peter L. Forey, Brian G. Gardiner, and Colin Patterson. 1981. Syst. Zool. 30:378–384. [Review].

Jarvik, E. 1986. The origin of the Amphibia. Pp. 1–24 *in* Roček, Z. (ed.), *Studies in Herpetology*. Prague: Charles Univ.

Jessen, H. 1968. *Moythomasia nitida* Gross und *M.* cf. *striata* Gross, devonische Palaeonisciden aus dem Oberen Plattenkalk der Bergisch-Gladbach–Paffrather Mulde (Rheinisches Schiefergebirge). Palaeontogr. A, 128:87–114.

Jurgens, J. D. 1971. The morphology of the nasal region of Amphibia and its bearing on the phylogeny of the group. Ann. Univ. Stellenbosch, Ser. A, 46(2):1–146.

Kesteven, H. L. 1950. The origin of the tetrapods. Proc. R. Soc. Victoria, new ser., 59:93–138.

Kulczycki, J. 1960. *Porolepis* (Crossopterygii) from the Lower Devonian of the Holy Cross Mountains. Acta Palaeontol. Polonica, 5:65–104.

Lapage, E. O. 1928a. The septomaxillary. I. In the Amphibia Urodela. J. Morphol. Physiol., 45(2):441–471.

Lapage, E. O. 1928b. The septomaxillary of the Amphibia Anura and of the Reptilia. II. J. Morphol. Physiol., 46(2):399–425.

Lebedev, O. A. 1984. First discovery of a Devonian tetrapod vertebrate in USSR. Dok. Akad. Nauk SSSR, 278:1470–1473. [In Russian].

Leonardi, G. 1982. Descoberta e pegada de um anfibio Devoniano no Paraná. Ciencias Terra, 1982:36–37.

Leonardi, G. 1983. *Notopus petri* nov. gen., nov. sp.: une empreinte d'amphibien de Dévonien au Paraná (Brésil). Geobios, 16:233–239.

Long, J. A. 1985. The structure and relationships of a new osteolepiform fish from the Late Devonian of Victoria, Australia. Alcheringa, 9:1–22.

Maisey, J. G. 1984. Higher elasmobranch phylogeny and biostratigraphy. Zool. J. Linn. Soc., 82:33–54.

Maisey, J. G. 1986. Heads and tails: a chordate phylogeny. Cladistics, 2:201–256.

Medvedeva, I. M. 1959. The nasolacrimal duct and its connection with the lacrimal and septomaxillary bones in *Ranodon sibiricus*. Dok. Akad. Nauk SSSR, 128:789–792. [In Russian].

Medvedeva, I. M. 1975. Olfactory organ in amphibians and its phylogenetic significance. Tr. Zool. Inst. Akad. Nauk SSSR, 58:1–174. [In Russian].

Miles, R. S. 1977. Dipnoan (lungfish) skulls and the relationships of the group: a study based on new species from the Devonian of Australia. J. Linn. Soc. London, Zool., 61:1–328.

Northcutt, R. G. 1987. Lungfish neural characters and their bearing on sarcopterygian phylogeny. Pp. 277–297 *in* Bemis, W. E., W. W. Burggren, and N. E. Kemp (eds.), *The Biology and Evolution of Lungfishes*. J. Morphol., Suppl. 1.

Panchen, A. L. 1967. The nostrils of choanate fishes and early tetrapods. Biol. Rev. Cambridge Philos. Soc., 42:374–420.

Panchen, A. L. 1985. On the amphibian *Crassigyrinus scoticus* Watson from the Carboniferous of Scotland. Philos. Trans. R. Soc. London, Ser. B, 309:505–568.

Panchen, A. L., and T. R. Smithson. 1987. Character diagnosis, fossils and the origin of tetrapods. Biol. Rev. Cambridge Philos. Soc., 62:341–438.

Pearson, D. M., and T. S. Westoll. 1979. The Devonian actinopterygian *Cheirolepis* Agassiz. Trans. R. Soc. Edinburgh, 70:337–399.

Remane, A. 1952. Die Grundlagen des natürlichen Systems, der vergleichenden Anatomie und der Phylogenetik. Leipzig: Akad. Verlagsges. Geest & Portig.

Roček, Z. 1986. An "intracranial joint" in frogs. Pp. 49–54 *in* Roček, Z. (ed.), *Studies in Herpetology*. Prague: Charles Univ.

Rosen, D. E., P. L. Forey, B. G. Gardiner, and C. Patterson. 1981. Lungfishes, tetrapods, paleontology and plesiomorphy. Bull. Am. Mus. Nat. Hist., 167:159–276.

Saint-Aubain, M. L. de. 1981. Amphibian limb ontogeny and its bearing on the phylogeny of the group. Z. Zool. Syst. Evolutionsforsch., 19:175–194.

Säve-Söderbergh, G. 1932. Preliminary note on Devonian stegocephalians from East Greenland. Medd. Groenland, 94:1–107.

Sawin, H. J. 1941. The cranial anatomy of *Eryops megacephalus*. Bull. Mus. Comp. Zool. Harvard College, 88:407–463.

Schmalhausen, I. I. 1958. Nasolacrimal duct and septomaxillae of Urodela. Zool. Zh., 37:570–583. [In Russian].

Schmalhausen, I. I. 1968. *The Origin of Terrestrial Vertebrates.* New York and London: Academic Press.

Schultze, H.-P. 1969. Die Faltenzähne der rhipidistiiden Crossopterygier, der Tetrapoden und der Actinopterygier-Gattung *Lepisosteus*, nebst einer Beschreibung der Zahnstruktur von *Onychodus* (struniiformer Crossopterygier). Palaeontogr. Ital., 65:63–136.

Schultze, H.-P. 1970. Folded teeth and the monophyletic origin of tetrapods. Am. Mus. Novit., 2408:1–10.

Schultze, H.-P. 1974. Osteolepidide Rhipidistia (Pisces) aus dem Pennsylvanian von Illinois/USA. Neues. Jahrb. Geol. Palaeontol. Abh., 146:29–50.

Schultze, H.-P. 1977a. The origin of the tetrapod limb within the rhipidistian fishes. Pp. 541–544 *in* Hecht, M. K., P. C. Goody, and B. M. Hecht (eds.), *Major Patterns in Vertebrate Evolution.* New York and London: Plenum Press.

Schultze, H.-P. 1977b. Ausgangsform und Entwicklung der rhombischen Schuppen der Osteichthyes (Pisces). Palaeontol. Z., 51:152–168.

Schultze, H.-P. 1981. Hennig und der Ursprung der Tetrapoda. Palaeontol. Z., 55:71–86.

Schultze, H.-P. 1987. Dipnoans as sarcopterygians. Pp. 39–74 *in* Bemis, W. E., W. W. Burggren, and N. E. Kemp (eds.), *The Biology and Evolution of Lungfishes.* J. Morphol., Suppl. 1.

Schultze, H.-P., and M. Arsenault. 1985. The panderichthyid fish *Elpistostege:* A close relative of tetrapods? Palaeontology, 28:293–309.

Shubin, N. H., and P. Alberch. 1986. A morphogenetic approach to the origin and basic organization of the tetrapod limb. Evol. Biol., 20:319–387.

Steiner, H. 1934. Ueber die embryonale Hand- und Fuss-Skelett-Anlage bei den Crocodiliern, sowie über ihre Beziehungen zur Vogel-Flügelanlage und zur ursprünglichen Tetrapoden-Extremität. Rev. Suisse Zool., 41:383–396.

Steiner, H. 1935. Beiträge zur Gliedmassentheorie: Die Entwicklung des Chiropterygium aus dem Ichthyopterygium. Rev. Suisse Zool., 42:715–729.

Szarski, H. 1962. The origin of the Amphibia. Q. Rev. Biol., 37:189–241.

Szarski, H. 1977. Sarcopterygii and the origin of tetrapods. Pp. 517–540 *in* Hecht, M. K., P. C. Goody, and B. M. Hecht (eds.), *Major Patterns in Vertebrate Evolution.* New York and London: Plenum Press.

Visser, M. H. C. 1963. The cranial morphology of Ichthyophis glutinosus (Linné) and Ichthyophis monochrous (Bleeker). Ann. Univ. Stellenbosch, Ser. A, 38(3):67–102.

Vorobyeva, E. I. 1977a. Morphology and nature of evolution of crossopterygian fishes. Tr. Paleontol. Inst. Akad. Nauk SSSR, 163:1–239. [In Russian].

Vorobyeva, E. I. 1977b. Evolutionary modifications of the teeth structure in the Palaeozoic Crossopterygii. J. Palaeontol. Soc. India, 20:16–20.

Vorobyeva, E. I. 1980. Phylogenetic methods in paleontology. Ichthyologia Beograd, 12:83–91.

Vorobyeva, E. I. 1984. Some procedural problems in the study of tetrapod origins. Proc. Linn. Soc. New South Wales, 107:409–418.

Vorobyeva, E. I. 1985. On the evolution of cranial structures in crossopterygians and tetrapods. Pp. 123–133 *in* Duncker, H.-R., and G. Fleischer (eds.), *Functional Morphology in Vertebrates.* Stuttgart and New York: Gustav Fischer.

Vorobyeva, E. I. 1986. The current state of the problem of amphibian origin. Pp. 25–28 *in* Roček, Z. (ed.), *Studies in Herpetology.* Prague: Charles Univ.

Warren, A. A., R. Jupp, and R. Bolton. 1986. Earliest tetrapod trackway. Alcheringa, 10:183–186.

Warren, J. W., and N. A. Wakefield. 1972. Trackways of tetrapod vertebrates from the Upper Devonian of Victoria, Australia. Nature, London, 238:469–470.

Westoll, T. S. 1943. The origin of the primitive tetrapod limb. Proc. R. Soc. London, Ser. B, 131:373–393.

Whiteaves, J. F. 1881. On some remarkable fossil fishes from the Devonian rocks of Scaumenac Bay, P.Q., with description of a new genus and three new species. Can. Naturalist, (2)10:27–35.

Wiedersheim, R. 1892. Das Gliedmassenskelett der Wirbelthiere mit besonderer Berücksichtigung des Schulter- und Beckengürtels bei Fischen, Amphibien und Reptilien. Jena: Gustav Fischer.

Worobjewa, E. 1973. Einige Besonderheiten im Schädelbau von *Panderichthys rhombolepis* (GROSS) (Pisces, Crossopterygii). Palaeontogr. A, 142:221–229.

Worobjewa, E. 1975. Formenvielfalt und Verwandtschaftsbeziehungen der Osteolepidida (Crossopterygii, Pisces). Palaeontol. Z., 49:44–55.

3 Description and Systematics of Panderichthyid Fishes with Comments on Their Relationship to Tetrapods

Emilia Vorobyeva and Hans-Peter Schultze

The origin of tetrapods has been debated vigorously in phylogenetic discussions. Part of the disagreement stems from the nature of the material that different authors have examined and analyzed. The conclusions of some were derived primarily from analyses of Recent forms (e.g., Szarski, 1962; Parsons and Williams, 1963; Løvtrup, 1977; Medvedeva, 1975; Lebedkina, 1979). Other authors emphasized fossil forms (e.g., Westoll, 1938, 1943; Romer, 1956; Worobjewa, 1975b; Shishkin, 1975; Tatarinov, 1976; Vorobyeva, 1977a), and some, both fossil and Recent forms (e.g., Jarvik, 1942, 1972, 1975; Schmalhausen, 1968; Lombard and Bolt, 1979; Rosen et al., 1981; Bolt and Lombard, 1985; Vorobyeva, 1985; Smirnov and Vorobyeva, 1986). However, the kind of material analyzed is less critical than the phylogenetic approach adopted. Jarvik (1960, 1964), using a "static ancestor-descendent" approach, related tetrapods to rhipidistian crossopterygians (osteolepiforms and porolepiforms, respectively). In contrast, Rosen et al. (1981) based their conclusion that dipnoans are the closest relatives of living tetrapods on a cladistic analysis. These authors maintained (contra Westoll 1938, 1943; Romer, 1956) that one cannot seek the ancestor of tetrapods in a specific osteolepiform rhipidistian. Subsequent studies have demonstrated that cladistic analyses intended to solve the same problem can produce quite different results (cf. Schultze, 1981, 1987; Maisey, 1986; Panchen and Smithson, 1987; Chang, this volume; Schultze, this volume). This variation suggests that the choice of taxa analyzed and the selection and interpretation of characters are more critical than the phylogenetic philosophy to the analytical results. Nonetheless, there exist various schools of phylogenetic interpretation (Rage and Janvier, 1982; Vorobyeva, 1986) that have had substantial influence on the development of hypotheses explaining the origin of tetrapods.

The Stockholm School (Vorobyeva, 1986) postulated a polyphyletic origin of tetrapods (Jarvik, 1942, 1960, 1964, 1972, 1980, 1986) in which urodeles were derived from porolepiforms ([Urodela + Porolepiformes] = Urodelomorpha of Jarvik), and anurans and all other tetrapods (Eutetrapoda) from the osteolepiforms. The fundamental assumption of this hypothesis is that structural types have remained stable (i.e, not evolved) since the Devonian. Thus, all basic vertebrate morphotypes would have originated more than 400 million years ago in a pre-Devonian adaptive radiation of organisms (Jarvik, 1960, 1964). Particular morphological features of urodeles are attributed to porolepiforms, and those of anurans to osteolepiforms. This line of reasoning involves a retrospective transposition of features of Recent forms into fishes that lived more than 400 million years ago and requires loose interpretations of homology (Rage and Janvier, 1982).

The Stockholm School opposes the monophyletic origin of tetrapods from osteolepiforms as proposed by the Anglo-American School of Westoll and Romer. The monophyletic origin of tetrapods and dipnoans from a common ancestor was proposed by the Transformed Cladist School (Rosen et al., 1981). Transformed cladists base their opposition to the Stockholm and Anglo-American schools of phylogenetic interpretation on Hennig's (1950, 1966) phylogenetic systematics. Nonetheless, transformed cladists, like disciples of the Stockholm School, employ a retrospective approach, because they base their phylogenetic evaluations largely on analysis of features of Recent taxa, but the key taxon in their postulated relationship of dipnoans and tetrapods is a Devonian dipnoan. If the sister-group of Recent tetrapods is the Dipnoi, then the same also must be true of fossil tetrapods. The analysis of Rosen et al. (1981) is restricted to the distribution of patterns without consideration of evolutionary processes; thus, convergences, for example, are ignored (Forey, 1987).

The Soviet School (Vorobyeva, 1986) reinstated use of the triple approach of Haeckel, in which data from comparative anatomy and embryology are combined with paleontology. This method emphasizes careful interpretation of anatomical features and their ontogenetic origin, as well as their occurrence and distribution in the fossil record (Schmalhausen, 1968). Functional, ecological, and physiological attributes are included in the evaluation of characters.

In this chapter, we accept Schmalhausen's (1968) conclusions regarding the origin of tetrapods (cf. Schultze, this volume) as follows. (1) The Tetrapoda is monophyletic (Gaffney, 1979). (2) The Tetrapoda is related closely to osteolepiform rhipidistians and specifically to panderichthyids (Vorobyeva, 1962; Schultze, 1969, 1970; Worobjewa, 1973; Schultze and Arsenault, 1985). Further, we accept Schultze's (1987) classification of osteichthyans. We deal with a restricted number of Rhipidistia, to the exclusion of the Onychodontiformes; thus, the Rhipidistia includes Porolepiformes and Choanata, and the Choanata is composed of Osteolepiformes, Panderichthyida, and Tetrapoda. Within tetrapods, we treat two groups—the Ichthyostegalia and all other tetrapods, the Neotetrapoda (Gaffney, 1979). The position of the Rhizodontida is unresolved within the rhipidistians.

The family Panderichthyidae (Vorobyeva and Lyarskaya, 1968) is composed of two genera, *Panderichthys* Gross 1941 and *Elpistostege* Westoll 1938; a third genus, *Obruchevichthys*, was included provisionally by Vorobyeva (1977a). Because the latter is represented only by fragments of the lower jaw, *Obruchevichthys* is not considered here. Complete skeletons of *Panderichthys* are known (Vorobyeva, 1980); in addition, there are two partial skulls and one postcranial specimen questionably assigned to *Elpistostege* (Schultze and Arsenault, 1985). The first known skull of *Elpistostege* was thought to be that of an amphibian (Westoll, 1938); however, examination of additional specimens revealed it to be a panderichthyid fish (Worobjewa, 1973; Schultze and Arsenault, 1985). Examination of the snout of *Elpistostege* reveals that the skulls of *Elpistostege* and *Panderichthys* are nearly identical (Figs. 1–2).

Herein we redescribe *Panderichthys* and compare it with *Elpistostege*. A new order, Panderichthyida, is erected and described. We demonstrate that many features support a sister-group relationship between the Panderichthyida and the Tetrapoda. Further, we suggest that the panderichthyids are related to osteolepiforms and that tetrapod features evolved independently and in parallel within the osteolepiform ancestor of tetrapods as well as in other osteolepiforms.

MATERIALS AND METHODS

The following description of *Panderichthys* is based on three specimens we examined in 1986. Museum acronyms are as follows: MHNM = Musée d'Histoire naturelle de Miguasha, Quebec, Canada; PIN = Paleontological Museum, Academy of Sciences, Moscow, USSR.

P. rhombolepis. PIN 3547/26: head of specimen (Figs. 2, 5, 9, 10, 12, 14–15); PIN 3549/19: shoulder girdle (Figs. 16–17).

P. stolbovi. PIN 54/169: partial snout (Figs. 6–8), previously published

Figs. 1–2. Skull roofs of panderichthyids. (1) *Elpistostege watsoni* Westoll, 1938 (MHNM 538)—Escuminac Formation, Frasnian, Late Devonian, shore of river Ristigouche at Miguasha, Province of Quebec, Canada. (2) *Panderichthys rhombolepis* (Gross, 1930) (PIN 3547/26)—Gauja-beds, Frasnian, Late Devonian; Lode, Latvia, USSR.

and figured by Vorobyeva (1960:Figs. 1–3; 1962:Fig. 31A, B, Pl. 19:Figs. 1a–d; 1971:Figs. 1, 2, 6, 8, Pl. 8:Figs. 2a–e; 1977a:Figs. 7B, 8F) and by Vorobyeva and Obruchev (1964:Fig. 32, Pl. 3:Fig. 9).

The snout of *P. stolbovi* had previously been prepared and cut obliquely to the midline. Clay was removed from specimens of *P. rhombolepis* in 1986 in order to reveal the supracoracoid foramen and the glenoid canal in the shoulder girdle and the right jaw articulation, palate, and left nasal capsule.

Photographs were taken with a Nikkormat camera. Enlarged prints served as the basis for the illustrations, although in the course of preparation, each drawing was compared closely with the actual specimen.

SYSTEMATICS OF PANDERICHTHYID FISHES

Redescription of *Panderichthys* Gross 1941

Skull

Snout and skull table. The snout is curved ventrally such that it forms a rostrum with the mouth subterminal. The position of the single pair of external nostrils (Fig. 3) is marginal. Each nostril is surrounded by the

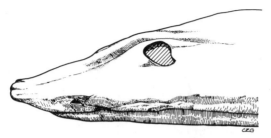

Fig. 3. Lateral view of skull of *Panderichthys rhombolepis* (Gross, 1930) (PIN 3547/18)—Gauja-beds, Frasnian, Late Devonian; Lode, Latvia, USSR. (After Vorobyeva, 1986:Fig. 1.)

anterior tectal dorsally and the narrow lateral rostral ventrally, and is separated from the margin of the upper jaw by a narrow posterior process of the premaxilla (Figs. 6, 12–14). A mosaic of rostral bones covers the snout (Figs. 4–5). A separate median rostral carries the ethmoidal sensory commissure. The nasal and postrostral series are short and do not reach the level of the orbits. A pair of frontals extends anteriorly from the level of the orbits and separates the nasal and postrostral series from the parietals. The narrow anterior portions of the parietals lie between the orbits. The paired postparietals are fused to an unpaired median bone (Fig. 5) or not (Fig. 4).

The orbits are located high on the flat skull roof (Figs. 4–5). The prefrontal (homologous to the posterior tectal of osteolepiforms) and postfrontal (homologous to supraorbital of osteolepiforms) border the orbit anteriorly and dorsally, respectively. The postfrontal forms a ridge medial to the orbit, the "eyebrow" of Schultze and Arsenault (1985). The bone articulates with the intertemporal posteriorly in *Panderichthys* (Worobjewa, 1973:Fig. 1A), whereas the intertemporal is fused with the postorbital in *Elpistostege*. A small parietal foramen is located posterior to the orbits.

There is no indication of an external intracranial joint between anterior and posterior dermal shields. The supratemporal and tabular lie lateral to the postparietals. The tabular is separated from the cheek region by the long spiracular slit, which extends forward toward the posterolateral corner of the supratemporal. The supratemporal and intertemporal articulate with the postorbital of the cheek region. There is no lateral kinesis between skull roof and cheek region as in osteolepiforms. The extrascapulars are small relative to the length and large size of the skull roof; the lateral extrascapulars overlap the median extrascapular.

A large lacrimal lies between the orbit and the anterior tectal, and it separates the prefrontal (i.e., posterior tectal) from the jugal. The cheek region is composed of a long jugal, a long postorbital that extends to the spiracular slit, a large squamosal, a quadratojugal, and a preopercular. According to Vorobyeva (1969), the squamosal articulates with the maxilla, thereby separating the jugal from the quadratojugal. Although this

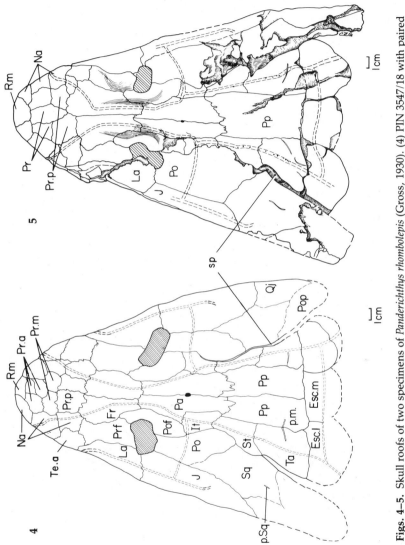

Figs. 4–5. Skull roofs of two specimens of *Panderichthys rhombolepis* (Gross, 1930). (4) PIN 3547/18 with paired postparietals. (5) PIN 3547/26 with unpaired postparietal. Both specimens from Gauja-beds, Frasnian, Late Devonian; Lode, Latvia, USSR. See Appendix I for key to abbreviations.

73

seems to be true of the left side of specimen PIN 3547/26 (Fig. 5), in some other specimens (e.g., PIN 3547/18; Figs. 4, 37), the jugal articulates with the quadratojugal, thereby separating the squamosal from the maxilla (as in tetrapods). The maxilla is narrow and does not extend as far posteriorly as does the jugal. The quadratojugal is the posterior component of the upper jaw.

Nasal capsule. The elongate, narrow nasal capsule lies adjacent to the lateral ethmoidal wall (Fig. 8). Because the capsule is only one-quarter of the width of the ethmoid, the paired capsules are separated widely (Figs. 7–8). The posterior wall of the nasal capsule is unossified; thus, there is a large ventrolateral opening, and a fenestra endochoanalis cannot be identified. There is only one external narial opening (Vorobyeva, 1986; contra Vorobyeva, 1962 and Worobjewa, 1973); the "posterior external nasal opening" of Worobjewa (1973:Fig. 2A) is a pore of the infraorbital canal. The external nasal opening enters the lower half of the nasal capsule. The internal nasal opening (i.e., choana) extends from the level of the posterior margin of the external nasal opening posteriorly to the level of the dermopalatine fangs (Figs. 9, 14). The length and posterior extent of the exochoanal opening in *P. rhombolepis* (Figs. 13–14) suggest that the nasal capsule extended posteriorly beyond the ethmoid. However, its proximity to the orbit cannot be estimated because the posterior wall of the nasal capsule is not ossified.

The dermintermedial process of the lateral rostral, the anterior tectal process, and the crista intermedia are present in the nasal capsule (Figs. 13–14); a medial recess is absent (Worobjewa, 1973:Fig. 3). The dermintermedial process extends rostrocaudally along the lateral wall of the nasal capsule; together with the anterior tectal process, it divides the lateral nasal capsule into dorsal and ventral chambers. The crista intermedia extends horizontally to unite with the medial wall of the nasal capsule and form a bridge over the nasobasal canal; the lamina inferior cannot be distinguished from the lamina superior. The nasobasal canal and inferior recess are present in the floor of the nasal capsule. The profundus nerve exits the thick medial ethmoidal wall via a single foramen and divides into many small branches within each nasal capsule. The rami exit each nasal capsule through openings in the anteromedial wall. The opening of the olfactory canal lies ventral to the ethmoidal foramen for the profundus nerve.

Palate. The palatal region of the snout has a large anterior palatal pit that accommodates the anterior fangs of the dentary (Figs. 10, 12–13). A premaxillary palatal plate is absent. The vertical tooth-bearing ridge of each vomer articulates with its counterpart medially. An intervomerine

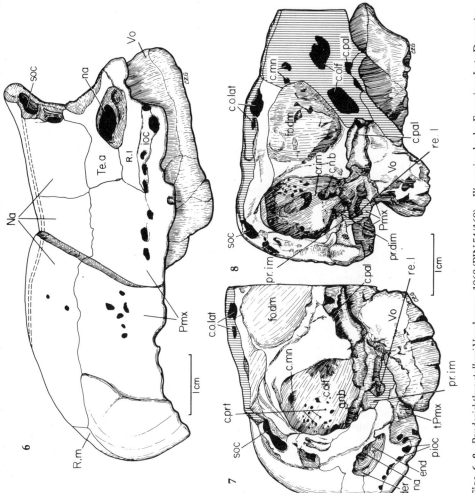

Figs. 6–8. *Panderichthys stolbovi* Vorobyeva, 1960 (PIN 54/169)—Il'menian beds, Frasnian, Late Devonian; Syas River, Stolbov, Leningrad district, USSR. (6) Snout in lateral view. (7–8) Posterior and posteromedial views of nasal capsule. See Appendix I for key to abbreviations.

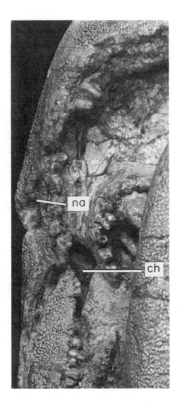

Fig. 9. Choana and external nasal opening in *Panderichthys rhombolepis* (Gross, 1930) (PIN 3547/26)—Gauja-beds, Frasnian, Late Devonian; Lode, Latvia, USSR. Cf. Figure 14. See Appendix I for key to abbreviations.

canal (Worobjewa, 1973:225), which occasionally is paired, traverses the vomerine ridge. A long keyhole-shaped choana is present in the buccal region of the palate; it is bordered by the premaxilla, maxilla, vomer, and dermopalatine (Figs. 9, 14). The wide anterior part of the choana lies between the vomer and premaxilla, whereas the elongate posterior portion lies between the dermopalatine and maxilla. The narrow parasphenoid terminates anteriorly at the level of the dentigerous ridges of the vomers; its anterior portion is flanked by the posterior vomerine processes. The parasphenoid is divided into an expanded dorsal portion and a tooth-bearing plate, which are connected by a narrow septum. The dermopalatines are posterolateral to the vomers, but the marginal tooth row of the dermopalatine is separated from the tooth-bearing ridge of the vomer by a smooth pitlike area (Worobjewa, 1973:227, Pl. 36:Fig. 4; Worobjewa, 1975a:Fig. 2; Rosen et al., 1981:Fig. 16B). The pit lies above the fang borne on the first coronoid of the lower jaw; thus, the fang does not fit into the choana as postulated by Rosen et al. (1981). The dermopalatine fang, in turn, fits into a pit between first and second coronoid of the mandible; thus, when the jaws are occluded, the dermopalatine

10   11

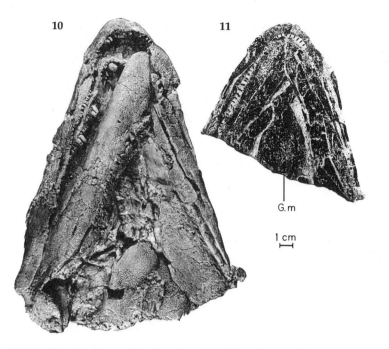

G. m

1 cm

Figs. 10–11. Ventral views of heads of panderichthyids. (10) *Panderichthys rhombolepis* (Gross, 1930) (PIN 3547/26)—Gauja-beds, Frasnian, Late Devonian; Lode, Latvia, USSR. Cf. Figure 12. (11) *Elpistostege watsoni* Westoll, 1938 (MHNM 538)—Escuminac Formation, Frasnian, Late Devonian; shore of river Ristigouche at Miguasha, Province of Quebec, Canada. See Appendix I for key to abbreviations.

fang would lie posterolateral to the fangs of the first coronoid. The third and fourth fangs of the ectopterygoid are accommodated by the second and third coronoid pits, respectively. The uninterrupted marginal tooth row extends from the dermopalatine to the ectopterygoid. An interpterygoid vacuity is absent owing to the presence of an entopterygoid between the dermopalatine, ectopterygoid, and parasphenoid.

Suspensorium. Medially, the palatoquadrate articulates synchondrotically with the ethmoidal region. A trigeminal notch and otic process may be present but were not observed. The posterior end of the right palatoquadrate was prepared on specimen PIN 3547/26 (Fig. 15). The double-headed quadrate articulates with the saddle-shaped glenoid surface of the articular. An elongate, flat hyomandibular lies in a depression on the dorsomedial side of the palatoquadrate. The size of the hyomandibular is moderate, it is straight rather than arched and lacks a foramen. The rough, unfinished anterior end of the bone indicates coverage by cartilage. A short stylohyal, which terminates far anteriorly to the jaw

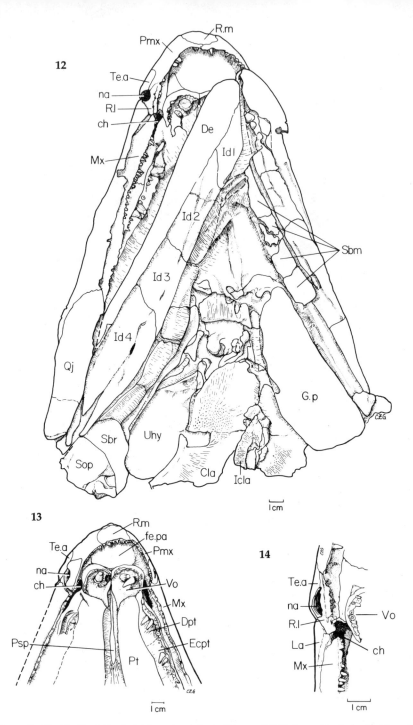

Figs. 12–14. Skull of *Panderichthys rhombolepis* (Gross, 1930) from Gauja-beds, Frasnian, Late Devonian; Lode, Latvia, USSR. (12) Ventral view of PIN 3547/26. Cf. Figure 7. (13) Anterior palate of PIN 3547/18. (14) Choana and external nasal opening of PIN 3547/26. Cf. Figure 7. See Appendix I for key to abbreviations.

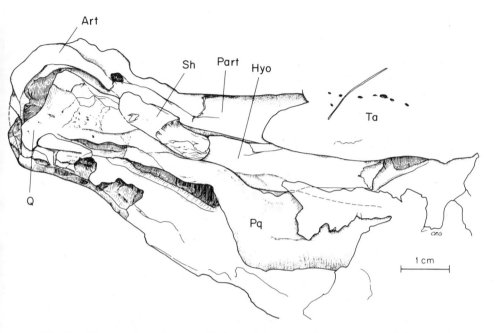

Fig. 15. Palatoquadrate, hyomandibula, stylohyal, and jaw articulation in oblique dorsal view of right side of skull of *Panderichthys rhombolepis* (Gross, 1930) (PIN 3547/26)—Gauja-beds, Frasnian, Late Devonian; Lode, Latvia, USSR. See Appendix I for key to abbreviations.

articulation, lies posteriorly adjacent to the hyomandibular. The sty-lohyal is thin anteriorly and thick posteriorly.

Lower jaw. The lower jaw (Figs. 10, 12) is typical of rhipidistians in being composed of a tooth-bearing dentary, four infradentaries, a prear-ticular on the lingual side, and three coronoids and an adsymphysial plate dorsally. The mandibular canal opens on the lateral surface through a single row of pores (Gross, 1941:Fig. 19B). The dentary bears small teeth, but a single fang lies close to the symphysis (Gross, 1941:Fig. 19A). The prearticular extends to the symphysis, where it is overlain by the adsymphysial plate (Vorobyeva, 1962:Fig. 50, Pl. 18:Fig. 3B). The adsym-physial plate bears small teeth or fangs. There are three well-developed coronoid pits—one anterior to the first coronoid, a second between Coronoids 1 and 2, and the third between Coronoids 2 and 3; these pits accommodate the fangs of the vomer, dermopalatine, and ectopterygoid, respectively.

Dentition. The teeth have polyplocodont structure with little or no attachment bone between the folds (Fig. 22; Schultze, 1969; Vorobyeva, 1977b).

Opercular Series

The opercular series is composed of extrascapulars, an opercular, subopercular, submandibulobranchiostegal, a long submandibular series of seven plates, and gulars (Vorobyeva, 1980:Fig. 2), as in osteolepiforms. The median gular is exceptionally large, and the lateral gulars shorter and wider than those in osteolepiforms. The median gular extends 40% the length of the lower jaw and comprises over 60% of the length of the principal gulars. The principal gulars of either side may articulate with one another for a short distance, or may be separated completely by the clavicles.

Postcranial Skeleton

Dermal shoulder girdle. The dermal shoulder girdle has a relatively narrow outer exposure; it is connected to the skull roof via the anocleithrum, supracleithrum, and posttemporal. The posttemporal articulates with the lateral and median extrascapulars. The abutment of the anocleithrum against the cleithrum is marked by a ridge that extends along the inner side of the anocleithrum. The ventral portion of the cleithrum is a short, anteroventrally directed ridged process that is overlapped by the clavicle (Fig. 16). Anterodorsally, the ascending process of the clavicle is overlapped by a rectangular part of the cleithrum. The dorsal ascending process of the clavicle fits into a groove between the cleithrum and scapulocoracoid. The clavicle has a broad ventral extension, and the

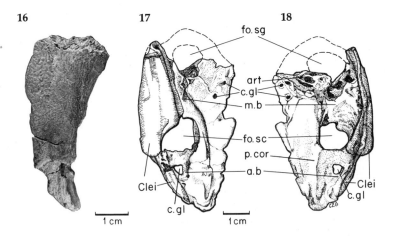

Figs. 16–18. Shoulder girdle of *Panderichthys rhombolepis* (Gross, 1930) (PIN 3547/19)—Gauja-beds, Frasnian, Late Devonian; Lode, Latvia, USSR. (16) Cleithrum in anterolateral view. (17) Anterodorsal view. (18) Posteroventral view. See Appendix I for key to abbreviations.

paired clavicles have an extensive medial articulation. The small inter-clavicle lies posteromedian to the anterior articulation of the clavicles (Figs. 10, 12). The clavicles extend between, and are overlapped ante-riorly by, the principal gulars.

Endoskeletal shoulder girdle. The endoskeletal shoulder girdle (Figs. 17–18) is formed by a massive scapulocoracoid. The three scapulocoracoid buttresses (i.e., anterior, middle, and posterior) converge to form a single extensive articulation with the cleithrum. The extensive coracoid plate is perforated by a large supracoracoid foramen. Because the posterior but-tress is not preserved, the posterior margin of the supraglenoid foramen was reconstructed. The supraglenoid and supracoracoid foramina do not extend to the cleithrum as they do in *Eusthenopteron* (Jarvik, 1980:Figs. 100, 165A). Instead, cartilage bone separates them from the inner side of the cleithrum as in *Ichthyostega* (Jarvik 1980:Fig. 165B). Two small canals perforate the scapulocoracoid medial to the glenoid fossa; the latter is preserved only partially in the specimen illustrated. Anterior to the su-pracoracoid foramen lies another smaller foramen. The head of the hu-merus fits into a shallow groove, the glenoid fossa, on the scapulocora-coid. The exact shape of the glenoid fossa cannot be determined because its borders are not preserved.

Fins. Pectoral and pelvic fins are located in an extreme ventral position (Fig. 19). The pelvic fins are much smaller than the pectoral fins and lie posterior, close to the caudal fin. The endoskeleton of the pectoral fin is of the uniserial archipterygial type (Worobjewa, 1975a:Fig. 3A). The long (about half the length of the pectoral endoskeleton) humerus has a well-developed deltoid crest, but no entepicondylar (postaxial) process. The

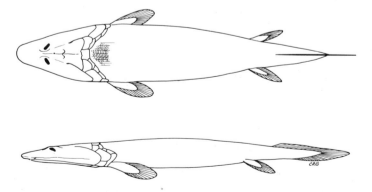

Fig. 19. Reconstruction of *Panderichthys rhombolepis* (Gross, 1930) in dorsal and lateral views.

radius and ulna articulate with the distal end of the humerus. The radius is as long as the humerus and has a well-developed external ridge on its ventral side. The ulna is three-quarters to one-half the length of the humerus. Two bones, the intermedium and ulnar plate, articulate with the distal end of the ulna. The intermedium lies between the radius and ulnar plate. The latter is large and includes the distal fin elements (i.e., ulnare and carpals). The bases of articulated and distally branched lepidotrichia overlie the distal portion of the radius, intermedium, and ulnar plate. Anal and dorsal fins are absent; a fin fold surrounds the tail to form a diphycercal tail. The lepidotrichia of all fins are articulated distally and branched.

Axial skeleton. The vertebrae (Fig. 20) are composed of broad intercentra and neural arches that are attached to the dorsal border of the intercentra; pleurocentra are absent. A median groove for the dorsal aorta and a lateral branch for the intermetameric artery are visible on the ventral side of each intercentrum (Vorobyeva and Tsessarskii, 1986:Fig. 3B). The neural arches do not meet dorsally, and the neural spines are paired (Vorobyeva and Tsessarskii, 1986:Fig. 3E). A large broad rib attaches to both the intercentrum and neural arch (Vorobyeva and Tsessarskii, 1986); the abdominal ribs are short (i.e., equal in length to the neural arch and spine).

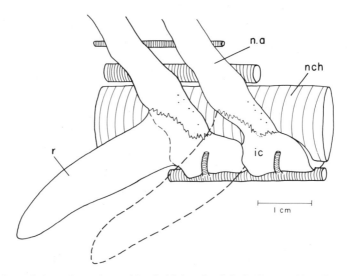

Fig. 20. Lateral view of vertebrae of *Panderichthys rhombolepis* (Gross, 1930) (PIN 3547/19)— Gauja-beds, Frasnian, Late Devonian; Lode, Latvia, USSR. See Appendix I for key to abbreviations. (After Vorobyeva and Tsessarskii, 1986:Fig. 3D.)

Scales. The rhombic scales lack a cosmine cover and are unexpectedly small relative to the large size of the organism. Their bony surface is tuberculate and ridged like the skull bones (Vorobyeva, 1962:Figs. 29A–D, Pls. 21–23). The superficial sculpturing of the scales is comparable to the ornamentation of the bones of labyrinthodonts.

Comparison of *Panderichthys* and *Elpistostege*

Similarities

Elpistostege resembles *Panderichthys* in features of the parietal shield (Worobjewa, 1973) and snout (Schultze and Arsenault, 1985). The snout of *Elpistostege watsoni* is similar in shape and size to that of *Panderichthys rhombolepis* (Figs. 1–2); in both species, the snout is prominent and the mouth subterminal (Figs. 10–11). The orbits are located high on the skull roof. The presence of "eyebrows" medial to the orbits results in a depression in the skull roof in which the suture between frontals and parietals and the parietal opening are located. Because the external nasal opening is located ventrally close to the margin of upper jaw, it is not visible in dorsal aspect. Most snout bones are arranged in pairs (paired postrostrals and nasals); exceptions include the unpaired median rostral and, when present, the unpaired median postrostral (Fig. 5). The pair of frontals is flanked by prefrontals and lies anterior to the postfrontals, which form the "eyebrows" along the dorsomedial margin of the orbits. The lower jaw is shallow in both taxa and bears a wide medial portion flanked by a narrow series of submandibulars. Both genera possess a large median gular (cf. Fig. 11, *E. watsoni* with *P. rhombolepis*, Vorobyeva, 1980:Fig. 2) and labyrinthodont-like bony ornamentation on the skull and scales. The ornamentation in *P. rhombolepis* is tuberculate, whereas that of *P. bystrowi* (Vorobyeva, 1962) is composed of a network of ridges similar to that of labyrinthodonts. These genera are alike and differ from all others known in possessing neural arches and intercentra that surround the notochord ventrally. Pleurocentra have not been discovered.

Differences

There are distinct differences between *Elpistostege* and *Panderichthys* (Schultze and Arsenault, 1985). The orbits are round in *Elpistostege* and elongated in *Panderichthys*, and the posterior margin of the skull roof is invaginated in *Elpistostege* and straight in *Panderichthys*. The genera differ in the shapes of numerous elements of the skull roof and in at least one proportion of the skull roof. The frontals are 29% of skull-roof length in *Elpistostege;* the posterior margin of the frontal lies just anterior to the level of the orbits, and the posterior part of the bone is half the width of the anterior part. In contrast, in *Panderichthys* the frontals are 20% of the skull-roof length; the bones that extend between the orbits diminish in width more gradually. The prefrontal is larger and much narrower in *Elpistostege* (width about 20% of length) than in *Panderichthys* (width 40–50% of length). The lacrimal of *Elpistostege* is unusually long; it forms an anterolateral flange above the contact between premaxilla and maxilla posterior to the nasal opening. Such a flange is absent in *Panderichthys*. The lacrimal reaches the postorbital in some specimens of both genera (cf. Schultze and Arsenault, 1985:Fig. 5A, *Elpistostege*, with Fig. 5, *Panderichthys*, in this chapter), and excludes the jugal from the margin of the orbit. These differences are correlated with the longer snout of *Elpistostege*; the ratio of the distance between the anterior border of the orbit and the tip of the snout to the distance between the posterior border of the orbit and the posterior margin of the parietal is twice as great in *Elpistostege* (3.5 vs. 1.8 in *Panderichthys*). In *Panderichthys*, the postorbital part of the posterior shield is longer than in *Elpistostege*; thus, the parietals and postorbitals are longer, and the postorbital portion of the postfrontal is longer and broader. The shapes of the parietals also differ. The width of the parietal is more or less uniform in *Panderichthys*. However, in *Elpistostege*, the anterior interorbital portion of the parietal is narrow; postorbitally, there is a narrow lateral expansion, and posteriorly the bone is broad. A separate intertemporal is absent through loss or fusion with the postorbital in *Elpistostege*.

Taxonomic Conclusions

On the basis of the material described herein, we deem panderichthyids to be sufficiently distinct from both osteolepiforms and tetrapods (cf. discussion of relationships, below) to justify the erection of a new order[1] to accommodate them.

Panderichthyida nov. ord.

Diagnosis. Choanate fishes that possess the following characters, some of which are shared with primitive tetrapods, and some of which are reversals (i.e., also occur in Porolepiformes). Skull roof flat; orbits dorsal, closely placed; external nasal opening ventral, close to the margin of the upper jaw. Nasal capsules relatively small and widely separated. Paired frontals sutured closely with parietals between the orbits; external intracranial joint not developed; extratemporal absent. The spiracular slit extending posteriorly to form squamosal embayment (= "otic notch"). Jugal meeting quadratojugal to separate squamosal from maxilla. Laby-

[1]The ordinal name may be preoccupied by Panderichthyida Vorobyeva, 1989. This name appeared in an article (in Russian) titled "Panderichthyida—New Order of Paleozoic Crossopterygian Fishes (Rhipidistia)," Dokl. Akad. Nauk SSSR 306, no. 1 (1989): 188–189.

rinthodont plicidentine. No median fins except caudal fin fold, which may include the second dorsal fin. Paired appendages developed as fins with uniserial internal skeleton. Body covered with rhombic bony scales.

The following features are synapomorphies of the order: (1) Median rostral separated from premaxilla; (2) paired posterior postrostrals; (3) large median gular (occupying 40% of the length of lower jaw); (4) lateral recess in nasal capsule; and (5) subterminal mouth (= prominent snout).

Content. The order contains a single family, Panderichthyidae Vorobyeva 1968, comprising three genera and five species.

1. *Panderichthys* Gross 1941: *P. rhombolepis* (Gross 1930); *P. bystrowi* Gross 1941; *P. stolbovi* Vorobyeva 1960
2. *Elpistostege* Westoll 1938: *E. watsoni* Westoll 1938
3. *?Obruchevichthys* Vorobyeva 1977: *?O. gracilis* Vorobyeva 1977

Remarks. Vorobyeva (in Vorobyeva and Lyarskaya, 1968) erected the family Panderichthyidae for the genus *Panderichthys.* She considered the Panderichthyidae to be one of the families of the Osteolepiformes (Worobjewa, 1975b) not closely related to tetrapods, whereas Schultze (1969, 1970, 1987), Worobjewa (1973), and Schultze and Arsenault (1985) placed *Panderichthys* or the Panderichthyidae as the closest relatives of the tetrapods. Long (1985) also removed the panderichthyids from the osteolepiforms.

Some of the features of the Panderichthyida may have resulted from preservation of juvenile structures. Thus, relative to a juvenile osteolepiform (Schultze, 1984), the short postorbital region, marginal position of the naris, narrow interorbital region, and presence of unossified postnasal wall and floor in the nasal capsule are indicative of fetalization. On the other hand, *Panderichthys* has an elongated cheek region as in osteolepiforms. The lengthening of the preorbital region is also the result of late ontogenetic development (also shown for temnospondyls by Shishkin, 1973:165), and not preservation of juvenile structures.

DISCUSSION OF RELATIONSHIPS OF PANDERICHTHYIDA

Panderichthyida and Osteolepiformes

Long (1985) proposed four advanced characters to unite the Osteolepiformes (i.e., the Osteolepididae and Eusthenopteridae): (1) unique cheek complex with enlarged lacrimal, and postorbital and jugal of equal size; (2) barlike preopercular in steeply inclined position at posterior margin of cheek; (3) well-developed dermal anocleithrum; and (4) large basal scutes on paired and unpaired fins. He excluded the Rhizodontida (which he placed closer to the porolepiforms) from the Osteolepiformes and admitted that the panderichthyids did not fit his definition of osteolepiforms.

Nevertheless, panderichthyids share many features with osteolepiforms. Among these are the shape and composition of the lower jaw, wide separation of nasal capsules, the shapes and relationship of the opercular and subopercular, the composition of the exoskeletal shoulder girdle, and possession of rhombic scales, polyplocodont teeth, and numerous bones on the snout; however, all these features are primitive for rhipidistians. In addition, the following features characterize the Choanata (i.e., Osteolepiformes + Panderichthyida + Tetrapoda): possession of one external nasal opening, choana, adsymphysial tooth plate, undivided palatal recess, connection between vomers and long parasphenoid, lateral rostral ventral to nasal opening, large lacrimal, high number of narrow submandibulars, median gular, ball-and-socket articulation of pectoral appendage, and dorsal position of ribs. The adsymphysial tooth plate, submandibular, and median gular are lost in tetrapods. The presence of a submandibulobranchiostegal, a median gular, and the overlap of the lateral over the median extrascapular are features shared by Osteolepiformes and Panderichthyida.

The rhombic scale with a broad-based peg of panderichthyids is a primitive sarcopterygian feature (Schultze, 1977; Vorobyeva, 1977a). However, the loss of cosmine and presence of bony ornamentation are advanced features that occur in many lineages of sarcopterygians (Schultze, 1977). The bony ornamentation (especially that of the skull bones of *Panderichthys stolbovi* and *P. bystrowi*, Vorobyeva, 1962) is similar to that of labyrinthodonts.

The lower jaw of panderichthyids is a conservative structure that preserves primitive features. The broad and comparatively low anterior end of the jaw, the presence of three coronoids separated by well-developed pits for the upper fangs, a Meckelian bone extending to the symphysis (Gross, 1941:Figs. 17–19), and long prearticulars are characters also found in osteolepiforms and porolepiforms. Panderichthyids have an adsymphysial tooth plate, as do osteolepiforms; porolepiforms have a parasymphysial tooth spiral in a similar position which is distinct in structure and position from the adsymphysial plate of panderichthyids and osteolepiforms. In *P. rhombolepis*, the adsymphysial plate bears a shagreen of small teeth, as in osteolepiforms, whereas it is developed into a coronoid-like plate with fangs in *P. bystrowi* (Vorobyeva, 1962) and *Obruchevichthys gracilis* (Vorobyeva, 1977a:Figs. 15D, 46A).

Possession of polyplocodont teeth (Fig. 21) often is cited as a synapomorphy of the Choanata (Schultze, 1969, 1970; Vorobyeva, 1977b), but Jessen (1980) described the same tooth structure in *Powichthys*, a primitive member of Porolepiformes or a primitive rhipidistian (Young-olepiformes, Gardiner, 1984). Therefore, the structure must be interpreted as a primitive feature for rhipidistians. The teeth of *Panderichthys* closely resemble those of primitive tetrapods (Figs. 22–26) in that bone of attachment does not reach far into the folded dentine and the teeth are labyrinthodont.

Another primitive feature of panderichthyids is the wide separation of the relatively small and poorly differentiated nasal capsules (Vorobyeva, 1971; Worobjewa, 1973). In this respect, panderichthyids are similar to porolepiforms. The structure of the nasal capsules may be explained by the posteroventral position of the external nasal opening (a derived feature) and the absence or weak ossification of the nasal walls—both characters are shared with tetrapods. In their possession of a rostrocaudal extension of the dermintermedial process, an ororostral groove, large ventrolateral fenestra, and posterior position of the opening for the nervus olfactorius, panderichthyids differ from osteolepiforms. Some of these features are similar to those in porolepiforms; panderichthyids are either more primitive than osteolepiforms or secondarily simplified in these characteristics.

The skull roof and cheek of panderichthyids (Figs. 31, 37) combine primitive and advanced features. The changes from the osteolepiform pattern (Figs. 30, 38) can be explained by (1) elongation of the preorbital, jugal, and specifically the prenasal region, (2) the marginal position of the external nares, (3) the dorsal position of the orbits, and (4) shortening of the posterior shield. The lengthening of the lacrimal and prefrontal (= posterior tectal of osteolepiforms) is related to the lengthening of the snout; the reduction of the external portion of the lateral rostral is related to the marginal position of the external nares. Posterior extension of the jugal separates the squamosal from the maxilla.

The presence of numerous rostral, postrostral, and nasal elements is primitive, although the number of elements is reduced compared with those osteolepiforms and porolepiforms in which the bone mosaic of the snout is visible. The presence of a pair of frontals in panderichthyids changes the whole appearance of the region. The rhipidistian posterior tectal and supraorbital(s) are altered to approach the shape and position of the prefrontal and postfrontal of tetrapods, respectively.

The position of the parietal foramen posterior to the level of the orbits and the shortening of the posterior shield occur in some osteolepiforms also. The ratio of the anterior to posterior shield lengths is about 2:1 in *Eusthenopteron* (Jarvik, 1948:Fig. 16B), *Platycephalichthys*, and *Jarvikina* (Vorobyeva 1977a:Fig. 42), whereas the posterior shield is nearly equal in length to the anterior shield in primitive osteolepiforms (Jarvik, 1948:Ta-

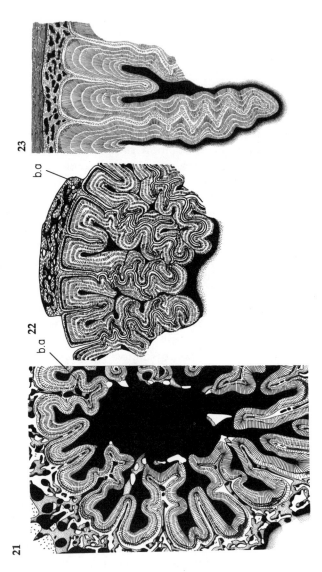

88

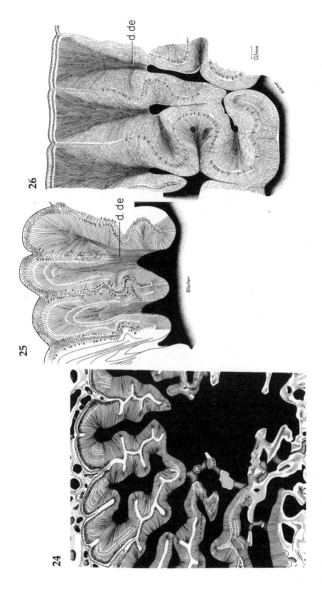

Figs. 21–26. Plicidentine in Choanata. (21) Polyplocodont plicidentine in osteolepiform *Eusthenopteron foordi* (after Schultze, 1969:Fig. 10). (22–26) Labyrinthodont plicidentine: (22) Panderichthyid *Panderichthys rhombolepis* (after Pander, 1860:Pl. G, Fig. 7); (23) Temnospondyl *Loxomma allmanni* (after Schultze, 1969:Fig. 15A); (24) *Ichthyostega* sp., cross section at base of tooth (after Schultze, 1969: Fig. 14); (25) *Ichthyostega* sp., higher cross section with "dark dentine" (after Schultze, 1969:Fig. 13); (26) Anthracosaur *Eogyrinus attheyi* with "dark dentine" (after Schultze, 1969:Fig. 15B). See Appendix I for key to abbreviations.

bles 3, 5, 7, 9). Nevertheless, panderichthyids with a ratio between 2.7:1 to 2.9:1 are similar to Ichthyostegalia (3:1 to 4.5:1, Säve-Söderbergh, 1932:Figs. 2, 15) and loxommatid tetrapods (3.1:1, *Colosteus*, Hook, 1983:Fig. 1; 4.2:1, *Greererpeton*, Smithson, 1982:Fig. 7). The loss of the extratemporal and postspiracular and the absence of an external intracranial joint clearly distinguish the posterior skull of panderichthyids from that of osteolepiforms. A unique feature of panderichthyids is the deep spiracular slit.

The opercular series of *Panderichthys* is similar to that of osteolepiforms (Figs. 37–38). The topographic and size relationships of the opercular and subopercular are characteristic of osteichthyan fishes, whereas Porolepiformes are specialized in the high dorsal position of opercular and subopercular. Branchiostegals are missing in osteolepiforms and panderichthyids; instead, a submandibulobranchiostegal is present and occupies the space between the opercular and submandibular series. The presence of the submandibulobranchiostegal and the overlap of the lateral extrascapulars over the median extrascapular are features shared by osteolepiforms and panderichthyids. They could be interpreted as synapomorphies of both groups, but we prefer to interpret them as synapomorphies for all Choanata that are lost secondarily in tetrapods.

An important feature that unites the Choanata is the presence of only one external nasal opening combined with the presence of a palatal opening (choana) that is surrounded by the premaxilla, maxilla, vomer, and dermopalatine, and located between the outer (maxilla, premaxilla) and inner (vomer, palatine) dental arcade. In contrast to osteolepiforms, in panderichthyids the palatal choanal opening is elongated and the nariochoanal lamina is narrow because the external nasal opening lies near it. There can be no doubt about the homology of this opening between both groups and tetrapods. In tetrapods, the opening is surrounded by the same bones, except in ichthyostegids, in which the premaxilla is excluded from the margin of the choana. Rosen et al. (1981) argued that this opening in panderichthyids is a pit to accept the first coronoid fang of the lower jaw. However, in its natural position the coronoid fang of the lower jaw of specimen PIN 3547/19 fits into the space between the fangs of the vomer and dermopalatine medial to the choana where the lateral toothed ridge is interrupted and a shallow depression formed (cf. pit in Fig. 14 of Rosen et al. 1981).

In the palatal region, the lack of dermal covering of the large anterior palatal recess is a primitive feature of Choanata. A median endocranial ridge separates the posterior part of the palatal recess in *Panderichthys stolbovi* (Worobjewa, 1975b:Fig. 2E; Vorobyeva, 1977a:Fig. 3E); this may be comparable to the division of the palatal recess in primitive osteolepiforms and porolepiforms (Worobjewa, 1975b). Such a structure is absent in *P. rhombolepis* (Figs. 12, 13). Behind the palatal recess, the vomerine tooth ridges of each side meet one another. This feature must be interpreted as a parallel development in advanced osteolepiforms (Worobjewa, 1975b), as also may be the case for the lateral and posterior processes of the vomer.

The presence of only one medial ethmoidal articulation of the palatoquadrate may be a primitive feature of Choanata. However, the palatoquadrate is immobile and synchondrotically united to the ethmoid, as in advanced Late Devonian osteolepiforms such as *Eusthenopteron, Jarvikina,* and *Platycephalichthys* (Vorobyeva, 1977a). The articulation is ligamentous in earlier osteolepiforms (e.g., *Thursius estonicus, Osteolepis macrolepidota, Megapomus markovskyi*) or movable in other osteolepiforms (e.g., *Megalichthys nitidus, Megistolepis klementzi, Gyroptychius pauli*).

The Choanata have a different form of articulation between scapulocoracoid and first, proximal pectoral endoskeletal element (i.e., humerus) than do other fishes (Schultze, 1987). The humerus forms the ball that resides in the socket of the scapulocoracoid. In other features of the postcranial skeleton, panderichthyids are distinct from osteolepiforms. They lack median fins except the caudal fin fold. The pelvic fins are located far posteriorly—at about the level of the anal fin in osteolepiforms. Basal scutes, a typical feature of osteolepiforms, are absent at the base of the paired fins. The fusion of the distal elements in the endoskeleton of the pectoral fin (Worobjewa, 1975a) may be another feature distinguishing panderichthyids from osteolepiforms.

The composition of the vertebrae varies tremendously from Porolepiformes to tetrapods and within each group (Andrews and Westoll, 1970; Vorobyeva and Tsessarskii, 1986). Only three elements take part in the formation of these vertebrae—the basiventral (= intercentrum), basidorsal (= neural arch), and interdorsal (= pleurocentrum) arcocentrum. Primitively, the pleurocentrum is small, but it is enlarged in some osteolepiforms and tetrapods (gastrocentrous vertebrae). *Panderichthys* and *Elpistostege,* like *Crassigyrinus,* lack pleurocentra (Fig. 20). The notochord is unconstricted in these forms, even though the intercentrum, neural arch, and rib fuse to form one unit in *Panderichthys.* The weak ossification of the vertebrae in *Crassigyrinus* was considered to be a degeneration by Panchen (1985), but it is comparable to that in Panderichthyida. Taking the weak vertebrae of panderichthyids and the small pleurocentrum in some osteolepiforms into account, one can argue that rhachitomous vertebrae developed independently in osteolepiforms and tetrapods, owing to the plasticity of their morphogenesis (Vorobyeva and Tsessarskii, 1986).

In conclusion, comparison of the distribution of characters in Panderichthyida and Osteolepiformes shows that the characters common to both groups are either primitive features or features of the larger unit, Choanata (Fig. 39). Of the four synapomorphies for Osteolepiformes (Long, 1985), only the barlike preopercular and large basal scute seem to remain. Postorbital and jugal bones of nearly equal size occur in *Porolepis* (similar to the osteolepiform *Beelarongia*, Long, 1987) and may be present in some panderichthyids, but most panderichthyids tend toward enlargement of the jugal, as do tetrapods. There is a morphocline of increasing size of the lacrimal; thus, the lacrimal is relatively larger in panderichthyids, Ichthyostegalia, and primitive tetrapods than in osteolepiforms. The preopercular lies at the posterior margin of the cheek in panderichthyids and ichthyostegids; however, in the latter (except *Acanthostega*) it has an oblique orientation. The external exposure of the anocleithrum may be similar in panderichthyids and osteolepiforms. Three other features (the submandibulobranchiostegal, the presence of a median gular, and the overlap of lateral over median extrascapular) are common to panderichthyids, osteolepiforms, and rhizodonts, whereas in porolepiforms, actinistians, and dipnoans the median extrascapula overlaps the lateral one. These must be interpreted as synapomorphies for the Choanata to be consistent with the large number of features that are common to panderichthyids and tetrapods (see below). Unfortunately, they cannot be examined in tetrapods because they are lost.

Panderichthyida and Rhizodontida

Our knowledge of the Rhizodontida is limited and based mainly on publications by Andrews (1972, 1973, 1985). The pattern of the posterior skull roof (separate supratemporal), the overlap of the lateral extrascapulars over the median one, the presence of a median gular, and the uniserial archipterygium are features that rhizodontids share with osteolepiforms and panderichthyids. The articulation between humerus and scapulocoracoid is another choanate character that unites rhizodontids with choanates; the rounded head of the humerus lies in a glenoid fossa (Andrews and Westoll, 1970) in contrast to the situation in porolepiforms and other sarcopterygians (Schultze, 1987), in which the humerus rotates around a rounded head of the scapulocoracoid.

Although the polyplocodont structure of the teeth is well developed in rhizodontids, osteolepiforms, panderichthyids, and tetrapods, this feature must be considered to be primitive because it also occurs in young-olepiforms. The "dark dentine" utilized by Panchen (1985) to characterize Anthracosauroidea occurs in rhizodontids (Schultze, 1969:Pl. 12, Fig. 3). Rhizodonts are more similar to panderichthyids than to osteolepiforms in possessing small pelvic fins and a symmetrical caudal fin. At present, we know of no features that would ally the rhizodontids more closely to one of these groups than the other; nonetheless, the aforementioned features would place them within the Choanata (Fig. 39). However, the presence of two external nares (as reconstructed by Andrews [1985]) is a primitive feature for rhipidistians and argues against such placement. If the Rhizodontida possesses two external nares, its phylogenetic position would lie between the Porolepiformes and Choanata. The characters they share with the Choanata would no longer be unique to the Choanata.

Panderichthyida and Tetrapoda

The earliest tetrapods are known from the Late Devonian. Whole skeletons of *Ichthyostega* have been described from the Late Devonian of Greenland (Säve-Söderbergh, 1932; Jarvik, 1952, 1980), whereas the second genus, *Acanthostega*, of the Ichthyostegalia is known only from the skull (Jarvik, 1952) and a little postcranial material (Clack, 1988). Recently, the fragmentary remains of a third Late Devonian genus, *Tulerpeton*, were described from central Russia (Lebedev, 1984, 1985). Monographic descriptions of early members of various new tetrapod groups are available: Temnospondyli—*Greererpeton* (Smithson, 1982); *Colosteus* (Hook, 1983); Batrachosauria—*Eoherpeton* (Smithson, 1985); *Proterogyrinus* (Holmes, 1984); *Anthracosaurus* (Panchen, 1977; Clack, 1987). *Crassigyrinus*, the amphibian recently described by Panchen (1985), deserves special consideration because some of the features that Panchen used to distinguish the genus from anthracosaurs also occur in panderichthyids (Schultze and Arsenault, 1985).

All early tetrapods except *Tulerpeton* (which possesses an ano- and supracleithrum) lack an opercular apparatus—i.e., the connection of the shoulder girdle with the skull roof—and fins. Each possesses typical tetrapod features such as dactylic extremities and a pelvic girdle fixed to the vertebral column (see Gaffney, 1979, for monophyly of Tetrapoda). In addition, they share the features listed below with panderichthyids (Fig. 39).

1. The skull roof is flat (Figs. 33–34, 36–37) relative to that of the typical fish and osteolepiform skull (Fig. 38).

2. The dorsally placed orbits are narrowly separated (Figs. 27–29, 31–32); the orbits are lateral in osteolepiforms.

3. The external naris (Figs. 33–34, 36–37) is located near the margin of the upper jaw (Panchen, 1967). Thus, the external naris and choana are close to each other, and both are visible in ventral view.

4. The presence of paired frontals is unique to panderichthyids and tetrapods (Figs. 27–29, 31–32). There are only two bones lateral to the frontals—the prefrontal and postfrontal; these are named the posterior tectal and supraorbital, respectively, in osteolepiforms.

5. The external intracranial joint is absent (Figs. 27–29, 31–32). Little is

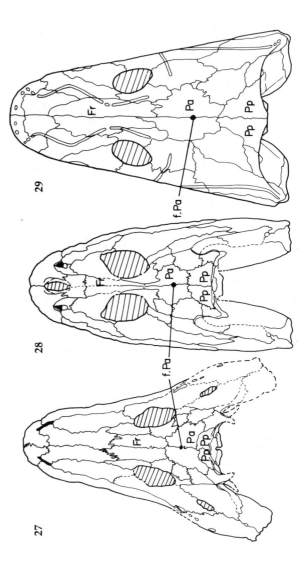

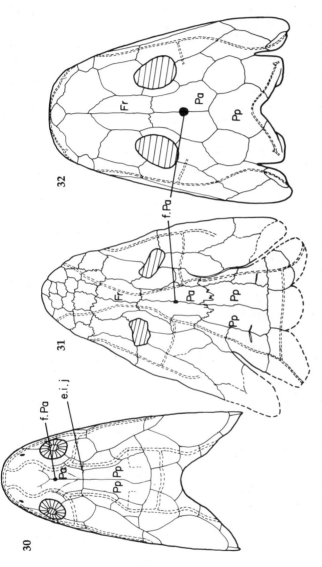

Figs. 27–32. The skull roofs of Choanata. (27) Anthracosauridae: *Anthracosaurus russelli* (after Panchen, 1977:Fig. 4). (28) *Crassigyrinus scoticus* (after Panchen, 1985:Fig. 7). (29) Temnospondyli: *Greererpeton burkemorani* (after Smithson, 1982:Fig. 7). (30) Osteolepiformes: *Osteolepis macrolepidotus* (after Jarvik, 1948:Fig. 32B). (31) Panderichthyida: *Panderichthys rhombolepis*; compare with Figure 8. (32) Ichthyostegalia: *Ichthyostega* sp. (after Jarvik, 1952:Fig. 35B). See Appendix I for key to abbreviations.

known about the endocranial intracranial joint in these groups. Neo-tetrapods lack the joint, but vestiges of it may be represented in embryonic stages (Roček, 1986; Schultze, 1987). Jarvik (1980) argued that *Ichthyostega* possessed an endocranial intracranial joint. The condition of this character is unknown in panderichthyids.

6. The postparietal shield is short in panderichthyids but longer than that of tetrapods (Figs. 27–29, 31–32). The parietal has a more posterior position in panderichthyids than it does in osteolepiforms; the bone extends between the orbits, but its main portion lies posterior the orbits, as it does in early tetrapods. The parietal foramen lies posterior to the level of the orbits. The Ichthyostegalia (Fig. 32) often are characterized as having an unpaired postparietal (Gaffney, 1979); however, in *Acanthostega* the postparietals are paired. In *Panderichthys,* pairing of the postparietal (Fig. 5) is individually variable.

7. The Tetrapoda (Figs. 33–36) is distinct from the Osteolepiformes (Fig. 38) in the composition of the cheek region; the jugal and quadratojugal articulate, thereby separating the squamosal from the maxilla (Panchen, 1985). In *Panderichthys,* the relationship between these bones varies. Whereas *Elpistostege* has the osteolepiform condition, one specimen of *Panderichthys rhombolepis* (Figs. 4, 37) has the tetrapod condition. In another specimen of *P. rhombolepis* (PIN 3547/26), the squamosal reaches the maxilla on the left side but not the right. This feature unites panderichthyids and tetrapods, and is not exclusive to tetrapods (contra Panchen and Smithson, 1987).

8. The teeth of panderichthyids and early tetrapods have an advanced polyplocodont structure (Figs. 22–26). The bone of attachment extends only a short distance into the folds of dentine, and the central axes of the folds undulate. The labyrinthodont structure of the teeth of *Panderichthys* and *Ichthyostega* is nearly indistinguishable (Schultze, 1969, 1970). The dentine folds of these genera retain side branches—a feature of polyplocodont teeth (Fig. 21) that is absent in advanced labyrinthodont teeth.

9. Panderichthyids lack median fins (Fig. 19) except for a caudal fin fold like that of *Ichthyostega.* The dorsal portion of this fin fold extends farther forward than the ventral portion, and it is possible that the second dorsal fin is incorporated into the caudal fin fold, an interpretation suggested for *Ichthyostega* by Jarvik (1980).

10. Attachment of ribs to the neural arch and intercentrum occurs in panderichthyids and the Ichthyostegalia. The small pleurocentrum of Ichthyostegalia is unknown in panderichthyids. The composition of the vertebrae of some osteolepiforms (i.e., *Osteolepis,* Andrews and Westoll, 1970; Andrews, 1977) more closely resembles that of certain neotetrapods than that of panderichthyids and ichthyostegids. In our opinion, this resemblance represents parallel evolution (Vorobyeva and Tsessarskii, 1986). Support for such an interpretation comes from *Crassigyrinus,*

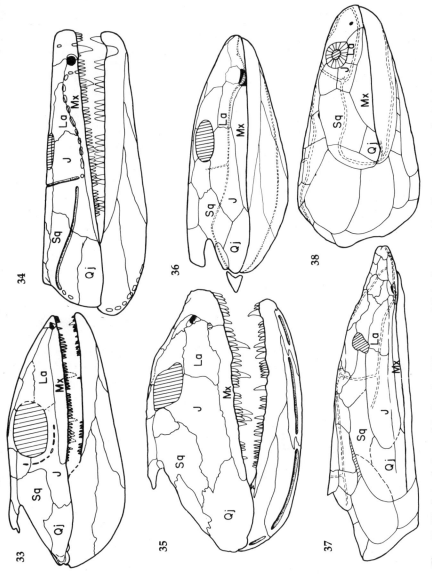

Figs. 33–38. Lateral views of the heads of Choanata. (33) Anthracosauridae: *Proterogyrinus scheelii* (after Holmes, 1984:Figs. 3a, 15a). (34) Temnospondyli: *Greererpeton burkemorani* (after Smithson, 1982:Fig. 9). (35) *Crassigyrinus scoticus* (after Panchen, 1985:Fig. 9). (36) Ichthyostegalia: *Ichthyostega* sp. (after Jarvik, 1952:Fig. 35A). (37) Panderichthyida: *Panderichthys rhombolepis*. (38) Osteolepiformes: *Osteolepis macrolepidotus* (after Jarvik, 1948:Fig. 32A). See Appendix I for key to abbreviations.

which possesses vertebrae similar to those of panderichthyids; thus, an intercentrum and neural arch are present, but the pleurocentrum is absent.

Crassigyrinus (Figs. 28, 35) is an unusual form that possesses a combination of tetrapod, panderichthyid, and osteolepiform features. Panchen (1985) placed *Crassigyrinus* within anthracosauroids on the basis of five shared features. The first, the presence of a consolidated braincase, seems to be a synapomorphy for all Neotetrapoda. The second feature is the presence of characteristic dermal ornament. The third, possession of a tabular horn, "may not seem a very compelling character. . . . Taken in combination with the next character [absence of posttemporal fossae] . . . the whole is quite unique" (Panchen, 1985:552). The fifth character, "dark dentine," is a primitive feature that occurs in *Ichthyostega* (Fig. 25; cf. Schultze, 1969:Pl. 13, Fig. 2b) and in the rhizodont *Strepsodus* (Schultze, 1969:Pl. 12, Fig. 3). The "dark dentine" results from the rapid retreat of peripheral odontoblasts toward the pulp cavity. This phenomenon is observed only in thin sections made at the proper level. Thus, Figure 3 of Plate 7 (Schultze, 1969) shows indications of "dark dentine" in *Panderichthys* that could be well developed at a slightly higher level. In any case, "dark dentine" is not a feature uniting anthracosauroids.

Two of the four autapomorphies of *Crassigyrinus* are features shared with panderichthyids—i.e., the presence of constricted parietals and frontals in the interorbital region, and a deep cheek below the orbit. There are many other characters that contraindicate the placement of *Crassigyrinus* within Neotetrapoda—e.g., the presence of preopercular, an undivided open palatal recess, anterior projection of notochord into basioccipital, the position of external nasal opening, and the distribution of foramina on the humerus. The grooves described by Panchen (1973) on the jugal are exaggerated (deep and narrow) components of the ornament. The lateral-line canal is limited to a superficial groove on the lower jaw. There are characters that Panchen (1985) interpreted as being degenerate—e.g., the lack of occipital condyle, the presence of weak vertebrae composed only of an intercentrum and neural arch, and ribs not clearly bicipital. Some characters, such as the wide separation of the external naris from the margin of upper jaw (reversal for *Crassigyrinus*) and the lack of reduction of the pars facialis of the lateral rostral, are primitive among tetrapods.

The composition of the lower jaw is stable in rhipidistians and lower tetrapods, consisting of four infradentaries and three coronoids. Coronoids are lost within tetrapods. Primitively, coronoids bear fangs that are bordered laterally by a row of small teeth. This arrangement changes in tetrapods; fangs are absent, and only a lateral row of small teeth is present. *Crassigyrinus* has a shagreen of small teeth on the coronoids comparable to that of anthracosauroids; the larger teeth are similar to the fangs in osteolepiforms and panderichthyids. The fangs of the upper jaw fit into large coronoid pits between the coronoids in rhipidistians. These pits and the dermopalatine and ectopterygoid fangs are absent in tetrapods. Like other early tetrapods, *Crassigyrinus* possesses fangs in the upper jaw on the dermopalatine and ectopterygoid; like osteolepiforms, it has small teeth on the pterygoid. To accommodate the tips of the palatal fangs in the lower jaw, *Crassigyrinus* has a row of pits lateral to the coronoids on the medial side of the dentary. Thus, a primitive upper jaw structure is combined with specialization of the lower jaw of *Crassigyrinus*.

It seems that the presence of so many primitive characters in *Crassigyrinus*, combined with the lack of many tetrapod synapomorphies, would place *Crassigyrinus* close to the base of tetrapod phylogeny, perhaps even below Ichthyostegalia. *Crassigyrinus* lacks four of the seven cited synapomorphies for Neotetrapoda in our cladogram (Fig. 39); the remaining three are not known or uncertain in *Crassigyrinus*. The genus resembles *Ichthyostega* in having a preopercular, a palatal recess not covered by bone, and a notochord entering the braincase; all these are primitive features. In addition, *Crassigyrinus* lacks some tetrapod synapomorphies involving the pelvic region and external nasal opening. The presence of only 11 of the 14 synapomorphies of tetrapoda would place *Crassigyrinus* below the branching between Ichthyostegalia and Neotetrapoda. Panchen (1985) dismissed some of Gaffney's (1979) tetrapod characters; in so doing he was able to support Watson's (1929) proposal that *Crassigyrinus* is an anthracosauroid.

A flat skull roof uniting panderichthyids and tetrapods is correlated with the dorsal position of the orbits. Lengthening of the preorbital region is correlated with the formation of frontals. The shortening of the postorbital region in connection with a lengthening of the preorbital region may require loss of the dermal intracranial joint. In addition, some of the features indicate correlation with environmental factors. Some Recent actinopterygians that leave the water in the intertidal zone possess dorsally placed eyes for subareal observation. The "eyebrows" in panderichthyids are indicative of an elevated position of the eyes in the dorsally placed orbits. The ventral position of the external nostrils also could be related to an amphibious life style.

In conclusion, the number of common features shared by tetrapods and panderichthyids is surprisingly large (Fig. 39) and includes four features that Gaffney (1979) used to justify the monophyly of tetrapods. Anterior and posterior shields of the skull roof are sutured (part of Character 1 of Gaffney, 1979; the situation in the endocranium is unknown). There is a clear tendency in tetrapods toward the shortening of the otico-occipital region (Character 2 of Gaffney, 1979); although pan-

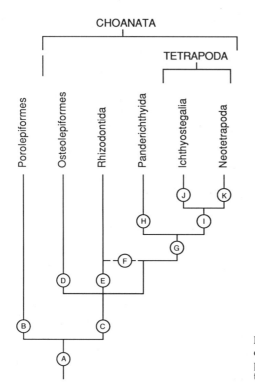

Fig. 39. Interrelationships of rhipidistian crossopterygians. See Appendix II for explanation of distribution of characters across Nodes A–K.

derichthyids have not attained the same degree of shortening found in tetrapods, the difference is only gradational. There is some ambiguity about the "otic notch" (Character 3 of Gaffney, 1979). Gaffney (1979) defined the "otic notch" as "a V-shaped cleft formed at the posterior margin of the skull between the tabular and squamosal" (= squamosal embayment of Godfrey et al., 1987), whereas Panchen (1985) associated the term with a restored tympanum. The opening of the spiracular cavity in rhipidistians has the same position as the squamosal embayment in labyrinthodonts (Shishkin, 1973:188; Smirnov and Vorobyeva, 1986). The spiracular canal opens lateral to the anterior part of the tabular and posterolateral to the supratemporal in osteolepiforms (Jarvik, 1980:Figs. 122, 144A) and in panderichthyids (Figs. 2, 4–5). In osteolepiforms, the posterior continuation is covered, whereas it is open in panderichthyids and tetrapods. The squamosal embayment has the same relative position in *Crassigyrinus* and neotetrapods (Figs. 27–28), whereas it is shortened in ichthyostegids (Fig. 32). *Greererpeton* (Fig. 29) and *Colosteus* (Hook, 1983:Fig. 1A) lack a squamosal embayment. The latter forms and the ichthyostegids have no lateral kinetism between the skull roof and cheek region, as is found in osteolepiforms (Vorobyeva, 1977a) and anthracosaurs (Holmes, 1984). There is also no movement possible among the supratemporal, intertemporal, and cheek in panderichthyids. It is difficult for Panchen (1985:514) to establish the presence of lateral kinetism in *Crassigyrinus* because the skull material is either incomplete or badly crushed. *Crassigyrinus* differs in the arrangement of the bones in the posterior skull roof from anthracosauroids in that it shows the primitive osteolepiform arrangement—i.e., tabular and supratemporal lateral to postparietal, and no contact between the tabular and the parietal.

Ossified ribs (Character 11 of Gaffney, 1979) also occur in panderichthyids. They are large relative to those of osteolepiforms, but not compared with tetrapods. Finally, the status of Gaffney's Character 8 (pectoral skeleton free from skull) is uncertain because Lebedev (1984:1471) described an anocleithrum and supracleithrum in a Late Devonian tetrapod, *Tulerpeton*.

The characters common to panderichthyids and tetrapods, and the close resemblance of *Crassigyrinus* to panderichthyids clearly indicate a close sister-group relationship between panderichthyids and tetrapods.

CONCLUSIONS

Panderichthys and *Elpistostege* are the two best known members of the Panderichthyida. Although they are similar in shape and proportions of the skull roof, there are clear differences (the shape of cranial elements—orbit, prefrontal, lacrimal, and parietal; presence vs. absence of an intertemporal; the posterior margin of the skull roof) that justify the separation of the two genera. The large median gular and the subterminal mouth are the most readily recognizable synapomorphies of panderichthyids.

The panderichthyids show a combination of porolepiform, osteolepiform, and tetrapod features that we could interpret as a case of mosaic development ("Watson's rule," de Beer, 1954). This means that panderichthyids retain primitive characters in common with porolepiforms and osteolepiforms—i.e., they are rhipidistians or choanates, respectively—and they share other advanced characters with tetrapods, which makes them the sister-group of the latter. Three characters (the presence of submandibulobranchiostegal, a large median gular, and an overlap of lateral over median extrascapular) could be used to unite panderichthyids with osteolepiforms. These bones are lost in tetrapods, so it is not possible to demonstrate that these features are part of the character set of the Choanata. Nevertheless, the numerous characters common to panderichthyids and tetrapods make it more parsimonious to support such an interpretation.

Although many characters unite panderichthyids and tetrapods (Fig. 39), features such as extrascapulars, opercular series with gular plates

and submandibulars, and paired fins allow us to recognize the panderichthyids as fishes. However, ichthyostegids possess so-called fish
features such as fin rays (lepidotrichia) and fin-ray supports in the caudal
fin, and preopercular, subopercular, dermal scales, and lateral-line canals
inside the bones. *Tulerpeton*, a tetrapod from the Late Devonian of the
USSR, possesses still more of these features (e.g., anocleithrum and
supracleithrum); therefore, it is possible that the connection between
skull roof and shoulder girdle still exists or that the gap between skull
and girdle is narrow. Like *Tulerpeton*, another tetrapod, *Crassigyrinus*, is
placed within the anthracosauroids (Panchen, 1985), although it shares
many features that are primitive for tetrapods with panderichthyids. It
even retains osteolepiform features (e.g., the nature of the external nares
and the shape of humerus). It seems more reasonable to place *Crassigyrinus* at the base of the tetrapods immediately above or below the
ichthyostegid level because it has not acquired all tetrapod synapomorphies. The similarities between *Crassigyrinus* and panderichthyids are
another strong indication that the panderichthyids are related closely to
tetrapods.

Acknowledgments Most of this chapter was written during a three-
month visit by Dr. E. Vorobyeva to Lawrence, Kansas, in 1986. This visit
was supported by the Interacademy Exchange Program between the
National Academy of Sciences of the United States, Washington, D.C.,
and the Academy of Sciences of the USSR. The drawings were prepared
by Colleen Z. Gregoire and Donna Stevens (Museum of Natural History,
The University of Kansas). The English was improved by Linda Trueb
and J. Chorn (Museum of Natural History, Lawrence, Kansas). J. A. Clack
(University Museum of Zoology, Cambridge, England) and Linda Trueb
(Museum of Natural History, The University of Kansas) reviewed the
manuscript carefully. Dr. J. K. Ingham, Hunterian Museum, The University, Glasgow, Scotland, supplied us with a colored photograph of the
holotype of *Crassigyrinus scoticus*. Jan Elder (Word Processing Service,
Division of Biology, The University of Kansas) typed the manuscript. We
extend our deepest thanks to all these persons who made possible the
completion of this essay.

APPENDIX I. Anatomical Abbreviations Used in
the Figures

a.b = anterior buttress; Art = articular; art = articular (glenoid) fossa
b.a = bone of attachment
c.gl = glenoid canal; ch = choana; Cla = clavicle; Clei = cleithrum; c.mn = canal
    for ramus medialis of nervus profundus; c.n.b = nasobasal canal; c.o.lat =

canal for ramus lateralis of nervus ophthalmicus; c.olf = canal for nervus olfactorius; c.pal = canal for palatine branch of nervus facialis; c.prt = canals for twigs of medial terminal branch of nervus profundus; cr.im = crista intermedia

d.de = "dark dentine"; De = dentary; der = dermal process of anterior tectal; Dpt = dermopalatine

Ecpt = ectopterygoid; e.i.j = external intracranial joint; end = endochondral bone; Esc.l, m = lateral, median extrascapula

fe.pa = palatal recess; fo.dm = dorsomedial depression in posterior face of postnasal wall; fo.sc = supracoracoid foramen; fo.sg = supraglenoid foramen; f. Pa = parietal foramen; Fr = frontal

G.m = median gular; g.n.b = nasobasal groove; G.p = principal gular

Hyo = hyomandibular

ic = intercentrum; icla = interclavicle; Id 1,2,3,4 = infradentary 1,2,3,4; ioc = infraorbital canal; It = intertemporal

J = jugal

La = lacrimal

m.b = middle buttress; Mx = maxilla

Na = nasal; na = external nasal opening; n.a = neural arch; nch = notochord

Pa = parietal; Part = prearticular; p.cor = coracoid plate; p.ioc = pores of the infraorbital canal; p.m. = middle pitline; Pmx = premaxilla; Po = postorbital; Pof = postfrontal; Pop = preopercular; Pp = postparietal; Pq = palatoquadrate; Pr = postrostral; Pr.a, m, p = anterior, median, posterior postrostral; pr.dim = dermintermedial process of lateral rostral; Prf = prefrontal; pr.im = intermedial process; Psp = parasphenoid; p.Sq = pitline of squamosal; Pt = pterygoid (endopterygoid)

Q = quadrate portion of the palatoquadrate; Qj = quadratojugal

r = rib; R.l, m = lateral, medial rostral; re.l = recessus lateralis

Sbm = submandibular; Sbr = preoperculosubmandibular; Sh = stylohyal; soc = supraorbital canal; Sop = subopercular; sp = spiracular slit; Sq = squamosal; St = supratemporal

Ta = tabular; Te.a = anterior tectal; t.Pmx = cross section of premaxillary teeth

Uhy = urohyal

Vo = vomer

APPENDIX II. Distribution of Character-States in Rhipidistian Crossopterygians (Fig. 39)

Node A: Rhipidistia excluding onychodonts (see Schultze, 1987). (1) Extratemporal lateral to tabular; (2) many narrow submandibular bones; (3) fenestra ventrolateralis, covered by vomer in porolepiforms; (4) four infradentaries; (5) three coronoids; (6) plicidentine, simply folded, polyplocodont being primitive; (7) supraorbital canal joining cephalic division of main lateral line.

Node B: Porolepiformes. (8) Intertemporal fused with parietal; (9) supratemporal fused with postparietal; (10) nariodal bone between anterior and posterior external nasal opening; (11) two or three squamosals versus only one; (12) prespiracular bone in upper cheek region; (13) preoperculosubmandibular, an additional bone at posterior margin of the cheek region; (14) high, dorsally placed opercular; (15) short anterior margin of opercular; (16) dorsolaterally placed "triangular" subopercular; (17) parietal foramen closed; (18) dendrodont tooth structure (Youngolepiformes, which have polyplocodont teeth, not included here); (19) teeth of dentary not reaching symphysis; (20) parasymphysial tooth spiral on symphysis; (21) olfactory bulbs ventral to cerebral hemisphere; (22) paired buccohypophysial canal.

Node C: Choanata. (23) One external nasal opening; (24) choana between premaxilla, maxilla, vomer, and dermopalatine; (25) median rostral fused with premaxilla; (26) few unpaired or paired postrostrals; (27) lateral extracapulars overlapping median extracapular; (28) lateral rostral ventral to external naris; (29) large lacrimal; (30) narrow preopercular bone at posterior margin of cheek; (31) submandibulobranchiostegal present, but no branchiostegals; (32) seven submandibulars; (33) median gular always present; (34) superior, inferior, and often median recess present in nasal capsule; (35) crista intermedia present; (36) nasobasal canal present; (37) palatal recess unpaired and without bone cover; (38) posterior process on vomer in most osteolepiforms; (39) long parasphenoid extending below otico-occipital region; (40) ossified dorsal ribs; (41) two-headed articulation of ribs; (42) uniserial archipterygium in pectoral fin; (43) tetrapod articulation between humerus and scapulocoracoid.

Node D: Osteolepiformes. (44) Steeply inclined preopercular at posterior margin of cheek region; (45) basal scutes on all fins; (46) large, ornamented anocleithrum.

Node E: Trichotomy of Rhizodontida, Osteolepiformes, and other Choanata. (47) Lateral extracapulars overlapping median extracapula; (48) median gular present; (49) uniserial archipterygium in pectoral fin.

Node F: Rhizodontida above Osteolepiformes. (50) Lateral line diverting from tabular and supratemporal into elongated postparietal; (51) reversed overlap between cleithrum and clavicle.

Node G: Panderichthyida and Tetrapoda. (52) Flat skull roof; (53) closely placed of orbits; (54) marginal position of external naris; (55) paired frontals in front of parietals and orbits; (56) external intracranial joint lost (intimate suturing between anterior and posterior portion of skull roof); (57) parietals extending anteriorly between orbits, but main portion with parietal opening behind orbits; (58) extratemporal absent; (59) long spiracular fenestra extended posteriorly to form squamosal embayment; (60) jugal meeting quadratojugal (squamosal separated from maxilla; not in all panderichthyid specimens, as far as known); (61) obliquely placed preopercular at posterior margin of cheek; (62) lateral wall and posterior floor of nasal capsule not ossified; (63) nasal capsule extending posteriorly beyond ethmoid region; (64) labyrinthodont plicidentine; (65) large dorsal ribs directed ventrally; (66) no median fins (except caudal fin fold).

Node H: Panderichthyida. (67) Median rostral separated from premaxilla (reversal; also in Porolepiformes); (68) paired posterior postrostrals; (69) large median gular; (70) lateral recess in nasal capsule (reversal; also in Porolepiformes); (71) subterminal mouth.

Node I: Tetrapoda (cf. Gaffney, 1979). (72) One pair of nasals meeting in midline; (73) posterior postrostrals lost; (74) lateral rostral transformed to septomaxilla (not in *Crassigyrinus*); (75) extrascapulars lost; (76) opercular and submandibulo-branchiostegal lost; (77) submandibulars lost; (78) gulars lost; (79) otico-occipital region of skull longitudinally compacted; (80) stapes; (81) fenestra ovalis; (82) presence of phalanges in fore and hind limbs combined with ankle and wrist joints; (83) pectoral girdle free from skull roof (opercular, supracleithrum and anocleithrum, posttemporal lost, except *Tulerpeton*); (84) iliac blade of pelvis attached to vertebral column (except in *Crassigyrinus*); (85) well-developed ischiac ramus of pelvis (except in *Crassigyrinus*, where the ischium is dermal [!] after Panchen, 1985).

Node J: Ichthyostegalia. (86) Premaxilla excluded from margin of choana (paralleled by some neotetrapods); (87) intertemporal fused with supratemporal; (88) short parasphenoid between and behind pterygoids; (89) adsymphysial plate incorporated into coronoid series as anterior coronoids (after Jarvik, 1980).

Node K: Neotetrapoda. (90) Preopercular lost (except in *Crassigyrinus*) and subopercular lost; (91) palatal recess covered by bone or divided by bony bridge (not in *Crassigyrinus*); (92) anterior coronoid (possibly including adsymphysial plate) lost; (93) ethmosphenoid and otico-occipital portions of braincase fused in adults; (94) notochord excluded from braincase in adults (except *Crassigyrinus*); (95) lateral-line system absent or in grooves; (96) median bony fin supports and lepidotrichia lost.

Literature Cited

Andrews, S. M. 1972. The shoulder girdle of "*Eogyrinus*." Pp. 35–48 *in* Joysey, K. A., and T. S. Kemp (eds.), *Studies in Vertebrate Evolution*. Edinburgh: Oliver and Boyd.

Andrews, S. M. 1973. Interrelationships of crossopterygians. Pp. 137–177 *in* Greenwood, P. H., R. S. Miles, and C. Patterson (eds.), *Interrelationships of Fishes*. London: Academic Press.

Andrews, S. M. 1977. The axial skeleton of the coelacanth, *Latimeria*. Pp. 271–288 *in* Andrews, S. M., R. S. Miles, and A. D. Walker (eds.), *Problems in Vertebrate Evolution*. Linn. Soc. Symp. Ser. 4.

Andrews, S. M. 1985. Rhizodont crossopterygian fish from the Dinantian of Foulden, Berwickshire, Scotland, with a re-evaluation of this group. Trans. R. Soc. Edinburgh Earth Sci., 76:67–95.

Andrews, S. M., and T. S. Westoll. 1970. The postcranial skeleton of rhipidistian fishes excluding *Eusthenopteron*. Trans. R. Soc. Edinburgh, 68(12):391–489.

Bolt, J. R., and R. E. Lombard. 1985. Evolution of the amphibian tympanic ear and the origin of frogs. Biol. J. Linn. Soc. 24:83–99.

Clack, J. A. 1987. Two new specimens of *Anthracosaurus* (Amphibia: Anthracosauria) from the Northumberland Coal Measures. Palaeontology, 30:15–26.

Clack, J. A. 1988. New material of the early tetrapod *Acanthostega* from the Upper Devonian of East Greenland. Palaeontology, 31:699–724.

de Beer, G. R. 1954. *Archaeopteryx* and evolution. Adv. Sci., 11:160–170.

Forey, P. 1987. Relationships of lungfishes. Pp. 75–91 *in* Bemis, W. E., W. W. Burggren, and N. E. Kemp (eds.), *The Biology and Evolution of Lungfishes*. J. Morphol., Suppl. 1.

Gaffney, E. S. 1979. Tetrapod monophyly: a phylogenetic analysis. Bull. Carnegie Mus. Nat. Hist., 13:92–105.

Gardiner, B. 1984. The relationships of the palaeoniscid fishes, a review based on new specimens of *Mimia* and *Moythomasia* from the Upper Devonian of Western Australia. Bull. Br. Mus. (Nat. Hist.) Geol., 37(4):173–428.

Godfrey, S. J., A. R. Fiorillo, and R. L. Carroll. 1987. A newly discovered skull of the temnospondyl amphibian *Dendrerpeton acadianum* Owen. Can. J. Earth Sci., 24:796–805.

Gross, W. 1941. Über den Unterkiefer einiger devonischer Crossopterygier. Abh. Preuss. Akad. Wiss. Math. Naturwiss. Kl., 1941:3–51.

Hennig, W. 1950. *Grundzüge einer Theorie der phylogenetischen Systematik.* Berlin: Deutscher Zentralverlag.

Hennig, W. 1966. *Phylogenetic Systematics.* (Translated by R. Zangerl.) Urbana: Univ. Illinois Press.

Holmes, R. 1984. The Carboniferous amphibian *Proterogyrinus scheeli* Romer, and the early evolution of tetrapods. Philos. Trans. R. Soc. London, Ser. B., 306:431–527.

Hook, R. W. 1983. *Colosteus scutellatus* (Newberry), a primitive temnnospondyl amphibian from the Middle Pennsylvanian of Linton, Ohio. Am. Mus. Novit., 2770:1–41.

Jarvik, E. 1942. On the structure of the snout of crossopterygians and lower gnathostomes in general. Zool. Bidr. Uppsala, 21:235–675.

Jarvik, E. 1948. On the morphology and taxonomy of the Middle Devonian osteolepid fishes of Scotland. K. Sven. VetenskAkad. Handl., (3)25:1–301.

Jarvik, E. 1952. On the fish-like tail in the ichthyostegid stegocephalians with descriptions of a new stegocephalian and a new crossopterygian from the Upper Devonian of East Greenland. Medd. Groenland, 114:1–90.

Jarvik, E. 1960. *Théories de l'évolution des vertébrés reconsidérées la lumière des récentes découvertes sur les vertébrés inférieurs.* Paris: Masson and Cie.

Jarvik, E. 1964. Specializations in early vertebrates. Ann. Soc. R. Zool. Belgique, 94(1):11–95.

Jarvik, E. 1972. Middle and Upper Devonian Porolepiformes from East Greenland with special reference to *Glyptolepis groenlandica* n. sp. and a discussion on the structure of the head in the Porolepiformes. Medd. Groenland, 187:1–307.

Jarvik, E. 1975. On the saccus endolymphaticus and adjoining structures in osteolepiforms, anurans and urodeles. Colloq. Int. C. N. R. S., 218:191–211.

Jarvik, E. 1980. *Basic Structure and Evolution of Vertebrates.* Vol. 1. New York and London: Academic Press.

Jarvik, E. 1986. The origin of the Amphibia. Pp. 1–24 in Roček, Z. (ed.), *Studies in Herpetology.* Prague: Charles Univ.

Jessen, H. 1980. Lower Devonian Porolepiforms from the Canadian Arctic with special reference to *Powichthys thorsteinssoni* Jessen. Palaeontogr. A, 167:80–214.

Lebedev, O. A. 1984. First discovery of a Devonian tetrapod vertebrate in USSR. Dok. Akad. Nauk SSSR, 278:1470–1473. [In Russian].

Lebedev, O. A. 1985. The first tetrapod: description and occurrence. Priroda Moscow, 1985 (11):26–26. [In Russian].

Lebedkina, N. S. 1979. *The Evolution of the Amphibian Skull. On the Problem of Morphological Integration.* Moscow: NAUKA. [In Russian].

Lombard, R. E., and J. R. Bolt. 1979. Evolution of the tetrapod ear: an analysis and reinterpretation. Biol. J. Linn. Soc., 11(1):19–76.

Long, J. A. 1985. The structure and relationships of a new osteolepiform fish from the Late Devonian of Victoria, Australia. Alcheringa, 9:1–22.

Long, J. A. 1987. An unusual osteolepiform fish from the Late Devonian of Victoria, Australia. Palaeontology, 30:839–462.

Løvtrup, S. 1977. *The Phylogeny of the Vertebrata.* London: John Wiley.

Maisey, J. 1986. Heads and tails: a chordate phylogeny. Cladistics, 2:201–256.

Medvedeva, I. M. 1975. The olfactory organ in amphibians and its phylogenetic significance. Tr. Zool. Inst. Akad. Nauk SSSR, 58:1–174. [In Russian, English summary].

Panchen, A. L. 1967. The nostrils of choanate fishes and early tetrapods. Biol. Rev. Cambridge Philos. Soc., 42:374–420.

Panchen, A. L. 1973. *Crassigyrinus scoticus* Watson, a primitive amphibian from the Lower Carboniferous of Scotland. Paleontology, 16:179–193.

Panchen, A. L. 1977. On *Anthracosaurus russelli* Huxley (Amphibia: Labyrinthodontia) and the family Anthracosauridae. Philos. Trans. R. Soc. London, Ser. B, 279:447–512.

Panchen, A. L. 1985. On the amphibian *Crassigyrinus scoticus* Watson from the Carboniferous of Scotland. Philos. Trans. R. Soc. London, Ser. B, 309:505–568.

Panchen, A. L., and T. R. Smithson. 1987. Character diagnosis, fossils and the origin of tetrapods. Biol. Rev. Cambridge Philos. Soc., 62:341–438.

Pander, C. 1860. *Über die Saurodipterinen, Dendrodonten, Glyptolepiden und Cheirolepiden des devonischen Systems.* St. Petersburg: Kaiserl. Akad. Wiss.

Parsons, T. S., and E. E. Williams. 1963. The relationships of the modern Amphibia: a re-examination. Q. Rev. Biol., 38(1):26–53.

Rage, J.-C., and P. Janvier. 1982. Le problème de la monophylie des amphibiens actuels, à la lumière des nouvelles données sur les affinités des tétrapodes. Geobios, Mem. Spec., 6:65–83.

Roček, Z. 1986. An "intracranial joint" in frogs. Pp. 49–54 *in* Roček, Z. (ed.), *Studies in Herpetology.* Prague: Charles Univ.

Romer, A. S. 1956. The early evolution of land vertebrates. Proc. Am. Philos. Soc., 100:157–167.

Rosen, D. E., P. L. Forey, B. G. Gardiner, and C. Patterson. 1981. Lungfishes, tetrapods, paleontology and plesiomorphy. Bull. Am. Mus. Nat. Hist., 167:159–276.

Säve-Söderbergh, G. 1932. Preliminary note on Devonian stegocephalians from East Greenland. Medd. Groenland, 94:1–107.

Schmalhausen, I. I. 1968. *The Origin of Terrestrial Vertebrates.* New York and London: Academic Press.

Schultze, H.-P. 1969. Die Faltenzähne der rhipidistiiden Crossopterygier, der Tetrapoden und der Actinopterygier-Gattung *Lepisosteus*, nebst einer Beschreibung der Zahnstruktur von *Onychodus* (struniiformer Crossopterygier). Palaeontogr. Ital., 65:63–136.

Schultze, H.-P. 1970. Folded Teeth and the monophyletic origin of tetrapods. Am. Mus. Novitat., 2408:1–10.

Schultze, H.-P. 1977. Ausgangsform und Entwicklung der rhombischen Schuppen der Osteichthyes (Pisces). Palaeontol. Z., 51:152–168.

Schultze, H.-P. 1981. Hennig und der Ursprung der Tetrapoda. Palaeontol. Z., 55:71–86.

Schultze, H.-P. 1984. Juvenile specimens of *Eusthenopteron foordi* Whiteaves, 1881 (osteolepiform rhipidistian, Pisces) from the Upper Devonian of Miguasha, Quebec, Canada. J. Vert. Paleontol., 4:1–16.

Schultze, H.-P. 1987. Dipnoans as Sarcopterygians. Pp. 39–74 *in* Bemis, W. E., W. W. Burggren, and N. E. Kemp (eds.), *The Biology and Evolution of Lungfishes.* J. Morphol., Suppl. 1.

Schultze, H.-P., and M. Arsenault. 1985. The panderichthyid fish *Elpistostege:* a close relative of tetrapods? Palaeontology, 28:293–309.

Shishkin, M. A. 1973. Morphology of early amphibians and problems of evolution of lower tetrapods. Tr. Paleontol. Inst. Akad. Nauk SSSR, 137:1–260. [In Russian].

Shishkin, M. A. 1975. Labyrinthodont middle ear and some problems of amniote evolution. Coll. Int. C. N. R. S., 218:337–348.

Smirnov, S. V., and E. I. Vorobyeva. 1986. The sound-conducting apparatus of anurans and urodeles and the problem of their origin. Pp. 156–179 *in* Vorobyeva, E. I., and N. S. Lebedkina (eds.), *Morphology and Evolution of Animals*. Moscow: NAUKA. [In Russian].

Smithson, T. R. 1982. The cranial morphology of *Greererpeton burkemorani* Romer (Amphibia: Temnospondyli). J. Linn. Soc. London Zool., 76:29–90.

Smithson, T. R. 1985. The morphology and relationships of the Carboniferous amphibian *Eoherpeton watsoni* Panchen. J. Linn. Soc. London Zool., 85:317–410.

Szarski, H. 1962. The origin of the Amphibia. Q. Rev. Biol., 37:189–241.

Tatarinov, L. P. 1976. *The Morphological Evolution of the Theriodonts and General Problems of Phylogenetics*. Moscow: NAUKA. [In Russian].

Vorobyeva, E. I. 1960. New data on *Panderichthys*, a crossopterygian genus from the Devonian of the USSR. Paleontol. Zh., 1960:87–96. [In Russian].

Vorobyeva, E. I. 1962. Rhizodont Crossopterygii of the main Devonian field of the USSR. Tr. Paleontol. Inst. Akad. Nauk SSSR, 94:1–139. [In Russian].

Vorobyeva, E. I. 1969. Zur Morphologie der Wangenplatte von *Panderichthys rhombolepis* (Gross). Eesti NSV Tead. Akad. Toim. Keem. Geol., 18(3):255–258. [In Russian; Estonian and German summaries].

Vorobyeva, E. I. 1971. The ethmoid region of *Panderichthys* and some problems of the cranial morphology of crossopterygians. Pp. 142–159 *in* Obruchev, D. V., and V. N. Shimanskii (eds.), *Current Problems in Palaeontology*. Tr. Paleontol. Inst. Akad. Nauk SSSR, Vol. 130. [In Russian].

Vorobyeva, E. I. 1977a. Morphology and nature of evolution of crossopterygian fishes. Tr. Paleontol. Inst. Akad. Nauk SSSR, 163:1–239. [In Russian].

Vorobyeva, E. I. 1977b. Evolutionary modifications of the teeth structure in the Palaeozoic Crossopterygii. J. Palaeontol. Soc. India, 20:16–20.

Vorobyeva, E. I. 1980. Observations on two rhipidistian fishes from the Upper Devonian of Lode, Latvia. J. Linn. Soc. London Zool., 70:191–201.

Vorobyeva, E. I. 1985. On the evolution of cranial structures in crossopterygians and tetrapods. Pp. 123–133 *in* Duncker, H.-R., and G. Fleischer (eds.), *Functional Morphology in Vertebrates*. Stuttgart and New York: Gustav Fischer.

Vorobyeva, E. I. 1986. The current state of the problem of amphibian origin. Pp. 25–28 *in* Roček, Z. (ed.), *Studies in Herpetology*. Prague: Charles Univ.

Vorobyeva, E. I., and L. A. Lyarskaya. 1968. Crossopterygian and dipnoan remains from the Amata beds of Latvia and their burial. Pp. 71–86 *in* Obruchev, D. V. (ed.), *Ocherki po filogenii i sistematike iskopaemykh ryb i bezcheliustnykh*. Moscow: Akad. "NAUKA" SSSR. [In Russian].

Vorobyeva, E. I., and D. V. Obruchev. 1964. Subclass Sarcopterygii. Pp. 268–321 *in* Orlov, Y. A. (ed.), *Osnovy Paleontologii*. Vol. 2. Agnatha, Pisces (D. V. Obruchev, ed.). Moscow: Izdatel'stvo "NAUKA" SSSR. [In Russian].

Vorobyeva, E. I., and A. A. Tsessarskii. 1986. The evolution of vertebrae in rhipidistians. Zh. Obshch. Biol., 67:735–747. [In Russian].

Watson, D. M. S. 1929. The Carboniferous Amphibia of Scotland. Paleontol. Hung., 1:219–252.

Westoll, T. S. 1938. Ancestry of the tetrapods. Nature, 141:127.

Westoll, T. S. 1943. The origin of the tetrapods. Biol. Rev. Cambridge Philos. Soc., 18:78–98.

Worobjewa, E. 1973. Einige Besonderheiten im Schädelbau von *Panderichthys rhombolepis* (Gross) (Pisces, Crossopterygii). Palaeontogr., A, 142:221–29.

Worobjewa, E. 1975a. Bemerkungen zu *Panderichthys rhombolepis* (Gross) aus Lode in Lettland (Gauja Schichten, Oberdevon). N. Jahrb. Geol. Palaeontol. Monatsh., 1975 (H.5):315–320.

Worobjewa, E. 1975b. Formenvielfalt und Verwandtschaftsbeziehungen der Osteolepidida (Crossopterygii, Pisces). Palaeontol. Z., 49:44–55.

4 The Early Tetrapods: Classification and the Shapes of Cladograms

A. L. Panchen

Recent discussion of the relationships of early tetrapods, the first land vertebrates, can be treated as a paradigm case of current controversies of taxonomic theory and practice in paleontology. The past two decades have witnessed the rise of what originally was known as "phylogenetic systematics," following the publication of the English version (1966) of Hennig's (1950) book of that title. Most systematists who have been participants in, or witnesses of, the often bitter controversy surrounding the birth and development of cladistic methodology probably would admit, however grudgingly, that the battles have resulted in at least partial victory for the cladist cause—i.e., cladistics is the best current taxonomic technique.

One should beware, however, of the "best in the field" fallacy. The best current methodology, like the best current theory, is not synonymous with the definitive or perfect one. The best may have serious shortcomings that may or may not be overcome in the future. If complacency and dogmatism take over, then they certainly will not be overcome. By taking as an example this systematics of the Tetrapoda, I intend to illustrate what I believe to be a serious shortcoming of cladistic methodology, particularly as used by vertebrate paleontologists.

Consideration of the relationships of early tetrapods, or any other major taxon, poses two categories of problems—first, the extrinsic relationships of the group, and second, its intrinsic relationships (i.e., the problems of classifying its members). In previous publications (Panchen,

1980, 1982, 1985; Panchen and Smithson, 1987), I have tried to say something about both. In our recent long review, Smithson and I addressed the problem of the extrinsic relationships of the Tetrapoda and concluded that the approach of some current practitioners of cladistics to the fossil record actually invalidated their results. (Schultze [1987] reached a similar conclusion.) In this chapter, I hope to give a similar demonstration using the classification of the earliest tetrapods as an example. Further, I hope to demonstrate in much the same spirit as that of Gould's (1977) splendid essay "Eternal Metaphors of Palaeontology" that many of the ideas inherent in cladistics and particularly those that generate its shortcomings are of ancient origin. Dogmatic cladists, like all of us, are not just heirs to the past; they are constrained, even incapacitated, by it.

THE CLASSIFICATION OF EARLY TETRAPODS

The Shapes of Cladograms

A few years ago, Gaffney presented two exercises in the cladistic classification of tetrapods, with particular reference to the "lower" tetrapods. The first (Gaffney, 1979a) is an attempt to demonstrate the monophyletic status of the taxon Tetrapoda and, further, to show within this taxon a sister-group relationship between *Ichthyostega* and a monophyletic group termed by him the "Neotetrapoda." *Ichthyostega* Säve-Söderbergh (1932) is a generic name applied to some of the Devonian tetrapods found in East Greenland. Säve-Söderbergh originally described several species, but more recent reconstructions by Jarvik (the latest in Jarvik, 1980) always have been attributed to *Ichthyostega* sp. Reconstructions of the external appearance of the skull, pectoral girdle, humerus, radius and ulna, pelvic girdle and limb, isolated vertebrae, ribs, and characteristic "fish-like-tail" with median fin-rays were presented by Jarvik (1980:218–244), together with a schematic reconstruction of the whole skeleton. Additionally, the skull of a second genus, *Acanthostega*, was described (Jarvik, 1952; Clack, 1988), and a third, *Ichthyostegopsis*, was named by Säve-Söderbergh. Unfortunately, all the East Greenland material is dissociated; thus, the association of cranial and postcranial elements as *Ichthyostega* is not absolutely certain. Despite all this, *Ichthyostega* has won the status of *the* primitive tetrapod, and is so regarded by Gaffney.

Gaffney's (1979b) second presentation of a tetrapod phylogeny is used as an example in a methodological essay on cladistics. It was set out in the form of a cladogram, together with a list of the characters used in its construction, and in a series of three-taxon cladograms claimed to be its constituent parts. In his first review, Gaffney (1979a) primarily was con-

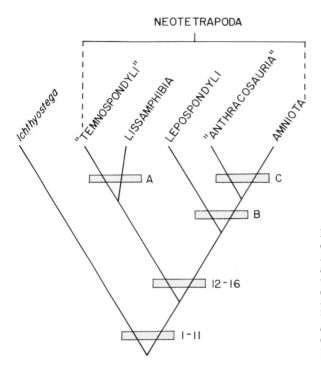

Fig. 1. "Hennigian comb" type of cladogram to show relationship of *Ichthyostega* to the "Neotetrapoda" and suggest possible relationships within the latter. Numbers refer to Gaffney's tetrapod and "neotetrapod" synapomorphies; "A," "B," and "C" refer to character distributions to corroborate the remaining grouping. Cf. Figure 2. (Adapted from Gaffney, 1979a.)

cerned with the monophyletic status of the Tetrapoda and his "Neotetrapoda," but he did provide a tentative arrangement of groups within the "Neotetrapoda" (Fig. 1). In Gaffney 1979b, however, the author provided a more complete division of the latter supported by a series of synapomorphies (Fig. 2). There are important differences between the corresponding regions of these two cladograms. The position and rank of the Temnospondyli plus Lissamphibia and the Lepospondyli are transposed, and the Anthracosauria is absent in the second cladogram. Important as these differences undoubtedly are, they are not my immediate concern.

I have cited Gaffney's two cladograms at this stage not to criticize the content, which might be regarded as having been superseded by subsequent work (Smithson, 1985; Bolt and Lombard, 1985; Panchen and Smithson, 1988), but to draw attention to their *shape*. This type of cladogram, which I dubbed (perhaps frivolously) a "Hennigian comb" (Panchen, 1982), has come to dominate cladistic methodology among paleontologists, with unfortunate results. In such a cladogram, the terminal taxa (e.g., Porolepiformes, Osteolepiformes, *Ichthyostega* et al. in Fig. 2) are arranged in a series such that each taxon is the sister-group of all taxa to its (conventional) right. Thus, the porolepiforms are the sister-group of

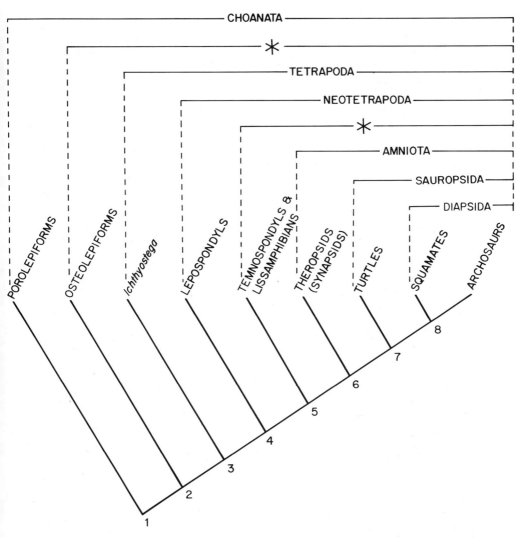

Fig. 2. "Hennigian comb" type of cladogram to show relationships of tetrapods and "rhipidistian" fishes. Asterisks refer to unnamed groups; numbers at nodes refer to Gaffney's list of synapomorphies. Cf. tetrapod arrangement with that of Figure 1. (Adapted from Gaffney, 1979b.)

the osteolepiforms *plus* all tetrapods, osteolepiforms the sister-group of all tetrapods (i.e., *Ichthyostega* plus "neotetrapods"), and so forth. As a result, the taxonomic rank of each terminal taxon descends from left to right. If porolepiforms constitute a taxon of, for example, the category subclass (to use the correct, if pedantic, formulation; Simpson, 1961), then osteolepiforms are a taxon of the infraclass category; *Ichthyostega* represents a superorder, and, finally, the amniotes a much lower rank. There is a reasonable convention for reducing the plethora of ranks that otherwise would result from the introduction of fossils into a classification (Patterson and Rosen, 1977, and see below), but it does not affect the principle.

Two further points need to be made about Gaffney's cladograms. I noted (Panchen, 1982:313) that in his methodological demonstration Gaffney (1979b) does not characterize the terminal taxa of his cladogram. I then went on to suggest that this is symptomatic of the "Hennigian comb" mentality in which the form of the comb takes precedence over the validity of the taxa that constitute it. Perhaps this was unfair, because in Gaffney's (1979a) other cladogram the terminal taxa are characterized. Nonetheless, there is a feature of both to which I must draw attention. As we see below, the principal justification for incorporating fossils in a Hennigian comb is that the fossil taxa should be placed within a preexisting cladogram of living forms. This necessitates specification of an extant taxon as the sister-group to the whole taxon represented by the comb. Gaffney does this for neither cladogram. In both cases, the "plesiomorph sister-group" arguably is the Dipnoi (lungfish) (Panchen and Smithson, 1987).

In contrast to the comb configuration, and again, concerning the classification of the "lower" tetrapods, we can cite Smithson's (1985) cladogram. It is a more recent study, and one based on a more detailed analysis of the fossil tetrapods involved, but, again, it is the shape that I wish to emphasize (Fig. 3). Elements of the "Hennigian comb" remain, and the comparison is rendered more difficult because Smithson introduced groups into the scheme that were absent in both of Gaffney's cladograms (e.g., Diadectomorpha, Loxommatoidea—mentioned by Gaffney [1979a]—and Palaeostegalia). Further, the "Lepospondyli," no

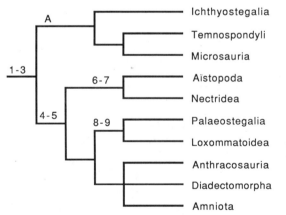

Fig. 3. "A phylogeny of the Amphibia"—a divergent cladogram based on sister-group pairing of terminal taxa. "A" refers to his discussion of relationships of extant Amphibia to the three fossil groups; numbers refer to key synapomorphies. Cf. the form and content of this figure to those of Figures 1 and 2. (Adapted from Smithson, 1985.)

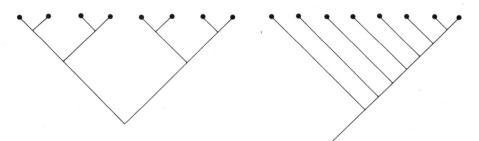

Fig. 4. The two extreme forms of cladogram shown for eight hypothetical terminal taxa. *Left,* symmetrical cladogram that gives the minimum number of ranks (r) per number of terminal taxa (*n*) (r = 1 + log$_2$ *n*). *Right,* "Hennigian comb" cladogram as often produced by paleontologists, giving the maximum number of ranks (r = *n*). (Adapted from Panchen, 1982.)

longer regarded as a valid taxon, was split into its constituent groups— i.e., the Nectridea, Aïstopoda, and Microsauria. Nevertheless, Smithson proposed a primary dichotomy at the root of the cladogram that separated Ichthyostegalia (essentially *Ichthyostega*) plus temnospondyls and microsaurs from the remaining tetrapods. Furthermore, it should be noted that no taxon in Smithson's cladogram occupies a unique rank, whereas each one (with the possible exception of temnospondyls and lissamphibians) does in Gaffney's cladograms. However, Gaffney noted that the Temnospondyli may be a paraphyletic taxon, incomplete without the extant Amphibia. Other authors (Bolt; Trueb and Cloutier) in this volume discuss the relationships of the living Amphibia.

Therefore, Smithson's cladogram approaches the other extreme in shape, that of the completely symmetrical cladogram (Fig. 4, *left*) in which the number of ranks for any given number of terminal taxa is minimal and every taxon at every rank has a sister-group that terminates in, and is terminated by, the same number of dichotomies.

The Significance of *Crassigyrinus scoticus*

Crassigyrinus scoticus Watson (1929) is a large tetrapod known from two mid-Carboniferous localities, both not far from Edinburgh, Scotland. The importance of *Crassigyrinus* to our theme is that, despite its much later date, the fossil seems to rival *Ichthyostega* in the primitive, fishlike nature of its skeleton. On the other hand, it possesses a series of characters that seem to unite it with a more advanced group of early tetrapods, the anthracosaurs. The holotype was collected in the 1850's, apparently by Hugh Miller, but not described until Watson's publication. It consists of the right side from snout to jaw articulation of a massive amphibian skull on a block of ironstone that almost certainly comes from the Gilmerton

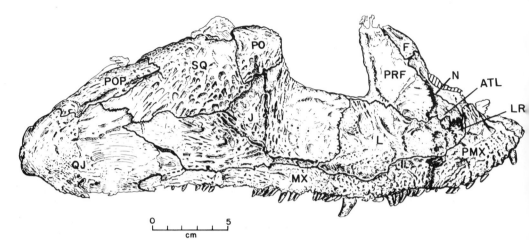

Fig. 5. *Crassigyrinus scoticus* Watson, holotype as preserved. ATL = anterior tectal; F = frontal; J = jugal; L = lacrimal; LR = lateral rostral (septomaxilla); MX = maxilla; N = nasal; PMX = premaxilla; PO = postorbital; POP = preopercular; PRF = prefrontal; QJ = quadratojugal; SQ = squamosal. (Adapted from Panchen, 1973.)

Ironstone of Edinburgh (Fig. 5). The horizon is Viséan, uppermost Lower Carboniferous (Panchen, 1973, 1985). A complete lower jaw also is known from Gilmerton. The great find, however, was made by Stanley Wood, the eminent Scottish collector, at an opencast coal site near Cowdenbeath, Fife (Andrews et al., 1977). This find consisted of a skull and most of the skeleton of a specimen of *Crassigyrinus*. The skull is virtually complete, though badly disrupted. Although disarticulated, the skeleton is little dispersed and consists of most of the presacral vertebral column and ribs, pectoral girdle and parts of the forelimb, the ischium of the pelvic girdle, and numerous belly scales. Nothing is known of the tail, but subsequent to my redescription, another *Crassigyrinus* specimen consisting of most of the pelvic girdle and hind limb turned up in Wood's Cowdenbeath collection; this material is now described (Panchen and Smithson, 1990). Based on our considerable knowledge of the fore and hind limb of *Crassigyrinus*, we can assert that it is a tetrapod, contrary to the opinion of some of our colleagues. The horizon of the Dora bonebed at Cowdenbeath from which both of Wood's specimens and some other fragments originated is basal Upper Carboniferous (Namurian A).

It was clear even from the holotype (Fig. 5) that *Crassigyrinus* is a primitive amphibian. It has a large preopercular bone, otherwise recorded among tetrapods only in *Ichthyostega* and its contemporary *Acanthostega* (Jarvik, 1980). The preopercular is a normal ossification of the cheek region in the sarcopterygian fishes that are related most closely to the tetrapods.

Proportionally, the preopercular of *Crassigyrinus* is considerably larger than that of either *Ichthyostega* or *Acanthostega*. In other respects also the features of the side of the skull seem intermediate between those of *Ichthyostega* and *Eusthenopteron* (Figs. 6–8). The latter is a Devonian fish representing the Osteolepiformes. Traditionally, the Osteolepiformes has been regarded as the group from which the tetrapods arose (and thus paraphyletic); however, now it is recognized generally to be the sister-group of tetrapods (Panchen and Smithson, 1987), despite claims on behalf of the Dipnoi (Gardiner, 1980; Rosen et al., 1981; Forey, 1987). The position of the jugal-lacrimal suture in *Crassigyrinus* relative to the orbit is nearer the fish condition than that of *Ichthyostega*; thus, in the former there is only a short jugal-prefrontal contact, in contrast to the longer contact in *Ichthyostega*. Furthermore, the pattern of the junction of the infraorbital lateral-line canal and the jugal canal on the jugal bone in *Crassigyrinus* (Fig. 12) seems to be similar to that of *Eusthenopteron* and other osteolepiforms, but different from that of *Ichthyostega*. More important, however, is the fact that the arrangement of bones around the external nostril of *Crassigyrinus* is the same as that of *Eusthenopteron*. This was less than certain in the holotype but is corroborated by the Cowdenbeath skull. In *Eusthenopteron*, the nostril is located well up on the side of the snout and bordered above by the anterior tectal bone, which sutures posteriorly with the prefrontal. Below the nostril is the lateral rostral, a homologue of the tetrapod septomaxilla (Panchen, 1967). The condition in *Crassigyrinus* is similar in all respects.

In *Ichthyostega*, the anterior tectal is retained, but the nostril is marginal; this condition is derived with respect to that of bony fish (Jarvik, 1980; Panchen, 1964, 1967) unless one accepts the dipnoan-tetrapod link of Rosen et al. (1981). Correlated with this, the lateral rostral of *Ichthyostega* is reduced to a small, tubular ossification that carries the infraorbital canal forward to the premaxilla. Despite my erroneous reconstruction in the anthracosaur *Palaeoherpeton* ("*Palaeogyrinus*"; Panchen, 1964), no other tetrapod is recorded with an anterior tectal.

The Cowdenbeath skeleton of *Crassigyrinus* has other features that may be primitive (Panchen, 1985). At the back of the braincase, the basioccipital bone is not developed into a formed occipital condyle. In *Ichthyostega*, the basioccipital consists largely of a bony tube (Jarvik, 1980) in which the notochord may have extended forward under the posterior moiety of the braincase. This is the condition in *Eusthenopteron*, but other bony fish have a formed condyle. Thus, the *Eusthenopteron-Ichthyostega* condition may be paedomorphic rather than primitive; a similar condition occurs in some fossil lungfish (Miles, 1977; Panchen and Smithson, 1987).

The postcranial skeleton of *Crassigyrinus* seems remarkably primitive but may be degenerate. The vertebrae have poorly ossified neural arches,

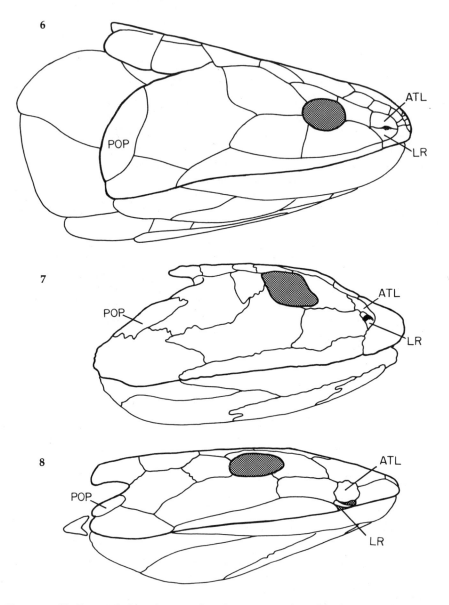

Figs. 6–8. Skulls in right lateral view reduced to same quadrate length. (6) *Eusthenopteron* after Jarvik. (7) *Crassigyrinus* holotype. (8) *Ichthyostega* after Jarvik. Abbreviations as in Figure 5. (Adapted from Panchen, 1973.)

which occur as unconsolidated bilateral halves. There is only one central vertebral element—a husklike, horseshoe-shaped centrum, which often shows signs of recent fusion of the two sides ventrally (Figs. 9–10). Most startling is the contact between successive neural archs which seems merely to have been represented by ill-formed facets anteriorly, with no sign of corresponding posterior facets. The strongly buttressed zygopophyses that are characteristic of tetrapods are not developed. Similarly, there is no clear transverse process for rib articulation on the neural arch. Correlated with the latter, the heads of the ribs lack well-developed capitular and tubercular facets for articulation with the centrum and neural arch, respectively. Nonetheless, the ribs are well-developed, curved cylinders that are strongly reminiscent of those of anthracosaurs. In all of these features, the axial skeleton of *Crassigyrinus* seems more primitive or degenerate than that of *Ichthyostega*. Like the rest of the forelimb, the humerus of *Crassigyrinus* is extraordinarily small, being some 35 mm long in an animal of an estimated length of about 2 m. The humerus is of particular interest because it possesses a series of foramina that, among tetrapods, otherwise are known only in *Ichthyostega* (Panchen, 1985; Jarvik, 1980); these foramina could be derived easily from the condition in *Eusthenopteron* (Andrews and Westoll, 1970). The hind limb of *Crassigyrinus* is considerably larger than the forelimb.

There are a number of important respects in which *Ichthyostega* may be regarded as more primitive than *Crassigyrinus*. *Ichthyostega* is thought to have a small triangular bone that lies free just behind the angle of the jaw. Jarvik identified this bone as the subopercular—a remnant of the gill-

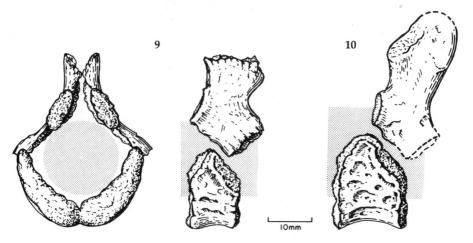

Figs. 9–10. *Crassigyrinus* vertebrae. (9) Midtrunk in anterior and lateral views. (10) Posterior trunk in lateral view. Stipple pattern indicates minimum notochordal diameter. (Adapted from Panchen, 1985.)

cover series that otherwise is unknown in tetrapods. Also, the lateral-line canals in the dermal skull roof of *Ichthyostega* take the form of canals (sensu stricto) that run through the substance of the bone, whereas those of *Crassigyrinus*, and other anamniotic tetrapods in which they are developed, take the form of open grooves or sulci. Although scarcely described, the braincase of *Ichthyostega* seems to be divided into two moieties—an anterior ethmosphenoid region and a posterior otico-occipital one—such that the investing dermal parasphenoid bone is confined to the region in front of the junction. This is the osteolepiform condition in both respects; in contrast, in other tetrapods (including *Crassigyrinus*) the parasphenoid extends posteriorly beneath the otic region, nearly reaching the craniocervical joint represented by the occipital condyle.

In the discussion following my description of *Crassigyrinus* in Panchen, 1985, I pointed out that there was an alternate interpretation of the condition of the braincase if the Dipnoi were regarded as the sister-group of the Tetrapoda, as was done by Rosen et al. (1981). The Dipnoi possess a consolidated braincase and an extended parasphenoid; thus, if they were used as the out-group with which tetrapod characters were polarized, *Crassigyrinus* would represent the plesiomorphic condition. The condition of the braincase is not the only character that is equivocal in this way, and although I concur with Jarvik (1981), Schultze (1981, 1986), Holmes (1985), and many other colleagues in my belief that Rosen et al. (1981) are wrong (cf. Panchen and Smithson, 1987), the general phenomenon resulting from dispute about the out-group does raise disturbing questions. I considered these in the discussion referred to above (Panchen, 1985); the disputed characters listed in Table 1 are derived from this paper. Briefly, the dilemma is as follows. In order to resolve the sister-group relationships of the Tetrapoda with either the Osteolepiformes or the Dipnoi, the apparent osteolepiform-tetrapod synapomorphies and

Table 1. Contrasting characters of *Ichthyostega* and *Crassigyrinus* of controversial polarity

Ichthyostega	*Crassigyrinus*
Braincase divided by suture (O)	Brain consolidated (D)
Parasphenoid underlying ethmoid region only (O)	Parasphenoid underlying whole braincase (D)
Course of infraorbital lateral-line canal as in dipnoans (D)	Course of infraorbital lateral-line canal as in osteolepiforms (O)
External nostril marginal (D)	External nostril not near margin of jaw (O)
Pars facialis of lateral rostral reduced (D)	Pars facialis not reduced (O)
"Fused" postparietals (D)	Separate postparietals (O)

Abbreviations: D = plesiomorphic character if Dipnoi are assumed to be the sister-group of tetrapods; O = plesiomorphic character if Osteolepiformes are assumed to be the sister-group of tetrapods.

apparent dipnoan-tetrapod synapomorphies must be evaluated against one another, using the principle of parsimony as the ultimate arbiter. However, these same putative synapomorphies can be established only after one of the two groups (i.e., osteolepiforms or dipnoans) has been selected as the sister-group of the tetrapods so that one can determine which of each alternate character state is the more primitive one for the Tetrapoda. Few arguments can be more obviously circular.

My concern herein is to comment on the position of *Crassigyrinus* within the tetrapods. In cladistics, each valid taxon (down to the species level) has (1) autapomorphies that define it uniquely, and (2) synapomorphies that ally it uniquely with another taxon, its sister-group. *Crassigyrinus* certainly has both. The enormous rhomboidal orbits and narrow interorbital region, and the extreme depth of the skull roof compared with the lower jaw (Figs. 11, 12) are diagnostic features within very primitive tetrapods. The minute forelimb also is distinctive. However, I think that *Crassigyrinus* shares a number of valid synapomorphies with the Anthracosauroideae (sensu Smithson, 1985), (i.e., the Anthracosauria, if that group is defined to exclude the reptile-like Seymouriamorpha of the Permian; Panchen, 1970). As discussed in detail by Panchen (1985:550–553), there are four anthracosauroid synapomorphies: (1) characteristic dermal ornament; (2) the presence of tabular horns extending back from posterior corners of the skull table, and each consisting of an ornamented dorsal area that is continuous with the external surface of the tabular, and a deeper, unornamented process extending farther back; (3) the absence of paired posttemporal fossae in the occiput—features that are developed in other early tetrapods and that are the homologues of the fossae bridgei of fish such as *Eusthenopteron* (Figs. 13–15); and (4) histology of the teeth that is seen best in transverse section at the base of the crown in large palatal tusks.

The tooth histology divides into two component characters. In his monograph on the histology of the "labyrinthodont" teeth of (inter alia) primitive tetrapods and their fish relatives, Schultze (1969) described a morphocline within temnospondyls and anthracosaurs in which the infolded primary dentine lost its short side branches but became more tortuous. The conditions in *Ichthyostega* and the osteolepidid osteolepiform *Panderichthys* are similar and provide good forerunners to the series. The advanced condition (i.e., "labyrinthodont" sensu stricto, Schultze) was attained by late and advanced temnospondyls and, surprisingly, also by Coal Measure anthracosauroids such as "*Eogyrinus*" (*Pholiderpeton*, fide Clack, 1987) and *Anthracosaurus*. Although the infoldings are less tortuous, the tusks of *Crassigyrinus* are similarly advanced. The second component character is more diagnostic. Atthey (1876) first noted and illustrated the characteristic triangular areas of "dark dentine" situated

11

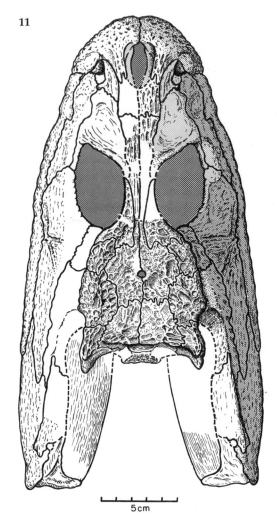

5 cm

Figs. 11–12. *Crassigyrinus scoticus,* composite reconstruction of the skull. (11) Dorsal view. (12) Lateral view. (Adapted from Panchen, 1985.)

between the primary infoldings and characterized by the presence of dentine tubules oriented radially, rather than perpendicularly, to the primary folds (Fig. 16). The contemporary Coal Measure loxommatids (of disputed relationship) have teeth little advanced over those of *Ichthyostega* and lack "dark dentine" (Fig. 17). *Crassigyrinus* has "dark dentine" indistinguishable from that of *"Eogyrinus"* and *Anthracosaurus* (Fig. 18).

Thus, *Crassigyrinus* seems to me to represent the sister-group of the Anthracosauroideae, although one character might appear to refute such

2

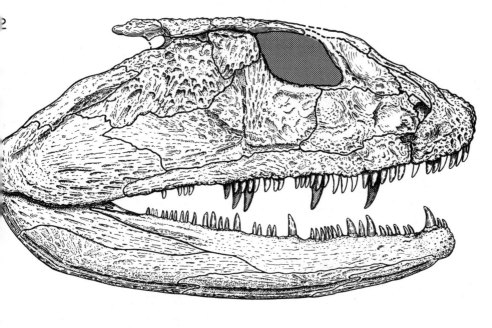

10 cm

relationship. At the back of the anthracosauroid skull table, the tabular
bone has a suture with the parietal; this is an apparent synapomorphy
shared with the Seymouriamorpha, and doubtfully with the Nectridea
and Aïstopoda. In contrast, in the temnospondyls tabular-parietal contact is excluded by a suture between the postparietals and supratem-
porals. The distinction was emphasized first by Säve-Söderbergh (1935).
Crassigyrinus has the temnospondyl condition (Panchen, 1980), but this
does not negate my conclusion, because that condition undoubtedly is
plesiomorphic, approaching the configuration in *Eusthenopteron* (Panchen, 1964). Bolt et al. (1988) provided a brief description of another early
tetrapod from the Mississippian of Iowa; like *Crassigyrinus*, this organism
seems to be related to the anthracosauroids, but it has a postparietal-
supratemporal suture and anomolous vertebrae. The authors associated
it taxonomically both with *Crassigyrinus* and the anthracosaurs.

Apart from its great intrinsic interest, *Crassigyrinus* seems important to
me in showing that there must have been an early dichotomy in the
cladogenesis of tetrapods, with *Ichthyostega* on one ramus and *Crassigyrinus* and the Anthracosauroideae on the other. In personal (and
friendly) discussion with a number of colleagues, I was rather surprised
to find that a number of them viewed my ideas on the relationships of
Crassigyrinus with less than fervent enthusiasm. Their alternate hypothe-

13

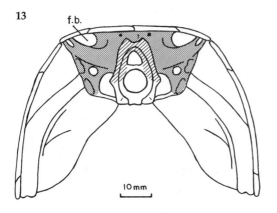

14

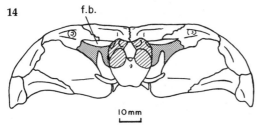

15

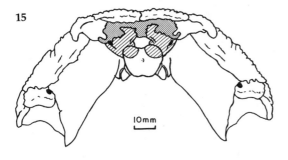

Figs. 13–15. Skulls in occipital view. (13) *Eusthenopteron foordi* (adapted from Smithson and Thomson, 1982). (14) *Greererpeton burkemorani* (after Smithson, 1982); note exoccipital–skull-roof contact. (15) *Palaeoherpeton decorum* (after Panchen, 1970); note absence of posttemporal fossae. Otic capsules shown in stippled pattern. Exoccipital bones hatched. f.b. = posttemporal fossa (fossa bridgei).

ses can be summarized as follows. (1) *Crassigyrinus* is so primitive that it must be a fish. Given our knowledge of the limbs and girdles, this hypothesis is not viable. (2) *Crassigyrinus* is related more closely to loxommatids than it is to anthracosaurs; nevertheless, *Crassigyrinus*, loxommatids, anthracosaurs, and their putative relatives together with amniotes form a monophyletic taxon that is the sister-group of the Nectridea and Aïstopoda. Together, these taxa are the sister-group of *Ichthyostega* et al. This view (Smithson, 1985:Fig. 3) is the only published alternative of which I am aware. (3) *Crassigyrinus* is a loxommatid. I can see no merit in this unpublished proposal. (4) *Crassigyrinus* is a "stem-group" tetrapod that is slightly advanced relative to *Ichthyostega*, but more primitive than

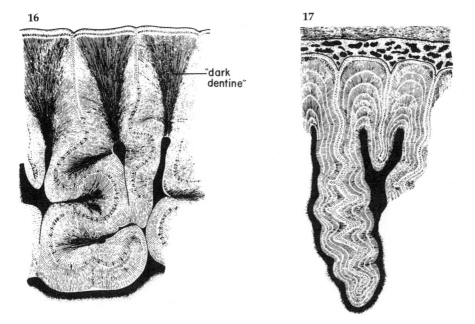

Figs. 16–17. Palatal tusks in transverse section (after Schultze, 1969). (16) "*Eogyrinus*" (*Pholiderpeton*) *attheyi* (after Atthey, 1876). (17) *Megalocephalus pachycephalus,* a loxommatid (after Embleton and Atthey, 1874).

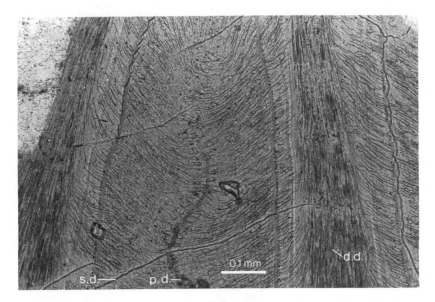

Fig. 18. Palatal tusk of *Crassigyrinus* to show dentine tubules. p.d. = primary dentine; s.d. = secondary dentine; d.d. = "dark dentine." (After Panchen, 1985.)

125

any other tetrapod taxon. The very primitiveness of *Crassigyrinus* precludes its close relationship to anthracosaurs, which usually are assumed to be close to amniotes.

Hypotheses 2 and 4 are of some interest. The former is a matter of interpretation of the morphology of *Crassigyrinus*—an issue that I do not propose to discuss here. However, the latter is directly pertinent to my theme. In Gaffney's (1979a) cladogram (Fig. 1), we see that he distinguished the Neotetrapoda as the apomorphous sister-group of *Ichthyostega*. The autapomorphies that Gaffney cited to characterize the "Neotetrapoda" are as follows: (1) persistent notochord excluded from braincase in the adult; the parachordal (otico-occipital) region formed without significant contribution from the notochord; (2) preopercular and opercular (= subopercular) ossifications absent; (3) median bony fin supports with lepidotrichia absent; (4) the lateral-line system absent or in grooves; not enclosed in bony canals; (5) ethmosphenoid and parachordal (i.e., otico-occipital) portions of the braincase solidly fused in the adult, not separated by suture.

If one compares the features of *Crassigyrinus*, as a candidate for the "Neotetrapoda," with this series of characters, *Crassigyrinus* seems to fall between *Ichthyostega* and the "Neotetrapoda." (1) The notochord does not extend significantly under the braincase, but the condition of the basioccipital shows signs of the (presumed) primitive condition. (2) The subopercular is lost, but a large preopercular is retained. (3) The condition or presence of a median bony fin is unknown. (4) The lateral-line system is in grooves. (5) The braincase is fused solidly. It should be noted that of these skull characters, only the large preopercular (plus the condition of the bones around the external nostrils) is unequivocally primitive. Taken together with what is presumed to be the primitive condition of the postcranial skeleton, however, *Crassigyrinus* can be placed in only one position in Gaffney's "Hennigian-comb" cladogram—between *Ichthyostega* and the "Neotetrapoda," as the sister-group of the latter. It is evident (at least to me) that obvious features that *Crassigyrinus* shares with anthracosauroids are precluded from consideration by the configuration of the cladogram, because, rightly or wrongly, the anthracosaurs (anthracosauroids plus seymouriamorphs) are considered more "advanced" in their position as the sister-group of amniotes.

The Temnospondyli and Microsauria

The same philosophy is evident in Gaffney's (1979a) placement of the Temnospondyli plus Lissamphibia as the most primitive of "neotetrapods." They have all the neotetrapod synapomorphies but lack the advanced features that accumulate as one moves toward the amniote condition. Thus, the attitude taken in construction of a "Hennigian comb" is not Hennig's (1966) precept of looking for the sister-group, in this case a sister-group of the Temnospondyli (including the "Lissamphibia," if it is concluded that it is part of the same clade). Instead, the Temnospondyli has been placed in a sort of multicharacter morphocline of all tetrapods.

This approach can be contrasted with that of Smithson (1985). Both Gaffney and Smithson would agree that the taxon Temnospondyli is difficult to characterize, hence its appearance in quotation marks in Gaffney (1979a). If the Temnospondyli is interpreted to include the extant amphibians, then the impedance-matching middle ear (Lombard and Bolt, 1979; Bolt and Lombard, 1985) characterized by a pair of otic notches and a slender stapes in fossil forms may define a group within the Temnospondyli (sensu lato). Such a group would exclude only primitive notchless forms, notably the colosteids (Panchen, 1975; Smithson, 1982; Hook, 1983). Smithson (1985) suggested that a condition of the tarsus, in which the intermedium articulates only with the fibula, may be an autapomorphy of the Temnospondyli. Be that as it may, there is an important feature of the braincase and skull roof pointed out by Smithson (1982, 1985). In most early tetrapods, the back of the braincase is attached to the underside of the skull roof by contact of the otic capsule; the tabulars are lateral to the posttemporal fossae, and the postparietals mesial to the fossae. However, in temnospondyls there is a secondary mesial contact of the paired exoccipitals with the postparietals (Fig. 14). This condition also is known in all adequately preserved microsaurs except *Pantylus* (Romer, 1969; Carroll and Gaskill, 1978). The exoccipital-postparietal contact undoubtedly is derived, because *Eusthenopteron* shows the primitive condition (Fig. 13). Thus, the derived condition is a persuasive synapomorphy of temnospondyls and microsaurs, although it also may characterize nectrideans (R. W. Hook, pers. comm.). Furthermore, the nature of the basal articulation and the reduction of the digits in the hand from five to a maximum of four also may unite these two groups. Disconcertingly, however, the Tuditanomorpha, one of the two major groups into which the microsaurs are divided, includes several taxa in which the amniote condition of the tarsus (i.e., with astragalus and calcaneum) is developed (Carroll and Gaskill, 1978). Furthermore, there is no tuditanomorph that possesses a tarsus that is sufficiently well preserved to show the primitive condition and thus suggest that the character is convergent or homoplastic (Figs. 19–21).

Obviously, there is room for a difference of opinion on the immediate extrinsic relationships of the microsaurs. However, I would urge that one should seek as their sister-group a taxon of the same rank and not attempt to place them in a sequence of descending ranks in a "Hennigian comb."

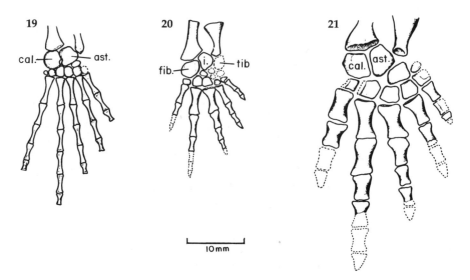

Figs. 19–21. Feet of tuditanomorph microsaurs. (19) *Tuditanus*. (20) *Saxonerpeton*. (21) *Pantylus*. Note resemblance of "fibulare" and "intermedium" of *Saxonerpeton* to calcaneum and astragalus of *Pantylus*. cal. = calcaneum; ast. = astragalus; fib. = "fibulare"; i. = "intermedium"; tib. = restored tibiale. (Adapted from Carroll and Gaskell, 1978.)

THE RATIONALE OF THE "HENNIGIAN COMB"

It is evident that I believe that it is undesirable to apply cladistic methodology with the presupposition that a "Hennigian comb" will result, because such an arrangement may obscure the real relationships of the taxa being classified and thus lead to an erroneous result. Nevertheless, it is only fair to point out that there is a rationale behind this type of classification; I shall attempt to set out its history within the development of cladistic methodology. Conceptually, the "Hennigian comb" seems to arise from Hennig's (1966:59, 207) "deviation rule." In many pairs of sister-species, one member of the pair often is more derived (i.e., has more autapomorphies) than the other. This, Hennig reasoned, is because speciation normally is allopatric and usually occurs by the separation of a peripheral isolate from the main species stock. Following speciation, the isolate would become the "apomorph sister-group" and the relatively unchanged main stock the "plesiomorph sister-group." A succession of such speciation events would produce a clade (Fig. 22) of which the intrinsic classification would be a "Hennigian comb" (Fig. 23).

The second conceptual thread arises from early attempts to incorporate fossils in a cladogram that includes extant taxa. In the early stages of the development of cladistics, there was some confusion as to whether the vertical dimension of a cladogram (in which the "terminal taxa" are

Fig. 22. Hypothetical phylogeny of an evolving clade with a series of speciation events, each taking place by allopatric speciation following isolation of a peripheral population (vertical axis representing time). The isolate becomes the "apomorph sister-group," while the remnant parental stock becomes the "plesiomorph sister-group." Thus, the Parental Stock "P" gives rise to Species "A" (plesiomorph sister-group) and Species "A'" (apomorph sister-group); "A" gives rise to "B" and "B'," etc. Note the counterintuitive effect that the plesiomorph sister-group contributes in each case to the evolving phyletic line (P–A–B–C–D). Cf. Hennig's figures (1966:Figs. 14–15).

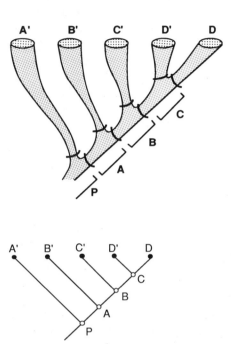

Fig. 23. "True" phylogenetic cladogram (i.e., "tree") corresponding to the phylogeny in Figure 22. Terminal taxa are represented by solid (i.e., closed) circles; ancestors of successive pairs of sister-groups (which would be hypothetical if characterized by cladistic analysis) are represented by open circles. Note "Hennigian-comb" form.

arranged horizontally along the top) represented time (e.g., Brundin, 1968). However, even if a cladogram is interpreted to be a direct representation of phylogeny (or, rather, cladogenesis), it soon became clear that cladistic analysis could yield only the sequence of speciation events leading to extant species, not the timing of the events. A horizontal line joining two or more nodes on a cladogram does not indicate that the speciation events represented by those nodes took place simultaneously.

Once this was accepted, Schaeffer et al. (1972) were able to draw the rational conclusion that all terminal taxa, including fossils, must be treated together and that all the subterminal nodes on the cladogram represented hypothetical ancestral forms. They further suggested that a cladogram was a pattern of character distribution rather than a direct representation of phylogeny, thereby preparing for the distinction between *cladograms* and *trees*, as these terms currently are used by cladists (e.g., Eldredge, 1979). Thus, a tree is a hypothesis of phylogeny derived from a cladogram; it does not necessarily have the same topology as the cladogram. A number of trees are compatible with any given cladogram (Nelson and Platnick, 1981:Ch. 3, but see Wiley, 1981), and a cladogram represents "not . . . the order of branching sister-groups, but the order of emergence of unique derived characters, whether or not the development of these characters happens to coincide with speciation events" (Hull, 1980). Paralleling these developments, and establishing the "Hennigian comb" as an outcome of methodology rather than of speculation about cladogenesis, was the development of Hennig's concept of the *stem-group* (Hennig, 1969, 1981).

If we select a Recent group that is agreed, without reference to fossils, to be monophyletic, then all those fossils that can be assigned to the extant subgroups of the monophyletic taxon are members of that taxon. Thus, using as an example the Amphibia, if it is agreed that the Anura, Urodela (Caudata), and the Apoda are related more closely to one another than any one is to an extant outside group (e.g., lungfish, or a reptilian group), then the taxon Amphibia is monophyletic. Any fossil that shares at least some of the apomorphies of *either* anurans, urodeles, or apodans is a member of what Jefferies (1979) aptly termed the *crown-group* of the Amphibia. The stem-group contains all those fossil taxa that possess some, but not all, of the apomorphies that characterize the crown-group Amphibia. Thus, it might be decided that all but the most primitive temnospondyls share the features of an impedance-matching middle ear, tympanum, transducing stapes, and the unique characteristics set out by Lombard and Bolt (1979) and Bolt and Lombard (1985) that are primitive for all extant Amphibia. In that case, those temnospondyls would constitute the stem-group (always paraphyletic by definition) of the Amphibia. The evidence that the ancestors of urodeles and apodans possessed a tympanum, and so forth, is equivocal (Bolt and Lombard, 1985; Bolt, this volume; Trueb and Cloutier, this volume). However, it is important to note that if it is concluded that they did not, then extant amphibians (i.e., "Lissamphibia") would still not be polyphyletic in cladist terms (contrary to the terminology of much recent discussion) unless they fail the test of closer relationship to one another than to any other extant group. The taxon Amphibia merely would have more fossil members; for example, different groups of microsaurs might constitute the stem-groups of the Apoda (Carroll and Currie, 1975) and of the Urodela (Carroll and Holmes, 1980).

Thus, the primary development of the "Hennigian comb" was a cladogram consisting of a crown-group (Amphibia, sensu stricto) and a stem-group (Temnospondyli, or those temnospondyls with an otic notch). The method of classification using this technique was explained by Patterson and Rosen (1977), and their discussion allows us to draw attention to another feature of the method. Fossils can be classified only with respect to a preexisting classification of extant taxa (Patterson, 1981), and Hennig's "deviation rule" comes into play again. In Patterson and Rosen's example, they are concerned to place a series of fossils in relationship to

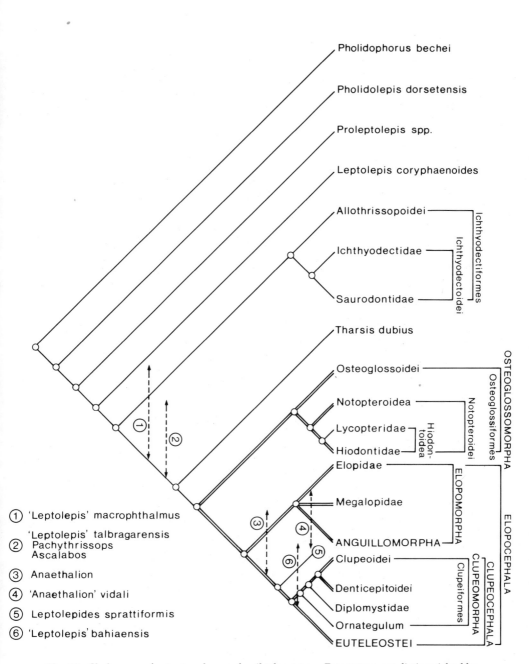

Fig. 24. Cladogram of extant and some fossil teleost taxa. Extant taxa are distinguished by double lines; numbers in circles and their double arrowhead lines give range of possible placements of fossil teleosts *incertae sedis*. See Fig. 25. (From Patterson and Rosen, 1977, courtesy of the American Museum of Natural History.)

INFRACLASS Neopterygii
DIVISION Ginglymodi
DIVISION Halecostomi
Halecostomi *incertae sedis* †"Semionotidae," †"Macrosemiidae," †"Oligopleuridae"
plesion †*Dapedium*
SUBDIVISION Halecomorphi
Halecomorphi *incertae sedis* †"Parasemionotidae," †"Caturidae"
   ORDER Amiiformes
      plesion †*Ospia*
      plesion †*Caturus*
      plesion †*Liodesmus*
      plesion †Sinamiidae
      FAMILY Amiidae
SUBDIVISION Teleostei
Teleostei *incertae sedis* †*Catervariolus*, †*Ceramurus*, †*Galkinia*, †*Ligulella*, †*Majokia*, †*Paleola-brus*, †Pleuropholidae, †"Pholidophoridae," †"Leptolepididae," †*Pachy-thrissops*, †*Ascalabos*, †Crossognathidae

plesion †Pachycormidae
plesion †Aspidorhynchidae
plesion †*Ichthyokentema*

plesion †*Pholidophorus bechei*
plesion †*Pholidolepis dorsetensis*
plesion †*Proleptolepis*
plesion †*Leptolepis coryphaenoides*
plesion †Ichthyodectiformes
   SUBORDER †Allothrissopoidei, new
      FAMILY †Allothrissopidae, new
   SUBORDER †Ichthyodectoidei
      FAMILY †Ichthyodectidae
      FAMILY †Saurodontidae

plesion †*Tharsis dubius*

SUPERCOHORT Osteoglossomorpha
  ORDER Osteoglossiformes
    SUBORDER Osteoglossoidei
    SUBORDER Notopteroidei
      SUPERFAMILY Hiodontoidea
        plesion †Lycopteridae
        FAMILY Hiodontidae
      SUPERFAMILY Notopteroidea
SUPERCOHORT Elopocephala
Elopocephala *incertae sedis* †*Anaethalion*
  COHORT Elopomorpha
  Elopomorpha *incertae sedis* †*"Anaethalion" vidali*
    ORDER Elopiformes
    ORDER Megalopiformes, new
    ORDER Anguilliformes
      SUBORDER Albuloidei
      SUBORDER Anguilloidei
  COHORT Clupeocephala
  Clupeocephala *incertae sedis* †*"Leptolepis" bahiaensis*
    plesion †*Leptolepides sprattiformis*
    SUBCOHORT Clupeomorpha
      ORDER Clupeiformes
        plesion †*Ornategulum*
        plesion †Diplomystidae
        SUBORDER Denticipitoidei
        SUBORDER Clupeoidei
    SUBCOHORT Euteleostei

Fig. 25. Classification of bony fish included in the Neopterygii by Patterson and Rosen (1977). Division Ginglymodi includes the extant *Lepisosteus* and is the sister-group of the Halecostomi. The latter is divided into Halecomorphi (including *Amia*) and Teleostei (sensu lato). Dagger denotes extinct taxon. The boxed part corresponds to the cladogram depicted in Figure 24. (From Patterson and Rosen, 1977, courtesy of the American Museum of Natural History.)

133

the teleost fishes. The (extant) "plesiomorph sister-group" of the Teleostei is the Halecomorphi represented by *Amia*. Their principal concern is, then, to arrange a series of fossil taxa, some consisting of single species, in a *scala naturae* of cumulative teleost characters from the halecomorph-teleost node until the crown-group of teleosts is reached. However, as noted above, Gaffney (1979a,b) and many other paleontological cladists fail to specify the extant plesiomorph group.

To avoid the plethora of ranks that otherwise would be involved in the written classification, one must make each stem-group taxon rankless, labeling it merely a *plesion* (Figs. 24–25). It is assumed that as the series of plesions ascends toward the crown-group teleosts, each plesion will have at least one more teleost apomorphy than its predecessor. Thus, if the apomorphies of the crown-group teleosts are "a + b + c + d . . . n," the plesions will have these characters in the following fashion: Plesion 1— "a"; Plesion 2—"a + b"; Plesion 3—"a + b + c"; and so forth. Of course, each plesion will have its own autapomorphies as well. There seems to me, however, no justification for the methodological assumption that character distributions should occur this way, or if the cladogram is interpreted as phylogeny, that the evolution of the teleosts (or the early tetrapods, or any group) occurred in such a way as to produce the *scala naturae* of the nodes of the stem-group.

Nevertheless, the "Hennigian comb" seems to have become the normal pattern among paleontologists and perhaps is reinforced, even in cladograms of exclusively extant groups, by computer algorithms used in classification. Whether phenetic or cladistic, such algorithms group by finding the most closely associated pair of taxa, treating them as a single taxon, and searching for its sister-group, and so on. This, of course, is an oversimplification, but if the program operated purely in this way, a "Hennigian comb" would result. I believe, however, that other factors are involved, and that these are revealing of the psychology of paleontological cladists.

STUFENREIHEN AND THE "TREE OF PORPHYRY"

In a series of papers in the early part of this century, Abel (1911, 1924, 1929) considered the nature of the fossil record in the reconstruction of phylogeny. Abel introduced the concept of the *Stufenreihe* ("step-series"), which corresponds exactly to the terminal taxa of a "Hennigian comb," if the latter represents phylogeny. Paleontologists long had realized that it was improbable that any simultaneously morphological and chronological series of fossils (i.e., a morphocline that also is a chronocline) known was a genuine ancestor-descendant sequence even at the species level, let alone the individual level. Thus, the series would be interpreted as a sequence of species, each divergent by a speciation event from a member of the true but unknown, and perhaps unknowable, ancestor-descendant series. The latter series was termed the *Ahnenreihe* ("ancestor-series") by Abel. Thus, whereas the members of the *Stufenreihe* represent the terminal taxa of a "Hennigian comb" (given that the latter represents phylogeny), the members of the *Ahnenreihe* correspond to the nodes of the comb, each of which represents the hypothetical common ancestor of one *Stufenreihe* member plus all subsequent members of both series. The only difference between an *Ahnenreihe-Stufenreihe* series and a "Hennigian comb" is that the former usually is arranged on an axis of geological time.

Abel's concepts were adopted by Simpson (1953), from whom Figure 26 is taken. The *Ahnenreihe* "a–b–c–d–e" is reconstructed from the *Stufenreihe* "A–B–C–D–E." In Simpson's diagram, both series represent sequences of species, or rather populations. The diagram is presented in the context of his discussion on adaptive zones; thus, each evolutionary change from an *Ahnenreihe*-member to a *Stufenreihe*-member (e.g., "a" to "A") represents the adaptive radiation into the corresponding zone. In Simpson's diagram, the line passing through the stages of the *Ahnenreihe* represents the course of evolution (anagenesis or phyletic evolution, in this case), but I would suggest that it further embodies two much older concepts.

As Patterson (e.g., 1981) pointed out on a number of occasions, one result, probably unintended by Darwin, of the publication of *On the*

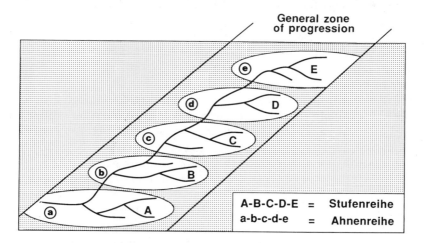

Fig. 26. Abel's concept of the *Stufenreihe* as used by Simpson (1953). The *Ahnenreihe* "a–b–c–d–e" represents phyletic evolution. The *Stufenreihe* members A–E represent evolution into a succession of adaptive zones (clear cells).

Origin of Species . . . (Darwin, 1859) was a changeover among systematists from a search for a pattern of the relationships of species to a search for ancestor-descendant sequences. Frequently, such searches involved supraspecific groups rather than species; thus, "some member of Taxon A gave rise to Taxon B," and so forth. Patterson cited Thomas Henry Huxley's (1870) discussion of the phylogeny of horses. Another feature of the post-*Origin* nineteenth century and early twentieth century was the number of paleontologists who accepted and promulgated orthogenetic or finalist theories of evolutionary mechanism. Persons such as Hyatt, Lang, Cope, and Osborn believed that their ancestor-descendant series of fossils (or, if they were more sophisticated, their reconstructed *Ahnenreihen*) were the result of predetermined internal evolutionary mechanisms. In Gould's (1977) terminology, they were "directionalist/internalist evolutionists."

I suggest, however, that the underlying concept for such people is an ancient one. It is the concept of the *scala naturae* or "Great Chain of Being" (Lovejoy, 1936). As phylogeny, it goes back not to Darwin or Wallace but to Lamarck; as pattern it is no doubt pre-ancient Greek in origin. The concept has two components; the first is attributed to Aristotle and the second to Plato. The first is that all organisms, or sometimes all created things, or (as in early Lamarck) all animals, can be arranged in a linear series of increasing perfection, or read from top to bottom, in a linear series of degradation. Man (at least of natural phenomena) always was placed at the top of the series. The second component was the concept of plenitude—i.e., that the *scala naturae*, if completely known, would have no gaps. It would form a series of insensibly small steps (cf. Darwin's gradualism), because in "the best of all possible worlds," everything that could exist, does exist. The *scala naturae* and the concept of plenitude appear in the work of seventeenth- and eighteenth-century philosophers such as Spinoza, Locke, and notably Leibniz, but also in the natural history of Buffon (1749) and Bonnet (1764), whose *scala* (or "Échelle des êtres") is reproduced as a diagram by Ritterbush (1964; cf. Fig. 27 herein).

Lamarck's initial theory of evolution was to turn the *scala naturae* into an escalator. There was an innate tendency for spontaneously generated organisms at the bottom of the series to evolve up the *scala* so that their descendants eventually reached the status of mankind (Lamarck, 1802). Alongside this idea, he developed the so-called theory of the inheritance of acquired characters to explain deviations from the ancestor-descendant sequence from "Monas" to man (Lamarck, 1809, 1815).

Both Darwin (1859) and Wallace (1855, 1858) attempted to repudiate the idea of phylogeny as a *scala naturae* (both using the metaphor of an irregularly branching tree), but the *scala* dies hard. It appears in the orthogenetic theories of post-*Origin* paleontologists, in Abel's recon-

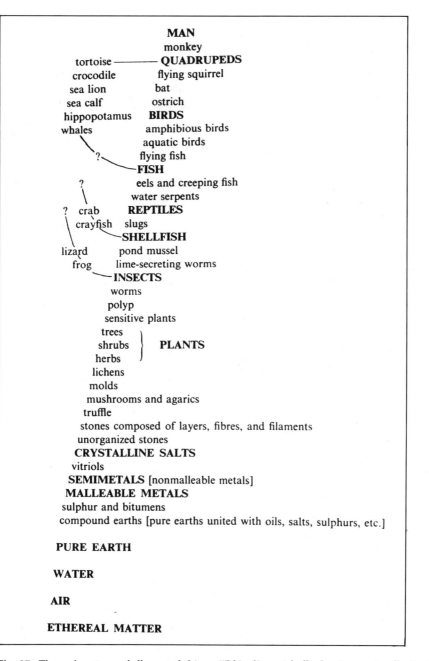

```
                        MAN
                       monkey
        tortoise ———— QUADRUPEDS
       crocodile       flying squirrel
       sea lion        bat
       sea calf        ostrich
     hippopotamus    BIRDS
     whales           amphibious birds
          \           aquatic birds
           ?           flying fish
            ——FISH
        ?            eels and creeping fish
         \           water serpents
     ?  crab      REPTILES
       \ crayfish   slugs
        \           ——SHELLFISH
     lizard      pond mussel
       frog      lime-secreting worms
         ——INSECTS
          worms
          polyp
          sensitive plants
          trees    ⎫
          shrubs   ⎬   PLANTS
          herbs    ⎭
          lichens
          molds
           mushrooms and agarics
          truffle
        stones composed of layers, fibres, and filaments
      unorganized stones
      CRYSTALLINE SALTS
     vitriols
      SEMIMETALS [nonmalleable metals]
      MALLEABLE METALS
     sulphur and bitumens
      compound earths [pure earths united with oils, salts, sulphurs, etc.]

   PURE EARTH

   WATER

   AIR

   ETHEREAL MATTER
```

Fig. 27. The *scala naturae* of all created things ("Idée d'une échelle des êtres naturelles") of Bonnet (1764). (From Philip C. Ritterbush, *Overtures to Biology: The Speculations of Eighteenth-Century Naturalists* [New Haven: Yale University Press, 1964], copyright © 1964 by Yale University, reprinted by permission.)

structed *Ahnenreihen*, and, as I want to urge here, in the presumption of cladists that fossils will fall into a neat *Stufenreihe* to give a tidy stem-group to their extant crown-group, thereby yielding a "Hennigian comb" founded on successive plesions.

But the "Hennigian comb" itself has a pedigree as a pattern of classification rather than phylogeny that may well be as long as that of the *scala naturae*. It was known from the Middle Ages on as the "Tree of Porphyry" after the third-century A.D. founder of the school of Neoplatonism. In that form it is a cladogram (in the sense of the transformed, rather than phylogenetic, cladist) in which an individual ("Socrates" in Fig. 28) is classified in a hierarchical series of dichotomously divided ranks. Read from the top, the "taxon" of most inclusive rank (the "Summum Genus"), down, each successive taxon is divided using "apomorph" features ("Differentia") into two subordinate taxa ("Subaltern genera") until the taxa of lowest rank ("Individua") are reached. It will be noticed that although the taxa at every rank are named along the axis leading to the "Individua," the alternate taxa at each rank are not named and are represented merely by "Differentia." It also will be noted that whereas "Socrates" is a monophyletic taxon, its "sister-group" ("Plato et al.") is paraphyletic, comprising all mankind except "Socrates." A case also can be made for saying that the unnamed sister-groups represented only by their "Differentia" are paraphyletic also. The "Tree of Porphyry" was applied to more than

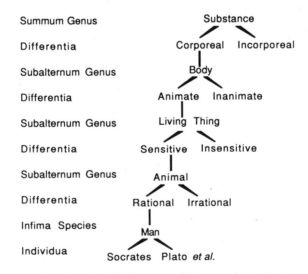

Fig. 28. The "Tree of Porphyry" with placement of the "taxon" ("Individuum") Socrates in a dichotomous classification of "Substance." "Genera" are supraspecific taxa at successive ranks. "Differentia" are apomorphies.

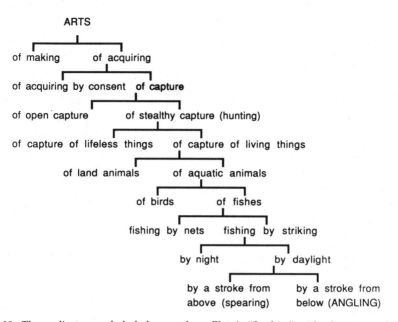

Fig. 29. The earliest recorded cladogram from Plato's "Sophist" with placement of "angling" in a dichotomous classification of the arts. Cf. Fig. 28.

the classification of creatures; in its earliest known manifestation in Plato's "Sophist" (Fig. 29), the activity of angling is classified similarly.

I do not claim that cladists who produce "Hennigian combs" have the "Tree of Porphyry" in mind, if only for the very good reason that most of them have never heard of it. I do claim, however, that the attitudes of mind inherent in both the concept of the *scala naturae* and that of the "Tree of Porphyry" are the same as those that predispose some cladists to produce "Hennigian combs." Darwin reminded himself continually not to refer to organisms as "higher" and "lower," and just as often broke his own resolution (Ospovat, 1981). We all have difficulty in not regarding the human being as the pinnacle of evolution and in not seeing any sequence of organisms (particularly fossils, and even more emphatically, vertebrate fossils) as an ascending or progressive series. The *scala naturae*, the orthogenetic series, the *Ahnenreihe*, and the axis of a phylogenetic "Hennigian comb" are "sisters under the skin." Similarly, the "Hennigian comb" as pure classification seems to have the same motivation as the "Tree of Porphyry." The aim is to place the taxon of the lowest rank, the "Individuum," in a dichotomous hierarchy; it is significant that the "plesions" in the "Tree of Porphyry" are not even named, as it is significant that the autapomorphies of the plesions in a stem-group tend not to be emphasized.

CONCLUSION

I began this review by discussing the relationships of Paleozoic tetrapods and then attempted to use that particular case, and especially the placement of *Crassigyrinus* and the temnospondyls, to illustrate the most common taxonomic method of paleontological cladists. Then I suggested that their prejudices extend back in time from an expected pattern of character distribution (the "Hennigian comb") for transformed cladists, through an expected pattern of phylogeny for Hennig and his untransformed disciples, to the *Stufenreihe* of Abel and Simpson. The conceptual pedigree then goes back to the orthogenetic series (and mechanisms) of Lamarck and some post-*Origin* paleontologists. But always there is the tendency to regard all phylogeny as inferred from the *scala naturae*—a "temporalizing of the Chain of Being," to use Lovejoy's phrase. Also, the usual attempt to arrange fossils as a series of plesions in a stem-group shows that a demonstration of the phylogeny of the crown-group has been given priority over an interest in the fossils themselves. For transformed cladists, this corresponds to a too-exclusive interest in placing the crown-group in a linear hierarchy, in the same way that the "Tree of Porphyry" is more concerned with classifying the "Individua" than exploring the properties of the "Subaltern genera."

To revert to the particular and perhaps the personal, I believe that classification of early, nonamniote tetrapods according to the precepts of the "Tree of Porphyry" almost certainly will yield an invalid result, and that my conclusion is highlighted by the nature of the fascinating and enigmatic Carboniferous tetrapod *Crassigyrinus scoticus*. I believe that *Crassigyrinus* is an anthracosauroid relative, albeit an extraordinarily primitive one. If I am correct, it signals the fact that the phylogeny of tetrapods was marked by an early division into two major rami. One ramus later contained, in addition to *Crassigyrinus*, the anthracosaurs, and possibly the amniotes and those nonamniote tetrapods (i.e., seymouriamorphs and diadectomorphs) traditionally associated with them. The other ramus probably contained, in addition to "*Ichthyostega* sp." (a rival to *Crassigyrinus* in primitiveness), the Temnospondyli, "Lissamphibia," and (if Smithson [1982, 1985] is correct) the microsaurs as well. This conclusion, reached in essence after the description of *Crassigyrinus* (Panchen, 1985), is sufficiently important, but if I have been able to show in this review that current taxonomic practice is in need of reappraisal, I shall regard that as even more important.

Acknowledgments Like the other chapters in this volume, the present review was presented originally in The University of Kansas course Biology 888: Topics in Evolutionary Morphology. It also was written during a visit to the Museum of Natural History in the spring of 1986.

Drs. Linda Trueb and William E. Duellman were my hosts during an enjoyable and stimulating period. The contents of this chapter were discussed with them, Drs. H.-P. Schultze and Larry Martin, and all those graduate students with whom I did battle following my verbal presentation. Of the latter, Michael Gottfried helped me in many ways. Dr. Robert Hook and an anonymous colleague suggested significant improvements as referees of the original manuscript. I am grateful to Drs. Schultze and Jennifer Clack (Cambridge) for allowing me access to unpublished manuscripts. Judith Hughes, Departments of Adult Education and Philosophy, The University, Newcastle upon Tyne, allowed me to prepare Figures 28 and 29 from her own diagrams. My travel expenses to Kansas were defrayed partially by the Staff Travel Fund of the University of Newcastle upon Tyne.

Literature Cited

Abel, O. 1911. *Grundzüge der Paläobiologie der Wirbeltiere*. Stuttgart: Schweizerbart.

Abel, O. 1924. *Lehrbuch der Paläozoologie*. 2nd ed. Jena: G. Fischer.

Abel, O. 1929. *Paläobiologie und Stammesgeschichte*. Jena: G. Fischer.

Andrews, S. M., M. A. E. Brown, A. L. Panchen, and S. P. Wood. 1977. Discovery of amphibians in the Namurian (Upper Carboniferous) of Fife. Nature, London, 265:529–532.

Andrews, S. M., and T. S. Westoll. 1970. The postcranial skeleton of *Eusthenopteron foordi* Whiteaves. Trans. R. Soc. Edinburgh, 68:207–329.

Atthey, T. 1876. On *Anthracosaurus russelli* Huxley. Ann. Mag. Nat. Hist., (4)18:146–167.

Bolt, J. R., and R. E. Lombard. 1985. Evolution of the amphibian tympanic ear and the origin of frogs. Biol. J. Linn. Soc., 24:83–99.

Bolt, J. R., R. M. McKay, B. J. Witzke, and M. P. McAdams. 1988. A new tetrapod locality in the Mississippian (Lower Carboniferous) of Iowa. Nature, London, 333:768–770.

Bonnet, C. 1764. *Contemplation de la nature*. 2 vols. 1st ed. Amsterdam: Marc-Michel Rey.

Brundin, L. 1968. Application of phylogenetic principles in systematics and evolutionary theory. Pp. 473–495 *in* Ørvig, T. (ed.), *Current Problems of Lower Vertebrate Phylogeny*. Nobel Symposium 4.

Buffon, G. L. L. 1749–1804. *Histoire naturelle, générale et particulière, avec la description du Cabinet du Roi*. 44 vols. plus Atlas. Paris: Imprimerie Royale, puis Plasson.

Carroll, R. L., and P. J. Currie. 1975. Microsaurs as possible apodan ancestors. J. Linn. Soc. London Zool., 57:229–247.

Carroll, R. L., and P. Gaskill. 1978. The order Microsauria. Mem. Am. Philos. Soc., 126:1–211.

Carroll, R. L., and R. Holmes. 1980. The skull and jaw musculature as guides to the ancestry of salamanders. J. Linn. Soc. London Zool., 68:1–40.

Clack, J. A. 1987. *Pholiderpeton scutigerum* Huxley; an amphibian from the Yorkshire Coal Measures. Philos. Trans. R. Soc., Ser. B, 318:1–107.

Clack, J. A. 1988. New material of the early tetrapod *Acanthostega* from the Upper Devonian of East Greenland. Palaeontology, 31:699–724.

Darwin, C. R. 1859. *On the Origin of Species by Means of Natural Selection, or the Preservation of Favoured Races in the Struggle for Life.* London: John Murray.

Eldredge, N. 1979. Cladism and common sense. Pp. 165–198 *in* Cracraft, J., and N. Eldredge (eds.), *Phylogenetic Analysis and Paleontology.* New York: Columbia Univ. Press.

Embleton, D., and T. Atthey. 1874. On the skull and some other bones of *Loxomma allmanni.* Ann. Mag. Nat. Hist., (4)14:38–63.

Forey, P. L. 1987. Relationships of lungfishes. Pp. 75–91 *in* Bemis, W. E., W. W. Burggren, and N. E. Kemp (eds.), *The Biology and Evolution of Lungfishes.* J. Morphol., Suppl. 1.

Gaffney, E. S. 1979a. Tetrapod monophyly: a phylogenetic analysis. Bull. Carnegie Mus. Nat. Hist., 13:92–105.

Gaffney, E. S. 1979b. An introduction to the logic of phylogeny reconstruction. Pp. 79–111 *in* Cracraft, J., and N. Eldredge (eds.), *Phylogenetic Analysis and Paleontology.* New York: Columbia Univ. Press.

Gardiner, B. G. 1980. Tetrapod ancestry: a reappraisal. Pp. 177–185 *in* Panchen, A. L. (ed.), *The Terrestrial Environment and the Origin of Land Vertebrates.* London: Academic Press.

Gould, S. J. 1977. Eternal metaphors of palaeontology. Pp. 1–26 *in* Hallam, A. (ed.), *Patterns of Evolution.* Amsterdam: Elsevier Scientific Publishing.

Hennig, W. 1950. *Grundzüge einer Theorie der phylogenetischen Systematik.* Berlin: Deutscher Zentralverlag.

Hennig, W. 1966. *Phylogenetic Systematics.* Urbana: Univ. of Illinois Press.

Hennig, W. 1969. *Die Stammesgeschichte der Insekten.* Frankfurt a M.: Kramer.

Hennig, W. 1981. *Insect Phylogeny.* [Edited and translated by A. C. Pont]. Chicester, N.Y.: J. Wiley.

Holmes, E. B. 1985. Are lungfishes the sister group of tetrapods? Biol. J. Linn. Soc., 25:379–397.

Hook, R. W. 1983. *Colosteus scutellatus* (Newberry), a primitive temnospondyl from the Middle Pennsylvanian of Linton, Ohio. Am. Mus. Novit., 2770:1–41.

Hull, D. L. 1980. The limits of cladism. Syst. Zool., 28:416–440.

Huxley, T. H. 1870. The anniversary address of the President. Q. J. Geol. Soc. London, 18:xi–ixiv.

Jarvik, E. 1952. On the fish-like tail in the ichthyostegid stegocephalians. Medd. Groenland, 114(12):1–90.

Jarvik, E. 1980. *Basic Structure and Evolution of Vertebrates.* Vol. 1. London: Academic Press.

Jarvik, E. 1981. Lungfishes, tetrapods, paleontology and plesiomorphy. By D. E. Rosen, P. L. Forey, B. G. Gardiner, and C. Patterson. 1981. Syst. Zool., 30:378–384. [Review].

Jefferies, R. P. S. 1979. The origin of the chordates—a methodological essay. Pp. 443–477 *in* House, M. R. (ed.), *The Origin of Major Invertebrate Groups.* London: Academic Press.

Lamarck, J.-B. P. A. de M. 1802. *Recherches sur l'organisation des corps vivans, et particulièrement sur son origine, . . . donné dans le Muséum national d'Histoire Naturelle, l'an X de la République.* Paris: Maillard.

Lamarck, J.-B. P. A. de M. 1809. *Philosophie zoologique, ou exposition des considérations relatives à l'histoire naturelle des animaux. . . .* Paris: Dentu.

Lamarck, J.-B. P. A. de M. 1815–1822. *Histoire naturelle des animaux sans vertèbres, présentant les caractères généraux et particuliers de ces animaux. . . .* 7 vols. Paris: Verdière.

Lombard, R. E., and J. R. Bolt. 1979. Evolution of the tetrapod middle ear: an analysis and reinterpretation. Biol. J. Linn. Soc., 11:19–76.

Lovejoy, A. O. 1936. *The Great Chain of Being.* Cambridge: Harvard Univ. Press.

Miles, R. S. 1977. Dipnoan (lungfish) skulls and the relationships of the group: a study based on new species from the Devonian of Australia. J. Linn. Soc. London Zool., 61:1–328.

Milner, A. R., T. R. Smithson, A. C. Milner, M. I. Coates, and W. D. I. Rolfe. 1986. The search for early tetrapods. Mod. Geol., 10:1–28.

Nelson, G., and N. I. Platnick. 1981. *Systematics and Biogeography: Cladistics and Vicariance.* New York: Columbia Univ. Press.

Ospovat, D. 1981. *The Development of Darwin's Theory.* Cambridge: Cambridge Univ. Press.

Panchen, A. L. 1964. The cranial anatomy of two Coal Measure anthracosaurs. Philos. Trans. R. Soc., Ser. B, 247:593–637.

Panchen, A. L. 1967. The nostrils of choanate fishes and early tetrapods. Biol. Rev. Cambridge Philos. Soc., 42:374–420.

Panchen, A. L. 1970. *Anthracosauria. Teil 5A. Handbuch der Paläontologie.* Stuttgart: Gustav Fischer.

Panchen, A. L. 1973. On *Crassigyrinus scoticus* Watson, a primitive amphibian from the Lower Carboniferous of Scotland. Palaeontology, 16:179–193.

Panchen, A. L. 1975. A new genus and species of anthracosaur amphibian from the Lower Carboniferous of Scotland and the status of *Pholidogaster pisciformis* Huxley. Philos. Trans. R. Soc., Ser. B, 269:581–640.

Panchen, A. L. 1980. The origin and relationships of the anthracosaur amphibia from the late Palaeozoic. Pp. 319–350 *in* Panchen, A. L. (ed.), *The Terrestrial Environment and the Origin of Land Vertebrates.* London: Academic Press.

Panchen, A. L. 1982. The use of parsimony in testing phylogenetic hypotheses. J. Linn. Soc. London Zool., 74:305–328.

Panchen, A. L. 1985. On the amphibian *Crassigyrinus scoticus* Watson from the Carboniferous of Scotland. Philos. Trans. R. Soc., Ser. B, 309:505–568.

Panchen, A. L., and T. R. Smithson. 1987. Character diagnosis, fossils, and the origin of tetrapods. Biol. Rev. Cambridge Philos. Soc., 62:341–438.

Panchen, A. L., and T. R. Smithson. 1988. The relationships of the earliest tetrapods. Pp. 1–32 *in* Benton, M. J. (ed)., *The Phylogeny and Classification of the Tetrapods.* Vol. 1. *Amphibians, Reptiles, Birds.* Spec. Vol. No. 35 A. Oxford: Clarendon Press.

Panchen, A. L., and T. R. Smithson. 1990. The pelvic girdle and hind limb of *Crassigyrinus scoticus* (Lydekker) from the Scottish Carboniferous and the origin of the tetrapod pelvic skeleton. Trans. R. Soc. Edinburgh (Earth Sci.), 81:31–44.

Patterson, C. 1981. Significance of fossils in determining evolutionary relationships. Ann. Rev. Ecol. Syst., 12:195–223.

Patterson, C., and D. E. Rosen. 1977. Review of ichthyodectiform and other Mesozoic teleost fishes and the theory and practice of classifying fossils. Bull. Am. Mus. Nat. Hist., 158:81–172.

Ritterbush, P. C. 1964. *Overtures to Biology: The Speculations of Eighteenth Century Naturalists.* New Haven: Yale Univ. Press.

Romer, A. S. 1969. The cranial anatomy of the Permian amphibian *Pantylus.* Breviora, 314:1–37.

Rosen, D. E., P. L. Forey, B. G. Gardiner, and C. Patterson. 1981. Lungfishes, tetrapods, paleontology and plesiomorphy. Bull. Am. Mus. Nat. Hist., 167:159–276.

Säve-Söderbergh, G. 1932. Preliminary note on Devonian stegocephalians from East Greenland. Medd. Groenland, 94(7):1–107.

Säve-Söderbergh, G. 1935. On the dermal bones of the head in labyrinthodont stego-cephalians and primitive Reptilia. Medd. Groenland, 98(3):1–211.

Schaeffer, B., M. K. Hecht, and N. Eldredge. 1972. Phylogeny and paleontology. Evol. Biol., 6:31–46.

Schultze, H.-P. 1969. Die Faltenzähne der rhipidistiiden Crossopterygier, der Tetra-poden und der Actinopterygier-Gattung *Lepisosteus*. Palaeontogr. Ital., 65(n.s.35): 59–137.

Schultze, H.-P. 1981. Hennig und der Ursprung der Tetrapoda. Palaeontol. Z., 55:71–86.

Schultze, H.-P. 1987. Dipnoans as sarcopterygians. Pp. 39–74 *in* Bemis, W. E., W. W. Burggren, and N. E. Kemp (eds.), *The Biology and Evolution of Lungfishes*. J. Morphol., Suppl. 1.

Simpson, G. G. 1953. *The Major Features of Evolution*. New York: Columbia Univ. Press.

Simpson, G. G. 1961. *Principles of Animal Taxonomy*. New York: Columbia Univ. Press.

Smithson, T. R. 1982. The cranial morphology of *Greererpeton burkemorani* Romer (Amphibia: Temnospondyli). J. Linn. Soc. London Zool., 76:29–90.

Smithson, T. R. 1985. The morphology and relationships of the Carboniferous am-phibian *Eoherpeton watsoni* Panchen. J. Linn. Soc. London Zool., 85:317–410.

Smithson, T. R., and K. S. Thomson. 1982. The hyomandibular of *Eusthenopteron foordi* Whiteaves (Pisces: Crossopterygii) and the early evolution of the tetrapod stapes. J. Linn. Soc. London Zool., 74:93–103.

Wallace, A. R. 1855. On the law which has regulated the introduction of new species. Ann. Mag. Nat. Hist., (2)16:184–196.

Wallace, A. R. 1858. On the tendency of varieties to depart indefinitely from the original type. J. Proc. Linn. Soc. London Zool., 3:53–62.

Watson, D. M. S. 1929. The Carboniferous Amphibia of Scotland. Palaeontol. Hun-garica, 1:219–252.

Wiley, E. O. 1981. *Phylogenetics: The Theory and Practice of Phylogenetic Systematics*. New York: John Wiley and Sons.

5 The Lungfish, the Coelacanth, and the Cow Revisited

P. L. Forey, B. G. Gardiner, and C. Patterson

In this chapter we comment on the preceding four chapters, written by H.-P. Schultze, E. Vorobyeva, Chang M.-M., and A. L. Panchen. These essays review the state of affairs in the current reevaluation of theories on the origin of tetrapods and the relationships between tetrapods and sarcopterygian fishes. The first sign of this reevaluation and of the controversy associated with it was an exchange in *Nature* (Gardiner et al., 1979; Halstead et al., 1979) titled "The Salmon, the Lungfish and the Cow." The title appropriately emphasized members of three extant groups (Actinopterygii, Dipnoi, Tetrapoda), although the controversy and the ensuing reevaluation of tetrapod origins have been dominated by paleontologists and by concern with fossils. To begin with, extant groups are appropriate because the problems that fossils should help to solve concern three main topics—i.e., how extant groups are interrelated, how those groups acquired their characters, and how old they are.

DISPUTED INTERPRETATIONS OF CHARACTERS OF FOSSILS

As for the interrelationships of extant groups, the authors of the four preceding chapters each propose solutions, summarized in Figures 1–4. Because there is no disagreement on the characters of the living members of these groups, the differences between the cladograms in these figures

145

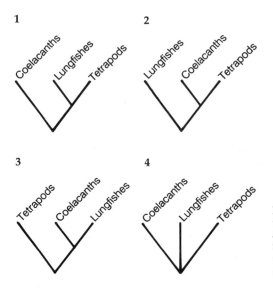

Figs. 1–4. Cladograms of three extant taxa proposed by (1) Panchen and Smithson (1987:Fig. 20), (2) Schultze (1987), (3) Chang (this volume), and (4) Panchen and Smithson (1987:438).

must rest on differences in interpretation of the characters of fossils. We comment on a few of these characters. Schultze (this volume) concentrates on two features, the nasal region and the fins.

Nasal Region

Although our homologization of the tetrapod choana with the dipnoan excurrent nostril (Rosen et al., 1981) has been criticized by most subsequent commentators, two points generally are now accepted. First, porolepiforms, which have been restored with three narial openings—two external and one internal (Jarvik, 1972)—have no choana (but see Jarvik, 1981); thus, no sarcopterygian is known to have three nares. Second, the tetrapod choana is homologous to the excurrent nostril of fishes and is not neomorphic, as was thought (following Jarvik, 1942, and Panchen, 1967) before our review. Two points deserve comment here, the nares in *Diabolepis* (Chang and Yu, 1984, 1987) and the primitive condition of the nares in tetrapods.

We accept *Diabolepis* as a lungfish, the sister of all others, owing to the features cited by Chang and Yu (1984) and the structure of the toothplates (Smith and Chang, 1990). Schultze (this volume) writes that the discovery of *Diabolepis* renders our homologization of the tetrapod choana with the dipnoan excurrent nostril obsolete, because both its nostrils are outside the premaxilla and maxilla. We dispute this for two reasons. First, the maxilla and the form and position of the excurrent nostril remain unknown in *Diabolepis*; Chang (this volume) infers that the maxilla was either toothless or absent. Second, the excurrent nostril of *Diabolepis* must have lain behind, rather than external to, the premaxilla (Chang and Yu, 1984:Fig. 2D). The bones bounding that nostril could well include the vomer (which is known) and the dermopalatine (which is not known), as in tetrapods. Because, as Chang and Yu said, the entire posterior part of the premaxilla was "situated inside the mouth cavity," all that can reasonably be inferred about the excurrent nostril of *Diabolepis* is that it was close to the margin of mouth. For this reason, the primitive condition of the nares in tetrapods is relevant.

There are two competing hypotheses on the primitive configuration of the nares in tetrapods. In 1981 (Rosen et al., 1981:196), we accepted the condition in *Ichthyostega* as primitive. In this taxon, the choana and external naris are close together so that both are visible in a ventral view of the skull; the external naris lies at the margin of the mouth, separated from the choana only by the inturned proximal end of the maxilla. Panchen (1985, this volume) and Panchen and Smithson (1987:376) proposed the condition in *Crassigyrinus* as an alternate and preferable primitive tetrapod configuration.

In *Crassigyrinus*, the external nostril is high on the side of the snout, bordered dorsally by the anterior tectal and separated from the maxilla by the lateral rostral and by contact between the lacrimal and premaxilla. As Panchen said, his restoration of *Crassigyrinus* resembles *Eusthenopteron* in these features. But our knowledge of the external naris in *Crassigyrinus* rests on one specimen, the holotype (Panchen, 1973). In that specimen, there are two potential external nostrils, one in the position just described and the other in a pit or deep concavity in the dorsal part of the suture between the maxilla and premaxilla (Panchen, 1973:185), close to the margin of the mouth. Regarding the possibility that the latter pit is the external naris, Panchen (1985:511) wrote, "I rejected that interpretation in 1963 [1973 intended] and do so now." In 1973 (p. 186), he wrote that the area he regarded as the external naris "is disrupted and difficult to interpret which is particularly unfortunate as, by analogy with *Eusthenopteron*, the nostril . . . should be situated in that region." Panchen went on to say that this putative nostril is "entirely occluded by bone" and interpreted that bone and the fragments surrounding it by analogy with *Eusthenopteron*. The second specimen of *Crassigyrinus* (Panchen, 1985) does not show the region of the suture between the premaxilla and maxilla, and of the bones surrounding the putative nostril in the holotype, it shows only the displaced anterior tectal. Therefore, it is hardly accurate for Panchen to repeat his claim (1985:Table 2; this volume:Table 1) that the polarity of this character (position of external nostril) is an example of the circular argument involved in resolving the sister-group of the Tetrapoda, for in this case circularity entered much earlier when

Panchen chose (in 1973) to interpret *Crassigyrinus* by using *Eusthenopteron* as a model.

The problem here is not the dilemma that Panchen attributed to "paleontological cladists," but a much older one—i.e., difficulty in interpreting unique or imperfectly preserved fossils. That this problem is still with us is evident from the fact that Chang (this volume) finds that "a choana has not yet been identified unequivocally in any 'rhipidistian,' " whereas Schultze (this volume) finds that "a choana can be documented at least in *Eusthenopteron* and *Panderichthys*." Both Chang and Schultze have seen the same fossils, but what is demonstrable to one is equivocal to the other. In just the same way, Schultze's (this volume) reading of Chang's (this volume) essay leads him to conclude that she "convincingly demonstrated that . . . *Diabolepis* [does] not possess a choana," whereas our reading of her essay and observation of her specimens leads to no such conclusion. And our reading of Vorobyeva and Schultze's account of *Panderichthys* (this volume) fails to convince us of the existence of a choana, because in the endoskeleton the postnasal wall is unossified, and in the exoskeleton the principal difference between the "fenestra exochoanalis" in specimens of *Panderichthys* (Vorobyeva and Schultze, this volume:Figs. 6–8) and *Eusthenopteron* (Rosen et al., 1981:Fig. 14) is that the slitlike opening is widened anteriorly in *Panderichthys* to produce a keyhole shape. This difference is explicable to us by the other difference between *Eusthenopteron* and *Panderichthys* in this region—the fact that in *Eusthenopteron* the fang of the first coronoid was received in a pit in the dermopalatine posteromedial to the "choana," whereas in *Panderichthys* that fang was received by a smooth area on the anterolateral margin of the dermopalatine which interrupts the marginal tooth row between the dermopalatine and vomer. Thus, we think that *Panderichthys* brings nothing really new to the choana problem and agree with Chang that a choana has yet to be demonstrated unequivocally in "rhipidistians."

Paired Fins

Schultze (this volume) raises the issue of the form of the joint between the humerus and the pectoral girdle in tetrapods and sarcopterygians. In tetrapods, there is a socket in the girdle that receives the ball-like proximal surface of the humerus, whereas in fishes, the situation is reversed, with the ball on the girdle. The latter pattern occurs in Recent coelacanths and lungfishes, and as evidence that it also occurred in Devonian lungfishes Schultze (1987:54) cited Janvier's (1980:Fig. 12A) illustration of the endoskeletal girdle in *Chirodipterus australis*. But as the text accompanying Janvier's figure states, it is based on a "freehand sketch" by one of us (P.L.F.); in fact, the specimen (BMNH P.52570) is lightly ossified, with a hollow glenoid bordered by thin perichondral bone and containing a core of irregular endochondral spicules from which no one could determine whether the intact articular surface was originally concave or convex. (See Rosen et al., 1981:caption to Fig. 40.) Janvier's purpose in illustrating this girdle of *Chirodipterus* was to point out the remarkable resemblance between it and osteolepiform girdles, and this similarity extends to the form of the glenoid (Rosen et al., 1981:Fig. 40). We remain ignorant of the form of the fully ossified glenoid and of the head of the humerus in Devonian dipnoans, but it seems to us improbable that girdles and glenoids so similar in form as those of *Eusthenopteron* and *Chirodipterus* (Rosen et al., 1981:Fig. 40) should have exhibited fundamentally different types of articulation. But in the absence of information on the endo-skeleton of the paired fins in primitive lungfishes, speculation on this and other matters is futile. In Appendix I, we have taken the form of the articulation from the Recent lungfishes in which there is no ambiguity about morphology.

Dermal Bones of the Skull

Comments on skull-roof patterns in the preceding chapters are restricted to comparisons between tetrapods and osteolepiforms or pan-derichthyids. Nowhere do Schultze, Vorobyeva and Schultze, or Panchen mention dipnoans or the observation that the rosette of five bones over the otico-occipital region in *Ichthyostega* is similar to that in dipnoans (Rosen et al., 1981). Instead, Schultze and Arsenault (1985) and Vorobyeva and Schultze (this volume) argue that the pattern of the dermal bones on the skull roof is directly comparable in *Panderichthys* and tetrapods, asserting that such bones as the tabular, supratemporal, and intertemporal of tetrapods are homologous to similarly positioned bones in *Panderichthys* and osteolepiforms (*Osteolepis*, *Eusthenopteron*). However, this view, though superficially attractive, has problems of its own.

There is no intertemporal in stem-group tetrapods (*Ichthyostega*, *Acanthostega*) or in loxommatids (other than *Loxomma*) or primitive temnospondyls. In the latter two groups, the intertemporal (which never is associated with the otic sensory canal) is minute in *Loxomma* (Beaumont, 1977) and occurs only in advanced temnospondyls such as the late Carboniferous *Dendrerpeton* and *Eugyrinus* and the Permian *Trimerorhachis* and *Edops*. In the more cladistically derived anthracosaurs and seymourians (Gardiner, 1983), the intertemporal similarly never is associated with the otic sensory canal, and in *Paleoherpeton*, it is associated with the supraorbital canal. Primitive tetrapods (*Ichthyostega*, *Acanthostega*, loxommatids, primitive temnospondyls) are characterized by a skull roof lacking an intertemporal, having the tabular separated from the parietal by sutural contact between the postparietal and supratemporal, and by a supratemporal having strong sutural junctions with the postparietal, parietal, and postorbital (Säve-Söderbergh, 1932:Fig. 11; Panchen, 1985:Fig. 24). We therefore agree with Stensiö (1947:93) that it is not possible to homologize the osteolepid (or panderichthyid) dermosphenotic with the tetrapod intertemporal. Similarly, the supratemporal of *Ichthyostega* and *Acanthostega* (J. A. Clack, pers. comm.), in which the sensory canals pass through the centers of ossification, is not associated with the otic sensory canal, nor is it in loxommatids or primitive temnospondyls (e.g., *Greererpeton, Eryops*), anthracosaurs (including *Crassigyrinus*), or aistopods. Nevertheless, the supratemporal is grooved by the otic canal in secondarily aquatic Triassic temnospondyls such as *Lyrocephalus, Trematosaurus, Aphaneramma,* and *Benthosaurus,* and also in *Neldasaurus, Rhinesuchus, Batrachosaurus, Trimerorhachis,* and *Eugyrinus.* In *Mastodonsaurus* and *Peltostega,* the otic canal ran over the squamosal. These sensory canal grooves presumably are not tied to centers of ossification, but the groove for the otic canal on the supratemporal of fossil amphibians never meets the groove for the supratemporal (occipital) commissure. Because the supratemporal of stem-group and primitive tetrapods is not a canal bone, it is doubtful whether it can be equated with the intertemporal of osteolepiforms and panderichthyids, as Schultze and Arsenault (1985) originally proposed.

In primitive tetrapods, the tube or groove for the otic sensory canal always ends short of the tabular (on the supratemporal in stem-group tetrapods or the squamosal in derived forms) and thus never joins the transverse tube or groove for the supratemporal (occipital) commissural canal. The latter canal rarely crosses the postparietals to meet its fellow in the midline; the only recorded instances are Jarvik's (1952:Fig. 35) reconstructions of *Ichthyostega,* Bystrow and Efremov's (1940:Fig. 18D) reconstruction of *Lonchorhynchus,* and Säve-Söderbergh's (1937) *Benthosaurus.* In most fossil amphibians, the groove for the commissure is confined to the tabulars. Thus, the tetrapod tabular, a bone primitively associated with the transverse commissural canal, is doubtfully homologous to the osteolepiform and panderichthyid supratemporal, a canal bone associated with the posterior part of the otic canal.

RELATIONSHIPS OF RECENT GROUPS

The characters on which Chang (this volume) and Schultze (1987) related *Latimeria* with dipnoans and with tetrapods, respectively (Figs. 2–3), are as follows.

Characters relating coelacanths and lungfishes (Chang, this volume:Fig. 15—Characters 7, 9–11):

1. Intracranial joint
2. Anocleithrum between supracleithrum and cleithrum (Schultze's Character 26, characterizing sarcopterygians)
3. Median extrascapular
4. Form of hyomandibular facet (extensive, straddling jugular vein at high angle)

Characters relating coelacanths and tetrapods (Schultze 1987:Fig. 12—Characters 32–34):
1. Form of head of hyomandibular (double-headed)
2. Intracranial joint
3. Three extrascapulars

There is an obvious symmetry between these two lists. An intracranial joint is known in no lungfish or tetrapod, and the difference between Chang's and Schultze's hypotheses rests on assumed loss of the joint in tetrapods (Schultze) and in lungfishes (Chang). Therefore, this character invokes a posteriori justification rather than characterization of groups. However, we note that in *Diabolepis* the otic and occipital portions of the braincase are missing in the only recorded specimen in which they would be expected (Chang and Yu, 1984:Fig. 2D), implying the possibility of a joint in the most primitive dipnoan.

The hyomandibular facet is known among dipnoans only in Devonian forms, and in primitive tetrapods the "hyomandibular facet" is the fenestra ovalis (perhaps accompanied by a facet adjoining it posteriorly, as implied in *Greererpeton* by Smithson, 1982). For Chang, the important feature is the form of the facet and its relation to the jugular vein, whereas for Schultze the important feature is its division (to match the double articular head of the hyomandibular). No known tetrapod has a divided hyomandibular head, whereas such a hyomandibular is known in the Devonian dipnoans *Griphognathus*, *Chirodipterus*, and *Stomiahykus* (inferred from the form of the hyomandibular facet in the latter); Schultze (1987:64) selected the undivided hyomandibular facet of *Dipnorhynchus* as morphotypic for dipnoans, but the element of a posteriori justification seems evident again here. As for the extrascapulars, Chang asserted that the median member of the series (Bone A in dipnoans) is the essential feature, whereas for Schultze, it is the presence of three bones in the series. No tetrapod has extrascapulars, and in the case of dipnoans there has been argument over whether five extrascapulars (rather than three) ever occur. Those who think that there are only three call the series of bones behind the skull roof Z-A-Z (e.g., Graham-Smith, 1978:Fig. 21) or Z-G-A-G-Z (e.g., Schultze, 1969:Fig. 16; Miles, 1977:Figs. 111, 116), where G is not a sensory canal bone. Those who think that there are five extrascapulars call the series Z-H-A-H-Z (e.g., Campbell and Barwick,

1982a:Fig. 9; Schultze and Campbell, 1987:Figs. 1e, 2; Schultze, 1987:Figs. 10C, 13B). Campbell and Barwick (1982a:524) summarized the evidence for the existence of Bone H; it is recorded in one specimen of *Chirodipterus australis* (lying in front of the large G usual in that species) but contains no sensory canal. That it was primitive for dipnoans and primitively contained a sensory canal remains hypothetical, as shown in Campbell and Barwick's (1982a) Figure 9a–d of alternate hypothetical extrascapular series in *Dipnorhynchus*. Schultze and Campbell (1987:Figs. 1e, 2) chose one of those hypotheses and extended it to *Uranolophus*, but that choice does not make the character any less hypothetical.

Thus, the grounds so far proposed for relating coelacanths to tetrapods (Schultze, 1987) or to dipnoans (Chang, this volume) are anything but secure. Similarly, the characters that we (Rosen et al., 1981) found relating dipnoans to tetrapods have been destructively criticized (e.g., Holmes, 1985; Panchen and Smithson, 1987:387; Schultze, 1987:69) or disregarded because they concern features not preserved in fossils. The arguments for all three possible dichotomous arrangements (Figs. 1–3) are sadly tenuous and evidently rely more on wishful thinking than on demonstrable shared characters. In the hope of reducing the role of wishful thinking, or the search for characters that support one's favored theory, we have tried a parsimony analysis of a set of characters for the three extant sarcopterygians groups and various fossils.

PAUP ANALYSIS

Among the variety of theories proposed for the interrelationships of sarcopterygian fishes and tetrapods during the past 20 years, some have been concerned only with living taxa (von Wahlert, 1968; Løvtrup, 1977), whereas others implicitly (e.g., Wiley, 1979) or explicitly (Forey, 1980; Rosen et al., 1981; Schultze, 1981; Janvier, 1986; Schultze, 1987; Panchen and Smithson, 1987; Chang, this volume) have included fossils. Among the latter, the range of fossils or fossil groups included varies considerably. These facts alone mean that it is difficult to compare the resulting classifications, but all have one refreshing common element in being formulated clearly, without the confusing inclusion of ancestor-descendant relationships.

Theories proposed by Schultze (1987), Rosen et al. (1981), Panchen and Smithson (1987), and Chang (this volume) are the most complete in their coverage and use similar data. The inclusion of many fossil taxa has meant that the data are primarily skeletal or can reasonably be inferred from the skeleton. Those four theories embrace many of the issues and characters that have played a part in discussion of the origin of tetrapods.

Understandably, each has stressed or played down or ignored certain similarities, such that the possible tree topologies have been restricted. Discussion of competing theories usually stresses the unsuitability of certain characters, substituting others and weighting characters. No doubt these practices will continue. However, the consequences of using unweighted character distributions has never been explored. And it is here that we find one application for computer-generated trees.

We ran some of the characters used in discussions through the PAUP program (Swofford, 1985). We do not claim freedom from subjectivity in character selection, but this type of analysis does reveal novel topologies and quantifies differences between topologies, thereby sparking new discussion.

Forty-three characters have been coded for nine taxa, including actinopterygians as the out-group (Appendix I). These characters are listed in the following section together with reasons for their inclusion and initial polarity assessments. The nine terminal taxa are regarded herein as monophyletic (although the monophyly of a few of them has been doubted on occasion by others). A few comments on the taxa are necessary. *Panderichthys* is used here instead of Panderichthyidae. There are differences between *Panderichthys* and *Elpistostege* (Schultze and Arsenault, 1985; Vorobyeva and Schultze, this volume) that affect coding. Thus, the intertemporal is scored as present in *Panderichthys* (Character 3), yet it is absent in *Elpistostege*. *Eusthenopteron* is used here, whereas other authors have used osteolepiforms on the assumption that they are monophyletic. Long (1985a) attempted to justify this but seemed uncertain as to which taxa should be included (cf. Forey 1987:81). Schultze (1987) regarded *Panderichthys* as an osteolepiform but now (Vorobyeva and Schultze, this volume) treats it as a separate taxon, more closely related to tetrapods. *Panderichthys* is treated here as a separate terminal taxon, but this still leaves osteolepiforms as problematic. When coding characters such as aspects of cosmine, we have used the condition in *Eusthenopteron* (which does not have cosmine) but are aware that many "osteolepiforms" do have this tissue. Similarly, the condition of the skull roofing bones is different in osteolepids and *Eusthenopteron* in that many osteolepids appear to have a mosaic of snout bones beneath the cosmine covered snout (cf. Character 4).

We have included, as separate taxa, *Diabolepis* and *Youngolepis* because the relationships of these taxa have been discussed or ignored. We can see no reason, however, to separate *Youngolepis* and *Powichthys* (except at most to regard them as sister-taxa), as Panchen and Smithson (1987:Fig. 20) did, separated by 12 derived characters. Onychodonts and rhizodonts have been omitted because of lack of data.

In collating characters, we have tried, where possible, to include those

that have been mentioned in previous discussions. Some characters that
we have used here (e.g., those numbered 15, 17, 36, 40) turn out to
characterize a group Sarcopterygii, and because we choose to root the
tree at actinopterygians, these may seem to be redundant to our basic
problem of the interrelationships of sarcopterygians. But we are aware
that not all authors consider sarcopterygians to be monophyletic (e.g.,
von Wahlert, 1968; Wiley, 1979), and these characters may assume a
greater importance when evaluating these theories against others such as
those shown in Figures 6–11. In some cases, we have reformulated them
(e.g., cheek characters). There are additional characters. Panchen and
Smithson (1987:421) used a variety of soft anatomical characters, and we
could have done the same, but some of our critics (e.g., Schultze, 1987)
suggested that this would bias the results. So for this exercise we restrict
ourselves to skeletal characters.

Characters Used

The following is a list of characters used and, where relevant, a justification for the polarity assumed. 0 = primitive. 1, 2, . . . n = derived
states. Where characters used by other workers are reexpressed, the
reasons are given.

Skull Roof

Character 1: The dermosphenotic forming part of the skull roof (1). The
dermosphenotic is identified as the bone containing the bend of the otic
canal as it turns to pass into the infraorbital canal. Sometimes (*Eusthenopteron, Panderichthys*) the bone also contains the junction with the supraorbital canal. That bone was called intertemporal by Schultze (1987) to bring
skull roof bone terminology into line with that of tetrapods. Because of
the canal pattern, we assume that the dermosphenotic is "fused" with
the parietals (frontals) in porolepiforms and, thus, is part of the skull
roof. The hinge of the cheek to the skull roof occurs ventral to the
dermosphenotic. In actinopterygians, the dermosphenotic is primitively
part of the cheek with the hinge dorsal to the bone. A dermosphenotic,
identified in this way, is absent in coelacanths. The actinopterygian condition is assumed to be primitive, but only because actinopterygians are
regarded here as the functional out-group.

Character 2: Many supraorbitals or tectals (1). Actinopterygians primitively have no supraorbitals or tectals (Gardiner and Schaeffer, 1989).
Supraorbitals develop several times within actinopterygians. *Eusthenopteron, Panderichthys*, and tetrapods have, at most, three bones in this
series (although the position of sutures varies, leading Jarvik [1948] to
construct his maximum number of bones in an osteolepid). The other

pattern is a large number of tectals or supraorbitals, which is represented in *Youngolepis*, *Diabolepis*, coelacanths, and lungfishes.

Character 3: The intertemporal absent (1). The absence of this bone is used here as a derived character. Polarity is assessed from actinopterygians in which ontogenetic evidence indicates that the bone may fuse with the parietal (postparietal) (*Polypterus*) or supratemporal (forming the dermopterotic, most actinopterygians). The intertemporal is assumed to be represented in *Powichthys* and *Diabolepis* by one or several bones lying lateral to the anterior end of the postparietals. We assume that lungfishes also have the equivalent of the intertemporal. The intertemporal is missing in *Elpistostege* (Schultze and Arsenault, 1985:298, 300), in which it is thought to have fused across the old joint with the postorbital or with the parietal. *Panderichthys*, however, has an intertemporal.

Character 4: The number of large bones anterior to the dermal intracranial joint. Irrespective of whether the joint is open or closed, it is thought possible to locate it by taking the bone bearing the anterior pit line as the parietal. (It may or may not also embrace the pineal opening.) The conditions are: $0 =$ one large pair; $1 =$ two large pairs; $2 =$ many. (This condition is always associated with the anterior mosaic.)

Character 5: Immobilization of the dermal portion of the intracranial joint (1). This character is included here because it often is discussed in theories of relationships of sarcopterygians and is used by Vorobyeva and Schultze (this volume) as a synapomorphy linking *Panderichthys* and tetrapods. Our assumption is that the joint was primitively open and mobile in sarcopterygians, as implied by the distribution of an open joint among these fishes. In *Diabolepis* and *Youngolepis*, the endocranium is divided completely or partially.

Character 6: Rostral tubuli (1). See Chang (this volume).

Character 7: Extrascapular overlap. Sarcopterygians primitively have an extrascapular series, which we assume included a median element (in contrast to actinopterygians and placoderms, in which the bones are primitively paired). In osteolepiforms, onychodonts, rhizodontiforms, porolepiforms, and coelacanths, there are primitively three bones. Lungfishes are claimed to have five (Schultze, 1987), but there are problems with this interpretation (see above). There is a secondary increase in the number of extrascapulars in coelacanths. Within sarcopterygians, the overlap relations between adjacent extrascapulars was noted by Jarvik (1972), who pointed out that porolepiforms have medial overlap, whereas osteolepiforms have lateral overlap. It is not possible to determine which is the plesiomorphic condition. In this analysis, the character-states are designated as: $0 =$ no overlap; $1 =$ medial; $2 =$ lateral. The character is run unordered.

Character 8: Interruption of the infraorbital canal at the anterior (incurrent) nostril (1). In actinopterygians, *Eusthenopteron,* and porolepiforms, the infraorbital canal is uninterrupted beneath the anterior nostril, and this is assumed to be the primitive condition (0). *Panderichthys* has this primitive condition (Vorobyeva and Schultze, this volume), although the path of the ethmoid commissure is unknown. In *Latimeria,* the infraorbital canal ends immediately behind the anterior nostril, and the ethmoid commissure begins immediately in front (Hensel, 1986); thus, the canal is interrupted. In dipnoans, the infraorbital canal ends at the anterior nostril, and there is no ethmoid commissure. The condition of the infraorbital canal and the ethmoid commissure in *Diabolepis* is unknown; Chang and Yu (1984) speculated that an ethmoid commissure was present in the suture between the premaxilla and snout (see below, Character 10). The anterior nostril lies at the mouth margin and no maxilla has been found. There can be only a narrow space between nostril and mouth, and an interrupted canal has been scored for *Diabolepis,* although there remains the possibility that the canal ended at the nostril and there was no ethmoid commissure. In tetrapods, the conditions are likewise not entirely clear (see above). In many primitive tetrapods, the anterior nostril lies at the mouth margin so that the canal, if present, would be interrupted. Jarvik (1952), on the basis of one specimen, however, restored a canal running beneath the nostril in a tubular lateral rostral. Rosen et al. (1981) doubted the existence of this bone and the contained canal. There is clearly a need for confirmation of the conditions in *Ichthyostega.* Schultze (1987:69) suggested that the "postulated interruption lies just above the choana." And Panchen and Smithson (1987:375) suggested that "a hypothesis that the infraorbital canal has been breached could still be maintained," following observations by Rage and Janvier (1982). We score the tetrapod infraorbital canal as interrupted beneath the anterior nostril.

Character 9: An anterior commissure between the supraorbital canals (1). The configuration of the supraorbital canal among osteichthyans varies. In actinopterygians, the canal primitively ends blindly between the nostrils (e.g., *Polypterus, Mimia, Moythomasia*). In *Eusthenopteron,* porolepiforms, and coelacanths the canal continues above both narial openings to join with the ethmoid commissure. This is probably a sarcopterygian character. An anterior commissure between left and right supraorbital canals is certainly a derived condition and is seen in coelacanths and Recent lungfishes. Fossil lungfishes probably also have this commissure (e.g., *Chirodipterus,* Miles 1977:Fig. 118D), although in many the anterior extent of the supraorbital canal has not been traced. It is however, noticeable that in most lungfishes the supraorbital canals curve medially as they pass forward.

Character 10: The ethmoid commissure taking a sutural course (1). It is assumed here that association between sensory canals and centers of ossification is a primitive osteichthyan condition; any deviation from this is derived. The condition in *Diabolepis* is in doubt, but Chang and Yu's (1984) suggestion of a sutural course is accepted.

Character 11: A processus dermintermedius present (1). This character and the polarity assessment is based on its use by Panchen and Smithson (1987). The process usually is assumed to be the homologue of the tetrapod septomaxilla. In addition to its distribution as recorded in the published literature, it also is recorded as present in porolepiforms (*Glyptolepis*; Chang, in press).

Neurocranium

Character 12: A ventrolateral fenestra in the floor of nasal capsule (1). This is a character used by Schultze (1987:66) to link porolepiforms, osteolepiforms, *Panderichthys*, and tetrapods. Its distribution is recorded here.

Character 13: A fossa autopalatina present (1). Gardiner (1984).

Character 14: The position of $V_{2, 3}$ with respect to the intracranial joint. This was discussed by Bjerring (1973) and Schultze (1987). Bjerring argued that the intracranial joint is nonhomologous in *Eusthenopteron* and coelacanths because the main trigeminal trunk leaves the neurocranium in different positions in a once completely segmented head. Here, the position of the nerve is simply noted as within the joint (0) or within the oticooccipital (1).

Histology

Character 15: Enamel in dermal skeleton (1).

Character 16: The cosmine with enamel-lined pore canals (1).

Cheek

Character 17: The cheek containing at least one squamosal bearing the jugal sensory canal (1).

Character 18: The number and pattern of cheek bones. These features have been used by various authors, usually to suggest relationship between osteolepiforms and tetrapods. Schultze (1987) argued that a pattern of seven cheek bones (maxilla, lacrimal, jugal, postorbital, squamosal, preoperculum, quadratojugal) is derived within sarcopterygians and is evidence of relationship between osteolepiforms, *Panderichthys*, and tetrapods. Rosen et al. (1981:Figs. 43–44) argued that this pattern probably was plesiomorphic for sarcopterygians. One of the reasons for Rosen et al.'s assessment was that onychodonts, considered to be primitive sarcopterygians, have a seven-bone pattern. Because Schultze (1987:Fig. 12) placed onychodonts as cladistically plesiomorphic to porolepiforms, which do not have the seven-bone arrangement, the plesiomorphic assessment becomes more likely.

Panchen and Smithson (1987) continued to accept the significance of the seven-bone cheek pattern but also adopted an additional stand by recognizing that osteolepiforms and tetrapods differ in the contact relationships between squamosal and maxilla. These bones meet in the osteolepiform cheek and exclude contact between the jugal and quadratojugal—a contact that occurs in tetrapods. Panchen and Smithson (1987:362) considered that these patterns were independently derived from the porolepiform pattern in which there are more bones in the cheek, and they did not use the character at any node in their synapomorphy scheme. Thus, for these authors, the similarity between osteolepiforms and tetrapods is homoplastic. Their general hypothesis of loss of bones from the cheek within sarcopterygians does not match their tree topology (Panchen and Smithson, 1987:Fig. 20) because coelacanths, which have the lowest number of bones in the cheek, are placed as the cladistically most primitive sarcopterygians, followed by lungfishes, which primitively have the highest number of cheek bones. Thus, there are problems with the number of bones forming the cheek, problems that are accentuated when we consider the osteolepiforms described by Long (1985a, 1987), in which there are supernumerary bones in the cheek. Additionally, to anticipate a deduction from the result of this analysis (which resolves *Youngolepis* as the sister of porolepiforms; cf. Jessen [1980] on *Powichthys*), the osteolepiform-type cheek may be primitive for sarcopterygians.

For these reasons, we have not included details of the cheek pattern in this analysis but, instead, point out another more obvious feature—the fact that the cheek is elongate and its hind margin slopes backward in most sarcopterygians and tetrapods but is short with an upright hind margin in lungfishes and coelacanths. The latter condition is considered derived because the long cheek and sloping hind margin is also primitive in actinopterygians.

Palate, Hyoid Arch, and Branchial Arches

Character 19: Palatal bite (1). It is assumed here that the primitive osteichthyan bite principally involved teeth on the premaxilla, maxilla, and dentary. In many sarcopterygians, these are supplemented by large teeth (fangs) on the coronoid series that occlude with fangs on the palatal bones (vomer, dermopalatines, ectopterygoids). A more derived condition is termed herein a palatal bite and exclusively involves the coronoids and prearticular in the lower jaw and the dermopalatine or ectopterygoids and pterygoids in the upper jaw. Lungfishes and coelacanths both show this feature, which is no doubt associated with reduction or loss of the dentary and maxilla.

Character 20: Double-headed hyomandibular (1). The hyomandibular is deduced as double-headed in *Youngolepis* by the shape of the facet on the braincase; see above on lungfishes.

Character 21: Short hyomandibular, free from the palate (1). The primitive osteichthyan condition is taken to be a long hyomandibular that is closely applied to the posterior margin of the palate. In coelacanths, lungfishes, and tetrapods, the hyomandibular is short (exceedingly so in modern lungfishes) and may (coelacanths and tetrapods) lie completely free from the palate, or lie free dorsally and brace the palate ventrally (primitive lungfishes fide Miles, 1977). In tetrapods (Carroll, 1980; Smithson and Thomson, 1982), the hyomandibular or stapes is thought to have primitively braced the palate and/or cheek, a rather similar situation to that in lungfishes.

Character 22: Upright jaw suspension (1). A palatoquadrate that has a marked posteroventral slope down to the jaw articulation is considered primitive, as seen in primitive actinopterygians, porolepiforms, osteolepiforms, and tetrapods, and in *Panderichthys*. We have scored this as a character different from Character 18 because the palate is known in *Diabolepis* (Chang, in press), but the cheek remains unknown.

Character 23: The posterior end of the ceratohyal expanded and reaching far dorsally (1). The primitive condition is assumed herein to be a ceratohyal that is in series with the hyomandibular (sometimes mediated by an interhyal), as in actinopterygians and *Eusthenopteron*.

Character 24: Reduction of the dorsal elements (pharyngobranchials and epibranchials) of the gill arches.

Character 25: Reduction or loss of hypobranchials (1).

Character 26: The last gill arch articulating with the penultimate (1). Rosen et al. (1981).

Lower Jaw

Features of the lower jaw have been used on many occasions to support the theory that osteolepiforms, porolepiforms, *Panderichthys*, and tetrapods are related (e.g., Westoll, 1943; Holmes, 1985; Schultze, 1987). Indeed, Schultze (1987:40) noted difficulty in distinguishing the rhipidistian lower jaw from that of primitive tetrapods. The issue was discussed by Forey (1987:81–82). Our assumption is that the long jaw with a long dentary, a series of infradentaries, three or four coronoids, and a series of submandibulars is primitive for sarcopterygians.

Coelacanths are uniquely specialized in having (1) a modified coronoid series, divided into an anterior series separated by a "diastema" from a posterior and modified enlarged coronoid; (2) only two infradentaries (splenial and angular); and (3) no submandibulars. These details are autapomorphous for coelacanths and are not included in this analysis. Primitive lungfishes have a series of infradentaries, but these have been given different names (Jarvik, 1967) implying nonhomology, which Schultze (1987) and Schultze and Campbell (1987) accepted. That view is not accepted herein because these bones have the same topographic relationships and association with the mandibular canal as in other sarcopterygians.

Character 27: Short dentary (1). In scoring this character, we assume that primitive lungfishes have a dentary (Rosen et al., 1981; Chang and Yu, 1984; cf. Campbell and Barwick, 1987).

Character 28: Labial pit (1). This is an area located laterally on the dentary and/or angular in lungfishes (Miles, 1977) and coelacanths; the pit is devoid of ornament and usually is excavated. It receives the lip fold in modern lungfishes and coelacanths, and is variably developed but universal in both groups.

Character 29: The angular as the dominant infradentary, bearing both the mandibular sensory canal and the oral canal (lungfishes) or pit line (coelacanths), which primitively extends forward onto the dentary (1).

Character 30: The loss of teeth at the anterior end of the dentary (1).

Character 31: Coronoid series reduced to the anterior quarter of the jaw (1). The primitive condition is assumed to be a continuous series of coronoids extending the greater length of the jaw medial to the dentary. Coelacanths are uniquely specialized in that the homologue of the most posterior coronoid is modified as a very distinctive element that stands above the oral margin (see above). However, as in lungfishes, the more typically shaped coronoids are restricted to the anterior quarter of the jaw. Therefore, there are two derived conditions within the coronoid series of coelacanths. There is reduction of primitive-shaped coronoids, a condition that we have coded, and there is modification of a posterior coronoid, a condition that is unique to coelacanths and one that we have not coded.

Teeth

Character 32: Folded teeth (1). The presence of folded teeth is regarded as derived, in comparison with the condition in primitive actinopterygians. Discrimination between the different types of folded teeth is not considered here because the dendrodont type is seen only in porolepiforms and commonly is regarded as an autapomorphy of that group. Folded teeth are included here because they figure in many discussions of tetrapod origins. Folded teeth are indicated on the parasphenoid of *Diabolepis*. It is acknowledged that some derived actinopterygians also have folded teeth (cf. Rosen et al., 1981).

Character 33: Radiate tooth arrangement (1).

Girdles and Paired Fins

Character 34: Sunken anocleithrum (1). See Rosen et al. (1981:218). It is assumed herein that the anocleithrum of *Panderichthys* is not sunken because Vorobyeva and Schultze (this volume) said that it is not overlapped by the cleithrum.

Character 35: Imperforate scapulocoracoid (1). It is assumed that the tripartite scapulocoracoid with large fenestrae (supraglenoid and supracoracoid foramina) is primitive for osteichthyans. This is the condition in actinopterygians (Gardiner, 1984; Jessen, 1972). The condition for *Youngolepis* has been taken from Chang (in press).

Character 36: Metapterygial fin (1). See Rosen et al., 1981:202. Although the fin endoskeleton of *Youngolepis* is unknown, the glenoid facet suggests that it articulated with a single basal element (Chang, in press).

Character 37: Paired fins elongate and composed of many endoskeletal segments (1). The primitive sarcopterygian condition for this character is assumed to be a fin composed of three or four segments, such as is seen in *Eusthenopteron, Panderichthys,* and coelacanths.

Character 38: Paired fins with a ball-and-socket joint between fin and girdle. Here, there appear to be three conditions: 0 = no ball-and-socket joint; 1 = ball on girdle, socket on fin; 2 = ball on basal element of fin (humerus), socket on girdle. These are two alternatives; thus, this character was run unordered. Porolepiforms, as represented by *Glyptolepis,* and *Youngolepis* are considered to lack a ball-and-socket joint (Ahlberg, 1989; Chang, in press); in both, the glenoid on the scapulocoracoid is flat, horizontal, and straplike. This may be a synapomorphy of *Youngolepis* + porolepiforms. Lungfishes are scored 1, following the condition in living forms (see Paired Fins, above).

Character 39: The orientation of the pectoral and pelvic fins differing by 180° (1). This character occurs in lungfishes and coelacanths. In lungfishes it is produced by rotation during ontogeny (Rosen et al., 1981); in coelacanths it is deduced from the fins in unborn young which are mirror images. Ahlberg (1989) described porolepiform pectoral and pelvic fins as having the same orientation—i.e., as not having mirror images.

Character 40: The basal element of the pectoral fin (humerus) with processes for muscle attachment (1). Much has been made of the processes on the postaxial side of the humerus of *Eusthenopteron* and their presumed homology with those on the humerus of primitive tetrapods. Exact process-to-process correspondence seems difficult to establish, but the fact that there are processes is important in terms of the potentiality for muscle attachment. It turns out that such processes have a wide distribution.

Character 41: A screw-shaped glenoid (1).

Character 42: The basal element (humerus) with entepicondylar foramen.

Character 43: Pectoral and pelvic fins of equal size (1). This character is regarded as derived because actinopterygians have fins of unequal size (0). We regard the condition in actinopterygians, in which the pelvic girdle may have been phylogenetically derived from the fin endoskeleton (Rosen et al., 1981), as a unique condition.

Method

The data matrix is given in Appendix I. Features that are not applicable in certain taxa are indicated. Thus, overlap of the extrascapulars is inapplicable in tetrapods in which these bones are absent. Missing data are indicated by question marks (and there is much missing data for *Youngolepis* and *Diabolepis*). The computer program treats "not applicable" and "missing" as equivalent, meaning that, as a tree is being built, alternate possible characters states (i.e., 0, 1; or 0, 1, 2) are considered. Sometimes this can have nonsensical consequences, such as the deduction that the extrascapulars of tetrapods overlap laterally. When such situations arise in derived taxa, there usually is no problem. But there are difficulties when a character is correctly coded as inapplicable in the out-group (in this case, actinopterygians) but is treated by the program as "missing." For example, Character 16 (pore canals of cosmine enamel-lined) is deduced as present in actinopterygians because this is the most parsimonious estimation of the character in this particular taxon set. There are additional problems with this character in this data set because the character also was scored as not applicable in *Eusthenopteron*, yet other "osteolepiforms" have cosmine and could have been scored 0 for this character.

Another source of ambiguity arises when the coding for certain kinds of transformation is considered. For example, extrascapular overlap was coded: 0 = no overlap; 1 = medial overlap; 2 = lateral overlap. Although the character was run unordered, meaning that any permutation of transformation was possible, it could be argued that only $0 \rightarrow 1$ and $0 \rightarrow 2$, or $1 \rightarrow 0$ and $2 \rightarrow 0$ are real possibilities. In the event, it turns out that the most parsimonious solution regards one of the overlap relations as a subset of the other, i.e., the assemblage showing one of the overlap relations is paraphyletic. The alternative would have been to split this character into two (lateral overlap versus any other condition; medial overlap versus any other condition), but there are problems with this method as well (Pimentel and Riggins, 1987). In this data set, three characters have been coded in this manner, and we discuss the consequences under "Results."

Various runs were undertaken. The principal run used the branch-and-bound option, the technique finding the shortest tree or trees. Multistate characters were run unordered. Actinopterygians were used to root the tree. We felt justified in this rooting because all theories compared here accepted this taxon as the out-group. As a check, we tried Lundberg rooting; this gave no difference in topology.

In a supplementary run, *Youngolepis*, *Diabolepis*, and *Panderichthys* were deleted, as three relatively poorly known taxa, in order to find out how far missing data affect the result.

Further runs were done using tree topologies published by other authors and optimizing our characters against their trees. Two pieces of information can be gathered from this exercise—the comparative lengths of the trees, and the character transformations predicted. In this set of runs, certain deviations from the original published trees had to be made, chiefly because the same taxa were not included in the original theory. Thus, *Diabolepis* was added to all published tree topologies, as the sister-group of lungfishes, although *Diabolepis* was not considered by Schultze (1987) or Panchen and Smithson (1987). *Panderichthys* was added as the sister of tetrapods in the theories proposed by Chang (this volume) and Panchen and Smithson (1987). In the theory of Rosen et al. (1981), *Youngolepis* and *Diabolepis* were added as the respective sisters of porolepiforms and lungfishes, and the trichotomy that Rosen et al. showed between coelacanths, porolepiforms, and lungfishes + tetrapods was arbitrarily resolved by (1) placing porolepiforms as the sister-group of tetrapods, and (2) placing coelacanths as the sister-group of tetrapods. This was done because the program will accept only fully dichotomous user trees.

Results

The principal run resulted in four equally parsimonious trees of length 64. One of the possibilities is shown in Figure 5, together with the variant part of the tree as insets. This solution involved *Eusthenopteron*, *Panderichthys*, and *Youngolepis* + porolepiforms. The solution suggested in the main figure is a trichotomy between *Eusthenopteron*, *Panderichthys*, and the remaining sarcopterygians. This arrangement was the result of there being no characters optimized on this particular tree that placed *Eusthenopteron* as cladistically more derived than *Panderichthys*.

This trichotomy is the solution that is favored here. But it is not the only parsimonious solution. Another possibility placed *Panderichthys* as cladistically more derived than *Eusthenopteron*. But this solution arose as a result of the prediction by the tree-building program that Characters 24, 25, and 26 (missing data for *Panderichthys*) were in fact present in the "1" state. In another of the possibilities, *Eusthenopteron* and *Panderichthys* were considered sister-groups; this grouping was based on three characters: 2 (1–0), a secondary loss of supraorbitals as a parallelism with tetrapods; 7 (2–1); 34 (1–0), an anocleithrum that secondarily became

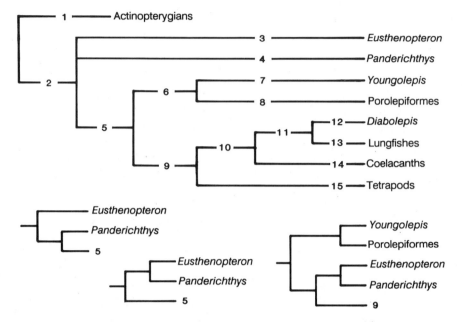

Fig. 5. One of the four equally parsimonious trees produced as a result of PAUP analysis of the data in Appendix I. The consistency index is 0.703; the length is 64 steps and the *F*-value 2.267. Inset, *below,* are the three other possibilities, with the numbers 5 and 9 referring to numbered legs in the main tree. The characters assigned to each numbered leg in the main tree are listed below, with reversals coded "−" and parallelisms coded "="; where relevant, states of characters or character transformations of multistate characters are shown in parentheses. (See text for discussion.) Leg 1: "nonapplicable" features of actinopterygians in Table 1 were resolved as 4(1), 5(1), 14(1). Leg 2:1, 7(0–1), 11, 15, 17, 20, 32, 36, 38(0–2), 40, 41. Leg 3: 4(1–0), = −5, =42. Leg 4: none. Leg 5: 2, 7(1–2), 10, −14, 24, 25, 34. Leg 6: =4(1–2), =6, =13, 30, =35, =37, 38(2–0), −41. Leg 7: =3, −10, 14. Leg 8: none. Leg 9: 8, 21, 23, 43. Leg 10: = −5, 9, −11, 18, 19, 22, 27, 28, 29, 31, −32, 38(2–1), 39. Leg 11: =4(1–2), =6, 7(2–0), =13, 33, =37. Leg 12: −12, 32. Leg 13: none. Leg 14: =3, = −5, =35, −41. Leg 15: −2, =42.

superficial and ornamented. None of these characters is particularly convincing. Characters suggesting that *Eusthenopteron* and *Panderichthys* are more derived than porolepiforms + *Youngolepis* include 4 (2–1), a reduction in the number of large bones in front of the intracranial joint; 38 (0–2), acquisition of a ball-and-socket joint in which the ball is located on the humerus, subsequently lost, and further transformed; 41, a screw-shaped glenoid, subsequently lost in coelacanths. With the possible exception of the last-mentioned character, these seem unconvincing.

All four trees were identical in other respects. All placed coelacanths as the sister of *Diabolepis* + lungfishes in a group with 11 synapomorphies, mainly features of the upright suspensorium, lower jaw, and paired fins that show opposite rotation. The idea that, among living animals, lung-fishes and coelacanths are sister-groups has not been explored seriously.

However, there are features of the brain such as similar development of a septum ependymale dividing the cerebral hemispheres, a superficial isthmal nucleus, and a thickened dorsal thalamus (Northcutt, 1987), and the hypertrophied development of snout electroreceptor cells that support this grouping (Forey, 1988). This tree implies that the processus dermintermedius, rated here as a sarcopterygian feature, has been lost in this combined group. The tree also implies that folded teeth, rated here as another sarcopterygian feature, were lost in this combined group and subsequently regained in *Diabolepis*. In reality, it may be more reasonable to postulate that folded teeth have been lost or modified independently in coelacanths and in lungfishes.

The conclusion that tetrapods are the sister of coelacanths + (*Diabolepis* + lungfishes) is founded on four synapomorphies (8, 21, 23, 43) that concern interruption of the infraorbital canal, the hyoid arch, and the paired fins of equal size.

The grouping of porolepiforms + *Youngolepis* with this combined group (tetrapods (coelacanths (*Diabolepis* + lungfishes))) as cladistically more derived than either *Panderichthys* or *Eusthenopteron* is supported by six synapomorphies involving the composition of the skull roof, features of the gill arches, and the anocleithrum. The gill arch characters may reflect a basic shift in the mechanism for ventilation. This node is also identified by features that are subsequently lost or changed, such as the position of the intracranial joint in relation to the trigeminal and the path of the ethmoid commissure (inapplicable in lungfishes and tetrapods).

The preferred solution offered here suggests that some of the characters traditionally used to link rhipidistians (or osteolepiforms) with tetrapods are plesiomorphic sarcopterygian features that have been lost or modified in lungfishes and coelacanths and, in some cases, porolepiforms as well. Characters falling into this category, and scored as apomorphic in this analysis, are the folded teeth, the septomaxilla or processus dermintermedius, and, in one solution, the entipecondylar foramen. Other characters, not used here but interpretable in the same way, would be a cheek pattern of seven bones, the long shallow lower jaw, and vertebral centra containing intercentra and pleurocentra (see Rosen et al., 1981, for discussion). We have some confidence that these characters are placed at the correct hierarchical level because of the phylogenetic positions of *Youngolepis* and *Diabolepis*. Between them, they show several of these features, yet they are related to derived sarcopterygians. In addition to the characters uniting *Youngolepis* to porolepiforms, as shown in this analysis *Youngolepis* is similar to some porolepiforms in showing three enlarged sensory pores along the length of the mandibular canal, widely separated vomers, downwardly pointing basipterygoid process (Gardiner, 1984), and a similarly shaped cleithrum (Chang, in press). *Diabolepis* shares features with lungfishes such as a

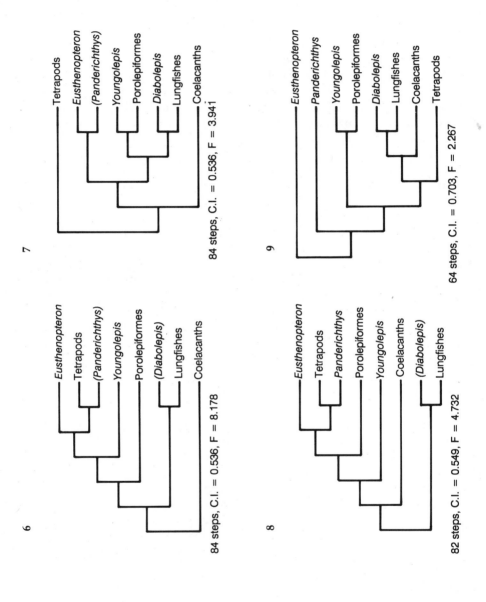

7

84 steps, C.I. = 0.536, F = 3.941

9

64 steps, C.I. = 0.703, F = 2.267

6

84 steps, C.I. = 0.536, F = 8.178

8

82 steps, C.I. = 0.549, F = 4.732

166

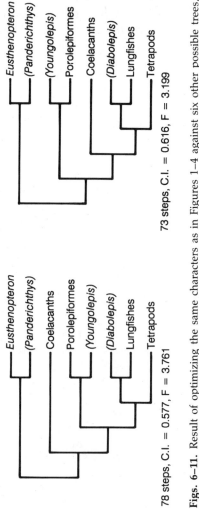

10

11

Eusthenopteron
(Panderichthys)
Coelacanths
Porolepiformes
(Youngolepis)
(Diabolepis)
Lungfishes
Tetrapods

78 steps, C.I. = 0.577, F = 3.761

Eusthenopteron
(Panderichthys)
(Youngolepis)
Porolepiformes
Coelacanths
(Diabolepis)
Lungfishes
Tetrapods

73 steps, C.I. = 0.616, F = 3.199

Figs. 6–11. Result of optimizing the same characters as in Figures 1–4 against six other possible trees, showing the lengths and consistency indices. (See text for discussion.) (6) Panchen and Smithson (1987). (7) Chang (this volume). (8) Schultze (1987). (9) Preferred solution (Fig. 3). (10–11) Alternate resolutions of Rosen et al. (1981:Fig. 62). C.I. = consistency index; F = F-value.

167

similar bone configuration on the rear half of the skull roof (I, B, and J bones), a palate with a dorsal palatal process (terminology of Campbell and Barwick, 1982b; pers. observ.), an inturned premaxilla (Chang and Yu, 1984).

It is worth noting that there is homoplasy between these derived sarcopterygians. The tree suggests, for instance, that a number of characters developed in parallel between porolepiforms + *Youngolepis* and lungfishes + *Diabolepis* (Characters 4, 6, 13, 37). The tree also implies parallelism between porolepiforms and coelacanths in Characters 3 and 5; Character 3, the disappearance of the intertemporal as an independent bone, was used by Andrews (1973) to link these two groups. The behavior of Character 5 may seem unusual because it suggests that the dermal intracranial joint became open independently in porolepiforms, coelacanths, and *Eusthenopteron* from a primitively immobile condition. But in reality this is no more parsimonious than assuming that it became immobile on three occasions. The parallelisms at this level reflect the trichotomy that Rosen et al. (1981) reached between coelacanths, porolepiforms, and choanates. That trichotomy is resolved in this tree by accepting similarity in lower jaw anatomy between lungfishes and coelacanths as homologous.

The characters used here were optimized against other published trees. Each gave a much longer tree (the numbers of steps are shown in Figs. 6–11) and inevitably lower consistency indices than the preferred tree. Those characters (8, 21, 23, 43) that in the preferred tree place tetrapods with coelacanths and lungfishes + *Diabolepis* must be assumed to have been gained several times (Schultze's and Chang's trees) or to have been gained, lost, then regained (Panchen and Smithson's tree). The gill arch characters supporting the grouping of porolepiforms + *Youngolepis* with (tetrapods (coelacanths (*Diabolepis* + lungfishes))) must be assumed to have been gained on several occasions. Homoplasy in these characters may be accepted or justified by future workers, but it cannot be ignored.

Finally, it is worth remarking that, irrespective of solutions involving fossil taxa, the distributions of characters among Recent taxa remain to be explained. There are many similarities in soft anatomy and biochemical features that are shared by lungfishes and tetrapods yet absent from coelacanths and actinopterygians. We see no reason to believe that what we deduce from the fossil record should be homology, whereas what we see in the modern world should be judged analogy.

Acknowledgments First, we are indebted to the late Donn Rosen for stimulating us to look at old problems in new ways; part of the foregoing is based on a short manuscript composed by Donn in the last year of his life. We also thank Chang Mee-Mann for showing us her specimens of *Diabolepis* and *Youngolepis*, and for allowing us to cite her work in press.

APPENDIX I. Summary of characters for nine osteichthyan taxa. N = nonapplicable character; ? = data not yet available for taxon.

Characters 1–22

Taxon	1	2	3	4	5	6	7	8	9	10	11	12	13	14	15	16	17	18	19	20	21	22
Actinopterygians	0	0	0	N	N	0	0	0	0	0	0	1	0	N	N	N	0	0	0	0	0	0
Eusthenopteron	1	0	0	0	0	0	1	0	0	0	1	1	0	1	1	N	1	0	0	1	0	0
Panderichthys	1	0	0	1	1	0	1	0	0	0	1	1	0	?	1	N	1	0	0	?	?	0
Porolepiforms	1	1	1	2	0	1	2	0	0	0	1	1	1	1	1	?	1	0	0	1	0	0
Youngolepis	?	1	0	?	1	1	2	?	0	1	1	1	1	0	1	?	1	0	0	1	0	?
Diabolepis	?	1	0	2	1	1	?	1	?	1	0	1	1	?	1	?	?	?	1	?	?	?
Lungfishes	1	1	1	2	1	1	0	1	1	N	0	1	N	N	1	1	1	1	1	1	1	1
Coelacanths	N	1	1	1	0	0	2	1	1	1	0	0	0	N	1	N	1	0	1	1	1	1
Tetrapods	?	0	0	1	1	0	N	1	0	N	1	1	0	N	1	N	1	0	0	N	1	0

Characters 23–43

Taxon	23	24	25	26	27	28	29	30	31	32	33	34	35	36	37	38	39	40	41	42	43
Actinopterygians	0	0	0	0	0	0	0	0	0	0	0	0	0	0	N	0	0	0	0	N	0
Eusthenopteron	0	0	0	0	0	0	0	0	0	1	0	0	0	1	0	2	0	1	1	1	0
Panderichthys	?	?	?	?	0	0	0	0	0	0	0	1	1	1	1	0	0	1	?	0	0
Porolepiforms	0	?	?	1	0	0	0	1	0	1	0	1	1	1	1	0	0	1	0	0	0
Youngolepis	?	?	?	?	0	1	0	1	0	1	1	?	1	1	?	0	?	?	0	?	?
Diabolepis	?	?	?	?	1	1	?	0	?	1	1	?	?	?	?	?	?	?	0	?	?
Lungfishes	1	1	1	1	1	1	1	0	1	0	1	1	0	1	0	1	1	1	0	0	1
Coelacanths	1	1	1	1	1	1	1	0	0	0	0	1	1	1	1	1	1	1	0	0	1
Tetrapods	1	1	1	1	0	0	0	0	0	1	0	N	0	1	1	2	0	1	1	1	1

169

Literature Cited

Ahlberg, P. 1989. Paired fin skeletons and relationships of the fossil group Por-
olepiformes (Osteichthyes: Sarcopterygii). J. Linn. Soc. London Zool., 96:119–166.
Andrews, S. M. 1973. Interrelationships of crossopterygians. Pp. 137–177 *in* Green-
wood, P. H., R. S. Miles, and C. Patterson (eds.), *Interrelationships of Fishes*. London:
Academic Press.
Beaumont, E. H. 1977. Cranial morphology of the Loxommatidae (Amphibia: Laby-
rinthodontia). Philos. Trans. R. Soc. London, Ser. B, 280:29–101.
Bjerring, H. C. 1973. Relationships of coelacanthiforms. Pp. 179–205 *in* Greenwood,
P. H., R. S. Miles, and C. Patterson (eds.), *Interrelationships of Fishes*. London:
Academic Press.
Bystrow, A. P., and J. A. Efremov. 1940. *Benthosuchus sushkini* Efr.—a labyrinthodont
from the Eotriassic of Sharjenga River. Tr. Paleontol. Inst. Akad. Nauk SSSR, 10:1–
152.
Campbell, K. S. W., and R. E. Barwick. 1982a. A new species of the lungfish *Dip-
norhynchus* from New South Wales. Palaeontology, 25:509–527.
Campbell, K. S. W., and R. E. Barwick. 1982b. The neurocranium of the primitive
dipnoan *Dipnorhynchus sussmilchi* (Etheridge). J. Vert. Paleontol., 2:286–327.
Campbell, K. S. W., and R. E. Barwick. 1987. Paleozoic lungfishes—a review. Pp. 93–
131 *in* Bemis, W. E., W. W. Burggren, and N. E. Kemp (eds.), *The Biology and
Evolution of Lungfishes*. J. Morph., Suppl. 1.
Carroll, R. L. 1980. The hyomandibular as a supporting element in the skull of
primitive tetrapods. Pp. 293–317 *in* Panchen, A. L. (ed.), *The Terrestrial Environment
and the Origin of Land Vertebrates*. Syst. Assoc. Spec. Vol. No. 15. London: Academic
Press.
Chang M.-M. In press. Head exoskeleton and shoulder girdle of *Youngolepis*. *In* Chang
M.-M., K. S. W. Campbell, and H.-P. Schultze (eds.), *Proceedings of the Fifth Sym-
posium on Early Vertebrate Studies and Related Problems in Evolutionary Biology*. Beijing:
Oceanography Publishing House.
Chang M.-M., and X. B. Yu. 1984. Structure and phylogenetic significance of *Diabolich-
thys speratus* gen. et sp. nov., a new dipnoan-like form from the Lower Devonian of
Eastern Yunnan, China. Proc. Linn. Soc. New South Wales, 107:171–184.
Chang M.-M., and X. B. Yu. 1987. A *nomen novum* for *Diabolichthys* Chang et Yu, 1984.
Vertebr. PalAsiatica 25:79.
Forey, P. L. 1980. *Latimeria:* a paradoxical fish. Proc. R. Soc. London, Ser. B, 208:369–
384.
Forey, P. L. 1987. Relationships of lungfishes. Pp. 75–91 *in* Bemis, W. E., W. W.
Burggren, and N. E. Kemp (eds.), *The Biology and Evolution of Lungfishes*. J. Morph.,
Suppl. 1.
Forey, P. L. 1988. Golden Jubilee for the coelacanth *Latimeria chalumnae*. Nature,
London, 336:727–732.
Gardiner, B. G. 1983. Gnathostome vertebrae and the classification of the Amphibia. J.
Linn. Soc. London Zool., 79:1–59.
Gardiner, B. G. 1984. The relationships of the palaeoniscid fishes, a review based on
new specimens of *Mimia* and *Moythomasia* from the Upper Devonian of Western
Australia. Bull. Br. Mus. Nat. Hist. Geol., 37:173–428.
Gardiner, B. G., P. Janvier, C. Patterson, P. L. Forey, P. H. Greenwood, R. S. Miles,
and R. P. S. Jefferies. 1979. The salmon, the lungfish and the cow: a reply. Nature,
London, 277:175–176.
Gardiner, B. G., and B. Schaeffer. 1989. Interrelationships of lower actinopterygian
fishes. J. Linn. Soc. London, 97:135–187.

Graham-Smith, W. 1978. On the lateral lines and dermal bones in the parietal region of some crossopterygian and dipnoan fishes. Philos. Trans. R. Soc. London, Ser. B, 282:41–105.

Halstead, L. B., E. I. White, and G. T. MacIntyre. 1979. [Letter]. Nature, London, 277:176.

Hensel, K. 1986. Morphologie et interprétation des canaux et canalicules sensoriels céphaliques de *Latimeria chalumnae* Smith, 1939 (Osteichthyes, Crossopterygii, Coelacanthiformes). Bull. Mus. Nat. Hist. Nat. Paris, (4)A8:379–407.

Holmes, E. B. 1985. Are lungfishes the sister group of tetrapods? Biol. J. Linn. Soc. London, 25:379–397.

Janvier, P. 1980. Osteolepid remains from the Devonian of the Middle East, with particular reference to the endoskeletal shoulder girdle. Pp. 223–254 *in* Panchen, A. L. (ed.), *The Terrestrial Environment and the Origin of Land Vertebrates*. Syst. Assoc. Spec. Vol. No. 15. London: Academic Press.

Janvier, P. 1986. Les nouvelles conceptions de la phylogénie et de la classification des "agnathes" et des sarcopterygiens. Oceanis, 12:123–138.

Jarvik, E. 1942. On the structure of the snout of crossopterygians and lower gnathostomes in general. Zool. Bidr. Uppsala, 21:235–675.

Jarvik, E. 1948. On the morphology and taxonomy of the Middle Devonian osteolepid fishes of Scotland. K. Svenska VetenskAkad. Handl., (3)25,3:1–301.

Jarvik, E. 1952. On the fish-like tail in the ichthyostegid stegocephalians with descriptions of a new stegocephalian and a new crossopterygian from the Upper Devonian of East Greenland. Medd. Groenland, 114(12):1–90.

Jarvik, E. 1967. On the structure of the lower jaw in dipnoans: with a description of an early Devonian dipnoan from Canada, *Melanognathus canadensis* gen. et sp. nov. J. Linn. Soc. London Zool., 47:155–183.

Jarvik, E. 1972. Middle and Upper Devonian Porolepiformes from East Greenland with special reference to *Glyptolepis groenlandica* n.sp. and a discussion of the structure of the head in the Porolepiformes. Medd. Groenland, 187(2):1–307.

Jarvik, E. 1981. Lungfishes, tetrapods, paleontology, and plesiomorphy.—Donn E. Rosen, Peter L. Forey, Brian G. Gardiner, and Colin Patterson. 1981. Syst. Zool., 30:378–384. [Review].

Jessen, H. 1972. Schultergürtel und Pectoralflossen bei Actinopterygiern. Fossils Strata, 1:1–101.

Jessen, H. 1980. Lower Devonian Porolepiformes from the Canadian Arctic with special reference to *Powichthys thorsteinssoni* Jessen. Palaeontographica, Ser. A, 167:180–214.

Long, J. A. 1985a. The structure and relationships of a new osteolepiform fish from the Late Devonian of Victoria, Australia. Alcheringa, 9:1–22.

Long, J. A. 1985b. New information on the head and shoulder girdle of *Canowindra grossi* Thomson, from the Upper Devonian Mandagery Limestone, New South Wales. Rec. Aust. Mus., 37:91–99.

Long, J. A. 1987. An unusual osteolepiform fish from the Late Devonian of Victoria, Australia. Palaeontology, 30:839–852.

Løvtrup, S. 1977. *The Phylogeny of Vertebrata*. London: John Wiley.

Miles, R. S. 1977. Dipnoan (lungfish) skulls and the relationships of the group: a study based on new species from the Devonian of Australia. J. Linn. Soc. London Zool., 61:1–318.

Northcutt, R. G. 1987. Lungfish neural characters and their bearing on sarcopterygian phylogeny. Pp. 277–297 *in* Bemis, W. E., W. W. Burggren, and N. E. Kemp (eds.), *The Biology and Evolution of Lungfishes*. J. Morph., Suppl. 1.

Panchen, A. L. 1967. The nostrils of choanate fishes and early tetrapods. Biol. Rev. Cambridge Philos. Soc., 42:374–420.

Panchen, A. L. 1973. On *Crassigyrinus scoticus* Watson, a primitive amphibian from the Lower Carboniferous of Scotland. Palaeontology, 16:179–193.

Panchen, A. L. 1985. On the amphibian *Crassigyrinus scoticus* Watson from the Carboniferous of Scotland. Philos. Trans. R. Soc. London, Ser. B, 309:505–568.

Panchen, A. L., and T. R. Smithson. 1987. Character diagnosis, fossils and the origin of tetrapods. Biol. Rev. Cambridge Philos. Soc., 62:341–438.

Pimentel, R. A., and R. Riggins. 1987. The nature of cladistic data. Cladistics, 3:201–209.

Rage, J.-C., and P. Janvier. 1982. Le problème de la monophylie des amphibiens actuels, à la lumière des nouvelles données sur les affinités des tétrapodes. Geobios, Mem. Sp., 6:65–83.

Rosen, D. E., P. L. Forey, B. G. Gardiner, and C. Patterson. 1981. Lungfishes, tetrapods, paleontology, and plesiomorphy. Bull. Am. Mus. Nat. Hist., 167:159–276.

Säve-Söderbergh, G. 1932. Preliminary note on Devonian stegocephalians from East Greenland. Medd. Groenland, 94(7):1–107.

Säve-Söderbergh, G. 1937. On the dermal skulls of *Lyrocephalus, Aphaneramma,* and *Benthosaurus,* labyrinthodonts from the Triassic of Spitsbergen and Russia. Bull. Geol. Inst. Univ. Uppsala, 27:189–208.

Schultze, H.-P. 1969. Die Faltenzähne der rhipidistiiden Crossopterygier, der Tetrapoden und der Actinopterygier–Gattung *Lepisosteus,* nebst einer Beschreibung der Zähnstruktur von *Onychodus* (struniiformer Crossopterygier). Palaeontogr. Ital., 65:63–136.

Schultze, H.-P. 1981. Hennig und der Ursprung der Tetrapoda. Palaeontol. Z., 55:71–86.

Schultze, H.-P. 1987. Dipnoans as sarcopterygians. Pp. 39–74 *in* Bemis, W. E., W. W. Burggren, and N. E. Kemp (eds.), *The Biology and Evolution of Lungfishes.* J. Morph., Suppl. 1.

Schultze, H.-P., and M. Arsenault. 1985. The panderichthyd fish *Elpistostege:* a close relative of tetrapods? Palaeontology, 28:293–309.

Schultze, H.-P., and K. S. W. Campbell. 1987. Characterization of the Dipnoi, a monophyletic group. Pp. 25–37 *in* Bemis, W. E., W. W. Burggren, and N. E. Kemp (eds.), *The Biology and Evolution of Lungfishes.* J. Morph., Suppl. 1.

Smithson, T. R. 1982. The cranial morphology of *Greererpeton burkemorani* Romer (Amphibia: Temnospondyli). J. Linn. Soc. London Zool., 76:29–90.

Smith, M. M., and Chang M.-M. 1990. The dentition of *Diabolepis speratus* Chang and Yu, with further consideration of its relationships and the primitive dipnoan dentition. J. Vert. Paleontol., 10:420–433.

Smithson, T. R., and K. S. Thomson. 1982. The hyomandibular of *Eusthenopteron foordi* Whiteaves (Pisces: Crossopterygii) and the early evolution of the tetrapod stapes. J. Linn. Soc. London Zool., 74:93–103.

Stensiö, E. A. 1947. The sensory lines and dermal bones of the cheek in fishes and amphibians. K. Svenska VetenskAkad. Handl., (3)24,3:1–195.

Swofford, D. L. 1985. Phylogenetic Analysis Using Parsimony, Version 2.4. Champaign, Ill.: distributed by the author.

Von Wahlert, G. 1968. *Latimeria* und die Geschichte der Wirbelthiere. Eine evolutionsbiologische Untersuchung. Stuttgart: G. Fischer.

Westoll, T. S. 1943. The origin of the tetrapods. Biol. Rev. Cambridge Philos. Soc., 18:78–98.

Wiley, E. O. 1979. Ventral gill arch muscles and the interrelationships of gnathostomes, with a new classification of the Vertebrata. J. Linn. Soc. London Zool., 67:149–179.

Section II

AMPHIBIANS

6 Toward an Understanding of the Amphibians: Two Centuries of Systematic History

Linda Trueb and Richard Cloutier

In a review of a recent two-volume work on tetrapod phylogeny and classification edited by Benton (1988), Gingerich (1988:628) pointed out, "Revolutions in science are like waves crashing on a shore. They arise quietly, gain energy and eventually break thunderously, spraying showers of new perspective that alter our view of the shoreline forever." The wave to which Gingerich refers is cladistics, but to this we also could add the data provided by molecular biology; the shoreline, in the case at hand, are relationships among fossil and Recent amphibians. The analogy clearly addresses change, which can be evaluated only in the context of history—a topic elegantly and lucidly discussed by Patterson (1987). As a preface to the two chapters that follow (by Bolt and by Trueb and Cloutier), we present a brief overview of the history of amphibian systematics as it has developed during the past 233 years.

PRE-DARWINIAN AND EARLY POST-DARWINIAN PERIODS

The origin(s) and relationships of both Recent and fossil amphibians have occasioned long-lived and spirited controversies among lower vertebrate paleontologists and neontologists. The arguments, as numerous as they are diverse, emanate from early schemes of classification that begin with Linnaeus (1758); the reader is referred to Gadow's (1901) and

175

Lescure's (1986) useful reviews of these classifications. It might be argued that the work of pre-Darwinian biologists has little bearing on the history of thought of phylogenetic relationships among modern amphibians. However, the taxonomic arrangements proposed by these workers reflect their perceptions of the relevance of phenetic similarities and dissimilarities among living groups of organisms. Before the development of the comparative method, amphibians and reptiles variously were grouped together, but beginning in the 1830's the distinctiveness of these two groups was reflected in their taxonomic separation. Thus, Haeckel's (1866) subdivision of the Amphibia into the Phractamphibia (= Labyrinthodontia + caecilians[1]) and Lissamphibia (= salamanders[2] + anurans[3]) marks the onset of a new era—one in which classifications included fossil as well as Recent amphibians, and one in which biologists sought to define the relationships within and between these groups. For the purposes of this review, Haeckel and Gadow are especially important. By extrapolation, one can assume that Haeckel (1866) considered modern amphibians to represent two phenetic or evolutionary assemblages (i.e., salamanders and anurans vs. labyrinthodonts and caecilians). Gadow (1901) provided 15 morphological features to characterize the Amphibia and allocated them to Haeckel's two subclasses—the Phractamphibia to accommodate fossil taxa, and the Lissamphibia, which included all Recent taxa. Gadow discussed the descent of amphibians from either the Crossopterygii or Dipnoi; although he is not explicit, we assume that Gadow was the first to consider all amphibians to have had a common ancestor (Appendix I).

THE FIRST HALF OF THE TWENTIETH CENTURY

Although Szarski (1962), Parsons and Williams (1963), Rage and Janvier (1982), and Milner (1988), among others, have presented substantive reviews of the various schools of thought on the origin(s) and relationships of amphibians during the first half of this century, some additional observations are relevant. The predominant theme seems to have been the search for fossil ancestor(s) to the modern groups. This led the majority of biologists to hypothesize multiple origins for Recent amphibians on the basis of anatomical similarities between fossil and living taxa (e.g., Jarvik, 1942, 1955, 1960, 1965, 1972, 1975, 1980–1981, 1986; Gregory

---

[1]Apoda or Gymnophiona.
[2]Urodela or Caudata.
[3]Anura, Acaudata, or Salientia.

et al., 1956; Noble, 1931; Romer, 1945; Eaton, 1959), or developmental evidence (e.g., Wintrebert, 1910, 1922; Holmgren, 1933, 1939, 1949a,b, 1952; Säve-Söderbergh, 1934, 1936). As shown in Appendix I, the majority of authors during this period seems to have proposed a diphyletic origin of the Amphibia, but as a point of clarity it should be noted that most did not consider caecilians in their studies. Moreover, there was little or no consensus as to the identities of ancestral stocks and their derivatives (Appendices II–IV).

Relatively few workers during the first half of the twentieth century seriously considered the possibility that the Lissamphibia was, in fact, a natural group derived from a common ancestor. Among these were two paleontologists—Kesteven (1950) on the basis of similarities between fish and amphibian skulls, and Kulczycki (1960) on the basis of an alternate interpretation of Jarvik's and his own observations on the rostral anatomy of *Porolepis* (Porolepiformes: Rhipidistia) and salamanders. Two anatomists, Pusey (1943) and Stephenson (1951a,b, 1955), noted similarities between the cranial anatomy of salamanders and primitive anurans that led them to suggest that these groups shared a common origin. A seminal but seldom cited paper of this period was Parker's (1956) short account on viviparous caecilians and amphibian phylogeny in which the author observed the pedicellate nature of the teeth of anurans, salamanders, and foetal caecilians. The structure and distribution of pedicellate teeth subsequently was investigated extensively by Parsons and Williams (1962). Quite independently of one another, two important papers appeared in the early 1960's—those by Szarski (1962) and Parsons and Williams (1963); taking somewhat different approaches, these authors critically reviewed the accumulated anatomical and developmental evidence and concluded that the Lissamphibia was, indeed, a monophyletic assemblage.

The impact of the papers by Szarski (1962) and Parsons and Williams (1963) lies less in their conclusions than in their methodological attitudes. Both emphasized the importance of integrating data derived from soft and hard anatomy, and from fossil and living taxa into analyses of relationships, and both were concerned with the logic involved in the construction of arguments. Szarski stressed the importance of using homologous characters in systematic studies and pointed out potential fallacies of then commonly accepted principles such as the "Biogenetic Law" and "the rule of the unspecialized" (= Commonality Rule). Parsons and Williams were rigorous in their distinction between primitive and derived features, and they emphasized the need to utilize a diversity of characters to minimize the effect of homoplasy. The work of these authors together with the broad acceptance of the Hennigian school of phylogenetics in the early 1970's profoundly affected the future research of most biologists concerned with lissamphibian relationships.

POST-HENNIGIAN POINTS OF VIEW

Research in the decade following the publication of Szarski's and Parsons and Williams's works marked a transition from a typological to a more holistic attitude. During this period, the only new proposal of a polyphyletic origin of living amphibians was that of Brundin (1966), and this was in his classic monograph on chironomid midges; the primary significance of this work to amphibian systematics is Brundin's application of Hennigian methodology. Among the paleoanatomists, Stensiö (1963) and Bertmar (1968) marshaled evidence from the central nervous system and the cranial anatomy of dipnoans, respectively, to support Jarvik's thesis of separate origins of anurans and salamanders from osteolepiform and porolepiform rhipidistians. The most significant contributions to the hypothesis of lissamphibian monophyly during this period are those of Estes, and Salthe and Kaplan. In his paper on fossil salamanders and their origins, Estes (1965) suggested that the dissorophid temnospondyls are the sister-group of the lissamphibians—an idea that subsequently has been perpetuated (Appendix III). Salthe and Kaplan (1966) brought data from molecular biology (micro-complement fixation studies to compare the structures of homologous proteins) to bear on the question of the origin of living amphibians, and they concluded that the Lissamphibia derive from a common ancestor. This work foreshadowed the diversity of investigations that were to follow.

Beginning in the 1970's, an eclectic array of research avenues were pursued in attempts to resolve the definition of the Amphibia and the question of the origin of the living representatives. For the most part, these efforts led authors to conclude that the Lissamphibia is monophyletic (Appendix I) and that the group is related most closely to dissorophids (Appendix III). For example, Schultze (1970) investigated the structure of the dentine of the teeth of rhipidistians and fossil and Recent lower tetrapods, and determined that among rhipidistians, only osteolepiforms possessed folded teeth (plicidentine) from which the folded teeth of fossil amphibians might have been derived. He also pointed out that the divided teeth characteristic of Recent amphibians and the dissorophoid *Doleserpeton* are absent in rhipidistians and other early tetrapods. Jurgens (1971), in an extensive survey of the internal and external anatomy of the nasal regions of living amphibians, concluded that the extant orders are related closely and form a monophyletic group. Morescalchi (1973) arrivéd at the same conclusion based on karyological criteria (i.e., the presence of high chromosome numbers and various acrocentrics and microchromosomes among primitive representatives of each order).

A Reconsideration of Ontogenetic Data

The renewed interest during the 1970's and 1980's in the relationship of ontogeny and phylogeny, particularly with respect to the role of heterochrony as an evolutionary mechanism, is reflected in several papers of the period. Thus, Bolt (1977b) postulated a paedomorphic origin of lissamphibians from dissorophoids and subsequently (1979) suggested progenesis as the most logical mechanism for that origin—an idea also espoused by Milner (1988). Bolt's most forceful arguments are based on reproductive strategy, and the similar structure of the teeth in a juvenile dissorophoid *Doleserpeton* and an adult dissorophoid *Amphibamus grandiceps*. Saint-Aubain (1981) reviewed amphibian limb ontogeny and rejected Holmgren's (1939) and Jarvik's (1965) contentions that limb development in anurans and salamanders was indicative of a diphyletic origin. A number of developmental biologists recently have described developmental features that might be useful phylogenetic characters (e.g., the presence or absence of blocks to polyspermy, the method of mesoderm formation, the source of origin of primordial germ cells). The reader is referred to Hanken's (1986) excellent review and commentary on the value and utility of these characters, which he considered equivocal, or of limited value owing to the restricted number of taxa for which information currently is available.

The Influence of Phylogenetic Methodology

Doubtless the single most significant force brought to bear on the resolution of amphibian systematics during the past 15 years is the Hennigian school of phylogenetics in combination with the availability of computer software for analysis of massive, and otherwise unmanageable, data sets. The cladistic method requires rigorous definition of characters and assessment of the polarity of their evolutionary change (i.e., plesiomorphic vs. apomorphic) based on the distribution of the characters in the group of organisms being analyzed and their presumed closest relatives (i.e., out-group comparison). It encourages incorporation of as many characters as possible into the analysis. Assemblages of taxa are identified by the possession of shared, derived characters (i.e., synapomorphies), and the preferred patterns of relationships are those with the least number of evolutionary steps (i.e., the most parsimonious tree[s]). The importance of this analytical method is that it produces hypotheses of relationships that can be examined critically. Although not directly relevant to the question of lissamphibian relationships, two papers are particularly significant because of their effects on contemporary

research. The first is that of Gaffney (1979), in which the author demonstrated the monophyly of the Tetrapoda based on 11 synapomorphies
and further diagnosed cladistically the Neotetrapoda (= [["Anthracosauria" + Amniota] + Lepospondyli] + ["Temnospondyli" + Lissam-
phibia]) on the basis of five synapomorphies. Despite the fact that Gaffney's character set has been modified by some subsequent workers, his
hypothesis of tetrapod monophyly remains unchallenged. The second
and more controversial publication is that by Rosen et al. (1981), in which
the authors proposed on the basis of 20 characters that the Dipnoi is the
sister-group of Tetrapoda rather than crossopterygians (represented by
Actinistia, Porolepiformes, and the osteolepiform *Eusthenopteron foordi* in
their phylogenetic tree). Although many biologists disagree with the
nontraditional phylogenetic scheme proposed by Rosen et al., the authors' methodology is sound; thus, critics must concentrate on the characters that Rosen et al. selected, the interpretation of the character-states
in fossil and Recent taxa, and the evaluation of the polarity of these
characters (cf. Holmes, 1985; Schultze, 1981, 1987; Forey et al., this volume). There is little doubt in our minds that the paper by Rosen et al. has
been the single greatest stimulus to renewed, methodologically sound
research on the morphology and systematics of both fossil and Recent
lower tetrapods since Szarski's (1962) and Parsons and Williams's (1963)
publications.

The Status and Results of Contemporary Systematic Research

Methodologically, contemporary (i.e., 1975–1988) contributions dealing with the origin(s) and relationships of amphibians are either phylogenetic (i.e., cladistic) or nonphylogenetic in their approach.

Nonphylogenetic Investigations

Among the nonphylogenetic types are phenetics and traditional evolutionary taxonomy. Pheneticists (i.e., numerical taxonomists) seek to
construct a nonhistorical classification based on similarities among organisms, whereas evolutionary systematists hypothesize historical classifications based on overall similarity. Examples of the latter are Carroll and
Currie (1975) and Carroll and Holmes (1980), who proposed the origins of
caecilians and salamanders from two different groups of microsaurs, and
Milner et al. (1986), who derived anurans and salamanders from temnospondyls and caecilians from microsaurs.

Phylogenetic Investigations: Idealized versus Pragmatic

Phylogenetic systematics differs from phenetics and evolutionary systematics in that it attempts to unravel the historical scheme of relationships among taxa on the basis of the presence of shared, derived characters. There are two sorts of approaches to phylogenetic systematics—one "idealistic" and the other more "pragmatic." The original, idealistic phylogenetics model is based on analysis of "idealized" (i.e., nonhomoplastic), nested sets of characters and, usually, organisms at a higher taxonomic level than the species; parsimony of characters is emphasized in preference to character congruence and parsimony of the entire system (i.e., character-states and taxa). The more pragmatic and rigorous pursuit of phylogenetic systematics is characterized by the use of computer algorithms to analyze larger data sets, wherein the data are derived from actual specimens.

By far, the majority of studies involving amphibian systematics falls into the "idealized" category. Furthermore, all phylogenetic studies can be divided into (1) those addressing the higher classification of tetrapods, and (2) those primarily concerned with the classification of extant groups of amphibians.

General tetrapod classifications. Two of the most provocative publications dealing with the higher classification of tetrapods are those of Gardiner (1982a, 1983). In 1982, Gardiner formulated a new phylogeny and a cladistic classification of the tetrapods. Included was a preliminary analysis of the interrelationships of amphibians in which [Aïstopoda + Nectridea] is the sister-group of lissamphibians. In his 1983 paper, Gardiner modified his original arrangement such that the Nectridea is the sister-group of the Lissamphibia on the basis of one cranial character and three vertebral features. Furthermore, [Aïstopoda + Adelogyrinidae] is the sister-group of [Nectridea + Lissamphibia]. Indirectly, Gardiner's work discredited Carroll and Currie's (1975) and Carroll and Holmes's (1980) hypotheses relating microsaurs to lissamphibians, because Gardiner (1983) demonstrated on the basis of vertebral characters that anthracosaurs, seymourians, and microsaurs are amniotes rather than amphibians. In another study of fossil taxa, Smithson (1985) advocated a diphyletic origin of living amphibians on the basis of the conclusions of Carroll and Currie (1975) and Carroll and Holmes (1980). Although Smithson attempted to implement a cladistic approach, his conclusions are weakened significantly by his a priori acceptance of Carroll and Currie's and Carroll and Holmes's hypotheses of amphibian-microsaur-temnospondyl relationships.

Panchen and Smithson (1988) is the most recent example of an "idealized" phylogenetic study of amphibians. In their paper on the relationships of the earliest tetrapods, Panchen and Smithson (1988) concluded that tetrapods comprise two monophyletic groups—the Amniota and the extant Amphibia. The Amniota are included in the Reptiliomorph Clade composed of Anthracosauroideae, *Crassigyrinus*, Seymouriamorpha, and Diadectomorpha, whereas the [Temnospondyli + Lissamphibia] is included in the Batrachomorph Clade composed of the Ichthyostegidae, Nectridea, Colosteidae, and Microsauria. The authors hypothesized the Microsauria to be the sister-group of the [Temnospondyli + Lissamphibia], and the Batrachomorph Clade to be the sistergroup of all other tetrapods; the Reptiliomorph and Batrachomorph clades are placed in the Division Amphibia (fide Milner, 1988).

Two landmark pragmatic phylogenetic studies involving amphibians are Gauthier et al. (1988a,b). In their paper on amniote phylogeny, Gauthier et al. (1988a) demonstrated that analysis of data derived from fossil and Recent taxa resulted in a phylogeny that was different than one derived from separate data sets, and one that provided a better fit to the stratigraphic record; the Lissamphibia was treated as an unresolved polytomy in this paper. In a second study on the early evolution of the Amniota, Gauthier et al. (1988b) hypothesized that the Amphibia is a monophyletic taxon that combines the extinct and paraphyletic Temnospondyli with the extant and monophyletic Lissamphibia, and that excludes the Anthracosauria, Batrachosauria, and Cotylosauria (contra Panchen and Smithson, 1988).

The status and relationships of lissamphibians. A variety of authors have adopted an "idealized" phylogenetic approach to the analysis of lissamphibian relationships. Only three of these authors proposed a polyphyletic origin of extant amphibians—Løvtrup, Rage, and Smithson. Løvtrup (1985) listed 12 lissamphibian synapomorphies from Gardiner (1982a) and Rage and Janvier (1982) and casually dismissed them with the following statement (p. 468): "However, most of the characters may just as well be plesiomorphic ones in the taxon Tetrapoda and, in any case, they seem to be of trifling significance." He then proposed salamanders and caecilians to be independent monophyletic taxa, with caecilians being the sister-group of amniotes.

In their phylogenetic analysis, Rage and Janvier (1982) used the characters of Parsons and Williams (1963) as a foundation and supplemented this data base with a variety of features described in more recent literature (e.g., the presence of spermatophores in salamanders, the differences in the formation of the choana in living amphibians, variation in the structure of the periotic system of the ear), and one character that had escaped attention—the method of visual accommodation in living amphibians. On the basis of these data, the authors attempted to support the monophyly of each living amphibian order, concluded that salamanders and anurans are sister-groups, and listed features that diagnose the Lissamphibia as a monophyletic assemblage. Rage and Janvier (1982) noted that the monophyly of the Dissorophoidea is not supported (as previously suggested by Bolt, 1974); nonetheless, they proposed that this taxon could represent the "stem-group" of the Lissamphibia. Rage and Janvier's paper probably is the most comprehensive review of characters in fossil and extant taxa and literature relating to amphibian relationships to appear since Parsons and Williams (1963) and Szarski (1962), and the reader is referred to it for further details.

Rage apparently was influenced by the work of Gardiner (1982a; see discussion below), because in his 1985 paper on the origin and phylogeny of amphibians he postulated quite a different phylogenetic arrangement for the amphibians. He proposed two possible hypotheses for the interrelationships of the extant orders. In both, salamanders plus anurans form the clade Paratoidia (sensu Gardiner, 1982a), and caecilians are the sister-group of the Paratoidia. In the first hypothesis, the salamanders are considered monophyletic and the entire clade represents the sister-group of anurans. In the second, the salamanders are considered to be paraphyletic, and one lineage represents the sister-group of anurans. Moreover, noting the presence of pedicellate teeth in dissorophoids and the possible presence of an operculum in the dissorophoid *Actiobates*, Rage (1985) tentatively included the dissorophoids in the Lissamphibia and postulated two possible phylogenetic arrangements. Thus, the Dissorophoidea may be the sister-group or the stem-group of the Paratoidia. In either case, caecilians are considered to be the sister-group of Paratoidia plus Dissorophoidea.

Another contribution that should be mentioned is the posthumous publication of Hennig (1983) on the evolutionary history of chordates. In this work, the author listed and discussed characters that he considered lissamphibian synapomorphies, as well as synapomorphies of each of the orders. Hennig's brief treatment is largely superseded by literature that apparently appeared after the manuscript had been prepared and before its publication.

Milner (1988) recanted his earlier (Milner et al., 1986) position that Recent amphibians constitute a paraphyletic group (see above); he concluded that the Lissamphibia is a monophyletic assemblage and that the Temnospondyli is a grade that represents the stem-group from which lissamphibians arose by a process of progenetic miniaturization. Milner hypothesized that among Recent amphibians, caudates are the sister-group of anurans, and caecilians the sister-group of both of the latter (contra Milner et al., 1986).

The foregoing studies were based on relatively unrestricted but primarily osteological data sets. There is a smaller body of literature that reports results garnered from restricted data sets such as particular anatomical systems or subcellular investigations. For example, based on their exhaustive anatomical investigations of the ear in tetrapods (Lombard and Bolt, 1979) and in amphibians (Bolt and Lombard, 1985), Lombard and Bolt are cautious, but positive, in their assertion that the Lissamphibia is a monophyletic assemblage derived from a temnospondyl ancestor. Although it is largely unexplored, there is some evidence indicating lissamphibian monophyly derived from neural anatomical studies at macro (Northcutt, 1987) and molecular levels (Franz et al., 1981). Recently, molecular biologists working with amino acid and nucleotide sequence data have hypothesized genealogical relationships within vertebrates (e.g., Goodman et al., 1987; Hillis and Dixon, 1989); many earlier papers are reviewed by Bishop and Friday (1988). Although these studies undeniably are interesting, at this time they can be applied only to living organisms, which have been represented by a very limited number of taxa. For example, among amphibians, only salamanders and anurans have been sampled—viz., *Ambystoma mexicanum, Amphiuma means, Taricha granulosa, Xenopus laevis, Rana catesbeiana,* and *R. esculenta.* With respect to the question of the monophyly or paraphyly of lissamphibians, the results of these studies are equivocal and depend upon the proteins examined and the method of their analysis.

The chapters that follow in this section represent two different approaches to the resolution of lissamphibian relationships. The first, by Bolt, is, in his own words, "mainly a reconsideration of observations and analyses that have been in the literature since 1959." Bolt analyzed skeletal characters of fossil and Recent amphibians, and concluded that lissamphibians are monophyletic and the sister-group of dissorophid labyrinthodonts. The second, by Trueb and Cloutier, adopts a different strategy to investigate the interrelationships among living and fossil amphibian groups. Trueb and Cloutier performed a tandem cladistic analysis of osteological data (58 characters of living and fossil taxa) and nonosteological data (72 characters of living taxa) using different sets of out-groups for each analysis. Trueb and Cloutier concluded that the monophyletic Lissamphibia is a group of dissorophoid temnospondyls and, further, that within the Dissorophoidea the Permian "branchiosaurid" *Apateon* is the sister-group of the Lissamphibia.

APPENDIX I. Hypotheses of the origins and relationships of the three orders of Recent amphibians

Monophyletic	Diphyletic	Polyphyletic
Gadow, 1901	[4]Haekel, 1866	Herre, 1935, 1964
[1]Pusey, 1943	[?]Wintrebert, 1910, 1922	[2,3]Jarvik, 1942, 1955, 1960,
Kesteven, 1950	[4]Noble, 1931	1965, 1972, 1975, 1980–
[1]Stephenson, 1951a,b, 1955	Holmgren, 1933, 1939,	1981, 1986
Parker, 1956	1949a,b, 1952[3]	Gregory, 1965
Williams, 1959	Säve-Söderbergh, 1934,	Gregory et al., 1956
Kulczycki, 1960	1936	Brundin, 1966
Szarski, 1962, 1977	[3]Romer, 1945	[?]Carroll & Currie, 1975
Parsons & Williams, 1962,	Colbert, 1955	Løvtrup, 1977 (part), 1985
1963	[3]Lehman, 1956, 1968, 1975	Carroll & Gaskill, 1978
Thomson, 1962, 1964a,b,	[3]von Huene, 1956	Carroll & Holmes, 1980
1967a,b, 1968	[2]Schmalhausen, 1958a,b,	Rage, 1985 (part)
Remane, 1964	1959	Carroll, 1988
Romer, 1964, 1968	[4]Eaton, 1959	
[1]Salthe & Kaplan, 1966	Stensiö, 1963	
Estes, 1965	[5]Reig, 1964	
[1]Cox, 1967	[2]Gross, 1964	
Hecht, 1969	Bertmar, 1968[1]	
Colbert, 1969	Bjerring, 1975, 1984	
Schultze, 1970	[2]Nieuwkoop & Satasurya,	
Jurgens, 1971	1976, 1979, 1983	
Morescalchi, 1973	[5]Løvtrup, 1977	
Bolt, 1977b, 1979	[?]Smithson, 1985	
Lombard & Bolt, 1979	Milner et al., 1986	
Gaffney 1979	[2]Bishop and Friday, 1987	
[1]Saint-Aubain, 1981, 1985	(part)	
[1]Franz et al., 1981		
[5]Gardiner, 1982a,b, 1983		
[1]Milner, 1982		
[5]Rage & Janvier, 1982		
Hennig, 1983		
[5]Rage, 1985 (part)		
Bolt & Lombard, 1985		
Duellman & Trueb, 1986		
Duellman, 1988		
[1]Goodman et al., 1987		
[1]Bishop & Friday, 1987		
(part)		
Northcutt, 1987		
Milner, 1988		

APPENDIX I.—*continued*

Monophyletic	Diphyletic	Polyphyletic
Panchen & Smithson, 1988 Gauthier et al., 1988b Bolt, 1990		

Note: Where a footnote number or symbol appears before an author's name, the note applies to each publication listed for that author; a footnote that follows a particular date applies only to the flagged publication.

?Placement in Diphyletic vs. Polyphyletic category questionable because authors were not explicit.

[1]Monophyly is assumed despite the fact that only anurans and salamanders are discussed in these papers.

[2]Proposed independent origins for anurans and salamanders without considering caecilians.

[3]Assumed salamanders and caecilians to constitute a monophyletic assemblage independent of anurans.

[4]Assumed salamanders and anurans to constitute a monophyletic assemblage independent of caecilians.

[5]Assumed salamanders and anurans to be sister-groups.

APPENDIX II. Proposed ancestral or stem-stocks of Recent amphibians by authors who hypothesized polyphyletic origins of the groups

Ancestral or stem-stock	Derivative	Authors
Unspecified, but different groups of fishes	Salamanders; Caecilians; Anurans	Herre, 1935
Porolepiform fishes	Salamanders	Jarvik, 1942, 1955, 1960, 1965, 1972, 1975, 1980–1981, 1986; Bertmar, 1968
	Lepospondyls + Salamanders + Caecilians	von Huene, 1956
Dipnoans	Salamanders	Holmgren, 1933, 1939, 1949a,b, 1952; Säve-Söderbergh, 1934, 1936; Lehman, 1956, 1959, 1968, 1975
Osteolepiform fishes	Anurans	Jarvik, 1942, 1955, 1960, 1965, 1972, 1975, 1980–1981, 1986; Bertmar, 1968
Rhachitomi (Ichthyostegalia)	Anurans	Schmalhausen, 1968
Nectridea	Salamanders	Gregory et al., 1956; Gregory, 1965
Stegocephalians	Anurans	Wintrebert, 1922
Labyrinthodonts	Salamanders + Anurans	Eaton, 1959
	Anurans	Colbert, 1955; Schmalhausen, 1958b; Gross, 1964
Lepospondyl labyrinthodont	Caecilians	Noble, 1931; Eaton, 1959
Lepospondyli[1]	Salamanders	Gross, 1964
Lepospondyl Stegocephalae	Salamanders + Caecilians	Schmalhausen, 1958b
Microsaurs[2]	Caecilians	Gregory et al., 1956; Reig, 1964; Gregory, 1965
	Salamanders + Caecilians	Schmalhausen, 1968
Microsaurs	Salamanders	Carroll & Holmes, 1980; Colbert, 1955
Hapsidopareiontidae or Ostodolepidae (Microsauria)	Salamanders	Carroll & Gaskill, 1978
Rhynchonkos (Microsauria)	Caecilians	Carroll & Currie, 1975
Lysorophus tricarinatus (Microsauria)	Salamanders Caecilians	Wintrebert, 1922 Colbert, 1955
"Phyllospondyl" labyrinthodonts	Salamanders + Anurans	Noble, 1931; Reig, 1964
Branchiosauria	Salamanders	Moodie, 1908

[1]Included Nectridea and Microsauria in Lepospondyli.
[2]Considered microsaurs to be members of the Lepospondyli.

APPENDIX III. Proposed sister-groups, or ancestral or stem-stocks of Recent
amphibians by authors who hypothesized a monophyletic origin of the group

Relationship	Authors
Ancestral or stem-stock	
Dipnoi and/or Crossopterygi	Gadow, 1901
Embolomeri	Kesteven, 1950
Rhipidistia	Szarski, 1962; Thomson, 1964a,b
"Single amphibian stock"	Thomson, 1967, 1968
Ancestral Permian amphibian	Salthe & Kaplan, 1966
Labyrinthodontia or Osteolepiformes	Schultze, 1970
Gymnarthrid microsaurs	Cox, 1967
"Stegocephalian groups"	Thomson, 1964a
Sister-group	
Aïstopoda + Nectridea	Gardiner, 1982
Nectridea	Gardiner, 1983
Temnospondyli	Duellman, 1988; Lombard & Bolt, 1988; Milner, 1988
"Temnospondyli"	Gaffney, 1979
Dissorophoidea	Bolt, 1974, 1977a,b
Doleserpetontidae	Bolt, 1969
Dissorophidae	Estes, 1965; Bolt, 1979, 1980; Rage & Janvier, 1982; Rage, 1985; Bolt & Lombard, 1985; Duellman & Trueb, 1986

APPENDIX IV. Proposed sister-groups of Recent amphibians by authors who
hypothesized polyphyletic origins of the groups

Sister-group	Recent amphibian groups	Authors
Microsaurs	[Salamanders + Caecilians]	Smithson, 1985
	Caecilians	Milner et al., 1986
Temnospondylii	Anurans	Smithson, 1985
	[Anurans + Salamanders]	Milner et al., 1986
Dissorophidae	Anurans	Bolt, this volume
[*Hapsidopareion* + [Trematopidae + [Dissorophidae + Anurans]]]	[Salamanders + Caecilians]	Bolt, this volume
Amniotes	Caecilians	Løvtrup, 1977, 1985
Caecilians + Amniotes	Anurans	Løvtrup, 1985
[Anurans + [Caecilians + Amniotes]]	Salamanders	Løvtrup, 1985
Salamanders	Anurans	Løvtrup, 1977
[Anurans + Amniotes]	Salamanders	Brundin, 1966
[Crossopterygii + [Actinopterygii + Brachiopterygii]]	[Salamanders + Anurans + Amniotes]	Brundin, 1966
Dissorophoid taxa	Salamanders, Anurans, or [Salamanders + Anurans]	Rage, 1985
[Salamanders + Anurans + Dissorophoid taxa]	Caecilians	Rage, 1985

Literature Cited

Benton, M. J. (ed.). 1988. *The Phylogeny and Classification of the Tetrapods*. Vol. 1: *Amphibians, Reptiles, Birds*. Syst. Assoc. Spec. Vol. No. 35 A. Oxford: Clarendon Press.

Bertmar, G. 1968. Lungfish phylogeny. Pp. 259–283 *in* Ørvig, T. (ed.), *Current Problems of Lower Vertebrate Phylogeny*. Stockholm: Almqvist and Wiksell.

Bishop, M. J., and A. E. Friday. 1988. Estimating the interrelationships of tetrapod groups on the basis of molecular sequence data. Pp. 33–58 *in* Benton, M. J. (ed.), *The Phylogeny and Classification of the Tetrapods*. Vol. 1: *Amphibians, Reptiles, Birds*. Syst. Assoc. Spec. Vol. No. 35 A. Oxford: Clarendon Press.

Bjerring, H. C. 1975. Contribution à la connaissance de la neuroépiphyse chez les Urodèles et leurs ancêtres Porolépiformes. Colloq. Int. C. N. R. S., 218:231–256.

Bjerring, H. C. 1984. Major anatomical steps toward craniotedness: a heterodox view based largely on embryological data. J. Vert. Paleontol., 4:17–29.

Bolt, J. R. 1969. Lissamphibian origins: possible protolissamphibian from the Lower Permian of Oklahoma. Science, 166:888–891.

Bolt, J. R. 1974. A trematopsid skull from the Lower Permian, and analysis of some characters of the dissorophoid (Amphibia: Labyrinthodontia) otic notch. Fieldiana Geol., 30(3):67–79.

Bolt, J. R. 1977a. *Cacops* (Amphibia: Labyrinthodontia) from the Fort Sill locality, Lower Permian of Oklahoma. Fieldiana Geol., 37(3):61–73.

Bolt, J. R. 1977b. Dissorophoid relationships and ontogeny, and the origin of the Lissamphibia. J. Paleontol., 51:235–249.

Bolt, J. R. 1979. *Amphibamus grandiceps* as a juvenile dissorophid: evidence and implications. Pp. 529–563 *in* Nitecki, M. H. (ed.), *Mazon Creek Fossils*. New York: Academic Press.

Bolt, J. R. 1980. New tetrapods with bicuspid teeth from the Fort Sill Locality (Lower Permian, Oklahoma). Neues Jahrb. Geol. Palaeontol. Monatsh., 8:449–459.

Bolt, J. R., and R. E. Lombard. 1985. Evolution of the amphibian tympanic ear and the origin of frogs. Biol. J. Linn. Soc., 24:83–99.

Brundin, L. 1966. Transantarctic relationships and their significance, as evidenced by chironomid midges. K. Sven. Vetenskapakad. Handl., 11:1–472.

Carroll, R. L. 1988. *Vertebrate Paleontology and Evolution*. New York: W. H. Freeman.

Carroll, R. L., and P. J. Currie. 1975. Microsaurs as possible apodan ancestors. J. Linn. Soc. London Zool., 57:229–247.

Carroll, R. L., and P. Gaskill. 1978. The Order Microsauria. Mem. Am. Philos. Soc., 126:1–211.

Carroll, R. L., and R. Holmes. 1980. The skull and jaw musculature as guides to the ancestry of salamanders. J. Linn. Soc. London Zool., 68(1):1–40.

Colbert, E. H. 1955. *Evolution of the Vertebrates*. New York: J. Wiley.

Colbert, E. H. 1969. *Evolution of the Vertebrates. A History of the Backboned Animals Through Time*. 2nd ed. New York: J. Wiley.

Cox, C. B. 1967. Cutaneous respiration and the origin of the modern Amphibia. Proc. Linn. Soc. London, 178:37–47.

Duellman, W. E. 1988. Evolutionary relationships of the Amphibia. Pp. 13–34 *in* Fritzsch, B., M. J. Ryan, W. Wilczynski, T. E. Hetherington, and W. Walkowiak (eds.), *The Evolution of the Amphibian Auditory System*. New York: John Wiley and Sons.

Duellman, W. E., and L. Trueb. 1986. *Biology of Amphibians*. New York: McGraw-Hill.

Eaton, T. H. 1959. The ancestry of modern Amphibia: a review of the evidence. Univ. Kansas Publ. Mus. Nat. Hist., 12:155–180.

Estes, R. 1965. Fossil salamanders and salamander origins. Am. Zool., 5:319–334.

Franz, T., T. V. Waehneldt, V. Neuhoff, and K. Wächtler. 1981. Central nervous system myelin proteins and glycoproteins in vertebrates: a phylogenetic study. Brain Res., 226:245–258.

Gadow, H. 1901. Amphibia and reptiles. *The Cambridge Natural History.* Vol. 8. London: Macmillan.

Gaffney, E. S. 1979. Tetrapod monophyly: a phylogenetic analysis. Bull. Carnegie Mus. Nat. Hist., 13:92–105.

Gardiner, B. G. 1982a. Tetrapod classification. J. Linn. Soc. London Zool., 74:207–232.

Gardiner, B. G. 1982b. Mammals, birds, and mammal-like reptiles. Pp. 11–17 *in* Jayakar, S. D., and L. Zanta (eds.), *Evolution and the Genetics of Populations.* Suppl. Atti Assoc. Genet. Ital., Vol. 29.

Gardiner, B. G. 1983. Gnathostome vertebrae and the classification of the Amphibia. J. Linn. Soc. London Zool., 79:1–59.

Gauthier, J. A., A. G. Kluge, and T. Rowe. 1988a. The early evolution of the Amniota. Pp. 103–155 *in* Benton, M. J. (ed.), *The Phylogeny and Classification of the Tetrapods.* Vol. 1: *Amphibians, Reptiles, Birds.* Syst. Assoc. Spec. Vol. No. 35 A. Oxford: Clarendon Press.

Gauthier, J. A., A. G. Kluge, and T. Rowe. 1988b. Amniote phylogeny and the importance of fossils. Cladistics, 4:105–209.

Gingerich, P. D. 1988. Cladistic futures. Nature, London, 336:268. [Book review].

Goodman, M., M. M. Miyamoto, and J. Czelusniak. 1987. Pattern and process in vertebrate phylogeny revealed by coevolution of molecules and morphologies. Pp. 141–176 *in* Patterson, C. (ed.), *Molecules and Morphology in Evolution: Conflict or Compromise?* London: Cambridge Univ. Press.

Gregory, J. T. 1965. Microsaurs and the origin of captorhinomorph reptiles. Am. Zool., 5:277–286.

Gregory, J. T., F. E. Peabody, and L. I. Price. 1956. Revision of the Gymnarthridae American Permian microsaurs. Bull. Peabody Mus. Nat. Hist. Yale, 10:1–77.

Gross, W. 1964. Polyphyletische Stämme im System der Wirbeltiere? Zool. Anz., 173(1):1–22.

Haeckel, E. 1866. *Generelle Morphologie der Organismen.* 2 vols. Berlin: Reimer.

Hanken, J. 1986. Developmental evidence for amphibian origins. Pp. 389–417 *in* Hecht, M. K., B. Wallace, and G. T. Prance (eds.), *Evolutionary Biology.* Vol. 20. New York: Plenum Press.

Hecht, M. K. 1969. The living lower tetrapods: their interrelationships and phylogenetic position. Ann. New York Acad. Sci., 167(1):74–79.

Hennig, W. 1983. *Stammesgeschichte der Chordaten.* Berlin: Paul Parey.

Herre, W. 1935. Die Schwanzlurche der mittel-eocänen (oberlutetischen) Braunkohle des Geiseltales und die Phylogenie der Urodelen unter Einschluss der fossilen Formen. Zoologica, 33(87):1–85.

Herre, W. 1964. Zum Abstammungsproblem von Amphibien und Tylopoden sowie über Parallelbildungen und zur Polyphyliefrage. Zool. Anz., 173:66–98.

Hillis, D. M., and M. T. Dixon. 1989. Vertebrate phylogeny: evidence from 28S ribosomal DNA sequences. Pp. 355–367 *in* Fernholm, B., K. Bremer, and H. Jörnvall (eds.), *The Hierarchy of Life.* Stockholm: Elsevier Science Publishers B.V. (Biomedical Division).

Holmes, E. B. 1985. Are lungfishes the sister group of tetrapods? Biol. J. Linn. Soc. London, 25:379–397.

Holmgren, N. 1933. On the origin of the tetrapod limb. Acta Zool. Stockholm, 14:185–295.

Holmgren, N. 1939. Contribution to the question of the origin of the tetrapod limb. Acta Zool. Stockholm, 20:89–124.

Holmgren, N. 1949a. Contributions to the question of the origin of tetrapods. Acta Zool. Stockholm, 30:459–484.

Holmgren, N. 1949b. On the tetrapod limb problem—again. Acta Zool. Stockholm, 30:485–508.

Holmgren, N. 1952. An embryological analysis of the mammalian carpus and its bearing upon the question of the origin of the tetrapod limb. Acta Zool. Stockholm, 33:1–115.

Jarvik, E. 1942. On the structure of the snout of crossopterygians and lower gnathostomes in general. Zool. Bijdr., 21:235–675.

Jarvik, E. 1955. The oldest tetrapods and their forerunners. Sci. Monthly, 80:141–154.

Jarvik, E. 1960. Théories de l'évolution des vertébrés reconsidérées à la lumière des récentes découvertes sur les vertébrés inférieurs. Paris: Masson.

Jarvik, E. 1965. The origin of girdles and paired fins. Israel J. Zool., 14:141–172.

Jarvik, E. 1972. Middle and Upper Devonian Porolepiformes from East Greenland with special reference to *Glyptolepis groenlandica* n. sp. and a discussion on the structure of the head in the Porolepiformes. Medd. Groenland, 187:1–307.

Jarvik, E. 1975. On the saccus endolymphaticus and adjacent structure in osteolepiforms, anurans and urodeles. Colloq. Int. C. N. R. S., 218:191–211.

Jarvik, E. 1980–1981. *Basic Structure and Evolution of Vertebrates*. 2 vols. New York: Academic Press.

Jarvik, E. 1986. The origin of the Amphibia. Pp. 1–24 *in* Roček, Z. (ed.), *Studies in Herpetology*. Prague: Charles Univ.

Jurgens, J. D. 1971. The morphology of the nasal region of Amphibia and its bearing on the phylogeny of the group. Ann. Univ. Stellenbosch, 46(Ser. A., No. 2):1–146.

Kesteven, H. L. 1950. The origin of the tetrapods. Proc. R. Soc. Victoria, 59:93–138.

Kulczycki, J. 1960. *Porolepis* (Crossopterygii) from the Lower Devonian of the Holy Cross Mountains. Acta Paleontol. Polonica, 5:65–106.

Lehman, J.-P. 1956. L'évolution des dipneustes et l'origine des urodèles. Colloq. Int. C. N. R. S., 60:69–76.

Lehman, J.-P. 1959. Les dipneustes du Dévonien supérieur du Groenland. Medd. Groenland, 160:1–58.

Lehman, J.-P. 1968. Remarques concernant la phylogénie des amphibiens. Pp. 307–315 *in* Ørvig, T. (ed.), *Current Problems of Lower Vertebrate Phylogeny*. Stockholm: Almqvist and Wiksell.

Lehman, J.-P. 1975. Quelques réflexions sur la phylogénie des vertébrés inférieurs. Colloq. Int. C. N. R. S., 218:257–264.

Lescure, J. 1986. Histoire de la classification des cécilies. Pp. 11–19 *in* Delsol, M., J. Flatin, and J. Lescure (eds.), *Biologie des amphibiens: Quelques mises au point des connaissances actuelles sur l'ordre des gymnophiones*. Mem. Soc. Zool. France, 43:1–177.

Linnaeus, C. 1758. *Systema Naturae*. 10th ed. Vol. 1, Pt. 1. Uppsala.

Lombard, R. E., and J. R. Bolt. 1979. Evolution of the tetrapod ear: an analysis and reinterpretation. Biol. J. Linn. Soc., 11:19–76.

Lombard, R. E., and J. R. Bolt. 1988. Evolution of the stapes in Paleozoic tetrapods. Conservative and radical hypotheses. Pp. 37–67 *in* Fritzsch, B. (ed.), *The Evolution of the Amphibian Auditory System*. New York: John Wiley and Sons.

Løvtrup, S. 1977. *The Phylogeny of Vertebrata*. New York: John Wiley and Sons.

Løvtrup, S. 1985. On the classification of the taxon Tetrapoda. Syst. Zool., 34(4):463–470.

Milner, A. R. 1982. Small temnospondyl amphibians from the Middle Pennsylvanian of Illinois. Palaeontology, 25:635–664.

Milner, A. R. 1988. The relationships and origin of living amphibians. Pp. 59–102 *in* Benton, M. J. (ed.), *The Phylogeny and Classification of the Tetrapods*. Vol. 1: *Amphibians, Reptiles, Birds*. Syst. Assoc. Spec. Vol. No. 35 A. Oxford: Clarendon Press.

Milner, A. R., T. R. Smithson, A. C. Milner, M. I. Coates, and W. D. I. Rolfe. 1986. The search for early tetrapods. Modern Geol., 10:1–28.

Moodie, R. L. 1908. The ancestry of the caudate Amphibia. Am. Nat., 42:361–373.

Morescalchi, A. 1973. Amphibia. Pp. 233–348 in Chiarelli, A. B., and E. Capanna (eds.), Cytotaxonomy and Vertebrate Evolution. New York: Academic Press.

Nieuwkoop, P. D., and L. A. Satasurya. 1976. Embryological evidence for a possible polyphyletic origin of recent amphibians. J. Embryol. Exp. Morphol., 35:159–167.

Nieuwkoop, P. D., and L. A. Satasurya. 1979. Primordial Germ Cells in the Chordates: Embryogenesis and Phylogenesis. Cambridge: Cambridge Univ. Press.

Nieuwkoop, P. D., and L. A. Satasurya. 1983. Some problems in the development and evolution of the chordates. Pp. 123–135 in Goodsin, B. C., N. Holder, and C. C. Wylie (eds.), Development and Evolution. Cambridge: Cambridge Univ. Press.

Noble, G. K. 1931. The Biology of the Amphibia. New York: McGraw-Hill.

Northcutt, R. G. 1987. Lungfish neural characters and their bearing on sarcopterygian phylogeny. Pp. 277–297 in Bemis, W. E., W. W. Burggren, and N. E. Kemp (eds.), The Biology and Evolution of Lungfishes. J. Morphol., Suppl. 1.

Panchen, A. L., and T. R. Smithson. 1988. The relationships of the earliest tetrapods. Pp. 1–32 in Benton, M. J. (ed.), The Phylogeny and Classification of the Tetrapods. Vol. 1: Amphibians, Reptiles, Birds. Syst. Assoc. Spec. Vol. No. 35 A. Oxford: Clarendon Press.

Parker, H. W. 1956. Viviparous caecilians and amphibian phylogeny. Nature, London, 178:250–252.

Parsons, T. S., and E. E. Williams. 1962. The teeth of Amphibia and their relation to amphibian phylogeny. J. Morphol., 110:375–390.

Parsons, T. S., and E. E. Williams. 1963. The relationships of the modern Amphibia: a re-examination. Q. Rev. Biol., 38:26–53.

Patterson, C. (ed.) 1987. Molecules and Morphology in Evolution: Conflict or Compromise? London: Cambridge Univ. Press.

Pusey, H. K. 1943. On the head of the leiopelmatid frog, Ascaphus truei. I. The chondrocranium, jaws, arches, and muscles of a partly-grown larva. Q. J. Microsc. Sci., 84:105–185.

Rage, J.-C. 1985. Origine et phylogénie des amphibiens. Bull. Soc. Herp. Fr., 34:1–19.

Rage, J.-C., and P. Janvier. 1982. Le problème de la monophylie des amphibiens actuels, à la lumière des nouvelles données sur les affinités des tétrapodes. Geobios, 6:65–83.

Reig, O. A. 1964. El problema del origen monofilético o polifilético de los amfibios, con consideraciones sobre las relaciones entre anuros, urodelos y ápodos. Ameghiniana, 3(7):191–211.

Remane, A. 1964. Das Problem Monophylie-Polyphylie mit besonderer Berücksichtigung der Phylogenie der Tetrapoden. Zool. Anz., 173:22–49.

Romer, A. S. 1945. Vertebrate Paleontology. 2nd ed. Chicago: Univ. Chicago Press.

Romer, A. S. 1964. The skeleton of the Lower Carboniferous labyrinthodont Pholidogaster pisciformis. Bull. Mus. Comp. Zool. Harvard, 131(6):129–159.

Romer, A. S. 1968. Notes and Comments on Vertebrate Paleontology. Chicago: Univ. Chicago Press.

Rosen, D., P. L. Forey, B. G. Gardiner, and C. Patterson. 1981. Lungfishes, tetrapods, paleontology, and plesiomorphy. Bull. Am. Mus. Nat. Hist., 167:161–275.

Saint-Aubain, M. L. de. 1981. Amphibian limb ontogeny and its bearing on the phylogeny of the group. Z. Zool. Syst. Evolutionsforsch., 19:175–194.

Saint-Aubain, M. L. de. 1985. Blood flow patterns in the respiratory systems in larval and adult amphibians: functional morphology and phylogenetic significance. Z. Zool. Syst. Evolutionsforsch., 23:229–240.

Salthe, S. N., and N. O. Kaplan. 1966. Immunology and rates of enzyme evolution in the Amphibia in relation to the origin of certain taxa. Evolution, 20(4):603–616.

Säve-Söderbergh, G. 1934. Some points of view concerning the evolution of vertebrates and the classification of this group. Arch. Zool., 26A(17):1–20.

Säve-Söderbergh, G. 1936. On the morphology of Triassic stegocephalians from Spitzbergen, and the interpretation of the endocranium in the Labyrinthodonta. K. Sven. Vetenskapakad. Handl., Ser. 3, 16(1):1–181.

Schmalhausen, I. I. 1958a. Die Entstehung der Amphibien im Verlauf der Erdgeschichte. Naturwiss. Beitr. Sowjetwiss., 9:941–961.

Schmalhausen, I. I. 1958b. Istoriya proiskhozhdeniya amfibii. Izv. Akad. Nauk SSSR Ser. Biol., 1:39–58. [In Russian].

Schmalhausen, I. I. 1959. Concerning monophyletism and polyphyletism in relation to the problem of the origin of land vertebrates. Byull. Mosk. Ova. Ispy. Prir. Otd. Biol., 64(4):15–33. [In Russian].

Schmalhausen, I. I. 1968. The Origin of Terrestrial Vertebrates. New York: Academic Press.

Schultze, H.-P. 1970. Folded teeth and the monophyletic origin of tetrapods. Am. Mus. Novit., 2408:1–10.

Schultze, H.-P. 1981. Hennig und der Ursprung der Tetrapoda. Palaeontol. Z., 55(1): 71–86.

Schultze, H.-P. 1987. Dipnoans as sarcopterygians. Pp. 39–74 in Bemis, W. E., W. W. Burggren, and N. E. Kemp (eds.), The Biology and Evolution of Lungfishes. J. Morphol., Suppl. 1.

Smithson, T. R. 1985. The morphology and relationships of the Carboniferous amphibian Eoherpeton watsoni Panchen. J. Linn. Soc. London Zool., 85:317–410.

Stensiö, E. 1963. The brain and the cranial nerves in fossil craniate vertebrates. Skr. Nor. Vidensk. Akad. Oslo, 13:5–120.

Stephenson, N. G. 1951a. On the development of the chondrocranium and visceral arches of Leiopelma archeyi. Trans. Zool. Soc. London, 27:203–253.

Stephenson, N. G. 1951b. Observations on the development of the amphicoelous frogs, Leiopelma and Ascaphus. J. Linn. Soc. London Zool., 42:18–28.

Stephenson, N. G. 1955. On the development of the frog Leiopelma hochstetteri Fitzinger. Proc. Zool. Soc. London, 124:785–795.

Szarski, H. 1962. The origin of the Amphibia. Q. Rev. Biol., 37:189–241.

Szarski, H. 1977. Sarcopterygii and the origin of tetrapods. Pp. 517–540 in Hecht, M. K., P. C. Goody, and B. M. Hecht (eds.), Major Patterns in Vertebrate Evolution. New York: Plenum Press.

Thomson, K. 1962. Rhipidistian classification in relation to the origin of the tetrapods. Breviora Mus. Comp. Zool., 177:1–12.

Thomson, K. 1964a. The ancestry of the tetrapods. Sci. Prog., 52(207): 451–459.

Thomson, K. 1964b. The comparative anatomy of the snout in rhipidistian fishes. Bull. Mus. Comp. Zool. Harvard, 131(10):313–357.

Thomson, K. 1967a. Notes on the relationships of the rhipidistian fishes and the ancestry of the tetrapods. J. Paleontol., 41(3):660–674.

Thomson, K. 1967b. Mechanisms of intracranial kinetics in fossil rhipidistian fishes (Crossopterygii) and their relatives. J. Linn. Soc. Zool., 46:223–253.

Thomson, K. 1968. A critical review of certain aspects of the diphyletic theory of tetrapod relationships. Nobel Symp., 4:285–305.

Von Huene, F. 1956. Paläontologie und Phylogenie der niederen Tetrapoden. Jena: Fischer.

Williams, E. E. 1959. Gadow's arcualia and the development of tetrapod vertebrae. Q. Rev. Biol., 34:1–32.

Wintrebert, P. 1910. Sur le déterminisme de la métamorphose chez les batraciens. XVIII. L'origine des urodèles. C. R. Seances Soc. Biol. Paris, 69:172–174.

Wintrebert, P. 1922. L'évolution de l'appareil ptérygopalatin chez les Salamandridae. Bull. Soc. Zool. France, 47:208–215.

7 Lissamphibian Origins

John R. Bolt

"Lissamphibia" is a convenient collective term for the three orders of living amphibians—Anura, Caudata, and Gymnophiona. The Lissamphibia may or may not be a monophyletic group; at present there is not a strong consensus regarding lissamphibian relationships, either inter se or between one or another lissamphibian group and any taxon of extinct amphibians. However, there is a majority view among paleontologists and herpetologists that the Lissamphibia *may* be a monophyletic group. The current respectability of lissamphibian monophyly is relatively recent and largely the result of a series of papers by Williams (1959) and by Parsons and Williams (1962, 1963), particularly the latter. Both these and other recent papers on lissamphibian relationships are mainly concerned with reevaluation of long-available data dealing with both fossil and Recent amphibians; the most important new data bearing on lissamphibian relationships involve comparisons of the dentition between fossil and Recent amphibians (Bolt, 1969, and other papers cited below). This chapter, too, reconsiders observations and analyses that have been in the literature since 1959.

Such reconsideration might be expected in a review paper, but there is a further reason (other than paucity of new observations) why it is appropriate now. By today's standards, some of the recent discussion of lissamphibian relationships is analytically imprecise and indeed irrelevant. The evolution of views on lissamphibian relationships since 1959 has been paralleled by a rather less gradual and considerably more strife-

194

ridden change in the theory and practice of systematics. Specifically, most systematists now employ some variety of cladistic analysis as the basis for phylogenetic hypotheses, and the vocabulary of cladism has become a sort of lingua franca for systematics. I believe that a review of the recent literature on lissamphibian relationships in cladistic terms will improve clarity of discussion and foster clearer recognition of the issues involved.

Description and analysis will be confined to skeletal characters because these are the only ones that can generally be compared among fossil and extant animals. The analysis of nonskeletal characters in extant lissamphibians could, of course, be restated in cladistic terms, but the result would leave unchanged the conclusions reached by Parsons and Williams (1963). Recent discussion of lissamphibian origins has centered on interpretation of data from fossils, and a cladistically based reexamination of the evidence is most needed here.

THE LISSAMPHIBIAN FOSSIL RECORD

Numerous fossils document a high diversity of nonlissamphibian amphibians in the late Paleozoic and the Triassic, with only one representative of the Paleozoic-Triassic groups persisting into the Jurassic (Warren and Hutchinson, 1983). The Paleozoic-Triassic groups do not converge morphologically on the Lissamphibia; if anything, the latest representatives are among the least similar to lissamphibians. The first undoubted lissamphibians appear in the Jurassic, although the record of Jurassic lissamphibians is extremely sparse. The post-Jurassic record is better, and by the end of the Miocene, at least one representative of nearly every frog and salamander family (most of which are still extant) has been identified (Estes and Reig, 1973; Estes, 1981). The number of species represented, however, is small, and much of the material is fragmentary or otherwise poorly preserved. So far, lissamphibian fossils have contributed little toward elucidating either the relationships among the living groups or their relationships to nonlissamphibian tetrapods.

The fossil record of Gymnophiona is confined to a single vertebra from the Upper Paleocene of Brazil (Estes and Wake, 1973). The earliest undoubted salamanders on record are middle (*Albanerpeton*) and late Jurassic (*Comonecturoides, Karaurus*) (see Estes, 1981, and references therein). Ivachnenko (1978) described as a larval urodele a small, poorly preserved specimen from the late Triassic of Uzbekhistan. As described, this specimen cannot be identified definitely as a urodele (Estes, 1981). The earliest known unequivocal fossil frog is *Vieraella*, from the early Jurassic of Argentina (Estes and Reig, 1973). The early Triassic *Triadobatrachus* from

Madagascar was described as a frog ancestor by Piveteau (1937); its status has been debated ever since. The unique type specimen is preserved as an impression of part of the skeleton. Interpretation is further complicated by the possibility that the specimen represents a transforming larva. Hecht (1962) concluded that *Triadobatrachus* is "an enigma." Griffiths (1963) considered it an ancestral frog, and Estes and Reig (1973), after a careful restudy based on a latex cast provided by Piveteau, concluded that *Triadobatrachus* "has too many similarities to frogs to be dismissed as convergent" (p. 41). I agree with this assessment. If *Triadobatrachus* is included in the Anura as a primitive sister-group, then we must conclude that anuran origins are in the early Triassic or late Paleozoic.

CHARACTERS AND CHARACTER ANALYSIS

Methods, and Types of Characters Considered

Some definitions and procedures used in this chapter should be stated explicitly, as follows. (1) "A taxon is monophyletic when all of its members have an ancestor in common that is not common to any other known taxon" (Gaffney, 1979:89). Similar formulations are available in many other publications, including Eldredge and Cracraft (1980) and Wiley (1981). (2) I will analyze characters as either primitive (= plesiomorphic) or derived (= apomorphic). Only derived characters are useful in constructing cladograms. (3) Out-group comparison constitutes the basis for all decisions on character polarity. (4) The basis for choice among cladograms is parsimony. The most parsimonious cladogram for any set of characters is the one that requires the smallest number of evolutionary steps.

Following the character analysis below, cladograms are presented for seven taxa and a maximum of nine characters. Cladogram length (in terms of the number of evolutionary steps) was determined using Version 1.01 of the MacClade program written by W. Maddison of Harvard University. MacClade is designed primarily to compute length of user-specified cladograms. Its automatic routines for finding the most parsimonious (i.e., shortest) cladogram are limited to local branch-swapping, unlike more sophisticated programs such as PAUP, and there is no assurance that the program will automatically find the shortest cladogram. This is not a disadvantage for such a small data set, because the user can quickly and easily construct cladograms that test all likely arrangements. All tests were performed with equally weighted characters that were assumed to be reversible for purposes of analysis.

No living tetrapod group is an appropriate out-group for lissamphibians, which generally are considered to be the sister-group of all other living tetrapods. The primitive sister-group of lissamphibians can be assumed to be some taxon of nonlissamphibian fossil amphibians. For the sake of brevity, "fossil amphibians" are equated herein with "non-lissamphibian fossil amphibians." For present purposes, it is unimportant that "fossil amphibians" is a nonmonophyletic group. The same comment applies to the "labyrinthodonts" and possibly to the "lepospondyls," both of which are traditionally recognized groups within "fossil amphibians." The sister-group of fossil amphibians and of all other tetrapods is assumed to be the osteolepiform crossopterygians (see Schultze and Arsenault, 1985).

Some group of fossil amphibians therefore would be the appropriate out-group for analyzing characters shared by lissamphibians and fossil amphibians. Unfortunately, the interrelationships of fossil amphibians still are poorly understood, and there is no published set of analyzed characters of these groups that is either sufficiently detailed or corroborated to permit the choice of an out-group. Thus, under the circumstances, it is difficult to resolve the character conflicts that occur when similar derived characters are found in groups that generally are considered not to be sister-groups. This chapter avoids the use of such ambiguous characters for the discussion of relationships. The only derived characters that will be considered in forming a hypothesis of relationships are those that pass a uniqueness test—i.e., they (1) are not found in osteolepiforms and (2) are uniquely shared by two or more lissamphibian groups, or by one or more lissamphibian groups with a fossil amphibian group that is presumed to be monophyletic. This operating rule should be recalled in reading the "comments" sections below.

I have tried to identify and discuss all such uniquely shared characters that appear in published discussions of lissamphibian origins. I have included only published discussions from 1959 to the present, because Eaton (1959), Williams (1959), and Parsons and Williams (1962, 1963) reviewed the literature before that date. In addition, I discuss in detail characters used by Carroll and Currie (1975) and Carroll and Holmes (1980) that purportedly support hypotheses of a nonmonophyletic Lissamphibia.

The first 11 characters discussed in this section pass the uniqueness test. They are presented in a standard format that consists of (1) the character number and name, (2) a statement of the character, and (3) comments on this and similar characters, including distribution. Distribution of the formally analyzed characters is summarized in Table 1. Vertebral characters are discussed separately and less formally. Despite the great importance historically accorded to vertebral characters as indicators of amphibian relationships, none appears to be useful in this

Table 1. Results of analysis of Characters 1–11 (0 = primitive,[1] 1 = derived)

Taxon	Character 1 SKF	2 MTP	3 MTCF	4 PTP	5 PTCF	6 MB	7 DQP	8 ON	9 CE	10 ASPD	11 OO
Primitive[1]	0	0	0	0	0	0	0	0	0	0	0
Hapsidopareion	0	0	0	0	0	0	0	?	1	?	0
Trematopsidae	0	0	0	0	0	0	1	1	0	?	0
Dissorophidae	0	1	1	1	1	0	1	1	0	?	0
Anura	1	1	1	1	1	1	1	1	0	1	1
Caudata	1	1	1	1	1	1	0	0	1	1	1
Gymnophiona	1	1	1	1	1	1	0	0	0	1	0

Abbreviations: ASPD = amphibian special periotic duct; CE = cheek emargination; DQP = dorsal quadrate process; MB = mentomeckelian bones; MTCF = marginal-tooth crown form; MTP = marginal-tooth pedicely; ON = otic notch; OO = otic operculum; PTCF = palatal-tooth crown form; PTP = palatal-tooth pedicely; SKF = skull fenestration.
[1] The primitive condition is that found in rhipidistians. It is also present in the majority of Paleozoic tetrapods of all major groups.

198

chapter. However, I do present new information on vertebral morphology in a key group of fossil amphibians. Finally, the two groups of characters used by Carroll and Currie (1975) and Carroll and Holmes (1980) are examined. With one exception, these characters do not meet the uniqueness test and therefore are not employed in constructing the cladograms figured herein.

Uniquely Derived Characters

Character 1: Skull fenestration. Absence of the supratemporal, intertemporal, tabular, postparietal, postfrontal, postorbital, and jugal.

Comment: A much more primitive (for tetrapods) complement of cranial roofing bones is seen in, for example, an early primitive temnospondyl such as *Greererpeton* (Smithson, 1982) or an anthracosaur such as *Proterogyrinus* (Holmes, 1984). Loss of a few skull roofing bones frequently is seen in Paleozoic and Triassic amphibian groups; bones of the temporal region (e.g., intertemporal and supratemporal) are especially susceptible to loss, but reduction never approaches the lissamphibian condition. A number of authors, including Parsons and Williams (1963), have suggested that extensive fenestration of the skull is a lissamphibian synapomorphy. This is surely a derived condition no matter what outgroup one chooses, because it is unique among tetrapods.

Character 2: Marginal-tooth pedicely. Presence of pedicely in marginal teeth at some ontogenetic stage.

Comment under Character 5.

Character 3: Marginal-tooth crown form. Marginal teeth bicuspid, with labial and lingual cuspules, at some ontogenetic stage.

Comment under Character 5.

Character 4: Palatal-tooth pedicely. Presence of pedicely in palatal teeth at some ontogenetic stage.

Comment under Character 5.

Character 5: Palatal-tooth crown form. Palatal teeth bicuspid at some ontogenetic stage.

Comment: Parsons and Williams (1962) were the first to emphasize the fact that lissamphibians are unique among living tetrapods in having pedicellate teeth—i.e., there is an uncalcified or weakly calcified annular zone within the tooth itself. In a mostly literature-based survey of living and fossil vertebrates, they found that this type of pedicely had been reported elsewhere only in pleuronectid teleosts, and there they considered the evidence for it equivocal. Bolt (1969) suggested that crown form is equally as characteristic of lissamphibian teeth as pedicely, and that lissamphibian teeth also should be characterized as bicuspid, with labial and lingual cuspules. Although a number of lissamphibians have other

tooth morphologies, the parsimonious interpretation that such pedicellate, bicuspid teeth are primitive for each of the lissamphibian orders
seems generally accepted. Pedicely and bicuspid crown form can be
treated as separate characters because they are not invariably found
together in lissamphibians or in nonlissamphibian fossil amphibians.
Similarly, tooth size, position, and arrangement (as in multiple rows) can
be considered as separate characters.

Parsons and Williams (1962) did not distinguish among palatal and
marginal teeth, and, in fact, palatal and marginal teeth of lissamphibians
are commonly of very similar size and morphology. Size variation is more
pronounced among fossil amphibians. In labyrinthodonts, for instance
(Fig. 1), one can generally distinguish large palatal fangs (usually found
on vomers, palatines, ectopterygoids, and the dentary near the symphysis) from the smaller marginal teeth, and both fangs and marginal
teeth from the relatively tiny denticles, which can occur on any or all of
the palatal bones as well as on various bones of the lower jaw. A similar

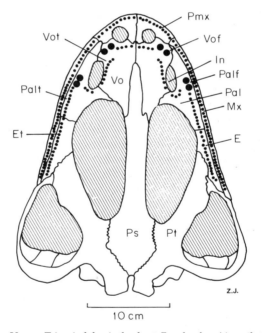

Fig. 1. Skull of the Upper Triassic labyrinthodont *Eupelor fraasi* in palatal view (redrawn
from Case, 1931). Described by Case as *Buettneria perfecta;* identification in this paper
follows Chowdhury (1965). E = ectopterygoid; Et = ectopterygoid teeth; In = internal
narial opening; Mx = maxilla; Pal = palatine; Palf = palatine fangs; Palt = palatine teeth;
Pmx = premaxilla; Ps = parasphenoid; Pt = pterygoid; Vo = vomer; Vof = vomerine fangs;
Vot = vomerine teeth.

dentitional pattern occurs in osteolepiforms (e.g., Jarvik, 1954) and thus can be considered primitive for tetrapods.

Bolt (1969) reported finding lissamphibian-like pedicellate, bicuspid marginal teeth in *Doleserpeton annectens*, a small labyrinthodont amphibian (Dissorophidae; superfamily Dissorophoidea) from the early Permian of Oklahoma. The two dissorophoid families that frequently are mentioned in this chapter—Dissorophidae and Trematopsidae—will be considered as monophyletic for present purposes, although their defining synapomorphies are still under discussion. Unlike the marginal teeth in most labyrinthodonts, those of *Doleserpeton* resemble lissamphibian teeth in lacking labyrinthine structure, which simply may be the result of their small size. The palatal dentition of *Doleserpeton* also is remarkable; non-labyrinthine, pedicellate, presumably bicuspid teeth of the same size as those in the marginal dentition are borne in short rows on the vomer, palatine, and the symphyseal area of the dentary (Figs. 2–3) in the usual locations of the large fang teeth seen in most other labyrinthodonts. The ectopterygoid was absent, perhaps owing to immaturity, in the material studied. The lissamphibian type of bicuspid, pedicellate tooth subsequently was reported in three other small dissorophid labyrinthodonts, *Amphibamus grandiceps* (middle Pennsylvanian), *Tersomius texensis*, and "cf. *Broiliellus* sp." (both from the early Permian) (Bolt, 1977, 1979). In the case of *A. grandiceps*, marginal teeth are bicuspid and pedicellate. The size and location of palatine and vomerine teeth resemble those of *Doleserpeton*, although pedicely and crown morphology could not be determined. As in *Doleserpeton*, there are no large palatal fang teeth; the symphyseal area of the dentary could not be observed. The marginal dentition of *T. texensis* is like that of *Doleserpeton*. The palatal dentition could not be fully described, especially in smaller (*Doleserpeton*-sized) specimens, but the short rows of small "fang" teeth seen in *Doleserpeton* were not found (see Bolt, 1977, for details). *Tersomius texensis*, however, showed variation in marginal tooth crown morphology (degree of "bicuspidity") correlated with skull size, and thus, presumably with maturity. Finally, bicuspid marginal and fang teeth that may or may not be pedicellate were found in a unique specimen identified as "cf. *Broiliellus* sp." It is significant that this specimen was the largest (snout-postparietal length 78 mm) of the small dissorophids found to have bicuspid teeth. The teeth were not so markedly bicuspid as those of other species represented by smaller skulls, and the single fang tooth preserved (apparently of an alternately replacing pair) was both weakly bicuspid and markedly larger than the marginal teeth.

On the basis of the size-correlated dental variation observed and of the dentitional ontogeny of salamanders, Bolt (1977, 1979) proposed that the

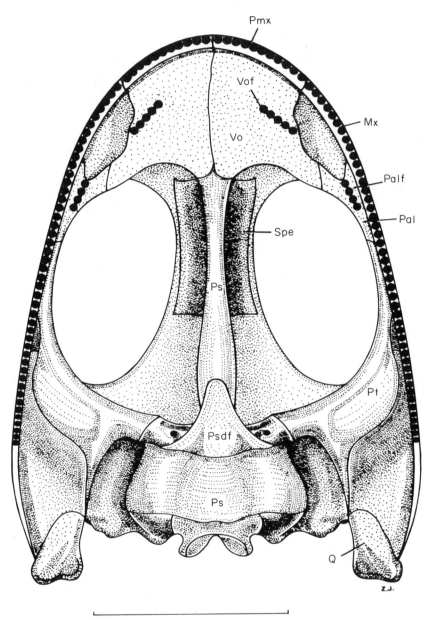

Fig. 2. Skull of the Lower Permian dissorophid labyrinthodont *Doleserpeton annectens* in palatal view (redrawn from Bolt, 1969). Psdf = area of parasphenoidal denticle field; Q = quadrate; Spe = sphenethmoid. See Figure 1 for key to other abbreviations. Note that the fang-tooth positions are occupied by teeth the same size as the marginal teeth, arranged in short rows.

202

Fig. 3. *Doleserpeton annectens*, scanning electron micrograph of symphyseal area of left dentary in medial view. Shows bases (pedicels) of two symphyseal fangs, with space for three more teeth. Note that the fang-teeth are the same size as marginal teeth and arranged in a short row.

1.0 mm

lissamphibian-like dentition of dissorophids was ontogenetically transient. Hypothetically, larval nonpedicellate monocuspids were replaced by larval pedicellate monocuspids. Around the time of metamorphosis, these were replaced by juvenile pedicellate bicuspids, which in turn were replaced gradually by adult monocuspid teeth with the usual labyrinthine structure. Under this interpretation, the dentition of adult lissamphibians is paedomorphic.

Other than lissamphibians and some dissorophids, no other tetrapod group is known to possess either bicuspid teeth of lissamphibian type (labial and lingual cuspules) or pedicellate teeth. The closest resemblance to lissamphibian-like bicuspid teeth is seen in the microsaur *Carrolla craddocki* from the early Permian of Texas (Langston and Olson, 1986). Marginal teeth of *Carrolla* are strongly bicuspid, but the cuspules lie on a mesiodistal rather than labiolingual axis. In addition, they are not visibly pedicellate, and there appear to be no palatal teeth of any kind (although the palate is imperfectly preserved).

Despite these data on distribution of Characters 2–5, some workers seem reluctant to accept these dentitional characters as possible lissamphibian synapomorphies. For instance, Carroll and Holmes (1980) noted that the frequent absence of pedicellate teeth in Mesozoic fossil salamanders could be due to the fact that many of them are neotenic. They go on to say (p. 36) that "the fossil record is very incomplete and biased toward the preservation of aquatic salamanders [which are frequently neotenic—JRB], but does not contradict an interpretation that pedicellate teeth evolved within the urodeles." However, the fossil record as it has been analyzed to date also does not *support* such an interpretation.

Jarvik (1981:222) dismissed the significance of bicuspid, pedicellate teeth, saying, "It is of course not possible to base any safe conclusions as to relationship on a single character and considering the justified objections raised by Lehman (1968) and the facts that such hinged teeth occur also in other groups of vertebrates, whereas many extant and obviously also several fossil 'lissamphibians' . . . have non-pedicellate teeth it is evident that hinged teeth cannot be a reliable 'lissamphibian' character either (as to bicuspid teeth in various fishes see Bystrow, 1939, fig. 8)."

Lehman's objections seem to rest on a misunderstanding of parsimony and of dentitional ontogeny in the living amphibians, and on the fact that crown morphology and histology are not completely identical among them. The significance of the latter differences is unclear at present, although there is no question that much remains to be learned regarding morphology and development of the teeth in lissamphibians. Jarvik cited no new evidence for his contention about the occurrence of "hinged" teeth in other groups and thus did not refute the conclusions of Parsons and Williams (1962). The crossopterygian teeth figured by Bystrow are better described as "hooked" than as "bicuspid." They do not closely resemble lissamphibian teeth.

Character 6: Mentomeckelian bones. Mentomeckelian bones present in the symphyseal region of the lower jaw.

Comment: Mentomeckelian bones are known in each of the lissamphibian orders and in no other amphibians. Carroll and Currie (1975) considered the mentomeckelian bones to have different functional roles in anurans and caecilians and so questioned their homology on this basis. However, given their similar position and origin (from Meckel's cartilage), there is no a priori reason to deny homology.

Character 7: Dorsal quadrate process. Presence of a dorsal process on the quadrate.

Comment: The dorsal quadrate process is found in dissorophids and trematopsids. In both of these dissorophoid families, the dorsal quadrate process lay within an otic notch formed from dermal skull elements and presumably suspended part of the tympanum. Both frogs and dissorophoids possess or possessed an otic notch and tympanum, as discussed in Lombard and Bolt (1979) and Bolt and Lombard (1985); lack of the tympanum within anurans is considered derived. Bolt and Lombard (1985) argued that the dorsal process of the quadrate is the homologue of the anuran tympanic annulus. Suspension of the tympanum on a cartilaginous tympanic annulus is a frog synapomorphy.

Character 8: Otic notch. Otic notch present as a concavity in the cheek or lateral part of the dorsal skull roof, toward which the distal end of the stapes is directed (from Lombard and Bolt, 1988).

Comment: This character occurs in frogs and dissorophoids, as well as the great majority of nondissorophoid labyrinthodonts. It is (primitively) absent in all microsaurs and other lepospondyls, and its absence in salamanders and caecilians is also scored as primitive.

Character 9: Cheek emargination. Maxilla and suspensorium separated by a gap, and quadratojugal small or absent.

Comment: Several genera in the microsaur family Hapsidopareiontidae have developed this type of emargination (Carroll and Holmes, 1980). It is undoubtedly derived, as noted by Carroll and Holmes, and is either rare or unique among fossil amphibians. If found among other fossil amphibian groups at all, it occurs only in the lysorophoids, which at present are not considered microsaurs (Carroll and Gaskill, 1978). *Lysorophus* indeed has a gap between the suspensorium and maxilla, which presumably arose in large part by shortening of the maxilla (see figures in Bolt and Wassersug, 1975). The quadratojugal in *Lysorophus* is either absent or relatively large, depending on interpretation of the questionable homologies of the temporal region. Most paleontologists would consider the highly derived (and nearly limbless) lysorophoids poor candidates for the sister-group even of caecilians. Therefore, this character can be accepted provisionally as a unique synapomorphy of hapsidopareiontid microsaurs and salamanders.

Character 10: Amphibian special periotic duct. Special periotic duct present, passing posterior to the lagena and to the vertical axis of the otic labyrinth.

Comment: The relationship of the special periotic duct to the vertical axis of the otic capsule was discussed by Lombard and Bolt (1979 and references therein). In amphibians, the special periotic duct connects the periotic cistern (associated with the fenestra ovalis) with the periotic sac by passing *posterior* to the lagena and to the vertical axis of the otic labyrinth. In amniotes, the duct passes *anterior* to the lagena and vertical axis. No intermediates between these two conditions are known, and the presumption is that each arose independently from a primitive (osteolepiform) condition in which there was no special duct. Neither condition can be argued to be primitive for tetrapods. This character is an exception to the expressed intent to consider only features that can be directly compared among fossil and extant amphibians. Data permitting such a comparison are not at present available. However, they may be obtainable because the course of the special periotic duct is known to be reflected in the otic capsule wall in a number of animals, including salamanders (Lombard, 1977). This character could be important in tracing the origin of amniotes, as well.

Character 11: Otic operculum. An otic operculum is present, lying within the fenestra ovalis and posterior to the stapedial footplate.

Comment: The otic operculum is often referred to as an "otic opercular bone," but it is cartilaginous rather than ossified, and this may be its primitive condition. It is known only from lissamphibians and is sometimes cited as a lissamphibian synapomorphy (Parsons and Williams, 1963), although its presence in caecilians is unconfirmed. An otic operculum has not been demonstrated in any fossil amphibian, despite the "accessory otic bones" described in some microsaurs by Carroll and

Gaskill (1978) and the "operculum" described by Eaton (1973) in the dissorophid *Actiobates*. Careful examination of the literature and of many of the specimens described by these authors provides no grounds for believing that the structures involved are homologous to the otic operculum of lissamphibians.

Vertebral Characters

Although not useful herein, characters of the vertebral centrum long have played an important role in the systematics of fossil amphibians. The lissamphibian centrum never has fit well into the paleontological scheme because the fact that it is a single ossification makes its homology uncertain at best. Until recently, an embryological approach to determining vertebral homologies for living species was widespread, based on the general theory of vertebral ontogeny propounded by Gadow around the turn of the century. A number of paleontologists even attempted to interpret the ossified centra of fossil amphibians and other fossil vertebrates in terms of the Gadowian "arcualia" of which they supposedly were composed. Gadow's theory was discarded by most comparative morphologists after a detailed and highly critical review of the subject by Williams (1959), who offered a radically simplified synthesis of vertebral ontogeny. As far as centrum homologies are concerned, Williams's approach was basically the traditional paleontological one: "The history of tetrapod vertebrae is to be understood quite simply in terms of the three separate elements that we find in the earliest Amphibia and in the fishes most closely related to them: the *neurapophysis*, the *pleurocentrum*, and the *hypocentrum* [= intercentrum]" (p. 25). Williams was able to use paleontological evidence to determine the homologies of the bi- or tripartite centrum found in most tetrapod groups. But in the absence of a fossil sister-group for lissamphibians, paleontology offered no clue to the homologies of the monospondylous lissamphibian centrum. For the lissamphibians, therefore, Williams returned to an embryological criterion of centrum homology (development in an intersegmental position). This led him into a series of increasingly speculative suggestions regarding vertebral embryology, function, and centrum homology in lissamphibians and in the extinct lepospondyl amphibians of the Paleozoic, the centra of which resemble those of lissamphibians in being monospondylous. This incongruity was noted by Wake (1970), who also concluded that embryology was not helpful in determining the homologies of the lissamphibian centrum, a conclusion later reinforced by Wake and Lawson (1973). Recently, Gardiner (1983) revived a modified version of the arcualia theory and made several novel propositions regarding centrum homologies in lissamphibians. As yet, there have been few published comments on his work (but see Wake and Wake, 1986), but I see no serious challenge to Wake's (1970:51) conclusion that "to discuss homologies of centra in living groups without fossil relatives is fruitless."

Although lissamphibian vertebral centrum homology cannot be determined at present, Parsons and Williams (1963:33) discussed one other vertebral character that merits comment here: "The first vertebra in modern Amphibia is very similarly modified in all three orders to articulate with the two occipital condyles. . . . It is, however, strikingly similar to the atlas of the microsaur *Euryodus* figured by Gregory, Peabody, and Price (1956)." This general type of atlas vertebra is found in a number of fossil amphibian groups, the microsaur mentioned by Gregory et al. being only one example. Therefore, it fails the uniqueness test, but it should be pointed out that the atlas of *Doleserpeton* is similar to that of lissamphibians in having a single centrum (most likely the pleurocentrum in this case) and in approaching a bicondylar condition (Figs. 4–5).

Doleserpeton is unusual for a rhachitome in having "reptilian" vertebral centra that combine a single cylindrical pleurocentrum as the principal element with a small ventral intercentrum. Therefore, it is parsimonious to suppose that the single, completely ossified atlantal centrum represents a modified pleurocentrum. Identification of the atlantal centrum as an intercentrum would imply a more radical change in relative size of the ossified pleurocentrum versus intercentrum compared with the trunk vertebrae than I am aware of in any other early amphibian. The fate of the atlantal intercentrum is uncertain because the atlas has not been exposed in articulation with a skull. Both *Amphibamus grandiceps* and *Tersomius texensis* apparently have vertebral centra similar, though not necessarily identical, to those of *Doleserpeton* (Daly, 1973; Bolt, 1977, 1979). As far as known, however, other dissorophids have "normal" rhachitomous centra, with a relatively larger ventral intercentrum and relatively smaller,

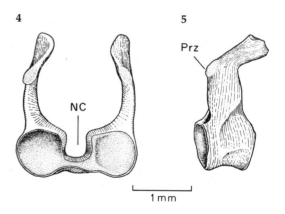

4   **5**

Figs. 4–5. *Doleserpeton annectens*, atlas vertebra. (4) Anterior view. (5) Lateral view. NC = notochordal canal; Prz = prezygapophysis.

NC

Prz

1 mm

paired pleurocentra posterodorsally. The *Doleserpeton*-style centrum, like the teeth, may represent an ontogenetically transient stage (Bolt, 1979).

Characters Supporting Alternative Hypotheses of Lissamphibian Relationships

Two recent and commonly cited papers on lissamphibian relationships propose an origin of salamanders and caecilians from the Paleozoic microsaurs, which despite their name are small, generally "lepospondylous" amphibians (Carroll and Currie, 1975; Carroll and Holmes, 1980). These papers contain much morphological detail, presented in noncladistic terms. I believe that rephrasing in cladistic terms what I see as the main arguments of these papers will contribute to clarity of discussion.

Carroll and Currie (1975) on Caecilian Ancestry

Characters discussed by Carroll and Currie (1975) as indicating possible microsaur-lissamphibian relationship include some that are widespread among microsaurs, and some that are peculiar to *Rhynchonkos (Goniorhynchus) stovalli*. The latter was emphasized because (p. 236) "much more specific similarities with apodans can be seen in the latest known microsaur, *Goniorhynchus*." Carroll and Currie considered five characters to be most significant—(1) temporal fenestration, (2) palatal structure, (3) palatal dentition, (4) braincase composition, and (5) mandibular dentition. The statement of each character below represents my interpretation of their more diffuse discussion.

Character 1: Temporal fenestration. The number of dermal bones in the temporal region of both microsaurs and caecilians is reduced considerably from the primitive condition, although reduction is much more extensive in the caecilian skull.

Comment: A serious criticism of this character is its corollary, explicitly adopted by Carroll and Currie (1975), that the solidly roofed skull of caecilians is primitive rather than derived. This interpretation is denied by recent work on caecilians, on the basis of cladistic analysis (Nussbaum, 1977, 1983) and ontogeny (Wake and Hanken, 1982). Further, reduction of the temporal series is not much more extensive in microsaurs, including *Rhynchonkos*, than in many other fossil amphibian groups, and microsaurs retain five of the seven bones listed under uniquely derived Character 1 (temporal fenestration). Absence of the intertemporal and supratemporal does not constitute a unique synapomorphy with any of the lissamphibian orders. The intertemporal is absent in numerous fossil amphibian groups, and the absence of both intertemporal and supratemporal is shared at least with advanced aïstopods, lysorophoids, and adelogyrinids, if the temporal-bone homologies proposed by Carroll and Gaskill (1978) for the latter two groups are accepted.

Character 2: Palatal structure. Microsaurs and caecilians retain the primitive complement of dermal bones in the palate, including the ectopterygoid, and an open basicranial joint.

Comment: The condition of the ectopterygoid bone in caecilians has been debated. Carroll and Currie (1975, see further discussion below), following Marcus et al. (1935), considered that an ectopterygoid is present in at least some caecilians. Wake and Hanken (1982) found no evidence for it in their developmental study of a single caeciliid species. They suggested that the description by Marcus et al., which is based on developmental study of one species of a different caeciliid genus, may not be adequate, and that resolution of the differences between their findings and those of Marcus et al. must await further study. This seems the only reasonable course; therefore, the presence or absence of the ectopterygoid is not a useful character at this time. Aside from this, if we accept their observations, Carroll and Currie are correct in identifying the resemblances in palatal structure as primitive.

Character 3: Palatal dentition. The presence on the palate of a row of teeth parallel to the marginal dentition. These have different relationships to the internal naris—medial to it in *Rhynchonkos,* and lateral to it in caecilians. Carroll and Currie (1975) attributed this difference to "the much smaller size of the nasal capsule" in microsaurs (p. 237).

Comment: Many fossil amphibians have a row of palatal teeth disposed like that of *Rhynchonkos*—this character has evolved independently a number of times, on almost any hypothesis of early amphibian interrelationships (see Fig. 1). The size of the nasal capsule in microsaurs is undetermined, and probably undeterminable.

Character 4: Braincase composition. Rhynchonkos and caecilians are unique among amphibians in having "an extensive pleurosphenoid [that] joins the otic-occipital portion of the braincase with the sphenethmoid" (Carroll and Currie, 1975:237).

Comment: The pleurosphenoid is an ossification in the pila antotica (see Goodrich, 1930). In fossil material, it perforce is defined operationally by position alone. Any ossification in the lateral wall of the braincase, lying between the sphenethmoid in front and the otic capsules behind, commonly is identified as a pleurosphenoid. By this definition, the pleurosphenoid of microsaurs is not unique. For instance, the braincase in the labyrinthodonts "*Eogyrinus*" (Panchen, 1972) and *Loxomma* (Beaumont, 1977) is ossified in this area, as is the braincase of the osteolepiform crossopterygian *Eusthenopteron* (Jarvik, 1954). Thus, presence of a "pleurosphenoid" ossification of some sort may be primitive. It is questionable whether caecilians possess a pleurosphenoid, in any case. Wake and Hanken (1982) failed to find a separate pleurosphenoid ossification center in their study of *Dermophis*.

Character 5: Mandibular dentition. Rhynchonkos has a row of teeth on the coronoid, parallel to the dentary tooth row. *Rhynchonkos* is "the only microsaur, and in fact the only Paleozoic amphibian known to have a single row of teeth in this position" (Carroll and Currie, 1975:241).

Comment: Carroll and Currie noted that in caecilians the inner row of mandibular teeth is borne on an area "indistinguishable from the dentary." The homology of the bone to which this inner tooth row is attached is unclear from the literature. In any case, the situation in *Rhynchonkos* is not unique among fossil amphibians. (Uniqueness among Paleozoic fossil amphibians, claimed by Carroll and Currie, is irrelevant in the present context.) A number of labyrinthodonts have a single row of coronoid teeth in this position, including the Paleozoic *Ichthyostega* (Jarvik, 1980:Fig. 174) and *Colosteus* (Hook, 1983), as do several Triassic species (Jupp and Warren, 1986).

The five characters considered most significant by Carroll and Currie (1975) thus are either primitive, convergent, or indeterminate. The other microsaur-caecilian resemblances they cited amount to little more than a demonstration of the fact that *any* sufficiently primitive amphibian can be supposed to be ancestral to caecilians or to any one of the other lissamphibian orders. Nothing in their account justifies a hypothesis that microsaurs and caecilians are sister-groups. This may become the preferred hypothesis as a result of future work, but at present, their catalog of resemblances has no more than heuristic value.

Carroll and Holmes (1980) on Salamander Ancestry

Carroll and Holmes (1980) presented seven cranial characters that "are considered of particular importance in primitive salamanders, and would be expected in the immediate ancestor of the group" (p. 17). In addition, they provided a discussion of the evolution of adductor jaw musculature and its relevance to the ancestry of salamanders. The seven "expected" ancestral-salamander characters (quoted below) are considered here using Carroll and Holmes's numbers.

Character 1: "Emargination of the cheek" by loss or reduction of the quadratojugal, so that there is always a gap between the quadrate and maxilla.

Comment: This character is dealt with above (as Character 9). It appears to pass the uniqueness test.

Character 2: "Jaw suspensorium formed by quadrate, squamosal and pterygoid. Pterygoid movable on the basicranial articulation and squamosal hinged to otic capsule and parietal."

Comment: This character consists of three statements, which may be considered separately. Thus: (A) "Jaw suspensorium formed by quadrate, squamosal and pterygoid." Carroll and Holmes (1980) did not discuss this character further, nor did they indicate its distribution within microsaurs. In the majority of fossil amphibians, including many microsaurs and such primitive forms as the labyrinthodont *Greererpeton* (Smithson, 1982), these bones are part of a suspensorium that includes in addition a quadratojugal in contact anteriorly with other skull bones (i.e., an unfenestrated cheek) and, in very primitive species, an ossified epipterygoid that is absent in almost all fossil amphibians. Therefore, this statement reduces to cheek fenestration via loss or reduction of the quadratojugal, which is discussed as Character 1 above. (B) "Pterygoid movable on the basicranial articulation." This is primitive for tetrapods. (C) "Squamosal hinged to otic capsule and parietal." Carroll and Holmes determined the existence of this condition by "manipulation of dissected specimens and observations of dried skulls" of salamanders, apparently hynobiids. The suggested movement (1–2 mm) is mediolateral, and they thought that anteroposterior movement was definitely precluded. Some sort of suspensorial mobility is certainly suggested by the gross morphology of many salamander skulls. However, their cited observations are inadequate to establish either its existence or its status as a primitive feature of urodeles. A quite different sort of movement (anteroposterior) was proposed by Eaton (1933; not cited by Carroll and Holmes) in ambystomatid salamanders.

Character 3: "Otic capsule made up of distinct opisthotic and prootic."

Comment: As Carroll and Holmes (1980) noted, this is likely primitive. Their claim that the otic capsule of dissorophids is formed "primarily by a single large ossification" (p. 17) is in error. Both prootic and opisthotic ossify as separate bones in *Doleserpeton* (Bolt, 1969). This could be true of other dissorophids as well if, as in *Doleserpeton*, the two bones were separate only in immature individuals.

Character 4: "Occipital condyle posterior to jaw articulation."

Comment: Carroll and Holmes (1980) noted that this condition is found in several families of microsaurs; they are undoubtedly correct in considering it derived. It is also found in several other fossil amphibian groups, including lysorophoids, aïstopods (see Lund, 1978), and brachyopid labyrinthodonts (e.g., Welles and Estes, 1969).

Character 5: "Double occipital condyle, loss of the basioccipital [ossification]."

Comment: This character (actually two characters) undoubtedly is derived, as Carroll and Holmes (1980) noted. They did not comment further on the condyle, although some of their figures show various microsaurs with what might be described as an incipiently double condyle. A double occipital condyle occurs in a number of other fossil amphibian groups, including nectrideans (e.g., *Diploceraspis*—Beerbower, 1963) and various labyrinthodonts (such as brachyopids—see above). Microsaurs are said to retain a basioccipital ossification (p. 18), which is a primitive character.

Character 6: "Absence of a tympanum."

Comment: This is a primitive character for tetrapods (cf. Lombard and Bolt, 1979).

Character 7: "Stapes with a large foot-plate and forming a structural link between the braincase and the cheek."

Comment: Carroll and Holmes (1980) characterized the microsaur stapes as having "a broad foot-plate, and a short stem, extending toward the squamosal or quadrate" (p. 18). Presence of an ossified shaft that extends toward the quadrate is primitive for tetrapods (Lombard and Bolt, 1979, 1988). A footplate with a single proximal head is a derived character that occurs in most labyrinthodont groups (Lombard and Bolt, 1988) and in aïstopods (McGinnis, 1967).

On present evidence, only one of these seven cranial characters (cheek emargination) is a synapomorphy unique to some or all of the lissamphibian orders and only one fossil amphibian group. The rest are duplicated in other groups, or are just primitive, some at a high level—i.e., the absence of a tympanum is primitive for tetrapods. Characters of Carroll and Holmes's (1980) second major category, those identified in the adductor jaw musculature, remain to be discussed.

Carroll and Holmes (1980) did not examine the jaw musculature of caecilians. Considering only frogs and salamanders, they found "consistent major differences between the patterns of the adductor jaw musculature" in the two groups (p. 28). They believed that "the distinctive patterns of the jaw musculature in frogs and salamanders could have evolved independently from a more primitive condition, but neither specialized pattern is likely to have evolved from the other" (p. 28). If this is accepted, the problem of frog-salamander relationships reduces to: Does the pattern of adductor jaw musculature indicate a unique synapomorphy between salamanders and one nonlissamphibian group, and between frogs and a different nonlissamphibian group?

This question can be answered only if the pattern of adductor jaw musculature in fossil amphibians is determinable to the necessary level of detail. Carroll and Holmes (1980) figured reconstructions of jaw musculature in both microsaurs and labyrinthodonts, but they cited no specific anatomical markers that would constrain such reconstructions. In the case of one of their reconstructions (*Doleserpeton*), there are *no* such markers (pers. observ.). Their reconstructions are constrained only by the shape of the subtemporal fossa. This sort of reconstruction is inadequate; plausible alternatives that do not suppose the presence of the distinctive "frog" or "salamander" patterns cannot be refuted by the available osteological data. As employed by Carroll and Holmes (1980), patterns of adductor jaw musculature are irrelevant to the question of lissamphibian origins.

HYPOTHESIS OF LISSAMPHIBIAN RELATIONSHIPS AND PROBLEMS OF CHARACTER ANALYSIS

Lissamphibian Relationships

The cladogram in Figure 6 was generated using nine of the 11 characters in Table 1. Characters 10 and 11 (amphibian special periotic duct; otic operculum) may prove difficult to determine in fossils and are not used in the analysis because their known distribution (restricted to lissamphibians) may be misleading. Figure 6 represents the single shortest (12 steps) cladogram for Characters 1–9, when rooted as shown. Lissamphi-

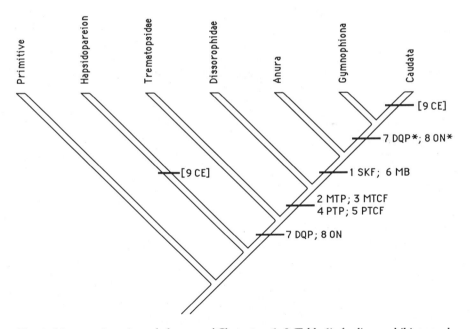

Fig. 6. Most parsimonious cladogram of Characters 1–9 (Table 1), for lissamphibians and fossil amphibians. See Table 1 for key to character numbers and abbreviations. Characters are mapped onto cladogram to show point of first appearance in each clade, or point at which a character is lost. Convergent evolution of characters in separate clades is indicated by square brackets. Loss of a character (i.e., convergence on the primitive condition) is indicated by an asterisk.

bian monophyly and a sister-group relationship to dissorophids are unambiguously indicated.

The same result follows if the dentitional characters are recoded into a single character. In the character analysis above the traditional one or two dentitional characters were subdivided into four. Because these all have the same distribution, it might be argued that subdivision in this case amounts to positive weighting of dentitional characters. (However, I maintain that subdivision could justifiably have been carried further, by basing additional characters on the short symphyseal tooth row of *Dole-serpeton*.) Table 2 shows the dentitional characters combined into one, which may be stated as follows: Character 2. Dentition. Marginal and palatal dentition includes similar-sized teeth that are pedicellate and bicuspid, with labial and lingual cuspules. As before, Characters 10 and 11 are not considered. The resulting (and single shortest) cladogram has nine steps rather than 12 but otherwise is identical with that of Figure 6.

Whichever set of characters is used, that in Table 1 or Table 2, other cladograms are possible that are almost equally parsimonious. As might be expected from the recoding of four dentitional characters into one, this is especially true of the smaller character set of Table 2. Several 10-step cladograms are possible if one uses this data set; Figure 7 is an example. Thus, with this data set, lissamphibian monophyly is the most parsimonious hypothesis by the narrowest possible margin. Nonetheless, it *is* the most parsimonious, using all the characters that meet the uniqueness test. It is obviously desirable to have more such characters, and the small number of characters that presently support a hypothesis of lissam-

Table 2. Results of analysis of Characters 1–8, with dentitional characters combined into Character 2 (0 = primitive,[1] 1 = derived)

	Character							
Taxon	1 SKF	2 DTN	3 MB	4 DQP	5 ON	6 CE	7 ASPD	8 OO
Primitive[1]	0	0	0	0	0	0	0	0
Hapsidopareion	0	0	0	0	0	1	?	0
Trematopsidae	0	0	0	0	1	0	?	0
Dissorophidae	0	1	0	1	1	0	?	0
Anura	1	1	1	1	1	0	1	1
Caudata	1	1	1	0	0	1	1	1
Gymnophiona	1	1	1	0	0	0	1	0

Abbreviations: DTN = dentition; see Table 1 for other abbreviations.

[1]The primitive condition is that found in rhipidistians. It is also present in the majority of Paleozoic tetrapods of all major groups.

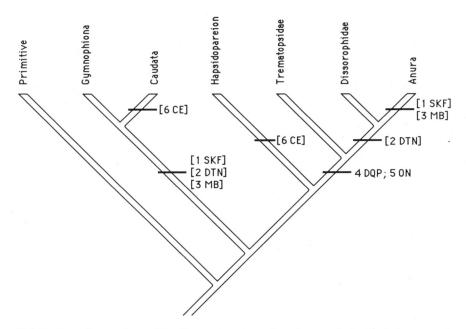

Fig. 7. One of several possible 10-step (most parsimonious = 9 steps) cladograms of Characters 1–6 (Table 2) for lissamphibians and fossil amphibians. See Tables 1 and 2 for keys to character numbers and abbreviations. Characters are mapped onto cladogram to show point of first appearance in each clade, or point at which a character is lost. Convergent evolution of characters in separate clades is indicated by square brackets.

phibian monophyly may be considered a problem. However, none of this should prevent us from exploring both other problems of character analysis and the implications of the hypothesis.

Problems of Character Analysis

The dentitional characters, as well as vertebral characters not employed in constructing cladograms, have been hypothesized to be ontogenetically transient, and this raises questions about their reliability that deserve some further discussion. Problems with the "ontogenetic" characters fall into two areas—i.e., determining their real distribution and assessing their phylogenetic significance within "dissorophids."

If we accept the suggested ontogenetic transience of the lissamphibian-type dentition and vertebral centrum within dissorophids, it is logical to ask whether such a developmental history was confined to dissorophids. It might, for instance, have characterized a much larger group of labyrinthodonts and be missed because the necessary developmental stages are not available. This issue could be resolved most convincingly by study of single-species developmental sequences that include both larval (as shown by external gills) and adult specimens of both dissorophid and nondissorophid groups. A few sequences that at least approach this ideal may be available, notably, *Amphibamus lyelli* and *Saurerpeton obtusum*. Both are from the middle Pennsylvanian locality of Linton, Ohio, and represent, respectively, the families Dissorophidae and Saurerpetontidae, which conventionally are classified in different superfamilies. Dentitional and vertebral ontogeny in them should be studied by techniques that are capable of revealing the necessary detail. However, it is a safe prediction that such developmental series will remain rare. Likely, we will be forced to continue to rely mostly on careful examination of tooth crown and vertebral morphology in whatever small specimens are available, to determine the distribution of lissamphibian-like teeth and centra.

"Weakly bicuspid," nonpedicellate marginal teeth occur in some small fossil tetrapods (whether reptiles or amphibians is uncertain) from the Early Permian Fort Sill fissure fills of Oklahoma (Bolt, 1980). However, except for *Doleserpeton*, strongly bicuspid or pedicellate teeth like those of lissamphibians do not occur among the 20 or more species of small tetrapods (many, perhaps most, of which are amphibians) at that locality. This is significant, although hardly conclusive, evidence that such teeth were not widespread among fossil amphibians.

If lissamphibian-like teeth and vertebral centra were not ontogenetically transient within dissorophids but instead represented the adult stage of a separate group of small labyrinthodonts related to dissorophids, their restricted distribution could be taken at face value. This makes it almost attractive to suppose that the ontogenetic hypothesis is wrong, a possibility raised recently by Milner (1982). Milner advanced evidence in favor of the proposition that *Amphibamus grandiceps* from the Mazon Creek localities and *A. lyelli* from Linton are separate species and thus do not form part of a single ontogenetic series as suggested by Bolt (1979). Acceptance of a specific distinction between these two species of *Amphibamus*, however, does not negate the strongest evidence favoring the ontogenetic hypothesis. That is the apparent correlation between skull size and degree of development of bicuspid crown form, plus the co-occurrence of both bicuspid and monocuspid crowns in some specimens. Taken together, these observations seem to indicate continuous variation in crown form from small to large dissorophids. Therefore, I believe that the ontogenetic hypothesis is preferable at this time as an explanation for the distribution of lissamphibian-like teeth and vertebrae within fossil amphibians, although more data would be highly desirable.

IMPLICATIONS OF LISSAMPHIBIAN MONOPHYLY

Keeping in mind the argument for lissamphibian paedomorphosis, one can see that the phylogenetic hypothesis represented by Figure 6 implies that characters of small (juvenile) dissorophids such as *Doleserpeton* and *Tersomius* are primitive for lissamphibians. It is difficult to say at present just how closely other juvenile dissorophids might resemble *Doleserpeton*, but for the sake of discussion it may be assumed that *Doleserpeton* and *Tersomius* are representative. The implications of this assumption for study both of the fossil record of lissamphibians and the development and morphology of the extant lissamphibian taxa then may be examined. Four categories of skeletal characters, two of which have received virtually no attention hitherto, acquire particular interest in view of the hypothesis of monophyly presented here. These are (1) the symphyseal dentition, (2) the palatal dentition, (3) the dorsal quadrate process-tympanic annulus, and (4) the vertebral centrum.

Symphyseal Dentition

A *Doleserpeton*-like symphyseal dentition should be present in primitive lissamphibians. It would consist of perhaps six or fewer pleurodont, bicuspid, pedicellate teeth borne on the dentary. Two of the three living groups have lost the symphyseal dentition. Its presence *on the dentary* in caecilians can reasonably be regarded as primitive, and the fossil record of caecilians should document an increase in symphyseal tooth number toward the condition in those extant species with an extensive inner mandibular tooth row.

Palatal Dentition

It can be predicted that, in primitive fossil lissamphibians, the palatal dentition included pleurodont, pedicellate, bicuspid teeth (with labial and lingual cuspules) arranged in clearly separated short rows—one row on the vomer and one on the palatine. The palatal rows were approximately parallel to the marginal tooth row, and replacement was from the lingual side. Available data unfortunately do not permit discussion of ectopterygoid teeth. Therefore, among extant lissamphibian species, a tooth row that is continuous from vomer to palatine is derived. This condition might be presumed to represent fusion of originally separate vomerine and palatine rows; evidence for intermediate conditions can be sought in the fossil record. In the absence of fossil evidence, interpretation of conditions in living lissamphibian species is difficult. For example, the presence of two separate palatal dental laminae at some ontogenetic stage could represent a primitive condition, as has been suggested for the salamander *Necturus* (Greven and Clemen, 1979). However, there are two "vomerine" laminae in some salamanders, and it is unclear whether these represent original palatine and vomerine laminae, or a subdivision of the vomerine lamina (see Greven and Clemen, 1979, 1985, and references therein).

These suggestions, in focusing attention on the history of the lissamphibian palatal dentition, are intended to be primarily heuristic. The actual history of that dentition might prove more complex, even if further work corroborates the phylogenetic hypothesis advanced here.

Dorsal Quadrate Process–Tympanic Annulus

Acceptance of lissamphibian monophyly implies that the ancestors of salamanders and caecilians must have greatly reduced the middle ear from the condition seen in *Doleserpeton* and other dissorophoids. The process included loss of the otic notch, tympanum, and dorsal quadrate process or its homologue, and reduction and reorientation of the stapes. As regards the stapes and otic notch, the predictions from the dissorophid condition are both obvious and, unfortunately, nonspecific. It would be interesting to learn that, for instance, the caecilian middle ear arose via reduction from a tympanic ear, or that the salamander stapes was derived from a dissorophid-like condition. However, neither of these findings would necessarily point to dissorophids or dissorophoids as the sister-group of the Lissamphibia or any one of its constituent orders, because a tympanum and a dissorophid-like stapes are synapomorphies at a much higher level than the Dissorophoidea (Lombard and Bolt, 1988).

On the other hand, the discovery of a dorsal quadrate process or its homologue, in caecilians or salamanders, would unequivocally imply a sister-group relationship with dissorophoids and would also constitute further evidence for lissamphibian monophyly. The search for evidence of a dorsal quadrate process in fossil lissamphibians should involve careful attention to preparation techniques, because the process may have been cartilaginous and be represented only by an impression. Whether cartilaginous or ossified, however, it will not necessarily be small; some of the larger dissorophids, such as *Cacops*, have a very prominent dorsal quadrate process that is relatively much larger than that of *Doleserpeton* or *Tersomius* (Bolt and Lombard, 1985).

Vertebral Centrum

With *Doleserpeton* and *Tersomius* representing the primitive condition, the lissamphibian vertebral centrum must be hypothesized as the homologue of the pleurocentrum. Given the traditional emphasis on centrum homologies in amphibian systematics, it is unnecessary to urge attention to vertebral morphology of fossil lissamphibians or to vertebral ontogeny of living species. Two less obvious points should be made, however. First, any remnant of the intercentrum in fossils might be represented only by the small impression of an originally cartilaginous structure. As in the case of the dorsal quadrate process, molding and casting technique could be crucial in study of vertebral impressions. Second, the former presence of an intercentrum would not necessarily be reflected in morphology of the pleurocentrum. This has been shown in several microsaurs (see Carroll and Gaskill, 1978), in which the intercentra are thin and lie superficial to the pleurocentra. If such intercentra are not preserved in situ, they can best be demonstrated in detailed impressions.

CONCLUSIONS

Cladistic analysis of lissamphibian relationships supports the hypotheses that (1) Lissamphibia is a monophyletic, likely paedomorphic group, and (2) the sister-group of Lissamphibia is the "labyrinthodont" family Dissorophidae. Such analysis does not support claims of close relationships between the Microsauria and either salamanders or caecilians. The most parsimonious hypothesis of lissamphibian relationships is supported by only a few characters, owing mainly to lack of data from fossil amphibian groups. The most promising types of additional skeletal characters for these groups appear to be the symphyseal and palatal dentition, the dorsal quadrate process–tympanic annulus, and the vertebral centrum. In examinations of fossil specimens for such characters, molding and casting techniques may prove at least as important as observation of preserved skeletal material.

Acknowledgments I am grateful to J. Hopson for a thoughtful review of an earlier version of this paper, to D. Berman and R. Hook for their helpful comments on the manuscript submitted for publication, and to R. Parshall and Z. Jastrzebski for drawings. The University of Kansas provided the travel funds for the symposium series in which this chapter is included. This work was supported in part by the Maurice Richardson Paleontological Fund at Field Museum and was made possible by NSF Collection Support Grants DEB 79-12482 and 82-06982 to W. D. Turnbull and myself.

Literature Cited

Beaumont, E. H. 1977. Cranial morphology of the Loxommatidae (Amphibia: Labyrinthodontia). Philos. Trans. R. Soc. London, Ser. B, 280:29–101.

Beerbower, J. R. 1963. Morphology, paleoecology and phylogeny of the Permo-Pennsylvanian amphibian *Diploceraspis*. Bull. Mus. Comp. Zool Harvard Univ., 130:31–108.

Bolt, J. R. 1969. Lissamphibian origins: possible protolissamphibian from the Lower Permian of Oklahoma. Science, 166:888–891.

Bolt, J. R. 1977. Dissorophoid relationships and ontogeny, and the origin of the Lissamphibia. J. Paleontol., 51:235–249.

Bolt, J. R. 1979. *Amphibamus grandiceps* as a juvenile dissorophid: evidence and implications. Pp. 529–563 *in* Nitecki, M. H. (ed.), *Mazon Creek Fossils*. New York: Academic Press.

Bolt, J. R. 1980. New tetrapods with bicuspid teeth from the Fort Sill locality (Lower Permian, Oklahoma). Neues Jahrb. Geol. Palaeontol. Monatsh., 1980(8):449–459.

Bolt, J. R., and R. E. Lombard. 1985. Evolution of the amphibian tympanic ear and the origin of frogs. Biol. J. Linn. Soc., 24:83–99.

Bolt, J. R., and R. J. Wassersug. 1975. Functional morphology of the skull in *Lysorophus*: a snake-like Paleozoic amphibian (Lepospondyli). Paleobiology, 1:320–332.

Bystrow, A. P. 1939. Zahnstruktur der Crossopterygier. Acta Zool., 20:283–338.

Carroll, R. L., and P. J. Currie. 1975. Microsaurs as possible apodan ancestors. J. Linn. Soc. London Zool., 57:229–247.

Carroll, R. L., and P. Gaskill. 1978. The Order Microsauria. Mem. Am. Philos. Soc., 126:1–211.

Carroll, R. L., and R. Holmes. 1980. The skull and jaw musculature as guides to the ancestry of salamanders. J. Linn. Soc. London Zool., 68:1–40.

Case, E. C. 1931. Description of a new species of *Buettneria* with a discussion of the braincase. Contrib. Mus. Paleontol. Univ. Michigan, 3:187–286.

Chowdhury, T. R. 1965. A new metoposaurid amphibian from the Upper Triassic Maleri Formation of central India. Phil. Tran. R. Soc. London, Ser. B, 250:1–52.

Daly, E. 1973. A Lower Permian vertebrate fauna from southern Oklahoma. J. Paleontol., 47:562–589.

Eaton, T. H., Jr. 1933. The occurrence of streptostyly in the Ambystomatidae. Univ. California Publ. Zool., 37:521–526.

Eaton, T. H., Jr. 1959. The ancestry of the modern Amphibia: a review of the evidence. Univ. Kansas Mus. Nat. Hist. Publ., 12:155–180.

Eaton, T. H., Jr. 1973. A Pennsylvanian dissorophid amphibian from Kansas. Univ. Kansas Mus. Nat. Hist. Occas. Pap., 14:1–8.

Eldredge, N., and J. Cracraft. 1980. *Phylogenetic Patterns and the Evolutionary Process*. New York: Columbia Univ. Press.

Estes, R. 1981. Teil 2. Gymnophiona, Caudata. *Handbuch der Paläoherpetologie*. Stuttgart: Gustav Fischer Verlag.

Estes, R., and O. Reig. 1973. The early fossil record of frogs: a review of the evidence. Pp. 11–63 *in* Vial, J. L. (ed.), *Evolutionary Biology of the Anurans*. Columbia: Univ. Missouri Press.

Estes, R., and M. H. Wake. 1973. The first fossil record of caecilian amphibians. Nature, London, 239:228–231.

Gaffney, E. S. 1979. Tetrapod monophyly: a phylogenetic analysis. Bull. Carnegie Mus. Nat. Hist., 13:92–105.

Gardiner, B. G. 1983. Gnathostome vertebrae and the classification of the Amphibia. J. Linn. Soc. London Zool., 79:1–59.

Goodrich, E. S. 1930. *Studies on the Structure and Development of Vertebrates*. London: Macmillan.

Gregory, J. T., F. E. Peabody, and L. I. Price. 1956. Revision of the Gymnarthridae: American Permian microsaurs. Peabody Mus. Nat. Hist. Yale Univ. Bull., 10:1–77.

Greven, H., and G. Clemen. 1979. Morphological studies on the mouth cavity of urodeles. IV. The teeth of the upper jaw and palate in *Necturus maculosus* (Rafinesque) (Proteidae: Amphibia). Arch. Histol. Japan (Niigata, Japan), 42:445–457.

Greven, H., and G. Clemen. 1985. Morphological studies on the mouth cavity of urodeles. VIII. The teeth of the upper jaw and the palate in two *Hynobius*-species (Hynobiidae: Amphibia). Z. Zool. Syst. Evolutionsforsch, 23:136–147.

Griffiths, I. 1963. The phylogeny of the Salientia. Biol. Rev. Cambridge Philos. Soc., 38:241–292.

Hecht, M. K. 1962. A reevaluation of the early history of the frogs. Part I. Syst. Zool., 11:39–44.

Holmes, R. B. 1984. The Carboniferous amphibian *Proterogyrinus scheelei* Romer and the early evolution of tetrapods. Phil. Tran. R. Soc. London, Ser. B, 306:431–524.

Hook, R. W. 1983. *Colosteus scutellatus* (Newberry), a primitive temnospondyl amphibian from the Middle Pennsylvanian of Linton, Ohio. Am. Mus. Nat. Hist. Novit., 2770:1–41.

Ivachnenko, M. F. 1978. Urodelens from the Triassic and Jurassic of Soviet Central Asia. Paleontol. Zh., 12:362–368.

Jarvik, E. 1954. On the visceral skeleton in *Eusthenopteron* with a discussion of the parasphenoid and palatoquadrate in fishes. K. Sven. Vetenskapsakad. Handl., 5:1–104.

Jarvik, E. 1980. *Basic Structure and Evolution of Vertebrates*. Vol. 1. London: Academic Press.

Jarvik, E. 1981. *Basic Structure and Evolution of Vertebrates*. Vol. 2. London: Academic Press.

Jupp, R., and A. A. Warren. 1986. The mandibles of the Triassic temnospondyl amphibians. Alcheringa, 10:99–124.

Langston, W., Jr., and E. C. Olson. 1986. *Carrolla craddocki*, a new genus and species of microsaur from the Lower Permian of Texas. Pearce-Sellards Ser. Texas Mem. Mus., 43:1–20.

Lehman, J.-P. 1968. Remarques concernant la phylogénie des amphibiens. Nobel Symposium, 4th, Stockholm, 1967, Proceedings, 1968:307–315.

Lombard, R. E. 1977. Comparative morphology of the inner ear in salamanders (Caudata: Amphibia). Pp. 1–143 *in* Hecht, M. K., and F. S. Szalay (eds.), *Contributions to Vertebrate Evolution*. Vol. 2. Basel: S. Karger.

Lombard, R. E., and J. R. Bolt. 1979. Evolution of the tetrapod ear: an analysis and reinterpretation. Biol. J. Linn. Soc. London, 11:19–76.

Lombard, R. E., and J. R. Bolt. 1988. Evolution of the stapes in Paleozoic tetrapods: conservative and radical hypotheses. Pp. 37–67 *in* Fritzsch, B., M. J. Ryan, W. Wilczynski, T. E. Hetherington, and W. Walkoviac (eds.), *The Evolution of the Amphibian Auditory System*. New York: John Wiley and Sons.

Lund, R. 1978. Anatomy and relationships of the Family Phlegethontiidae (Amphibia, Aïstopoda). Ann. Carnegie Mus., 47:53–79.

Marcus, H., E. Stimmelmayr, and G. Porsch. 1935. Beiträge zur Kenntnis der Gymnophionen. XXV. Die Ossifikation des Hypogeophisschädels. Morpholog. Jahrb., 76:375–420.

McGinnis, H. J. 1967. The osteology of *Phlegethontia*, a Carboniferous and Permian aïstopod amphibian. Univ. California Publ. Geol. Sci., 71:1–49.

Milner, A. R. 1982. Small temnospondyl amphibians from the Middle Pennsylvanian of Illinois. Palaeontology, 25:635–664.

Nussbaum, R. A. 1977. Rhinatrematidae: a new family of caecilians (Amphibia: Gymnophiona). Occ. Pap. Mus. Zool. Univ. Michigan, 682:1–30.

Nussbaum, R. A. 1983. The evolution of a unique jaw-closing mechanism in caecilians (Amphibia: Gymnophiona) and its bearing on caecilian ancestry. J. Zool. London, 199:545–554.

Panchen, A. L. 1972. The skull and skeleton of *Eogyrinus attheyi* Watson (Amphibia: Labyrinthodontia). Philos. Tran. R. Soc. London, Ser. B, 263:279–326.

Parsons, T. S., and E. E. Williams. 1962. The teeth of Amphibia and their relation to amphibian phylogeny. J. Morphol., 110:375–389.

Parsons, T. S., and E. E. Williams. 1963. The relationships of the modern Amphibia: a re-examination. Q. Rev. Biol., 38:26–53.

Piveteau, J. 1937. Un amphibien du Trias Inférieur. Essai sur l'origine et l'évolution des amphibiens anoures. Ann. Paleontol., 26:135–177.

Schultze, H.-P., and M. Arsenault. 1985. The panderichthyid fish *Elpistostege:* a close relative of tetrapods? Palaeontology, 28:293–309.

Smithson, T. R. 1982. The cranial morphology of *Greererpeton burkemorani* Romer (Amphibia: Temnospondyli). J. Linn. Soc. London Zool., 76:29–90.

Wake, D. B. 1970. Aspects of vertebral evolution in the modern Amphibia. Forma et Functio, 3:33–60.

Wake, D. B., and R. Lawson. 1973. Developmental and adult morphology of the vertebral column in the plethodontid salamander *Eurycea bislineata,* with comments on vertebral evolution in the Amphibia. J. Morphol., 139:251–300.

Wake, D. B., and M. H. Wake. 1986. On the development of vertebrae in gymnophione amphibians. Mem. Soc. Zool. France, 43:67–70.

Wake, M. H., and J. Hanken. 1982. Development of the skull of *Dermophis mexicanus* (Amphibia: Gymnophiona), with comments on skull kinesis and amphibian relationships. J. Morphol., 173:203–223.

Warren, A. A., and M. N. Hutchinson. 1983. The last labyrinthodont? A new brachyopoid (Amphibia, Temnospondyli) from the early Jurassic Evergreen formation of Queensland, Australia. Philos. Trans. R. Soc. London, Ser. B, 303:1–62.

Welles, S. P., and R. Estes. 1969. *Hadrokkosaurus bradyi* from the Upper Moenkopi Formation of Arizona. Univ. California Publ. Geol. Sci., 84:1–56.

Wiley, E. O. 1981. *Phylogenetics: The Theory and Practice of Phylogenetic Systematics.* New York: John Wiley and Sons.

Williams, E. E. 1959. Gadow's arcualia and the development of tetrapod vertebrae. Q. Rev. Biol., 34:1–32.

8 A Phylogenetic Investigation of the Inter- and Intrarelationships of the Lissamphibia (Amphibia: Temnospondyli)

Linda Trueb and Richard Cloutier

The roots of this chapter are many and far-reaching. One of us (L.T.) has been involved in the study of anuran osteology for the past quarter century and more recently has addressed the soft-anatomical and osteological features of the other living amphibians as part of a project that culminated in the publication of a book on amphibian biology (Duellman and Trueb, 1986). The process of accumulating and attempting to distill information for that volume revealed the distressing fact that there are a great many critical gaps in our knowledge about Recent taxa, let alone fossil forms. On the other hand, there existed data that had not been considered in assessments of relationships of living and fossil groups. Moreover, a refined cladistic analysis of amphibians had yet to be undertaken. These facts provided the impetus for the first author to accept an invitation to present a modest paper on the monophyly and interrelationships of living amphibians as part of the seminar series described in the Preface.

The reality of attempting to prepare a manuscript was a humbling and enlightening experience for the first author. If significant progress were to be made in analysis of the relationships of living amphibians, it became apparent that one would have to bridge the gap to fossil taxa. Once this step was taken, "levels of universality" became an issue. Thus, the seemingly simple question of the monophyly and interrelationships of living amphibians evolved into a complex series of analyses involving neontological and paleontological data derived from actinopterygians,

223

actinistians, dipnoans, amphibians, and amniotes—clearly more than one person could handle with confidence and expertise. This chapter represents a truly cooperative effort from the combined neontological (L.T.) and paleontological and methodological (R.C.) perspectives of the authors. We do not profess to have solved all the problems to our satisfaction. There remain many missing data and unresolved questions of homology. Nonetheless, we present an hypothesis with the intention and hope that our results will be challenged and thereby stimulate further research efforts into relationships of this fascinating group of lower tetrapods.

In the first section (Materials and Methods), we provide an explanation of our methodology, character determination and coding, rationale for ingroup and out-group selection, notes dealing with nomenclature, and definitions of specialized terms used in the text. A section follows on the phylogenetic status of the various taxa that were analyzed. The Analysis of Characters contains the descriptions of 72 soft-anatomical characters (Data Set I) and 58 osteological features (Data Set II). Data matrices are presented in Appendices I and II, and a list of specimens examined specifically for this study appears in Appendix III. We caution the reader that our descriptions are, of necessity, brief. Citations are limited to those kinds of information that are not readily accessible or secondarily cited in general reference books (e.g., comparative anatomy texts) or summary volumes such as Duellman and Trueb (1986) or Carroll (1988). Much of the primary literature consulted for specific taxa (e.g., *Latimeria*) is cited in the section concerned with the status of the taxa analyzed. The cladograms that resulted from our analyses are presented in the Results, which is followed by our taxonomic conclusions and consideration of some of the more interesting implications of the phylogenetic arrangements in the Discussion and Conclusions.

MATERIALS AND METHODS

General Methodology

This study is based on phylogenetic principles and methodology throughout (sensu Hennig, 1966). The analyses in this study were conducted using PAUP (Phylogenetic Analysis Using Parsimony) software (version 2.4) of David L. Swofford as implemented by the Academic Computing Center at The University of Kansas. Operationally, parsimony (defined to be the minimum number of evolutionary steps to construct a tree[s]—i.e., the one or more topologies with the highest consistency index) was used to determine interrelationships among monophyletic units and was the primary criterion used to select among phylogenetic hypotheses (or trees).

This chapter investigates two questions—(1) the monophyly of lissamphibians (i.e., Anura, Caudata, and Gymnophiona), and (2) the interrelationships of these taxa with fossil amphibians. Traditionally, amphibian systematists have concentrated on either fossil or Recent taxa, but not both simultaneously (but see Gauthier et al., 1988a). The reasons for this are several. Few systematists are equally competent in the areas of neontology and paleontology. The nature of fossils usually precludes deriving soft-anatomical data (in the broad sense) from them; therefore, relevant comparisons about soft anatomy, biochemical, and ontogenetic features rarely can be made between the two groups. As pointed out by Maisey (1988) and Gauthier et al. (1988a), neontological systematists are disadvantaged because they are much less likely to discover phylogenetically intermediate taxa than are paleontologists; moreover, data derived from Recent taxa lack a geological time component useful in the determination of minimum dates of taxonomic divergence, as well as biogeographic and other evolutionary events (Paul, 1982; Grande, 1985). Maisey (1986) also mentioned that paleontologists who exclude fossil taxa from their analyses risk ascribing too much significance to putative synapomorphies of ancient higher taxa.

It is obvious that in order to obtain the best estimation of phylogeny, one should seek to maximize the numbers of relevant taxa and characters included in the analysis. Moreover, as Hillis (1987) and Kluge (1989) pointed out, the best phylogenetic reconstruction is derived from analysis of combined data sets. In the present analysis, we deal with two kinds of characters (i.e., soft-anatomical and osteological) as well as two kinds of taxa (i.e., extinct and extant). As a consequence, several methodological problems arise (Table 1). For example, osteological characters can be coded for both extant and extinct taxa, but, in general, soft-anatomical characters can be coded only for extant taxa. In order to polarize soft-

Table 1. Kinds of specimens analyzed with respect to relative levels of universality and limiting methodological factors

Kinds of specimens	Data	Number of taxa	Out-groups	Relative level of universality
Living and fossil	Hard (*reduced*)	Maximum	Less remote	Lower
Fossil	Hard (*reduced*)	*Reduced intermediate*	Less remote	Lower
Living	Hard and soft (maximum)	*Reduced*	*Most remote*	Highest

Note: Factors limiting the analyses are in italics.

anatomical characters, one must resort to remote extant out-groups; in contrast, osteological characters can be coded from more closely related extinct out-groups. Consequently, the use of soft-anatomical features is a limiting factor on the analysis in terms of the availability of comparable structures and the restricted number of out-groups that, phylogenetically, are related only remotely to the in-groups. Thus, if one combines data sets (when dealing with a high level of universality encompassing extant and extinct taxa), the use of soft-anatomical characters imposes a constraint on the analysis that potentially can reduce accuracy in interpretation of osteological characters and determination of the polarity of these characters (Table 1).

Patterson (1987), Hillis (1987), and Kluge (1989) emphasized that phylogenetic reconstruction should treat congruence between different kinds of taxa (e.g., extinct vs. extant), as well as congruence among different kinds of characters (e.g., soft-anatomical vs. osteological, morphological vs. biochemical). Kluge (1989) pointed out that character congruence (i.e., patterns among characters) is more important than taxonomic congruence (i.e., patterns among taxa). Both issues must be considered with respect to the interrelationships of lissamphibians.

To maximize the information content of our analysis and to have a better understanding and interpretation of character congruence with respect to the lissamphibians, we have performed a "tandem analysis" at "high" (fossil and extant osteichthyans) and "low" (fossil and extant amphibians) levels of universality (Tables 1–2). Thus, living and fossil taxa were analyzed separately (or in tandem) to assess character congruence. Subsequently, taxonomic congruence was assessed. In this way we were able to contrast character congruence and taxonomic congruence (Fig. 1). First, character congruence was investigated within individual analyses—i.e., among extant taxa the congruence of (1) soft-anatomical and

Table 2. Three analyses involved in the tandem analysis of the interrelationships of the Lissamphibia

Analyses	Characters	Taxa	Out-groups
I	Soft	Living	Actinistia, Actinopterygii
II	Soft and osteological	Living	Actinistia, Actinopterygii
III	Osteological	Fossil and living	"Temnospondyls," Loxommatoidea

Note: The order of out-groups corresponds to first and second taxonomic out-groups, respectively.

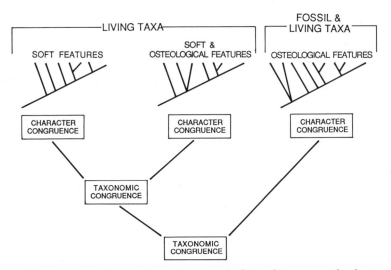

Fig. 1. The structure of the tandem analysis, and relationships among the three analyses performed (see Table 2) and the levels of taxonomic and character congruence investigated. Analysis I involved living taxa and soft features, Analysis II living taxa and soft and osteological features, and Analysis III fossil and living taxa and osteological features.

osteological characters, and (2) soft-anatomical characters, and among fossil taxa, the congruence of osteological characters. Second, congruence between taxonomic patterns derived from soft-anatomical and combined soft-anatomical and osteological data sets of living taxa was examined. Third and last, taxonomic congruence between living and fossil taxa was examined.

Taxa Set I represents only living organisms, whereas Taxa Set II represents both fossil and living organisms. Three different sets of characters were employed to analyze the two groups of taxa. Data Set I consists of 72 soft-morphological features, whereas Data Set II comprises 58 osteological characters. A third data set represents a combination of Data Set I and a subset of Data Set II (i.e., 72 soft-anatomical features plus 50 osteological characters). Each data set was analyzed with different sets of outgroups.

Consensus trees (strict and Adams) were used to summarize the topology of numerous equally parsimonious trees (Adams, 1972; Nelson, 1979). Analysis of consensus trees was accomplished by CONTREE (Consensus Tree Program) of David L. Swofford. Unless otherwise stated, the PAUP analyses were executed with the following set of commands: ROOT = ANCESTOR, SWAP = GLOBAL, MAXTREE = 1000, OPTION = FARRIS, and MULPARS ON. The BANDB (Branch and Bound) option was used for Analyses I and II because of the reduced numbers of OTUs (Operational Taxonomic Units). The ANCESTOR was coded as "9" for unpolarized·characters.

Character Determination

All characters are unweighted and considered to be independent of one another. Interdependence of characters or the occurrence of complex characters are inferred from the phylogenetic pattern(s) that emerged (Kluge and Farris, 1969). The observed characters were coded as binary-character transformation series whenever possible. Multiple-state characters were run unordered (i.e., without assumption of a transformation series), because transformation series can be inferred objectively only after the pattern is determined. Characters and character-states are defined below (see Analysis of Characters).

Nonosteological characters (e.g., biochemical, chromosomal, developmental, and soft-anatomical features) were selected primarily from the literature and are referred to herein as "soft-morphological" or "soft-anatomical" characters. Thus, we surveyed general texts, systematic works, and basic morphological treatises. In the case of conflicting or contradictory information, specimens were examined to confirm the state of a character; determination of character-states of fossil taxa was based largely on examination of specimens. One suite of osteological characters was chosen a priori without examination of specimens. These included the presence or absence of elements, and the existence of sutures or fusions between bones. Because we do not consider loss of elements to be of any less evolutionary significance or phylogenetic value than appearance of novelties, absence of characters is accorded equivalent weight to presence of characters contra Hecht and Edwards (1976, 1977) and Hecht (1976). Other osteological characters became apparent after observation of the specimens. For example, bony processes may be present or absent, and the same elements may have various orientations and configurations.

In no case was character coding reconstructed in the data sets. Missing characters are coded by various letters in the data sets to specify their type, although all such characters were coded as 9 in the data sets run on PAUP. N denotes "nonapplicable," where the organism does not possess the morphological structure analyzed or the character does not apply owing to a logical conflict with other character definition(s). For example, if the character describes the presence or absence of contact between Bone A and B, and Bone B is not present in the taxon, the character would be coded as N. U denotes "unclear" and is used to designate taxa in which the state of the character could not be ascertained owing to the state of preservation of the organism.

Character Coding

The character codes reported in the data sets were determined in several ways. When possible, we examined specimens (see Appendix III: Specimens Examined). Otherwise, we relied on published illustrations of specimens or detailed descriptions available in the literature. Much of the information is summarized in general reference books such as comparative anatomy texts, or in the case of living amphibians, in Duellman and Trueb, 1986. Such information is considered generally available and no specific citations have been made to a source. The sources of more specialized, recent, or less widely disseminated information are cited. Unless specified otherwise, character coding corresponds to the condition found in adult rather than subadult organisms, in which there might be ontogenetic variation in the character-states. The state of maturity of fossils was determined, insofar as possible, on the basis of the relative size and degree of ossification of the specimen.

In order to use PAUP more efficiently, we reduced the number of OTUs. This was accomplished by constructing a hypothetical coding for some clades, such that conditions were coded for either the basal taxa(on) or the stem-species—i.e., relatively primitive member(s) or a hypothetical ancestor having the minimum number of apomorphic conditions that have been derived within the clade. Whenever possible, previous phylogenetic studies were used to determine the coding; theoretically, this procedure minimizes the possibility of coding an apomorphic condition derived within the clade as a "basal" condition of the clade (Cloutier and Ford, in prep.). For example, "hynobiid" and cryptobranchid salamanders represent the basal members of the Caudata (Cloutier, in prep.), whereas ascaphid and leiopelmatid frogs are the basal anuran families (fide Cannatella, 1985). Coding at the generic level represents the condition hypothesized for the clade on the basis of a consensus among the species. Coding at the familial level is based on the hypothesized plesiomorphic conditions in the monophyletic assemblage. Whenever the plesiomorphic and apomorphic conditions were found in a taxon, the plesiomorphic was used in the PAUP analysis unless additional phylogenetic information was available to justify coding the group with the apomorphic condition (the assumption being that a reversal had occurred within the in-group).

In Analysis II, osteological characters were coded on the basal taxa of each group having extant representatives. For example, within the Actinistia, *Latimeria chalumnae* was used to code the soft-anatomical characters, whereas the Late Devonian *Miguashaia bureaui* was used to code the osteological features.

Taxa (In-Group and Out-Group) Selection

To be consistent with the cladistic methodology, every higher taxonomic unit used in the analysis must be monophyletic (Wiley, 1981). Monophyly is evidenced by the possession of one or more unique shared-derived characters (i.e., synapomorphies) that occur at the base of the clade; it should be remembered, however, that derived members within a clade may possess conditions derived from the basal condition. A group of organisms that has been recognized as a taxonomic unit but that lacks a synapomorphy supporting its monophyly must be divided into clades (i.e., monophyletic units) that then represent lower taxonomic units. Supraspecific taxa are included in the data matrix; however, we agree with Wiley (1979a) that the species is the evolutionary unit.

Each group that has been considered in the literature to be a sister-group or ancestor of the Lissamphibia (see Appendices II–IV in Trueb and Cloutier, this volume), or parts of the Lissamphibia, has been treated in various analyses. The groups were analyzed at different levels of universality, thereby allowing us to justify the exclusion of most of them. A phylogenetic investigation of the major monophyletic groups of fossil lower Tetrapoda at a high level of universality was carried out systematically in order to justify the exclusion of previously proposed sister-group relationships and to support the selection of alternate sister-groups. Successive embedded analyses from higher to lower levels of universality were used to determine sister-group relationships from which out-group relationships were inferred.

Out-Group Comparison

Out-group comparison is used to polarize the characters. More than one out-group is necessary to polarize characters in order to reduce the possibility of incorrectly assuming that a synapomorphy of the first out-group is plesiomorphic (Maddison et al., 1984; Waltrous and Wheeler, 1981).

In order to polarize the characters and to infer phylogenetic interrelationships, we used two sets of out-groups. Out-group Set I involves living taxa with (1) soft-anatomical characters and (2) soft-anatomical and osteological features, whereas Out-group Set II includes fossil and living taxa and osteological characters. The choice of out-groups to polarize nonosteological characters is restricted to living taxa because of the information constraints encountered in dealing with fossils. This parallel, or tandem, analytical approach has been adopted in order to minimize the effects of homoplasy primarily owing to convergence in independent groups having living representatives. It also has the advantage of maximizing the accuracy of determination of character polarities by utilizing a greater number of out-groups that potentially may be closely related.

Nomenclature

A system of brackets, mathematical symbols $(+,/,-)$, and quotation marks is used to summarize phylogenetic relationships among groups. The phylogenetic information corresponds to the topology of the tree. For example, the synoptic cladistic statement "[A + [B + [C − D]]]" means Taxon A is the sister-group of the Sister Taxa B and C, with Taxon C being defined as all taxa traditionally included in Taxon C minus Taxon D. The slash (/) is used to represent equally parsimonious hypotheses of relationships. Taxa belonging to a clade in which the interrelationships are unresolved, thereby creating a polytomy, are listed with addition signs between each member; in this case, no brackets appear to delimit sister-group relationships other than the two clustering brackets that embrace the entire clade. Quotation marks enclose names of taxa that are considered to be paraphyletic as they presently are defined in the literature (Wiley, 1981).

Classification must be consistent with phylogeny (Wiley, 1981). Modification of existing classifications is made only if it is substantiated by well-corroborated branching patterns. Nomenclatorial rules follow Wiley (1979b).

Definitions

Throughout the text certain terms are used in specific ways. These terms along with our definitions of them are listed below.

Basal taxon: The basal taxon of a clade shares the synapomorphies that diagnose the clade but lacks the synapomorphies derived within the in-group.

Bracing bone: A bracing bone is one or more of a number of dermal palatal components (e.g., ectopterygoid, palatine, pterygoid) that singly or together act to support the upper jaw against the braincase.

Characterization (definition, diagnosis, and description): We use these terms as defined by de Queiroz (1987), Rowe (1987), and Cloutier (1991), wherein the characterization of a group comprises the definition, diagnosis, and description. A group is *defined* by the taxa of which it is composed (i.e., by its ancestor and all members sharing this common ancestor). A group is *diagnosed* by its synapomorphies. A group is *described* by a list of apomorphies, plesiomorphies, characters of uncertain polarities, and variable characters.

Endocranium: The endocranium is defined as those parts of the skull that are endochondral in origin. These include the rostral cartilages, planum antorbitale, maxillary cartilages, Meckel's cartilage, the palato-quadrate and its associated processes, the otic capsules, and the brain-case proper.

Exocranium: The exocranium includes all bones of dermal origin that invest the endocranial components.

Maxillary arcade: The maxillary arcade includes all components of the upper jaw—the premaxillae, maxillae, quadratojugals, and occasionally, the jugal.

Snout and Rostrum: The snout is that part of the skull anterior to the orbits, whereas the rostrum is restricted to that part of the skull including, and anterior to, the external nares.

Topological variant: Equally parsimonious trees (i.e., trees having the same number of steps and consistency index) based on the same data sets and taxa result from variation in the position of taxa and the distribution of characters. In many cases, the distributions of the taxa (i.e., topologies) on these trees are identical. Thus, equally parsimonious trees can be reduced to the minimum number of different topologies that do not account for variation in the patterns of character distribution. Topologically, the trees can be reduced further to a series of "topological variants"—i.e., subsets or components of the trees that represent the fundamental source(s) of topological variation.

THE PHYLOGENETIC STATUS OF VARIOUS TAXA ANALYZED

The taxa included as in-groups and out-groups in this study are listed in Table 3. If interrelationships among taxa are to be established, it is requisite that each taxon analyzed (in-group and out-group) be monophyletic. The monophyly of a taxon is determined by the presence of shared-derived character(s). For taxa that include fossil as well as living representatives, it is necessary to determine monophyly on the basis of osteological features that can be assessed in both groups. The phylogenetic status (i.e., monophyly vs. paraphyly) for each taxon included in the dual analyses is discussed, and opposing views are presented. Whenever possible, we have provided brief evaluations of synapomorphies that have been proposed in the literature to support a particular clade, and the representative member(s) used for coding is noted. Because this prelimi-

Table 3. Fossil and living amphibian taxa used in Analysis III

Subclass Labyrinthodontia
  Order *incertae sedis*
    Suborder *incertae sedis*
      Superfamily Loxommatoidea
        Family Loxommatidae: *Megalocephalus, Loxomma, Baphetes*
  Order Temnospondyli
    Suborder Rhachitomi
      Superfamily *incertae sedis*
        Family Dendrerpetontidae: *Dendrerpeton*
      Superfamily Edopoidea
        Family Edopidae: *Edops*
      Superfamily Eryopoidea
        Family Eryopidae: *Eryops*
      Superfamily Dissorophoidea
        Family Dissorophidae: *Dissorophus, Tersomius, Broiliellus, Amphibamus*
        Family Doleserpetontidae: *Doleserpeton*
        Family Micromelerpetontidae: *Micromelerpeton*
        Family Branchiosauridae: *Branchiosaurus, Apateon, Leptorophus, Schoenfelderpeton*
        Family Trematopidae: *Trematops, Actiobates*
      Superfamily Trimerorhachoidea
        Family Trimerorhachidae: Stem-species
Subclass Lissamphibia
  Superorder Salientia
    Order Proanura
        Family Protobatrachidae: *Triadobatrachus*
    Order Anura
      Suborder Archaeobatrachia
      Superfamily Discoglossoidea
        Family Leiopelmatidae: *Ascaphus*
  Superorder Urodela
    Order Caudata
      Suborder Karauroidea
        Family Karauridae: *Karaurus*
      Suborder Cryptobranchoidea
        Family Cryptobranchidae: *Cryptobranchus*
        Family Hynobiidae: Stem-species
      Suborder Salamandroidea
        Family Prosirenidae: *Albanerpeton*

Source: Classifications from Carroll, 1988; Duellman and Trueb, 1986; and Bolt, 1974b.

nary analysis is necessarily brief, criticisms of questionable characters are not elaborated upon extensively.

Lissamphibian In-Groups

The taxa included as in-groups are the Anura (= Salientia, = Ecaudata of some authors), Caudata (= Urodela of some authors), and Gymnophiona (= Apoda of some authors), and the "Kayenta fossil caecilian"

(Jenkins and Walsh, in prep.). The phylogenetic positions of *Albanerpeton* spp., *Karaurus sharovi* (Middle and late Jurassic salamanders), and *Triadobatrachus massinoti* (early Triassic anuran) among the three living orders of lissamphibians are problematic, because their phylogenetic positions depend on the taxonomic definition and diagnosis of the three living orders. Owing to the uncertainty of the phylogenetic definition of the living orders, each of these three genera was included as individual taxon in the analysis.

Anura

The monophyly of anurans has remained unquestioned (save for Roček, 1981), probably owing to their unique structure. Anurans have been characterized by a suite of readily recognized features, including the low number of presacral vertebrae, the lack of caudal vertebrae and the presence of a urostyle, the presence of hind limbs that are longer than the forelimbs, and modified pectoral and pelvic girdles, among other features. Three characters have been repeatedly considered as fundamental anuran features (Griffiths, 1963): (1) large frontoparietal, (2) anteriorly elongated ilium, and (3) elongation of the proximal tarsals.

Saint-Aubain (1981) based her phylogeny of the Lissamphibia on limb characters. Anuran synapomorphies included the following fusions: (1) ulnare and intermedium; (2) radiale with the proximal preaxial centrale; (3) Distal Carpale IV with the proximal postaxial centrale; (4) naviculare with distal preaxial centrale; and (5) radius and ulna. Despite the fact that this is a reasonably satisfactory suite of derived characters, some of them vary among living taxa (e.g., Characters 1, 2; Jarosova, 1973; Andersen, 1978) and exclude some fossils forms (e.g., *Notobatrachus;* Characters 1, 2) from the Anura.

Saint-Aubain (1985) reinvestigated the monophyly of the living Anura based on four synapomorphies related to the circulatory and respiratory systems. These characters are (1) the enlargement of the cutaneous artery of the ductus arteriosus, (2) the "disappearance" of the segment of ductus arteriosus between the cutaneous artery and the dorsal aorta, (3) the reduction of vascularization of larval skin, and (4) the increase in the complexity of the carotid labyrinth. However, Characters 1, 2, and 4 are relative statements with respect to the Caudata; therefore, the polarities of the characters are questionable.

A. R. Milner (1988) discussed the monophyly of the Salientia (i.e., [*Triadobatrachus* + Anura]) and the Anura. He listed four diagnostic features possessed by the Salientia—(1) the frontoparietal, (2) toothless dentary (except in the hylid, *Gastrotheca guentheri*), (3) 14 or fewer presacral vertebrae, and (4) elongated, anteriorly directed ilium.

We have accepted the phylogeny of archaeobatrachian anurans proposed by Cannatella (1985). Thus, the conditions found in the basal taxa Ascaphidae, Leiopelmatidae, Bombinatoridae, and Discoglossidae (fide Cannatella, 1985) were hypothesized to be plesiomorphic for the Anura. Unfortunately, for various nonosteological characters, information is available for only a limited number of taxa—often species belonging to derived families of neobatrachians such as the ranids.

Although *Triadobatrachus massinoti* from the early Triassic of Madagascar is thought to be closely related to anurans, its phylogenetic status is uncertain; it has been hypothesized to be the ancestor of the Anura (Piveteau, 1937; Watson, 1941; Griffiths, 1963), the link between dissorophids and primitive frogs (Carroll, 1988), the sister-group of the Anura (Estes and Reig, 1973; Boy, 1981), or a primitive anuran of uncertain phylogenetic position but closely related to the ancestor of the Anura (Roček, 1989; Rage and Roček, 1989). We have elected to treat *Triadobatrachus* as a separate taxon in this analysis and coded our characters on the basis of the detailed description provided by Rage and Roček (1989).

Caudata

Unlike anurans, salamanders are not easily characterized as a monophyletic assemblage, although previous workers have assumed their monophyly. Rage and Janvier (1982) considered the monophyly of the Caudata questionable because they did not find a shared-derived character unique to the group, although they suggested that the presence of a spermatophore[1] might be considered a caudate synapomorphy. However, male cryptobranchids do not deposit spermatophores, and the single "hynobiid" reported to deposit a spermatophore is *Ranodon sibiricus*, in which the female presumably deposits eggs on a spermatophore that previously was deposited by a male (Bannikov, 1958). The mode of sperm transfer in sirenids is unknown, but as pointed out by Hecht and Edwards (1977), the urogenital morphology of male sirenids does not suggest that they produce spermatophores. Thus, sperm transfer via spermatophores seems to represent a condition that is derived within salamanders, and not a feature that can be considered a synapomorphy of the order (Sever, 1987).

Estes (1981) listed a series of derived characters shared by the Caudata, as follow. (1) The otic notch is absent; this feature may be considered derived if it is accepted that urodeles arose from labyrinthodonts that possessed an otic notch. (2) Associated with loss of otic notch is the absence of a middle ear. (3) The stapes possesses a large footplate and (4)

[1]A gelatinous mass secreted by the cloacal glands of some male salamanders; the spermatophore consists of a cone of jelly with a sperm cap on top. The spermatophore is deposited by male salamanders; subsequently, the cap of the spermatophore is picked up by female salamanders of most families (Peters, 1964).

a small (i.e., short) stylus. (5) The m. adductor mandibulae internus superficialis extends far posterior, originating on the exoccipital (or in some cases, the cervical vertebra). (6) The m. adductor mandibulae posterior is small and poorly differentiated. (7) The postorbital, (8) jugal, (9) postfrontal, (10) postparietal, (11) tabular, (12) supratemporal, (13) supraoccipital, (14) basioccipital, (15) ectopterygoid, and (16) quadratojugal bones absent. (17) The m. interhyoideus posterior is present and probably functions in adduction of the suspensorium. (18) The vertebrae have bicipital rib-bearers. (19) The vertebrae bear basapophyses in many groups. Estes (1981) noted that in *Karaurus* the m. adductor mandibulae internus superficialis did not extend far posterior (Character 5), and that a quadratojugal (Character 9) may have been present in the latter taxon and is present in Group II salamandrids.[2]

Unfortunately, most of Estes' (1981) characters contain little information relevant to an assessment of the monophyly of salamanders. Characters 1, 9–12, 14, and 18 are plesiomorphic, and Characters 2–4, 7, 8, and 15 are shared with other lissamphibian groups. The quadratojugal (Character 16) is present in some "hynobiids" (Cloutier, in prep.). Character 5 may represent a synapomorphy for all salamanders with the exception of the upper Jurassic fossil *Karaurus sharovi* Ivachnenko, 1978. Because some characters seem to support the monophyly of the Caudata with the exclusion of *Karaurus*, this genus is incorporated in the analysis as an individual taxon.

A. R. Milner (1988) listed five synapomorphies of the Caudata: (1) bicapital rib-bearers and basapophyses present on vertebrae; (2) a tuberculum interglenoideum present; (3) a stapes bearing large footplate and short, slender stylus; (4) the Basibranchial I bearing radial horns in metamorphosed individuals; and (5) the possession of a co-ossified scapulocoracoid. A large footplate also is known in the Kayenta fossil caecilian (Cloutier, pers. observ.), whereas a short stylus occurs in primitive fossil anurans. The radial "horns" referred to in Character 4 correspond to the hypohyals, the presence of which is not unique to the Caudata; as described, the condition refers to advanced salamanders (Cloutier, in prep.). D. Wake (1979) noted first that all living and fossil salamanders possess a tuberculum interglenoideum on the atlas, thereby providing one character that, potentially, could be assessed in both living and fossil taxa.

M. Wake (1968, 1979) and Saint-Aubain (1985) provided additional characters uniting living salamanders. M. Wake (1968, 1979) reported regional differentiation in the spermatogenic activity of the testes of

---

[2]According to Estes (1981) and Naylor (1978), Group II salamandrids are primitive newts consisting of the fossil genera *Brachycormus*, *Chelotriton*, and *Palaeopleurodeles*, and the Recent genera *Pleurodeles* and *Tylototriton*.

salamanders. Thus, within different lobes of the testis, development of spermatocytes is not necessarily equivalent. Moreover, in some salamanders, additional lobes form on the testes as the animals mature. In contrast, in anurans and caecilians, new lobes do not develop with age, and there is not interlobular differentiation in spermatogenic activity. Saint-Aubain (1981, 1985) proposed that the reduction of Distal Carpal V and the prepollex, and the union of Gill Arches II–IV to form a single aortic arch that joins the dorsal aorta were characters common to all salamanders.

The basal members of the Caudata are the Cryptobranchidae and "hynobiids" (fide Herre, 1935; Edwards, 1976; Hecht and Edwards, 1976; Cloutier, in prep.).

Fox and Naylor (1982) proposed the recognition of a new order possibly referable to the Lissamphibia—the Allocaudata. Allocaudata was created to accommodate the fossil genus *Albanerpeton*, which previously had been considered to represent a prosirenid salamander (Estes and Hoffstetter, 1976).

Gymnophiona

The monophyly of caecilians never has been questioned. A. R. Milner (1988) diagnosed the Gymnophiona by eight synapomorphies: (1) 95–285 trunk vertebrae; (2) the absence of limbs and girdles; (3) a skull composed of several compound ossifications; (4) the phallodeum present; (5) a tentacle present; (6) the possession of more than 200 intersegmental, subcutaneous lymph hearts; (7) an interhyoideus muscle modified as a jaw levator; and (8) a subvertebralis muscle with a ventral layer. According to Nussbaum (1977, 1979), the Rhinatrematidae represents a primitive family of caecilians, whereas Walsh (1987) proposed a basal trichotomy including the Ichthyophiidae, Rhinatrematidae, and a clade composed of the remaining caecilian families. In the analysis, the condition found in the Rhinatrematidae and Ichthyophiidae was used to characterize the Gymnophiona. Recently, a Lower Jurassic caecilian was discovered from the Kayenta Formation of Arizona (Walsh and Jenkins, pers. comm.). Although the description has not been published yet, Walsh and Jenkins permitted us to use this organism in our analysis. We refer to this fossil as the "Kayenta caecilian."

Set I Out-Groups

The out-groups for Set I were chosen on the basis of a previous cladistic analysis of fossil and living Sarcopterygii (Cloutier, in prep.) that supports the general topology of the living taxa proposed by Maisey (1986) and Panchen and Smithson (1987). Although we agree with Maisey's

interpretation of the phylogenetic interrelationships of the sarcopterygians, it should not be assumed that we accept all of his synapomorphies.
The relationships of the living taxa are summarized as follows:

[Actinopterygii + [Actinistia + [Dipnoi + [Amphibia + Amniota]]]]

Actinistia

The Actinistia (= Coelacanthiformes) is considered a monophyletic
group (Middle Devonian–Recent) (Andrews, 1973; Forey, 1981; Lund
and Lund, 1985; Panchen and Smithson, 1987; Cloutier, 1991). The Actinistia is represented by a single living species, *Latimeria chalumnae*.
Millot and Anthony (1958, 1965) and Millot et al. (1978) were the primary
references for anatomical descriptions of this coelacanth. (See Schultze
and Cloutier, 1991, for review of morphological literature.) Coding for
osteological characters is based on the basal taxon of the Actinistia—
Miguashaia bureaui (Schultze, 1973; Cloutier, 1991).

Actinopterygii

The monophyly of the Actinopterygii was proposed by Gardiner (1984)
and Lauder and Liem (1983). The characters coded for this clade are
based primarily on the conditions found in *Polypterus*, or *Amia* and *Lepisosteus*. The Teleostei has little bearing on the basal actinopterygian
condition, because cladistically, teleosts are embedded in the actinopterygian clade (Lauder and Liem, 1983; Gardiner, 1984). *Moythomasia*,
Mimia, and *Cheirolepis* are considered to be the fossil basal taxa of the
actinopterygians (Gardiner, 1984).

Amniota

The monophyly of the Amniota has been corroborated in numerous
studies (e.g., Gardiner, 1982a,b; Løvtrup, 1977; Gauthier et al., 1988a,b).
We relied on the diagnosis of the Amniota provided by Gauthier et al.
(1988a,b). As defined by them, the Amniota is restricted to the most
recent common ancestor of extant mammals and reptiles, representing
two major groups—the Synapsida (including Mammalia), and the Reptilia (including Chelonia, Lepidosauria, and Archosauria). Gauthier et al.
(1988a,b) diagnosed the Amniota by the following characters: (1) the
intertemporal absent in taxa in which the postorbital contacts the parietal
and supratemporal; (2) the squamosal entering the posttemporal fenestra; (3) exoccipitals in contact at the occipital condyle; (4) the occipital
condyle hemispherical and well ossified; (5) a caniniform maxillary tooth
present; (6) the second cervical vertebra (axis) sloping anterodorsally; (7)
the cleithrum reduced, not capping the scapula; (8) two coracoid ossification centers present in the scapulocoracoid; (9) an astragalus present; (10)

bony dorsal scales absent; (10) gastralia overlapping midventrally and often fusing to form a small V; (11) the presence of three nasal turbinals; (12) the penis single with erectile tissue; (13) the ciliary body of the eye bearing ciliary processes; (14) the presence of cartilage canals in the epiphyses of long bones; (15) the presence of a sinus cavernosus in the dura mater of the brain; (16) the production of ornithuric acid in benzoic acid metabolism; and (17) the presence of a horny caruncle in embryos.

Dipnoi

The monophyly of the fossil and living Dipnoi is corroborated by a suite of synapomorphies (see Maisey, 1986; Schultze and Campbell, 1987). There are six species of living dipnoans: *Neoceratodus forsteri*, *Lepidosiren paradoxa*, *Protopterus annectens*, *P. dolloi*, *P. aethiopicus*, *P. amphibius*). Among living taxa, the Dipnoi is the sister-group of the Tetrapoda; however, among fossil taxa, the Panderichthyiformes is the sister-group of the Tetrapoda (Schultze, 1987; Cloutier, in prep.). Among the fossil Dipnoi, *Diabolepis speratus* and *Uranolophus wyomingensis* are hypothesized to be basal taxa (Chang and Yu, 1984; Maisey, 1986). The condition coded for the soft anatomy of the Dipnoi corresponds to the stem-species coding inferred from the phylogeny proposed by Miles (1977) and reanalyzed by Marshall (1987). The relationships at the generic level for the three living genera are as follows:

[*Neoceratodus* + [*Protopterus* + *Lepidosiren*]]

Set II Out-Groups

It is not worthwhile to use "osteolepiforms" as the closest sister-group of Lissamphibia, because phylogenetically the taxon is far removed from the origin of the Lissamphibia, and were it to be used as a closely related group, intermediate fossil forms would be overlooked. Because our preliminary analysis supports the hypothesis that the lissamphibian groups are more closely related to other groups of temnospondyls than to any other groups of tetrapods, we have disregarded the hypotheses of Gardiner (1982a, 1983) supporting the Nectridea and Aïstopoda, and Carroll and Currie (1975) and Carroll and Gaskill (1978) supporting the Microsauria. The Loxommatoidea is hypothesized as the sister-group of the clade, including the temnospondyl and lissamphibian taxa. If monophyletic, the Microsauria is related more closely to amniote groups than to any group of amphibians. The Lepospondyli is paraphyletic as defined presently—i.e., it includes Nectridea, Aïstopoda, and Microsauria. However, on the basis of vertebral characters, the Lepospondyli might represent a monophyletic assemblage if it is composed only of the Nectridea and Aïstopoda. The Lepospondyli either is the sister-group of [Loxom-matoidea + Temnospondyli + Lissamphibia], or it is part of a trichotomy with the latter clade and the [Anthracosauria + Microsauria + Amniota].

The taxa that have been discussed in relation to the origin of lissamphi-bian groups are listed in Table 3. Only amphibian taxa were considered; microsaurs were excluded owing to their controversial alliance with ei-ther the Amphibia or the Amniota. Generally, the Dissorophidae has been recognized as the sister-group of the Lissamphibia. This hypothesis relies primarily on the presence of bicuspid, pedicellate teeth (Bolt, 1979) and assumes that the family represents a natural grouping.

By selecting out-groups in addition to the dissorophoids, we are able to (1) investigate if there are other group(s) more closely related to the lissamphibians, (2) provide a second out-group, and (3) assess the phy-logenetic status (i.e., monophyly or paraphyly) of the dissorophoids. Rather than using the Temnospondyli as an individual out-group (which includes the Dissorophoidea), we analyzed the group at the superfamilial level.

Microsauria

There are several hypotheses of microsaur relationships. The Micro-sauria has been proposed to be the sister-group of Gymnophiona (Walsh, 1987) or the ancestor of the Lissamphibia (or parts of the Lissamphibia; Carroll, 1988; Carroll and Currie, 1981). *Lysorophus* has been suggested as the ancestor of the Caudata (Williston, 1916; Wintrebert, 1922) or of the Gymnophiona (Nussbaum, 1983), and more recently, Walsh (1987) has proposed [Lysorophia + Aïstopoda] as the sister-group of the [Caudata + Anura]. Detailed descriptions of *Lysorophus* are given by Bolt and Wassersug (1975) and Wellstead (1985). Three microsaur genera were examined—*Rhynchonkos* (= "*Goniorhynchus*"), *Hapsidopareion*, and *Lysoro-phus*.

Lepospondyli [Nectridea + Aïstopoda]

Gardiner (1982a, 1983) considered the Nectridea to be closely related to the Lissamphibia on the basis of the presence of four characters: (1) the large squamosal contacts the parietal; (2) haemal arches fused to cen-trum; (3) bifurcated, membrane bone ribbanders; and (4) condyles medi-ally directed and having a notochordal pit. The coding for the clade Nectridea corresponds to the hypothesized coding of the stem-species according to the phylogeny proposed by A. C. Milner (1980).

Gardiner (1983) proposed the following topology for the relationships of the Lissamphibia with closely related Amphibia:

[[Aïstopoda + Nectridea] + [Gymnophiona + [Caudata + Anura]]]

Loxommatoidea

Beaumont (1977) surveyed the cranial morphology of this group and demonstrated its similarity to Ichthyostegalia and Temnospondyli. Panchen (1980) considered the Loxommatoidea to be the sister-group of the Anthracosauria, whereas Gardiner (1982a, 1983) hypothesized that it was the sister-group of all tetrapods except the Ichthyostegidae (= Ichthyostegalia). According to Smithson (1985), the Loxommatoidea is the sister-group to *Crassigyrinus*, but the relationship of this clade to either the Anthracosauria or Amniota is unclear. Herein, the Loxommatoidea is hypothesized to be the sister-group of the Temnospondyli. The L-shaped orbit supports the monophyly of the group. However, further study may justify its inclusion within the Temnospondyli. Because the interrelationships within the Loxommatoidea are unclear, a consensus of three genera (*Megalocephalus*, *Loxomma*, and *Baphetes*) is used.

Temnospondyli

The interrelationships among the Temnospondyli have been studied cladistically by Boy (1981), Godfrey et al. (1987), Smithson (1985), and Warren and Black (1985). Gardiner (1983) considered the Temnospondyli to be the sister-group of [Amniota + Amphibia] (where Amphibia = [[Adelogyrinidae + Aïstopoda] + [Nectridea + [Apoda + [Anura + Caudata]]]]). This sister-group relationship is supported by four characters (Gardiner's Characters 20–23): (1) the internasal bone absent, (2) labyrinthodont teeth present, (3) the apical fossa small, vomers meeting premaxillae anteriorly, and (4) the infraorbital canal looped over the lacrimal and continuous over the jugal and lacrimal. We consider Characters 2 and 3 to be plesiomorphic.

Some authors consider the Temnospondyli to be paraphyletic (e.g., Gaffney, 1979; Smithson, 1982, 1985; Bolt and Lombard, 1985); however, Gardiner (1983) and Godfrey et al. (1987) proposed that it is monophyletic. Gardiner (1983) based his assumption of monophyly on five characters (his Characters 1–5): (1) the basipterygoid region of the parasphenoid sutured to the pterygoid, (2) development of the interpterygoid vacuity, (3) the stapes projects dorsally, sutured to the parasphenoid, (4) the cleithrum narrow, and (5) interdorsals ossified, articulating with neural arches. Godfrey et al. (1987) listed four characters as temnospondyl synapomorphies—possession of: (1) postparietal-exoccipital contact, (2) interpterygoid vacuities, (3) a small uncinate process on the ribs, and (4) a single, undivided iliac blade. Godfrey et al. (1987) further proposed the following scheme of interrelationships for the major groups of Temnospondyli:

[*Greererpeton* + [*Caerorhachis* + [*Dendrerpeton* + [[Trimerorhachoidea + Edopoidea] + [Dissorophoidea + Eryopoidea]]]]]

Because the interrelationships among temnospondyl groups are controversial, subunits of this order are included in the analysis.

Dendrerpeton

The Carboniferous genus *Dendrerpeton* is included as an individual taxon because its phylogenetic position is questionable (see A. R. Milner, 1980; Godfrey et al., 1987). The generic coding is based on two species, *D. acadianum* and *D. rugosum* (Carroll, 1967; A. R. Milner, 1980; Godfrey et al., 1987). Berman et al. (1985:30) considered *Dendrerpeton* to be an "unspecialized primitive temnospondyl generally considered near the ancestral stock of the dissorophids." However, Godfrey et al. (1987) favored the following phylogenetic scheme:

[*Dendrerpeton* + [Trimerorhachoidea + Edopoidea] + [Dissorophoidea + Eryopoidea]]

Dissorophoidea

The taxonomic status of "dissorophids" in the question of the origin of the Lissamphibia has been taken in a broad sense. Often the family Dissorophidae is mentioned when reference should have been made to the order Dissorophoidea. The superfamily Dissorophoidea is composed of Branchiosauridae (4 genera), Dissorophidae (19 genera), Doleserpetontidae (1 genus), Micromelerpetontidae (3 genera), and Trematopidae (4 genera) (Olson, 1941; Bolt, 1969, 1974a,b,c; Boy, 1972; Carroll, 1988). According to Godfrey et al. (1987), the monophyly of the Dissorophoidea is supported by four characters (their Characters 33–36): (1) the palatine exposed along orbital margin, (2) the quadrate bearing a posterodorsal process, (3) the otic notch greatly expanded, and (4) supratympanic flange present. Character 1 is found in other groups (Saurerpetontidae and Trimerorhachoidea) as they specify.

As traditionally defined, the Late Carboniferous–Early Triassic "Dissorophidae" is not monophyletic (Rage and Janvier, 1982; Bolt, this volume). Carroll (1964) and DeMar (1968) proposed a phylogeny of dissorophids, assuming *Amphibamus* to be the "ancestor" of most of dissorophids; however, the monophyly of the group was not corroborated, and there is no theoretical rationale supporting either the recognition of an ancestor or their phylogenetic scenario.

The Late Carboniferous–Late Permian Branchiosauridae is represented in the analysis by four genera: *Apateon, Branchiosaurus, Leptorophus,* and *Schoenfelderpeton.* Romer (1966, 1968) considered the "branchiosaurs" to be simply "larval and growth stages of contemporary rhachitomes" found in association in the fossil record. However, Boy (1972, 1974, 1987) showed that the branchiosaurids represent valid species with ontogenetic series available and proposed the inclusion of the Branchiosauridae within the Dissorophoidea.

Dissorophoid taxa were selected such that each family of the superfamily is represented by at least one taxon; however, not all dissorophoid genera were included in the analysis. Choice of genera was based on two criteria—(1) those genera that have been considered in the literature with respect to the origin of the Lissamphibia (cf. Trueb and Cloutier, this volume), and (2) those for which either detailed morphological descriptions exist or specimens were available for examination. The coding for *Amphibamus* is derived from observations made on *A. grandiceps* and *A. lyelli. Apateon* is represented by a single species, *A. pedestris. Broiliellus* is based on *B. texensis, B. brevis,* and "cf. *B.* sp." sensu Bolt (1977a, 1979), and *Dissorophus* on *D. angustus* and *D. multicinctus.*

"Edopoidea"

A. R. Milner (1980) and Godfrey et al. (1987) considered the Edopoidea to be a monophyletic group based on two characters (their Characters 30–31): (1) the presence of an ornamented septomaxilla incorporated into the skull roof, and (2) the presence of a prefrontal-jugal suture that excludes the lacrimal from the orbit. However, the preservation and the identification of the septomaxilla in many fossil groups is unclear. The prefrontal-jugal suture also is known in *Crassigyrinus* (Palaeostegalia), some Anthracosauria, and the Seymouriamorpha. Therefore, the monophyly of the Edopoidea is questionable. *Edops* was used as an individual taxon that belongs within the Edopoidea without the assumption that the conditions found in this genus are actually representative of all five genera contained in the family.

"Eryopoidea"

The monophyly of the "Eryopoidea" is not supported (Godfrey et al., 1987). For the purpose of this analysis, we will include in the taxonomic grouping "Eryopoidea" the Eryopidae, the Archegosauridae, and most stereospondyls. This arrangement follows the phylogenetic relationships proposed by Gardiner (1983). A close relationship between "Eryopoidea" (sensu stricto) and "Stereospondyli" was suggested by Romer (1966). The phylogenetic status of the "Stereospondyli" is unclear (Carroll, 1988); Gardiner (1983) and Warren and Black (1985) proposed the monophyly of most stereospondyl taxa. For convenience, the species *Eryops megacephalus* is included to represent this grouping of taxa.

Trimerorhachoidea

Hook (1983) included three families in the Trimerorhachoidea—Eugyrinidae, Saurerpetontidae, and Trimerorhachidae. The phylogenetic status of the superfamily is questionable. Coldiron (1978) considered the Trimerorhachoidea to be paraphyletic, but Chorn (1985) supported the monophyly of the taxon on the basis of five characters proposed by him and others (Chase, 1965; Godfrey et al., 1987). The characters are as follows: (1) the presence of a broad cultriform process of the parasphenoid; (2) the possession of a "modest" retroarticular process; (3) the presence of a reduced otic notch (or squamosal embayment); (4) the presence of a short antorbital region; and (5) the presence of anterior palatal fenestrae and symphyseal tusks. Character 1 is plesiomorphic; Characters 2, 3, and 4 are highly subjective in their definition. The condition coded for the Trimerorhachoidea is based on the stem-species hypothesized from the cladogram proposed by Chorn (1985).

ANALYSIS OF CHARACTERS

Data Set I: Soft-Anatomical Characters

Skin and Skin Derivatives

Character 1: Dermis, attachment of (DER). The deepest layer of vertebrate skin is the stratum compactum [= stratum reticulare] of the dermis. Separating the latter from the underlying deep fascia, aponeuroses, or periosteum is the connective tissue hypodermis, which is bound either loosely or firmly to the dermis above.

0 = Dermis and hypodermis firmly and uniformly connected.

1 = Dermis and hypodermis not uniformly connected over body surface.

Character 2: Myoepithelial cells (MYO). Myoepithelial [= basal] cells are slender, spindlelike cells that lie between mucous cells and the basement membrane. They are presumed to act as smooth-muscle cells and to facilitate the movement of glandular secretions into excretory ducts.

0 = Absence of a distinct myoepithelium.

1 = Presence of a distinct myoepithelium.

Character 3: Leydig cells (LEY). Leydig cells are epidermal cells found only in Recent larval anurans and salamanders; the cells disappear at metamorphosis. These cells are not in contact with the skin surface, nor do they release any secretion onto the epidermal surface. Kresja (1979) proposed that the mucuslike content of the Leydig cells might be released into extracellular compartments and thereby serve as an internal fluid reserve beneath the epidermis.

0 = Absence of Leydig cells.
1 = Presence of Leydig cells.

Character 4: Stratum corneum (COR). The stratum corneum is the superficial layer of the epidermis, the outermost layer of the skin. When present, the stratum corneum consists of one or more layers of dead or dying horny cells.

0 = Absence of stratum corneum.
1 = Presence of stratum corneum.

Character 5: Mucous cuticle (MUC). The extracellular mucous cuticle is composed of mucopolysaccarides that are secreted from, or through, the plasma membrane of each apical epidermal cell; thus, the cuticle covers the skin of organisms that possess it (Kresja, 1979).

0 = Presence of a mucous cuticle.
1 = Absence of a mucous cuticle.

Character 6: Dermal folds (FLD). The presence of dermal folds in the skin that reflect body segmentation seems to characterize only two groups of Recent amphibians.

0 = Absence of dermal body folds correlated with body segmentation.
1 = Presence of dermal body folds correlated with body segmentation.

Character 7: Iridophores (IRD). Iridophores [= guanophores] are bright-colored dermal chromatophores composed of purines. They are located in reflecting platelets and produce a silvery color.

0 = Presence of iridophores.
1 = Absence of iridophores.

Character 8: Lateral-line system (LAT). Lateral-line systems are present in nearly all anamniotes. The lateral-line system is composed of two kinds of organs—mechanoreceptive neuromasts and electroreceptive ampullae—in all anamniotes except anurans, which lack ampullae (cf. Fritzsch and Wake, 1986, and Fritzsch and Wahnschaffe, 1983, for literature summaries).

0 = Presence of ampullae.
1 = Absence of ampullae.

Character 9: Lorenzinian ampullae (LOR). Two kinds of lateral-line ampullae currently are distinguished—Lorenzinian ampullae and so-called teleostean ampullae (Bullock et al., 1982; Munz et al., 1982). The apical surface of a Lorenzinian ampulla, an electroreceptor organ, bears a kinocilium, or a kinocilium along with a few microvilli, whereas the apical surface of the teleostean ampullae bears only microvilli. According to Fritzsch and Wahnschaffe (1983), all vertebrate ampullary receptors are homologous, and Northcutt (1986) confirmed that primitively the ampullae bear both a kinocilium and microvilli, and that loss of one or the other structure represents a derived state. Northcutt (1986) provided the distribution of this character for most taxa analyzed.

0 = Presence of kinocilium and microvilli on apical surface of ampulla.

1 = Absence of either kinocilium or microvilli on apical surface of ampulla.

Glands

Character 10: Granular glands (GRA). Granular, or so-called poison, glands are dermal glands that produce a protective apocrine secretion.

0 = Presence of granular glands.

1 = Absence of granular glands.

Character 11: Harderian gland (HAR). Tetrapods typically have a Harderian gland associated with the anterior part of the eye. The gland secretes an oily liquid that usually functions to lubricate the eye.

0 = Absence of Harderian gland.

1 = Presence of Harderian gland.

Character 12: Intermaxillary glands (IMX). A complex of mucous glands located between the nasal capsules and/or ventral and ventromedial to them. The glands open into the roof of the mouth via one or more openings anterior to the vomerine teeth. Mucous glands in this position are known variously as *Zungendrüsen, Rachendrüsen, Choanendrüsen,* and *glandulae vomerales anteriores* (Parsons and Williams, 1963; Webb, 1951; Du Plessis, 1945). We assume that these glandular structures are homologous on the basis of topographical evidence.

0 = Intermaxillary glands absent.

1 = Intermaxillary glands present.

Character 13: Adrenal gland, association with kidney (AD1). The adrenal gland is an endocrine structure that produces steroid hormones, and adrenaline [= epinephrine] and noradrenalin [= norepinephrine]. The steroid hormones are produced by steroidogenic cells, whereas adrenalin and noradrenalin are produced by chromaffin cells. The adrenal glands may lie adjacent to the kidneys or their tissues may be embedded within the kidney.

0 = Adrenal tissues embedded within, or closely associated with, kidney and not protruding into coelom.

1 = Adrenal gland discrete and protruding into coelom.

Character 14: Adrenal gland, separation of steroidogenic and chromaffic tissues (AD2). The two principle tissue components of the adrenal gland (i.e., steroidogenic and chromaffin cells) may be intermingled or separate. The conditions in salamanders and anurans were described by Grassi Milano and Accordi (1986).

0 = Steroidogenic and chromaffin cells separate.

1 = Steroidogenic and chromaffin cells intermingled.

General Sensory Organs

Character 15: Tentacle and choanenschleimbeutel (TEN). Caecilians are unique among vertebrates in their possession of a retractile tentacular apparatus with a chemosensory function (Duellman and Trueb, 1986). Associated with their chemosensory apparatus is the choanenschleimbeutel, a ventral chamber of the cavum principale, that lies adjacent to, but separate from, the vomeronasal organ.

0 = Tentacle and choanenschleimbeutel absent.

1 = Tentacle and choanenschleimbeutel present.

Character 16: Vomeronasal organ (VOM). The vomeronasal [= Jacobson's] organ is a specialized area of olfactory epithelium, the primary function of which seems to be receiving olfactory sensations from the buccal cavity.

0 = Vomeronasal organ absent.

1 = Vomeronasal organ present.

Character 17: Visual accommodation I (VA1). Visual accommodation is accomplished either by movement of the lens (anamniotes) or deformation of the lens (amniotes). In those vertebrates that move the lens, the lens is moved either proximally (i.e., backward) or distally (i.e., forward) to accommodate distance vision by means of an interocular muscle(s) of mesodermal origin—the m. retractor or m. protractor lentis. In those organisms in which visual accommodation is accomplished by deformation of the lens, the ciliary body is enlarged around the lens; within the body are circular muscle fibers. Contraction of the muscle deforms the lens into a more rounded shape for close vision.

0 = Lens moved by interocular muscle(s).

1 = Lens moved by ciliary muscle.

Character 18: Visual accommodation II (VA2). See description in Character 17.

0 = Visual accommodation by movement of lens proximally.

1 = Visual accommodation by movement of lens distally.

Character 19: Green rods (ROD). Among tetrapods, salamanders and anurans are unique in having specialized retinal receptor cells known as green rods. These cells absorb wavelengths of 432 nm and have short outer segments and long, thin myoids.

0 = Absence of green rods.

1 = Presence of green rods.

Character 20: Tympanum (TYM). The tympanum is a thin membrane that acts as a resonator for the ossicles of the middle ear. This character was used by Lombard and Bolt (1979) and Bolt and Lombard (1985).

0 = Tympanum absent.

1 = Tympanum present.

Character 21: Tympanic cavity (CAV). The tympanic cavity is defined as an enclosed cranial cavity that lies lateral to the auditory capsule and that is derived from outpocketings of the first two visceral pouches. This character was used by Lombard and Bolt (1979) and Bolt and Lombard (1985).

0 = Tympanic cavity absent.
1 = Tympanic cavity present.

Character 22: Eustachian tube (EUS). The eustachian tube connects the middle-ear cavity with the pharynx.

0 = Eustachian tube absent.
1 = Eustachian tube present.

Character 23: Papilla amphibiorum (PAP). The papilla amphibiorum is a patch or patches of neuroepithelium embedded in the labyrinth wall of the inner ear. The papilla amphibiorum may consist of a single or a double patch of hair cells. Caecilians, salamanders, and some primitive anurans possess one patch of hair cells, whereas most anurans have two (Lewis et al., 1985). If two contiguous patches of hair cells are present, each is innervated by a separate branchlet of the auditory nerve and each has two populations of oppositely polarized hair cells. Fritzsch and Wake (1988) hypothesized that the papilla amphibiorum represents a part of the papilla neglecta that was shifted into the sacculus, and that its presence is an amphibian synapomorphy.

0 = Absence of papilla amphibiorum.
1 = Presence of papilla amphibiorum.

Character 24: Papilla neglecta (PAN). When present, this neuroepithelial patch (also known as the utricular papilla [Wever, 1985]) is located either adjacent to the utriculosaccular foramen or in the utriculus (in caecilians). The distribution of this character has been assessed by Lewis et al. (1985) and Fritzsch and Wake (1988).

0 = Presence of papilla neglecta.
1 = Absence of papilla neglecta.

Character 25: Perilymphatic cistern (CIS). The perilymphatic duct of the inner ear may be expanded into a perilymphatic cistern.

0 = Perilymphatic cistern absent.
1 = Perilymphatic cistern present.

Character 26: Sacculus (SAC). The labyrinth of the inner ear is composed of the sacculus, lagena, and utricle, which are associated with the three semicircular canals. The perilymphatic duct may be associated with either the sacculus or the lagena. This character was not polarized.

0 = Perilymphatic duct associated with the lagena.
1 = Perilymphatic duct associated with the sacculus.

Character 27: Round window (RND): The fenestra rotunda [= round window], lies in the posterolateral wall of the neurocranium and opens into the inner ear.

0 = Round window absent.
1 = Round window present.

Character 28: Operculum (OPC). The operculum [= opercularis auris of Lombard and Bolt, 1979] lies in the fenestra ovalis posterior to the footplate of the stapes and is connected to the shoulder girdle by the opercularis muscle (Hetherington et al., 1986). Kunkel (1912) reported the presence of a cartilaginous element in a similar position to the operculum in the turtle *Emys*, but no muscle is associated with the element.

0 = Opercular element not associated with proximal stapes.
1 = Opercular element associated with proximal stapes.

Muscles

Character 29: Smooth narial muscles (NAR). The opening and closing of the external naris in salamanders and caecilians is controlled by three smooth muscles (cf. Duellman and Trueb, 1986, for explanation).

0 = Absence of intrinsic narial musculature.
1 = Presence of intrinsic narial musculature.

Circulatory System

Character 30: Sinus venosus, size (SV1). The sinus venosus opens into the dorsal side of the right atrium via the sinoatrial opening.

0 = Sinus venosus large.
1 = Sinus venosus reduced.

Character 31: Sinus venosus, position (SV2). When present, the sinus venosus may be positioned medially or to the left.

0 = Medial position.
1 = Left position.

Character 32: Sinus venosus, subdivision of (SV3). The sinus venosus opens in the right atrium. The sinus venosus in some groups is divided into right and left portions. This separation could either be absent or present; if present, it is the result of an indentation of the wall (e.g., caecilians), of the presence of a valve(s) (e.g., caecilians), or of constrictions of the inner wall of the sinus (e.g., anurans).

0 = Sinus venosus not subdivided.
1 = Sinus venosus subdivided.
2 = Sinus venosus partially divided.

Character 33: Atrium, size (ATR). The heart is divided into a number of chambers—one sinus venosus, one or two atria, one or two ventricles, and one conus arteriosus. When two atria are present and separated by the interatrial septum, the relative size of the left and right atria varies.

0 = Right atrium larger than left.
1 = Right atrium smaller than left.
2 = Right and left atria equal in size.

Character 34: Interatrial septum (IAT). The two atria of the heart are separated by a wall—the interatrial septum. The interatrial septum could be incomplete, fenestrate, or complete. In Dipnoi, the atria are separated partially by a septum termed the pulmonaris fold, which arises from a deformation of the atrial wall (Burggren and Johansen, 1987). This condition is considered to be an incomplete interatrial septum.

0 = Incomplete interatrial septum.
1 = Complete interatrial septum.
2 = Fenestrate interatrial septum.

Character 35: Ventricle, subdivision of (VEN). The ventricle may be divided into right and left halves by an interventricular septum, which varies in the completeness of its development.

0 = Subdivision absent.
1 = Subdivision present.

Character 36: Carotid labyrinth (CAR). This dense network of blood vessels is situated midstream at the bifurcation of the common carotid artery into the internal and external carotid arteries in those organisms in which it occurs (Toews et al., 1982).

0 = Carotid labyrinth absent.
1 = Carotid labyrinth present.

Character 37: Cutaneous branches of ductus arteriosus (CUT). When present, these branches of the ductus arteriosus supply the skin (Saint-Aubain, 1985).

0 = Cutaneous branches absent.
1 = Cutaneous branches present.

Character 38: Aortic Arch II (AII). Primitively within the gnathostomes, there are six pairs of aortic arches, each one of which is associated with a branchial arch. Aortic Arch I (mandibular) is lost in all adult vertebrates, and Aortic Arch II (hyoid) is retained only in some. Aortic Arches III–VI are associated with Gill Arches 1–4. Originally, the second aortic arch lies between the spiracle and Gill Slit I.

0 = Aortic Arch II present.
1 = Aortic Arch II absent.

Character 39: Aortic Arch V (A–V). If present in the adult, the fifth aortic arch lies between Gill Arches III and IV.

0 = Aortic Arch V present.
1 = Aortic Arch V absent.

Character 40: Aortic Arch VI (AVI). If present in the adult, the sixth aortic arch lies posterior to Gill Arch IV.

0 = Aortic Arch VI present.
1 = Aortic Arch VI absent.

Respiratory System

Character 41: Choana, development of (CHO). Embryologically, the choana originates in one of two ways. (See Schultze, 1987, for osteological definition, and Bertmar, 1968, and Panchen, 1967, for embryological definition of the choana.) It may form via a connection between two evaginations—

a posterior evagination from the nasal sac that joins with an anterior evagination of the endodermic archenteron posteriorly. In the alternate embryological scheme, there is a connection established between the nasal sac and the ectoderm of the stomodeum anterior to the archenteron (Rage and Janvier, 1982).

  0 = Choanal tube opens in stomodeum.
  1 = Choanal tube opens in archenteron.

Character 42: Tracheal cartilages (TRA). The trachea is a tube extending from the larynx or esophagus to the bronchi. When the trachea is present, it may or may not be supported by ringlike tracheal cartilages. Among anurans, only the pipids (a derived group fide Cannatella, 1985) possess cartilaginous support of the trachea; therefore, anurans have not been coded as possessing tracheal cartilages.

  0 = Tracheal cartilages absent.
  1 = Tracheal cartilages present.

Urogenital System

Character 43: Pronephros (PRO). Different types of kidneys have been identified in the past according to their embryological origin or formation—opisthonephric, mesonephric, metanephric, and pronephric. However, each one of these conditions represents a complex of characters. Thus, we divide the complex into a series of separate characters (Characters 44–48). The pronephros is the first and most anterior part of the vertebrate kidney. In most taxa, it is transitory, but in some it is retained in the adult, in which the organ may or may not be functional.

  0 = Pronephros retained in adult.
  1 = Pronephros absent in adult.

Character 44: Utilization of the pronephros for sperm transport (SPM). The pronephros originally has an urinary function. In addition, in a few taxa, the pronephros may be used for sperm transport.

  0 = Pronephros used for sperm transport.
  1 = Pronephros not used for sperm transport.

Character 45: Modification of the pronephros (MOD). If the anterior part of the kidney is utilized for sperm transport, the kidney structure may remain unmodified or it may be altered by reduction or loss of glomeruli from the kidney tubules.

  0 = Pronephros not modified for sperm transport.
  1 = Pronephros modified in sperm transport.

Character 46: Pronephric units, arrangement (ARR). Functional nephric units that develop from anterior nephrostomes are found in embryos and free-living larvae of some vertebrates and in the adults of a few (Fraser, 1927). When present, these nephric units may or may not retain a serial arrangement. This character was not polarized.

  0 = Serial arrangement of pronephric nephrostomes in adult.

1 = Random arrangement of pronephric nephrostomes in adult.

Character 47: Pronephric units, integrity of walls (INT). If functional pronephric nephrostomes are retained in an adult vertebrate, the walls of the units may break down partially or remain intact. This character was unpolarized.

0 = Pronephric nephrostome walls retained.

1 = Pronephric nephrostome walls partially broken down.

Character 48: Ciliated pronephric tubule segment (CIL). In the pronephros and adult kidneys of some vertebrates, individual tubules may bear a ciliated funnel opening from the coelom to the tubular lumen or tubular blood supply. These ciliated segments of the tubule are considered homologous to those of the mesonephros (Olsen, 1977).

0 = Ciliated pronephric tubule segment present.

1 = Ciliated pronephric tubule segment absent.

Character 49: Testis, connection to kidney (CON). The vertebrate testis develops medially adjacent to the kidney. If the testis is associated structurally with the kidney, both sperm and urine are transported to the cloaca via the archinephric duct. If the testis is independent of the kidney, the archinephric duct [= vas deferens] transports sperm only, whereas kidney products are moved to the cloaca via a separate duct, the ureter.

0 = Testis connected by a series of ducts to the anterior part of the kidney.

1 = Testis connected by a series of ducts to the posterior part of the kidney.

2 = Testis independent of kidney.

Character 50: Fat bodies associated with gonads (FAT). Fat bodies are composed of adipose tissue and are associated dorsally, medially, or ventrally with the gonads (of both sexes) of some vertebrates. This character was used by Noble (1931) and Parsons and Williams (1963). It has been reported that in lissamphibians the fat bodies originate from the germinal ridge; however, little is known about the condition in other groups.

0 = Fat bodies originating from germinal ridge absent.

1 = Fat bodies originating from germinal ridge present.

Character 51: Externally lobed testes (LOB). Among vertebrates, the surface of the testis may be either uniform or lobulate. The number of lobes may increase with the age of the organism (e.g., salamanders) or may remain constant throughout life (e.g., caecilians).

0 = Surface of testes smooth and uniform.

1 = Surface of testes lobulate.

Character 52: Site of spermatogenesis (SIT). There are two patterns of spermatogenesis among vertebrates—tubular and cystic. In the tubular pattern, the spermatogenic portions of the testis contain permanent populations of primordial germ cells, whereas in the cystic pattern, the permanent population of germ cells lies outside the spermatogenic crypts.

0 = Site of spermatogenesis separate from location of permanent populations of primordial germ cells.

1 = Site of spermatogenesis coincident with location of permanent populations of primordial germ cells.

Nervous System

Character 53: Mauthner cells (MAU). In some lower vertebrates, there is a pair of giant cells, Mauthner cells, that lie in the floor of the medulla oblongata and that have axons that extend the entire length of the spinal cord. The cell bodies are associated with the acoustico-lateralis centers. It is presumed that localized cord reflexes may regulate undulatory movement common to the organisms having Mauthner cells.

0 = Presence of Mauthner cells.

1 = Absence of Mauthner cells.

Character 54: Superior olivary nucleus (OLI). The superior olivary nucleus is a cell group located in the tegmental basal plate of the medulla oblongata. Its presence in amphibians has been correlated with the transformation of part of the central lateral-line apparatus into an acoustic system (Kuhlenbeck, 1975).

0 = Absence of superior olivary nucleus.

1 = Presence of superior olivary nucleus.

Character 55: Cranial Nerve XII (XII). In amniotes and actinistians, the last cranial nerve is considered to be the hypoglossal (XII), which exits the skull via a foramen. Some vertebrates lack a twelfth nerve exiting from the occiput of the skull, and the first spinal nerve seems to represent a primordial composite of fibers from the first and second spinal nerves and the hypoglossal of other vertebrates.

0 = Cranial foramen for C.N. XII present.

1 = Cranial foramen for C.N. XII absent.

Character 56: First spinal nerve (SPI). The first spinal nerve of Recent amphibians seems to represent a composite of the first two spinal nerves plus C.N. XII (see Character 55). In salamanders and caecilians, this nerve exits from the spinal cord through a foramen in the cervical vertebra, whereas in anurans, it exits via an intervertebral foramen between the cervical and second presacral vertebra. In the lamprey, the spinal nerve exits the spinal cord via an intervertebral foramen. Because of ambiguity of the character-states of the out-groups, this character is not polarized.

0 = First spinal nerve exiting spinal cord via vertebral foramen.

1 = First spinal nerve exiting spinal cord via intervertebral foramen.

Character 57: Development of corpus cerebelli and auricular subdivisions of the cerebellum (CC1). The vertebrate cerebellum develops from the dorsal plate of the rostral part of the hindbrain. The cerebellum is subdivided into two areas—the medial corpus cerebelli or transverse cerebellar plate, and the paired lateral auricular subdivision. The relative development of these two areas may be subequal or about the same.

0 = Degree of development of corpus cerebelli and auricular subdivisions of cerebellum about equal.

1 = Marked difference in the degree of development of corpus cerebelli and auricular subdivision of cerebellum.

Character 58: Relative development of corpus cerebelli and auricular subdivisions of the cerebellum (CC2). If there is a marked difference in the degrees of development of these two regions of the cerebellum, one of two situations seems to prevail. Either the auricular subdivisions are exceedingly small relative to the corpus cerebelli, or the auricular subdivisions are large and the corpus cerebelli is small. This character was not polarized.

0 = Auricular subdivision small and corpus cerebelli of normal size.

1 = Auricular subdivision large and corpus cerebelli smaller than usual.

Character 59: Purkinje cells, arrangement (PU1). Purkinje cells are located in the cerebellar cortex of all vertebrates except agnathans; the cells are arranged in a layer between the molecular (superficial) and granular (internal) layers of the cortex. The organization of these cells varies from a seemingly random, irregular arrangement to a regular arrangement.

0 = Purkinje cells partly or wholly irregular in their arrangement.

1 = Purkinje cells arranged regularly.

Character 60: Purkinje cells, presence of cerebellar nuclei (PU2). Purkinje cell neurites represent the "output" channels of the cerebellar cortex that communicate with deep cerebellar cells. In some vertebrates, the cells are not organized into distinct cell masses, whereas in others, the cells are organized into defined cerebellar nuclei.

0 = Absence of cerebellar nuclei.

1 = Presence of cerebellar nuclei.

Character 61: Torus semicircularis, differentiation (TOR). The dorsal alar plate of the mesencephalon of vertebrates tends to be differentiated into two zones—the tectum mesencephali s. tectum opticum, and the torus semicircularis. The former receives optic tract fibers. The torus semicircularis protrudes to varying degrees into the ventricular lumen ventral to the tectum opticum and contains the griseum of the nucleus mesencephali lateralis. The torus semicircularis is delimited variably from the tectum opticum by the sulcus lateralis mesencephali internus (and externus in anurans) of the ventricular wall, and from the basal plate by the sulcus limitans.

0 = Torus semicircularis differentiated from optic tectum but not markedly protuberant into ventricular lumen of mesencephalon.

1 = Torus semicircularis not clearly distinguishable from optic tectum and not protruding into ventricular lumen.

2 = Torus semicircularis differentiated from optic tectum and forming a prominent protrusion into the ventricular lumen of mesencephalon.

Character 62: Nucleus lentiformis mesencephali (LEN). In the boundary zone between the mesencephalon and diencephalon, there are a number of mesencephalic and diencephalic grisea that represent the nuclei of the pretectal region or the "pretectal complex." One of these is the nucleus lentiformis mesencephali. It is formed by a more or less distinct rostrobasal derivative of the tectum and probably receives optic input (Kuhlenbeck, 1975).

0 = Nucleus lentiformis mesencephali poorly defined.

1 = Nucleus lentiformis well defined.

Character 63: Nucleus radicis mesencephalicae, arrangement of cell masses (RAD). The nucleus radicis mesencephalicae trigemini is a derivative of the mesencephalic alar plate. These proprioceptive neurons seem to be related to the muscles of mastication, with the afferent neurites of the cells joined to the mandibular nerve and the intracerebral efferent neurites connected to the motor nucleus of the trigeminal nerve (Kuhlenbeck, 1975).

0 = Cell masses scattered and disorganized.

1 = Cell masses distinctly stratified.

Character 64: Basal plate of mesencephalon, regional differentiation (BAS). The basal plate of the mesencephalon may display two griseal systems depending upon the level of organization of this part of the brain (Kuhlenbeck, 1975).

0 = Presence of regional differentiation of basal plate of mesencephalon.

1 = Absence of regional differentiation of basal plate of mesencephalon.

Character 65: Oculomotor nucleus, differentiation (OCU). If differentiated, the nuclei of the oculomotor and trochlear nerves are located in the tegmentum or motor zone of the basal plate of the mesencephalon.

0 = Oculomotor nucleus undifferentiated.

1 = Oculomotor nucleus differentiated.

Character 66: Nucleus of trochlear nerve, differentiation (TRO). The nucleus of the trochlear nerve may be undifferentiated or differentiated.

0 = Trochlear nucleus undifferentiated.

1 = Trochlear nucleus differentiated.

Miscellaneous

Character 67: Gastrocoel formation (GAS). The formation of the gastrocoel results from the invagination of the outside wall of the egg. This invagination may originate ventrolaterally (i.e., from the vegetal hemisphere), or dorsally.

0 = Dorsal invagination.

1 = Ventrolateral invagination.

Character 68: Mesotocin (MES). This neurohypophysial hormone differs from oxytocin (found in actinopterygians and holocephalians) by the substitution of one amino acid; thus, LEU is replaced by ILE (Acher et al., 1972).

0 = Absence of mesotocin.

1 = Presence of mesotocin.

Character 69: Bile salts (BIL). Bufol, a derivative of 5-alpha cyprinol, is a bile salt that characterizes some taxa (Løvtrup, 1977; Haslewood, 1967, 1968).

0 = Absence of bufol.

1 = Presence of bufol.

Character 70: Lens protein D1 (LP1). Biochemical studies of the lenses of the eyes of vertebrates (Manski et al., 1967a,b,c; Løvtrup, 1977) have revealed a variety of proteins (e.g., AC1–3; AG1–6; CH1; DI1–3; GN1–5; PA1–4; PL1, 4, and 6; TE1–2).

0 = Absence of lens protein D1.

1 = Presence of lens protein D1.

Character 71: Lens protein D2 (LP2). See Character 70.

0 = Absence of lens protein D2.

1 = Presence of lens protein D2.

Character 72: Lens protein D3 (LP3). See Character 70.

0 = Absence of lens protein D3.

1 = Presence of lens protein D3.

Data Set II: Osteological Features

Character polarities correspond to those inferred from temnospondyl out-groups. The character polarities of nine characters were changed in the analysis of living taxa because of the condition found in actinistians and actinopterygians, the first and second out-groups, respectively. Characters for which polarities were changed are so indicated in their descriptions. Parenthetical numbers refer to the number of the character in Analysis II (Appendix I).

Cranial Characters

Character 1: Orbit size (ORB; 73). The size of the orbit in the adult is determined by the configuration of the endocranium (i.e., the length between the planum antorbitale and auditory capsule vs. the width between maxilla and neurocranium) together with the numbers, shapes, and sizes of circumorbital dermal bones (e.g., maxilla, nasal, lacrimal, squamosal, jugal, parietal, frontal, and pre- and postfrontal). If the orbit constitutes only a small fraction of the skull (e.g., apodans), it is characterized as small. A moderate-sized orbit is approximately one-fourth or one-fifth the length of the skull, whereas a large orbit is one-third or more the total skull length. Frequently, there is an ontogenetic change in the relative size of the orbit; the coding is based on the adult or the largest available individual of a given taxon.

0 = Orbit small.

1 = Orbit moderate.

2 = Orbit large.

Character 2: External naris size (EXN; 74). The size of the narial opening is defined as the ovoid space bounded by the maxilla, premaxilla, nasal, and lacrimal and/or septomaxilla. If the long axis of the narial opening in the adult is one-fifth or less the width of the snout at the midlength of the opening, the size is small. Openings approximately one-quarter the width of the snout are moderate, and those greater than one-third the width are large. Because the size of the external naris is dependent on the condition of surrounding bones, Characters 2 (EXN) and 23 (PNR) could be associated.

0 = External naris small.

1 = External naris moderate.

2 = External naris large.

Character 3: Pars dorsalis of premaxilla (PDP). When present, the pars dorsalis [= alary process; = ascending process] is the dorsal rostral component of the premaxilla, which generally serves as an abutment for internal nasal cartilages in Recent taxa. The pars dorsalis usually is narrower than the width of the pars dentalis. Although always dorsal to the pars dentalis, the pars dorsalis varies greatly in its orientation, relative size, and association with the posteriorly adjacent nasal bones.

0 = Premaxilla bearing pars dorsalis.

1 = Premaxilla lacking pars dorsalis.

Character 4: Pars palatina of premaxilla (PPP; 75). If present, the pars palatina is an edentate palatal shelf along the lingual margin of the pars dentalis of the premaxilla. The polarity of this character was reversed in Analysis II.

0 = Premaxilla bearing pars palatina.

1 = Premaxilla lacking pars palatina.

Character 5: Maxilla and ectopterygoid, association between (MET). The maxilla is the primary component of the upper jaw. The ectopterygoid is a dermal palatal element that, if present, articulates with the posterior margin of the palatine and lies between the maxilla and pterygoid. The medial margin of the ectopterygoid always articulates with the pterygoid;

however, the lateral margin may be free or may bear a complete articulation with the maxilla. Character 5 may be associated with Character 10.

0 = Articulation present between maxilla and ectopterygoid.

1 = Articulation absent between maxilla and ectopterygoid.

Character 6: Maxillary arcade, articulation with suspensorium (MSP; 76). The maxillary arcade includes all the components that constitute the ventral margin of the upper jaw (usually premaxillae, maxillae, and quadratojugals; occasionally the squamosal is involved also). If complete, the upper jaw is connected posteriorly to the palatoquadrate of the suspensorium. If incomplete, the posterior end of the arcade is separated from the suspensorium.

0 = Maxillary arcade complete.

1 = Maxillary arcade incomplete.

Character 7: Palatine (PAL; 77). This dermal palatal bone variably braces the maxilla against the neurocranium directly and/or via the ectopterygoid (if present) and pterygoid.

0 = Palatine present in adults.

1 = Palatine absent in adults.

Character 8: Palatine, orbital component (POB). The configuration, position, and articulations of the palatine are highly variable. Frequently, the bone lies in the anterior part of the orbit or bears a posterior process that extends into the orbit such that the entire bone, or part of it, is exposed laterally. Eaton (1973) described this variation in terms of the palatine's forming (or not forming) part of the lower rim of the orbit. Bolt (1974b) discussed the distribution of this character among early tetrapods, especially dissorophoids.

0 = Palatine not exposed laterally.

1 = Palatine exposed laterally.

Character 9: Palatine, anterolateral process (PAP). In some taxa, the anterior end of the palatine is bifurcated to form anterolateral and anteromedial processes. The former articulates with the lingual margin of the maxilla, whereas the latter articulates with the vomer.

0 = Anterolateral process of palatine absent.

1 = Anterolateral process of palatine present.

Character 10: Palatine, articulation with maxilla (PMX). Among those taxa that bear a palatine with an anterolateral process, the palatine has two different associations with the maxilla. It may articulate throughout its length with the lingual margin of the maxilla, or the palatine may diverge posteromedially from the maxilla such that the only point of articulation with upper jaw is the anterolateral process.

0 = Palatine articulates with maxilla posterior to anterolateral process.

1 = Palatine articulates with maxilla only at anterolateral process.

Character 11: Lacrimal (LAC; 78). When present, this dermal bone lies on the lateral aspect of the snout between the nasal and pars facialis of the maxilla. It forms the anteroventral margin of the orbit, and, in the absence of an external septomaxilla, forms the posterior margin of the opening for the external naris.

0 = Lacrimal present.

1 = Lacrimal absent.

Character 12: Prefrontal (PFR; 79). When present, the prefrontal is a dermal bone that forms part of the anteromedial margin of the orbit. It lies lateral to the frontal and posterior to the nasal. The polarity of this character is reversed in Analysis II.

0 = Prefrontal present.

1 = Prefrontal absent.

Character 13: Postfrontal (POF; 80). If present, this dermal roofing bone lies lateral to the frontal and parietal, medial to the postorbital, and forms the posteromedial margin of the orbit. The polarity of this character is reversed in Analysis II.

0 = Postfrontal present.

1 = Postfrontal absent.

Character 14: Prefrontal-postfrontal articulation (PPF; 81). The prefrontal and postfrontal vary significantly in their sizes and relative contributions to the supraorbital margin. The posterior end of the prefrontal may articulate with the anterior end of the postfrontal lateral to the frontal at the dorsomedial margin of the orbit, in which case they form the entire medial edge of the orbit. Alternately, the bones are separated from one another, and the frontal contributes to the circumorbital series.

0 = Prefrontal-postfrontal articulation present.

1 = Prefrontal-postfrontal articulation absent.

Character 15: Frontal and parietal (FPR; 82). In anurans, the frontal is fused with the parietal to form the frontoparietal bone. In most anurans, the frontoparietal forms from one center of ossification. However, in the early ontogeny of a few species (e.g., *Rana*), it has been claimed that there are two centers of ossification that are topographically coincident with the positions of the frontal and parietal bones of other vertebrates (Lebedkina, 1968).

0 = Frontal and parietal separate.

1 = Frontal and parietal fused to form frontoparietal.

Character 16: Exposure of the parietal organ (PAR; 83). Primitively, there is a foramen in the skull roof for the parietal organ. In adults of some taxa, the skull is completely roofed and lacks a parietal foramen, whereas in others, there is a medial fenestra between the paired frontals and/or parietals. The presence of such a fenestra is considered to be a separate character-state despite the fact that the homology of this condition is not clear.

0 = Parietal foramen present.

1 = Parietal foramen absent; parietal organ covered by dermal bone.

2 = Parietal foramen absent; parietal organ not covered by bone.

Character 17: Postparietal (PPR; 84). If present, the paired postparietals are dermal roofing bones that lie posteriorly adjacent to the parietals and form the medial occiput of the skull.

0 = Postparietal present.

1 = Postparietal absent.

Character 18: Intertemporal (ITP; 85). The paired intertemporals are dermal roofing bones that, if present, lie between the parietals and postorbitals, posterior to the postfrontal and anterior to the supratemporal.

0 = Intertemporal present.

1 = Intertemporal absent.

Character 19: Supratemporal (STP; 86). Each member of this pair of posterior dermal roofing bones is located posterior to the orbit between the squamosal laterally and the parietal medially.

0 = Supratemporal present.

1 = Supratemporal absent.

Character 20: Tabular (TAB; 87). When present, the paired tabulars are the most posterolateral of the series of dermal bones in the skull roof. They lie laterally adjacent to the postparietals (if present) and form the lateral occiput of the skull. The polarity of this character is reversed in Analysis II.

0 = Tabular present.

1 = Tabular absent.

Character 21: Postorbital (POB; 88). If present, the postorbital is one of a series of dermal bones that surrounds the orbit. It lies lateral or ventrolateral to the postfrontal and forms part or all of the posterior margin of the orbit.

0 = Postorbital present.

1 = Postorbital absent.

Character 22: Jugal (JUG; 89). The jugal is a dermal cheek bone that forms the ventral margin of the orbit when it is present. The jugal lies dorsal to the maxillary arcade and articulates with the squamosal posteriorly.

0 = Jugal present.

1 = Jugal absent.

Character 23: Prefrontal, association with opening for external naris (PNR; 90). In some taxa, the anterior limit of the prefrontal lies posterior to the opening for the external naris and is separated from it by the lacrimal and nasal. In such organisms, the nasal and lacrimal articulate with one another, and the nasal forms the dorsal margin of the opening of the external naris, whereas the lacrimal forms its posterior edge. In other taxa, the nasal and lacrimal are separated by the prefrontal, which then forms the posterodorsal margin of the external narial opening. The polarity of this character is reversed in Analysis II.

0 = Prefrontal not associated with margin of opening for external naris.

1 = Prefrontal forming part of margin of opening for external naris.

Character 24: Position of jaw articulation relative to posterior limits of endocranium (JAW; 91). There is considerable variation in the position of the jaw articulation relative to the position of the otic capsules and the posterior limits of the neurocranium medially. In the adult, the articulation may lie posterior to the endocranium, or at a level more or less coincident with the posterior limits of the endocranium. If the articulation lies medially adjacent to the otic capsule or at its anterolateral corner, its position is considered anterior.

0 = Jaw articulation posterior to endocranium.

1 = Jaw articulation anterior to posterior limits of endocranium.

2 = Jaw articulation coincident with posterior limits of endocranium.

Character 25: Squamosal embayment (SQE; 92). The squamosal embayment is defined as a distinct dorsolateral emargination at the posterior end of the skull roof in the sense of Godfrey et al. (1987) and includes the "otic notch" of other authors (cf. Bolt, this volume). The size and shape of the embayment is a function of the orientation of the palatoquadrate and the posterior extent of the dorsolateral skull table. In organisms possessing a squamosal embayment, the long axis of the palatoquadrate is inclined such that the dorsal end lies anterior to the ventral end. Because the palatoquadrate is covered by the squamosal laterally and posterolaterally, this bone forms the ventral margin of the squamosal embayment. The dorsal margin is composed variably of the otic ramus of the squamosal and/or a temporal series of dermal roofing bones (e.g., intertemporal, supratemporal, tabular). The polarity of this character is reversed in Analysis II.

0 = Squamosal embayment present.

1 = Squamosal embayment absent.

Character 26: Parasphenoid, relationship to neurocranium (PSH; 93). The parasphenoid is a median, unpaired palatal investing bone. Usually it is discrete; however, in some taxa the parasphenoid becomes indistinguishably incorporated into endocranial ossifications.

0 = Parasphenoid separate from neurocranial ossifications.

1 = Parasphenoid incorporated into neurocranial ossifications.

Character 27: Parasphenoid shape (PRS; 94). The parasphenoid may have a rectangular or triangular configuration. In the latter case, the parasphenoid is widest posteriorly and gradually narrows into the cultriform process anteriorly, whereas in the former case, the posterior portion narrows abruptly to form the cultriform process anterior to the otic capsules.

0 = Parasphenoid with rectangular corpus.

1 = Parasphenoid with triangular corpus.

Character 28: Pterygoid, orientation of quadrate ramus (QRO; 95). The pterygoid is a dermal palatal bone that is involved with bracing the maxilla and suspensorium against the braincase. Generally, it is tripartite, bearing anterior (= palatine), medial, and quadrate (= posterior) rami, the relative lengths and orientations of which vary. The quadrate ramus articulates with the medial palatoquadrate and may be directed posteriorly, posterolaterally, or laterally with respect to the longitudinal axis of the skull. This character can be associated with the relative position of the jaw articulation (Character 24).

0 = Quadrate ramus with posterior orientation.

1 = Quadrate ramus with posterolateral orientation.

2 = Quadrate ramus with lateral orientation.

Character 29: Pterygoid, relative length of quadrate ramus (QRL; 96). The length of the quadrate ramus varies depending on the configuration of the braincase with respect to the position of the jaw articulation. Thus, the distance from the otic capsule to the palatoquadrate and the length of the quadrate ramus may be long or short.

0 = Quadrate ramus long.

1 = Quadrate ramus short.

Character 30: Pterygoid, relative width of quadrate ramus (QRW). When viewed in ventral aspect, that portion of the quadrate ramus that is visible may appear as a narrow flange or a relatively robust (i.e., wide) strut.

0 = Quadrate ramus narrow in ventral aspect.

1 = Quadrate ramus robust in ventral aspect.

Character 31: Pterygoid, temporal flange (PTF). In some taxa, the pterygoid bears a flange that extends toward the temporal fossa from the posterolateral margin of the bone in the area between the quadrate ramus of the pterygoid and the maxilla.

0 = Pterygoid flange present.

1 = Pterygoid flange absent.

Character 32: Pterygoid, anterior ramus articulating with maxilla (ARM; 97). The anterior (or palatine) ramus of the pterygoid extends anterolateral from the otic capsule. Its distal articulation(s) is highly variable depending on the configuration of the skull and the number and position of palatal bones. In some taxa, the ramus articulates along the medial margin of the maxilla in the orbital region. Character 32 may be associated with Character 37.

0 = Anterior ramus does not articulate with maxilla.

1 = Anterior ramus articulates with maxilla.

Character 33: Pterygoid, anterior ramus articulating with palatine (ARP; 98). Depending on the position of the palatine and the length of the anterior pterygoid ramus, the ramus may articulate with the palatine along its medial or posterolateral margin.

0 = Anterior ramus articulates with palatine.

1 = Anterior ramus does not articulate with palatine.

Character 34: Pterygoid, anterior ramus articulating with vomer (ARV; 99). In taxa with large pterygoids and vomers, the anterior ramus of the pterygoid may articulate with the posterior margin of the vomer. The polarity of this character is reversed in Analysis II.

0 = Anterior ramus articulates with vomer.

1 = Anterior ramus does not articulate with vomer.

Character 35: Pterygoid, anterior ramus free (ARA; 100). Occasionally, owing to reduction in the length of the anterior ramus of the pterygoid and/or the presence of an incomplete maxillary arcade, the ramus may lack any bony articulation.

0 = Anterior ramus bears one or more bony articulations.

1 = Anterior ramus lacks any bony articulation.

Character 36: Pterygoid, medial articulation between anterior rami (AMA; 101). In some taxa, the anterior rami of the pterygoids bear a medial articulation with one another posterior to the vomers and anterior to the parasphenoid. The polarity of this character is reversed in Analysis II.

0 = Anterior rami with medial articulation.

1 = Anterior rami separated and lacking medial articulation.

Character 37: Ectopterygoid (EPT; 102). If present, this dermal palatal element is located medial to the maxilla and posterior to the palatine. It articulates with the palatine and pterygoid, and may or may not articulate with the maxilla.

0 = Ectopterygoid present.

1 = Ectopterygoid absent.

Character 38: Stapedial foramen (STF; 103). The proximal part of the stylus of the stapes may bear a foramen for passage of the facial (i.e., stapedial) blood vessels and/or the hyomandibular trunk of Cranial Nerve VII (facialis).

0 = Stapedial foramen present.

1 = Stapedial foramen absent.

Character 39: Articular (ART; 104). Meckel's cartilage may be ossified at the articulation of the mandible with the palatoquadrate cartilage to form the articular bone.

0 = Articular present as an independent bone in adult.

1 = Discrete articular bone absent in adult.

Character 40: Prearticular (PRT; 105). This dermal element sheaths the lingual surface of Meckel's cartilage anterior to the articular bone (if present) and forms the dorsal margin of the prearticular fossa.

0 = Prearticular present as independent bone in adult.

1 = Prearticular absent in adult.

Character 41: Surangular (SUR; 106). When present, this dermal element lies on the lateral surface of the mandible between the anterior dentary and the posterior articular (if an articular is present). The surangular is located dorsal to the angular.

0 = Surangular present as independent bone in adult.

1 = Surangular absent in adult.

Character 42: Splenial (SPL; 107). If present, the splenial(s) is a lingual dermal element of the mandible that lies ventral to the dentary. The polarity of this character is reversed in Analysis II.

0 = Splenial(s) present as independent bone(s) in adult.

1 = Splenial(s) absent in adult.

Character 43: Retroarticular process of the mandible (RAR; 108). In a few taxa, the mandible may bear an extension posterior to the articulation of the lower jaw with the palatoquadrate. The extension usually is curved posterodorsally and is termed a retroarticular process.

0 = Retroarticular process absent.

1 = Retroarticular process present.

Pectoral Girdle and Forelimb

Character 44: Interclavicle (ICL; 109). The interclavicle is a ventromedial dermal element of the pectoral girdle that, if present, lies between the paired clavicles. This character is considered to be a lissamphibian synapomorphy by A. R. Milner (1988).

0 = Interclavicle present.

1 = Interclavicle absent.

Character 45: Cleithrum (CLE; 110). When present, this dermal component of the pectoral girdle invests the leading edge of the scapula and articulates with the distal end of the clavicle.

0 = Cleithrum present.

1 = Cleithrum absent.

Character 46: Clavicle (CLA; 111). When present, the clavicle is a dermal element of the pectoral girdle that articulates with the proximal end of the scapula. Ventromedially, the clavicles either form a symphysis with one another or articulate with an interclavicle. The clavicles may be reduced, slim elements or may be expanded to be the predominant elements of the pectoral girdle.

0 = Clavicles are the predominant components of the pectoral girdle.

1 = Clavicles reduced to slender elements along leading edge of ventromedial part of pectoral girdle.

2 = Clavicles absent.

Character 47: Supinator process of humerus (SPH; 112). This process of the main shaft of the humerus lies proximolateral to the radial condyle and provides a surface for the attachment of a muscle that presumably facilitated rotation of the distal forelimb. A. R. Milner (1988) used this character to characterize a clade including lissamphibians within the dissorophoids.

0 = Supinator process of humerus present.
1 = Supinator process of humerus absent.

Character 48: Radius and ulna (R&U; 113). These skeletal elements of the forearm usually are separate; however, in some taxa they are fused.

0 = Radius and ulna not fused.
1 = Radius and ulna partially or wholly fused.

Pelvic Girdle

Character 49: Ilium, shape (ILM; 114). The dorsal part of the pelvic girdle, which is connected to the axial skeleton, is the ilium. When present, the ilium may be short and club-shaped, or elongated and slender.

0 = Ilium short and club-shaped.
1 = Ilium long and slender.

Fore- and Hind Limbs

Character 50: Limbs (LIM; 115). Limbs are defined as pectoral and pelvic appendages consisting of a humerus, radius, ulna and manus, and femur, tibia, fibula and pes, respectively.

0 = Limbs present.
1 = Limbs absent.

Axial Column

Character 51: Tuberculum interglenoideum (TIG; 116). If present, this structure is a well-developed anteroventral process of the atlas that projects into the foramen magnum and articulates with the cranium.

0 = Tuberculum interglenoideum absent.
1 = Tuberculum interglenoideum present.

Character 52: Urostyle (URO; 117). Caudal vertebrae are defined as those elements of tetrapods located posterior to the sacrum; their numbers vary, and they may be present or absent. The urostyle [= coccyx] is a specialized, rodlike axial structure that lies posterior to the sacrum and is formed developmentally by fusion of caudal vertebrae.

0 = Caudal vertebrae unmodified.
1 = Caudal vertebrae modified into urostyle.

Character 53: Articulation of ribs (RIB; 118). In most organisms, the proximal end of the rib articulates with the vertebral column via two heads, which may or may not be fused. However, in a few taxa, the proximal rib is unicapitate and articulates with modified transverse processes of vertebrae. *Triadobatrachus* is coded as having unicapitate ribs because only the ribs of the atlas are bicapitate (Estes and Reig, 1973; Rage and Roček, 1989). The polarity of this character is reversed in Analysis II.

0 = Bicapitate ribs present.

1 = Unicapitate ribs present.

Character 54: Axial dermal armor (ADA). These segmental dermal os-sifications lie dorsal to, and sometimes are fused with, the neural spines of the vertebrae (DeMar, 1966).

0 = Axial dermal armor absent.

1 = Axial dermal armor present.

Scales and Dentition

Character 55: Scales of dermal origin (SCA; 119). When present, scales are composite tissue structures that generally consist of varying combina-tions of an outer, ectodermally derived, acellular enamel layer [= ga-noine of Actinopterygii], a mesodermally derived middle layer of den-tine, and a fused inner (i.e., basal) layer of dermal bone. The later may be of one or more types including cementum, "isopedine," among others.

0 = Scales of dermal origin present.

1 = Scales of dermal origin absent.

Character 56: Pedicellate teeth (PED; 120). When present, pedicellate teeth are characterized by division into two parts—a basal dentine pedicel and a distal dentine crown—with a fibrous connection (Parsons and Wil-liams, 1962). The division between the pedicel and crown lies in the orthodentine (Schultze, 1970). Bolt (1977a, 1979) described the presence of pedicellate teeth in *Tersomius, Doleserpeton,* and *Amphibamus grandiceps.* Bolt (this volume) provides a review of this character. *Apateon* is consid-ered to have pedicellate teeth based on illustrations of Boy (1978:Fig. 5, 1987:Fig. 27).

0 = Pedicellate teeth absent.

1 = Pedicellate teeth present.

Character 57: Cusping of teeth (CSP; 121). The distal portion of the tooth may be monocuspid, or subdivided to produce two (bicuspid) or more (multicuspid) prominences or points. Boy (1972:Fig. 11, 1987:Fig. 6) illus-trated the multicuspid teeth of *Apateon, Leptorophus, Schoenfelderpeton, Branchiosaurus,* and *Micromelerpeton.* These denticules are similar to the fetal teeth of caecilians (M. Wake, 1976). A taxon is coded as 1 if the apomorphic condition is present at some ontogenetic stage.

0 = Teeth monocuspid.

1 = Teeth bicuspid or multicuspid.

Character 58: Tooth replacement (TRP; 122). Among vertebrates bearing teeth, tooth replacement occurs in vertical waves that travel along the rami of the jaws. Thus, as a tooth is shed owing to resorption or a loosening of its connection with the jaw bone, it is replaced by a new tooth that has formed from a tooth germ in the tissue above (or below) the mature tooth. However, in Recent amphibians, replacement teeth form lingual to mature teeth and move peripherally to replace them.

0 = Vertical tooth replacement.

1 = Mediolateral tooth replacement.

RESULTS

Relationships among Living Taxa

The relationships among living sarcopterygians (i.e., Actinistia, dipnoans, apodans, salamanders, anurans, and amniotes) were analyzed using two different data sets. The first (Analysis I) was composed of 72 soft-anatomical characters (Data Set I, Character 1–72; Appendix I). The second (Analysis II) was based on a total of 122 characters—soft-anatomical Characters 1–72 plus 50 osteological features (i.e., Data Set I + part of Data Set II; Appendix I). In the case of hard characters, two numbers separated by a slash appear in parentheses; the first number refers to the character number in the 122-character analysis (see Appendix I), whereas the second refers to the character's number in the description of characters in Data Set II. In each case, the description of the cladogram proceeds from the most inclusive to the least inclusive nodes of the tree. In both Analyses I and II, the hypothetical ANCESTOR was reconstructed based on the Actinistia (first out-group) and the Actinopterygii (second outgroup). In the case of equally parsimonious trees, each node is described by the total possible number of characters and the total number of characters in common to all trees. The total number of characters corresponds to the sum of all different character-states occurring at a given node, independent of the number of trees in which the character is present. The second descriptor is more conservative and reflects only the number of character-states common to a given node. Only the common characters would be described to diagnose each node. The distribution of the characters reflects the option selected in the analysis; equally parsimonious patterns of individual characters are not reported.

Tree 1: Soft-Morphological Characters

When rooted on the hypothetical ANCESTOR, six equally parsimonious trees, each with 112 steps (105 steps without autapomorphies) and a Consistency Index of 0.670 (C.I. = 0.648, with autapomorphies deleted) were obtained. Six different tree topologies that can be reduced to two topological variants were obtained (Fig. 2). The first topological variant involves the relative positions of the Actinistia and Actinopterygii,

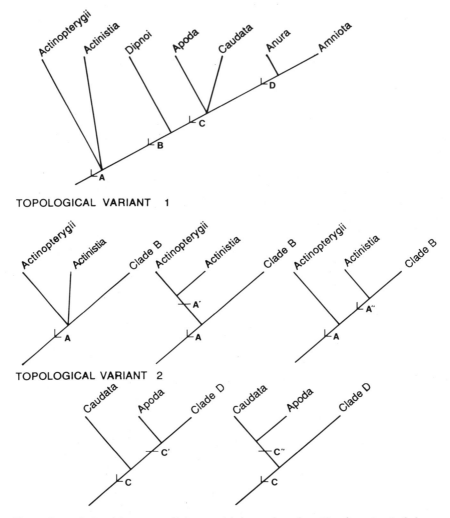

Fig. 2. Interrelationships among living osteichthyans based on 72 soft-anatomical characters in Analysis I. Strict/Adams consensus tree based on six equally parsimonious trees at
112 steps and a Consistency Index of 0.670 (105 steps and C.I. = 0.648, with autapomorphies deleted). The Mickevich Consensus Information Index is 0.667. The different topologies of Variants 1 and 2 are illustrated and discussed in the text.

whereas the second topological variant concerns the relative positions of
the Caudata and Apoda. The topologies of the strict and Adams consensus trees are identical (Fig. 2) (Mickevich C.I. = 0.667):

[Actinopterygii + Actinistia + [Dipnoi + [Apoda + Caudata + [Anura +
Amniota]]]]

The principal points of interest in this cladistic arrangement are that modern amphibians are shown to be paraphyletic with respect to the Amniota, and dipnoans are the living sister-group of tetrapods.

Node A: [*Actinopterygii* + *Actinistia* + [*Dipnoi* + [*Apoda* + *Caudata* + [*Anura* + *Amniota*]]]]. No character that corroborates the monophyly of Osteichthyes has been included in the analysis.

Node B: [*Dipnoi* + [*Apoda* + *Caudata* + [*Anura* + *Amniota*]]]. Dipnoans are shown to be the sister-group of tetrapods on the basis of 11 synapomorphies at Node B. Only one feature is a unique shared-derived character—the absence of the second aortic arch (38). Of the remaining 10 characters, only three homoplastic features are common to the six topologies: presence of an interventricular septum in the heart (35; reversed in anurans and caudates); absence of a cranial foramen for the passage of C.N. XII (55; reversed in amniotes); and presence of the bile salt, bufol (69; absent in amniotes and presence or absence unknown in caecilians).

Node C: [*Apoda* + *Caudata* + [*Anura* + *Amniota*]]. There are 29 possible synapomorphies at Node C (i.e., the Tetrapoda) of Tree 1; of these, 19 are common to the six topologies. Among the congruent characters, eight are unique shared-derived features. All taxa possess a stratum corneum (4), Harderian gland associated with the eye (11), intermaxillary glands (12), a discrete adrenal gland that protrudes into the coelom (13), vomeronasal organ (16), a visual accommodation by movement of lens distally (18; not applicable in amniotes and apodans), a tympanic cavity (21), a complete (34.1) or fenestrate (34.2) interatrial septum, and Purkinje cells that are organized into distinct nuclei in the cerebellum (60; unknown in apodans).

There are an additional 11 homoplasies at Node C. The conditions of seven of these are reversed in amniotes (3, 19, 23, 28, 37, 50, 67). The characters present are as follows: Leydig cells (3; unknown in caecilians), absence of either kinocilium or microvilli on apical surface of Lorenzinian ampullae (9; reversed in apodans), green rods in the eye (19; unknown in apodans), a papilla amphibiorum in the ear (23), round window (27; reversed in apodans), an opercular element associated with the proximal stapes (28), cutaneous branches of the truncus arteriosus (37; unknown in apodans), tracheal cartilages (42; reversed in anurans), fat bodies (50), and formation of the gastrocoel from a ventrolateral invagination (67).

Node D: [*Anura* + *Amniota*]. There are 19 possible synapomorphies at Node D of Tree 1; of these, 11 are common to the six topologies. Among the congruent characters, seven are unique shared-derived features. In each taxon ampullae are absent in the lateral-line system (8), and a tympanum (20) and eustachian tube (22) are present. In addition, the interatrial septum is complete (34.1), and ciliated pronephric tubule segments are absent (48; unknown in actinistians). The site of spermatogenesis is coincident with the location of permanent populations of primordial germ cells (52; unknown in actinistians and dipnoans). The torus semicircularis is not clearly distinguishable from the optic tectum and does not protrude into the ventricular lumen (61.1); however, the other derived state (61.2) occurs in apodans.

There are four problematic characters at Node D. Three (26, 56, 58) are unpolarized, and Character 63 is homoplastic. The perilymphatic duct is associated with the sacculus (26.1) in the Caudata and Apoda, whereas it is associated with the lagena (26.0) in amniotes; its condition is unknown in the Anura. The state of Character 56, the exit of first spinal nerve via an intervertebral foramen, is unknown in actinopterygians, dipnoans, and amniotes. With respect to the relative development of the corpus cerebelli and auricular subdivisions of the cerebellum (58), we know that apodans have a condition different from that of anurans and amniotes; however, the polarity of the character is unknown. The possession of distinctly stratified cell masses in the nucleus radicis mesencephalicae (63; unknown in actinistians) is homoplastic with respect to the Actinopterygii.

Apoda. There are 13 possible synapomorphies for apodans in Tree 1; of these, 10 are common to the six topologies. Apodans are diagnosed by five synapomorphies: the possession of laminophores (and absence of iridophores) (7), a tentacle (15), a heart having a right atrium that is smaller than the left (33.1), a serial arrangement of pronephric nephrostomes in the adult (46), and nephrostomes that retain the integrity of the nephrostome walls (47).

Of the remaining five characters, three are reversals. The reversals include the presence of kinocilia and microvilli in the apical surface of the Lorenzinian ampulla (9; unclear in *Latimeria*) and the absence of a round window (27). Another reversal involves the structure of the kidney; apodans lack modification of the pronephros for sperm transport (45). Two additional characters are homoplastic with respect to dipnoans. In both dipnoans and apodans, the basal plate of the mesencephalon is not differentiated regionally (64), but the trochlear nucleus is differentiated (66).

Caudata. There are nine possible synapomorphies for caudates in Tree 1; of these, five are common to the six topologies. Salamanders have only a single unique feature, the sinistral position of the sinus venosus of the heart (31). Three homoplasies characterizing the caudates also occur in anurans. A papilla neglecta is absent (24), a carotid labyrinth is present (36), and during development, the choanal tube opens into the archenteron (41). There is one homoplastic reversal in caudates—the absence of subdivision of the ventricle (35).

Anura. There are nine possible synapomorphies for anurans in Tree 1; of these, six are common to the six topologies. Anurans possess only a single synapomorphy—Character 1, the absence of a uniform and firm connection between the dermis and hypodermis. Three homoplasies, as noted above, also occur in salamanders—the absence of a papilla neglecta (24), the possession of a carotid labyrinth (36), and the choanal tube opening into the archenteron during development (41). One homoplastic reversal occurs in the Anura, the absence of tracheal cartilages (42), which is a reversal from the derived state at Node C.

Amniota. There are 19 possible synapomorphies for amniotes in Tree 1; of these, 16 are common to the six topologies. Four characters (10, 17, 33, 40) are unique to amniotes. The first is the absence of granular glands (10; unknown in dipnoans). The second (17) involves the evolution of the ciliary muscle that focuses the eye by deformation of the lens, in contrast to the system in the anamniotes in which the lens is moved either proximally or distally via the m. protractor or retractor lentis. The third (33) is the possession of equal-sized atria, and the fourth (40), the absence of Aortic Arch VI.

Of the 12 homoplastic characters, 10 are reversals, seven of which reverse from the derived conditions of Node C. These include the absence of the following structures: Leydig cells (3; unknown in apodans), green rods (19; unknown in apodans), papilla amphibiorum (23), an opercular element associated with the proximal stapes (28), cutaneous branches of the ductus arteriosus (37; unknown in apodans), and fat bodies (50; variable within amniotes). And last, the gastrocoel results from a dorsal invagination of the outer wall of the egg (67). Three of the homoplastic characters reverse from the derived conditions at Node B. These include the presence of a cranial foramen for C.N. XII (55) and the absence of the liver bile salt bufol (69; unknown in apodans) and lens protein D3 (72; unknown in apodans and actinistians). There are two additional homoplastic characters; as in actinistians and dipnoans, a pronephros is absent in adult amniotes (43), and as in actinistians, Mauthner cells are absent (53).

Topological variants in Tree 1. There are two topological variants in the first tree. The first involves Actinopterygii, Actinistia, and Clade B, whereas the second involves Caudata, Apoda, and Clade D (Fig. 2).

First topological variant: There are three topologies associated with this variant, which is located at Node A (Fig. 2). In none of the six trees is Node A supported by any characters.

One topology is as follows:

[Actinistia + Actinopterygii + Clade B]

In the two trees in which this arrangement occurs, this trichotomy is unsupported. However, in addition to the four characters described above common to each of the six topologies, Clade B is diagnosed by seven additional characters, of which four are unique shared-derived characters (14, 68, 70, 71). Thus, these organisms have intermingled steroidogenic and chromaffin cells (14) and possess mesotocin (68) and lens proteins D1 and D2 (70, 71). The three remaining features reverse in more advanced taxa. Thus, the modification of the pronephros for sperm transport (45) reverses in apodans, and the elimination of the pronephros for sperm transport (44) reverses at Node D. The presence of lens protein D3 (72) reverses in amniotes.

The second typology is as follows:

[[Actinistia + Actinopterygii] + Clade B]

In the two trees in which this arrangement occurs, [Actinistia + Actinopterygii] is supported by two characters. The nucleus lentiformis is well defined (62), and the cell masses of the nucleus radicis mesencephalicae are distinctly stratified (63); these characters are homoplastic with respect to Nodes C and D, respectively. The diagnosis of Clade B is identical to that of the first topology described above.

The third topology is as follows:

[Actinopterygii + [Actinistia + Clade B]]

In the two trees in which this arrangement occurs, Clade A'' [Actinistia + Clade B] is corroborated by a suite of 10 characters, seven of which were used to diagnose Clade B in the first two topologies described (14, 44, 45, 68, 70, 71, 72). The three additional homoplastic features reverse in more advanced taxa. These are the possession of a perilymphatic cistern (25) and the differentiation of the oculomotor nucleus (65), which reverse in dipnoans. The last feature, the absence of the pronephros in the adult (43), reverses at Node C and then is regained in amniotes. In this topology, Clade B is diagnosed only by the four characters in common to the six trees—35, 38, 55, and 69.

Second topological variant: There are two topologies associated with this variant, which is located at Node C (Fig. 2). This topological variant influences the diagnoses of Nodes C and D. At Node C of the following two topologies, the variation of Characters 25, 43, and 65 is associated with the third topology of Topological Variant 1 discussed above.

The first topology is as follows:

[Caudata + [Apoda + Clade D]]

which occurs in three trees. Thus, these organisms possess dermal body folds correlated with body segmentation (6), intrinsic narial musculature (29), and lobulate testes (51). Clade C' [Apoda + Clade D] is diagnosed by 10 characters, six of which are unique shared-derived features. Included are the possession of a reduced (30), subdivided (32.1) sinus venosus, and the absence of Aortic Arch V (39). In addition, a superior olivary nucleus is present (54), and there is a marked difference in the degree of development of the corpus cerebelli and auricular subdivision of the cerebellum (57). Last, the torus semicircularis is differentiated from the optic tectum and forms a prominent protrusion into the ventricular lumen of the mesencephalon (61.2). Of the remaining four characters, two—the presence of a distinct myoepithelium (2) and the absence of a mucous cuticle (5)—reverse in anurans. The two final characters—regular arrangement of Purkinje cells (59) and well-defined nucleus lentiformis (62)—are homoplastic with respect to actinopterygians.

In addition to the 11 common characters diagnostic of Node D, this topology bears five more characters at this node. Three are reversals from Node C—6, 29, 51 (described above). The single reversal from Node B is Character 44—utilization of the pronephros for sperm transport. The last character, independence of the testis and kidney (49.2), is homoplastic with respect to *Latimeria*.

The second topology is as follows:

[[Caudata + Apoda] + Clade D]

which occurs in three trees. In addition to the 19 characters in common among all six trees for Node C (cf. description above), there are four others in common to these three trees. These are Characters 30, 32, 54, and 57, each of which reverses in Caudata. Thus, these organisms possess a reduced (30), subdivided (32.1) sinus venosus, a superior olivary nucleus (54), and a marked difference in the degree of development of the corpus cerebelli and auricular subdivision of the cerebellum (57). Clade C'' [Caudata + Apoda] is supported by five characters, four of which are unique shared-derived features. Caudates and apodans possess dermal body folds correlated with body segmentation (6), intrinsic narial musculature (29), a pronephros that is not used for sperm transport (44), and lobulate testes (51). The fifth feature is a reversal from Node B—the testis is connected by a series of ducts to the anterior part of the kidney (49.0).

In addition to the 11 characters in common among the six trees, Clade D [Anura + Amniota] is diagnosed by three additional characters in this topology. The absence of Aortic Arch V (39) is a unique shared-derived character. The last two features are homoplastic with respect to actinopterygians; anurans and amniotes possess regularly arranged Purkinje cells (59) and a well-defined nucleus lentiformis (62).

Tree 2: Soft-Morphological and Osteological Characters

Only one tree (Fig. 3) with 184 steps and a Consistency Index of 0.707 (C.I. = 0.660 with autapomorphies deleted) was generated in this analysis. Because this analysis was conducted at a high level of universality, we have had to deal with remote out-groups in order to include amniotes and dipnoans. Accordingly, the plesiomorphic condition of the characters (i.e., the hypothetical ANCESTOR) was determined by a double comparison with the out-groups Actinopterygii and Actinistia, and this resulted in a reversal in the polarities of 10 of the 50 osteological characters. Characters 75/4, 79/12, 80/13, 87/20, 90/23, 92/25, 99/34, 101/36, 107/42, and 118/53 are flagged throughout the following discussion of Tree 2 with an asterisk to remind the reader of the reversal. The topology of Tree 2 is represented as follows:

[Actinopterygii + [Actinistia + [Dipnoi + [[Apoda + [Caudata + Anura]] + Amniota]]]]

In Tree 2, in contrast to Tree 1, living amphibians are shown to be a monophyletic assemblage in which the Amniota is the living sister-group. Within the amphibians, the apodans are the sister-group to the salamanders and anurans.

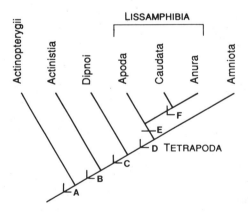

Fig. 3. Interrelationships among living osteichthyans based on 122 soft-anatomical and osteological characters in Analysis II. The single tree generated by PAUP has 184 steps and a Consistency Index of 0.707 (C.I. = 0.660 with autapomorphies deleted). See text for explanation of Character Nodes A–F.

Node A: [*Actinopterygii* + [*Actinistia* + [*Dipnoi* + [[*Apoda* + [*Caudata* + *Anura*]] + *Amniota*]]]]. As in Tree 1, Node A lacks any uniquely shared-derived characters; however, there are two homoplasies. The nucleus lentiformis is well defined (62); this character is reversed in dipnoans and caudates. The orbit is moderate-sized (73.1). This character reverses to the plesiomorphic state at Node C, and its alternate state (large orbit) occurs at Node F.

Node B: [*Actinistia* + [*Dipnoi* + [[*Apoda* + [*Caudata* + *Anura*]] + *Amniota*]]]. This group is supported rather weakly by four synapomorphies at Node B. The steroidogenic and chromaffin cells of the adrenal gland are intermingled (14), but the condition of this character is unknown in actinistians and dipnoans. The states of the remaining three synapomorphies are unknown in actinistians and apodans; these include the presence of the neurohypophysial hormone mesotocin (68), as well as lens proteins D1 and D2 (70, 71).

Of the 10 homoplasies at Node B, two are reversed to the plesiomorphic condition at Node E for the lissamphibians—viz., the absence of a pronephros in the adult (43), and the presence of a tabular (*87/20). Six other characters show single reversals elsewhere on the tree. Among these are the presence of a perilymphatic cistern (25; reversed in dipnoans), pronephros not used for sperm transport (44; reversed in anurans), differentiation of the oculomotor nucleus (65; reversed in dipnoans), the presence of lens protein D3 (72; unknown in actinistians and apodans, and reversed in amniotes), and the presence of the following bones—prefrontal (*79/12; reversed in anurans), postfrontal (*80/13; reversed at Node F), splenial (*107/42; reversed in anurans). The last homoplasy, the independence of the testes and kidney (49.2) is an unordered character. Reversals to the plesiomorphic state occur at Node E; the alternate state of the character (49.1) occurs in dipnoans and anurans.

Node C: [*Dipnoi* + [[*Apoda* + [*Caudata* + *Anura*]] + *Amniota*]]. Dipnoans and tetrapods are characterized by one shared-derived character in addition to Character 38 of Tree 1—the absence of an intertemporal bone (85/18; not applicable for dipnoans).

Whereas 10 homoplastic features characterized dipnoans and tetrapods in Tree 1, here there are eight plus one homoplastic reversal. Two of the homoplasies are duplicated in Tree 1—Characters 55 and 69 (both reversed in amniotes). Another three are reversed at Node F. These are the presence of a distinct myoepithelium (2; unknown in actinopterygians and dipnoans), a subdivided ventricle (35), and a triangular parasphenoid (94/27). Two other features are reversed at Node E—anterior ramus of pterygoid not articulating with vomer (*99/34), and the presence of a medial articulation between the anterior rami of the pterygoids (*101/36). The final homoplasy is the coincidence of the jaw articulation with the posterior limits of the endocranium (91/24). This is an unordered character, and the same condition occurs in anurans, whereas the alternate state (24.1) characterizes Node E.

The single homoplastic reversal involves unordered Character 73/1. Dipnoans and tetrapods have a small orbit (1.0), whereas the ancestor of the osteichthyan clade has a moderate-sized orbit (1.1), and anurans and salamanders (Node F) are characterized by a large orbit (1.2).

Node D: [[*Apoda* + [*Caudata* + *Anura*]] + *Amniota*]. The tetrapods have nine unique shared-derived features in Tree 2. The suites of synapomorphies in Trees 1 (Node C) and 2 (Node D), are the same except for the presence of an additional synapomorphy in Tree 2—visual accommodation by distal movement of the lens (18); both trees lack any unique shared-derived osteological features at these nodes.

Whereas Tree 1 has 21 homoplasies at this node, there are 13 in Tree 2, three of which are duplicates of those in Tree 1 (9, 27, 42). The derived states of six characters are reversed in the Caudata—viz., the possession of a reduced sinus venosus (30), subdivision of the sinus venosus (32), superior olivary nucleus (54), a marked difference in the degree of development of corpus cerebelli and auricular subdivision of cerebellum (57), the regular arrangement of Purkinje cells (59; unknown in apodans, and derived in actinopterygians), and the absence of Aortic Arch V (39). The derived states of three other tetrapod characters are reversed in apodans; among the latter are the absence of either kinocilia or microvilli on the apical surface of the ampulla (9), the presence of a round window in the ear (27), and the absence of scales of dermal origin (119/55). Two tetrapod synapomorphies are reversed in the Anura; tetrapods possess tracheal cartilages (42) and unicapitate ribs (118/53; variable in the Caudata). A single tetrapod character is reversed at Node F [Caudata + Anura]; tetrapods lack a mucous cuticle (5).

The distribution of the character-states of the remaining two homoplastic characters is somewhat more complex. Tetrapods possess a fenestrate interatrial septum (34.2); an alternate state (complete interatrial septum; 34.1) occurs in anurans and amniotes. The final homoplasy is also an unordered character—the possession of a torus semicircularis that is not clearly distinguished from the optic tectum and does not protrude into the ventricular lumen (61.1). Apodans possess the alternate state of this character (61.2), and the caudates are characterized by the plesiomorphic state (61.0).

Node E: [*Apoda* + [*Caudata* + *Anura*]]. Lissamphibians are diagnosed by a total of 29 synapomorphies at Node E in Tree 2, of which 17 are unique shared-derived characters. Eight are soft-anatomical features that include the presence of Leydig cells (3; unknown in apodans), green rods in the eye (19; unknown in caecilians), a papilla amphibiorum in the ear (23), an opercular element associated with the proximal stapes (28), cutaneous branches of the ductus arteriosus (37; unknown in caecilians), and fat bodies that originate from the germinal ridge associated with the gonads (50). In addition, the gastrocoel is formed by a ventrolateral invagination of the outside wall of the egg (67; unknown in caecilians). The remaining nine characters are osteological. The premaxilla has a pars palatina (*75/4), and the quadrate ramus of the pterygoid has a posterolateral orientation (*95/28). Several elements are absent; among these are the postparietal (84/17), supratemporal (86/19), jugal (89/22), cleithrum (110/45), and clavicle (*111/46; alternate state of reduced clavicles [111.1] in anurans). The final three unique shared-derived characters are the possession of pedicellate (120/56) and bicuspid (121/57) teeth that are replaced mediolaterally (122/58).

There are 12 homoplastic characters at Node E. Three are reversed in the Anura—viz., the presence of dermal body folds correlated with body segmentation (6), intrinsic narial musculature (29), and lobulate testes (51). Another character (104/39), the absence of a discrete articular bone, is reversed in the Caudata. Five other characters represent reversals from derived features at Nodes B and C. Among the former are the retention of the pronephros in adult tetrapods (43), the connection of the testis to the anterior part of the kidney by a series of ducts (49; alternate state 49.1 in anurans), and the absence of a tabular (87/20). The two characters reversed from their derived states at Node C are anterior rami of pterygoids lacking articulation with the vomer (99/34) and being medially separated from one another (101/36). Of the three remaining characters, the absence of an interclavicle (109/44) is homoplastic with respect to the Actinistia. The other two are unordered characters. Tetrapods lack a parietal foramen and the parietal organ is covered by dermal bone (83/16.2; alternate state 83/16.1 in apodans), and they possess a jaw articulation that lies forward to the posterior limits of the endocranium (91/24.1; alternate state 91/24.2 in anurans).

Node F: [*Caudata + Anura*]. A suite of 15 synapomorphies diagnoses this clade at Node F; nine are unique shared-derived characters. Of the latter, four involve soft-anatomical features—the absence of a papilla neglecta (24), the possession of a carotid labyrinth (36) and a choanal tube that opens into the archenteron during development (41), and modification of the pronephros for sperm transport (45). The remaining five unique shared-derived characters are osteological and include possession of a large orbit (73/1.2) and moderate-sized external naris (74/2). Postorbital (88/21) and surangular (106/41) bones are absent, and the anterior ramus of the pterygoid does not articulate with the palatine (98/33).

Of the six homoplastic characters, one—the absence of an ectopterygoid bone (102/37)—is homoplastic with respect to dipnoans, whereas the other five represent reversals of derived characters at preceding nodes on the tree. One feature is reversed from its condition at Node B—caudates and anurans lack a postfrontal (*80/13). Three are reversed from their conditions at Node C. Thus, these taxa lack a distinct myoepithelium (2) and subdivided ventricle (35) and possess a parasphenoid with a rectangular corpus (94/27). There is a single reversal from Node D involving the presence of a mucous cuticle (5).

Apoda. Apodans are diagnosed by a total of 18 synapomorphies, 12 of which are unique shared-derived characters, in Tree 2. Five of the 12 unique shared-derived characters (7, 15, 33.1, 46, 47) are identical to those obtained in Analysis I. Of the additional features obtained in this analysis, two are soft-anatomical features—viz., auricular subdivision of the cerebellum markedly large and corpus cerebelli small (58), and torus semicircularis differentiated from the optic tectum and forming a prominent protrusion into the ventricular lumen of the mesencephalon (61.2). The remaining five unique shared-derived features are osteological. In caecilians, the prefrontal bone does not form part of the margin of the opening for the external naris (*90/23), and the parasphenoid is coossified with the neurocranium (93/26). The quadrate ramus of the pterygoid is short (96/29); a retroarticular process is present (108/43), and limbs are absent (115/50).

There are six homoplastic characters. Three are reversals from derived conditions at Node D (tetrapods). These are the presence of kinocilia and microvilli on the apical surface of the Lorenzinian ampulla (9; unclear in actinistians), the absence of a round window in the ear (27), and the presence of scales of dermal origin (119/55). Two apodan characters are homoplastic with respect to dipnoans—viz., the lack of regional differentiation of the basal plate of the mesencephalon (64), and the presence of differentiation of the trochlear nucleus (66). The final character involves the absence of a parietal foramen with the parietal organ covered by dermal bone (83/16); this feature of apodans is homoplastic with respect to actinistians.

Caudata. This group is diagnosed by 13 synapomorphies on Tree 2; there are an additional three unique shared-derived characters in addition to Character 31 discussed above in Tree 1. The group is characterized by having an incomplete maxillary arcade (76/6), a pterygoid having a free anterior ramus (*100/35), and a tuberculum interglenoideum (116/51).

Caudates possess nine homoplastic reversals on Tree 2. One, which also is reversed in dipnoans, is a reversal of a derived character at Node A—poor definition of the nucleus lentiformis mesencephali (62). Seven others represent reversals of derived, soft-anatomical characters at Node D (tetrapods). Included are a large, undivided sinus venosus (30, 32), the presence of Aortic Arch V (39), the absence of the superior olivary nucleus (54), equally developed corpus cerebelli and auricular subdivisions of the cerebellum (57), irregularly arranged Purkinje cells (59), and the possession of a differentiated torus semicircularis that is not markedly protuberant into the ventricular lumen of the mesencephalon (61). The final homoplastic character is a reversal from Node E—the possession of an independent articular bone in the adult (104/39).

Anura. In Tree 1, anurans were characterized by only one unique shared-derived character—the absence of firmly attached skin (1)—among nine synapomorphies. However, in Tree 2, there is a total of 30 synapomorphies that includes an additional 12 unique shared-derived characters. One is a soft-anatomical character—the exit of the first spinal nerve from the spinal cord via an intervertebral foramen (56). Absence predominates among the remaining 10 features. Thus, anurans lack a palatine (77/7), lacrimal (78/11), stapedial foramen (103/38), and prearticular (105/40). In anurans, the caudal vertebrae are modified to form a urostyle (117/52), and the frontal and parietal are fused to form one bone, the frontoparietal (82/15). In contrast to other living lower vertebrates, in anurans a distinct squamosal embayment is present (*92/25), and the anterior ramus of the pterygoid articulates with the maxilla (97/32). Other unique anuran features include fusion of the radius and ulna (113/48), the presence of a long, slender ilium (114/49), and reduction of the clavicles to slender elements that lie along the leading edge of the ventromedial part of the pectoral girdle (111.1/46).

Whereas there are three homoplasies and one homoplastic reversal for anurans in Tree 1, there are totals of nine homoplasies and eight homoplastic reversals for this taxon in Tree 2. Eight of the homoplasies also occur in the Amniota. These include the following soft-anatomical features: The absence of electroreceptive ampullae in the lateral-line system (8) and ciliated pronephric tubule segments (48; unknown in actinistians), and the presence of a tympanum (20), eustachian tube (22), and complete interatrial septum (34). Also, in both amniotes and anurans, the site of spermatogenesis is coincident with the location of the permanent populations of primordial germ cells (52; unknown in dipnoans and actinistians). Character 63 (unknown in actinistians), distinct stratification of the cell masses of the nucleus radicis mesencephalicae, also occurs in amniotes and actinopterygians. The single osteological homoplastic character present in anurans and amniotes also occurs in dipnoans—

viz., coincidence of the jaw articulation with the posterior limits of the endocranium (91.2). The ninth homoplastic character is the connection of the testis to the posterior part of the kidney by a series of ducts (49.1), a feature also found in dipnoans.

The eight remaining synapomorphies are reversals. Three are reversals from Node B; anurans utilize the pronephros for sperm transport (44) and lack prefrontal (*79/12) and splenial (*107/42) bones. There are two reversals from Node D; anurans lack tracheal cartilages (42) and possess unicapitate ribs (*118/53). Finally, there are three reversals from Node E; thus, anurans lack dermal body folds correlated with body segmentation (6) and intrinsic narial musculature (29), and they have smooth, rather than lobulate, testes (51).

Amniota. The diagnosis of the Amniota in Tree 2 is markedly different than that in Tree 1. There is a total of 18 synapomorphies, of which six features are unique shared-derived characters. Three of the latter (10, 17, 40) are the same as in Tree 1. The other three are association of the perilymphatic duct with the sacculus (26), and the presence of an articulation between the prefrontal and postfrontal (81/14) and supinator process on the humerus (112/47).

Of the nine homoplastic characters, seven also occur in the Anura. Included are the absence of electroreceptive ampullae (8) and ciliated pronephric tubule segment (48), and the presence of a tympanum (20), eustachian tube (22), and complete interatrial septum (34.1). The site of spermatogenesis is coincident with the location of permanent populations of primordial germ cells (52), and the cell masses of the nucleus radicis mesencephalicae are distinctly stratified (63), as they also are in actinopterygians. The eighth character, the absence of Mauthner cells (53), is homoplastic with respect to *Latimeria*, and the ninth, possession of equal-sized atria (33.2), is homoplastic with respect to apodans.

There are three homoplastic reversals on Tree 2. With respect to Node B, Character 72—the absence of lens protein D3—represents a reversal in the Amniota. The other two characters are reversals from Node C. Thus, in amniotes, there is a foramen for Cranial Nerve XII (55), and the bile salt bufol is absent (69).

Relationships between Dissorophoid and Lissamphibian Taxa, and among Lissamphibians

Tree 3A: Consensus Tree for Dissorophoids, Including Lissamphibians

This analysis is based on 22 taxa and 58 osteological characters (Data Set II), including a total of 64 apomorphic character-states. A consensus of temnospondyls (excluding the Dissorophoidea) was used as the first out-group for the Dissorophoidea. The temnospondyl out-group taxa included Edopoidea, Trimerorhachoidea, *Dendrerpeton*, and Eryopoidea. The second out-group used was the Loxommatoidea. Discussion focuses on dissorophoids (i.e., branchiosaurids, micromelerpetontids, trematopids, doleserpetontids, *Tersomius*, *Dissorophus*, and *Broiliellus*) and lissamphibians (i.e., Apoda, Caudata, *Karaurus*, *Albanerpeton*, *Triadobatrachus*, and Anura); however, nondissorophoid and nonlissamphibian temnospondyls also are included. The analysis for the temnospondyl taxa yielded 12 equally parsimonious trees, each with 104 steps and a Consistency Index of 0.615 (C.I. = 0.583 with autapomorphies deleted). The Adams and strict consensus trees based on these 12 trees have identical topologies (Fig. 4), and share a Mickevich Consensus Information Index of 0.645. The topologies of the cladograms obtained varied in the relationships of (1) the dissorophoid taxon *Dissorophus*, (2) the branchiosaurids *Branchiosaurus* and *Schoenfelderpeton*, and (2) the caudates and their allies (i.e., Caudata, *Albanerpeton*, and *Karaurus*). Thus, the topological variants are located at Nodes E, I, and R (Fig. 4). As above, explanation of the cladograms proceeds from the most inclusive nodes of the tree to the least inclusive nodes. Synapomorphies, homoplasies, and reversals are described for each node.

The primary significance of this cladistic arrangement is that the Micromelerpetontidae and "Branchiosauridae" are shown to be dissorophoids,

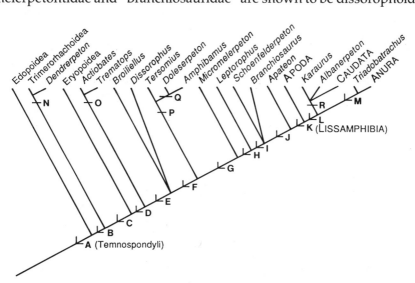

Fig. 4. Interrelationships among the Temnospondyli based on 58 osteological characters in Analysis III. The Strict/Adams consensus tree is based on 12 equally parsimonious trees of 104 steps and having a Mickevich Consensus Information Index of 0.645. The Consistency Index of the individual trees is 0.615. See text for explanation of Character Nodes A–R.

and the branchiosaurid *Apateon* is the sister-group of lissamphibians. The monophyletic lissamphibians are dissorophoids.

Node A: [*Edopoidea* + [[*Trimerorhachoidea* + Dendrerpeton] + [*Eryopoidea* + [*Trematopidae* + [Broiliellus + Dissorophus + [[Tersomius + *Doleserpetontidae*] + [*Micromelerpetontidae* + [Leptorophus + [Schoenfelderpeton + Branchiosaurus + [Apateon + [*Apoda* + [*Caudata* + [Triadobatrachus + *Anurans*]]]]]]]]]]]]]]. No character that corroborates the monophyly of the Temnospondyli has been included in the analysis.

Node B: [[*Trimerorhachoidea* + Dendrerpeton] + [*Eryopoidea* + [*Trematopidae* + [Broiliellus + Dissorophus + [[Tersomius + *Doleserpetontidae*] + [*Micromelerpetontidae* + [Leptorophus + [Schoenfelderpeton + Branchiosaurus + [Apateon + [*Apoda* + [*Caudata* + [Triadobatrachus + *Anurans*]]]]]]]]]]]]]. The trimerorhachoids and *Dendrerpeton* are shown to be the sister-group of the erypoids, dissorophoids, and lissamphibians on the basis of a single synapomorphy at Node B. The character, anterior rami of the pterygoid separated and lacking a medial articulation (36), is homoplastic, reversing in *Trematops*.

Clade N [Trimerorhachoidea + *Dendrerpeton*] is corroborated by two homoplastic characters. The first, possession of a moderate-sized orbit (1.1), is homoplastic with respect *Trematops* and *Broiliellus*. The second feature, the presence of an anterolateral process on the palatine (9), is homoplastic with respect to Node F and reverses at Node K.

Node C: [*Eryopoidea* + [*Trematopidae* + [Broiliellus + Dissorophus + [[Tersomius + *Doleserpetontidae*] + [*Micromelerpetontidae* + [Leptorophus + [Schoenfelderpeton + Branchiosaurus + [Apateon + [*Apoda* + [*Caudata* + [Triadobatrachus + *Anurans*]]]]]]]]]]]]. Node C is diagnosed by one unique shared-derived character—the absence of an intertemporal bone (18).

Node D: [*Trematopidae* + [Broiliellus + Dissorophus + [[Tersomius + *Doleserpetontidae*] + [*Micromelerpetontidae* + [Leptorophus + [Schoenfelderpeton + Branchiosaurus + [Apateon + [*Apoda* + [*Caudata* + [Triadobatrachus + *Anurans*]]]]]]]]]]]. Clade D includes dissorophoid and lissamphibian taxa and is diagnosed by four synapomorphies, of which only one is a unique shared-derived feature—viz., the possession of mediolateral tooth replacement (58). However, it should be noted that the condition of this character is unknown in most of the temnospondyls, and it is known only in doleserpetontids, some branchiosaurids, and lissamphibians. Consequently, we have only limited confidence in mediolateral tooth replacement as a shared-derived character of the dissorophoids and lissamphibians; it is possible that it is part of a character complex involving pedicellation of the teeth (56).

Three homoplastic features are found at this node. Dissorophoids and lissamphibians possess a large orbit (1.2); alternate states of this character occur in Clade N (1.1), *Trematops* (1.1), *Broiliellus* (1.1), and the Apoda (1.0). The palatine is exposed laterally (8); this character is reversed in *Actiobates* and at Node H. A prefrontal-postfrontal articulation is absent (14), but this feature is reversed at Node K and in *Amphibamus*.

The Trematopidae [*Actiobates* + *Trematops*] (Clade O of Fig. 4) is supported by four homoplastic characters. The possession of a large external naris (2.2) is homoplastic with respect to *Leptorophus*, and the absence of a pars palatina on the premaxilla (4) is homoplastic with respect to *Doleserpeton*. In the Trematopidae, the prefrontal forms part of the margin of the opening for the external naris (23), as it also does in *Albanerpeton* and Caudata, and the parasphenoid bears a triangular corpus (27), as it does in caecilians.

Node E: First topological variant. The first of three topological variants occurs at this node involving *Broiliellus*, *Dissorophus*, and Clade F and is expressed by two different topologies (Fig. 5). Of the five possible synapomorphies at Node E, three are common to the two different topologies. One of these features, the absence of an articulation between the anterior ramus of the pterygoid and the vomer (34), is a unique shared-derived character. The remaining two features—the absence of a

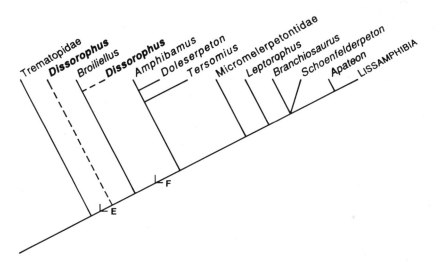

Fig. 5. The alternate positions of *Dissorophus* among the dissorophoids. This arrangement represents the First Topological Variant of Analysis III (see Node E of Fig. 4).

supinator process of the humerus (47) and the possession of bicuspid or multicuspid teeth (57)—are reversed in *Triadobatrachus* and urodeles (except Caudata), respectively.

One topology common to six trees is as follows:

[*Dissorophus* + [*Broiliellus* + Clade F]]

In addition to the three characters described above common to each of the 12 topologies, Node E is diagnosed by the presence of axial dermal armor (54)—a character that reverses at Clade F. [*Broiliellus* + Clade F] is diagnosed by having clavicles that are reduced to slender elements along the leading edge of the ventromedial part of the pectoral girdle (46.1). Alternate states of this character occur in apodans and urodeles (46.2) (unknown in *Albanerpeton*) and *Micromelerpeton* (46.0).

The alternate topology is as follows:

[[*Dissorophus* + *Broiliellus*] + Clade F]

In this arrangement, Node E is diagnosed by the three characters described (i.e., 34, 47, 57) in addition to Character 46.1. The sister-group relationship of *Dissorophus* and *Broiliellus* is corroborated by Character 38, the absence of a stapedial foramen, which is unknown in *Broiliellus* and homoplastic with respect to the Caudata and Salientia, and the presence of dermal axial armor (54).

Node F: [[Tersomius + *Doleserpetontidae*] + [*Micromelerpetontidae* + [Leptorophus + [Schoenfelderpeton + Branchiosaurus + [Apateon + [*Apoda* + [*Caudata* + [Triadobatrachus + Anurans]]]]]]]]. Depending upon which of the two arrangements is considered from the first topological variant described above, there are two possible characters diagnosing Node F. The feature common to all 12 trees is the possession of an anterolateral process on the palatine (9), which is homoplastic with respect to Clade N [Trimerorhachoidea + *Dendrerpeton*] and reverses at Node K. If the first arrangement of the first topological variant is investigated wherein *Dissorophus* and *Broiliellus* are *not* sister-groups, then the absence of dermal axial armor (54.0) also diagnoses Clade F.

Clade P [*Tersomius* + [*Doleserpeton* + *Amphibamus*]] is corroborated by one character, the presence of pedicellate teeth (56), which is homoplastic with respect to Node J. The Doleserpetontidae (Clade Q) is diagnosed weakly by a single feature, the presence of a moderate-sized external naris (2.1), which is homoplastic.

Node G: [*Micromelerpetontidae* + [Leptorophus + [Schoenfelderpeton + Branchiosaurus + [Apateon + [*Apoda* + [*Caudata* + [Triadobatrachus +

Anurans]]]]]]]]. Of the three synapomorphies that diagnose this node, two are unique shared-derived characters. These organisms have pterygoid bones that possess a robust quadrate ramus (30) and that lack a temporal flange (31). The third feature, the presence of a posterolaterally oriented quadrate ramus on the pterygoid (28), is homoplastic with respect to the Trimerorhachoidea.

Node H: [Leptorophus + [Schoenfelderpeton + Branchiosaurus + [Apateon + [*Apoda* + [*Caudata* + [Triadobatrachus + *Anurans*]]]]]]. There are four possible synapomorphies at this node, only one of which is a unique shared-derived character—viz., articulation of the palatine with the maxilla only at the anterolateral process of the palatine (10). One feature, the absence of lateral exposure of the palatine (8), is a reversal from Node D. The two remaining characters are homoplastic. These organisms possess a large narial opening (2.2), which is homoplastic with respect to Clade O and which reverses in Apoda and *Albanerpeton*. Further, they lack an articulation between the maxilla and ectopterygoid (5)—a feature that is reversed at Node K.

Node I: Second topological variant: [Schoenfelderpeton + Branchiosaurus + [Apateon + [*Apoda* + [*Caudata* + [Triadobatrachus + *Anurans*]]]]]. The second topological variant, which occurs at Node I, involves only the positions of *Schoenfelderpeton* and *Branchiosaurus* with respect to one another and Clade J. Because *Schoenfelderpeton* and *Branchiosaurus* have the same coding, no further resolution of this trichotomy is possible with our data set. Node I is corroborated by two synapomorphies, one of which is a unique shared-derived character—jaw articulation located anterior to the posterior limits of the endocranium (24.1); the alternate state (24.2) of this feature occurs in the Salientia and *Micromelerpeton*. The second synapomorphy is possession of a moderate-sized external narial opening (2.1), which is homoplastic with respect to the Doleserpetontidae.

Node J: [Apateon + [*Apoda* + [*Caudata* + [Triadobatrachus + *Anurans*]]]]. Only one possible synapomorphy occurs at this node—the presence of pedicellate teeth (56)—and this character is homoplastic with respect to Clade P, and reversed in *Albanerpeton*.

Node K: [*Apoda* + [*Caudata* + [Triadobatrachus + *Anurans*]]]. Lissamphibians are diagnosed by a suite of 14 synapomorphies, of which seven are unique shared-derived characters that involve the absences of bones—viz., the postparietal (17), supratemporal (19), tabular (20), postorbital (21), jugal (22), interclavicle (44), and cleithrum (45). Three other synapomorphies are reversals of derived character-states found at preceding nodes. Thus, in lissamphibians there is an articulation between the maxilla and ectopterygoid (5; reversed from Node H), the palatine lacks an anterolateral process (9; reversed from Node F), and an articulation is present between the prefrontal and postfrontal (14; reversed from Node D). Two other features are reversed within the lissamphibians; these are the absence of a squamosal embayment (25; reversed in the Salientia), and the absence of an articular bone in the adult (39; reversed in the Urodela). Lissamphibians lack a parietal foramen and the parietal organ is covered by dermal bone (16.1), but an alternate state of this character diagnoses the Salientia (16.2). Similarly, they lack clavicles (46.2), but the Salientia possesses reduced clavicles (46.1).

Node L: [*Caudata* + [Triadobatrachus + *Anurans*]]. There are nine possible synapomorphies at Node L (Batrachia); of these, eight are common to the 12 trees. Among the congruent characters, four are unique shared-derived features. All taxa lack a postfrontal (13), surangular (41), splenial (42), and scales of dermal origin (55). Among the congruent homoplastic characters are the absence of an articulation of the anterior pterygoid ramus with the palatine (33; homoplastic with respect to *Micromelerpeton*), the absence of an ectopterygoid (37; homoplastic with respect to *Doleserpeton*), and the absence of a stapedial foramen (38; homoplastic with respect to *Dissorophus* and reversed in *Karaurus*). The final synapomorphy of batrachians is the absence of a palatine in adults (7), a character that is reversed in *Triadobatrachus*.

The occurrence of the ninth synapomorphy—the prefrontal forming part of the margin of the opening for the external naris (23)—is absent at this node in the six trees associated with Topological Variant 3A described below.

Node M: [Triadobatrachus + *Anurans*]. There are nine possible synapomorphies at Node L (Salientia); of these, eight are common to the 12 trees. Among the congruent characters, four are unique shared-derived features. All salientians lack a lacrimal (11) and possess a frontoparietal bone (15), a long, slender ilium (49), and unicapitate ribs (53). Among the congruent homoplastic characters are the placement of the jaw articulation coincident with the posterior limits of the endocranium (24.2; homoplastic with respect to *Micromelerpeton*), and the possession of an anterior pterygoid ramus that articulates with the maxilla (32; homoplastic with respect to *Doleserpeton*). Character 46.1—the possession of reduced, slender clavicles—also is found in *Amphibamus, Apateon, Branchiosaurus, Broiliellus, Leptorophus,* and *Schoenfelderpeton*. The presence of a squamosal embayment (25) is a reversal of the derived character-state present at Node K.

The occurrence of the ninth synapomorphy—the absence of a parietal foramen and the parietal organ not covered by bone (16.2)—is absent at this node in the six trees associated with Topological Variant 3B described below.

Clade R: Third topological variant: caudates and their allies. Included in the urodeles are living salamanders, *Albanerpeton*, and *Karaurus*. Of the five possible synapomorphies diagnosing Clade R, four are common to the two different topologies and include two unique shared-derived features. The latter are an anterior ramus of the pterygoid lacking any bony articulation (35), and the presence of a tuberculum interglenoideum (51). The possession of an incomplete maxillary arcade (6) is homoplastic with respect to *Schoenfelderpeton*, and the presence of an independent articular bone in the adult (39) is a reversal of the derived character-state that occurs at Node K.

Topology 3A (Fig. 6) common to six trees is as follows:

[*Karaurus* + [*Albanerpeton* + Caudata]]

In addition to the four characters described above, Clade R' is diagnosed by the possession of monocuspid teeth (57.0). In this arrangement, Clade S' [*Albanerpeton* + Caudata] is diagnosed by a single homoplastic character—the prefrontal forms part of the margin of the opening for the external naris (23).

The alternate topology (3B; Fig. 7) is as follows:

[Caudata + [*Karaurus* + *Albanerpeton*]

In this arrangement, Clade R'' is diagnosed only by the four characters in common to both topologies, but the Clade S'' [*Karaurus* + *Albanerpeton*] is diagnosed by five synapomorphies. These include the absence of a parie-

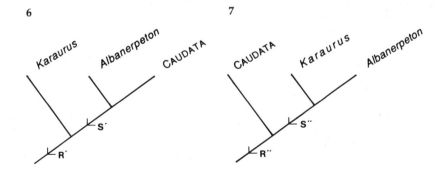

Figs. 6–7. The alternate interpretations of the interrelationships of the Urodela. These arrangements represent the two topologies of the Second Topological Variant of Analysis III (see Node R of Fig. 4). (6) Topology 3A. (7) Topology 3B.

tal foramen with the parietal organ covered by dermal bone (16.1), and lateral orientation of the quadrate ramus of the pterygoid (28.2). The remaining three characters are reversals; thus *Karaurus* and *Albanerpeton* have a stapedial foramen (38.0) and nonpedicellate (56.0), monocuspid (57.0) teeth.

Apoda. Of the eight possible synapomorphies diagnosing apodans, seven are shared by all trees. Of these, four features are unique shared-derived characters—the incorporation of the parasphenoid into endo-cranial ossifications (26), the possession of a short quadrate ramus of the pterygoid (29) and retroarticular process on the mandible (43), and the absence of limbs (50). Two other features are homoplastic reversals—viz., the possession of small orbits (1) and external nares (2). The seventh congruent synapomorphy, parasphenoid with a triangular corpus (27), is homoplastic with respect to Node O.

The occurrence of the eighth synapomorphy—the absence of a parietal foramen and the parietal organ covered by bone (16.2)—is absent at this node in the six trees associated with Topological Variant 3A described above.

Karaurus. In the six trees associated with Topological Variant 3A described above, this fossil is diagnosed by two synapomorphies, one of which is a unique shared-derived feature—lateral orientation of the quadrate ramus of the pterygoid (28.2; condition unknown in *Albaner-peton*). The presence of a stapedial foramen (38) is a homoplastic reversal. In the six trees associated with Topological Variant 3B, *Karaurus* is diagnosed by a single homoplastic reversal—viz., lack of association of the prefrontal with the margin of the opening for the external naris.

Albanerpeton. This poorly known taxon is characterized by one homoplastic reversal in all 12 trees—the possession of a small external narial opening (2). In the six trees associated with Topological Variant 3A, *Albanerpeton* also is diagnosed by the absence of pedicellate teeth (56).

Caudata. The monophyly of living salamanders is uncorroborated in the six trees associated with Topological Variant 3B described above; however, in Variant 3A, the Caudata is diagnosed by two synapomorphies, both of which are homoplastic. The first is the absence of a parietal foramen with the parietal organ not covered by bone (16.2; homoplastic with respect to the Anura), and the second is the possession of bicuspid teeth (57).

Triadobatrachus. This fossil sister-taxon to the anurans is character-ized by only three homoplastic reversals. The palatine (7), absent at Node

L, is present. It possesses the plesiomorphic features of a parietal foramen (16; reversed from the condition 16.2 derived at Node M) and a supinator process on the humerus (50; reversed from Node E).

Anura. Anurans are diagnosed by four unique shared-derived characters. The prefrontal (12) and prearticular (40) are absent, the radius and ulna are totally or partially fused (51), and the caudal vertebrae are modified to form a urostyle (52).

DISCUSSION

Relationships among Living Taxa

Analysis I: Soft Morphology

The results of the first analysis were unexpected. As shown in Figure 2, there is (1) a trichotomy among the Actinopterygii, Actinistia, and remaining Sarcopterygii, and (2) the Anura is shown to be the sister-group of the Amniota (thereby rendering living amphibians paraphyletic). However, evaluation of this cladogram suggests that the results may be biased in at least two respects. First, the data set (Appendix I, Characters 1–72) is less complete than we might have wished. A total of 31 (i.e., 6.2% of 504) codings was unknown (i.e., missing); most of them reflect the lack of available information for apodans. Second, the selection of characters was biased toward resolution of the relationships among advanced sarcopterygians (i.e., the tetrapods); doubtless, this accounts in part for the lack of resolution of relationships among actinopterygians, actinistians, and [Dipnoi + Tetrapoda]. However, it is well corroborated on the basis of other studies that actinistians are related more closely to tetrapods than are actinopterygians (Maisey, 1986; Schultze, 1987; Panchen and Smithson, 1987).

The explanation for the hypothesized sister-group relationship between anurans and amniotes was clarified in another analysis in which we forced the soft-morphology data set on the "conventional" topology (i.e., [Actinistia + [Dipnoi + [[Apoda + [Caudata + Anura]] + Amniota]]]; see Fig. 3). The resulting tree was only five steps longer than the six equally parsimonious trees of Analysis I (i.e., 117 steps and C.I. = 0.650, as compared with 112 steps and C.I. = 0.670 in Fig. 2). The primary difference among the six trees lies in the homoplasticity of the characters at Node C (Tetrapoda) in Tree 1 (Fig. 2). Of the 11 homoplastic characters at this node, seven are reversed in the Amniota. It is also instructive to examine the characters that unite anurans and amniotes at Node D in Tree 1, because many of these seem to represent novelties that commonly are associated with successful and efficient existence in terrestrial environments. Thus, the circulatory system is modified for more complete separation of arterial and venous blood. Ampullae are absent in the lateral-line system. The ear is modified for reception of air-borne sound waves, and a eustachian tube facilitates pressure adjustments. The male reproductive system has become considerably more complex, with (among other features) the site of spermatogenesis being coincident with the location of the permanent populations of primordial germ cells. Finally, the central nervous systems of anurans and amniotes are mark- edly more complex than those of lower sarcopterygians. Generally, in- creased complexity of the central nervous system has been associated with the increased importance and diversity of sensory and motor ac- tivities in terrestrial situations.

Analysis II: Soft Morphology and Osteological Features

The second analysis included a series of 50 osteological characters (Appendix I: Characters 73–122) in addition to the 72 soft morphological features employed in Analysis I. These osteological characters are a subset of the 58 characters used in Analysis III; the eight characters discarded represent five that showed no change in the seven taxa ana- lyzed (Characters 3, 8, 9, 31, and 54 of Appendix II) and three others for which there were insufficient data (Characters 5, 10, and 30 of Appendix II). Because the out-groups used for Analysis II were the Actinopterygii and Actinistia, the polarities of 10 of the osteological characters had to be changed; these are Characters 75/4, 79/12, 80/13, 87/20, 90/23, 92/25, 99/34, 101/36, 107/42, and 118/53 of Appendix I (numbers of characters in Appendix II and in Analysis of Characters in parentheses).

In general, the topology of the cladogram (Fig. 3) yielded by Analysis II is more conventional than that resulting from Analysis I. This topology is accepted in preference to the topology of Analysis I because this hypoth- esis is based on the congruence of a more complete and diversified data set. The Actinistia is shown to be the sister-group of advanced sarcop- terygians, and living amphibians constitute a monophyletic assemblage that is the sister-group of the Amniota. We believe that we have an- swered successfully Løvtrup's (1985:469) challenge in defense of his para- phyletic arrangement of living amphibians, wherein he stated that he believes "that this [i.e., his] classification can be seriously challenged only by means of data referring to the lower levels of organization."

The Dipnoi is shown to be the living sister-group of tetrapods, as in the first analysis. This is a consequence of the fact that we could deal only with living taxa if we wished to use soft-morphological features (limiting factor of the tandem analysis). Thus, owing to the exclusion of many fossil taxa, we were forced to work at an extraordinarily high level of universality. It is likely that the inclusion of the fossil taxa would have provided a more refined hypothesis of phylogenetic relationships, and one in which the Dipnoi was not the first sister-group of tetrapods. Furthermore, the results of Analysis II suggest that the discrimination of the clade [Anura + Amniota] in the first analysis may have been artificial; thus, the characters constituting the synapomorphies at Node C in Tree 1 may not be pertinent at this high level of universality.

The second analysis revealed that some of the characters that traditionally have been thought to represent synapomorphies of Lissamphibia have a different distribution among tetrapods. For example, the presence of intermaxillary glands (Character 12) is not a synapomorphy for the Lissamphibia, as proposed by Parsons and Williams (1963). Intermaxillary glands also seem to be present in amniotes (Webb, 1951; du Plessis, 1945). Granular glands (Character 10) are present in actinistians, and some actinopterygians in addition to the Lissamphibia; thus, this feature is not a synapomorphy of the latter group. Restriction of the presence of an operculum (Character 28) to the Lissamphibia also is suspect because the turtle *Emys* has a cartilaginous element in a similar position (Kunkel, 1912). Finally, the presence of green rods (Character 19) should be accepted as a lissamphibian synapomorphy with caution. Green rods occur in salamanders and anurans, but their presence has not been documented in caecilians. Thus, it is possible that this character is a synapomorphy of the Batrachia (= [Urodela + Salientia]) rather than the Lissamphibia.

Relationships between Dissorophoid and Lissamphibian Taxa, and among Lissamphibians

Analysis III

The third analysis is based on a suite of 58 osteological characters analyzed in 22 living and fossil temnospondyl taxa (Appendix II), to which the Loxommatoidea was hypothesized to be the sister-group. The results of this analysis (Tree 3; Fig. 4), like those of the second, indicate that the Lissamphibia is a monophyletic assemblage. Among temnospondyls, the Lissamphibia is more closely related to dissorophoid taxa (Fig. 8). The Dissorophoidea is redefined by the inclusion of previously recognized dissorophoid taxa plus the Lissamphibia. The Eryopoidea is considered to be the sister-group to Dissorophoidea, as suggested by Godfrey et al. (1987). It is particularly interesting that the "Branchiosauridae" is paraphyletic with respect to the Lissamphibia; thus, the dissorophids are not the lissamphibian sister-group, as has been postulated (cf. Trueb and Cloutier, this volume). Boy (1981) proposed that

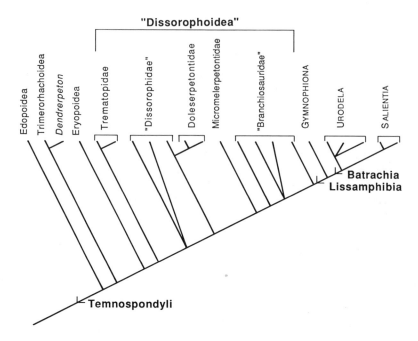

Fig. 8. Interrelationships of traditionally recognized suprageneric taxa of temnospondyls. The branching pattern is the same as that of Figure 4.

[Micromelerpetontidae + Branchiosauridae] was the sister-group of the remaining Dissorophoidea. However, Boy (1981) did not include the Lissamphibia within the Dissorophoidea.

For the most part, branchiosaurids have been dismissed rather casually from consideration as possible close relatives to living amphibians. For example, Romer (1966, 1968) did not recognize "branchiosaurids" as a natural group; he considered them to be larval temnospondyls and suggested that some of them might be related to *Eryops*. We were led to include "branchiosaurids" and micromelerpetontids in our analysis on the basis of Boy's work (1974). Boy studied ontogenetic series of the branchiosaurid genus *Apateon* and demonstrated that the genus was composed of several species, none of which represented larval stages of other temnospondyls. The monophyly of the [Micromelerpetontidae + ["branchiosaurids" + Lissamphibia]] is supported by three synapomorphies involving the structure of the pterygoid. One unique shared-derived character involving the articulation of the palatine and maxilla, and three homoplastic characters (i.e., size of the narial opening, lateral exposure of the palatine, and position of the jaw articulation) unite the [Branchiosauridae + Lissamphibia]. On the basis of osteological data, the Lissamphibia is diagnosed by 14 synapomorphies, seven of which are unique shared-derived characters and involve absences of elements. The possession of pedicellate, bicuspid teeth (Characters 56–57) is not a lissamphibian synapomorphy, as proposed by many previous authors. Instead, this is a homoplastic feature that is present in the Doleserpetontidae and *Apateon*; the condition is unclear or unknown in *Dissorophus*, most "branchiosaurids," and micromelerpetontids.

Like the Lissamphibia, the Batrachia (= [Urodela + Salientia]) and the Salientia (= [*Triadobatrachus* + Anura]) are characterized by a suite of characters that represent absences of cranial elements—the postfrontal, lacrimal, prefrontal, prearticular, ectopterygoid, surangular, palatine, and splenial bones. The absence of these elements seems to be associated with (1) a simplification of the mandible, and (2) a trend toward increased zygokrotaphy of the skull. It seems likely that many of these absences represent actual losses, rather than absences resulting from fusion of adjacent centers of ossification, for two reasons. First, the elements tend to become smaller as one progresses through the dissorophoid clade, and second, they are absent in the ontogenies of living taxa. Examples of the latter are the postparietal, tabular, supratemporal, interclavicle, cleithrum, postfrontal, postorbital, and palatine bones. The absence of other elements in adults of living amphibians such as a discrete ectopterygoid, surangular, and splenial may represent ontogenetic fusion of centers of ossification or resorption of early centers of ossification during development. Thus, the ectopterygoid apparently is incorporated into the maxillopalatine during the development of some caecilians (Marcus et al., 1935), and the center of ossification of the parietal may fuse with that of the frontal in some anurans (Griffiths, 1954; Lebedkina, 1968). In contrast, a bone such as the splenial is present as a discrete element in some salamander larvae (e.g., *Ambystoma*, *Batrachuperus*) but subsequently is lost through bone resorption at metamorphosis (Bonebrake and Brandon, 1971; Reilly, 1987; Cloutier, in ms.).

Among lissamphibians, the primitive status of gymnophionans is marked by their retention of many cranial elements that are lost in urodeles and salientians. At the same time, they possess many derived features associated with a semifossorial or fossorial mode of existence that characterizes all caecilians except the aquatic typhlonectids. Six of the seven synapomorphies diagnosing this clade involve cranial modifications of fossil and Recent species. The seventh synapomorphy, loss of limbs, must be restricted to the Apoda rather than the Gymnophiona as a whole, because the Kayenta fossil (Fig. 9) has limbs (Jenkins and Walsh, in prep.; Cloutier, pers. obser.).

In comparison with caecilians and anurans, most adult urodeles retain a relatively unspecialized morphotype suited to a generalized terrestrial mode of existence. Thus, salamanders possess attenuate bodies, tails,

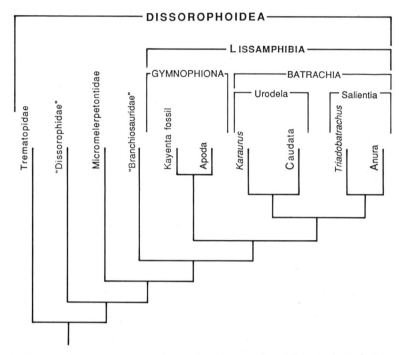

Fig. 9. Classification of the Dissorophoidea based on Analysis III. Note the inclusion of the Kayenta fossil gymnophionan, which was incorporated subsequent to the generation of the tree. See text for discussion of the proposed classification.

and, usually, fore- and hind limbs of equal length. The clade is character-ized uniquely by the presence of a four-faceted articulation between the skull and axial column (i.e., occipital condyles and tuberculum inter-glenoideum). In addition, the maxillary arcade is incomplete and lacks bony articulations with either the neurocranium or palatoquadrate. In this respect, the skulls of most salamanders are less robust than those of either caecilians or most anurans; possibly, the evident differences are correlated with the specialized feeding mechanisms of both of the latter taxa in contrast to the relatively simple mechanism of all salamanders except that of some highly derived bolitoglossines that have projectile tongues. On the other hand, the skull roof is less well developed than those of caecilians but retains more elements and is roofed more com-pletely than those of most anurans. Further, the pectoral girdle is re-duced by the absence of a clavicle and, therefore, apparently is weaker than those of the fossil sister-groups of the Lissamphibia, as well as the highly modified pectoral girdles of anurans.

In contrast to Naylor and Fox (1981), we do not recognize the order Allocaudata erected for *Albanerpeton*. In the third analysis, *Albanerpeton* is associated with *Karaurus* and the Caudata (Tree 3, Figs. 4, 6–7). However, we agree with Estes (1981) that this genus belongs within the clade Caudata and hypothesize that the topological variant of Tree 3 (Figs. 6–7) could be the result of inclusion of a derived taxon (i.e., *Albanerpeton*) with coding that was based on conditions in basal taxa. In addition, the conditions of some of the characters diagnosing the Caudata are unknown in *Albanerpeton*.

The Salientia (= [*Triadobatrachus* + Anura]) is diagnosed by the absence of the lacrimal bone and the fusion of the centers of ossification of the frontal and parietal. Thus, in terms of numbers of elements, salientians have the simplest skulls of the Lissamphibia. In addition, the pelvic girdle of salientians is specialized by the elongation and anterior (rather than dorsal) orientation of the ilium. The Anura is distinguished from the less derived *Triadobatrachus* by various specializations associated with saltatorial locomotion. Thus, the hind limbs are elongated, the radius and ulna are partially or wholly fused, discrete caudal vertebrae are fused to form a urostyle, the sacrum is modified, and the number of presacral vertebrae is reduced. On the basis of the synapomorphies possessed by anurans, we disagree with Roček's (1981) hypothesized diphyletic origin of the group.

CLASSIFICATION

The following classification of the Dissorophoidea is congruent with the phylogenetic hypothesis illustrated in Figure 9. The taxa are listed, with their authors and dates, according to phylogenetic sequence only, and thereby define clades. Until cladistic analyses of the remaining Amphibia have been completed, we refrain from designating hierarchical ranks to the names. We use Lissamphibia in its usual meaning to include the Gymnophiona, Urodela, and Salientia. Batrachia usually has been used in the same way, but we restrict it to include only the Urodela and Salientia. Batrachia is an available older name for Paratoidia Gardiner (1982a); we agree with A. R. Milner (1988) that Paratoidia is an inappropriate (and misspelled) name.

Tetrapoda
    Ichthyostegalia Säve-Söderbergh, 1932
    [Tetrapoda - Ichthyostegalia] Clade
        Palaeostegalia Panchen, 1973
            Crassigyrinus Watson, 1929
        [Amniota + Amphibia] Clade
            Amniota Haeckel, 1866

Amphibia Linnaeus, 1758
  Loxommatoidea Watson, 1917
  Temnospondyli Zittel, 1887–1890
    Edopoidea Romer, 1947
    [Temnospondyli - Edopoidea] Clade
      Trimerorhachoidea Clade
        Trimerorhachoidea Romer, 1947
        Dendrerpeton Owen, 1853
      [Eryopoidea + Dissorophoidea] Clade
        Eryopoidea Säve-Söderbergh 1935
        Dissorophoidea Bolt, 1969
          Trematopidae Williston, 1910
            Actiobates Eaton, 1973
            Trematops Williston, 1909
          [Dissorophoidea - Trematopidae] Clade
            "Dissorophidae" Williston, 1910
              Broiliellus Williston, 1914
              Dissorophus Cope, 1895
            Doleserpetontidae Clade
              Tersomius Case, 1911
              Doleserpetontidae Bolt, 1969
                Doleserpeton Bolt, 1969
                Amphibamus Cope 1865
              [Micromelerpetontidae + ["Branchiosauridae" +
              Lissamphibia]] Clade
                Micromelerpetontidae Boy, 1972
                ["Branchiosauridae" + Lissamphibia] Clade
                  "Branchiosauridae" Fritsch, 1879
                  Lissamphibia Haeckel, 1866
                    Gymnophiona Rafinesque-Schmaltz, 1814
                      (Kayenta fossil; Jenkins and Walsh, in prep.)
                      Apoda Oppel, 1811
                    Batrachia Brongniart, 1800
                      Urodela Rafinesque, 1815
                        Karaurus Ivachnenko, 1978
                        Caudata Scopoli, 1777
                      Salientia Laurenti, 1768
                        Triadobatrachus Kuhn, 1962
                        Anura Rafinesque, 1815

Acknowledgments Completion of this paper has proven to be a particularly taxing commitment for both authors. It would not have been
possible without the forbearance of our colleagues, who allowed us to
pillage their libraries and, on more than one occasion, overlooked aca-

demic obligations that we were remiss in filling. In this regard, we are especially indebted to William E. Duellman (Division of Herpetology) and H.-P. Schultze (Division of Paleontology), who also indulged us with translations of German literature. In addition to both of the latter, Darrel Frost and Linda Ford provided moral support and acute criticism throughout the course of this study; we acknowledge their contributions gratefully. Special thanks are extended to Arnold G. Kluge (The University of Michigan), Andrew Milner (Birkbeck College, London), and Jurgen Boy (Johannes Gutenberg–Universität Mainz), who critically reviewed previous drafts of this manuscript. For permission to examine specimens, we thank Robert Carroll (Redpath Museum, McGill University), Eugene Gaffney and John Maisey (American Museum of Natural History), David S Berman (Carnegie Museum of Natural History), and Robert Drewes (California Academy of Sciences). Various of our colleagues provided essential data. We particularly thank Carl F. Wellstead (University of Oklahoma) for information on *Lysorophus*, Denis Walsh (Cambridge University) and Farish Jenkins (Harvard University) for information on an undescribed fossil caecilian, and Brian Foreman (The University of Kansas) who provided pictures of *Actiobates* and branchiosaurids for our use. Research was supported by National Science Foundation Grant BSR 85-08470 (L.T.).

APPENDIX I. Data matrix of soft and hard characters: Characters 1–72 (Data Set I; Analysis I); Characters 1–122 (Data Set II; Analysis II). 0 = plesiomorphic state; 1, 2 = apomorphic states; ? = unknown; U = coding unclear; N = character not applicable. The characters ?, U, and N were coded as 9 in the analyzed matrix. Codes for abbreviation appear in descriptions of characters. Unpolarized characters are indicated in italics; characters for which the polarity has been changed between Analyses II and III are indicated in boldface.

Characters 1–17

Taxon	1 DER	2 MYO	3 LEY	4 COR	5 MUC	6 FLD	7 IRD	8 LAT	9 LOR	10 GRA	11 HAR	12 IMX	13 AD1	14 AD2	15 TEN	16 VOM	17 VA1
Actinistia	0	0	0	0	0	0	0	0	U	U	0	0	0	?	0	0	0
Actinopterygii	0	?	0	0	0	0	0	0	0/1	0/1	0	0	0	0	0	0	1
Amniota	0/1	1	0	1	1	0	0	1	1	1	1	1	1	1	1	1	1
Anura	1	0/1	1	1	0	0	0	1	1	0	1	1	1	1	0	1	0
Caudata	0	0/1	1	1	0	1	0	0	1/0	0	1	1	1	?	0	1	0
Dipnoi	0	?	0	0	0	0	0	0	0	?	0	0	0	?	0	0	0
Apoda	0	1	?	1	1	1	1	0	0	0	1	1	1	1	1	1	N

Characters 18–35

Taxon	18 VA2	19 ROD	20 TYM	21 CAV	22 EUS	23 PAP	24 PAN	25 CIS	26 SAC	27 RND	28 OPC	29 NAR	30 SV1	31 SV2	32 SV3	33 ATR	34 IAT	35 VEN
Actinistia	0	0	0	0	0	0	0	1	N	0	0	0	0	0	0	N	N	0
Actinopterygii	0	0	0	0	0	0	0	0	N	0	0	0	0	0	0	N	N	0
Amniota	N	0	1/0	1	1	0	0	1	0	1	0/1	0	1	N	N	N	1	1
Anura	1	1	1	1	1	1	1	1	?	1	1	0	0	0	1	0	0	0
Caudata	1	1	0	1	0	1	1	1	1	1	1	1	0	1	0	0	2	2
Dipnoi	0	0	0	0	0	0	0	0	N	0	0	0	0	0	0	0	0	0
Apoda	N	?	0	1	0	1	0	1	1	0	1	1	1	0	1	1	2	1

Characters 36–52

Taxon	36 CAR	37 CUT	38 AII	39 A-V	40 AVI	41 CHO	42 TRA	43 PRO	44 SPM	45 MOD	46 ARR	47 INT	48 CIL	49 CON	50 FAT	51 LOB	52 SIT
Actinistia	0	0	0	0	0	N	0	1	N	N	N	N	?	2	0	0	?
Actinopterygii	0	0	0	0	0	N	0	0	0	0	1	1	0	0	0	0	0
Amniota	0	0	1	1	1	0	1	1	N	N	N	N	1	2	0/1	0	1
Anura	1	1	1	1	0	1	0	0	0	1	1	1	0/1	1	1	1	1
Caudata	1	1	1	0	0	1	1	0	1/0	1	1	1	0	0	1	0	1
Dipnoi	0	0	1	0	0	0	0	1	N	N	N	N	0	0	0	1	?
Apoda	0	?	1	?	0	0	1	0	1	0	0	0	0	0	1	1	0

Characters 53–70

Taxon	53 MAU	54 OLI	55 XII	56 SP1	57 CC1	58 CC2	59 PU1	60 PU2	61 TOR	62 LEN	63 RAD	64 BAS	65 OCU	66 TRO	67 GAS	68 MES	69 BIL	70 LP1
Actinistia	1	N	0	1	0	N	0	0	0	?	?	0	1	0	0	?	0	?
Actinopterygii	0	0	0/1	?	0	N	1	0	0	1	1	0	0	0	0	0	0	0
Amniota	1	1	0	?	1	0	1	1	1	1	1	0	1	0	0	1	0	1
Anura	0	1	1	0	1	0	1	1	1	0	1	0	0	0	1	1	1	1
Caudata	0	0	1	1	0	N	0	1	0	0	0	0	0	0	1	1	1	1
Dipnoi	0	0	1	?	0	N	0	0	0	?	0	1	1	1	0	1	1	1
Apoda	0	1	1	1	1	1	?	?	2	?	0	1	1	1	1	?	?	?

APPENDIX I.—*continued*

Characters 71–87

Taxon	71 LP2	72 LP3	73 ORB	74 EXN	75 **PPP**	76 MSP	77 PAL	78 LAC	79 **PFR**	80 **POF**	81 *PPF*	82 *FPR*	83 PAR	84 PPR	85 ITP	86 STP	87 **TAB**
Actinistia	?	?	1	0	0	0	0	0	N	N	N	N	1	0	0	0	1
Actinopterygii	0	0	1	0	0	0	0	0	0	0	N	N	0	0	0	0	0
Amniota	1	0	0	0	0	0	0	0	1	1	0	0	0	0	1	0	1
Anura	1	1	2	1	1	0	1	1	0	0	N	1	2	1	1	1	0
Caudata	1	1	2	1	1	1	0	0	1	0	N	0	2	0	1	1	0
Dipnoi	1	1	0	0	N	0	0	0	N	N	N	N	0	0	N	N	N
Apoda	?	?	0	0	1	0	0	0	1	1	1	0	1	1	1	1	0

Characters 88–105

Taxon	88 POB	89 JUG	90 *PNR*	91 JAW	92 **SQE**	93 PSH	94 PRS	95 *QRO*	96 *QRL*	97 ARM	98 ARP	99 **ARV**	100 ARA	101 **AMA**	102 EPT	103 STF	104 ART	105 PRT
Actinistia	0	0	N	0	0	0	0	N	N	N	0	0	0	0	0	0	0	0
Actinopterygii	0	0	N	0	0	0	0	N	N	N	0	0	0	0	0	0	0	N
Amniota	0	0	*0*	2	0	0	1	0	0	0	0	1	0	1	0	0	0	1
Anura	1	1	N	2	1	0	0	1	0	1	N	0	1	0	1	1	1	1
Caudata	1	1	*0*	1	0	0	0	1	0	0	1	1	1	1	1	0	0	0
Dipnoi	N	0	N	2	0	1	1/0	N/1	N/1	N	N	1	0	1	0	0	0	N
Apoda	0	1	*1*	1	0	1	1	1	1	0	0	0	0	0	0	0	1	0

Characters 106–122

Taxon	106 SUR	107 **SPL**	108 RAR	109 ICL	110 CLE	111 CLA	112 SPH	113 R&U	114 ILM	115 LIM	116 TIG	117 URO	118 **RIB**	119 SCA	120 PED	121 CSP	122 TRP
Actinistia	0	1	0	1	0	0	N	N	N	0	0	0	N	0	0	0	0
Actinopterygii	0	0	0	0	0	0	N	N	N	0	0	0	0	0	0	0	0
Amniota	0	1	0	0	0	0	0	0	0	0	0	0	1	1	1	0	0
Anura	1	0	0	1	1	1	1	1	1	0	0	1	0	1	1	1	1
Caudata	1	1	0	1	1	2	1	0	0	0	1	0	1/0	1	1	1	1
Dipnoi	N	1	0	0	0	0	N	N	N	0	0	0	0	0	0	0	N
Apoda	0	1	1	1	1	2	N	N	N	1	0	0	1	0	1	1	1

301

APPENDIX II. Data matrix of osteological characters for temnospondyls, including lissamphibians: characters 1–58 (Data Set II; Analysis III). 0 = plesiomorphic state; 1, 2 = apomorphic states; ? = unknown; U = coding unclear; N = character not applicable. The characters ?, U, and N were coded as 9 in the analyzed matrix.

Taxon	1 ORB	2 EXN	3 PDP	4 PPP	5 MET	6 MSP	7 PAL	8 POB	9 PAP	10 PMX	11 LAC	12 PFR	13 POF	14 PPF	15 FPR
Actiobates	2	2	0	?	?	0	?	0	?	?	0	0	0	1	0
Albanerpeton	2	0	0	0	N	1	?	?	?	?	0	0	1	N	0
Amphibamus	2	1	0	0	0	0	0	1	1	0	0	0	0	0	0
Anura	2	1	0	0	N	0	1	N	N	N	1	1	1	N	1
Apateon	2	1	0	0	1	0/1	0	0	1	1	0	0	0	1/0	0
Branchiosaurus	2	1	0	?	1	0	0	0	1	1	0	0	0	1	0
Broiliellus	1	0	0	?	1	0	0	1	0	1	0	0	0	1	0
Caudata	2	1	0	0	N	1	1	N	N	N	0/1	0/1	1	N	0
Dendrerpeton	1	0	1	0	0	0	0	0	1	0	0	0	0	0	0
Dissorophus	2	0	0	?	0	0	0	1	?	?	0	0	0	1	0
Doleserpeton	2	1	1	1	?	0	0	1	1	**0**	0	0	0	1	0
Edops	0	0	0	0	0	0	0	0	0	N	0	0	0	0	0
Eryops	0	0	0	0	0	0	0	0	0	N	0	0	0	0	0
Apoda	0	0	0	0	N	0	1	N	0	N	0	0	0/1	0/1	0
Karaurus	2	1	0	0	N	1	1	N	N	N	0	0	1	N	0
Leptorophus	2	2	0	?	1	0	1	0	1	1	0	0	0	1	0
Micromelerpeton	2	0	0	?	0	0	0	1	1	0	0	0	0	1	0
Schoenfelderpeton	2	1	0	?	1	1	0	0	1	1	0	0	0	1	0
Tersomius	2	0	0	0	0	0	0	1	1	0	0	0	0	1	0
Trematops	1	2	0	1	0	0	0	1	1	N	0	0	0	0	0
Triadobatrachus	2	?	?	?	N	0	0	0	0	N	?	0	1	N	1
Trimerorhachis	1	0	0	0	0	0	0	0	1	0	0	0	0	0	0

Characters 1–15

Characters 16–29

Taxon	16 PAR	17 PPR	18 ITP	19 STP	20 TAB	21 POB	22 JUG	23 PNR	24 JAW	25 SQE	26 PSH	27 PRS	28 QRO	29 QRL
Actiobates	0	0	1	0	0	0	0	1	0	0	0	1	0	0
Albanerpeton	1	1	1	1	1	1	1	1	1	1	?	?	?	?
Amphibamus	0	0	1	0	0	0	0	0	0	0	0	0	0	0
Anura	2	1	1	1	1	1	1	N	2	0	0	0	1	0
Apateon	0	0	1	0	0	0	0	0	1	0	0	0	1	0
Branchiosaurus	0	0	1	0	0	0	0	0	0	0	0	0	1	0
Broiliellus	0	0	1	0	0	0	0	0	1	0	0	0	0	0
Caudata	2	1	1	1	1	1	1	1	0	1	0	0	1	0
Dendrerpeton	0	0	0	0	0	0	0	0	1	0	0	0	0	0
Dissorophus	U	0	1	0	0	0	0	0	U	0	0	0	0	0
Doleserpeton	0	0	1	0	0	0	0	0	0	0	0	0	0	0
Edops	0	0	0	0	0	0	0	0	0	0	0	0	U	0
Eryops	0	0	1	0	0	0	0	0	0	0	0	0	1	1
Apoda	1	1	1	1	1	1	1	0	1	1	1	1	1	0
Karaurus	1	1	1	1	1	1	1	0	1	1	0	0	2	0
Leptorophus	0	0	1	0	0	0	0	0	0	0	0	0	1	0
Micromelerpeton	0	0	1	0	0	0	0	0	2	0	0	0	1	0
Schoenfelderpeton	0	0	1	0	0	0	0	0	1	0	0	0	1	0
Tersomius	0	0	1	0	0	0	0	0	0	0	0	0	1	0
Trematops	0	0	1	1	0	0	1	1	0	0	0	1	0	0
Triadobatrachus	0	1	1	1	1	1	1	?	2	0	0	0	1	0
Trimerorhachis	0	0	0	0	0	0	0	0	0	0	0	0	1	0

APPENDIX II.—_continued_

							Characters 30–44								
Taxon	30 QRW	31 PTF	32 ARM	33 ARP	34 ARV	35 ARA	36 AMA	37 EPT	38 STF	39 ART	40 PRT	41 SUR	42 SPL	43 RAR	44 ICL
Actiobates	0	0	?	?	?	0	1	?	0	?	?	?	?	0	0
Albanerpeton	?	?	?	?	?	?	?	?	?	0	0	1	1	0	?
Amphibamus	0	0	0	0	1	0	1	0	0	0	0	0	1	0	0
Anura	1	1	1	N	1	0	1	1	1	1	1	1	0	0	1
Apateon	1	1	0	0	1	0	1	0	0	0	0	0	0	0	0
Branchiosaurus	1	1	0	0	1	0	1	0	0	0	0	?	?	0	0
Broiliellus	0	0	0	0	1	0	1	0	?	?	?	?	?	0	?
Caudata	1	1	0	1	0	1	1	1	1	0	0	1	1	0	1
Dendrerpeton	0	0	0	0	1	0	1	0	0	0	0	0	0	0	0
Dissorophus	0	0	?	?	0	0	1	0	1	?	?	0	?	0	0
Doleserpeton	0	0	1	0	1	0	1	1	0	?	?	0	?	0	0
Edops	0	0	0	0	0	0	0	0	0	?	?	?	?	0	?
Eryops	0	0	0	0	0	0	1	0	0	0	0	?	0	0	0
Apoda	U	1	0	0	1	0	1	1	0	1	?	0	0	1	1
Karaurus	1	1	0	?	1	1	1	1	1	?	?	1	1	0	1
Leptorophus	1	1	0	0	1	0	1	0	0	0	0	0	0	0	0
Micromelerpeton	1	1	0	1/0	1	0	1	0	0	0	0	0	0	0/1	0
Schoenfelderpeton	1	1	0	0	1	0	1	0	?	0	0	0	0	0	0
Tersomius	0	0	0	0	1	0	1	0	?	0	0	?	0	0	?
Trematops	0	0	9	0	0	0	0	0	?	0	?	?	?	0	0
Triadobatrachus	1	1	1	1	1	0	1	1	?	?	0	?	1	0	1
Trimerorhachis	?	?	0	?	?	?	?	0	?	0	0	0	0	0	0

Characters 45–58

Taxon	45 CLE	46 CLA	47 SPH	48 R&U	49 ILM	50 LIM	51 TIG	52 URO	53 RIB	54 ADA	55 SCA	56 PED	57 CSP	58 TRP
Actiobates	0	0	?	0	0	0	0	0	?	0	?	?	0	?
Albanerpeton	?	?	?	?	?	?	1	?	?	?	?	0	0	?
Amphibamus	0	1	1	0	0	0	0	0	0	0	0	1/0	1	1
Anura	1	1	1	1	1	0	0	1	1	0	1	1	1	1
Apateon	0	1	1	0	0	0	0	0	0	0	0	1/0	1	?
Branchiosaurus	0	1	1	0	0	0	?	0	0	1	0	0	1	1
Broiliellus	0	1	?	?	?	0	?	0	?	0	?	0	1	?
Caudata	1	2	1	0	0	0	1	0	0/1	0	1	1	1	1
Dendrerpeton	0	0	0	0	0	0	0	0	?	0	0	1	0	?
Dissorophus	0	0	1	0	0	0	0	0	0	1	?	0	?	?
Doleserpeton	0	?	?	0	?	0	?	?	?	?	?	1	1	1
Edops	?	?	0	0	0	0	0	0	?	?	0	0	0	?
Eryops	0	0	0	0	0	0	0	0	?	0	0	0	0	0
Apoda	1	2	N	N	N	1	0	0	0	0	0	1	1	1
Karaurus	1	2	1	0	0	0	1	0	0	0	1	?	0	?
Leptorophus	0	1	1	0	0	0	?	0	0	0	0	0	1	?
Micromelerpeton	0	0	?	0	0	0	?	0	0	0	0	0	0	?
Schoenfelderpeton	?	1	1	0	?	0	?	?	0	0	0	1	1	?
Tersomius	0	?	1	0	?	0	?	?	?	?	?	0	1	?
Trematops	0	0	0	0	0	0	0	0	?	0	?	1	1	?
Triadobatrachus	?	1	0	0	1	0	0	0	1	0	1	?	0	?
Trimerorhachis	?	0	0	0	0	0	0	0	?	0	?	0	0	?

305

APPENDIX III. Specimens Examined

AMNH = American Museum of Natural History; BM(NH) = British Museum (Natural History); CAS = California Academy of Sciences; FM = Field Museum of Natural History; KU = The University of Kansas Museum of Natural History (Division of Herpetology) and KUVP (Division of Vertebrate Paleontology); UMMZ = The University of Michigan Museum of Zoology.

Anura: *Ascaphus truei:* KU 153205–209.
Caudata: *Batrachuperus musteri:* CAS 147029, 147041 152088; KU 194381–87. *Hynobius nebulosus:* KU 151821, 151827. *H. leechi:* KU 38774, 152349. *Cryptobranchus alleghaniensis:* UMMZ 150644. *Salamandrella keyserlingii:* KU 204753, 204755.
Apoda: *Ichthyophis glutinosus:* KU 31289–93. *Epicrionops petersi:* KU 119397–402.
Branchiosauridae: *Apateon pedestris:* KUVP 60724–725. *Branchiosaurus salamandroides:* AMNH 2010 (electrocast of type).
Dendrerpeton: *Dendrerpeton acadianum:* AMNH 51 (peel of BM[NH] 4555), 53 (peel of BM[NH] 436), 23249 (peel of BM[NH] 4554).
Dissorophoidea: *Amphibamus grandiceps:* KUVP 80419 (cast of FM UC 2000). *A. lyelli:* KUVP 80422 (cast of AMNH 6841). *Broiliellus texensis:* AMNH 1824 (cast). *Dissorophus multicinctus:* AMNH 4343, 4376. *Tersomius mosesi:* KUVP 86657 (cast of UR 1251).
Edopoidea: *Edops craigi:* AMNH 7511 (cast).
Micromelerpetontidae: *Branchierpeton amblystomus:* AMNH 10457, 10481–482.
Microsauria: *Cardiocephalus sternbergi:* AMNH 4763a. *Lysorophus tricarinatus:* AMNH 4879, 23134, 23163. *L.* sp.: KUVP 49540–541.

Literature Cited

Acher, R., J. Chauvet, and M.-T. Chauvet. 1972. Phylogeny of the neurohypophysial hormones: two new active peptides isolated from a cartilaginous fish, *Squalus acanthias.* Eur. J. Biochem., 29:12–19.

Adams, E. N. III. 1972. Consensus techniques and the comparison of taxonomic trees. Syst. Zool., 21:390–397.

Anderson, M. L. 1978. The comparative myology and osteology of the carpus and tarsus of selected anurans. Doctoral dissertation. Lawrence: Univ. of Kansas.

Andrews, S. M. 1973. Interrelationships of crossopterygians. Pp. 138–177 *in* Greenwood, P. H., R. S. Miles, and C. Patterson (eds.), *Interrelationships of Fishes.* Zool. J. Linn. Soc. 53, Suppl. 1.

Bannikov, A. G. 1958. Die Biologie des Froschzahnmolches *Ranodon sibiricus* Kessler. Zool. Jahrb. Syst., 86:245–249.

Beaumont, E. H. 1977. Cranial morphology of the Loxommatidae (Amphibia: Labyrinthodontia). Philos. Tran. R. Soc. London, Ser. B, 280:29–101.

Berman, D. S, R. R. Reisz, and D. A. Eberth. 1985. *Ecolsonia cutlerensis,* an Early Permian dissorophid amphibian from the Cutler Formation of north-central New Mexico. New Mexico Bur. Mines Miner. Resour., Circ. 191:1–31.

Bertmar, G. 1968. Lungfish phylogeny. Pp. 259–283 *in* T. Ørvig (ed.), *Current Problems of Lower Vertebrate Phylogeny.* Stockholm: Almqvist and Wiksell.

Bolt, J. R. 1969. Lissamphibian origins: possible protolissamphibian from the Lower Permian of Oklahoma. Science, 166:888–891.

Bolt, J. R. 1974a. A trematopsid skull from the Lower Permian, and analysis of some characters of the dissorophoid (Amphibia: Labyrinthodontia) otic notch. Fieldiana Geol., 30(3):67–79.

Bolt, J. R. 1974b. Evolution and functional interpretation of some suture patterns in Paleozoic labyrinthodont amphibians and other lower tetrapods. J. Paleontol., 48(3):434–458.

Bolt, J. R. 1974c. Osteology, function, and evolution of the trematopsid (Amphibia: Labyrinthodontia) nasal region. Fieldiana Geol., 33:11–30.

Bolt, J. R. 1977a. Dissorophoid relationships and ontogeny, and the origin of the Lissamphibia. J. Paleontol., 51:235–249.

Bolt, J. R. 1977b. Cacops (Amphibia: Labyrinthodontia) from the Fort Sill Locality, Lower Permian of Oklahoma. Fieldiana Geol., 37(3):61–73.

Bolt, J. R. 1979. Amphibamus grandiceps as a juvenile dissorophid: evidence and implications. Pp. 529–563 in Nitecki, M. H. (ed.), Mazon Creek Fossils. New York: Academic Press.

Bolt, J. R., and R. E. Lombard. 1985. Evolution of the amphibian tympanic ear and the origin of frogs. Biol. J. Linn. Soc., 24:83–99.

Bolt, J. R., and R. J. Wassersug. 1975. Functional morphology of the skull in Lysorophus: a snake-like Paleozoic amphibian (Lepospondyli). Paleobiology, 1(3):320–332.

Bonebrake, J. E., and R. A. Brandon. 1971. Ontogeny of cranial ossification in the small-mouthed salamander, Ambystoma texanum (Matthes). J. Morphol., 133:189–204.

Boy, J. A. 1972. Die Branchiosaurier (Amphibia) des saarpfälzischen Rotliegenden (Perm, SW-Deutschland). Abh. Hess. L.-Amt. Bodenforsch., 65:1–137.

Boy, J. A. 1974. Die Larven der rhachitomen Amphibien (Amphibia: Temnospondyli; Karbon–Trias). Palaeontol. Z., 48:236–268.

Boy, J. A. 1978. Die Tetrapodenfauna (Amphibia, Reptilia) des saarpfälzischen Rotliegenden (Unter-Perm; SW-Deutschland). 1. Branchiosaurus. Mainzer Geowiss. Mitt., 7:27–76.

Boy, J. A. 1981. Zur Anwendung der Hennigschen Methode in der Wirbeltierpaläontologie. Palaeontol. Z., 55:87–107.

Boy, J. A. 1986. Studien über die Branchiosauridae (Amphibia: Temnospondyli). 1. Neue und wenig bekannte Arten aus dem mitteleuropäischen Rotliegenden (?oberstes Karbon bis unteres Perm). Palaeontol. Z., 60(1/2):131–166.

Boy, J. A. 1987. Studien über die Branchiosauridae (Amphibia: Temnospondyli; Ober-Karbon–Unter-Perm). 2. Systematische Übersicht. Neues. Jahrb. Geol. Palaeontol. Abh., 174(1):75–104.

Bullock, T. H., R. G. Northcutt, and D. A. Bodznick. 1982. Evolution of electroreception. Trends Neurosci., 5:50–53.

Burggren, W. W., and K. Johansen. 1987. Circulation and respiration in lungfishes (Dipnoi). Pp. 217–236 in Bemis, W. E., W. W. Burggren, N. E. Kemp (eds.), The Biology and Evolution of Lungfishes. J. Morphol., Suppl. 1.

Cannatella, D. C. 1985. A phylogeny of primitive frogs (archaeobatrachians). Doctoral dissertation. Lawrence: Univ. of Kansas.

Carroll, R. L. 1964. Early evolution of the dissorophid amphibians. Bull. Mus. Comp. Zool., 131:161–250.

Carroll, R. L. 1967. Labyrinthodonts from the Joggins Formation. J. Paleontol., 41:111–142.

Carroll, R. L. 1988. Vertebrate Paleontology and Evolution. New York: W. H. Freeman.

Carroll, R. L., and P. J. Currie. 1975. Microsaurs as possible apodan ancestors. J. Linn. Soc. London Zool., 57:229–247.

Carroll, R. L., and P. Gaskill. 1978. The Order Microsauria. Mem. Am. Philos. Soc., 126:1–211.

Chang M.-M., and X. Yu. 1984. Structure and phylogenetic significance of *Diabolich-thys speratus* gen. et sp. nov., a new dipnoan-like form from the Lower Devonian of eastern Yunnan, China. Proc. Linn. Soc. New South Wales, 107:171–184.

Chase, J. N. 1965. *Neldasaurus wrightae*, a new rhachitomous labyrinthodont from the Texas Lower Permian. Bull. Mus. Comp. Zool. Harvard, 133:153–225.

Chorn, J. 1984. A trimerorhachid amphibian from the Upper Pennsylvanian of Kansas. Doctoral dissertation. Lawrence: Univ. of Kansas.

Cloutier, R. 1991. Interrelationships of Palaeozoic actinistians: patterns and trends. Proceedings of the Fifth Symposium on Early Vertebrate Studies and Related Problems in Evolutionary Biology. Beijing: Oceanography Publishing House.

Cloutier, R. Manuscript submitted. Ontogenetic changes in the osteology of *Batrachuperus musteri* (Hynobiidae: Caudata). J. Morphol.

Coldiron, R. W. 1978. *Acroplous vorax* Hotton (Amphibia, Saurerpetontidae) restudied in light of new material. Am. Mus. Novit., 2662:1–27.

DeMar, R. E. 1966. The phylogenetic and functional implications of the armor of the Dissorophidae. Fieldiana Geol., 16(3):55–88.

DeMar, R. E. 1968. The Permian labyrinthodont amphibian *Dissorophus multicinctus*, and adaptations and phylogeny of the family Dissorophidae. J. Paleontol., 42(5): 1210–1242.

De Queiroz, K. 1987. Phylogenetic systematics of iguanine lizards. A comparative osteological study. Univ. California Publ. Zool., 118:1–203.

Duellman, W. E., and L. Trueb. 1986. *Biology of Amphibians.* New York: McGraw-Hill.

du Plessis, S. S. 1945. Cranial anatomy and ontogeny of the South African cordylid, Chamaesaura anguina. S. Afr. J. Sci., 41:245.

Eaton, T. H. 1941. The family Trematopsidae. J. Geol., 49:149–176.

Eaton, T. H. 1973. A Pennsylvanian dissorophid amphibian from Kansas. Occas. Pap. Mus. Nat. Hist. Univ. Kansas, 14:1–18.

Edwards, J. L. 1976. Spinal nerves and their bearing on salamander phylogeny. J. Morphol., 148:305–327.

Estes, R. 1981. Gymnophiona, Caudata. Handb. Palaeoherpetol., 2:1–115.

Estes, R., and R. Hoffstetter. 1976. Les urodèles du Miocène de La Grive-Saint-Alban (Isère, France). Bull. Mus. Nation. Hist. Nat. Paris, 3rd Ser., no. 398, Sciences de la Terre, 57:297–343.

Estes, R., and O. A. Reig. 1973. The early fossil records of frogs. A review of the evidence. Pp. 11–63 *in* Vial, J. L. (ed.), *Evolutionary Biology of the Anurans.* Columbia: Univ. of Missouri Press.

Forey, P. L. 1981. The coelacanth *Rhabdoderma* in the Carboniferous of the British Isles. Palaeontology, 24(1):203–229.

Fox, R. C., and B. G. Naylor. 1982. A reconsideration of the relationships of the fossil amphibian *Albanerpeton*. Can. J. Earth Sci., 19:118–128.

Fraser, E. A. 1927. Observations on the development of the pronephros of sturgeon, *Acipenser rubicundus*. Q. J. Microsc. Sci., 71:75–112.

Fritzsch, B., and U. Wahnschaffe. 1983. The electroreceptive ampullary organs of urodeles. Cell Tiss. Res., 229:483–503.

Fritzsch, B., and M. Wake. 1986. A note on the distribution of ampullary organs in gymnophiones. J. Herpetol., 20:90–93.

Fritzsch, B., and M. Wake. 1988. The inner ear of gymnophione amphibians and its nerve supply: a comparative study of regressive events in a complex sensory system (Amphibia, Gymnophiona). Zoomorphology, 108:201–217.

Gaffney, E. S. 1979. Tetrapod monophyly: a phylogenetic analysis. Bull. Carnegie Mus. Nat. Hist., 13:92–105.

Gardiner, B. G. 1982a. Tetrapod classification. J. Linn. Soc. London Zool., 74:207–232.

Gardiner, B. G. 1982b. Mammals, birds, and mammal-like reptiles. Pp. 11–17 *in* Jayakar, S. D., L. Zanta (eds.), *Evolution and the Genetics of Populations*. Suppl. Atti. Ass. Genet. Ital., Vol. 29.

Gardiner, B. G. 1983. Gnathostome vertebrae and the classification of the Amphibia. J. Linn. Soc. London Zool., 79:1–59.

Gardiner, B. G. 1984. The relationships of the palaeoniscid fishes, a review based on new specimens of *Mimia* and *Moythomasia* from the Upper Devonian of Western Australia. Bull. Br. Mus. Nat. Hist. Geol., 37(4):173–428.

Gauthier, J., A. G. Kluge, and T. Rowe. 1988a. Amniote phylogeny and the importance of fossils. Cladistics, 4:105–209.

Gauthier, J., A. G. Kluge, and T. Rowe. 1988b. The early evolution of the Amniota. Pp. 103–155 *in* Benton, M. J. (ed.), *The Phylogeny and Classification of the Tetrapods*. Vol. 1: *Amphibians, Reptiles, Birds*. Syst. Assoc. Spec. Vol. 35 A. Oxford: Clarendon Press.

Godfrey, S. J., A. R. Fiorillo, and R. L. Carroll. 1987. A newly discovered skull of the temnospondyl amphibian *Dendrerpeton acadianum* Owen. Can. J. Earth Sci., 24:796–805.

Grande, L. 1985. The use of paleontology in systematics and biogeography, and a time control refinement for historical biogeography. Paleobiology, 11(2):234–243.

Grassi Milano, E., and F. Accordi. 1986. Evolutionary trends in adrenal gland of anurans and urodeles. J. Morphol., 189:249–259.

Griffiths, I. 1954. On the nature of the fronto-parietal in Amphibia, Salientia. Proc. Zool. Soc. London, 123:781–792.

Griffiths, I. 1963. The phylogeny of the Salientia. Biol. Rev. Cambridge Philos. Soc., 38:241–292.

Haslewood, G. A. D. 1967. *Bile Salts*. London: Methuen.

Haslewood, G. A. D. 1968. Bile salt differences in relation to taxonomy and systematics. Pp. 159–172 *in* Hawkes, J. G. (ed)., *Chemotaxonomy and Serotaxonomy*. London: Academic Press.

Hecht, M. K. 1976. Phylogenetic inference and methodology as applied to the vertebrate record. Evol. Biol., 9:335–363.

Hecht, M. K., and J. Edwards. 1976. The determination of parallel or monophyletic relationships: the proteid salamanders—a test case. Am. Nat., 110:653–677.

Hecht, M. K., and J. Edwards. 1977. The methodology of phylogenetic inference above the species level. Pp. 3–51 *in* Hecht, M. K., P. C. Goody, and B. M. Hecht (eds.), *Major Patterns in Vertebrate Evolution*. NATO Advanced Studies Series A., vol. 14. New York: Plenum Press.

Hennig, W. 1966. *Phylogenetic systematics*. Urbana: Univ. of Illinois Press. [Translation].

Herre, W. 1935. Die Schwanzlurche der mittel-eocänen (oberlutetischen) Braunkohle des Geiseltales und die Phylogenie der Urodelen unter Einschluss der fossilen Formen. Zoologica, 33(87):1–85.

Hetherington, T. E., A. P. Jaslow, and R. E. Lombard. 1986. Comparative morphology of the amphibian opercularis system: I. General design features and functional interpretation. J. Morphol., 190:43–61.

Hillis, D. M. 1987. Molecular versus morphological approaches to systematics. Ann. Rev. Ecol. Syst., 18:23–42.

Hook, R. W. 1983. *Colosteus scutellatus* (Newberry), a primitive temnospondyl from the Middle Pennsylvanian of Linton, Ohio. Am. Mus. Novit., 2770:1–41.

Ivachnenko, M. F. 1978. Urodelans from the Triassic and Jurassic of Soviet Central Asia. Paleontol. J., 12(3):362–368.

Jarosova, J. 1973. The components of the carpus in Palaeobatrachus and their development in two related recent species. Casopis Narodmiho Muzea, odd. prirodovedny, 142:89–106.

Kluge, A. G. 1989. A concern for evidence and a phylogenetic hypothesis of relationships among *Epicrates* (Boidae, Serpentes). Syst. Zool., 38:7–25.

Kluge, A. G., and J. S. Farris. 1969. Quantitative phyletics and the evolution of anurans. Syst. Zool., 18:1–32.

Krejsa, R. J. 1979. The comparative anatomy of the integumental skeleton. Pp. 112–191 *in* Wake, M. H. (ed.), *Hyman's Comparative Vertebrate Anatomy.* 3rd ed. Chicago: Univ. of Chicago Press.

Kuhlenbeck, H. 1975. *The Central Nervous System of Vertebrates: Spinal Cord and Deuterencephalon.* Vol. 4. New York: S. Karger.

Kunkel, B. W. 1912. On a double fenestral structure in *Emys.* Anat. Record, 6:267–280.

Lauder, G. V., and K. F. Liem. 1983. The evolution and interrelationships of the actinopterygian fishes. Bull. Mus. Comp. Zool., 150:95–197.

Lebedkina, N. S. 1968. The development of bones in the skull roof of Amphibia. Pp. 317–329 *in* Ørvig, T. (ed.), *Current Problems of Lower Vertebrate Phylogeny.* Stockholm: Almqvist and Wiksell.

Lewis, E. R., E. L. Leverenz, and W. Bialek. 1985. The vertebrate inner ear. Boca Raton, Fla.: CRC Press.

Lombard, R. E., and J. R. Bolt. 1979. Evolution of the tetrapod ear: an analysis and reinterpretation. Biol. J. Linn. Soc., 11:19–76.

Løvtrup, S. 1977. *The Phylogeny of Vertebrata.* 2nd ed. New York: John Wiley and Sons.

Løvtrup, S. 1985. On the classification of the taxon Tetrapoda. Syst. Zool., 34(4):463–470.

Lund, R., and W. D. Lund. 1985. Coelacanths from the Bear Gulch Limestone (Namurian) of Montana and the evolution of the Coelacanthiformes. Bull. Carnegie Mus. Nat. Hist., 25:1–74.

Maddison, W. P., M. J. Donoghue, and D. R. Maddison. 1984. Outgroup analysis and parsimony. Syst. Zool., 33(1):83–103.

Maisey, J. G. 1986. Heads and tails: a chordate phylogeny. Cladistics, 2(3):210–256.

Maisey, J. G. 1988. Phylogeny of early vertebrate skeletal induction and ossification patterns. Pp. 1–36 *in* Hecht, M. K., B. Wallace, and G. T. Prance (eds.), *Evolutionary Biology.* Vol. 22. New York: Plenum Publ. Corp.

Manski, W., S. P. Halbert, T. Auerbach-Pascal, and P. Janvier. 1967a. On the use of antigenic relationships among species for the study of molecular evolution. I. The lens proteins of Agnatha and Chondrichthyes. Internat. Arch. Allergy, 31:38–56.

Manski, W., S. P. Halbert, and P. Janvier. 1967b. On the use of antigenic relationships among species for the study of molecular evolution. II. The lens proteins of the Choanichthyes and early Actinoptergyii. Internat. Arch. Allergy, 31:475–489.

Manski, W., S. P. Halbert, P. Janvier, and T. Auerbach-Pascal. 1967c. On the use of antigenic relationships among species for the study of molecular evolution. III. The lens proteins of the late Actinopterygii. Internat. Arch. Allergy, 31:529–545.

Marcus, H., O. Winsauer, and A. Hueber. 1935. Beiträge zur Kenntnis der Gymnophionen. XXV. Die Ossifikation des Hypogeophisschädels. Morphol. Jahrb., 76:375–420.

Marshall, C. R. 1987. Lungfish: phylogeny and parsimony. Pp. 151–162 *in* Bemis, W. E., W. W. Burggren, N. E. Kemp (eds.), *The Biology and Evolution of Lungfishes.* J. Morphol., Suppl. 1.

Miles, R. S. 1977. Dipnoan (lungfish) skulls and the relationships of the group: a study based on new species from the Devonian of Australia. J. Linn. Soc. London Zool., 61:1–328.

Millot, J., and J. Anthony. 1958. *Anatomie de* Latimeria chalumnae. Vol. I: *Squelette, muscles et formations de soutien.* Paris: C.N.R.F.

Millot, J., and J. Anthony. 1965. *Anatomie de* Latimeria chalumnae. Vol. II: *Système nerveux et organes des sens.* Paris: C.N.R.F.

Millot, J., J. Anthony, and D. Robineau. 1978. *Anatomie de* Latimeria chalumnae. Vol. III: *Appareil digestif . . . conclusions générales*. Paris: C.N.R.F.

Milner, A. C. 1980. A review of the Nectridea (Amphibia). Pp. 377–405 *in* Panchen, A. L. (ed.), *The Terrestrial Environment and the Origin of Land Vertebrates*. London: Academic Press.

Milner, A. R. 1980. The temnospondyl amphibian *Dendrerpeton* from the Upper Carboniferous of Ireland. Palaeontology, 23(1):125–141.

Milner, A. R. 1988. The relationships and origin of living amphibians. Pp. 59–102 *in* Benton, M. J. (ed.), *The Phylogeny and Classification of the Tetrapods*. Vol. 1: *Amphibians, Reptiles, Birds*. Syst. Assoc. Spec. Vol. No. 35 A. Oxford: Clarendon Press.

Munz, H., B. Class, and B. Fritzsch. 1982. Electrophysiological evidence of electroreception in the axolotl *Siredon mexicanum*. Neurosci. Lett., 28:107–111.

Naylor, B. G. 1978. The systematics of fossil and Recent salamanders (Amphibia: Caudata), with special reference to the vertebral column and trunk musculature. Doctoral dissertation. Alberta: Univ. of Alberta.

Nelson, G. 1979. Cladistic analysis and synthesis: principles and definitions, with a historical note on Adanson's "Familles des plantes" (1763–1764). Syst. Zool., 38:1–21.

Noble, G. K. 1931. *The Biology of the Amphibia*. New York: McGraw-Hill.

Northcutt, R. G. 1986. Electroreception in nonteleost bony fishes. Pp. 257–285 *in* Bullock, T. H., and W. Heiligenberg (eds.), *Electroreception*. New York: John Wiley and Sons.

Nussbaum, R. A. 1977. Rhinatrematidae: a new family of caecilians (Amphibia: Gymnophiona). Occ. Pap. Mus. Zool. Univ. Michigan, 682:1–30.

Nussbaum, R. A. 1979. The taxonomic status of the caecilian genus *Uraeotyphlus* Peters. Occ. Pap. Mus. Zool. Univ. Michigan, 687:1–20.

Nussbaum, R. A. 1983. The evolution of a unique dual jaw-closing mechanism in caecilians (Amphibia: Gymnophiona) and its bearing on caecilian ancestry. J. Zool. London, 199:545–554.

Olsen, I. D. 1977. The sensory receptors. Pp. 425–462 *in* Kluge, A. G. (ed.), *Chordate Structure and Function*. New York: Macmillan.

Panchen, A. 1967. The nostrils of choanate fishes and early tetrapods. Biol. Rev. Cambridge Philos. Soc., 42:374–420.

Panchen, A. 1980. The origin and relationships of the anthracosaur amphibia from the late Palaeozoic. Pp. 319–350 *in* Panchen, A. L. (ed.), *The Terrestrial Environment and the Origin of Land Vertebrates*. London: Academic Press.

Panchen, A., and T. R. Smithson. 1987. Character diagnosis, fossils and the origin of tetrapods. Biol. Rev. Cambridge Philos. Soc., 62:341–438.

Parsons, T. S., and E. E. Williams. 1962. The teeth of Amphibia and their relation to amphibian phylogeny. J. Morphol., 110:375–389.

Parsons, T. S., and E. E. Williams. 1963. The relationships of the modern Amphibia: a re-examination. Quart. Rev. Biol., 38:26–53.

Patterson, C. 1987. Introduction. Pp. 1–22 *in* Patterson, C. (ed.), *Molecules and Morphology in Evolution: Conflict or Compromise?* Cambridge: Cambridge Univ. Press.

Paul, C. R. C. 1982. The adequacy of the fossil record. Pp. 75–117 *in* Joysey, K. A., and A. E. Friday (eds.), *Problems of Phylogenetic Reconstruction*. Syst. Assoc. Spec. Vol. 21. London: Academic Press.

Peters, J. A. 1964. *Dictionary of Herpetology*. New York: Hafner.

Piveteau, J. 1937. Un amphibien du Trias inférieur, essai sur l'origine et évolution des amphibiens anoures. Ann. Paleontol., 26:135–177.

Rage, J.-C., and P. Janvier. 1982. Le problème de la monophylie des amphibiens actuels, à la lumière des nouvelles données sur les affinités des tétrapodes. Geobios, 6:65–83.

Rage, J.-C., and Z. Roček. 1989. Redescription of *Triadobatrachus massinoti* (Piveteau, 1936), an anuran amphibian from the Early Triassic. Paleontographica, Abt. A, 206:1–16.

Reilly, S. M. 1987. Ontogeny of the hyobranchial apparatus in the salamanders *Ambystoma talpoideum* (Ambystomatidae) and *Notophthalmus viridescens* (Salamandridae): the ecological morphology of two neotenic strategies. J. Morphol., 191:205–214.

Roček, Z. 1981. Cranial anatomy of frogs of the family Pelobatidae Stannius, 1856, with outlines of their phylogeny and systematics. Acta Univ. Carolinae Biol., 1980: 1–164.

Roček, Z. 1989. Developmental patterns of the ethmoidal region of the anuran skull. Pp. 412–415 *in* Splechtna, H., and H. Hilgers (eds.), *Trends in Vertebrate Morphology.* Proc. 2nd Int. Symp. Vert. Morphol. Fortschritte Zool., Vol. 35. New York: Gustav Fischer Verlag.

Romer, A. S. 1966. *Vertebrate Paleontology.* 3rd ed. Chicago: Univ. of Chicago Press.

Romer, A. S. 1968. *Notes and Comments on Vertebrate Paleontology.* Chicago: Univ. of Chicago Press.

Rowe, T. 1987. Definition and diagnosis in the phylogenetic system. Syst. Zool., 36(2):208–211.

Saint-Aubain, M. L. de. 1981. Amphibian limb ontogeny and its bearing on the phylogeny of the group. Z. Zool. Syst. Evolut.-forsch., 19:175–194.

Saint-Aubain, M. L. de. 1985. Blood flow patterns of the respiratory systems in larval and adult amphibians: functional morphology and phylogenetic significance. Z. Zool. Syst. Evolut.-forsch., 23:229–240.

Schultze, H.-P. 1970. Folded teeth and the monophyletic origin of tetrapods. Am. Mus. Novit., 2408:1–10.

Schultze, H.-P. 1973. Crossopterygier mit heterozerker Schwanzflosse aus dem Oberdevon Kanadas, nebst einer Beschreibung von Onychodontida-Resten aus dem Mitteldevon Spaniens und aus dem Karbon der USA. Palaeontographica, Ser. A, 143:188–208.

Schultze, H.-P. 1987. Dipnoans as sarcopterygians. Pp. 39–74 *in* Bemis, W. E., W. W. Burggren, N. E. Kemp (eds.), *The Biology and Evolution of Lungfishes.* J. Morphol., Suppl. 1.

Schultze, H.-P., and K. S. W. Campbell. 1987. Characterization of the Dipnoi, a monophyletic group. Pp. 25–38 *in* Bemis, W. E., W. W. Burggren, N. E. Kemp (eds.), *The Biology and Evolution of Lungfishes.* J. Morphol., Suppl. 1.

Schultze, H.-P., and R. Cloutier. 1991. Computed tomography and magnetic resonance imaging studies of *Latimeria chalumnae* (Sarcopterygii: Actinistia). *In* Musick, J. A., M. Bruton, and E. Balon (eds.), *The Evolution and Biology of Coelacanths.* Env. Biol. Fish. 26. In press.

Sever, D. M. 1987. *Hemidactylium scutatum* and the phylogeny of cloacal anatomy in female salamanders. Herpetologica, 43:105–116.

Smithson, T. R. 1982. The cranial morphology of *Greererpeton burkemorani* Romer (Amphibia: Temnospondyli). Zool. J. Linn. Soc., 76:29–90.

Smithson, T. R. 1985. The morphology and relationships of the Carboniferous amphibian *Eoherpeton watsoni* Panchen. J. Linn. Soc. London Zool., 85:317–410.

Toews, D., G. Shelton, and R. Boutilier. 1982. The amphibian carotid labyrinth: some anatomical and physiological relationships. Can. J. Zool., 60:1153–1160.

Wake, D. 1979. The endoskeleton: the comparative anatomy of the vertebral column and ribs. Pp. 192–237 *in* Wake, M. H. (ed.), *Hyman's Comparative Vertebrate Anatomy.* 3rd ed. Chicago: Univ. of Chicago Press.

Wake, M. 1968. Evolutionary morphology of the caecilian urogenital system. I. The gonads and the fat bodies. J. Morphol., 126(3):291–332.

Wake, M. 1976. The development and replacement of teeth in viviparous caecilians. J. Morphol., 148:33–64.

Wake, M. 1979. The urogenital system. Pp. 555–614 in Wake, M. H. (ed.), Hyman's Comparative Vertebrate Anatomy. 3rd ed. Chicago: Univ. of Chicago Press.

Walsh, D. M. 1987. Systematics of the caecilians (Amphibia: Gymnophiona). Doctoral dissertation. Montreal: McGill Univ.

Waltrous, L. E., and Q. D. Wheeler. 1981. The out-group comparison method of character analysis. Syst. Zool., 30(1):1–11.

Warren, A., and T. Black. 1985. A new rhytidosteid (Amphibia, Labyrinthodontia) from the early Triassic Arcadia Formation of Queensland, Australia, and the relationships of Triassic temnospondyls. J. Vert. Paleontol., 5(4):303–327.

Watson, D. M. S. 1941. The origin of frogs. Trans. R. Soc. Edinburgh, 60:195–231.

Webb, M. 1951. The cranial anatomy of the South African geckoes Palamtogecko rangei (Andersson), and Oedura karroica (Hewitt). Ann. Univ. Stellenbosch, 27(Sect. A, No. 5):131–165.

Wellstead, C. F. 1985. Taxonomic revision of the Permo-Carboniferous lepospondyl amphibian families Lysorophidae and Molgophidae. Doctoral dissertation. Montreal: McGill Univ.

Wever, E. G. 1985. The Amphibian Ear. Princeton: Princeton Univ. Press.

Wiley, E. O. 1979a. An annotated Linnean hierarchy, with comments on natural taxa and competing systems. Syst. Zool., 28(3):308–337.

Wiley, E. O. 1979b. Ventral gill arch muscles and the interrelationships of gnathostomes, with a new classification of the Vertebrata. J. Linn. Soc. London Zool., 67:149–179.

Wiley, E. O. 1981. Convex groups and consistent classifications. Syst. Bot., 6(4):346–358.

Williston, S. W. 1916. Synopsis of the American Permocarboniferous Tetrapoda. Contrib. Walker Mus., 1(9):193–236.

Wintrebert, P. 1922. L'évolution de l'appareil ptérygopalatin chez les Salamandridae. Bull. Soc. Zool. France, 47:208–215.

Section III

REPTILES

9 Amniote Phylogeny

Michael J. Benton

The amniotes (reptiles, birds, mammals), which arose during the Carboniferous, represent one of the most prominent vertebrate groups today. Their key innovation, the cleidoic ("closed") egg, has a semipermeable shell (either calcareous or leathery) that allows the embryo to develop outside the mother's body in its own pond of fluid. Water is retained by the shell, and the eggs can be laid on land, unlike the eggs of most amphibians. The cleidoic eggs of amniotes also contain extraembryonic membranes that function in respiration, feeding, and waste disposal.

The amniotes arose during the Carboniferous period. The oldest described forms date from the early part of the Pennsylvanian (Late Carboniferous), about 300 million years (Myr) ago (Carroll, this volume), although an older reptile from the Mississippian (Early Carboniferous, ca. 340 Myr) of Scotland has been reported (Smithson, 1989). Regrettably, cleidoic eggs have not been found in sediments older than the Early Permian (ca. 270 Myr), and the validity of that specimen has been questioned (Kirsch, 1979). However, the Carboniferous "reptiles" almost certainly are members of the Amniota, because the major amniote lineages arose in the Carboniferous, and they all share very similar egg characters that are unlikely to have arisen independently more than once. Carroll (this volume) gives other arguments in favor of this view.

The closest out-group of the Amniota currently is disputed, as are the

317

relationships of the major groups within Amniota. These two topics are reviewed here, and a fuller review is given in Benton, 1990.

THE OUT-GROUPS OF THE AMNIOTA

The problem of identifying the sister-group of the Amniota, or indeed the series of out-groups leading to that clade, generally has been tackled under the rubric "the origin of the reptiles." Most authors have accepted for some time that the most reptilelike "amphibians" are the anthracosaurs, or batrachosaurs, of the Carboniferous and Permian. Another amphibian group of that time, the Microsauria, is reptilelike in many respects, but the similarities probably are convergent (Carroll, this volume).

The groups typically classified as amphibians seem to fall into two major groupings (Panchen and Smithson, 1988; Milner, 1988; Panchen, this volume) that cut across the old split into labyrinthodonts and lepospondyls: a batrachomorph clade (i.e., "true" amphibians, including nectrideans, colosteids, microsaurs, "temnospondyls," and lissamphibians), and a reptiliomorph clade—i.e., those amphibians on the line to the reptiles, as well as all amniotes.

The key reptiliomorph taxa, according to the cladistic analysis of Panchen and Smithson (1988), are the Loxommatidae, *Crassigyrinus*, the Anthracosauroidea, Seymouriamorpha, Diadectomorpha, and Amniota. These taxa all share a basal articulation (where the braincase rotates against the palatal bones), a specialized retractor pit for the eye muscles on the basisphenoid (Panchen and Smithson, 1988), and vertebrae in which the pleurocentrum dominates and the intercentrum is reduced.

If the loxommatids and *Crassigyrinus* are reptiliomorphs, the group arose early in the Mississippian. *Crassigyrinus* (Fig. 1) possesses the reptiliomorph characters noted above, and others—such as the presence of a single convex occipital condyle with a convex atlas articulation, and five digits in the hand (typically four in "true" amphibians)—appear in the anthracosauroids (Fig. 2), seymouriamorphs (Fig. 3), diadectomorphs (Fig. 4), and amniotes.

The seymouriamorphs and diadectomorphs long have been regarded as the closest out-groups to the Amniota, or even as full-fledged reptiles (e.g., Romer, 1945; Heaton, 1980; Carroll, 1982, this volume). It is unclear whether the Seymouriamorpha is the sister-group of the Anthracosauroidea (Smithson, 1985), of the Diadectomorpha (Heaton, 1980; Fracasso, 1987), or of the Diadectomorpha and Amniota (Panchen, this volume), as shown in Figure 5. However, the diadectomorphs, such as *Diadectes*

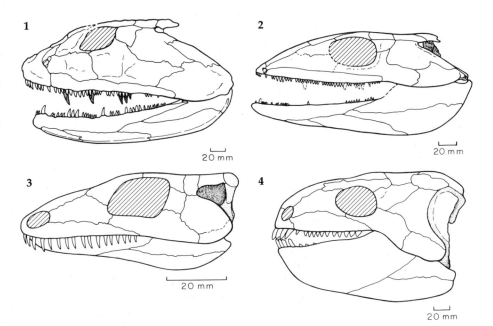

Figs. 1–4. Skulls, in lateral view, of reptiliomorph amphibians. (1) *Crassigyrinus*. (2) The anthracosauroid *Proterogyrinus*. (3) *Seymouria*. (4) *Diadectes*. (From various sources, after Carroll, 1987.)

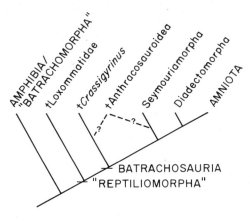

Fig. 5. The phylogeny of the reptiliomorph amphibians. (From information in Panchen and Smithson, 1988, and other sources.)

(Heaton, 1980) or *Limnoscelis* (Fracasso, 1987), are very amniote-like in many ways. Postulated diadectomorph-amniote synapomorphies include (Panchen and Smithson, 1988; Gauthier et al., 1988a) the following: (1) the pterygoid flange directed ventrally and often bearing teeth, (2) the convex occipital condyle fully developed, (3) postparietal and tabular bones exposed on occiput only, (4) the presence of at least two sacral vertebrae, and (5) the ?presence of an astragalus.[1]

THE RELATIONSHIPS OF LIVING AMNIOTE GROUPS

Morphological Data

Six monophyletic groups (clades) of living amniotes may be assessed for their mutual relationships: turtles (Chelonia or Testudinata), mammals (Mammalia), the tuatara (Sphenodontida), lizards and snakes (Squamata), crocodilians (Crocodylia), and birds (Aves). The turtles are diagnosed by their "shell," a carapace and plastron formed from bone and keratin, as well as other characters (Gaffney and Meylan, 1988). Mammals are diagnosed by possession of hair, mammary glands, and skeletal characters (Kemp, 1988; Rowe, 1988), and the squamates by their skin, paired copulatory organs, kinetic quadratic bone, and other modifications to the skull and skeleton (Evans, 1988; Rieppel, 1988). Crocodilians are diagnosed by their pneumatic posterior skull bones, ear lid, elongate wrist bone, modified pelvis, and numerous other features (Benton and Clark, 1988), and birds by their feathers, furcula (wishbone), fused lower leg bones, reduced tail, and wings (Cracraft, 1988).

A "standard" view of relationships presented by Gaffney (1980) united the tuatara, squamates, crocodilians, and birds as the Diapsida, and paired these with the mammals first and placed turtles as the out-group (Fig. 6). In this arrangement, the tuatara and the squamates form the Lepidosauria, and the crocodilians and the birds the Archosauria. Gaffney argued that the mammals and diapsids share a lower temporal fenestra and a Jacobson's organ in a ventromedial pocket in the roof of the mouth at some stage in ontogeny. The other "traditional" view is that the turtles and diapsids are sister-groups, and that mammals are the out-group to them (see below).

Gardiner (1982) proposed a rather revolutionary cladogram (Fig. 7) in which the Diapsida and the Archosauria were separated, and the birds were the sister-group of the mammals. He listed 28 postulated bird-mammal synapomorphies—i.e., shared characters of the brain case, brain, snout, vertebral column, circulatory system, glands, and physiology (both groups being endothermic)—and an additional 20 characters shared by turtles, crocodilians, birds, and mammals, but not by the tuatara and squamates.

---

[1]In Amniota and *Diadectes*, there is a single, medial element of the ankle formed by fusion of the tibiale, intermedium, and centrale IV of other forms; it is not clear whether the astragalus of *Diadectes* is homologous to that of amniotes.

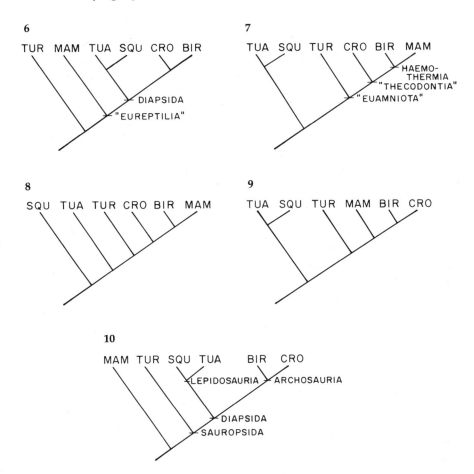

Figs. 6–10. The relationships of living amniotes. (6) Gaffney (1980). (7) Gardiner (1982). (8) Løvtrup (1985). (9) Gauthier et al. (1988b), based on living and extinct taxa only. (10) Gauthier et al. (1988b), based on living and extinct taxa. BIR = birds; CRO = crocodilians; MAM = mammals; SQU = squamates; TUA = tuatara; TUR = turtles.

Løvtrup (1985) proposed a third cladogram (Fig. 8), which also involved a breakup of the Diapsida and Archosauria, and the establishment of a bird-mammal clade, the Haemothermia. He also split up the Lepidosauria, making the squamates and the tuatara the most primitive amniote groups.

The views of Gardiner (1982) and Løvtrup (1985) were criticized vigorously by Benton (1985), Gauthier et al. (1988a,b), and Kemp (1988), all of whom found that many of the putative synapomorphies in support of the cladogram in Figures 7 and 8 were nonhomologous, ill-defined, or present in wider groups than at first proposed. The remaining postulated synapomorphies were said to be heavily outweighed by those in favor of a monophyletic Lepidosauria, Archosauria, and Diapsida, with the mammals and turtles as out-groups (Benton, 1985; Kemp, 1988).

Gauthier et al. (1988b) attempted a thorough analysis of the relationships of living amniotes on the basis of 109 characters. The resulting cladogram (Fig. 9) differs from Gardiner's (Fig. 7) only in the exchange of the positions of the mammals and crocodilians, respectively. However, the amount of agreement of the characters (i.e., their congruence as measured by the Consistency Index was low (C.I. = 0.674). Moreover, the addition of 25 fossil taxa to the analysis and the use of a total of 207 characters of the skull and skeleton yield a very different cladogram (Fig. 10) in which the integrity of the Archosauria and Diapsida is restored. The minimum of at least seven postulated synapomorphies shared by turtles and diapsids in this scheme far outweighs those proposed to unite mammals and diapsids in Figure 6. The turtle-diapsid synapomorphies are as follows: (1) the tabular small or absent, (2) the supratemporal small or absent, (3) a supraoccipital with anterior crista, (4) a suborbital fenestra or foramen in palate, (5) a simple coronoid, (6) the atlas centrum and axis intercentrum fused, and (7) the medial centrale of ankle absent. Gauthier et al. (1988b) argued that the fossils were crucial in establishing their final cladogram of amniote relationships (Fig. 10). Attempts to fit the fossil taxa into a cladogram based solely on modern taxa (e.g., Gardiner, 1982) lead to absurd problems of high levels of parallelism and reversal, as well as major stratigraphic anomalies.

Molecular Data

Independent lines of evidence for amniote phylogeny have been obtained from studies of amino acid sequences in proteins. These molecular studies are based on the Molecular Clock Hypothesis in some form or another—i.e, the idea that the primary structure of proteins changes in a clocklike, stochastic way. For any particular protein, a rate of substitution per Myr can be established, and this can be used to determine patterns of relationships among taxa; the more distantly related two taxa are, the more differences will be discovered between homologous proteins.

Molecular sequences from a variety of amniotes now are available for the following polypeptides: α- and β-parvalbumin, α- and β-hemoglobin, myoglobin, lens α-crystallin A, fibrinopeptides A and C, cytochrome c, and ribonuclease. These have given rise to a number of maximum-parsimony trees. Although the wider relationships of major tetrapod groups are still tentative because of the paucity of nonmammalian sequences, nearly every pairing of mammals, birds, crocodilians, lizards, snakes, turtles, and amphibians has been found (e.g., Goodman et al., 1985, 1987; Bishop and Friday, 1987, 1988). The arrangements derived from morphological and molecular data are presented below for comparison.

Myoglobin
[[[[Turtle] Lizard] Crocodilian] [Bird [Mammal]]]
or
[[[[Turtle] Crocodilian] Lizard] [Bird [Mammal]]]

β-*Hemoglobin*
[Snake [Crocodilian [Bird [Mammal]]]]
or
[Mammal [Snake [Crocodilian [Bird]]]]
or
[Snake [Mammal [Bird [Crocodilian]]]]

α-*Hemoglobin*
[Crocodilian [Bird [Mammal]]]

Lens α-*crystallin A*
[Mammal [Crocodilian [Lizard [Bird]]]]
or
[Mammal [Lizard [Crocodilian [Bird]]]]

Cytochrome c
[Bird [Snake [Mammal]]]

"Standard Morphological"
[[Turtle [[Lizard [Snake]] [Crocodilian [Bird]]]] Mammal]

The majority of these protein-based phylogenetic trees hypothesizes a sister-group relationship between birds and mammals, in apparent support of the morphological views of Gardiner (1982) and Løvtrup (1985). In addition, where relevant sequences are available, turtles often are associated with squamates (lizards and snakes) to form a clade separate from crocodilians or birds. Some authors have accepted these results at face value, whereas others have urged caution until more nonmammalian sequences become available. It has been noted that the relative difference in parsimony values between the most parsimonious tree or trees, and any of a large number of other patterns, often is very small. Further, the structures of some of the polypeptides, such as the hemoglobins and myoglobins, might be correlated functionally with, for example, the endothermy of birds and mammals, and some of their similarities might be convergent or the result of resistance to mutation (Bishop and Friday, 1988).

THE RELATIONSHIPS OF EARLY AMNIOTES

If the cladogram in Figure 10 is accepted as the best current solution, where do the extinct Carboniferous and Permian amniotes fit in?

Carboniferous Amniotes

Basically, there are three families of early reptiles known thus far from the Carboniferous—the Protorothyrididae, Petrolacosauridae, and Ophiacodontidae. Each of these has a skull pattern that seems to place it in a different major amniote lineage. Thus, *Paleothyris* (Fig. 11), a typical protorothyrid, has an *anapsid* skull; that is, there are no temporal openings behind the orbit. *Petrolacosaurus*, on the other hand, has a typical *diapsid* skull (Fig. 12), with two temporal openings, and *Ophiacodon* has a *synapsid* skull (Fig. 13), with only the lower temporal opening present.

The relationships of the Petrolacosauridae and Ophiacodontidae would seem to be clear, with the former being close to the origin of the great clade Diapsida, and the Ophiacodontidae being close to the origin of the Synapsida, which includes the mammals (Fig. 14). The anapsid skull of the protorothyrids is the primitive pattern for amniotes and also is characteristic of amphibians and fishes; thus, this character cannot be used as a synapomorphy to link protorothyrids with the only living anapsid amniotes, the turtles. In fact, the protorothyrids seem to be an

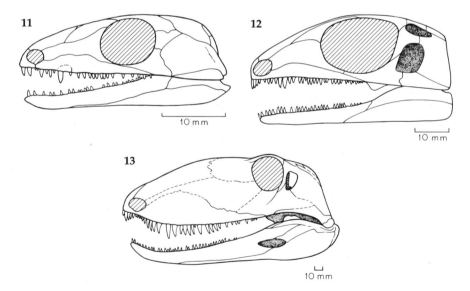

Figs. 11–13. Skulls, in lateral view, of Carboniferous amniotes. (11) The protorothyrid *Paleothyris*. (12) *Petrolacosaurus*. (13) *Ophiacodon*. (From various sources, after Carroll, 1987.)

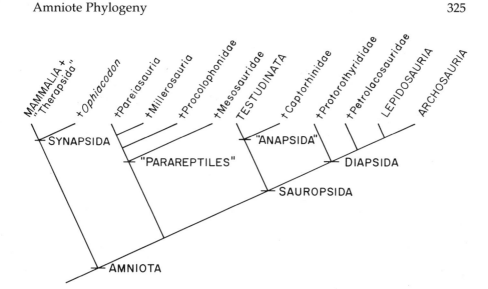

Fig. 14. The phylogeny of the major early amniote groups. (From information in Heaton and Reisz, 1986; Gaffney and Meylan, 1988; Gauthier et al., 1988a,b; and other sources.)

out-group of the diapsids on the basis of several postulated synapomorphies (Heaton and Reisz, 1986)—(1) a short postorbital region of the skull, (2) keels on the underside of the anterior presacral pleurocentra, (3) slender limbs, and (4) long, slender feet.

Permian Amniotes

The synapsids radiated extensively during the Permian, giving rise to numerous lineages of "pelycosaurs," of which *Ophiacodon* is an example in the Late Carboniferous—Early Permian and therapsids in the Late Permian. These are considered further by Hotton (this volume) and Hopson (this volume). The diapsids also radiated in the Late Permian, after an unusual and apparent gap during the Early Permian; their later evolution is considered herein by Carroll and Currie (this volume).

Several anapsid groups also had their heyday in the Permian, and they are more difficult to place in the phylogenetic scheme. These include the captorhinids (Fig. 15), millerettids (Fig. 16), procolophonids (Fig. 17), pareiasaurs (Fig. 18), and mesosaurs (Fig. 19). Hitherto, the captorhinids generally have been bracketed with the protorothyridids as the Captorhinomorpha, but their shared characters all seem to be plesiomorphous. The millerettids occasionally have been linked with the diapsids, or even with the lizards, but the evidence for this alliance is weak. The other three groups generally have been abandoned to a rag-bag group of basal reptiles, the "Cotylosauria," because they have no particular fea-

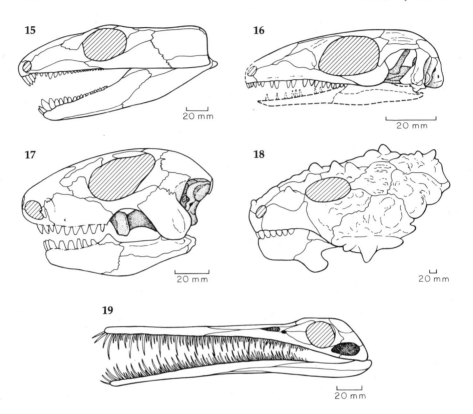

Figs. 15–19. Skulls, in lateral view, of anapsid Permian amniotes. (15) *Captorhinus*. (16) *Millerosaurus*. (17) *Procolophon*. (18) The pareiasaur *Pareiasaurus*. (19) *Mesosaurus*. (From various sources, after Romer, 1956, and Carroll, 1987.)

tures of the major amniote clades. Carroll (1982, this volume) argued that these five groups cannot be placed readily in a phylogenetic scheme, because they were given off piecemeal from a long-lived protorothyridid stock over a span of about 70 Myr extending from the Late Carboniferous to the Early Permian. He demonstrated that the postulated synapomorphies for any pairing of the five are matched by equally convincing shared derived characters for quite different patterns.

However, some modest progress has been made in attempts to disentangle the relationships of these groups. The captorhinids seem to be the sister-group of the Testudines (Fig. 14) on the basis of four skull characters (Gaffney and Meylan, 1988; Gauthier et al., 1988a): (1) the medial process of jugal absent, (2) the ectopterygoid absent, (3) the tabular absent, and (4) the foramen orbitonasale present.

The remaining four anapsid groups are brigaded tentatively as the "parareptiles" by Gauthier et al. (1988a). Pareiasaurs and millerettids are regarded as sister-groups because they share reduction in the size of premaxillary teeth (?),[2] fusion of the caudal ribs to the vertebrae (?), and the absence of the supinator process of the humerus. The procolophonids are the postulated out-group of these two groups (Fig. 14) on the basis of the position of the articulation in front of the occiput, and the loss of the caniniform maxillary teeth (cf. Figs. 16–18 with Figs. 11–13, 15). The "parareptiles" as a whole (Fig. 14) are diagnosed (Gauthier et al., 1988a) by the greatly swollen neural arches in trunk vertebrae, the fusion of the caudal ribs to the vertebrae (?reversed in procolophonids), and the loss of the supraglenoid foramen in the scapulocoracoid.

SUMMARY

The Amniota is a major vertebrate clade that includes reptiles, birds, and mammals. The amniotes arose in the Early Carboniferous, and their subsequent success probably is the result of their possession of the cleidoic egg, which allowed them to become fully terrestrial.

The out-groups of the Amniota include a series of reptiliomorph "amphibians" that acquired various reptilelike synapomorphies. These are best seen in the diadectomorphs, the postulated sister-group of the Amniota.

There has been much dispute over the relationships of living amniotes, occasioned by the fact that many soft-part anatomical, physiological, and molecular data seem to ally birds closely with mammals. However, the balance of evidence strongly favors a monophyletic Lepidosauria (tuatara, lizards, snakes), Archosauria (crocodilians, birds), Diapsida (lepidosaurs, archosaurs), and Sauropsida (turtles, diapsids), with the Synapsida (mammals plus extinct relatives) as the sister-group of the Sauropsida (reviewed in more detail in Benton, 1990).

The Carboniferous and Permian amniotes can be accommodated within this cladogram (Fig. 20), and it becomes clear that all three amniote lineages are present in the Pennsylvanian (the Diapsida with Petrolacosauridae, the Synapsida with Ophiacodontidae) or the Early Permian (the Anapsida [turtles, etc.] with Captorhinidae). Other Permian amniotes fall on the diapsid or synapsid line, or in a fourth postulated lineage, the "parareptiles," which died out in the Late Triassic (the last procolophonid). The relationships of the "parareptiles" still are problematic, because each of the four groups is quite distinctive, and yet none of them shows any convincing synapomorphies with another clade.

Acknowledgments I thank Una McCauley for typing the manuscript,

[2]? = polarity uncertain.

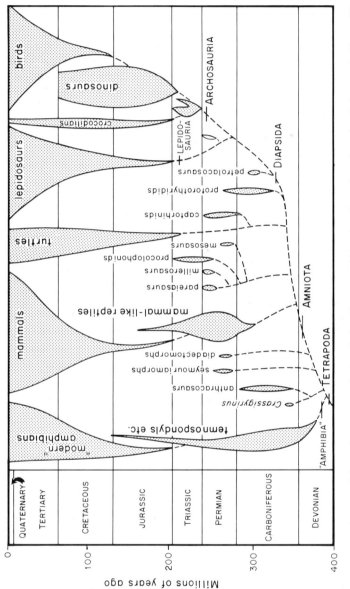

Fig. 20. Phylogeny of the early amniotes during Carboniferous and Permian times, based on the cladogram in Figure 14, with the addition of stratigraphic and diversity data from Carroll (1987) and other sources.

328

Hans-Peter Schultze for the invitation to write it, and Linda Trueb for her editorial work. This work was supported by S.E.R.C. grant GR/F 87912.

Literature Cited

Benton, M. J. 1985. Classification and phylogeny of the diapsid reptiles. Zool. J. Linn. Soc. London, 84:97–164.

Benton, M. J. 1990. Phylogeny of the major tetrapod groups: morphological data and divergence dates. J. Molec. Evol., 30:409–424.

Benton, M. J., and J. M. Clark. 1988. Archosaur phylogeny and the relationships of the Crocodylia. Pp. 295–338 *in* Benton, M. J. (ed.), *The Phylogeny and Classification of the Tetrapods*. Vol. 1: *Amphibians, Reptiles, Birds*. Syst. Assoc. Spec. Vol. No. 35 A. Oxford: Clarendon Press.

Bishop, M. J., and A. E. Friday. 1987. Tetrapod relationships; the molecular evidence. Pp. 123–139 *in* Patterson, C. (ed.), *Molecules and Morphology in Evolution: Conflict or Compromise?* Cambridge: Cambridge Univ. Press.

Bishop, M. J., and A. E. Friday. 1988. Estimating the interrelationships of tetrapod groups on the basis of molecular sequence data. Pp. 33–58 *in* Benton, M. J. (ed.), *The Phylogeny and Classification of the Tetrapods*. Vol. 1: *Amphibians, Reptiles, Birds*. Syst. Assoc. Spec. Vol. No. 35 A. Oxford: Clarendon Press.

Carroll, R. L. 1982. Early evolution of reptiles. Annu. Rev. Ecol. Syst., 13:87–109.

Carroll, R. L. 1987. *Vertebrate Paleontology and Evolution*. New York: W. H. Freeman.

Cracraft, J. 1988. The major clades of birds. Pp. 339–361 *in* Benton, M. J. (ed.), *The Phylogeny and Classification of the Tetrapods*. Vol. 1: *Amphibians, Reptiles, Birds*. Syst. Assoc. Spec. Vol. No. 35 A. Oxford: Clarendon Press.

Evans, S. E. 1988. The early history and relationships of the Diapsida. Pp. 221–260 *in* Benton, M. J. (ed.), *The Phylogeny and Classification of the Tetrapods*. Vol. 1: *Amphibians, Reptiles, Birds*. Syst. Assoc. Spec. Vol. No. 35 A. Oxford: Clarendon Press.

Fracasso, M. A. 1987. Braincase of *Limnoscelis paludis* Williston. Postilla, 201:1–22.

Gaffney, E. S. 1980. Phylogenetic relationships of the major groups of amniotes. Pp. 593–610 *in* Panchen, A. L. (ed.), *The Terrestrial Environment and the Origin of Land Vertebrates*. Syst. Assoc. Spec. Vol. 15. London: Academic Press.

Gaffney, E. S., and P. A. Meylan. 1988. A phylogeny of turtles. Pp. 157–219 *in* Benton, M. J. (ed.), *The Phylogeny and Classification of the Tetrapods*. Vol. 1: *Amphibians, Reptiles, Birds*. Syst. Assoc. Spec. Vol. No. 35 A. Oxford: Clarendon Press.

Gardiner, B. G. 1982. Tetrapod classification. Zool. J. Linn. Soc. London, 74:207–232.

Gauthier, J. A., A. G. Kluge, and T. Rowe. 1988a. The early evolution of the Amniota. Pp. 103–155 *in* Benton, M. J. (ed.), *The Phylogeny and Classification of the Tetrapods*. Vol. 1: *Amphibians, Reptiles, Birds*. Syst. Assoc. Spec. Vol. No. 35 A. Oxford: Clarendon Press.

Gauthier, J. A., A. G. Kluge, and T. Rowe. 1988b. Amniote phylogeny and the importance of fossils. Cladistics, 4:105–209.

Goodman, M., J. Czelusniak, and J. E. Beeber. 1985. Phylogeny of primates and other eutherian orders: a cladistic analysis using amino acid and nucleotide sequence data. Cladistics, 1:171–185.

Goodman, M., M. M. Miyamoto, and J. Czelusniak. 1987. Pattern and process in vertebrate phylogeny revealed by coevolution of molecules and morphologies. Pp. 141–176 *in* Patterson, C. (ed.), *Molecules and Morphology in Evolution: Conflict or Compromise?* Cambridge: Cambridge Univ. Press.

Heaton, M. J. 1980. The Cotylosauria: a reconsideration of a group of archaic tetrapods. Pp. 497–551 *in* Panchen, A. L. (ed.), *The Terrestrial Environment and the Origin of Land Vertebrates*. London: Academic Press.

Heaton, M. J., and R. R. Reisz. 1986. Phylogenetic relationships of captorhinomorph reptiles. Can. J. Earth Sci., 23:402–418.

Kemp, T. S. 1988. Haemothermia or Archosauria? the interrelationships of mammals, birds, and crocodiles. Zool. J. Linn. Soc. London, 92:67–104.

Kirsch, K. F. 1979. The oldest vertebrate egg? J. Paleontol., 53:1068–1084.

Løvtrup, S. 1985. On the classification of the taxon Tetrapoda. Syst. Zool., 34:463–470.

Milner, A. R. 1988. The relationships and origin of living amphibians. Pp. 59–102 *in* Benton, M. J. (ed.), *The Phylogeny and Classification of the Tetrapods*. Vol. 1: *Amphibians, Reptiles, Birds*. Syst. Assoc. Spec. Vol. No. 35 A. Oxford: Clarendon Press.

Panchen, A. L., and T. R. Smithson. 1988. The relationships of the earliest tetrapods. Pp. 1–32 *in* Benton, M. J. (ed.), *The Phylogeny and Classification of the Tetrapods*. Vol. 1: *Amphibians, Reptiles, Birds*. Syst. Assoc. Spec. Vol. No. 35 A. Oxford: Clarendon Press.

Rieppel, O. 1988. The classification of the Squamata. Pp. 261–293 *in* Benton, M. J. (ed.), *The Phylogeny and Classification of the Tetrapods*. Vol. 1: *Amphibians, Reptiles, Birds*. Syst. Assoc. Spec. Vol. No. 35 A. Oxford: Clarendon Press.

Romer, A. S. 1945. *Vertebrate Paleontology.* 2nd ed. Chicago: Univ. of Chicago Press.

Romer, A. S. 1956. *Osteology of the Reptiles.* Chicago: Univ. of Chicago Press.

Rowe, T. 1988. Definition, diagnosis and origin of Mammalia. J. Vert. Paleontol., 8:241–264.

Smithson, T. R. 1985. The morphology and relationships of the Carboniferous amphibian *Eoherpeton watsoni* Panchen. Zool. J. Linn. Soc. London, 85:317–410.

Smithson, T. R. 1989. The earliest known reptile. Nature, London, 342:676–678.

10 The Origin of Reptiles

Robert L. Carroll

One of the most important achievements in the history of vertebrates was the capacity to reproduce in the absence of standing water. The eggs of reptiles, birds, and mammals have specialized extraembryonic membranes that allow them to develop on land, in contrast to those of fishes and amphibians. Typically, the young are born or hatched as miniatures of the adults, and none has an aquatic larval stage.

Together, reptiles, birds, and mammals constitute a monophyletic group, the Amniota. Among amniotes, birds and mammals can be defined as monophyletic groups. Reptiles are a paraphyletic assemblage recognized by the absence of characters present in their descendants. The term *Reptilia* is used to include the turtles, crocodiles, lizards, and snakes of the modern fauna, and many groups of more primitive amniotes including the ancestors of both birds and mammals. Because the earliest known amniotes are osteologically similar to living reptiles, the origin of the amniotes is typically considered to be equivalent to the origin of reptiles.

Although the oldest known amniotes come from deposits more than 300 million years old, their skeletons are sufficiently similar to those of primitive living lizards that it would be possible to identify them as reptiles, even in the absence of any intervening fossils. In fact, the oldest known reptiles were recognized as such by their discoverer, Sir William Dawson, when they were first described in 1860.

First I describe the skeletal anatomy of these early reptiles and, on the basis of modern forms, postulate some broad aspects of their way of life. Then I discuss the problem of their relationship with other Paleozoic tetrapods in an effort to reconstruct earlier stages in the origin of reptilian characteristics.

THE OLDEST KNOWN REPTILES

The oldest fossil reptiles were collected from the sea cliffs of the little coal-mining town of Joggins, Nova Scotia (Carroll et al., 1972).[1] The Joggins locality is unusual in that all of the specimens are preserved within the stumps of upright lycopod trees. These grew in a succession of forests, each of which was inundated rapidly and partially covered by sediments. Subsequently, the water withdrew, leaving the partially decayed stumps still in place. The hollow trees served as traps for animals walking on the new land surface. The importance of this peculiar manner of deposition is that it preferentially preserved terrestrial animals. Except for a single fish scale, which might have been transported into the tree in some animal's stomach, there are no remains of animals of a strictly aquatic nature in this fauna. This is in contrast with the more typical coal-swamp deposits of Linton, Ohio, and Nyrany, Czechoslovakia, which are approximately contemporary in age with the Joggins locality, and which have produced hundreds of skeletons of fish and aquatic amphibians, but only a handful of reptiles (Milner, 1980; Hook and Ferm, 1985).

The scores of reptile specimens at Joggins support the contention that one of the most important characteristics of modern reptiles, a primarily terrestrial way of life, was manifest at the time of the first appearance of the group.

The bones of the earliest reptile from Joggins, *Hylonomus lyelli*, are scattered but individually well preserved; nearly the entire skeleton can be reconstructed (Fig. 1). Specimens of a similar genus, *Paleothyris*, from a second, slightly younger tree-stump locality, near Sydney, Nova Scotia, were collected by Romer in 1956. These remains are even more complete and well articulated, and they show every bone in the skeleton except for the end of the tail (Carroll, 1969a). The size and general proportions of these early reptiles are similar to those of *Sphenodon* and primitive modern lizards such as the iguanids. *Hylonomus* and other primitive amniotes have been assigned to the family Protorothyridae.

---

[1]Subsequent to submission of the manuscript for this chapter, notice of a much older amniote was published (Smithson, 1989). This preliminary study mentions few skeletal features that differ from those of previously described early amniotes but extends their range an additional 38 million years. Reference to specific features of its anatomy are made in a subsequent section.

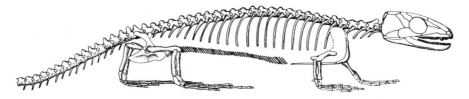

Fig. 1. Skeleton of the earliest known reptile, *Hylonomus lyelli,* from the Westphalian B of Joggins, Nova Scotia. Approximately two-thirds natural size.

Hylonomus and *Paleothyris* are recognized as primitive reptiles by comparison with other amniote groups known from the Carboniferous and Early Permian, including pelycosaurs, diapsids, and captorhinids. Through time, each of these groups shows similar character transformations involving size increase, loss and fusion of bones, and changes in proportions. With some possible exceptions to be discussed below, the characters exhibited by the early protorothyrids resemble the most primitive character states of the other early amniote groups. The problem of using out-group comparison for establishing polarities among early amniotes is also discussed in a later section of this chapter.

Fossils of *Hylonomus* and *Paleothyris* form a good basis for establishing the primitive skeletal pattern from which all other amniotes have evolved, as well as providing a framework for discussing some aspects of the way of life of primitive reptiles.

In comparison with more primitive Paleozoic tetrapods, the skull of protorothyrids is distinguished by its small size relative to the trunk, and the small size of the tabular, supratemporal, and postparietal bones, which are primarily occipital in position. The intertemporal is lost (Figs. 2–4). The occipital surface is distinguished by the presence of a large supraoccipital bone that linked the dermal bones of the skull roof with the back of the braincase.

The palate is distinguished from that of primitive amphibians by the loss of large palatine fangs and the development of a transverse flange on the pterygoid (Figs. 5–6). The pterygoid flange is one of the most important features by which even the earliest reptiles can be associated with their modern counterparts. In lizards and crocodiles, a comparable flange is associated with the origin of one of the most important of the jaw muscles, the pterygoideus. This muscle extends around beneath the back of the lower jaw. Its fibers are essentially horizontal, in contrast with the more vertical fibers of the other adductors.

The importance of the pterygoideus muscle rests not only in the overall increase in muscle mass, compared with that of primitive amphibians, but also in the differentiation of the adductor musculature into two major portions, fibers of which have significantly different angles of orienta-

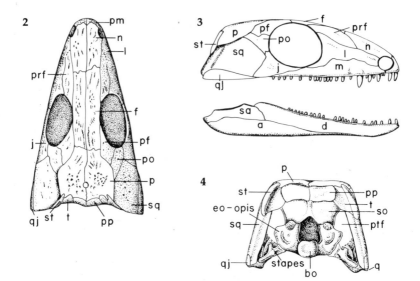

Figs. 2–4. Skull of the primitive amniote *Paleothyris* in (2) dorsal, (3) lateral, and (4) occipital views. See Appendix II for key to abbreviations.

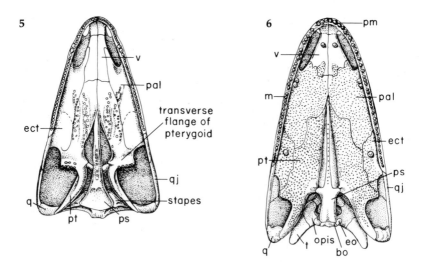

Figs. 5–6. Comparative views of the palate of (5) the primitive amniote *Paleothyris* and (6) the anthracosaur *Gephrostegus*, showing the different patterns of the subtemporal fenestra. *Paleothyris* is distinguished by the presence of the transverse flange of the pterygoid, from which originated the pterygoideus muscle. See Appendix II for key to abbreviations.

tion. Simple, straplike muscles such as the primitive adductors act most strongly when they are oriented at right angles to the bone being moved. In primitive amphibians, the adductor chamber is oriented obliquely, and the musculature must have extended posteriorly so that it was at right angles to the lower jaw when it was fully open. This, together with the presence of large palatal fangs, indicates that primitive labyrinthodonts were adapted, like the crocodile, for feeding on large prey. The adductors would have closed the jaw rapidly but could not have applied their maximum force when the jaws were nearly closed. Olson (1961) applied the term *kinetic inertial* to this mode of jaw closure. In the early reptiles, in contrast, the adductor chamber and the main adductor muscles had a more vertical orientation so that the major force was applied when the mouth was nearly closed. These forms lacked palatine fangs and had rows of smaller teeth on the palate. The pterygoideus probably acted at nearly right angles to the other adductors so that it could apply its greatest force when the mouth was wide open. Hence, the two muscles provided the most effective utilization of their force during different stages of jaw closure (Carroll, 1969b).

The pterygoideus muscle is a feature of modern reptile groups that have become adapted to a variety of diets and ways of manipulating food. The closest approximation of skull structure, proportions, and dentition to the pattern of early reptiles is seen in primitive lizards. As has been documented by Pough (1973), primitive iguanids, agamids, and geckos feed almost exclusively on insects. It seems likely that insects and other small arthropods also constituted a large part of the diet of primitive reptiles. This also corresponds well with their small body size.

The skull of early reptiles differs from that of lizards in lacking temporal openings, but this difference does not seem to affect the general distribution of the jaw muscles, nor the broader aspects of the mechanics of their feeding.

Another primitive feature of the skull of protorothyrids is the absence of the otic notch that in modern reptiles supports a tympanum for reception of airborne sound. It was long assumed that early reptiles also had a tympanum but that it was supported by soft tissue behind the skull. Because early reptiles had a stapes, they were assumed to have been capable of responding to airborne vibrations, although perhaps of lower frequencies than those to which modern amniotes respond.

Physiological work by Wever (1978) and his colleagues established the functional significance of the structural features of the modern reptilian middle ear. It consists of a large tympanum that activates a slim stapes supported in an air-filled chamber. This arrangement acts to magnify the force of airborne sound so that it can be transmitted through the much

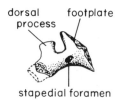

Fig. 7. Stapes of the earliest reptile, *Hylonomus*.

more dense fluid of the inner ear. In order to be mechanically effective, the tympanum has to have approximately 20 times the area of the footplate of the stapes.

In all early reptiles, the stapes is a massive structure with a huge footplate (Fig. 7). In order to receive sufficient force from airborne vibrations to activate the fluid in the inner ear, the tympanum would have had to be nearly as large as the entire skull. Furthermore, the stapes does not lie at a right angle to the skull surface; instead, it is oriented obliquely, further diminishing its efficiency in transmitting vibrations. Although early reptiles, like snakes and other squamates without a tympanum or middle-ear cavity, might have been sensitive to loud low-frequency sounds, there is no evidence that the middle-ear region functioned as an impedance-matching system as it does in most modern amniotes. The development of an impedance-matching ear almost certainly occurred separately in the ancestors of turtles, lizards, crocodiles, and mammals (Allin, 1975; Carroll, 1987).

Although the skeleton of protorothyrids is small and light, it is well ossified and includes a complete complement of carpals and tarsals (Figs. 8–9). In contrast to the larger labyrinthodonts that form the major elements of the Carboniferous fauna, the early reptiles presumably were relatively agile.

The structure of the cervical vertebrae is complex and would have allowed the skull to rotate and bend in any direction. As in the modern reptile *Sphenodon*, the atlas arch is a small paired structure that is linked to the back of the skull by the proatlas. The atlas centrum and the first two intercentra are separate ossifications, whereas the axis arch is large and fused to the pleurocentrum (Fig. 10).

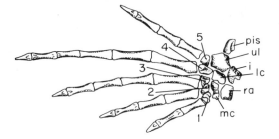

Fig. 8. Carpus and manus of the protorothyrid *Paleothyris* from the Westphalian D of Florence, Nova Scotia. See Appendix II for key to abbreviations.

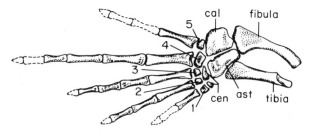

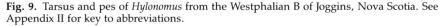

Fig. 9. Tarsus and pes of *Hylonomus* from the Westphalian B of Joggins, Nova Scotia. See Appendix II for key to abbreviations.

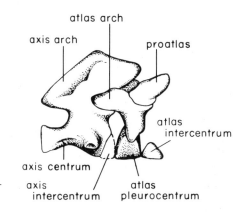

Fig. 10. Atlas-axis complex of *Hy-lonomus.*

Except for the absence of temporal openings and the primitive nature of the ear, most of the skeleton of early amniotes broadly resembles that of primitive modern reptiles, especially *Sphenodon*. What evidence is there of the most important feature of reptiles and other amniotes—i.e., their mode of reproduction? Did the early reptiles lay their eggs on land and undergo direct development, or did they deposit their eggs in the water and have a larval stage as do amphibians and fishes? There are no skeletal features that answer this question. Fossils of amniote eggs are known, but they are much younger than these Carboniferous fossils.

Extraembryonic membranes, which provide support and facilitate gas exchange and water retention, do not fossilize, and the subsequent development of a calcareous shell provides no direct evidence as to the time of origin of these membranes. The actual process of the evolution of extraembryonic membranes is subject to speculation. In 1968, Szarski proposed a model for their origin via an intermediate stage in which nonamniotic eggs were laid on land. This is a practice of many plethodontid salamanders.

There are no features of the skeleton that indicate that plethodontids lay their eggs on land, but the hatchlings are very large relative to the adult. The large size of the young may be necessary so that they can assume an essentially adult way of life immediately at hatching in a terrestrial environment. The maximum size of hatchlings is limited by the physical constraints of the anamniotic egg, which among modern frogs and salamanders rarely exceeds 10 mm in diameter. Presumably neither support nor exchange of respiratory gasses could occur in a larger egg without specialized membranes.

Reproductive modes of modern amphibians are discussed at length by Salthe and Mecham (1974) and reviewed by Duellman and Trueb (1986:Ch. 5). The largest salamander eggs are those of the plethodontid *Aneides lugubris*, which reach 7.4 mm in diameter. The largest anuran ovum, 12 mm in diameter, is that of the hylid frog *Gastrotheca cornuta*. Both of these species have direct development. Caecilians (apodans or gymnophionans) are exceptions in their egg size. The eggs of *Siphonops paulensis* and *Ichthyophis glutinosus* exceed 40 mm in diameter. The large size of their eggs may result from the presence of elastic fibers in the egg capsules and the early elaboration of large filamentous gills that serve for gas exchange, as do the extraembryonic membranes of amniotes. Caecilians seem to have evolved an alternative means of achieving large terrestrial eggs distinct from that of the amniotes.

There is a close correlation between egg size and body size among frogs and salamanders. Salamanders that lay eggs on land have the largest ratio of hatchling to adult size (Salthe and Mecham, 1974).

Reptiles have an even stronger correlation between the size of the egg and the adult, with the smallest body size corresponding to that of terrestrial plethodontid salamanders (Fig. 11) (Carroll, 1970). A similar size relationship presumably applied to the earliest reptiles and their immediate antecedents. This would indicate that the body size in reptilian ancestors that laid anamniotic eggs on land (which are necessarily less than 10 mm in diameter) would have been limited to approximately 100 mm in snout-vent length. This corresponds well with the smallest of the Joggins reptiles. Such small size is not enough to establish that they had, in fact, developed extraembryonic membranes by this time, but it is a necessary condition for them to have done so. Most lines of early reptiles show marked size increase in the later Carboniferous and Permian.

Small size might have affected other aspects of the biology of early amniotes. One of their most striking features is the high degree of ossification of the vertebrae, carpals, and tarsals. Because these structures typically ossify slowly and appear fully developed only in large adults of other early tetrapods, it is surprising that they ossify without intervening cartilage-filled gaps in very small early amniotes. This apparent paradox was also observed by Hanken (1985) in species of extremely small mod-

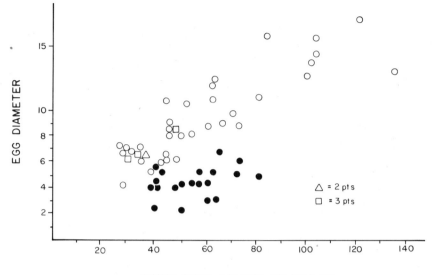

Fig. 11. Correlation between mean diameter of eggs and adult body size in geckos (*closed symbols*) and plethodontid salamanders (*open symbols*).

ern salamanders. He attributed this phenomenon to selection for premature ossification that serves to terminate growth at a small size. This explanation may apply to the early amniotes as well.

What selection pressure can account for the early termination of growth? The answer might be dietary specialization. In large reptiles such as crocodiles, the diet changes during growth; thus, hatchlings feed on small prey (largely insects), whereas adults feed on large prey. A similar dietary shift presumably occurred among Paleozoic tetrapods that were large as adults. The limited gape in the young individuals and the more vertical orientation of the space for the jaw muscles observed in the young of all Paleozoic tetrapods would have facilitated feeding upon small, agile prey. If such prey (e.g., early terrestrial arthropods, particularly insects) were abundant, selection may have led to the retention of juvenile size and proportions in one or more lineages of primitive tetrapods, including the ancestors of reptiles.

We are fortunate in having numerous well-preserved fossils of primitive reptiles that provide an excellent basis for determining the relationships of the various groups of early amniotes close to the time of their initial radiation (Carroll, 1982). On the other hand, we have no direct evidence of the amphibian group ancestral to the reptiles.

Because the earliest reptiles are clearly distinct from all other early tetrapods, their sister-group relationship among this assemblage is difficult to establish. If early labyrinthodonts and the rhipidistian fish are used as successive out-groups for establishing polarity, the earliest amniotes can be characterized by the following derived features: (1) small body size, (2) skull small relative to trunk, (3) loss of squamosal embayment of cheek, (4) loss of intertemporal bone, (5) presence of a large supraoccipital bone, (6) palate lacking large fangs, (7) small denticles on palate, arranged in three radiating rows, (8) presence of transverse flange on pterygoid, (9) limbs slender and well ossified, (10) spool-shaped pleurocentra strongly integrated with neural arches, (11) intercentra reduced to crescents, and (12) proximal tarsals integrated to form astragalus.

POSSIBLE SISTER-GROUPS OF THE AMNIOTES

In order to establish how these features evolved in the origin of amniotes, it is necessary to determine to which other group of Paleozoic tetrapods the amniotes have the closest affinities. No amphibians that can be identified as possible ancestors are known from deposits earlier than those yielding the oldest reptiles. Rather, we must consider which of a number of Late Carboniferous and Early Permian genera may be the most closely related sister-group.

Paleozoic amphibians generally have been classified in two large assemblages, the labyrinthodonts and the lepospondyls (Carroll, 1977). The labyrinthodonts are a primitive group whose skeletal anatomy can be derived directly from the pattern of rhipidistian fish. They are almost certainly a paraphyletic group in the sense of including the ancestry of all more advanced tetrapods.

Lepospondyls include a heterogenous assemblage of distinct lineages, united by the common presence of derived features of the skull and vertebral column. Their body proportions and other aspects of the skeleton are so different from one another, however, that there is considerable question as to whether the features they share are indicative of a common ancestry. Many of these features may be attributed instead to small size (Carroll, 1986b). The lepospondyls presumably evolved from one or more groups with an anatomical pattern generally comparable to that of the known labyrinthodonts.

Labyrinthodonts may be divided into three groups—(1) the primitive ichthyostegids, known only from the late Devonian of East Greenland; (2) the anthracosaurs; and (3) the temnospondyls. The lepospondyls include three major groups—(1) the limbless aïstopods; (2) the aquatic nectrideans; and (3) a heterogeneous group termed microsaurs that includes some lizardlike forms, others that were obligatorily aquatic, and some that may have been burrowers with long bodies and reduced limbs.

Among the lepospondyls, the microsaurs have frequently been suggested to be close to the ancestry of reptiles. Among the labyrinthodonts, three families—the Diadectidae, Tseajaiidae, and the Limnoscelida (grouped as the Diadectomorpha [Heaton, 1980])—show the greatest number of derived features in common with early amniotes. *Diadectes* and *Limnoscelis* were long classified as reptiles (Romer, 1956). Derived features of the vertebrae, including the dominance of the pleurocentra, and the pattern of bones of the skull table unite the diadectomorphs with the anthracosaurs among the labyrinthodonts.

Microsaurs and limnoscelids each share a number of derived features with early amniotes. The patterns of distribution of these characters are listed below.

Derived features shared with early amniotes by both limnoscelids and microsaurs (as exemplified by Tuditanus *Carroll and Baird, 1968).* (1) Loss of intertemporal bone, (2) loss of palatal fangs, (3) loss of squamosal embayment of cheek (vertical posterior wall of adductor chamber), and (4) spool-shaped vertebral centra.

Derived features shared with microsaurs, but not limnoscelids. (1) Loss of labyrinthine infolding of teeth, (2) fusion of tibiale, intermedium, and proximal centrale to form astragalus, (3) small body size, (4) skull small relative to trunk, and (5) limbs slender and well ossified.

Derived features shared with limnoscelids, but not microsaurs. (1) Large supraoccipital bone linking back of braincase to skull roof, (2) transverse flange of pterygoid, (3) specialized structure of atlas-axis complex.

Difference of uncertain polarity. Microsaurs have a four-toed manus, whereas anthracosaurs and reptiles have five toes.

Specializations of microsaurs absent in both anthracosaurs and reptiles. (1) Loss or fusion of one temporal bone. Reptiles retain both the tabular and supratemporal, but only one bone is present in this area in microsaurs. (2) Fusion of elements of anterior cervical vertebrae to form a single functional unit.

The number of characters supporting alternative hypotheses of relationship is not large. A simple parsimony argument is not sufficient to choose between the two alternative patterns of relationship.

This problem can be resolved *directly* only on the basis of fossil remains of the actual ancestors of reptiles. However, this question can be approached *indirectly* through an appraisal of evolutionary patterns among other Paleozoic tetrapods. Of these, changes in the cervical vertebrae may be the most informative.

Lepospondyls are characterized by the possession of spool-shaped vertebral centra, in contrast with the multipartite, crescentic central elements that characterize primitive labyrinthodonts (Carroll, 1989). Spool-shaped centra have also evolved among some derived labyrinthodonts (embolomeres and stereospondyls) and may have evolved separately among the ancestors of each of the lepospondyl groups. The pattern of the cervical vertebrae seems to have evolved divergently among labyrinthodonts and lepospondyls. In rhipidistian fish and the most primitive labyrinthodonts (i.e., ichthyostegids), the primary connection between the skull and the postcranial skeleton is maintained by the unrestricted notochord, which extends forward through the base of the occiput, nearly to the level of the pituitary. In most tetrapods, the notochord does not penetrate the occiput deeply, and support of the skull is provided by a bony articulation between the occiput and the first cervical vertebra. The nature of this articulation differs from group to group. In aïstopods, the circular margin of the most anterior centrum fits against the rim of a circular recess at the back of the occiput. In microsaurs and nectrideans, the anterior end of the first centrum is expanded laterally. In microsaurs, the exoccipitals and basioccipital form a broad strap-shaped surface into which the broad anterior surface of the first cervical fits. In nectrideans, the basioccipital is not involved in the articular surface, which is formed by the paired exoccipitals. In all three "lepospondyl" groups, the articulation is provided by a single, specialized cervical vertebra. Temnospondyls and anthracosaurs (including diadectomorphs) retain the multipartite pattern of the anterior vertebrae common to rhipidistians but specialize the centra and arches of the first two segments to form a multipartite atlas-axis complex that fits around a median occipital condyle (Fig. 12). Reptiles share with diadectomorphs advanced features of this structure, including the elaboration of the pleurocentrum of the second cervical vertebra at the expense of the intercentra and the fusion of the axis arch and centrum (Fig. 13). This pattern differs fundamentally from that of microsaurs (Fig. 14).

The patterns of the cervical vertebrae in diadectomorphs and reptiles, on one hand, and microsaurs, on the other, are both derived relative to that of primitive tetrapods. The patterns in the two groups are widely divergent, however; thus, it is improbable that either gave rise to the other. This is the primary reason for thinking that other similarities between microsaurs and reptiles evolved by convergence, rather than from a common ancestor.

There are also difficulties involved in accepting a common ancestry for reptiles and advanced anthracosaurs. Previously, I argued that the configuration of the occiput of anthracosaurs and amniotes indicates divergent patterns in the consolidations of the braincase and the dermal skull roof (Carroll, 1980). Recent descriptions of the early anthracosaur *Eoherpeton* (Smithson, 1985) and *Limnoscelis* (Fracasso, 1987) suggest the possibility of a transition between early anthracosaurs, limnoscelids, and primitive amniotes. *Eoherpeton* has a single ossification in the otic region; this ossification apparently includes the areas that are distinguished in

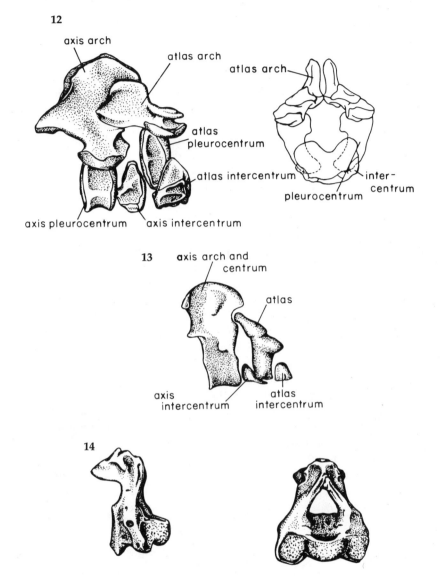

Figs. 12–14. Anterior cervical vertebrae of primitive tetrapods. (12) Lateral and anterior views of the atlas-axis complex of the anthracosaur *Proterogyrinus*. (Modified from Holmes, 1984.) (13) The atlas-axis complex of the cotylosaur *Tseajaia;* like that of reptiles (Fig. 10), the atlas pleurocentrum has fused to form a medial structure, and the axis arch and centrum have fused. (14) First cervical vertebra of the microsaur *Euryodus* in lateral and anterior views. In all "lepospondyls," the elements of the anterior cervical vertebra have fused to form a single functional unit. The first cervical of microsaurs has a broad, strap-shaped anterior surface that articulates with the occiput as the ball of a ball-and-socket joint. In contrast, the multipartite atlas-axis complex of primitive anthracosaurs, cotylosaurs, and reptiles fits around the occipital condyle in the role of the socket of a ball-and-socket joint. (Modified from Carroll and Gaskill, 1978.)

343

amniotes as three separate bones—the supraoccipital, opisthotic, and prootic (Fig. 15). This bone is attached firmly to the tabulars by what are referred to as the paroccipital processes, as in embolomeres. This is quite different from the pattern in early reptiles, in which a separate supraoccipital joins the otic capsules to the skull roof, and the opisthotic bears a paroccipital process that extends toward the cheek. In *Paleothyris*, the opisthotic appears incompletely ossified, but in early pelycosaurs, it makes contact with the ventral portion of the tabular. In early diapsids, the opisthotic extends toward the squamosal. In *Limnoscelis*, as restored by Fracasso (Fig. 16), the supraoccipital is distinguished as a plate of bone between the otic capsule and the skull table. The otic capsule is a recognizably separate ossification beneath it. The pattern in *Limnoscelis* suggests that the paroccipital processes of *Eoherpeton* and embolomeres are not strictly homologous to those of reptiles. The paroccipital process of *Eoherpeton* and embolomeres is a dorsal process from the supraoccipital portion of a unified otic bone. The paroccipital process of reptiles is an extension from the opisthotic that developed in a more ventral position. Evolution could proceed from one pattern to another without reversal or loss of functional continuity. This interpretation makes it possible to reconcile changes in the pattern of the occiput with the suite of derived characters that support a sister-group relationship between limnoscelids and amniotes.

Posttemporal fossae are absent in primitive anthracosaurs, but they are present in temnospondyls and reptiles. If reptiles evolved from animals like *Limnoscelis*, these openings must have been evolved within the descendants of anthracosaurs and therefore are not homologous to openings of this name in temnospondyls.

If limnoscelids and other diadectomorphs are the sister-group of amniotes, one might include both in a single larger taxonomic assemblage. It is difficult to justify application of the term *amniote* to this assemblage, however, because there is some evidence that both *Tseajaia* and *Lim-*

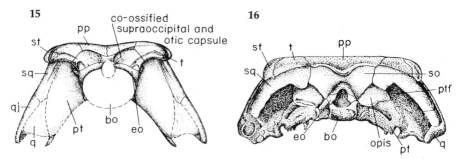

Figs. 15–16. Occiput of the anthracosaurs (15) *Eoherpeton* (modified from Smithson, 1985) and (16) *Limnoscelis* (modified from Fracasso, 1987). See Appendix II for key to abbreviations.

noscelis had lateral-line canal grooves; this indicates that these taxa had aquatic larval stages and, thus, were not amniotes reproductively. In fact, all anthracosaurs probably belong to the same monophyletic assemblage as the amniotes, although primitive anthracosaurs, including embolomeres, almost certainly were amphibians physiologically.

Eoherpeton, from the Visean and Namurian of Great Britain, provides a plausible model for the cranial anatomy of the early anthracosaur group that might have been ancestral to reptiles. The skull is large and typical of other early anthracosaurs except for the more vertical posterior margin of the cheek. The orientation of the cheek margin suggests that the major adductor muscles already had achieved a nearly vertical orientation. The palate is primitive in the retention of large fangs on the palatines and ectopterygoids. There is no evidence of the development of the transverse flange of the pterygoid—a feature that is associated with the elaboration of the pterygoideus muscle in early reptiles.

Even if anthracosaurs such as the limnoscelids can be demonstrated to be the sister-group of early amniotes, the absence of immediate ancestors leaves many aspects of the origin of amniotes unresolved. Of particular importance is the question of when the small size, characteristic of early amniotes, was achieved. Accepting that primitive tetrapods were relatively large, is the small size of the early amniotes a specialization of this lineage, in contrast with the much larger limnoscelids that are known in the later Carboniferous, or were the early relatives of *Limnoscelis* also small, and the later members of that group secondarily large?

If the common ancestors of limnoscelids and amniotes were small, the loss of palatine fangs and the elaboration of the pterygoideus musculature may be associated with small size and associated dietary specializations. If the ancestors of *Limnoscelis* were large throughout their evolution, such a correlation would not be acceptable.

If the large size of the known limnoscelids is a primitive feature of this group, it is unlikely that they achieved an amniotic mode of reproduction, but this question is more difficult to resolve if they, like the amniotes, had gone through a period of their evolution during which they were of much smaller body size. Neither of these problems can be resolved without the discovery of fossils of primitive limnoscelids and the group that was immediately ancestral to reptiles in the early Carboniferous, nor can the polarities of specific character-states that distinguish limnoscelids and early amniotes be established.

THE EARLY RADIATION OF AMNIOTES

In addition to the protorothyrid *Hylonomus*, two other amniote genera have been described from the Middle Pennsylvanian of Joggins, Nova

Scotia (Carroll, 1964). *Archerpeton* is known too incompletely for us to be certain of its ordinal affinities. *Protoclepsydrops* also is incompletely known, but characters of the humeri indicate affinities with a second major group—the pelycosaurs—the first of the mammal-like reptiles. Pelycosaurs were important and diverse elements in the slightly later tree-stump fauna from Florence, Nova Scotia (Reisz, 1972), and they continued to radiate in the Late Pennsylvanian and Early Permian. A third major amniote group, the Diapsida, appeared by the end of the Pennsylvanian (Reisz, 1981). By the base of the Permian, the family Captorhinidae (which has been suggested as the sister-group of turtles, Gaffney and McKenna, 1979) is recognizable (Clark and Carroll, 1973; Heaton, 1979). Mesosaurs, pareiasaurs, millerosaurs, and procolophonoids, none of which survived beyond the end of the Triassic, all appeared in the middle to late Permian. The interrelationships of these early amniote groups was discussed by Carroll (1982) and by Heaton and Reisz (1986). All agree on the close affinities of the protorothyrid genera and the diapsids. The other relationships are subject to dispute. Carroll argued that nearly all the skeletal characters of protorothyrids represent the primitive character states relative to those of all other groups of amniotes (Appendix I). Heaton and Reisz argued that several characters of the protorothyrids are more derived than those of stem amniotes. The most conspicuous of these contentious characters are the nature of the neural arches; the proportions of the limbs, manus, and pes; and the configuration of the supratemporal. Heaton and Reisz's arguments regarding the polarity of these characters apparently were based primarily on outgroup comparison with limnoscelids.

Although limnoscelids might have diverged from a common ancestor shared with the amniotes, Late Pennsylvanian and Early Permian limnoscelids form a dubious basis for establishing the polarity of individual characters found among Early Pennsylvanian amniotes. In the absence of the actual temporal ancestors of amniotes, is there any other way in which polarities of amniote characters can be judged? Character polarity can be established within several of the most important early amniote groups. The most primitive character state common to these groups is probably the character state of their immediate common ancestor.

In general, the greatest degree of similarity lies among the earliest and most primitive members of each of the early amniote lineages. The earliest diapsids, captorhinids, and pelycosaurs are most similar to the early protorothyrids (Carroll, 1986a). All groups diverge in time. In most groups, divergence in morphology is accompanied by increase in size. This is most conspicuous among the ophiacodont and sphenacodont pelycosaurs, and less so among the captorhinids. There is little change in either size or morphology among protorothyrids throughout their 30-million-year history. Diapsids are poorly known in the Upper Carboniferous and Early Permian but are morphologically diverse and larger in the later Permian and Early Triassic.

The pattern of each of the contentious traits may be viewed through time.

Neural Arches

Heaton and Reisz (1986) claimed that swollen neural arches are plesiomorphic for amniotes, and that steeply angled zygapophyses and narrow neural arches are derived. All protorothyrids and most pelycosaurs have narrow neural arches and steeply angled zygapophyses. One possible exception is *Varanosaurus*, the position of which within the Pelycosauria has been seriously questioned. This genus is known only in the Permian, and there is no evidence that its possession of swollen neural arches is primitive for pelycosaurs. None of the 12 genera of pelycosaurs described from the Pennsylvanian has swollen neural arches. The later Lower Permian pelycosaur *Cotylorhynchus* is extremely large and has relatively wide, flat zygapophyses, but the arches do not appear swollen because of the high spines.

Heaton and Reisz (1986) cited swollen neural arches as characteristic of procolophonoids, pareiasaurs, and captorhinids, but none of these groups appeared until the Permian. In both procolophonoids and captorhinids, the relative width of the neural arches increases with size, both ontogenetically and phylogenetically, thereby suggesting that this feature is at least partially size-related within these groups. It is extremely unlikely that the presence of wide neural arches was a primitive character of amniotes, to judge by their absence in all Carboniferous members of this group. It is much more logical to conclude that swollen neural arches were achieved separately in several lines of Permian reptiles and in the limnoscelids and other diadectomorphs. Olson (1976) attributed the elaboration of flattened zygapophyses and swollen neural arches in these groups to the reduction of tortion necessary to accommodate their large body size. Large pelycosaurs and temnospondyls achieved stability of the vertebral column through the evolution of long neural spines.

Broad Manus and Pes, and Short, Heavy Limbs

The proportions of the limbs, manus, and pes follow a similar pattern. The great girth and width of these structures in pareiasaurs and large captorhinids can be associated with their great body size. This is seen most clearly among the pelycosaurs, within which ophiacodonts, sphenacodonts, edaphosaurs, and caseids increase in size gradually through time. Concomitantly, there is a proportionate increase in the relative girth of the limb bones and the relative width of the manus and pes (Romer and Price, 1940; Stovall et al., 1966).

The early protorothyrids *Hylonomus* and *Paleothyris* have slender limbs and long manus and pes. The Westphalian D genus *Paleothyris* is more gracile than the earlier *Hylonomus*. *Hylonomus* may represent the most primitive condition, with divergent evolution leading to the more gracile later protorothyrids and the less gracile members of other lineages.

The Supratemporal

The primitive nature of the supratemporal is more difficult to establish. In primitive tetrapods and diadectomorphs, it is large. In anthracosaurs and *Limnoscelis*, it is in contact with the postorbital; this area is difficult to interpret in *Tseajaia*. In all described Pennsylvanian amniotes, the supratemporal is a narrow splint of bone that fits into the corner of the parietal. This area, however, is not adequately known in any Pennsylvanian pelycosaur. In diapsids and protorothyrids, the supratemporal does not make contact with the postorbital. In contrast, the postorbital and supratemporal articulate in most pelycosaurs, but it is uncertain whether this is a primitive condition for amniotes or derived among early pelycosaurs.

Heaton and Reisz (1986) cited the presence of a wide supratemporal in contact with the postorbital in procolophonids and pareiasaurs as evidence that this is a primitive feature of amniotes. However, the configuration of the remainder of the temporal region is highly specialized in these Middle and Upper Permian genera, and they constitute an unlikely basis for establishing the primitive condition for amniotes. Some pelycosaurs (notably, caseids and eothyridids) and millerosaurs also have a fairly broad supratemporal (Godfrey and Reisz, in prep.). Both pelycosaurs and millerosaurs have a large lateral temporal opening that certainly is a derived feature among amniotes. The close association of the supratemporal with the temporal opening suggests that the large size of the bone might serve to strengthen the surrounding area of the skull. The configuration of the supratemporal in Lower Permian pelycosaurs only hints at its possible form in Lower Pennsylvanian members of this group. Lower Permian pelycosaurs may retain the primitive pattern seen in limnoscelids, but this feature cannot be demonstrated by out-group comparison. It can only be substantiated by the condition in early Pennsylvanian pelycosaurs themselves.

Despite the continuing controversy about the primitive configuration of the supratemporal, an animal with an anatomical pattern very similar to the early protorothyrids *Hylonomus* and *Paleothyris* still seems a logical ancestral form for all later amniote groups. However, this can be confirmed with assurance only by the discovery of appropriate fossils from earlier in the Carboniferous.

A fossil 38 million years older than the Joggins reptiles has recently been discovered (Smithson, 1989). It has not yet been fully prepared, but several vertebrae are exposed that clearly show narrow neural arches, with the zygapophyses close to the midline. The bones of the limb are nearly as gracile as those of *Hylonomus*, rather than having short, thick shafts and large extremities, as was predicted for early amniotes by Heaton and Reisz. The digits are only slightly shorter, relative to the width of the pes, than those of *Hylonomus*. The parietal is very wide and may have reached the squamosal. The temporal region is poorly exposed, but neither the supratemporal nor the tabular appear to have been large.

This newly described specimen is primitive in having contact between the pre- and postfrontal bones so as to separate the frontal from the orbital margin, and in the very large size of the intercentra and the ventral scales. In all these features, the new specimen resembles primitive anthracosaurs. Smithson was unwilling to speculate as to the specific affinities of this specimen with other early amniotes, but it provides the best available evidence for establishing the polarity of character transformations within this assemblage.

Limnoscelids may be the closest known sister-group to the amniotes, but arguing from the standpoint of parsimony, there are far fewer character transformations necessary between *Hylonomus* and the early members of the other amniote groups than there are between *Limnoscelis* and any of the early amniotes. Thus, *Limnoscelis* provides a questionable basis for establishing the polarity of particular amniote characters.

On the basis of their relative time of appearance in the fossil record, pelycosaurs were the first group to diverge from the primitive amniote stock, followed by the diapsids and captorhinids. This sequence is not contradicted by the distribution of character states among these groups. The classification of early amniotes is certain to remain contentious because it is difficult to avoid recognition of paraphyletic groups.

Acknowledgments The research that forms the basis for this chapter has been supported by grants from the Natural Sciences and Engineering Research Council of Canada. The illustrations were drafted by Pamela Gaskill. I thank Dr. Tim Smithson for suggestions regarding the phylogenetic relationships of amniotes and early anthracosaurs. Other helpful suggestions were provided by Dr. Robert Holmes and the subsequent reviewers of the manuscript.

APPENDIX I. Primitive Character States Exhibited by Protorothyrid Reptiles

A. General features
    1. Small size: 100 mm in snout-vent length

    2. Skull small, limbs relatively short
    3. Habitus: terrestrial, "lacertoid," insectivorous

B. Cranial features
    4. No otic notch; stapes forming a mechanical link between the braincase and the lower cheek
    5. No temporal opening
    6. Supratemporal, tabular, and postparietal small, paired bones
    7. Supratemporal not contacting postorbital
    8. Supraoccipital a broad plate of bone, loosely attached to postparietal and tabular dorsally
    9. Opisthotic not reaching cheek or tabular
    10. Braincase movably articulating with palate
    11. Ectopterygoid retained; no palatal fenestra
    12. Jaw articulation at level of occipital condyle
    13. Approximately 40 pointed marginal teeth, including two canines

C. Postcranial features
    14. 25–32 presacral vertebrae
    15. Axis intercentrum not fused to atlas centrum
    16. Short neck
    17. ? 1 sacral vertebra
    18. Long tail
    19. Neural arch narrow and zygapophyses close to midline
    20. Transverse processes short
    21. No caudal autotomy
    22. Caudal ribs not fused to transverse processes; ends extending posteriorly to parallel tail
    23. Scapulocoracoid developing rapidly into a single ossification
    24. Slim, posteriorly directed iliac blade
    25. No pubic tubercle
    26. Humerus with supinator process
    27. Carpus consisting of pisiform, ulnare, intermedium, radiale, two centralia, and five distal carpals
    28. Metacarpals not overlapping; phalangeal count of 2, 3, 4, 5, 3
    29. Tarsus consisting of astragalus and calcaneum, two centralia, and five distal tarsals
    30. Metatarsal not overlapping; phalangeal count of 2, 3, 4, 5, 4

APPENDIX II. Anatomical Abbreviations Used in the Figures

a = angular; ast = astragalus
bo = basioccipital
cal = calcaneum; cen = centrale

d = dentary

ect = ectopterygoid; eo = exoccipital

f = frontal

i = intermedium

j = jugal

l = lacrimal; lc = lateral centrale

m = maxilla; mc = medial centrale

n = nasal

opis = opisthotic

p = parietal; pal = palatine; pf = postfrontal; pis = pisiform; pm = premaxilla; po = postorbital; pp = postparietal; prf = prefrontal; ps = parasphenoid; pt = pterygoid; ptf = posttemporal fossa

q = quadrate; qj = quadratojugal

ra = radiale

sa = surangular; so = supraoccipital; sq = squamosal; st = supratemporal

t = tabular

ul = ulnare

v = vomer

1–5 = distal carpals and tarsals

Literature Cited

Allin, E. F. 1975. Evolution of the mammalian middle ear. J. Morphol., 147:403–438.

Breckenridge, W. R., and S. Jayasinghe. 1979. Observations on the eggs and larvae of *Ichthyophis glutinosus*. Ceylon J. Sci. (Biol. Sci.), 13:187–202.

Carroll, R. L. 1964. The earliest reptiles. J. Linn. Soc. London Zool., 45:61–83.

Carroll, R. L. 1969a. A Middle Pennsylvanian captorhinomorph, and the interrelationships of primitive reptiles. J. Paleontol., 43:151–170.

Carroll, R. L. 1969b. Problems of the origin of reptiles. Biol. Rev. Cambridge Philos. Soc., 44:393–432.

Carroll, R. L. 1970. Quantitative aspects of the amphibian-reptilian transition. Forma Functio, 3:165–178.

Carroll, R. L. 1977. Patterns of amphibian evolution: an extended example of the incompleteness of the fossil record. Pp. 405–437 *in* Hallam, A. (ed.), *Patterns of Evolution*. Amsterdam: Elsevier Scientific Publishing.

Carroll, R. L. 1980. The hyomandibular as a supporting element in the skull of primitive tetrapods. Pp. 293–317 *in* Panchen, A. L. (ed.), *The Terrestrial Environment and the Origin of Land Vertebrates*. London: Academic Press.

Carroll, R. L. 1982. Early evolution of reptiles. Annu. Rev. Ecol. Syst., 13:87–109.

Carroll, R. L. 1986a. The skeletal anatomy and some aspects of the physiology of primitive reptiles. Pp. 25–45 *in* Hotton, N., III., P. D. MacLean, J. Jan, and E. C. Roth (eds.), *The Evolution and Ecology of Mammal-Like Reptiles*. Washington D.C.: Smithsonian Institution Press.

Carroll, R. L. 1986b. Developmental patterns and the origin of lepospondyls. Pp. 45–48 *in* Roček, Z. (ed.), *Studies in Herpetology*. Prague: Charles Univ.

Carroll, R. L. 1987. *Vertebrae Paleontology and Evolution*. New York: W. H. Freeman.

Carroll, R. L. 1989. Developmental aspects of lepospondyl vertebrae in Paleozoic tetrapods. Hist. Biol., 3:1–25.

Carroll, R. L., and D. Baird. 1968. The Carboniferous amphibian *Tuditanus [Eosauravus]* and the distinction between microsaurs and reptiles. Am. Mus. Novit., 2337:1–50.

Carroll, R. L., E. S. Belt, D. L. Dineley, D. Baird, and D. C. McGregor. 1972. Excursion A59: Vertebrate Paleontology of Eastern Canada. 24th Int. Geol. Congr., Ottawa.

Carroll, R. L., and P. Gaskill. 1978. The Order Microsauria. Mem. Am. Philos. Soc., 126:1–211.

Clark, J., and R. L. Carroll. 1973. Romeriid reptiles from the Lower Permian. Bull. Mus. Comp. Zool. Harv. Univ., 144:353–407.

Dawson, J. W. 1860. On a terrestrial mollusk, a chilognathous myriapod, and some new species of reptiles from the Coal-Formation of Novia Scotia. J. Geol. Soc. London, 16:268–277.

Duellman, W. E., and L. Trueb. 1986. *Biology of Amphibians*. New York: McGraw-Hill.

Fracasso, M. 1987. Braincase of *Limnoscelis paludis* Williston. Postilla, 201:1–22.

Gaffney, E. S., and M. C. McKenna. 1979. A Late Permian captorhinid from Rhodesia. Am. Mus. Novit., 2688:1–15.

Hanken, J. 1985. Morphological novelty in the limb skeleton accompanies miniaturization in salamanders. Science, 229:871–874.

Heaton, M. J. 1979. Cranial morphology of primitive captorhinid reptiles from the late Pennsylvanian and early Permian, Oklahoma and Texas. Bull. Oklahoma Geol. Surv., 127:1–84.

Heaton, M. J. 1980. The Cotylosauria: a reconsideration of a group of archaic tetrapods. Pp. 497–551 *in* Panchen, A. L. (ed.), *The Terrestrial Environment and the Origin of Land Vertebrates*. London: Academic Press.

Heaton, M. J., and R. R. Reisz. 1986. Phylogenetic relationships of captorhinomorph reptiles. Can. J. Earth Sci., 23(3):402–418.

Holmes, R. 1984. The Carboniferous amphibian *Proterogyrinus scheelei* Romer, and the early evolution of tetrapods. Philos. Trans. R. Soc. London, Ser. B, 306:431–527.

Hook, R. W., and J. C. Ferm. 1985. A depositional model for the Linton tetrapod assemblage (Westphalian D, Upper Carboniferous) and its palaeoenvironmental significance. Philos. Trans. R. Soc. London, Ser. B, 311:101–109.

Milner, A. R. 1980. The tetrapod assemblage from Nyrany, Czechoslovakia. Pp. 439–496 *in* Panchen, A. L. (ed.), *The Terrestrial Environment and the Origin of Land Vertebrates*. London: Academic Press.

Olson, E. C. 1961. Jaw mechanisms: rhipidistians, amphibians, reptiles. Am. Zool., 1:205–215.

Olson, E. C. 1976. The exploitation of land by early tetrapods. Pp. 1–30 *in* Bellairs, A. d'A., and C. B. Cox (eds.), *Morphology and Biology of Reptiles*. Symposium Series No. 3. Linn. Soc. London: Academic Press.

Pough, F. H. 1973. Lizard energetics and diet. Ecology, 54:837–844.

Reisz, R. R. 1972. Pelycosaurian reptiles from the Middle Pennsylvanian of North America. Bull. Mus. Comp. Zool. Harv. Univ., 144:27–62.

Reisz, R. R. 1981. A diapsid reptile from the Pennsylvanian of Kansas. Occas. Pap. Mus. Nat. Hist. Univ. Kansas, 7:1–74.

Romer, A. S. 1956. *Osteology of the Reptiles*. Chicago: Univ. of Chicago Press.

Romer, A. S., and L. I. Price. 1940. Review of the Pelycosauria. Spec. Pap. Geol. Soc. Am., 28:1–538.

Salthe, S. N., and J. S. Mecham. 1974. Reproductive and courtship patterns. Pp. 309–521 *in* Lofts, B. (ed.), *Physiology of the Amphibia*. Vol. II. New York: Academic Press.

Smithson, T. R. 1985. The morphology and relationships of the Carboniferous amphibian *Eoherpeton watsoni* Panchen. J. Linn. Soc. London Zool., 85:317–410.

Smithson, T. R. 1989. The earliest known reptile. Nature, London, 342:676–678.

Stoval, J. W., L. I. Price, and A. S. Romer. 1966. The postcranial skeleton of the giant Permian pelycosaur *Cotylorhynchus romeri*. Bull. Mus. Comp. Zool. Harv. Univ., 135:1–30.

Szarski, H. 1968. The origin of vertebrate foetal membranes. Evolution, 22:211–214.

Wever, E. G. 1978. *The Reptilian Ear*. Princeton: Princeton Univ. Press.

11 The Early Radiation of Diapsid Reptiles

Robert L. Carroll and Philip J. Currie

The Diapsida, represented in the modern fauna by *Sphenodon*, lizards, snakes, and crocodiles, encompasses one of the most wide-scale radiations in the history of vertebrates (Fig. 1). Diapsids first appeared in the late Carboniferous but remained rare and showed little diversity until the end of the Paleozoic. During the Late Permian and Triassic, they radiated dramatically, giving rise to a host of lineages including the dinosaurs, pterosaurs, and a variety of secondarily aquatic forms that dominated both the terrestrial and marine environments during the Mesozoic. Many diapsid groups became extinct by the end of the Cretaceous, but lizards and snakes continued to diversify throughout the Cenozoic. Birds are direct descendants of dinosaurs and thus also may be considered within the diapsid radiation. Analysis of any large-scale evolutionary process such as the origin and radiation of the diapsids requires an understanding of the nature of the relationships of the included taxa. Each of the major diapsid groups can be defined readily, but their interrelationships have been subject to continuing dispute. For the past 20 years, cladistics or phylogenetic systematics has been touted widely as an objective and testable procedure for establishing relationships. The emphasis on derived characters and the corresponding need to establish polarities and recognize truly homologous features have given methodological rigor to what Mayr and Simpson previously had considered more of an art than a science.

Despite the clear and logical methodology elaborated by Hennig (1966),

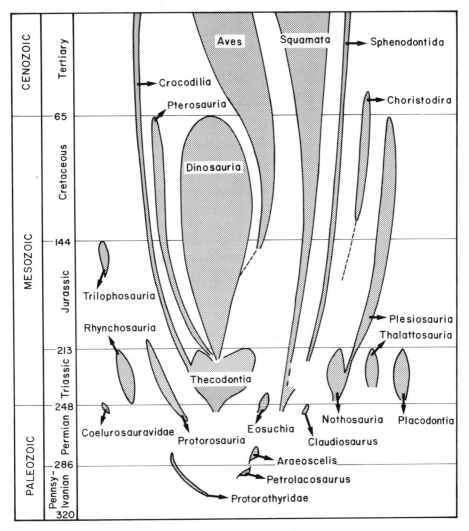

Fig. 1. Temporal distribution of major groups of diapsids.

the application of this methodology has led to some highly contentious phylogenetic conclusions. Examples include the suggestion by Rosen et al. (1981) that lungfish are the sister-group of tetrapods (criticized by Jarvik, 1981; Schultze, 1981; Holmes, 1985; Panchen and Smithson, 1987), Gardiner's proposal (1982) of a sister-group relationship between birds and mammals (disputed by Kemp, 1988; Gauthier et al., 1988), and Løv-trup's phylogeny of the vertebrates (1977). These examples, also noted by Ax (1987), suggest that there may be some general procedural problems in applying Hennig's methodology. These problems should be considered before an effort is made to apply this methodology to the diapsid radiation.

METHODS OF PHYLOGENETIC RECONSTRUCTION

There should be no conflict between the procedures of phylogenetic reconstruction elaborated by Hennig and those used by evolutionary systematists in the school of Simpson and Mayr, because ultimately both are based on fundamental processes of biological evolution. If evolutionary change proceeds in a consistent manner from primitive to derived character-states, the presence of unique derived characters in any two taxa demonstrates that they are more closely related to each other than either is to any other taxon that lacks these characters. Application of this principle requires knowledge of (1) the direction of evolution in order to differentiate derived from primitive characters, and (2) the homology of characters so that similarities resulting from convergence are not mistaken for similarities resulting from common ancestry.

The followers of Hennig and those in the school of Simpson and Mayr differ principally in the way that they establish polarity and homology. Evolutionary systematists rely heavily on the fossil record, whereas cladists emphasize the importance of the modern fauna in developing phylogenetic hypotheses. For an evolutionary systematist, and especially for most paleontologists, the fossil record provides the most direct means of establishing homologies and the polarity of evolutionary change. The direction of evolution can be determined from the relative time of appearance of changing character-states within a monophyletic lineage. Homology can be established on the basis of the presence of the character in question in the immediate common ancestor of the groups being studied. Both can be determined if actual ancestor-descendant relationships can be established. However, if we consider all of organic diversity, knowledge of the fossil record is rarely sufficiently complete to provide this information. Hennig's own research was concentrated on the phylogeny of insects, which have a very incomplete fossil record. Hence, he and his followers have attempted to establish procedures for establishing homology and polarity that are applicable to all groups, even if they lack an adequate fossil record.

Efforts to establish relationships without recourse to the fossil record have figured significantly in cladistic literature. Hennig and his followers have placed emphasis on establishing sister-group relationships, rather than on ancestor-descendant relationships. They also have developed a variety of methods for determining the polarity of evolutionary change and the homology of characters through the study of character distribution among essentially contemporary groups. Homology is judged by essentially probabilistic criteria. According to Patterson (1982), if a large number of other derived features support a sister-group relationship between two groups, then other derived features shared by the two groups are judged to be homologous as well. According to Wiley (1981: 138–139) "homologies can be treated as hypotheses which are tested by other hypotheses of homology and their associated phylogenetic hypotheses." Most of the derived characters found in common in closely related groups would be expected to be homologous, but this method is not capable of establishing which, if any, *specific* characters are homologous, *and* which may be evolved convergently within the groups in question. In contrast, evidence from the fossil record would be expected to be capable of establishing the homology or lack of homology in each character individually.

The most generally used method for establishing polarities in groups without a fossil record is out-group comparison (Kluge, 1977; Wiley, 1981). If particular character-states are common to both the group in question and other related groups (i.e., the out-groups), the character-state is probably primitive. If a character-state is exhibited in one group but is not observed in related groups, the less common occurrence probably represents the derived character-state.

The Importance of Fossils

Ironically, the elaboration of methods of establishing polarity and homology in groups without a fossil record has led to the downplaying of the significance of fossil evidence in groups that do have a significant fossil record.

Wiley (1981), Gaffney (1979), Patterson (1982), and especially Løvtrup (1985) argued that living, rather than fossil, representatives should be emphasized in phylogenetic analysis, even in groups that have a rich fossil record. The primary reason is that living members possess a host of characters that never can be studied in fossils. On the other hand, there are obvious evolutionary reasons why the fossil record should be of greater significance in establishing phylogenetic relationships than even the best known living species.

There is no simple correlation between elapsed time and the amount of evolutionary change, but in general, more changes are likely to have taken place in groups that share only a very distant ancestry than in those that have diverged from a common ancestor more recently. The number of features that result from convergence will almost inevitably increase during evolutionary history, whereas the number of shared characters resulting from common ancestry cannot increase, and may decrease, if synapomorphies are lost or altered. Hence, the likelihood that particular characters are homologous is progressively lower the more distant their common ancestors are in time. The older that groups are geologically, the more hazardous it is to assume that the characters their living members share in common are homologous, and the more difficult it is to establish their relationships on the basis of modern representatives.

It is clear from the writings of Hennig that he considered that the fossil record might provide the ultimate test for phylogenetic hypotheses. For example, he stated (1981:441): "In the preceding chapter I have shown that many of the prerequisites for a really satisfactory account of insect phylogeny are still lacking. The most obvious shortcoming is the almost complete absence of fossils from before the Lower Carboniferous, when a considerable number of insect groups must have arisen according to phylogenetic trees."

Although features of the soft anatomy and physiology may be valuable in establishing hypotheses of relationship in groups without an adequate fossil record, these features have the weakness that they cannot be tested in fossil forms. For example, one may argue that it is more probable that ancestral amniotes had evolved an effective feedback system involving stretch receptors in the axial and appendicular muscles than to assume that this system evolved separately in the ancestors of modern reptiles, birds, and mammals. On the other hand, the presence of this derived condition in modern members of these groups cannot be used to establish that they did share a common ancestor because it is impossible to test whether this system was actually present in any group of Paleozoic tetrapods.

The Prevalence of Convergence

Cladists have based their arguments for the suitability of using modern representatives of groups to establish relationships on the assumption that the majority of similar derived characters are the result of common ancestry and not convergence. According to Kluge (1984), there is no obvious biological basis for assuming that convergence is not a common phenomenon. Surprisingly, no cladists have attempted to demonstrate that convergence is a relatively rare phenomenon.

Because application of the rule of parsimony that governs cladistic analysis depends on the assumption that unique character changes are more common than convergent events, it would be impossible to establish the frequency of convergence through the study of living taxa if it involved more than 50% of the characters being analyzed. On the other hand, detailed knowledge of the fossil record should provide an objective basis for establishing the prevalence of convergence. If the most primitive members of two monophyletic groups are adequately known, it should be possible to establish most of the character changes that occur in their descendants. From them, the relative frequency of unique and convergent character changes can be tabulated. This never can be achieved on the basis of living groups.

Character Selection

Another reason that phylogenies proposed by some cladists may run counter to long-accepted phylogenetic hypotheses is the matter of character selection. Little attention has been paid to how particular characters and taxa are chosen for cladistic analysis. This problem is obscured in many works on Hennigian methodology by the use of hypothetical examples and the use of letters and numbers to represent characters and taxa. Study of the writings of some cladists suggests that the choice of characters is essentially arbitrary. Løvtrup (1985), giving his own examples and citing from other recent papers, listed 22 characters that unite birds and mammals, 14 that establish crocodiles as the sister-group of a combined taxon including birds and mammals, and 16 that unite chelonians and thecodontians. These are chosen from the nearly limitless number of characters exhibited by these groups. The choice of other characters has led previous authors to entirely different interpretations of the relationships of these groups. Clearly, there must be some objective criteria for selecting characters for analysis. One approach would be to use all characters known to change within an assemblage, or some random means of character selection that avoids bias produced by prior assumptions of phylogenetic relationships. This problem is less acute in the case of fossils in which it may be possible to use all characters of the skeletal system.

The problem of character selection strongly influences the application of the rule of parsimony. Parsimony, in the sense of choosing a simple rather than a complex solution for problems, is a general procedure in all science. It has been applied to systematics as a means of choosing between two or more alternate hypotheses of relationship. In general, the hypothesis selected is that which involves the fewest cases of convergence or reversal, or the smallest number of evolutionary steps of any sort. There is, however, no direct evidence that evolution is parsimonious, and convergence may be a common phenomenon. There is another problem with this procedure. It is valid only if the characters considered include all the characters that differ between the groups in question, or some nonbiased sample of these characters. One may say that two groups are related more closely to one another than either is to a third because the first groups are united by 27 shared-derived characters, as opposed to some smaller number supporting the alternate relationships; however, this conclusion is of little significance if there are hundreds of other characters that have not been considered. Parsimony is a practical means of choosing between alternate phylogenetic schemes only if the organisms are basically similar to one another and there is a relatively small number of differences involved, all of which can be analyzed.

It is also necessary to establish some objective means to select the taxa for analysis. This has been "solved" to some extent by cladists who choose to restrict their analysis to living forms (e.g., Løvtrup, 1985). Clearly, such an approach is not justified when the patterns of evolutionary change suggest that the *earliest* known (i.e., fossil) members of monophyletic groups would be much more informative in establishing the pattern of probable relationships.

For nearly 20 years the role of fossils in phylogenetic analysis has been denigrated by many followers of Hennig (but not by Hennig himself). There is now a growing awareness, as illustrated by recent papers by Maisey (1984), Novacek (1986), and Gauthier et al. (1988), that information from the fossil record can contribute substantially to phylogenetic analysis. The following discussion of the relationships of diapsids is intended to demonstrate how Hennig's methodology can be applied to a group with a significant fossil record. At the same time, it serves to test the assumption that convergence is a relatively rare phenomenon and demonstrates the degree to which the use of modern genera is appropriate in evaluating the relationships of the groups to which they belong.

Benton (1985) produced a classification of diapsids that provides an informative contrast to the approach that is attempted here. (See also Evans, 1984; Evans, 1988 and Laurin, 1991 were published too late for evaluation in this review.)

Methodology

In an effort to establish relationships among the major diapsid groups, a relatively simple list of procedures for phylogenetic analysis has been developed. It is intended to be as objective and as free from prior assumptions as possible. The procedures include the following: (1) establishment of the monophyly of the group in question, (2) establishment of the polarities of all characters that vary within the group, (3) recognition of subgroups on the basis of unique apomorphies, (4) determination of derived character-states present in the most primitive members of each monophyletic subgroup, and (5) determination of possible interrelationships on the basis of derived characters shared by the most primitive known members of each subgroup. It would be convenient if this list could be followed in numerical sequence, but, in fact, it has been necessary to work on all categories at once, moving back and forth as information accumulated.

The term *monophyly* is used in this paper to refer to the origin of a group, rather than its subsequent history. For example, the archosaurs are referred to as a monophyletic group on the assumption that they evolved from a single ancestor, without concern for the fact that the term *archosaur* usually is used to the exclusion of their probable descendants, the birds.

Establishment of the Monophyletic Nature of the Group

No dependable classification is possible in the absence of at least a general knowledge of other, related groups. The difficulty of classifying the placoderms, for example, can be attributed at least partially to the fact that their specific relationships to other jawed fish have not been determined (Denison, 1978; Miles and Young, 1977; Goujet, 1984; Gardiner, 1984, 1986). Fortunately, the relationship of diapsids to other early amniotes is well established (Reisz, 1981; Heaton and Reisz, 1986; Carroll, 1982). This knowledge assists in the selection of appropriate out-groups for establishing polarity so that truly unique and derived characters can be used to establish the monophyletic nature of the diapsid subgroups.

Early amniotes are distinguished by the following combination of characters that are derived relative to most other Paleozoic tetrapods: (1) Small adult body size, with highly ossified joint surfaces, carpals, and tarsals. (2) Tooth-bearing transverse flange of pterygoid; loss of fangs on palatal bones. (3) Absence of labyrinthine infolding of teeth. (4) Supraoccipital, a platelike ossification that is movable relative to the skull table. (5) Loss of intertemporal bone and reduction in size of supratemporal, tabular, and postparietal. (6) Stapes with short perforate stem, large footplate, and a dorsal process. (7) Pleurocentra cylindrical and forming the dominant element of the vertebrae, sutured or fused to the arches of all vertebrae except the atlas; intercentra reduced to small crescents but retained throughout the trunk region. (8) Proximal centrale, intermedium, and tibiale co-ossified to form astragalus. In these and other skeletal features, the early representatives of the Paleozoic amniote groups resemble one another closely and apparently diverged from a similar common ancestor not long before their first appearance in the early Pennsylvanian.

Diapsids, in turn, can be differentiated from all members of other early amniote groups by a few features that are derived relative to all other early tetrapods. The most conspicuous traits are the presence of dorsal and lateral temporal openings and a suborbital fenestra. These three cranial characters establish the monophyly of the diapsids. The temporal and suborbital openings may be lost or modified in later species, but all derived patterns can be traced to the primitive condition of the early diapsids.

Reisz (1981) identified some other, less definitive synapomorphies including the possession of well-developed posttemporal fenestrae, a relatively small skull, and a locked tibioastragalar joint. Benton's (1985) objection that these features do not occur in later diapsids or occur in turtles (which do not appear in the fossil record until the late Triassic) is irrelevant in establishing the monophyly of diapsids among early amniotes.

Benton (1985) listed five additional characters of the soft anatomy and development to support the monophyly of diapsids: (1) a Jacobson's organ that develops as a ventromedial outpocketing of the early embryonic nasal cavity; (2) olfactory bulbs anterior to the eyes and linked to the forebrain by the stalklike olfactory tract; (3) one or more nasal conchae in the cavum nasi proprium; (4) a "Huxley's foramen" at the distal end of the extracolumella, surrounded medially by the processus dorsalis and intercalary, and laterally by a laterohyal that links the intercalary to the distal extracolumellar plate; and (5) low levels of urea in the blood. Benton's choice of these characters is based entirely on their presence in living squamates, sphenodontids, birds, and crocodiles (none of which had differentiated in the late Paleozoic) and their absence in modern frogs, salamanders, mammals, and turtles.

It is extremely unlikely that an extracolumella was present in early diapsids, because the structure and function of the stapes is entirely different from that of any living diapsids. There is no evidence that early diapsids possessed either a tympanum or a middle-ear cavity. It is impossible to establish directly if any of the other soft anatomical and developmental characters cited by Benton were present in *early* diapsids (as defined by the nature of the skull) and not in other early amniotes. They are thus of no utility in testing any phylogenetic hypothesis involving Paleozoic genera.

Establishment of the Polarity of Each Character
That Varies within the Group

In order to establish relationships among the diapsids, it is necessary to establish the polarity of all characters that vary within the group. By dealing with *all* variable characters, one can achieve a level of objectivity that would not be possible if characters are selected a priori. A similar approach has been taken by Novacek (1986) in establishing a morphotype for primitive placental groups.

Determination of polarity is a serious problem in itself. Many criteria have been discussed, of which out-group comparison is usually stressed by cladists. To be useful in establishing polarity, the out-group must be a closely related member of the same larger monophyletic assemblage. In the case of the diapsids, the larger monophyletic group is the Amniota, within which protorothyrids and mammal-like reptiles are successively more distantly related sister-groups of diapsids (Heaton and Reisz, 1986).

The most informative out-group for establishing polarity would be the actual ancestors of the group in question. Most followers of Hennig have argued that it is impossible to establish ancestor-descendant relationships, but Wiley (1981:105–107) admitted that it is possible under certain circumstances. Ancestors may be recognized on the basis of the following criteria. An ancestor is a species that can be included within a monophyletic group on the basis of shared derived characters. In addition, the ancestor must (1) retain the primitive character-state for one or more of the morphoclines in which possible descendants possess a more derived condition, *and* (2) lack all apomorphies other than those encountered in the possible descendant forms. The position as putative ancestor can be falsified by the discovery or recognition of such apomorphies.

In the case of early diapsids, the protorothyrid *Paleothyris* (Carroll, 1969) appears to be close to the pattern expected for an ancestor. The only recognized autapomorphy that it possesses is the fusion of the axis intercentrum and the atlas centrum. In the earlier but less well-known genus *Hylonomus*, these bones remain separate. The axis intercentrum and atlas centrum are closely integrated but not fused in the early diapsids. To avoid the apparent circularity of these arguments, one can test the polarity of morphoclines in early diapsids further against the character-states in pelycosaurs and more primitive tetrapods.

Character-states that are present in both *Paleothyris* and early diapsids may be hypothesized as primitive for diapsids as a whole. The most primitive character-states known in diapsids are listed in Appendix I. Most of these are observed in *Petrolacosaurus*, the earliest known genus (Reisz, 1981). In a few characters (indicated in the appendix by an asterisk), *Petrolacosaurus* is derived relative to the state seen in some later diapsids, notably a second genus (*Apsisaurus*; Laurin, 1991) from the Lower Permian of Texas (see also Reisz, 1988). All character-states other than those listed in the left column of Appendix I may be considered derived relative to the condition in the earliest and most primitive diapsids.

Recognition of Subgroups on the Basis of Unique Apomorphies

Numerous subgroups among the early diapsids have been recognized by previous workers. In most cases, they can be diagnosed on the basis of clearly recognizable apomorphies that are unique for each group and widespread within them. Such features as the antorbital fenestra of archosaurs, the streptostylic quadrate of squamates, and the medial narial opening of rhynchosaurs are so clearly unique, obviously derived, and nearly universal in the groups under consideration that they can be used as a preliminary basis for the recognition of the groups. These traits then can be tested for their occurrence elsewhere in this assemblage or their conflict with other characters.

Unique derived features must be present in the earliest members of each group to be useful in their diagnosis. For example, most rhynchosaurs have multiple rows of cheek teeth, and most archosaurs have a lateral mandibular foramen, but these characters cannot be used to establish the monophyly of the groups because they are not present in the earliest known members as established by other criteria. The most conspicuous apomorphies possessed by the diapsid subgroups are listed in Table 1.

As discussed by Hennig (1981), the earliest and most primitive members of a group may have very few derived characters. One would expect to find some taxa in which the diagnostic characters of the group to which they probably belong are poorly developed if present at all. In the case of the diapsids, there are relatively few adequately known forms for which this is a problem. They will be discussed individually.

Most of the long-recognized groups of early diapsids can be defined readily by the autapomorphies listed in Table 1. This analysis has revealed one major change from previous usage. There are no synapomorphies that unite the genera that Romer (1956, 1966) included in the order Eosuchia. In fact, eosuchians must be divided into several distinct taxa. The groups that Romer (1966) classified as the suborders Choristodera and Thalattosauria are each characterized by autapomorphies that distinguish them clearly from all other major diapsid groups. Their phylogenetic positions remain uncertain. The genera that Romer grouped as the Prolacertiformes share unique derived features with *Protorosaurus*, *Tanystropheus*, and *Tanytrachelos* and are here termed the Protorosauria. Within Romer's Younginiformes, *Palaeagama*, *Paliguana*, and *Saurosternon* share unique derived characters with the modern lepidosaur groups (Carroll, 1975, 1977). *Heleosaurus* shares several derived features with archosaurs, and *Noteosuchus* with rhynchosaurs. The only genera that share a common anatomical pattern with *Youngina*, on which Broom (1914) based the Eosuchia, are *Acerodontosaurus*, *Kenyasaurus*, *Thadeosaurus*, *Hovasaurus*, and *Tangasaurus* (Currie, 1981a,b, 1982; Currie and Carroll, 1984). The term *Eosuchia* is here confined to these genera. Other recent authors have chosen to abandon the term *Eosuchia* because it has been used in reference to many genera that do not share a close common ancestry.

Most major groups of vertebrates show a similar natural division into a

Table 1. Autapomorphies of diapsid subgroups

Araeoscelida
1. Mammillary processes on neural spines of posterior cervical and anterior dorsal vertebrae
2. Conspicuous process for triceps muscle on posterior coracoid
3. Greatly enlarged lateral and distal pubic tubercles

Coelurosauravidae
1. Squamosal frill
2. Trunk ribs ossified in two segments, forming supports for large gliding membrane

Eosuchia (Younginiformes)
1. Medial centrale interposed between lateral centrale and third distal carpal

Squamata
1. Streptostylic quadrate supporting large tympanum

Sphenodontida
1. Acrodont dentition and particular pattern of limited tooth replacement

Nothosauria
1. Suborbital and interpterygoid vacuities closed with pterygoids meeting along midline as far posteriorly as the occipital condyle
2. Large unossified area between the transversely oriented ventral portions of the clavicles and interclavicle and the elongate, posteromedially directed coracoids

Plesiosauria
1. Pectoral and pelvic girdles greatly expanded ventrally
2. Similar, paddle-shaped hind and forelimbs
3. Ilium not attached to pubis

Protorosauria
1. At least seven very elongate cervical vertebrae
2. Cervical ribs extremely long and slender
3. Tympanum probably supported by squamosal and quadrotojugal

Rhynchosauria
1. Medial narial opening
2. Premaxillae forms overhanging beak
3. Ankylothecodont tooth attachment

Archosauria
1. Antorbital opening

Trilophosauridae
1. No lateral temporal opening
2. Loss of teeth in premaxilla and front of dentary
3. Cheek teeth laterally expanded

Choristodera
1. Elongation of snout with nasal bones fused at midline
2. Prefrontals meeting at midline, separating nasals from frontals
3. Temporal area greatly expanded laterally and extending well posterior to the occipital condyle
4. Internal nares extended posteriorly as grooves in the roof of the palate

Order Thalattosauria
1. Dorsal temporal opening much restricted or entirely closed
2. Jaw articulation far behind level of skull table and occipital condyle
3. Premaxillae elongated and reaching frontals; nasals displaced laterally

Placodontia
1. Trunk vertebrae deeply amphicoelus, but with long transverse processes
2. Closure of lateral temporal opening; quadratojugal forming much of cheek

number of large subgroups—e.g., the orders among placental mammals and Devonian placoderms, and the families of modern birds and bony fish. Presumably, this coherence of natural groups is associated with specific relationships to their physical and biological environments and is perpetuated further by genetic and developmental constraints (Carroll, 1986). For analysis, these monophyletic units may be of any size, from a single genus the phylogenetic position of which is in question, up to (in this study) an entire subclass, the Archosauria, the unity of which is clearly evident on the basis of one or more autapomorphies.

The recognition of such monophyletic subunits at an early stage in cladistic analysis is necessary for the identification of strictly homologous characters. Characters or character-states that evolve within a particular monophyletic group cannot be homologous to characters that evolve within a separate monophyletic group, no matter how similar they may be in structure and function.

For example, the tabular and postparietal are missing in all Recent diapsid groups. However, it can be demonstrated that these losses occurred separately at least twice, and perhaps three or more times within distinct monophyletic subgroups, and thus cannot be homologous traits. This procedure demonstrates that many character-states that seem similar have been achieved convergently. Therefore, it is difficult to establish a reliable phylogeny on the basis of a preliminary analysis of character distribution without first establishing the strict homology of the characters.

Determination of All the Derived Character-States Present in the Most Primitive Members of Each of the Monophyletic Subgroups

The *unique* derived characters that are present in the earliest members of each monophyletic group are of no significance in establishing relationships with other groups. However, each group may possess additional derived features that are shared by other groups which can provide evidence of relationships.

It would seem obvious that only the character-states expressed in the earliest members of each group can be used to establish relationships between groups. Characters evolved *within* a group are of no significance in demonstrating relationships with other groups. This guideline has been ignored by many cladists, however, who typically argue that living members of groups provide much more information than do fossils for establishing relationships. For example, Gardiner (1982) in his discussion of tetrapod phylogeny and Rosen et al. (1981) in their attempt to establish close relationships between lungfish and tetrapods make no effort to demonstrate that the characters that they used were present in the early members of the groups which they discuss. The use of call notes in modern lungfish and salamanders by Rosen et al. to demonstrate the affinities of Paleozoic labyrinthodonts and lungfish was notable.

The importance of establishing the earliest members of monophyletic groups was expressed clearly by Hennig (1981:34) "the task of phylogenetic research is to trace the history of 'modern groups' as far back into the past as possible: this can only be done if we assign to each 'modern group' all the fossils that belong to its ancestral line." This raises a very important procedural question: How do we recognize the earliest member of a monophyletic group? Because of the incompleteness of the fossil record, we have no way of being certain that a particular fossil represents the absolute earliest member of a group. Unless the fossil record of both the group in question and the group that is immediately ancestral to it are known in great detail, ever older "earliest" species might be postulated.

Although we cannot necessarily recognize the earliest member of a group, we can recognize the species with the most primitive suite of character-states. Such a species would belong to what Hennig (1981) referred to as a stem-group. This is an assemblage characterized by the possession of some, but not all, the apomorphies of the typical or modern members of a monophyletic assemblage. The earliest adequately known archosaur, *Chasmatosaurus*, and the earliest known squamate, *Paliguana*, for instance, can be recognized by only one or two of the many features that characterize later members of these groups.

The earliest known member of a group is not necessarily the most primitive morphologically, but in general, species that possess a majority of primitive character-states are also among the earliest forms to appear in the fossil record (Butler, 1982). On the other hand, the early appearance of a species should not, by itself, be taken as evidence for the primitive nature of all its expressed character-states. For example, coelurosauravids (Evans, 1982) are among the earliest diapsids, yet their greatly elongate ribs and peculiar elaboration of the squamosal certainly are not primitive character-states. Paul (1982) commented on the generally close correlation between stratigraphic level and phylogenetic position, and Maisey (1984) recently discussed the close correlation among Mesozoic chondrichthyans.

The derived characters present in the most primitive members of each of the monophyletic groups among the diapsids are listed in Appendices II–VII.

RELATIONSHIPS AMONG DIAPSIDS

Possible interrelationships can now be considered on the basis of derived characters that are shared by the most primitive known members of two or more of the monophyletic subgroups. This is, in fact, simply establishing monophyly at a new level. Again, the characters chosen can be tested against other features to determine congruence.

An important procedure is to begin with the most widespread features that define the largest number of included subgroups. All adequately known diapsids from the Upper Permian and later are united by the following derived characters: (1) strengthening of the temporal bars, (2) downgrowth of the parietals beneath the adductor jaw musculature, (3) exclusion of the lacrimal bone from the narial opening, (4) absence of caniniform maxillary teeth, (5) elimination of the primitive separation between the two coracoids, (6) distal condyles of femur level with each other, (7) caudal ribs straight and fused to centra, and (8) median row of gastralia fused at midline.

Several additional features cited by Benton (1985) as characterizing these diapsids are not present in all early members of this assemblage. For example, in the younginoid eosuchians, the quadrate is not emarginated, as he claimed, or notched posteriorly, nor is the stapes slender by comparison with modern diapsids that have an impedance-matching middle ear. Neither is the retroarticular process well developed, as would be expected in animals such as modern lizards in which this area contributes to support of a tympanum. *Chasmatosaurus*, the earliest adequately known archosaur, does not have a slender sigmoidal femur, nor is the proximal head inflected medially.

Araeoscelida

Two families, the Petrolacosauridae and Araeoscelidae, are more primitive in these features and can be recognized as a distinct monophyletic group by the presence of the autapomorphies cited in Appendix II. These families may be placed in a distinct order, the Araeoscelida (Reisz et al., 1984). Benton suggested that all other adequately known diapsids should be included in a separate taxon, the Neodiapsida.

Coelurosauravidae

Five groups of advanced diapsids are recognized by the Late Permian. Most genera belong to two large groups designated by Benton (1985) as the Lepidosauromorpha and the Archosauromorpha. The third is represented by two highly specialized genera, *Weigeltisaurus* from Europe and *Coelurosauravus* [*Daedalosaurus*] from Madagascar (Carroll, 1978; Evans, 1982; Evans and Haubold, 1987). Both are characterized most dramatically by the possession of greatly elongated ribs that almost certainly supported a gliding membrane, as in the modern lizard *Draco*. They lack the synapomorphies that distinguish the lepidosauromorphs and archosauromorphs and retain the following primitive features that are modified by the major groups of advanced diapsids: (1) ribs of atlas vertebra probably retained; (2) large cleithrum; and (3) well-ossified olecranon. Retention of these primitive features probably indicates that the coelurosauravids diverged from the ancestral diapsid lineage prior to the emergence of either the Lepidosauromorpha or the Archosauromorpha.

Lepidosauromorpha

Among the advanced diapsids, the distribution of derived character-states suggests the recognition of two large groups, including all the living genera, the Archosauromorpha and the Lepidosauromorpha. The position of several other groups is more difficult to ascertain. The derived features of these major groups are cited in Appendices III–IV. Benton provided similar character lists, but with little comment on their adaptive significance.

Characters can be used to establish phylogenies without any concern for their biological significance, but an understanding of evolutionary processes is certainly assisted by an appreciation of their function. The different structures of the limbs and girdles in advanced diapsid groups may be associated with divergent patterns of posture and locomotion. Most lepidosauromorphs are relatively small animals whose fossil members are broadly comparable to living lizards and *Sphenodon* in their skeletal anatomy. Presumably, they had a similar sprawling posture and a locomotary pattern based on sinusoidal movement of the trunk. This pattern accentuates that of more primitive tetrapods, but the lepidosauromorphs also are united by specific derived features of the skeleton. The most important of these is the presence of a large sternum (Figs. 2–5), which has specific areas of attachment for the distal ends of the anterior trunk ribs. The basic structure is very similar in modern lizards, *Sphenodon*, and the Permo-Triassic eosuchians (younginiforms). In all, the sternum is a massive median ventral plate, the anterior margins of which articulate with the posterior margin of the scapulocoracoids. Gray (1968) pointed out that the sternum of lizards functions to prevent posterior movement of the scapulocoracoid when the humerus is strongly retracted and forms a surface on which the coracoids can rotate. This rotation moves the glenoid in an extensive horizontal arc that enables the humerus to be extended much farther anteriorly than would be possible from its limited movement within the glenoid. Jenkins and Goslow (1983) provided detailed evidence from X-ray cinematography and electromyography of the range of movement between the coracoids and the sternum in living lizards, and its contribution to locomotion in this group.

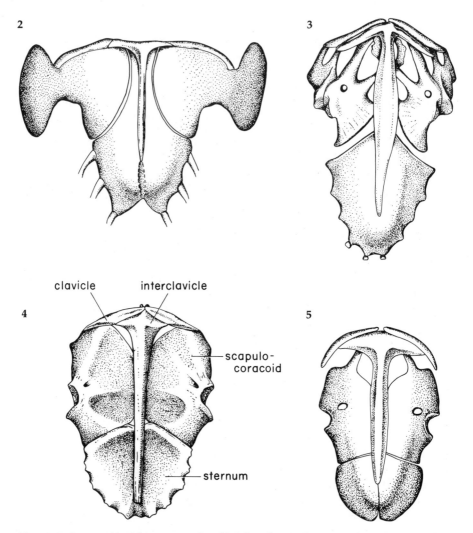

Figs. 2–5. Sterna of lepidosauromorphs. (2) *Sphenodon*. (3) *Iguana*. (4) *Hovasaurus*. (5) *Saurosternum*, attributed to the Paliguanidae. (Modified from Carroll, 1987.)

The term *sternum* has been applied to ventromedial ossifications in the thoracic region in birds, mammals, and archosauromorphs, but the structure and function of these elements differ significantly in each group and there is no evidence for their homology. There certainly is no evidence for the presence of a sternum in early members of the archosauromorph assemblage. Hence, the nature of the sternum in primitive lepidosauromorphs provides very strong evidence for the monophyly of this assemblage.

Possibly associated with the differences in their pattern of locomotion, the transverse processes of the trunk vertebrae are shorter and the rib heads single in lepidosauromorphs, whereas the transverse process of archosauromorphs tend to become more elongate than in primitive diapsids, and the ribs are clearly double headed.

The limbs of primitive lepidosauromorphs do not differ greatly from those of more primitive diapsids. However, the neck region is distinctive in having cervical vertebrae 3–5 noticeably shorter than those of the anterior trunk region.

The presence of accessory vertebral articulations in early lepidosauromorphs was noted by Benton (1985) and Currie (1982). Because these areas of articulation vary in position and their degree of expression increases with size, they are not convenient taxonomic indicators.

Eosuchians (Younginiformes)

Among the lepidosauromorphs, the eosuchians (younginiforms) retain a suite of primitive character-states relative to sphenodontids and squamates that suggest that they might occupy a position ancestral to the modern orders. All adequately known eosuchians possess at least one autapomorphy not present in modern lepidosaurs. The medial centrale is interposed between the lateral centrale and the third distal carpal (Fig. 6). This character seems to be sufficient to establish the monophyly of the known eosuchians, excluding the poorly known genus *Galesphyrus* (Carroll, 1976b), which retains the primitive pattern of the wrist in common with early diapsids, squamates, and sphenodontids. Other characters cited by Benton seem less important in establishing the monophyly of the eosuchians. The large size of the entepicondyle of the humerus is characteristic of tangasaurids but has not been demonstrated clearly in *Youngina*. The shape of the neural spines changes during ontogeny and in relationship to adult size, and thus is not a clear-cut difference that can be used for phylogenetic analysis.

Although primitive characters in themselves do not demonstrate the monophyly of a group, they are, in fact, the most conspicuous aspects of eosuchians. The group retains a complete lower temporal bar, and the quadrate, although exposed posteriorly, is not obviously emarginated for support of a tympanum. The retroarticular process is barely evident. The stapes remains massive and angled posteroventrally, as in primitive amniotes. The structure of the stapes, articular, and quadrate provides no evidence for an impedance-matching function of the middle ear.

Aside from the structure of the carpus, eosuchians have no conspicuous derived features of the skeleton that would preclude them from having been ancestral to sphenodontids and squamates.

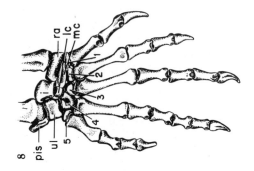

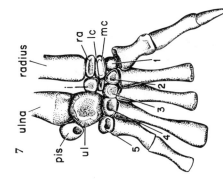

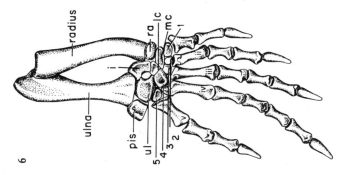

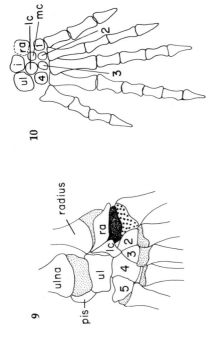

Figs. 6–10. Carpals of diapsids. (6) Lower forelimb of the eosuchian *Thadeosaurus*. (7) Carpus of the putative early lizard *Saurosternon*. (8) Carpus of *Sphenodon*. (9) Carpus of *Iguana*. Epiphyseal ossifications are identified by fine dots. Bone shown by coarse dots might be the epiphysis of the first digit or the first distal carpal. Dark element is the medial centrale. The intermedium is lost. (10) Carpus and manus of the archosauromorph *Protorosaurus*, restored on the basis of illustrations in Meyer (1856). See Appendix VIII for key to abbreviations. (Modified from Carroll, 1987.)

Lepidosauria

Benton (1985) compiled a substantial list of features that unite the Sphenodontida and Squamata within the Lepidosauria, but these are based primarily on the advanced members of these groups. He did not specifically consider the early members, which would be expected to demonstrate the most primitive character-states expressed within each group. Many of the features he cited might have been achieved independently within each group.

Sphenodontida

Benton distinguished sphenodontids on the basis of a large number of derived characters present in a variety of Triassic and Jurassic forms as well as the living genus *Sphenodon*. In fact, primitive character-states of many of these features are observed among the earliest genera, suggesting that the derived condition has been achieved within the group rather than being characteristic of the earliest representatives (Fraser, 1982, 1986). The following list retains the original numbering of Benton (1985):

1. "Lacrimal absent." This is true of all known sphenodontids.
2. "Parietals narrow and reduced to two nearly vertical back-to-back plates with ventro-lateral flanges that contact the supraoccipital." Broad parietals (a primitive condition) are present in the Triassic genera *Planocephalosaurus* and *Polysphenodon*, as well as in *Homeosaurus* from the late Jurassic.
3. "Supratemporal absent." A supratemporal was reported in *Clevosaurus* from the Upper Triassic (Robinson, 1973).
4. "Quadrate not emarginated." No early sphenodontid is known that definitely lacks an emarginated quadrate. In both *Clevosaurus* and *Planocephalosaurus*, the quadrate (perhaps incorporating the quadratojugal) bears a lateral conch that broadly resembles the structure that supports a tympanum in modern lizards. The likelihood of a tympanum being present is further indicated by the presence of a well-developed retroarticular process in these genera. *Sphenodon*, which has a long slender stapes like that of most lizards, might have secondarily lost the impedance-matching function of the middle ear (Wu, 1988).
5. "Teeth acrodont." In all early genera that are definitely identified as sphenodontids, most of the teeth are acrodont, although some genera have been reported to have pleurodont teeth (Fraser, 1986).
6. "One to three fused teeth on the premaxilla which are longer than the maxillary teeth and give a 'beaked' appearance to the skull." *Planocephalosaurus* has five small conical teeth in the premaxilla, closely resembling the pattern in primitive lepidosauromorphs.

7. "Tiny juvenile teeth at the front of the maxilla and dentary." These are evident in many early sphenodontids.
8. "Tooth replacement occurs by addition at the back of the maxilla and dentary." Apparently, this is true of all sphenodontids.
9. "A single row of large teeth on the palatine which are separated from the maxillary teeth by a deep groove." *Planocephalosaurus*, like more primitive diapsids, has several rows of palatal teeth, although the lateral palatine row is the largest.
10. "The dentary teeth fit tightly into the maxilla-palatine groove and the propalinal jaw action polishes the teeth and bone in a uniform way." Tooth occlusion in *Clevosaurus* and *Planocephalosaurus* occurs by vertical jaw movements. These genera do not show evidence for propalinal movement. Vertical tooth occlusion seems to be a feature of most, if not all, early sphenodontids. Propalinal movement is a specialization of the modern genus *Sphenodon*.
11. "No teeth on the palate except for the palatine row." (See number 9.)
12. "Splenial absent." This bone has not been reported in any genera, but it could easily be lost during fossilization. A splenial might have been retained in the closely related family Pleurosauridae (Carroll, 1985).
13. "Broad mandibular symphysis formed entirely by the dentary." Not a clear-cut feature.
14. "Dentary runs well back, forming most of the lateral side of the lower jaw." To this can be added the distinct feature of a high coronoid process of the dentary, supported medially by a well-developed coronoid bone.
15. "Large mandibular foramen bounded by the dentary and surangular." This is not illustrated in Fraser's (1986) reconstruction of *Planocephalosaurus*, but this foramen can be seen in his Figure 2 of Plate 70 of this genus. A notch for this foramen is also apparent in Fraser and Walkden's illustration of *Clevosaurus* (1983).

Several other derived features of early sphenodontids were noted by other authors. In common with most, if not all, other members of the order, the prefrontal of *Planocephalosaurus* extends to the palatine. *Planocephalosaurus* also has a specialized tail-break mechanism, as have other terrestrial sphenodontids. All sphenodonts and early pleurosaurs have a large quadratojugal foramen.

The polarity of another important feature of sphenodontids cannot be established on the basis of evidence currently available. *Sphenodon* has a well-developed lower temporal bar. This feature has long been accepted as primitive for sphenodontids. However, several early forms, including *Clevosaurus* and *Planocephalosaurus*, have an incomplete lower temporal bar. Two possibilities present themselves. The reduction of the lower temporal bar may be a specialization evolved within sphenodontids from a primitive diapsid condition. Alternatively, the open condition might have been retained from their immediate ancestors and the complete bar re-elaborated within some members of the group. Only direct evidence from the immediate ancestors of the group will answer this question with assurance.

In contrast with the condition in lizards, the reduction in the lower temporal bar in early sphenodontids is not associated with mobility of the quadrate, which remains firmly attached to the quadrate ramus of the pterygoid in *Planocephalosaurus* and *Clevosaurus*. Rieppel and Gronowski (1981) associated reduction of the lower temporal bar with elaboration of the lateral portion of the adductor jaw musculature. At least among the pleurosaurs, there seems to be a progressive reduction of the bar (Carroll, 1985). In *Sphenodon*, *Polysphenodon*, and *Brachyrhinodon*, the lower temporal bar is arched laterally away from the lower jaw. This seems an alternate way of accommodating an enlarged adductor musculature, thus suggesting that a complete lower bar was primitive for the group. Whiteside (1986), in contrast, presented other evidence that suggests that the temporal bar was incomplete in the earliest sphenodontids and their immediate ancestors.

Gephyrosauridae

Gephyrosaurus, from the Lower Jurassic of Great Britain, shares a number of derived features with both squamates and sphenodontids, thereby leaving little question of its assignment to the Lepidosauria (Evans, 1980, 1981).

Benton (1985) considered the following features to be indicative of a sister-group relationship of gephyrosaurids with the Squamata: (1) fused parietal and frontal, (2) reduced lower temporal bar, (3) quadrate notched with well-rounded conch, (4) articular fused to the prearticular, (5) all ribs holocephalous, and (6) sacral ribs fused indistinguishably to sacral vertebrae. Three of these features are also characteristic of early sphenodontids—the reduced lower temporal bar, the quadrate notch, and the fusion of the articular to the prearticular. Solid attachment of the sacral ribs to the centra is observed in mature eosuchians and cannot be considered a unique feature of any advanced group of lepidosaurs. Not all the ribs are holocephalous. Those in the cervical region, as in both sphenodontids and lizards, clearly are double headed. The trunk ribs are single headed in all three groups.

The frontals and/or the parietals are fused along the midline in a variety of squamates, but this is not the case in *Paliguana* (see below), which may be the most primitive lizard. The nature of the foot, the presence of a thyroid fenestra, and specialized caudal autonomy are shared with squamates and sphenodontids.

The presence of separate epiphyseal ossifications and a quadrate conch may be common to all lepidosaurs. The absence of a sternal ossification may be attributed to the nature of preservation.

No features of *Gephyrosaurus* are shared uniquely with the early squamates. However, two features—the contact of the prefrontal with the palatine and the elaboration of a row of large palatine denticles—are otherwise unique to the Sphenodontida, suggesting a sister-group relationship with that order. *Gephyrosaurus* may be a relict of an early stage in the evolution of sphenodontids, in which case the stem-group of the order might be characterized by many fewer derived characters than were listed by Benton.

Squamata

Benton (1985) recognized 23 derived characters of the skeleton as being unique to squamates. It is clear from his list that he has drawn most of the features from the modern lizard groups, known no earlier than the late Jurassic, by which time all the modern infraorders had differentiated. Hennig (1981) argued that it is important to trace taxonomic groups back to the earliest members, although the latter may be recognized by only few of the derived characters that distinguish living species.

The Late Permian or Early Triassic genus *Paliguana* is known only from an isolated skull (Carroll, 1977, 1988). It has several important characters that otherwise are unique to the modern lizard groups. These include (1) presence of a streptostylic quadrate, supported dorsally by the squamosal; (2) a distinctive lateral conch of quadrate, similar to the surface that in modern lizards supports a tympanum (early sphenodontids also have a conch, but it is formed partially by a vestige of the quadratojugal, which does not have this position in squamates); and (3) absence of the lower temporal bar. If *Paliguana* is a lizard, as suggested by these shared derived characters, the other cranial features cited by Benton must have evolved within the early squamates.

Saurosternon, known only from a headless skeleton from the Late Permian or Early Triassic of southern Africa, also was assigned to the squamates. The features that suggest this assignment—the small size of the intermedium, possible fenestration of the scapula, and fusion of the atlas neural arch—are not clearly visible in the specimen. The probable elaboration of separate epiphyseal ossification, the very short fifth metatarsal, and the development of a mesotarsal joint are characters shared by both squamates and sphenodontids. *Saurosternon* can be associated with the lepidosaurs with some assurance, although its assignment to the Squamata may be questioned.

If one accepts *Paliguana* as an early squamate and *Saurosternon* as an early lepidosaur, the following derived characters define the Lepidosauria: (1) an impedance-matching middle ear (evidenced by a lateral conch of the quadrate or quadrate-quadratojugal complex, a slender stapes, and a conspicuous retro-articular process), (2) separate epiphyseal ossifications, (3) short fifth metatarsal, and (4) incipiently mesotarsal foot joint.

If the presence of a lower temporal bar and the solid attachment of the quadrate were primitive features of sphenodontids, the conditions seen in *Paliguana* are derived and indicate that lizards and sphenodontids must have diverged by the Late Permian. If this is the case, the following derived features shared by advanced lizards and sphenodontids, but not by *Paliguana* (or *Sauropternon* and *Palaeagama*, if these three genera are closely related), must have been achieved by convergence: (1) postparietal and tabular lost, (2) thyroid fenestration of the pelvis, (3) fusion of the astragalus and calcaneum, (4) loss or fusion of the centrale pes, (5) loss of Distal Tarsals 1 and 5, and (6) hooking of the fifth metatarsal.

On the other hand, if the presence of a complete lower temporal bar in some sphenodontids is not primitive for that group but derived from a pattern like that of squamates, sphenodontids might have evolved from animals such as *Paliguana*. If this were the case, *Paliguana* and other members of the Paliguanidae may belong to a stem- or sister-group of advanced lepidosaurs. If this were true, the characters listed above might have been achieved within a single lineage that includes the immediate common ancestors of the sphenodontids and the modern lizard orders. Whiteside (1986) argued that lizards and sphenodontids might have diverged as late as the Middle Triassic. This problem can only be resolved by the discovery of well-preserved lepidosaurs from the Early Triassic.

Kuehneosauridae

Most authors have placed the kuehneosaurids among the primitive squamates. Benton (1985), in contrast, placed them in Neodiapsida *incertae sedis*, arguing that they share at least seven synapomorphies with archosauromorphs. This disagreement shows the difficulty of establishing relationships on the basis of characters the homology of which is not adequately known. Benton listed the following characters as synapomorphies with archosauromorphs: (1) absence of tabulars, (2) vertebrae nonnotochordal, (3) transverse processes of vertebrae projecting laterally, (4) absence of cleithrum, and (5) absence of entepicondylar foramen. All of these features are also encountered among lepidosauromorphs, although they are not found in the most primitive members of this group. All but the great width of the transverse processes are found in Upper Jurassic squamates. Another feature that is shared by kuehneosaurids

and some archosauromorphs and that is not seen in other lepidosauromorphs is the confluence of the external nares; this feature is known in
rhynchosaurs and crocodiles among the primitive archosauromorphs (as
well as in champsosaurs, of doubtful phylogenetic position). None of
these groups is likely to have a sister-group relationship with kuehneosaurs on the basis of the majority of other skeletal traits. The configuration of the other bones making up the nasal region in rhynchosaurs,
crocodiles, and champsosaurs makes it extremely unlikely that confluent
external narial openings are a homologous character. The great elongation of the transverse processes in kuehneosaurids is certainly related to
the elaboration of the ribs to support a gliding membrane and is no basis
for suggesting homology with the pattern in archosaurs. This condition
also is achieved separately in plesiosaurs (see below). Thus, doubt is cast
on all the characters used by Benton to suggest affinities between kuehneosaurids and archosauromorphs.

The problem of classifying the kuehneosaurids demonstrates how important it is to establish the specific homology of characters before they
can be used as evidence for phylogeny reconstruction. For example, the
tabular, postparietal, and supratemporal may be considered homologous
bones in all tetrapods, but their loss is not necessarily a homologous
process in different lineages. The absence of the tabular and postparietal
in modern squamates, sphenodontids, crocodiles, and kuehneosaurs
can be considered homologous only if that loss occurred in the immediate
common ancestors of all these groups. That can be established only on
the basis of the fossil record of probable common ancestors. Even the
presence of a large number of derived characters in common with other
groups, as in the case of the kuehneosaurids, can lead to contradictory
phylogenetic conclusions, only one of which could possibly be correct.
Clearly, parsimony is of no use in establishing relationships on the basis
of characters whose homology has not been established.

The most important character that might link kuehneosaurids with
lepidosauromorphs, the presence of a sternum, cannot be established in
Icarosaurus because of the position in which the specimen was preserved
(Colbert, 1970). It is doubtful that a sternum would be preserved in the
fissure-filling deposits from which material of *Kuehneosaurus* has been
described. None has been reported in *Planocephalosaurus*, for example, although this animal is unquestionably a sphenodontid. Sterna are known
in the articulated Upper Jurassic specimens of sphenodontids from
Solenhofen. The prefrontal does appear to enter the margin of the upper
temporal opening in *Kuehneosaurus*, as restored by Robinson (1962). As in
other lizards, the dorsal ribs are single headed. One undeniable squamate feature exhibited by *Kuehneosaurus* is the configuration of the quadrate, which suggests both streptostyly and support of a lizardlike tym-

panum. A quadrate of this pattern is known only among lepidosaurs, and never among archosauromorphs.

Sauropterygians

Two major groups of Mesozoic aquatic reptiles, the nothosaurs and plesiosaurs, are readily characterized by autapomorphies (Appendix IV). They can be united in a single large assemblage on the basis of the unique configuration of the shoulder girdle, in which the scapula is superficial to the clavicular blade—a reversal of the usual relationship between these bones. This supports the long-held view that they could be classified in a common group, the Sauropterygia, on the basis of the similar manner of specialization for aquatic locomotion, the presence of a dorsal temporal opening, and an emarginated cheek with a fixed quadrate.

The relationship of sauropterygians with other reptiles has long been controversial. They were once assigned to a separate subclass, the Synaptosauria (together with the placodonts). Kuhn-Schnyder (1962) argued that nothosaurs and plesiosaurs could be derived from primitive diapsids by the loss of the lower temporal bar. This idea was dismissed by Romer (1971) but was elaborated further by Carroll (1981). The problem of resolving these alternate hypotheses is that there are no clear-cut synapomorphies uniting sauropterygians with the major diapsid groups. Without earlier, more primitive fossils, their phylogenetic position cannot be established unequivocally.

Claudiosaurus (Carroll, 1981) from the Upper Permian of Madagascar combines features of primitive lepidosauromorphs and primitive sauropterygians. *Claudiosaurus* resembles eosuchians in a host of primitive features and specifically shares the derived lepidosauromorph features of the entrance of the postfrontal into the margin of the upper temporal opening and short transverse processes with holocephalous trunk ribs. *Claudiosaurus* also can be included within the Lepidosauromorpha on the basis of the presence of a large sternum. This bone is not calcified as in eosuchians and lepidosaurs; instead, it is preserved as an impression between the coracoids and the ventral gastralia in several well-preserved specimens. The carpus of *Claudiosaurus* resembles that of eosuchians, except for its retention of the primitive pattern of contact between the lateral central and the third distal carpal. Except for the elongation of the neck, a common feature in secondarily aquatic reptiles, there are no derived characters in common with any archosauromorphs. *Claudiosaurus* shares with nothosaurs and plesiosaurs the loss of the lower temporal bar and the restriction of the palatal vacuities.

At least some nothosaurs have a quadrate with an emarginated posterior margin, a character shared with lepidosaurs, but they lack the other features by which that group is recognized. The stapes is missing in plesiosaurs (Brown, 1981); there is no evidence for an impedance-matching middle ear in that group. If *Claudiosaurus* is a member of the sister-group of sauropterygians, an impedance-matching ear must be assumed to have evolved separately in nothosaurs and lepidosaurs. Separate epiphyseal ossifications are absent in *Claudiosaurus* and sauropterygians; they are unlikely to have been lost, as must be hypothesized if they shared a common ancestry with lepidosaurs. Sauropterygians, including *Claudiosaurus*, are thought to have diverged from the stem-group of lepidosauromorphs before either eosuchian or lepidosaurs.

Consideration of the sauropterygians reveals a methodological weakness in the procedures discussed here. Concentration on the character-states of the most primitive members of each group is based on the assumption that all relationships can be traced to the earliest members of major groups. This may be a common evolutionary pattern, to judge by the radiation of placental mammals and advanced teleosts in the Late Cretaceous and Early Tertiary and the radiation of placoderms and primitive chondrichthyans in the Late Silurian and Early Devonian. On the other hand, this methodology makes it difficult to recognize successive divergences within a particular assemblage. For example, two major groups of mammalian aquatic carnivores, the odobenoids and phocids, evolved from within separate advanced families of the Order Carnivora, long after the initial radiation of the order (Tedford, 1976). Plesiosaurs and nothosaurs can be placed within a larger monophyletic assemblage, the Sauropterygia, by cladistic analysis, but this does not reveal the specific nature of their relationships. Nothosaurs are characterized by unique derived features of the palate and pectoral girdle, but it is possible (or even probable) that these features have undergone evolutionary modification toward the pattern of plesiosaurs among taxa that retain many nothosaurian characters. This conclusion cannot be established using all nothosaurs for comparison with primitive plesiosaurs, but it can be detected if comparison is made with one particular genus among the nothosaurs, *Pistosaurus* (Sues, 1987a). This genus was long classified as a nothosaur on the basis of general skeletal similarities, but Edinger (1935) noted that the palate was not closed, as in typical nothosaurs, and that the nasal bones were much reduced. The pectoral girdle may be somewhat modified toward the pattern of plesiosaurs but retains definitive features of nothosaurs.

Pistosaurus may be considered a member of the stem-group of plesiosaurs with which it shares the loss of the entepicondylar foramen, the great reduction of the nasal bones, and the incipient development of large, platelike coracoids. All of these features are clearly derived relative to the most primitive nothosaurs, suggesting that *Pistosaurus*, and hence plesiosaurs as a group, evolved from within the nothosaurs, rather than having a typical sister-group relationship (divergent evolution from a common ancestral stock). It can be argued that the open nature of the palate in *Pistosaurus* and plesiosaurs is a primitive, rather than a derived, feature, which implies separation of these groups prior to the elaboration of the definitive nothosaur character states. In contrast, specialization of other aspects of the skull and the pectoral girdle implies divergence of plesiosaurs from a nothosaur pattern, rather than from a common ancestral stock.

If one follows the latter interpretation, nothosaurs would be considered a paraphyletic group in cladistic terminology. With or without the inclusion of *Pistosaurus*, nothosaurs nevertheless are a clearly defined group on the basis of specializations of the appendicular skeleton for aquatic locomotion that are distinct from the pattern of plesiosaurs (Carroll and Gaskill, 1985).

Archosauromorpha

With minor exceptions, those advanced diapsids that are not included in the Lepidosauromorpha can be included in a second large assemblage, the Archosauromorpha. The archosauromorphs embrace the archosaurs (represented during the Upper Permian and Lower Triassic by the proterosuchian thecodontians), rhynchosaurs, protorosaurs, and trilophosaurids (Figs. 11–14). In contrast with lepidosauromorphs, they are generally large and tend toward a more upright posture; the limbs would have moved in a more fore-and-aft direction. This is reflected initially in changes in the structure of the rear limb. All groups of archosauromorphs known in the Upper Permian and Lower Triassic are characterized by a foot structure in which the astragalus and calcaneum articulate with one another so that the feet can face more nearly forward throughout the stride. The fifth distal tarsal is lost, and the fifth metatarsal is inflected medially to articulate with the fourth distal tarsal (Figs. 15–18) (Characters 1–3).

Lepidosaurs also evolved a hooked fifth metatarsal, but in that group it is not known until the late Triassic and is associated with a much different pattern of the proximal tarsals, which become fused with one another. Benton (1985) pointed out that the fifth metatarsal of archosauromorphs lacks the plantar tubercles that characterize that bone in lepidosaurs. Goodrich (1916) noted that turtles also have a hooked fifth metatarsal, but it is clear from the distribution of many other derived traits that the hooking of the bone evolved separately in lepidosauromorphs, archosauromorphs, and Chelonia. Although similar in both structure and function and ultimately derived from the same element in primitive amniotes, the hooked fifth metatarsal should not be considered a homol-

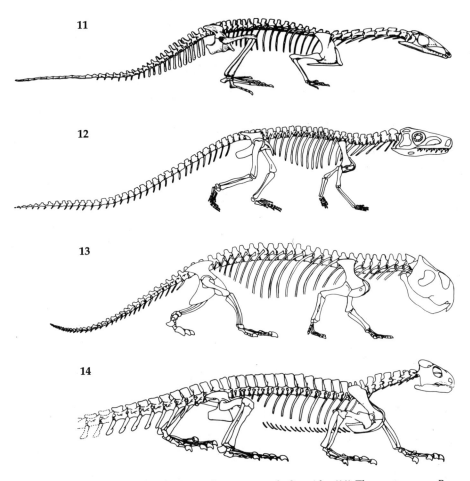

Figs. 11–14. Skeletons of primitive archosauromorph diapsids. (11) The protorosaur *Prolacerta*, approximately 1 m long (modified from Gow, 1975). (12) The archosaur *Euparkeria*, approximately 0.5 m long (modified from Ewer, 1965). (13) The rhynchosaur *Paradapedon*, approximately 1.5 m long (modified from Chatterjee, 1974). (14) *Trilophosaurus*, approximately 2 m long (modified from Gregory, 1945).

ogous structure in these three groups. Specialization of this bone also may have proceeded separately in squamates and sphenodontids.

Because of the basically different pattern of the skull and/or dentitions among early archosaurs, rhynchosaurs, protorosaurs, and trilophosaurids, it may be difficult to accept these features of the tarsus as truly synapomorphous. Despite the near identity of the tarsus in the three groups, it is possible that this pattern developed convergently. This hypothesis seems less likely, however, if we consider the character transformation within each of the groups. Although the earliest and otherwise

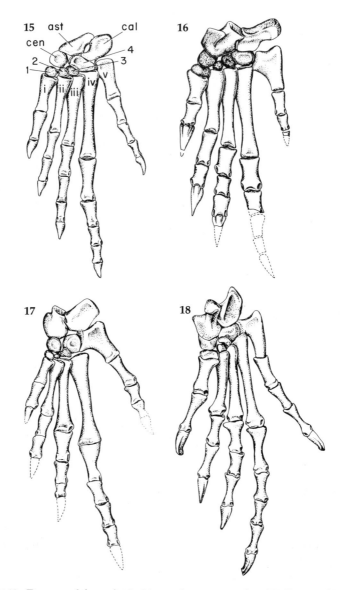

Figs. 15–18. Tarsus and foot of primitive archosauromorphs. (15) Tarsus of *Protorosaurus* (restored on the basis of illustrations in Meyer, 1856). (16) The primitive archosaur *Chasmatosaurus* (modified from Cruickshank, 1972). (17) The early rhynchosaur *Noteosuchus* (modified from Carroll, 1976a). (18) *Trilophosaurus* (modified from Gregory, 1945).

384

most primitive member of each of these groups exhibits a common pattern, later members of all adequately known groups show progressive divergence. Little more than a change in proportions and a loss of distal elements occur among the protorosaurs (Wild, 1974). The rhynchosaurs show a progressive integration of the centrale with the more proximal tarsals, and a partial fusion into a single unit (Carroll, 1976a). The archosaurs show a variety of derived patterns, but all could have evolved from that of the earliest genus (Brinkman, 1981).

Benton (1983, 1985) listed many other synapomorphies of early archosauromorphs. If one includes only the derived characters of the most primitive members of all the groups, the list continues as follows: (4) presence of a lateral tuber on the calcaneum (only incipiently developed in *Protorosaurus* from the Upper Permian), (5) loss of the tabular, (6) stapes without a foramen, (7) vertebrae not notochordal, (8) transverse process on dorsal vertebrae extends beyond level of zygapophyses, (9) cleithrum absent, (10) no entepicondylar foramen, and (11) loss of pisiform. Items 2, 3, 5, 6, 7, 9, and 10 also occur within the lepidosauromorph lineage but evolve there long after these features appear in the early archosauromorphs.

Benton (1985) listed the absence of a foramen in the carpus between the ulnare and intermedium as characteristic of archosauromorphs. However, a foramen between two carpal bones is retained in *Noteosuchus*, *Trilophosaurus*, and *Protorosaurus*. In the fossils of these genera, the elements are disarticulated; thus, it is not certain which bones are represented, but a perforating foramen definitely is retained. The surrounding bones were probably in a pattern little different from that of the eosuchian *Acerosodontosaurus* (Currie, 1980). The carpus is poorly ossified in *Chasmatosaurus* (the earliest adequately known archosaur), and none of the elements can be identified by their shape.

An item not included at this level by Benton, but which is probably common to all archosauromorphs, is the more extensive ossification of the braincase (Character 12). The basioccipital and basisphenoid are contiguous, whereas they are separated by an unossified gap in primitive diapsids, and the basisphenoid and prootic are united anterolateral to the dorsal sellae.

Benton suggested that the trilophosaurids, rhynchosaurs, archosaurs, and protorosaurs (Prolacertiformes) are progressively more derived groups within the archosauromorph assemblage. He considered trilophosaurids to be the sister-group of other archosauromorphs on the basis of the primitive configuration of the premaxilla and quadratojugal. Most of the cranial sutures were fused in the specimens examined by Gregory (1945) in his comprehensive description of *Trilophosaurus*. In material examined by Parks (1969), the premaxilla extends behind the narial opening to meet the nasal in the manner of other archosauromorphs. The configuration of the quadratojugal is equivocal. The postfrontal enters the margin of the upper temporal opening, a feature considered characteristic of lepidosauromorphs, but this also occurs in some early rynchosaurs.

Benton recognized a sister-group relationship between the protorosaurs and archosaurs because of the possession of the following features: (1) long snout and narrow skull, (2) nasal longer than frontal, (3) posttemporal fenestra small or absent, (4) recurved teeth, (5) extensive participation of the parasphenoid/basisphenoid in the side wall of the braincase, and (6) long, thin, tapering cervical ribs with two or three heads and an anterior dorsal process. None of these characters is convincing. The first two features, both aspects of skull proportion, do not greatly differ from those of *Youngina* and might be primitive for what Benton referred to as neodiapsids.

The posttemporal fossae (Character 3, above) in *Prolacerta*, the earliest member of these groups in which the occiput can be reconstructed (Camp, 1945; Robinson, 1962; Gow, 1975) are not significantly smaller than those of eosuchians and early lepidosaurs. The posttemporal fossae are small in *Chasmatosaurus* and other early archosaurs, but this feature cannot be used to unite this group with protorosaurs.

Recurved teeth (Character 4, above) are a common feature of carnivorous reptiles and are very strongly developed in *Heleosaurus* (Carroll, 1976c), which might be close to the base of the archosauromorph assemblage. Otherwise, the dentition is highly specialized in rhynchosaurs and trilophosaurids; this suggests that the condition in archosaurs and protorosaurs may be primitive for the Archosauromorpha, rather than a synapomorphy of these two groups. The nature of the braincase is not known in the Upper Permian genus *Protorosaurus*. The differences between the structure of the braincase in archosauromorphs and lepidosauromorphs is more significant than that between primitive members of the archosauromorph groups and is not considered to be an effective character for differentiation taxa at this level.

The ribs (Character 6, above) are distinctive in protorosaurs, but the configuration of the heads of the anterior ribs in the early archosaur *Chasmatosaurus* are quite similar to those of the lepidosauromorph *Claudiosaurus* and primitive nothosaurs (Carroll and Gaskill, 1985), champsosaurs, and thalattosaurs (see below). In all these groups, the configuration of the ribs may be associated with the great length of the cervical vertebrae and may not be indicative of close taxonomic affinities.

If *Chasmatosaurus* is accepted as the most primitive archosaur, the only adequately documented apomorphy of this group is the presence of an antorbital fenestra (Charig and Sues, 1976). The configuration of the orbit in the shape of an inverted triangle also occurs in the early genera *Euparkeria* and *Erythrosuchus* but might not be expressed clearly in smaller, more primitive genera. Benton listed the possession of a fifth trochanter on the femur, but this does not apply to *Chasmatosaurus*. The relatively small number of synapomorphies shared between *Chasmatosaurus* and later archosaurs is characteristic of members of a stem-group.

Chasmatosaurus is specialized in having a down-turned premaxilla. Benton cited this as a character shared with *Prolacerta*, but it is not so figured in any restorations of the latter genus, nor does this feature occur in the earlier protorosaur, *Protorosaurus*.

The haemal arches are long in both *Prolacerta* and *Chasmatosaurus*, and expanded distally. *Protorosaurus*, however, has haemal arches of normal proportions. Close affinities between *Chasmatosaurus* and *Prolacerta* to the exclusion of other protorosaurs would require that *Chasmatosaurus* had undergone several significant evolutionary reversals, including the redevelopment of a lower temporal bar and shortening of the length of the cervical vertebrae.

Benton suggested the alliance of *Malerisaurus* (Chatterjee, 1980, 1986) with the Protorosauria on the basis of the elongated cervical vertebrae and the presumed loss of the temporal bar. The nature of the lower temporal bar is uncertain, but the quadratojugal is illustrated as a large platelike bone, in sharp contrast with that of all adequately known protorosaurs. The cervical vertebrae, although somewhat elongated, actually resemble those of *Trilophosaurus*, rather than those of *Protorosaurus*. The centra are sharply angled, which would have resulted in a permanently arched, elevated neck. Chatterjee (1980) described the fifth metatarsal as not hooked, but the head is considerably expanded and may have functioned as a hooked element. *Malerisaurus* does not fit clearly with any of the better known groups of archosauromorphs and its specific affinities remain uncertain.

There is ample evidence for the union of the four major groups of archosauromorphs within a single, monophyletic assemblage, but much less evidence for their specific interrelationships. None of the groups is obviously close to the point of origin of any of the others. Most of the characters evident in the earliest members of these groups are autapomorphies, suggestive of rapid and marked divergence. A few characters can be used to support special affinities among these groups, but their significance is uncertain. Primitive archosaurs and protorosaurs have a median postparietal, rather than paired bones. This condition is unique among diapsids and may indicate that archosaurs and protorosaurs are related more closely to one another than either is to rhynchosaurs and trilophosaurids. Primitive archosaurs and protorosaurs also retain teeth on the transverse flange on the pterygoid. In both rhynchosaurs and

Trilophosaurus, the postfrontal enters the margin of the upper temporal opening. This feature is common to lepidosauromorphs, but the taxonomic significance of its appearance in the archosauromorphs is not known.

Heleosaurus, from the Upper Permian of southern Africa, was suggested as an archosaur ancestor by Carroll (1976c) on the basis of the presence of dermal armor of a pattern vaguely similar to that of thecodonts, and the nature of the teeth (laterally compressed, recurved, and serrated). Other features of archosauromorphs, as opposed to lepidosauromorphs, are the presence of six elongate cervical vertebrae and the absence of a sternum, which would almost certainly be evident in an eosuchian preserved in this manner.

Unfortunately, most of the definitive features of archosauromorphs cannot be determined. Neither the carpus or tarsus is present, nor is the dorsal portion of the skull, which would reveal the presence of an antorbital fenestra and dorsal process of the quadratojugal behind the ventral temporal opening. *Heleosaurus* is a plausible member of the sister-group of later archosauromorphs, but its phylogenetic position is too uncertain for it to be used to establish polarity of character transformation in that group.

Mesozoic Diapsids Not Related to Either the Lepidosauromorpha or the Archosauromorpha

All living diapsids and the majority of Mesozoic forms can be classified among either the Archosauromorpha or the Lepidosauromorpha. A few Mesozoic groups cannot be assigned at present to either taxon. They may have evolved directly from primitive stem diapsids or have diverged so greatly from one of the two major groups that their correct affinities cannot be established without additional fossils.

The Choristodera, Thalattosauria, and Placodontia all lack the key features by which either lepidosauromorphs or archosauromorphs are recognized. None shows either the specialization of the tarsus common to archosauromorphs or the presence of a sternum that is characteristic of lepidosauromorphs. Because both of these features are associated specifically with effective terrestrial locomotion, it is conceivable that they have been lost in these groups, all of which are secondarily aquatic. However, in the absence of other evidence to support their affinities with either of these groups, it seems more parsimonious to assume that these features are primitively absent. If this is the case, all of these groups may have evolved separately from the stem diapsids. Early members of all three groups also lack thyroid fenestration of the pelvis, indicating that they must have diverged from the base of the derived groups, if not from stem diapsids.

Choristodera

Champsosaurs (order Choristodera) were relatively common forms in the Upper Cretaceous and Lower Tertiary of North America (Erickson, 1972) and Europe (Sigogneau-Russell and Russell, 1978). Incomplete remains have also been reported from the Lower Cretaceous of Mongolia (Sigogneau-Russell and Efimov, 1984) and the Jurassic of Europe and North America (Evans, 1990). Most champsosaurs are similar in size and proportions to crocodiles and apparently had similarly semiaquatic habits. The Lower Cretaceous fossils contribute little to our understanding of their affinities. Aside from their autapomorphies, other derived traits of champsosaurs resemble those of both archosauromorphs and lepidosauromorphs (Appendix VII). Champsosaurs, however, lack any of the key features that define either group.

In addition to the derived features, it is important to consider some primitive features that indicate derivation from a relative primitive level of diapsid evolution. Champsosaurs lack the thyroid fenestra and retain denticles on the palate, including those on the transverse flange of the pterygoid, which are lost in all other diapsid groups before the Cretaceous. There is no retroarticular process, and the dermal elements of the shoulder girdle are massive and primitive. The radius resembles that of eosuchians, but the humerus is similar in shape to that of protorosaurs and lacks an enclosed ectepicondylar foramen in most genera. Most trunk ribs are holocephalous, but the anterior ribs are partially separated. Presumably this group has evolved separately from other diapsids since the Late Permian or Early Triassic.

If only the most primitive members of other major groups are compared, Cretaceous and Tertiary champsosaurs share most derived features with the archosauromorphs: (1) thecodont implantation of marginal teeth, (2) more than five cervical vertebrate, (3) loss of notochordal canal of vertebrate, (4) loss of trunk intercentra, (5) loss of cleithrum, (6) loss of entepicondylar foramen, (7) loss of pisiform, (8) three fused sacral vertebrae, and (9) hooked fifth metatarsal. On the other hand, all of these characters are also encountered among one or more advanced members of the lepidosauromorph assemblage (specifically, advanced squamates) except for the fusion of the sacral vertebrae.

Several features of champsosaurs are shared with other aquatic forms: (1) thecodont implantation of teeth, (2) long neck, (3) lack of fusion between centrum and neural arch, (4) reduced ossification of the carpals and tarsals, (5) pachyostotic ribs, and (6) short epipodials.

Champsosaurs demonstrate the difficulty of trying to establish relationships on the basis of particular isolated traits, rather than considering the organism as a whole. In the case of this group, it should be admitted that they are highly specialized diapsids with no obvious features that unite them with any particular ancestral group. The only way to ascertain their relationship with more certainty is to discover fossils of intermediate morphology that may link them with some particular group of more primitive diapsids.

Further evidence of the affinities of the Choristodera has recently been provided by the discovery of fossils from the Middle Jurassic of England belonging to the genus *Cteniogenys* (Evans, 1990). Remains of this genus are completely disarticulated and represent animals much smaller than the better known Cretaceous genera, but share with them 19 derived characters. *Cteniogenys* exhibits 10 derived characters in common with primitive archosauromorphs, but several of the derived characters shared by protorosaurs, rhynchosaurs, and archosaurs retain a more plesiomorphic condition, suggesting that the Choristodera were the first group to diverge from the base of the archosauromorph radiation.

Thalattosauria

The Thalattosauria encompasses an assemblage of secondarily aquatic genera from the Middle and Upper Triassic. They are united by striking specializations of the skull (Appendix VII) (Figs. 19–22). Three families are recognized, each represented by one or two genera: Thalattosauridae from the Upper Triassic of western North America (Merriam, 1905), Claraziidae (Peyer, 1936), and Askeptosauridae (Kuhn, 1952) from the Middle Triassic of Switzerland. The families are clearly distinct from one another and none is uniformly more primitive or close to the pattern expected of a common ancestor, although *Askeptosaurus* retains more primitive characters than either *Clarazia* or *Thalattosaurus*, which Rieppel (1987) considered to form the more derived sister-group.

All cranial features can be derived from those of either *Prolacerta* or *Youngina*, but the postcranial skeleton has character-states that are more primitive than those of early members of either the Archosauromorpha or Lepidosauromorpha. There is no evidence of archosauromorph specialization of the tarsus or fifth metatarsal, nor is there evidence for the presence of a lepidosauromorph sternum. The transverse processes of the trunk vertebrae are not as reduced as in early lepidosauromorphs, nor is the maxilla excluded from the margin of the external nares as in most archosauromorphs. *Clarazia* is primitive in retaining a solid, platelike pelvis, but a thyroid fenestra is developed in *Askeptosaurus*.

Placodontia

Placodonts have long been associated with the aquatic nothosaurs and plesiosaurs, on the basis of vaguely similar specializations toward an aquatic way of life and the presence of a conspicuous dorsal temporal opening. In a recent discussion of the skull of *Placodus*, Sues (1987b)

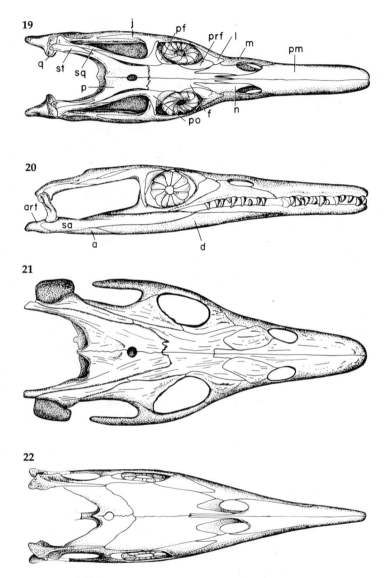

Fig. 19–22. Skulls of thalattosaurs. (19) Dorsal view of the skull of *Askeptosaurus* (modified from Kuhn-Schnyder, 1952). (20) Lateral view of *Askeptosaurus*. (21) Dorsal view of the skull of *Clarazia* (modified from Rieppel, 1987). (22) Dorsal view of the skull of *Thalattosaurus* (modified from Merriam, 1905).

pointed out that there are no specific shared derived characters that unite placodonts with sauropterygians. Rather, he placed them as Diapsida *incertae sedis*.

The most complete knowledge of the group is provided by *Placodus*. This is a relatively primitive genus within the order but already highly specialized relative to other diapsids (Drevermann, 1933). Some more primitive features are evident in *Helveticosaurus*, but this genus is not well known and is divergent in other features, notably the presence of more than 40 presacral vertebrae; thus, it is difficult to justify its use as an example of primitive placodonts (Peyer, 1955). A striking feature that unites *Helveticosaurus* with the better-known placodonts is the configuration of the vertebrae, which are unique among diapsids. As in some archosaurs and plesiosaurs, the transverse processes are greatly elongated, but they accompany centra that are conspicuously amphicoelous and notochordal. The pedicles of the arches are very high.

Helveticosaurus also resembles *Placodus* in possessing very long, spatulate anterior teeth, but it definitely lacks the large, flattened cheek teeth that otherwise characterize most placodonts. The shoulder girdle is unlike those of nothosaurs and plesiosaurs in the small size and oval configuration of the coracoids. In *Helveticosaurus*, the interclavicle is primitive in retaining a long, robust stem. Peyer (1955) reconstructed the relationships of the clavicle and interclavicle according to the pattern of *Placodus*, in which the interclavicle is largely superficial to the clavicular blades, but this cannot be unequivocally established from the specimen. The interclavicle has clearly demarcated recesses for the clavicular blades, as in primitive amniotes. Unfortunately, it is not possible to establish whether the bone is preserved in dorsal or ventral view. If in dorsal view, *Helveticosaurus* shows a derived condition, as in *Placodus* and nothosaurs, but, if the bone is seen in ventral view, the pattern resembles that of primitive reptiles (confirmed by Rieppel, 1989:133).

Placodonts do not have the reversed relationship of the base of the scapula and clavicular blade that characterizes nothosaurs and plesiosaurs. As illustrated by *Placodus*, placodonts have a number of derived features shared by both lepidosauromorphs and archosauromorphs, but placodonts cannot be associated convincingly with either group. The possession of numerous primitive character states suggests that placodonts might have diverged separately from the stem diapsids. They possess neither the specializations of the foot that are characteristic of archosauromorphs, nor any evidence of the sternum of lepidosauromorphs. There is little development of a thyroid fenestration, and the opisthotic is not suturally attached to the cheek. As in lepidosauromorphs, the postfrontal enters the margin of the upper temporal opening, but this also occurs in early rhynchosaurs.

It is conceivable that placodonts diverged from the lepidosauromorphs at about the level of *Claudiosaurus*, with which they share the restriction of the suborbital fenestra, and the extension of the pterygoids to the midline, closing the interpterygoid vacuities. The palate differs from that of nothosaurs in the expansion of the palatine at the expense of the pterygoid. The absence of a sternum occurs in other aquatic groups and is not strong evidence for affinities with sauropterygians.

Rieppel (1989) recently reviewed the anatomy and taxonomic position of *Helveticosaurus*. He concluded that there was no strong evidence of close affinity with the placodonts. Depending on how characters were analyzed, *Helveticosaurus* shares approximately the same number of derived states with lepidosauromorphs and archosauromorphs. Rieppel considered that classification of *Helveticosaurus* within the Archosauromorpha should be considered tentative for the time being. He concluded by stating: "The question must again be raised whether this rather high degree of character incongruence is due to an as yet unsatisfactory characterization of the two major subgroups of diapsid reptiles, or whether it does in fact reflect a high degree of convergence in early diapsid reptiles."

DISCUSSION

The relationships discussed in this paper are summarized in a cladogram (Fig. 23) keyed to the characters that support the affinities of the various groups. Despite the relatively complete knowledge of the skeleton in representatives of most of the groups, many relationships remain incompletely resolved. In particular, the affinities of the Choristodera, Thalattosauria, and Placodontia, and the interrelationships of the several archosauromorph groups cannot be convincingly established. In these cases, we lack sufficient knowledge of early members of these groups which might show characters lost in the descendants that would permit them to be placed in a specific phylogenetic position. As noted by Hennig (1981), the earliest members of derived groups may be differentiated by only rather trivial features that may be lost in their descendants.

The remaining diapsids may be grouped in three categories—the stem diapsids, including the Araeoscelida and coelurosauravids; the Archosauromorpha; and the Lepidosauromorpha. These conclusions do not differ greatly from those reached by Benton (1985), but they are accomplished by a much different approach to character analysis.

Perhaps the most important conclusion reached by this study is the apparent frequency of convergence. If the monophyletic groups have been identified correctly and the polarity of character transformations interpreted correctly, a great number of similar derived character-states

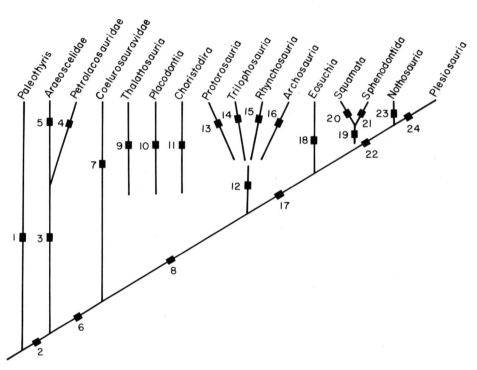

Fig. 23. Cladogram showing hypothesized relationships among major diapsid groups. Derived characters are as follows. (1) Fusion of atlas centrum and axis intercentrum. (2) Dorsal and lateral temporal openings and suborbital fenestra. (3) Six cervical vertebrae, 2–6 much longer than trunk centra; mammillary processes on neural spines; conspicuous triceps process on coracoid; enlarged lateral and distal public tubercles; and epipodials nearly equal in length to propodials. (4) Conspicuous ischiadic notch. (5) Nine cervials, 29 presacral vertebrae; fusion of axial intercentrum to atlantal centrum; cheek teeth enlarged; and ectepicondylar foramen. (6) Strengthening of temporal bars; downgrowth of the parietal beneath the adductor jaw musculature; exclusion of the lacrimal bone from the narial opening; absence of caniniform maxillary teeth; elimination of the primitive separation between the two coracoids; acetabulum rounded, without distinct supra-acetabular buttress; tibial and fibular condyles of femur in same plane; caudal ribs fused to centra; medial rows of gastralia fused at midline. (7) Trunk ribs ossified in two segments, forming support for gliding membrane; squamosal frill; trunk vertebrae elongate; and loss of dorsal intercentra. (8) Ribs of atlas vertebra lost (exceptions exist); cleithrum reduced; olecranon reduced; and caudal ribs extend straight laterally. (9) Dorsal temporal opening much restricted; premaxillae elongated and reaching frontals, nasals displaced laterally; posterior margin of skull table deeply emarginated; and occiput lying well forward relative to quadrate suspension. (10) Vertebrae with elongate transverse processes, but deeply amphicoelous centra; loss of lateral temporal opening; quadratojugal forms much of cheek; anterior teeth spatulate and procumbent; quadrate fitting into a socket formed by squamosal and quadratojugal; and prefrontal and postfrontal separate frontal from orbital margin. (11) Elongation of snout with nasal bones fused at midline; prefrontals meeting at midline separating nasals from frontals; temporal area greatly expanded laterally and extending well posterior to the occipital condyle; and internal nares extended posteriorly as grooves in the roof of the palate. (12) Astragalus and calcaneum articulating with one another; fifth distal tarsal lost; fifth metatarsal hooked; presence of a lateral tuber on the calcaneum; loss of tabular; stapes without a foramen; vertebrae not notochordal; transverse

394

evolved in separate monophyletic groups. This can be demonstrated most readily in well-established groups such as the Lepidosauromorpha and the Archosauromorpha. If genera such as the younginoids are recognized correctly as primitive lepidosauromorphs, and *Protorosaurus* and *Chasmatosaurus* are primitive archosauromorphs, then a host of characters recognized in later members of both groups must have evolved convergently. A conservative list includes the following: (1) closure of upper temporal opening, (2) loss of tabular, (3) loss of postparietal, (4) loss of supratemporal, (5) loss of pineal opening, (6) reduction or loss of lacrimal, (7) loss of teeth from transverse flange of pterygoid, (8) fusion of bones along the midline of the skull (premaxillae, nasals, frontals, parietals, and/or vomers), (9) reduction or loss of lower temporal bar, (10) extension of opisthotic to cheek, (11) development of thecodont implantation, (12) ossification of area of pleurosphenoid or laterosphenoid, (13) loss of stapedial foramen, (14) development of an impedance-matching middle ear, (15) elaboration of retroarticular process, (16) development of procoely, (17) loss of notochordal perforation, (18) elongation of neck, (19) loss of trunk intercentra, (20) elongation of transverse processes of trunk vertebrae, (21) loss of entepicondylar foramen, (22) development of thyroid fenestration, (23) loss or fusion of centrale in pes, (24) loss of Distal Tarsals 1 and 5, (25) hooking of fifth metatarsal, (26) reduction of phalangeal count, and (27) elaboration of dermal armor.

---

process on dorsal vertebrae extending beyond level of zygapophyses; cleithrum absent; no entepicondylar foramen; and loss of pisiform. (13) Seven very elongate cervical vertebrae; cervical ribs very thin; and loss of lower temporal bar. (14) No lateral temporal opening; loss of teeth in premaxilla and front of dentary; cheek teeth laterally expanded; extensive fusion of dermal bones in the adult; and cervical vertebrae angled and procoelous. (15) Ankylothecodont tooth implantation; median narial opening; and overhanging premaxillae. (16) Antorbital fenestrae. (17) Cervical Centra 3–5 shorter than trunk centra; reduced length of transverse processes of trunk vertebrae; trunk ribs holocephalous; and large, ossified sternum with which coracoids articulate. (18) Medial centrale interposed between lateral centrale and Distal Carpal 3. (19) Epiphyses elaborated as specialized joint surfaces; incipient mesotarsal joint; and lateral conch of quadrate supporting tympanum. (20) Quadrate streptostylic; loss of lower temporal bar; loss of ventral process of squamosal; and intermedium much reduced. (21) Acrodont attachment of at least some of the marginal teeth; enlargement of the lateral row of palatine teeth; precise occlusion of marginal teeth; tooth replacement of acrodont teeth occurs by addition at back of maxilla and dentary; dentary characterized by its great posterior extension, coronoid process and mandibular foramen; prefrontal forms solid articulation with dorsal surface of palatine; very large quadratojugal foramen; and prominent posterior process of ischium. (22) Ventral portion of scapula superficial to blade of clavicle; loss of lower temporal bar, without development of streptostyly; and reduction of palatal fenestra. (23) Suborbital and interpterygoid vacuities closed, with pterygoids meeting along midline as far posteriorly as the occipital condyle; and large unossified area between the transversely oriented ventral portions of the clavicle and the elongate, posteromedially directed coracoids. (24) Loss of stapes; pectoral and pelvic girdles greatly expanded ventrally; and hind and forelimbs with similar paddle shape.

These characters are all considered to exemplify convergence, because it would be difficult, if not impossible, to determine whether or not they evolved from the same or a distinct ancestral pattern if we knew only the descendant forms. Other characters might be cited as examples of convergence, but they are not really comparable in the context of the remainder of the anatomy. The medial narial openings of kuehneosaurids and crocodiles clearly are different structurally, as well as having evolved separately. The structures that are called sterna in dinosaurs, birds, mammals, and lepidosauromorphs are functionally and structurally distinct, as well as being nonhomologous, and it may be questioned as to whether they should be considered as examples of convergence at any level.

Many other cases of convergence appear to have occurred among the early diapsids, but their demonstration rests on hypotheses of relationships that are not so clearly established. If *Saurosternon* is a primitive squamate, many of the features of the postcranial skeleton that characterize Triassic and Jurassic lizards and sphenodontids must have evolved separately—development of thyroid fenestration, fusion of centrale, fusion of astragalus and calcaneum, loss of Distal Tarsals 1 and 5, and hooking of fifth metatarsal.

Whatever the specific relationships of the placodonts, Thalattosauria, and Choristodera, the presence of some primitive features suggest that they evolved from near the base of the radiation of advanced diapsids. This implies that many of their derived similarities also were achieved convergently.

Accepting that the conclusions regarding the monophyly of groups and the polarity of character change have been correctly assessed, we find that there are relatively few cases of reversals compared with the frequency of convergence. If the pisiform is actually lost in the ancestors of archosaurs, then it must have re-evolved in the earliest crocodiles. This bone may simply have been poorly ossified in early archosaurs, but there is no evidence for it elsewhere in this assemblage. Early crocodiles exhibit an antorbital fenestra, but it is lost in later forms, showing a reversal from the primitive archosaur condition. In contrast with most other early diapsids, *Araeoscelis* and *Trilophosaurus* lack a lateral temporal opening. If these genera are diapsids, as indicated by all other features of the skeleton, this character has reversed to a pre-diapsid pattern. Temporal openings also are closed among several groups of dinosaurs and within the Squamata, although this is accompanied by so many other changes in the skull that it may not be proper to think of these changes as reversals. If sauropterygians evolved from primitive lepidosauromorphs, the sternum must have been lost in the process—a reversal to a more primitive character state. Advanced archosaur groups loose the hooking of the fifth metatarsal that evolved in early archosauromorphs. The impedance-matching ear is lost in many modern lepidosauromorph groups.

It is difficult to quantify objectively the relative number of changes that can be referred to as convergences and reversals, but convergence appears to be much more common. There is no justification for presuming that convergences and reversals are equally likely events, as has been assumed by many authors who make use of numerical procedures for establishing phylogenies.

What is clearly evident is that convergence is a *much* more common phenomenon than is the origin of strictly unique features. If this is the case, statistical arguments, based on parsimony, are very unlikely to provide correct phylogenies in groups with long evolutionary histories.

The problem of phylogenetic analysis is not how to establish possible relationships from a host of characters whose possibility of convergence or uniqueness is difficult to establish, but how to recognize the specific homology of each character. Clearly this cannot be done statistically, but can be achieved only by a knowledge of the actual evolutionary history of the species in question. This can be established only on the basis of the fossil record.

SUMMARY AND CONCLUSIONS

Establishing relationships among diapsid reptiles provides an informative mode for extending cladistic methodology to groups with a good fossil record. Cladistic methods were elaborated initially to deal with groups without an adequate fossil record, but their basis in fundamental evolutionary processes should make them even more effective when applied to groups in which there is more direct evidence of the specific homology of traits and the polarity of character transformation.

Phylogenetic analysis of the early diapsids was undertaken employing the following procedures: (1) establishment of the monophyletic nature of the entire group; (2) establishment of the polarity of all characters that vary within the group; (3) recognition of subgroups on the basis of unique apomorphies; and (4) determination of the derived character-states present in the most primitive members of each of the monophyletic subgroups. Relationships were recognized on the basis of derived characters shared by the most primitive known members of each group.

This procedure differs significantly from that practiced by most cladists in that it emphasizes the significance of the earliest members of each group, rather than their living representatives, for establishing relationships. The earliest members of any group would be expected to share the greatest number of uniquely derived characters with their sister-group, and to show the least amount of convergence. The longer the period of time since the divergence of two groups, the greater the possibility for the loss of original synapomorphies and the accumulation of characters achieved by convergence.

Recognition of subgroups on the basis of unique derived characters at an early stage in analysis demonstrated that many derived characters have evolved convergently in two or more groups. Of 142 character transformations that can be systematically treated, 43 or approximately 30%, appear to have occurred uniquely in a single lineage. Ninety-nine, or approximately 70%, exhibit convergence. Approximately 20 additional characters exhibit less clear-cut or less well-established patterns of change (see Appendix I). These particular numbers are not significant, for they would change depending on how characters are defined and how exhaustively character changes are followed among more derived archosauromorphs and lepidosauromorphs. The general pattern is clear, however. Convergence, as evidenced by the fossil record, is a very important phenomenon among diapsids. It is not possible to establish reliable relationships on the basis of character distribution without knowing the homologous nature of the characters in question. Homology can be established on the basis of the fossil record, by the discovery of the character in the common ancestors of the groups in question. Homology of particular characters cannot be judged accurately on the basis of the distribution of other characters.

On the basis of analysis of character-states in the most primitive known members of each major subgroup of early diapsids, this assemblage may be divided into two major groups with living representatives and several entirely extinct groups. The archosauromorphs include the following subgroups: archosaurs, rhynchosaurs, protorosaurs, and trilophosaurids. Lepidosauromorphs include eosuchians, lepidosaurs (paliguanids, kuehneosaurids, advanced lizards and snakes, sphenodontids), and the sauropterygians (nothosaurs and plesiosaurs). The Araeoscelida is the sister-group of all other diapsids, and the coelurosauravids may belong to the sister-group of the common ancestors of archosauromorphs and lepidosauromorphs. The positions of the Choristodera, Thalattosauria, and Placodontia remain uncertain. They may have evolved separately from the stem diapsids.

Acknowledgments This reanalysis of the relationships among early diapsids depends primarily on recent studies of fossil material. We wish to acknowledge the help of the following individuals in making specimens available for study: J. P. Lehman, and Daniel Goujet (Institut de Paléontologie, Paris), Arthur Cruickshank and Cris Gow (Bernard Price Institute for Palaeontological Research, Johannesburg, South Africa), Thomas Barry and Michael Cluver (South African Museum), Rupert Wild (Staatliches

Museum für Naturkunde in Stuttgart), Hans Reiber and Olivier Rieppel (Paläontologisches Institut und Museum der Universität Zürich), Claude Germain (Port Manech, Brittany, France), and Hans-Peter Schultze (Museum of Natural History, The University of Kansas). Drs. Schultze, Rieppel, and Reisz, and Mr. Laurin provided information based on unpublished studies of early diapsids. Drs. Sues, Schultze, Rieppel, and Walsh read early drafts of the manuscript and offered many helpful suggestions. We thank Pamela Gaskill for her careful drafting of the illustrations. This work was supported by grants from the Natural Sciences and Engineering Research Council of Canada.

APPENDIX I. Distribution of character-states in diapsids. Primitive character-states shown by *Petrolacosaurus* (Reisz, 1981) and *Apsisaurus* (Laurin, 1991) (indicated by *). The protorothyrid *Paleothyris* is the primary basis of out-group comparison to establish polarity. Ar = Araeoscelida; Co = Coelurosauridae, Eo = Eosuchia; Sq = Squamata; Sph = Sphenodontida; No = Nothosauria; Ple = Plesiosauria; Pro = Protorosauria; Rhy = Rhynchosauria; Arch = Archosauria; Tri = Trilophosauridae; Ch = Choristodera; Th = Thalattosauria; Pla = Placondontia.

Primitive diapsid condition	Derived condition	Groups in which derived state is observed
SKULL		
1. Upper temporal opening	1a. Loss of upper temporal opening	Some Sq, some Arch, some Th, some Pla
2. Lateral temporal opening	2a. Loss of lateral temporal opening	*Araeoscelis*, Tri, some Arch
3. Suborbital fenestra	3a. Loss of suborbital fenestra	No
4. Temporal bars flattened in plane of skull	4a. Temporal bars rounded and thickened	All groups other than Ar
5. Parietal flat, without ventral processes medial to the upper temporal opening	5a. Ventral process medial to upper temporal opening	All groups other than Ar
6. Lower temporal bar complete	6a. Reduction or loss of lower	Co, Sq, some Sph, No, Ple, Pro
7. Retention of all bones present in the skull of primitive amniotes, including paired post-parietals, tabulars, and supratemporals	7a. Fusion of postparietal	Primitive Pro, primitive Arch
	7b. Loss of postparietal	Advanced Sq, Sph, No, Ple, advanced Pro, Rhy, advanced Arch, Tri, Ch, Th, Pla
	7c. Loss of tabular	Advanced Sq, Sph, No, Ple, Pro, Arch, Tri, Ch, Th, Pla

The appendixes have not been updated since they were originally submitted and corrected in 1988.

APPENDIX I.—*continued*

Primitive diapsid condition	Derived condition	Groups in which derived state is observed
	7d. Loss of supratemporal	Some Sq, advanced Sph, No, Ple, advanced Arch, Tri, Ch, Th, Pla
8. No midline fusion of premaxillae, nasals, frontals, parietals, or vomers	8a. Fusion of premaxillae	Advanced squamates, some Arch
	8b. Fusion of nasal	Some Sq, Ch, Pla
	8c. Loss of nasals	Some Sq, Ple
	8d. Fusion of frontals	Some Sq, some Sph, some No, some Arch
	8e. Fusion of parietals	Some Sq, some Sph, advanced Pro, some Rhy, some Arch
	8f. Fusion of vomers	Some Sq, some Ple, some Arch, Pla
9. Conspicuous pineal opening midway in length of parietal	9a. Closure of pineal opening	Some Sq, some Ple, some Rhy, some Arch, Tri, Ch
	9b. Pineal opening between parietal and frontal	Some Sq
10. Postfrontal does not form margin of upper temporal opening	10a. Postfrontal enters margin of upper temporal opening	Eo, Sq, Sph, No, Ple, primitive Rhy, Tri, Pla
11. Postfrontal and postorbital separate ossification	11a. Fusion of postfrontal and postorbital	Some Sq, some Ple, some Arch, some Ch
12. Orbital and narial openings round	12a. Changes in many ways that can not be systematically tabulated	
13. Narial opening paired, anterior position	13a. Medial narial opening	Some Sq, some Arch, Ch
	13b. Narial opening posterior	Some Sq, some Sph, No, Ple, some Arch, Th, Pla
14. No antorbital opening	14a. Antorbital opening	Arch
15. No external mandibular fenestra	15a. External mandibular fenestra	Sph, most Arch
16. Premaxilla not deflected ventrally	16a. Premaxilla deflected	*Chasmatosaurus*, Rhy
17. Premaxilla excluded from posterior margin of external nares by maxilla	17a. Premaxilla forms posterior margin of external nares	? some Sq, Pro, Rhy, Arch, Tri, Ch
18. Quadrate with slender dorsal process, not exposed posteriorly	18a. Quadrate with strong dorsal process, exposed posteriorly	Sq, Sph, No, Ple, Pro, Rhy, Arch, Tri, Th, Pla

APPENDIX I.—*continued*

Primitive diapsid condition	Derived condition	Groups in which derived state is observed
19. Quadrate firmly supported by pterygoid, squamosal, and quadratojugal	19a. Quadrate movable, streptostylic	Sq, ?Pro, ?Th
20. Posterior margin of cheek not embayed for support of tympanum	20a. Posterior margin of cheek embayed for support of tympanum	Sq, Sph, No, Pro, Rhy, Arch, Tri, Th, Pla
21. Occipital condyle in same transverse plane as jaw articulation	21a. Jaw articulation anterior to occipital condyle	Sq, Sph, some Arch
	21b. Jaw articulation posterior to occipital condyle	Some Sq, advanced No, Ple, some Rhy, some Arch, Tri, Ch, Th, Pla
22. Lacrimal bone extending from orbit to external nares	22a. Lacrimal bone not reaching narial opening	All groups other than Ar
	22b. Lacrimal bone lost	Some Sq, Sph, some No, Ple, some Arch, Pla
23. Quadratojugal essentially horizontal, not extending behind lateral temporal opening	23a. Quadratojugal extending behind lateral temporal opening	Pro, Rhy, Arch, ?Ch
	23b. Quadratojugal lost	Sq, Sph, most No, Ple, advanced Pro
24. Prefrontal does not extend to palatine	24a. Prefrontal reaches palatine	Sph, some Arch, Pla
25. Single row of conical marginal teeth	25a. Multiple rows of marginal teeth	Advanced Rhy, some Arch
	25b. Change of tooth shape	Too diverse to document
26. Tooth implantation subpleurodont	26a. Pleurodont	Advanced Sq
	26b. Acrodont	Some Sq, Sph
	26c. Thecodont implantation	Some Sq, No, Ple, Pro, Rhy, Arch, Tri, Ch, Th, Pla
	26d. Ankylothecodont	Rhy
	26e. Loss of teeth from premaxilla and/or end of dentary	Advanced Rhy, some Arch, Tri, some Pla
27. Tooth replacement	27a. Loss of regular replacement	Some Sq, most Sph
28. Approximately 5 teeth in premaxilla	28a. Increase or decrease	Too varied to tabulate
29. Approximately 35 teeth in maxilla	29a. Increase or decrease	Too varied to tabulate
30. Maxillary teeth of positions 6 and 7 enlarged canines	30a. No conspicuous canine teeth	All groups other than Ar

APPENDIX I.—*continued*

Primitive diapsid condition	Derived condition	Groups in which derived state is observed
31. Three rows of denticles radiating from area of basicranial articulation	31a. Loss of tooth row on transverse flange of pterygoid	Sq, Sph, No, Ple, advanced Pro, Rhy, advanced Arch, Tri, Ch, Th, Pla
	31b. Loss of all palatal teeth	Some Sq, No, Ple, Rhy, some Arch, Tri
	31c. Elaboration of teeth on palatine	Sph, Pla (in quite a different way)
32. Movable basicranial articulation	32a. Loss of basicranial articulation	No, Ple, some Arch
33. Epipterygoid a narrow vertical rod	33a. Epipterygoid widened to form a longitudinal plate	Some No
	33b. Epipterygoid lost	Some Sq, some Arch (not consistently described)
34. Unossified gap between basioccipital and basisphenoid	34a. Basioccipital contiguous with basisphenoid	Pro, Rhy, Arch, Tri, Ch
35. Basisphenoid not in contact with prootic	35a. Basisphenoid reaches prootic	Pro, Rhy, Arch, Tri (not consistently described)
36. Area of pleurosphenoid (or laterosphenoid) not ossified	36a. Pleurosphenoid ossified	Advanced Sq
	36b. Laterosphenoid ossified	Advanced Arch
37. Long tapering cultiform process	37a. Cultiform process short or eliminated	Some Sq, No, Ple, advanced Rhy, advanced Arch, Tri, Ch, Pla
38. No vidian canal	38a. Vidian canal formed between basisphenoid and parasphenoid	Advanced Sq, advanced Arch
39. No fenestra rotundum	39a. Fenestra rotundum	Advanced Sq, advanced Arch
40. Ossified portion of paroccipital process not reaching cheek or top of quadrate	40a. Ossified portion of opisthotic reaching cheek or top of quadrate	Advanced Sq, Sph, advanced No, Ple, Pro, Rhy, Arch, Tri, Ch, Th, Pla
41. Large perforate stapes with dorsal process and posteroventrally directed stem	41a. Stapes directed laterally	Sq, Sph, No, Pro, Rhy, Arch, Tri
	41b. Stapes without foramen	Some Sq, No, Pro, Rhy, Arch, Tri
	41c. Loss of stapes	Ple, ?Ch, ?Pla
LOWER JAW 42. Retention of two splenial bones and one coronoid	42a. Loss of one splenial	Most advanced diapsids (condition not systematically documented)

APPENDIX I.—*continued*

Primitive diapsid condition	Derived condition	Groups in which derived state is observed
	42b. Loss of both splenials	Some Sq, most if not all Sph
43. No conspicuous coronoid process	43a. Conspicuous coronoid process	Advanced Sq, Sph, some Arch, Pla
44. No retroarticular process	44a. Conspicuous retroarticular	Sq, Sph, No, Ple, Pro, Rhy, Arch, Tri
45. Ceratobranchial I the only ossified element of hyoid apparatus, a long, slender rod	45a. Part of corpus ossified	Crocodiles (condition not documented in most groups)
POSTCRANIAL SKELETON		
46. 25 or 26 presacral vertebrae	46a. Increased number	Some Ar, ?Co, some Sq, some Sph, No, Ple, some Arch, Th, some Pla
47. All vertebral centra amphicoelous and notochordal	47a. Reduced amphicoely to platycoely	Some Sq, No, Ple, Pro, Rhy, Arch, Tri, Ch, Th
	47b. Procoely	Advanced Sq, advanced Arch
48. Intercentra retained throughout trunk	48a. Loss of trunk intercentra	Most Sq, some Sph, No, Ple, Rhy, advanced Arch, Ch, Th, Pla
49. *Five cervical vertebrae	49a. Increased number of cervical vertebrae	Ar, some Sq, some Sph, No, Ple, Pro, Rhy, Arch, Tri, Ch, Th, Pla
50. *Cervical vertebrae 3–5 approximately length of the anterior trunk vertebrae	50a. Cervical vertebrae 3–5 shorter than anterior trunk vertebrae	Eo, primitive Sq, Sph, some Rhy, some Arch
	50b. Cervical vertebrae longer than anterior trunk vertebrae	Ar, some Sq, Pro, Arch, Tri
51. Intercentra of atlas and axis separate atlas pleurocentrum from ventral margin of ventral column	51a. Retained in primitive members of most derived groups. Pattern varied in advanced genera but not sufficiently well known for consistent analysis	
52. Proatlas paired	52a. Proatlas lost	Most advanced groups, but not crocodiles and *Sphenodon*. Not consistently described
53. Atlas arch paired and not fused to centrum	53a. Arch fused at midline	Sq
	53b. Arch fused to centrum	Some Arch
	53c. Arch lost	Some Sq, some Arch

APPENDIX I.—*continued*

Primitive diapsid condition	Derived condition	Groups in which derived state is observed
54. Axis intercentrum suturally attached, but not fused to atlas pleurocentrum	54a. Intercentrum fused to atlas pleurocentrum	Ar, some Arch (pattern not consistently described)
55. *20 or 21 trunk vertebrae	55a. Increased number	Some Ar, some Sq, some Sph, No, Ple, some Arch, Th, some Pla
	55b. Decreased number	*Petrolacosaurus, Claudiosaurus,* Pro, Tri, some Arch
56. *Neural spines slender and roughly rectangular in lateral view. Arch and centrum solidly attached	56a. Mammalary processes	Ar
	56b. A variety of other shapes the distribution of which cannot be tabulated conveniently	
	56c. Arches and centra not solidly attached	Some Sph, No, Ple, Ch, Pla
57. Transverse processes of trunk vertebrae extending slightly beyond level of zygaphophyses	57a. Shorter transverse processes	Eo, most Sq, Sph
	57b. Elongate transverse processes	Kuehneosauridae, some No, Ple, Pro, Rhy, Arch, Tri, Pla, some Sq
58. Two sacral vertebrae	58a. Reduction in number of sacral vertebrae	Some Sq
	58b. Additional sacral vertebrae	No, Ple, most Arch, Ch
59. 60–65 caudal vertebrae	59a. Increase or decrease in number	Not consistent enough to tabulate
60. No caudal autotomy	60a. Caudal autotomy	Most advanced Sq, most Sph
61. Haemal arches beginning behind third caudal centrum, slender and progressively reduced in size posteriorly	61a. Change in position of first haemal arch and change in shape	Too varied to tabulate conveniently
62. Ribs articulating with all presacral, sacral, and first 11–12 caudal vertebrae	62a. Loss of anterior cervical ribs	At least atlas rib lost in most groups other than Ar and Co, and some Arch
	62b. Fusion of cervical ribs to transverse processes	Advanced Arch, Pla, some Sq
	62c. Fusion of posterior trunk ribs to vertebrae, or loss	Some Sq, some Arch
	62d. Fusion of caudal ribs to transverse processes	All diapsids other than Ar

APPENDIX I.—*continued*

Primitive diapsid condition	Derived condition	Groups in which derived state is observed
63. Cervical ribs clearly double-headed	63a. Single-headed cervical ribs	Sq, advanced Ple
	63b. Addition of anterior process	No, Ple, Pro, Arch
64. Heads of trunk ribs separated by narrow constriction	64a. Single-headed trunk ribs	Eo, Sq, Sph, No, Ple, Th
	64b. Clearly double-headed trunk ribs	Some Arch
65. Sacral ribs not fused to either centrum or ilium, widely expanded distally	65a. Sacral ribs fused to centrum and/or ilium	Some Arch
	65b. Sacral ribs not expanded distally	Some Sq, some Sph, No, Ple
66. Caudal ribs sharply recurved to the rear	66a. Straight caudal ribs	All diapsids other than Ar; Co somewhat transitional
67. Cleithrum, clavicle, and interclavicle retain pattern of primitive amniotes	67a. Cleithrum lost	Sq, Sph, Not, Ple, Pro, Rhy, Arch, Tri, Ch, Th, Pla
	67b. Blade of clavicle medial to scapula	No, Ple
	67c. Interclavicle entirely superficial to blade of clavicles	No, Ple, Pla
	67d. Interclavicle absent	Some Sq, advanced Arch
68. No trace of sternal elements	68a. Large medial sternum with which coracoids articulate	Eo, Sq, Sph
	68b. Median rod	Crocodiles
	68c. Paired plates	Some dinosaurs
	68d. Median plate for origin of flight muscles	Pterosaurs
69. Both anterior and posterior coracoids	69a. A single coracoid	All diapsids except Ar
	69b. Scapula and coracoid slow to co-ossify or not co-ossified	Advanced Sq, some Sph, advanced Pro, Arch, No, Ple, Ch, Pla
70. *No conspicuous process for triceps muscle	70a. Conspicuous triceps process	Ar
71. Screw-shaped glenoid	71a. Open glenoid	Sq, Sph, No, Ple, Pro, Rhy, Arch, Tri, Ch, Th, Pla
72. Supraglenoid and coracoid foramina	72a. Loss of supraglenoid foramen	Pro, some Sq, Rhy, No, Ple, Ch, Th, Pla
	72b. Loss of coracoid foramen	Some Arch, some Sq
73. No separate epiphyseal ossification	73a. Epiphyseal ossification	Sq, Sph

APPENDIX I.—*continued*

Primitive diapsid condition	Derived condition	Groups in which derived state is observed
74. Humerus with entepicondylar foramen	74a. Loss of entepicondylar foramen	Advanced Sq, some No, Ple, Pro, Rhy, Arch, Tri, Ch, Th, Pla
75. Distinct supinator process	75a. Loss of supinator process	*Araeoscelis,* Co, Eo, Sq, Sph, No, Ple, Rhy, Arch, Tri, Th, Pla (not Ch)
76. Groove for radial nerve not enclosed by bone to form ectepicondylar foramen	76a. Enclosure of ectepicondylar foramen	*Araeoscelis,* Co, Eo, Sq, Sph, No
	76b. Loss of ectepicondylar foramen	Some Sq, No, Ple, Arch, Th, Pla
77. Ulna with clearly defined olecranon and sigmoid notch	77a. Reduction of olecranon and sigmoid notch	Eo, some Sq, some Sph, No, Ple, some Arch, Ch, Th, Pla
78. *Shaft of ulna and radius subequal in length, and shorter than humerus	78a. Ulna and radius subequal to humerus, or longer	Ar, some Sq, Pro, some Arch
	78b. Shaft of radius longer than that of ulna	Eo
79. All primitive elements of amniote carpus retained	79a. Loss of elements	Some Sq, some Sph, No, Ple, Pro, Rhy, Arch, Tri, Ch, Th, Pla
80. Perforating foramen surrounded by intermedium, ulnare, and lateral centrale	80a. Loss of perforating foramen	? Co, Sq, Sph, No, Ple, some Pro, some Rhy, Arch, Ch, Th, Pla
81. Medial centrale articulating with third distal tarsal	81a. Medial centrale interposed between lateral centrale and third distal carpal	Eo
82. Proximal ends of metacarpals slightly overlapping	82a. Changes too varied to categorize systematically	
83. Length of metacarpals increase from I–IV; Metacarpal V is intermediate in length between I and II	83a. Changes too varied to categorize systematically	
84. Phalangeal count 2-3-4-5-4	84a. Reduced phalangeal count	Some Sq, some No, some Pro, ?Rhy, Arch, ?Ch, Th
	84b. Hyperphalangy	Some Sq, some No, Ple
85. Unguals sharply pointed and laterally compressed	85a. Reduction and modification of unguals too varied to categorize systematically	

APPENDIX I.—*continued*

Primitive diapsid condition	Derived condition	Groups in which derived state is observed
86. Puboischiadic plate plate complete, without thyroid fenestra	86a. Thyroid fenestra	Advanced Sq, Sph, No, Ple, advanced Pro, advanced Rhy, advanced Arch, advanced Th
87. Narrow, posteriorly directed iliac blade	87a. Expansion of iliac blade	Co, advanced Rhy, advanced Arch, Tri
	87b. Reduction of iliac blade	Some Sq, No, Ple
88. Ilium in contact with pubis	88a. Ilium loses contact with pubis	Ple, some Arch
89. Supra-acetabular buttress retained	89a. Supra-acetabular buttress lost	All diapsids except Ar, and some No
90. Pubis perforated by obturator foramen	90a. Loss of obturator foramen	Some Arch, Ple
91. Femur with straight shaft and terminal head	91a. Sigmoidal shaft	Eo, Sq, Sph, Pro, Rhy, most Arch, Tri
	91b. Head displaced medially	Advanced Arch
92. Deep intertrochanteric fossa, internal trochanter, but no fourth trochanter	92a. Reduction and/or loss of intertrochanteric fossa and internal trochanter	Some Sq, Not, Ple, advanced Arch, advanced Pla
	92b. Fourth trochanter	Advanced Arch
93. Fibular condyle extending beyond tibial condyle	93a. Condyles in same plane	All diapsids other than Ar
94. *Tibia and fibular substantially shorter than femur	94a. Tibia and fibula as long or longer than femur	Ar, some Sq, Pro, some Arch
95. Astragalus and calcaneum meet along a straight, flat articulating surface that is indented by perforating foramen	95a. Fusion of astragalus and calcaneum	Advanced Sq, Sph
	95b. Specialized articulation between astragalus and calcaneum	Pro, Rhy, Arch, Tri
96. Separate centrale	96a. Fusion of centrale to astragalus	Advanced Sq, Sph
	96b. Loss of centrale	No, Ple, some Pro, advanced Arch, Ch, Th
97. Five distal tarsals	97a. Loss of distal tarsals	Most Sq, Sph, No, Ple, Pro, Rhy, Arch, Tri, Ch, Th, Pla
98. Proximal heads of metatarsals overlapping	98a. Changes too varied to categorize systematically	
99. Length of metatarsals increases from I to IV	99a. Changes too varied to categorize systematically	

APPENDIX I.—*continued*

Primitive diapsid condition	Derived condition	Groups in which derived state is observed
100. Metatarsal V approximately the length of II, not markedly shorter and not hooked	100a. Metatarsal V markedly shorter than II, but not hooked	Paliguanids, some Arch
	100b. Metatarsal V hooked	Advanced Sq, Sph, Pro, Rhy, Arch, Tri, Ch
101. Phalangeal count 2-3-4-5-4	101a. Phalangeal count reduced	Some Sq, some Not, some Arch
	101b. Hyperphalangy	Some No, Ple, some Sq
102. Fifth digit divergent	102a. Fifth digit absent	Some Sq, some Arch
103. Gastralia in chevron pattern, elements not fused at midline	103a. Gastralia fused at midline	All diapsids except Ar
	103b. Gastralia lost	Sq, Arch derivatives
104. No dermal armor	104a. Dermal armor	Some Eo, some Sq, some Arch, some Pla

APPENDIX II. Derived Characters of Early Diapsid Families

A. Synapomorphies Shared by Petrolacosauridae and Araeoscelidae

1. At least six cervical vertebrae; axis and more posterior cervicals much longer than trunk vertebrae
2. Mammillary processes on neural spines of posterior cervical and anterior dorsal vertebrae
3. Conspicuous process for triceps muscle on posterior coracoid
4. Greatly enlarged lateral and distal pubic tubercles
5. Epipodials nearly equal in length to propodials

B. Apomorphy of Petrolacosauridae

1. Conspicuous ischiadic notch

C. Apomorphies of Araeoscelidae

1. Nine cervical, 29 presacral vertebrae
2. Fusion of axial intercentrum to atlantal centrum
3. Secondary closure of lateral temporal opening

    4. Cheek teeth enlarged and reduced in number

    5. Enclosure of radial nerve to form ectepicondylar foramen

D. Synapomorphies of All Diapsids Other Than the Araeoscelida (these characters have been systematically omitted from the subsequent lists)

    1. Strengthening of the temporal bars

    2. Downgrowth of the parietals beneath the adductor jaw musculature

    3. Exclusion of the lacrimal bone from the narial opening

    4. Absence of caniniform maxillary teeth

    5. Elimination of the primitive separation between the two coracoids

    6. Acetabulum rounded, without distinct supra-acetabular buttress

    7. Tibial and fibular condyles of femur in same plane

    8. Caudal ribs fused to centra

    9. Medial rows of gastralia fused at midline

E. Synapomorphies of Coelurosauravidae

    1. Trunk ribs ossified in two segments, forming support for large gliding membrane

    2. Squamosal frill

    3. Trunk vertebrae elongate

    4. Loss of dorsal intercentra

APPENDIX III. Derived Features of Lepidosaurs and Eosuchians, Based on Character-States in the Most Plesiomorphic Genera

AA = autapomorphy of group; M = characters shared by advanced lizard groups and sphenodontids, but not by paliguanids; SL = synapomorphy of lepidosauromorphs; SS = synapomorphies shared by modern squamates and paliguanids; and SSS = synapomorphies shared by squamates, paliguanids, and sphenodontids.

Eosuchians (based on *Youngina* and *Thadeosaurus*)

    AA   Medial centrale interposed between lateral centrale and Distal Carpal 3

    AA   Reduced ossification of olecranon; radius may exceed shaft of ulna in length

    SL   Postfrontal entering margin of upper temporal opening

    SL   Cervical Centra 3–5 shorter than trunk centra

    SL   Reduced length of transverse processes of trunk vertebrae

    SL   Trunk ribs holocephalous

    SL   Large, ossified sternum with which coracoids articulate

    SL   Quadrate exposed posteriorly (no evidence of support of tympanum)

Paliguanidae (based on *Paliguana, Saurosternon,* and
Palaeagama)

SL	Postfrontal entering margin of upper temporal opening
SL	Cervical Centra 3–5 shorter than trunk centra
SL	Reduced length of transverse processes of trunk vertebrae
SL	Trunk ribs holocephalous
SL	Large ossified sternum with which coracoids articulate
SSS	Epiphyses elaborated as specialized joint surfaces
SSS	Incipient mesotarsal joint
SSS	Fifth metatarsal much shorter than fourth
SS	Quadrate streptostylic
SS	Loss of lower temporal bar
SS	Loss of ventral process of squamosal
SS	Lateral conch of quadrate supporting tympanum
SS	Intermedium much reduced

Sphenodontida (based on the most primitive pattern
of Upper Triassic genera)

AA	Acrodont attachment of at least some of the marginal teeth
AA	Enlargement of the lateral row of palatine teeth
AA	Precise occlusion of marginal teeth
AA	Tiny juvenile teeth at the front of the maxilla and dentary
AA	Tooth replacement of acrodont teeth occurring by addition at the back of the maxilla and dentary
AA	Lacrimal absent
AA	Dentary characterized by its great posterior extension, coronoid process, and mandibular foramen
AA	Prefrontal forming solid articulation with dorsal surface of palatine (condition in crocodiles and placodonts not considered homologous)
AA	Very large quadratojugal foramen
AA	Prominent posterior process of ischium
SL	Postfrontal enters margin of upper temporal opening
SL	Cervical Centra 3–5 shorter than trunk centra
SL	Reduced length of transverse processes of trunk vertebrae
SL	Trunk ribs holocephalous
SL	Large ossified sternum with which coracoids articulate (not known in Triassic sphenodontids, but preserved in Upper Jurassic genera)
SSS	Separate epiphyseal ossifications
SSS	Impedance-matching middle ear
M	Caudal centra specialized for autotomy
M	Thyroid fenestration of pelvis
M	Fusion of astragalus and calcaneum, probably incorporating centrale
M	Loss of Distal Tarsals 1 and 5
M	Fifth metatarsal short and hooked

Kuehneosauridae

> AA   Greatly elongated trunk ribs
> AA   Median narial opening (not considered homologous to condition in rhynchosaurs)
> AA   Scapular blade narrow
> SS   Quadrate streptostylic and emarginated posteriorly
> SL   Postfrontal entering margin of upper temporal opening
> SL   Cervical Centra 3–5 shorter than trunk centra
> SL   Holocephalous trunk ribs

Characters shared with some archosauromorphs and some lepidosauromorphs

1. Loss of tabular, supratemporal, and postparietal
2. Lower temporal bar lost
3. Conspicuous retroarticular process
4. Elongate transverse processes of trunk vertebrae
5. Vertebrae non-notochordal
6. Loss of entepicondylar foramen
7. Reduced ossification of carpus
8. Thyroid fenestration

Characters of advanced lizards

1. Pineal opening between frontals and parietal

APPENDIX IV. Derived Characteristics of Lizard Groups Known from the Upper Jurassic to the Present

(S) = character shared with sphenodontids; (P) = synapomorphies of paliguanids and modern lizard groups.

Skull

1. Loss of lower temporal bar and ventral ramus of squamosal (P)
2. Absence of fixed connection between quadrate and pterygoid (P)
3. Dorsal end of quadrate enlarged to articulate with squamosal (streptostyly) (P)
4. Quadrate embayed posteriorly for reception of tympanum (P)
5. Light, rodlike stapes (S)
6. Paroccipital process extending to top of quadrate (S)
7. Tabular and postparietal lost (S)
8. Supratemporal expanded distally, lateral to distal end of parietal
9. Transverse hinge between parietals and frontals; one or both bones fused at midline
10. Fenestra rotunda present
11. Vidian canal enclosed by parasphenoid and basisphenoid
12. Coronoid bone extended as long coronoid process (S)
13. Dentition pleurodont or acrodont, exceptionally thecodont (not subpleurodont)
14. Denticles lost from transverse flange of pterygoid (S)

Postcranial Skeleton

1. Except for some members of the Gekkota, centra procoelous; trunk intercentra lost
2. Cervical intercentra specialized as hypopophyses
3. Atlas arch fused medially (?P)
4. Tail-break mechanism, unless secondarily lost (S)
5. Scapulocoracoid with two to four fenestrae (?P)
6. Scapula and coracoid slow to co-ossify, compared with eosuchians (S)
7. Large sternum, the anterior edge of which forms a surface for rotation of the coracoid (S) (P)
8. Bony epiphyses of limb bones form specialized, articulating surfaces (S) (P)
9. Entepicondylar foramen of humerus lost
10. Intermedium of carpus reduced or lost (P)
11. Puboischiadic plate with large thyroid fenestra; no bony connection of pubis and ischium below fenestra (S)
12. Pubis outturned dorsally (S)
13. Astragalus (including centrale) fused to calcaneum (S)
14. Mesotarsal joint formed between proximal tarsals and fourth distal tarsal (S) (P)
15. Usually only one distal tarsal retained
16. Fifth metatarsal much shorter than fourth (S) (P)
17. Fifth metatarsal hooked (S)
18. Gastralia lost

APPENDIX V. Derived Features of Sauropterygians, Based on Character-States in the Most Plesiomorphic Genera

AA = unique derived feature of this group; SA = derived features shared with one or more groups of advanced diapsids; SAL = derived feature shared with some archosauromorphs and some lepidosauromorphs; SL = derived feature shared with lepidosauromorphs; SS = derived feature shared specifically with nothosaurs and plesiosaurs; and SSS = synapomorphy uniquely shared by nothosaurs and plesiosaurs.

Derived Characters of Nothosaurs

AA   Suborbital and interpterygoid vacuities closed; pterygoids meet along midline as far posteriorly as the occipital condyle

AA   Large unossified area between the transversely oriented ventral portions of the clavicle and the elongate, posteromedially directed coracoids

SA   Loss of postparietal, tabular, and supratemporal

SA   Quadrate embayed posteriorly

SA   Stapes reduced to a narrow rod
SA   Loss of palatal dentition
SA   Loss of transverse flange of pterygoid
SA   Thecodont implantation of marginal dentition
SA   Posterior position of external nares
SA   Loss of lower temporal bar without development of streptostyly
SA   Reduction or loss of lacrimal
SA   Conspicuous retroarticular process
SA   Loss of trunk intercentra
SA   Centra not notochordal
SA   Arches and centra not co-ossified
SA   Neck elongate (at least 17 cervical vertebrae)
SA   Three or more sacral vertebrae
SA   Cervical ribs with anterior process
SA   Trunk ribs single-headed
SA   Ribs variably pachyostotic
SA   Sacral ribs not expanded distally
SA   Articulating surfaces of limbs, girdles, carpals, tarsals poorly os-
          sified
SA   Loss of cleithrum
SSS   Ventral portion of scapula superficial to blade of clavicle
SA   Interclavicle superficial to clavicle
SA   Epipodials reduced
SA   Thyroid fenestra
SA   Blade of ilium much reduced

Derived Characters of Plesiosaurs (based on
superficial observation of Lower Jurassic genera,
never described in detail, and Brown, 1981, on
Upper Jurassic genera)

AA   Loss of stapes
AA   Pectoral and pelvic girdles greatly expanded ventrally
AA   Similar paddle shape of hind and forelimbs
AA   Ilium not attached to pubis (not considered homologous to condi-
          tion in crocodiles)
SA   Lower temporal bar lost
SA   Nasals lost
SA   Loss of palatal dentition
SA   Loss of transverse flange of pterygoid
SA   Thecodont implantation of marginal dentition
SA   Opisthotic extending to cheek
SA   Jaw articulation well below tooth row
SA   Conspicuous retroarticular process
SA   Vertebrae non-notochordal
SA   Arches and centra not co-ossified
SA   No trunk intercentra
SA   28 or more cervical vertebrae

SA   Great elongation of transverse processes of trunk vertebrae
SA   Three or more sacral vertebrae
SA   Single-headed trunk ribs
SA   Articulating surfaces of limbs, girdles, carpals, and tarsals poorly
        defined
SA   Loss of cleithrum
SSS   Ventral portion of scapula superficial to blade of clavicle
SA   Humerus lacking both ectepicondylar and entepicondylar fora-
        mina
SA   Epipodials reduced
SA   Hyperphalangy
SA   Thyroid fenestra
SA   Blade of ilium much reduced

Derived Characters of *Claudiosaurus*

SL   Postfrontal borders on upper temporal opening
SS   Loss of lower temporal bar without development of streptostyly
SS   Reduction of suborbital and interpterygoid vacuities
SAL   ? Loss of postparietal and tabular
SAL   Eight cervical vertebrae
SL   Short transverse processes on trunk ribs
SS   Partial integration of third sacral rib
SAL   Loss of cleithrum
SAL   Shoulder girdle displaced posteriorly
SL   Cartilaginous sternum
SS   Slight reduction in degree of ossification of carpals and tarsals

APPENDIX VI. Derived Characters of Archosauromorphs, Based on Character-States in the Most Plesiomorphic Genera

AA = autapomorphy of each group *within* archosauromorphs; and SA = synapomorphies uniting archosauromorphs.

Protorosauria (based on *Protorosaurus* and *Prolacerta*)

AA   Seven very elongate cervical vertebrae
AA   Cervical ribs very thin
AA   Loss of lower temporal bar
AA   ? Tympanum supported by quadrate, squamosal, and quadratojugal
        Postparietal median (also occurs in early archosaurs)
SA   Tooth implantation subthecodont or thecodont
SA   Premaxilla extending behind external nares
SA   High quadrate
SA   Quadratojugal behind lower temporal opening
SA   Elongate stapes without stapedial foramen

SA   Paroccipital process extends to top of quadrate
SA   Basioccipital and basisphenoid contiguous; basisphenoid reach-
        ing prootic
SA   Loss of postparietal and tabular (supratemporal retained)
SA   Prominent retroarticular process
SA   Vertebrae not notochordal
SA   Elongate transverse processes of trunk vertebrae
SA   Cleithrum absent
SA   Pisiform, radiale, and fifth distal carpal lost or slow to ossify
SA   Loss of entepicondylar foramen
SA   Astragalus and calcaneum articulate with one another
SA   Fifth distal tarsal lost
SA   Fifth metatarsal hooked

Rhynchosauria (based on *Noteosuchus*, *Howesia*, and *Mesosuchus*)

AA   Ankylothecodont tooth implantation
AA   Median narial opening
AA   Overhanging premaxillae (condition in *Chasmatosaurus* not con-
        sidered homologous)
        Postfrontal entering margin of upper temporal opening
SA   Premaxilla extends behind external nares
SA   High quadrate
SA   Quadratojugal behind lower temporal opening
SA   Elongate stapes without stapedial foramen
SA   Paroccipital process extending to top of quadrate
SA   Basioccipital and basisphenoid contiguous; basisphenoid reach-
        ing prootic
SA   Loss of postparietal and tabular (supratemporal retained)
SA   Prominent retroarticular process
SA   Vertebrae not notochordal
SA   Elongate transverse process of trunk vertebrae
SA   Cleithrum absent
SA   Pisiform, radiale, and fifth distal carpal lost or slow to ossify
SA   Loss of entepicondylar foramen
SA   Astragalus and calcaneum articulating
SA   Fifth distal tarsal lost
SA   Fifth metatarsal hooked

Archosaurs (based on *Proterosuchus* [*Chasmatosaurus*])

AA   Antorbital fenestrae
AA   Laterally compressed serrate teeth (also occurs in *Heleosaurus*)
        Postparietal median (also occurs in early Protorosaurs)
SA   Long neck; cervical vertebrae as long or longer than trunk ver-
        tebrae
SA   Thecodont implantation
SA   Premaxilla extending behind external nares

SA   High quadrate, presumably supporting large tympanum
SA   Quadratojugal behind lower temporal opening
SA   Paroccipital process extending to top of quadrate
SA   Basioccipital and basisphenoid contiguous; basisphenoid reach-
        ing prootic
SA   Tabular lost (supratemporal retained)
SA   Vertebrae not notochordal
SA   Elongate transverse processes
SA   Loss of cleithrum
SA   Loss of entepicondylar foramen
SA   Number of carpals reduced; ?pisiform lost
SA   Astragalus and calcaneum articulating with one another
SA   Heel of calcaneum directed posteriorly
SA   Fifth distal tarsal lost
SA   Fifth metatarsal hooked

Trilophosauridae (based on *Trilophosaurus*)

AA   No lateral temporal opening
AA   Loss of teeth in premaxilla and front of dentary
AA   Cheek teeth laterally expanded
AA   Extensive fusion of dermal bones in the adult
AA   Cervical vertebrae procoelous
SA   Tooth implantation thecodont
        Postfrontal entering margin of upper temporal opening
SA   Premaxilla excluding maxilla from narial opening
SA   High quadrate
        Pineal opening lost
SA   Elongate stapes without stapedial foramen
SA   Paroccipital process extending to top of quadrate
SA   Prominent retroarticular process
SA   Vertebrae not notochordal
SA   Elongate transverse processes
SA   Cleithrum absent
SA   Loss of entepicondylar foramen
SA   ?Pisiform lost; radiale and fifth distal carpal lost or slow to ossify
SA   Astragalus and calcaneum articulating with one another
SA   Calcaneal heel
SA   Fifth distal tarsal lost
SA   Fifth metatarsal hooked

Malerisaurus

AA   Articulating surfaces of posterior cervical centra angled at 50°–75°
        with long axis of centra to provide a fixed curvature of neck
AA   Parietal fused at midline
SA   Premaxilla extending behind external nares
SA   High quadrate

SA   Quadratojugal behind lower temporal opening
SA   Paroccipital process extending to top of quadrate
SA   Basioccipital and basisphenoid contiguous; basisphenoid reaching prootic
SA   Loss of postparietal, tabular, and supratemporal
SA   Prominent retroarticular process
SA   Nine cervical vertebrae; all but atlas elongate
SA   Vertebrae not notochordal
SA   Elongate transverse processes of trunk vertebrae
SA   Cleithrum absent
SA   Loss of entepicondylar foramen
      Head of fifth metatarsal not hooked, but widely expanded

APPENDIX VII. Derived Characters of Groups
Whose Phylogenetic Position Is Uncertain

AA = autapomorphies unique to this group; SA = derived similarities with some archosauromorphs; SL = derived similarities with some lepidosauromorphs; SAL = derived similarities with some archosauromorphs and some lepidosauromorphs.

Choristodera (based on the more primitive of
derived character-states described in *Champsosaurus*
and *Simoedosaurus*)

AA   Elongation of snout with nasal bones fused at midline; prefrontals meeting at midline separating nasals from frontals
AA   Temporal area greatly expanded laterally and extending well posterior to occipital condyle
AA   Internal nares extended posteriorly as grooves in the roof of palate
SAL   Loss of postparietal, supratemporal, tabular, and pineal foramen
SA   Basioccipital and basisphenoid form continuous floor of braincase
SAL   External nares medial in position
SAL   Extensive paroccipital process supported by quadrate
SAL   Extremely long jaw symphysis
SAL   Thecodont implantation of marginal teeth
SAL   Nine cervical vertebrae
SA   Three fused sacrals
SL   Reduced ossification of vertebrae, girdles, carpals, and tarsals
SAL   Vertebrae not notochordal
SA   Loss of trunk intercentra and cross-pieces of haemal arches
SL   Ribs pachyostotic
SAL   Loss of cleithrum
SAL   Loss of entepicondylar foramen
SAL   Cervical ribs with anterior processes

SAL   Femur much longer than humerus
SAL   Reduced phalangeal formula
SAL   No pisiform
SL   Short epipodials

Placodontia (based on the more primitive of derived
character states observed in *Placodus* and
Helveticosaurus)

AA   Vertebrae with elongate transverse processes, but deeply amphi-
      coelous centra
AA   Loss of lateral temporal opening; quadratojugal forming much of
      cheek
AA   Anterior teeth spatulate and procumbent
AA   Quadrate fitting into a socket formed by squamosal and quadrato-
      jugal
AA   Prefrontal and postfrontal separating frontal from orbital margin
SAL   Loss of postparietal, supratemporal, and tabular
SAL   Loss of lacrimal
SAL   Nasal fused at midline
SAL   External nares posterior in position (common to many aquatic
      groups)
SAL   Loss of teeth on pterygoid
SL   Postfrontal entering margin of upper temporal opening
SAL   Quadrate recessed posteriorly, suggesting support of tympanum
SAL   Prefrontal having pillarlike contact with palatine (crocodiles,
      sphenodontids, and the nothosaur *Simosaurus*)
SL   Closure of interpterygoid vacuity and suborbital fenestra reduced
      to a slit
      Contact between maxilla and ectopterygoid
SAL   Vomer fused
SL   Cervical centra shorter than those of trunk
SAL   Loss of trunk intercentra
SL   Single-headed trunk ribs
SAL   Loss of cleithrum
SAL   Loss of entepicondylar foramen
SL   Reduced ossification of ends of limb bones, girdles, tarsals, and
      carpals (common to many aquatic groups)

Thalattosauria (based primarily on *Askeptosaurus* and
Clarazia)

AA   Dorsal temporal opening much restricted or entirely closed
AA   Premaxillae elongated and reaching frontals; nasals displaced lat-
      erally
AA   Posterior margin of skull table deeply emarginated; occiput lying
      well forward relative to quadrate suspension
SAL   At least 7–8 cervical and 30–32 trunk vertebrae

SAL External nares posterior in position and close to midline
SAL Reduction of lower temporal bar
SAL Loss of tabular and postparietal, but not supratemporal
SA   Loss of trunk intercentra and cross-piece of haemal arches
SAL Cervical ribs with anterior process
SAL Loss of cleithrum
SAL Loss of entepicondylar foramen
SL   Reduced length of epipodials (common to many aquatic groups)
SA   Reduced ossification of carpals and tarsals (common to many
     aquatic groups)
SAL Loss of fifth distal tarsal
SAL Reduction in phalangeal count

APPENDIX VIII. Anatomical Abbreviations Used in the Figures

a = angular; art = articular; ast = astragalus

cal = calcaneum; cen = centrale

d = dentary

f = frontal

i = intermedium

j = jugal

l = lacrimal; lc = lateral centrale

m = maxilla; mc = medial centrale

n = nasal

p = parietal; pf = postfrontal; pis = pisiform; pm = premaxilla; po = postorbital; prf = prefrontal

q = quadrate

ra = radiale

sa = surangular; sq = squamosal; st = supratemporal

ul = ulnare

1–5 = distal carpals or tarsals

I–V = metacarpals or metatarsals

Literature Cited

Ax, P. 1987. *The Phylogenetic System*. New York: Wiley Interscience.

Benton, M. J. 1983. The Triassic reptile *Hyperodapedon* from Elgin: functional morphology and relationships. Philos. Trans. R. Soc. London, Ser. B, 302:605–717.

Benton, M. J. 1985. Classification and phylogeny of diapsid reptiles. Zool. J. Linn. Soc., 84(2):97–164.

Brinkman, D. 1981. The origin of the crocodiloid tarsi: the interrelationships of the thecodontian archosaurs. Breviora, 464:1–23.

Broom, R. 1914. A new thecodont reptile. Proc. Zool. Soc. London, 1072–1077.

Brown, D. S. 1981. The English Upper Jurassic Plesiosauroidea (Reptilia) and a review of the phylogeny and classification of the Plesiosauria. Bull. Br. Mus. (Nat. Hist.) Geol., 35(4):253–345.

Butler, P. M. 1982. Direction of evolution in the mammalian dentition. Pp. 235–244 _in_ Joysey, K. A., and A. E. Friday (eds.), _Problems of Phylogenetic Reconstruction._ London: Academic Press.

Camp, C. L. 1945. _Prolacerta_ and the protosaurian reptiles. Am. J. Sci., 243(1):17–32, 84–101.

Carroll, R. L. 1969. A Middle Pennsylvanian captorhinomorph, and the interrelationships of primitive reptiles. J. Paleontol., 43(1):151–170.

Carroll, R. L. 1975. Permo-Triassic "lizards" from the Karroo. Palaeontol. Afr., 18:71–87.

Carroll, R. L. 1976a. _Noteosuchus_—the oldest known rhynchosaur. Ann. S. Afr. Mus., 72(3):37–57.

Carroll, R. L. 1976b. _Galesphyrus capensis_, a younginid eosuchian from the _Cistecephalus_ Zone of South Africa. Ann. S. Afr. Mus., 72(4):59–68.

Carroll, R. L. 1976c. Eosuchians and the origin of archosaurs. Pp. 58–79 _in_ Churcher, C. S. (ed.), _Athlon—Essays on Palaeontology in Honour of Loris Shano Russell._ R. Ont. Mus. Life Sci. Misc. Publ.

Carroll, R. L. 1977. The origin of lizards. Pp. 359–396 _in_ Andrews, S. M., R. S. Miles, and A. D. Walker (eds.), _Problems in Vertebrate Evolution._ Linn. Symp. Ser. No. 4.

Carroll, R. L. 1978. Permo-Triassic "lizards" from the Karroo system. Part II. A gliding reptile from the Upper Permian of Madagascar. Palaeontol. Afr., 21:143–159.

Carroll, R. L. 1981. Plesiosaur ancestor from the Upper Permian of Madagascar. Philos. Trans. R. Soc. London, Ser. B, 293:315–383.

Carroll, R. L. 1982. Early evolution of reptiles. Annu. Rev. Ecol. Syst., 13:87–109.

Carroll, R. L. 1985. A pleurosaur from the Lower Jurassic and the taxonomic position of the Sphenodontia. Palaeontogr. Abt. A, 189:1–28.

Carroll, R. L. 1986. Physical and biological constraints on the pattern of vertebrate evolution. Geosc. Canada, 13:85–90.

Carroll, R. L. 1987. _Vertebrate Paleontology and Evolution._ New York: W. H. Freeman.

Carroll, R. L. 1988. Late Paleozoic and early Mesozoic lepidosauromorphs and their relation to lizard ancestry. Pp. 99–118 _in_ Estes, R., and Pregill, G. (eds.), _Phylogenetic Relationships of the Lizard Families._ Stanford: Stanford Univ. Press.

Carroll, R. L., and P. Gaskill. 1985. The nothosaur _Pachypleurosaurus_ and the origin of plesiosaurs. Philos. Trans. R. Soc. London, Ser. B, 309:343–393.

Charig, A. J., and H.-D. Sues. 1976. Thecodontia. Pp. 11–39 _in_ Kuhn, O. (ed.), _Handbuch der Paläoherpetologie._ Stuttgart: Gustav Fischer Verlag.

Chatterjee, S. 1974. A rhynchosaur from the Upper Triassic Maleri Formation of India. Philos. Trans. R. Soc. London, Ser. B, 267:209–261.

Chatterjee, S. 1980. _Malerisaurus_, a new eosuchian reptile from the later Triassic of India. Philos. Trans. R. Soc. London, Ser. B, 291:163–200.

Chatterjee, S. 1986. _Malerisaurus langstoni_, a new diapsid reptile from the Triassic of Texas. J. Vert. Paleontol., 6:297–312.

Colbert, E. H. 1970. The Triassic gliding reptile _Icarosaurus_. Bull. Am. Mus. Nat. Hist., 143(2):89–142.

Cruickshank, A. R. I. 1972. The proterosuchian thecodonts. Pp. 89–119 _in_ Joysey, K. A., and T. S. Kemp (eds.), _Studies in Vertebrate Evolution._ Edinburgh: Oliver & Boyd.

Currie, P. J. 1980. A new younginid (Reptilia: Eosuchia) from the Upper Permian of Madagascar. Can. J. Earth Sci., 17(4):500–511.

Currie, P. J. 1981a. The vertebrae of *Youngina* (Reptilia: Eosuchia). Can. J. Earth Sci., 18(4):815–818.

Currie, P. J. 1981b. *Hovasaurus boulei*, an aquatic eosuchian from the Upper Permian of Madagascar. Palaeontol. Afr., 24:99–168.

Currie, P. J. 1982. The osteology and relationships of *Tangasaurus mennelli* Haughton (Reptilia, Eosuchia). Ann. S. Afr. Mus., 86(8):247–265.

Currie, P. J., and R. L. Carroll. 1984. Ontogenetic changes in the eosuchian reptile *Thadeosaurus*. J. Vert. Paleontol., 4(1):68–84.

Denison, R. 1978. Placodermi. Pp. 1–128 *in Handbook of Paleoichthyology*. Vol. 2. Stuttgart: Gustav Fischer Verlag.

Drevermann, F. 1933. Das Skelett von *Placodus gigas* Agassiz im Senckenberg-Museum. Abh. Senckenb. Naturforsch. Ges., 38:319–364.

Edinger, T. 1935. *Pistosaurus*. Neues Jahrb. Mineral. Geol. Palaeontol., Beilage-Band, 74:321–359.

Erickson, B. R. 1972. The lepidosaurian reptile *Champosaurus* in North America. Monogr. Sci. Mus. Minnesota Paleontol., 1:1–91.

Evans, S. E. 1980. The skull of a new eosuchian reptile from the Lower Jurassic of South Wales. J. Linn. Soc. London Zool., 70:203–264.

Evans, S. E. 1981. The postcranial skeleton of the Lower Jurassic eosuchian *Gephyrosaurus bridensis*. J. Linn. Soc. London Zool., 73:81–116.

Evans, S. E. 1982. The gliding reptiles of the Upper Permian. J. Linn. Soc. London Zool., 76:97–123.

Evans, S. E. 1984. The classification of the lepidosauria. J. Linn. Soc. London Zool., 82:87–100.

Evans, S. E. 1988. The early history and relationships of the Diapsida. Pp. 221–260 *in* Benton, M. J. (ed.), *The Phylogeny and Classification of the Tetrapods*. Vol. 1: *Amphibians, Reptiles, Birds*. Syst. Assoc. Spec. Vol. No. 35 A. Oxford: Clarendon Press.

Evans, S. E. 1990. The skull of *Cteniogenys*, a christodere (Reptilia: Archosauromorpha) from the Middle Jurassic of Oxfordshire. Zoo. J. Linn. Soc., 99:205–237.

Evans, S. E., and H. Haubold. 1987. A review of the Upper Permian genera *Coelurosauravus*, *Weigeltisaurus* and *Gracilisaurus* (Reptilia: Diapsida). J. Linn. Soc. London Zool., 90:275–303.

Ewer, R. F. 1965. The anatomy of the thecodont reptile *Euparkeria capensis* Broom. Philos. Trans. R. Soc. London, Ser. B, 248:379–435.

Forey, P. L., and B. G. Gardiner. 1986. Observations on *Ctenurella* (Ptyctodontida) and the classification of placoderm fishes. J. Linn. Soc. London Zool., 86:43–74.

Fraser, N. C. 1982. A new rhynchocephalian from the British Upper Trias. Palaeontology, 24(4):709–725.

Fraser, N. C. 1986. New Triassic sphenodontids from south-west England and a review of their classification. Palaeontology, 29(1):165–186.

Fraser, N. C., and G. M. Walkden. 1983. The ecology of a later Triassic reptile assemblage from Gloucestershire, England. Palaeogeogr. Palaeoclimatol. Palaeoecol., 42:341–365.

Fraser, N. C., and G. M. Walkden. 1984. The postcranial skeleton of the Upper Triassic sphenodontid *Planocephalosaurus robinsonae*. Palaeontology, 27(3):575–595.

Gaffney, E. S. 1979. Tetrapod monophyly: a phylogenetic analysis. Bull. Carnegie Mus. Nat. Hist., 13:92–105.

Gardiner, B. G. 1982. Tetrapod classification. J. Linn. Soc. London Zool., 74:207–232.

Gardiner, B. G. 1984. The relationship of placoderms. J. Vert. Paleontol., 4:379–395.

Gauthier, J. A., A. G. Kluge, and T. Rowe. 1988. The early evolution of the Amniota. Pp. 103–155 in Benton, M. J. (ed.), *The Phylogeny and Classification of the Tetrapods.* Vol. 1: *Amphibians, Reptiles, Birds.* Syst. Assoc. Spec. Vol. No. 35 A. Oxford: Clarendon Press.

Goodrich, E. S. 1916. On the classification of the Reptilia. Proc. R. Soc. London, 89:261–276.

Goujet, D. F. 1984. Placoderm interrelationships: a new interpretation with a short review of placoderm classification. Proc. Linn. Soc. New South Wales, 107(3):211–243.

Gow, C. E. 1975. The morphology and relationships of *Youngina capensis* Broom and *Prolacerta broomi* Parrington. Palaeontol. Afr., 18:89–131.

Gray, J. 1968. *Animal Locomotion.* London: Weidenfeld and Nicolson.

Gregory, J. T. 1945. Osteology and relationships of *Trilophosaurus.* Univ. Texas Publ., 4401:263–359.

Heaton, M. J., and R. R. Reisz. 1986. Phylogenetic relationships of captorhinomorph reptiles. Can. J. Earth Sci., 23(3):402–418.

Hennig, W. 1966. *Phylogenetic Systematics.* Chicago: Univ. of Illinois Press. [Translation].

Hennig, W. 1981. *Insect Phylogeny.* (Edited and translated by A. C. Pont.) New York: John Wiley.

Holmes, E. B. 1985. Are lungfishes the sister group of tetrapods? Biol. J. Linn. Soc., 24:379–397.

Jarvik, E. 1981. Lungfishes, tetrapods, paleontology and plesiomorphy. D. E. Rosen, P. L. Forey, B. G. Gardiner, and C. Patterson. 1981. Syst. Zool., 30(3):378–384. [Review].

Jenkins, F. A., Jr., and G. E. Goslow. 1983. The functional anatomy of the savannah monitor lizard (*Varanus exanthematicus*). J. Morphol., 175:195–216.

Kemp, T. S. 1988. Haemothermia or Archosauria? The interrelationships of mammals, birds and crocodiles. J. Linn. Soc. London Zool., 92:67–104.

Kluge, A. G. 1977. *Chordate Structure and Function.* New York: Macmillan.

Kluge, A. G. 1984. The relevance of parsimony to phylogenetic inference. Pp. 24–38 in Duncan, T., and T. F. Stuessy (eds.), *Perspectives on the Reconstruction of Evolutionary History.* New York: Columbia Univ. Press.

Kuhn-Schnyder, E. 1952. Die Triasfauna der Tessiner Kalkalpen. XVII. *Askeptosaurus italicus* Nopcsa. Schweiz. Palaeontol. Abh., 69:1–82.

Kuhn-Schnyder, E. 1962. La position des nothosaurides dans le système des reptiles. Pp. 135–144 in Sur les problèmes actuels de paléontologie: Évolution des vertébrés. Colloq. Int. C. N. R. S., 104. Paris: C.N.R.S.

Laurin, M. 1991. The osteology of a Lower Permian eosuchian from from Texas and a review of diapsid phylogeny. J. Linn. Soc. London Zool., 101:59–95.

Løvtrup, S. 1977. *The Phylogeny of Vertebrata.* London: Wiley-Interscience.

Løvtrup, S. 1985. On the classification of the taxon Tetrapoda. Syst. Zool., 34(4):463–470.

Maisey, J. C. 1984. Higher elasmobranch phylogeny and biostratigraphy. J. Linn. Soc. London Zool., 82:33–54.

Merriam, J. C. 1905. The Thalattosauria, a group of marine reptiles from the Triassic of California. Mem. Cal. Aca. Sci., 5:1–52.

Meyer, H. von. 1856. Zur Fauna der Vorwelt. Saurier aus dem Kupferschiefer der Zechstein-Formation. Frankfurt-am-Main.

Miles, R. S., and G. C. Young. 1977. Placoderm interrelationships reconsidered in the light of new ptyctodontids from Gogo, Western Australia. Linn. Soc. Symp. Ser., 4:123–197.

Novacek, M. J. 1986. The skull of leptictid insectivorans and the higher-level classification of eutherian mammals. Bull. Am. Mus. Nat. Hist., 183:1–111.

Panchen, A. L., and T. R. Smithson. 1987. Character diagnosis, fossils and the origin of tetrapods. Biol. Rev. Cambridge Philos. Soc., 62:341–438.

Parks, P. 1969. Cranial anatomy and mastication of the Triassic reptile *Trilophosaurus*. Doctoral dissertation. Austin: Univ. of Texas.

Patterson, C. 1981. Significance of fossils in determining evolutionary relationships. Annu. Rev. Ecol. Syst., 12:195–223.

Patterson, C. 1982. Morphological characters and homology. Pp. 21–74 *in* Joysey, K. A., and A. E. Friday (eds.), *Problems of Phylogenetic Reconstruction*. London: Academic Press.

Paul, C. R. C. 1982. The adequacy of the fossil record. Pp. 75–117 *in* Joysey, K. A., and A. E. Friday (eds.), *Problems of Phylogenetic Reconstruction*. London: Academic Press.

Peyer, B. 1936. Die Triasfauna der Tessiner Kalkalpen. X. *Clarazia schinzi* nov. gen. nov. spec. XI. *Hescheleria rubeli* nov. gen. nov. spec. Abh. Schweizer. Palaeontol. Ges., 52:1–59; 53:1–48.

Peyer, B. 1955. Die Triasfauna der Tessiner Kalkalpen. XVIII. *Helveticosaurus zollingeri* n. g. n. sp. Abh. Schweizer. Palaeont. Ges., 146:1–50.

Reisz, R. R. 1981. A diapsid reptile from the Pennsylvanian of Kansas. Occas. Pap. Mus. Nat. Hist. Univ. Kansas, 7:1–74.

Reisz, R. R. 1988. Two small reptiles from a Late Pennsylvanian quarry near Hamilton, Kansas. Pp. 189–194 *in Hamilton Quarry Guidebook.*

Reisz, R. R., D. S. Berman, and D. Scott. 1984. The anatomy and relationships of the Lower Permian reptile *Araeoscelis*. J. Vert. Paleontol., 4(1):57–67.

Rieppel, O. 1987. *Clarazia* and *Hescheleria*: a re-investigation of two problematical reptiles from the Middle Triassic of Monte San Giorgio (Switzerland). Palaeontographica, 195:101–129.

Rieppel, O. 1989. *Helveticosaurus zollingeri* Peyer (Reptilia, Diapsida) skeletal paedomorphosis, functional anatomy and systematic affinities. Palaeontographica, A, 208:123–152.

Rieppel, O., and R. W. Gronowski. 1981. The loss of the lower temporal arcade in diapsid reptiles. J. Linn. Soc. London Zool., 72:203–217.

Robinson, P. L. 1962. Gliding lizards from the Upper Keuper of Great Britain. Proc. Geol. Soc. London, 1601:137–146.

Robinson, P. L. 1973. A problematic reptile from the British Upper Trias. J. Geol. Soc. London, 129:457–479.

Romer, A. S. 1956. *Osteology of the Reptiles*. Chicago: Univ. of Chicago Press.

Romer, A. S. 1966. *Vertebrate Paleontology*. 3rd ed. Chicago: Univ. of Chicago Press.

Romer, A. S. 1971. Unorthodoxies in reptilian phylogeny. Evolution, 25(1):103–112.

Rosen, D. E., P. L. Forey, B. G. Gardiner, and C. Patterson. 1981. Lungfishes, tetrapods, paleontology, and plesiomorphy. Bull. Am. Mus. Nat. Hist., 167(4):159–276.

Schultze, H.-P. 1981. Hennig und der Ursprung der Tetrapoda. Palaeontol. Z., 55(1):71–86.

Sigogneau-Russell, D., and M. Efimov. 1984. Un Choristodera (Eosuchia?) insolite du Crétacé Inférieur de Mongolie. Palaeontol. Z., 58(3/4):279–294.

Sigogneau-Russell, D., and D. E. Russell. 1978. Étude ostéologique du reptile *Simoedosaurus* (Choristodera). Ann. Paleontol. Vertebr., 64:1–84.

Sues, H.-D. 1987a. The postcranial skeleton of *Pistosaurus* and the interrelationships of the Sauropterygia (Diapsida). J. Linn. Soc. London Zool., 90:109–131.

Sues, H.-D. 1987b. On the skull of *Placodus gigas* and the relationship of the Placodontia. J. Vert. Paleontol., 7:138–144.

Tedford, R. H. 1976. Relationship of pinnipeds to other carnivores (Mammalia). Syst. Zool., 25(4):363–374.

Whiteside, D. I. 1986. The head skeleton of the Rhaetian Sphenodontid *Diphydontosaurus avonis* gen. et sp. nov. and the modernizing of a living fossil. Philos. Trans. R. Soc. London, Ser. B, 312:379–430.

Wild, R. 1974. Die Triasfauna der Tessiner Kalkalpen. 23. *Tanystropheus longobardicus* (Bassani) (New developments). Schweiz. Palaeontol. Abh., 95:1–162.

Wiley, E. O. 1981. *Phylogenetics*. New York: John Wiley.

Wu, X. 1988. The middle ear region of the Late Triassic–Early Jurassic sphenodontids from China. J. Vert. Paleontol., 8(3)Abstracts:30A.

Section IV

BIRDS

12 Perspectives on Avian Origins

Lawrence M. Witmer

Before the work of John H. Ostrom in the 1970's, few contemporary scientists were interested in the origin of birds. Although the belief that birds were descended from reptiles was nearly universal, little research was directed toward developing a more specific hypothesis of avian ancestry. Most workers accepted the authoritative treatment of Heilmann (1927) that suggested that small early archosaurs ("pseudosuchian thecodonts") were involved in some way with the origin of birds. Ostrom's discovery and description of the small, birdlike theropod *Deinonychus* (Ostrom, 1969) ignited both the scientific community and the public. In the succeeding years, there has been an unrelenting flow of literature on dinosaurs and their physiology, the remarkable diversity of other archosaurs, and the origin of birds.

This and the following three chapters are devoted to this last topic. John Ostrom restates his well-known advocacy for the relationship of birds to coelurosaurian theropod dinosaurs and answers his critics. Samuel Tarsitano provides a novel approach to investigating avian ancestry and discusses features that suggest to him that birds descended from thecodonts. Larry Martin provides a detailed description of the Jurassic bird *Archaeopteryx* and reasserts the validity of his hypothesis that birds are closely related to crocodilians.

The present chapter summarizes the historical development of ideas on bird origins. Upon reaching the modern era (post-1970), I focus on the diversity of current opinion and attempt to put some of the recent controversies in perspective. In all but a few cases, it is impossible to examine in detail the separate, but related, topic of the origin of avian flight (Padian, 1986). Although today we are in the period of liveliest discourse on the origin of flight, perhaps the most insightful historical review should await the passing of a few more years. Several papers have included short historical summaries of the debate on avian origins. Among the most important are those by Osborn (1900), Gregory (1916), Heilmann (1927), de Beer (1954), Ostrom (1976, 1985), Feduccia (1980), and Gauthier (1986). Furthermore, although his book is not dedicated to the subject, Desmond (1982) provided an engaging and well-researched account of the debate in mid-Victorian England. Rather than list all of the papers for or against a particular hypothesis (see de Beer, 1954, and Ostrom, 1976, 1985, for more comprehensive citations), I will examine some of the more important contributions in detail.

Three authors, more than anyone else, have shaped the debate on avian origins. T. H. Huxley (1868b, 1870a,b) developed the first well-articulated, specific argument for avian ancestry—the dinosaurian hypothesis. The second is Gerhard Heilmann (1927), whose monumental work *The Origin of Birds* held sway for 50 years. Its importance led Ostrom (1985:17) to exclaim rightly, "The impact of Heilmann's book cannot be exaggerated. On the question of bird origins, its impact has been second only to the original discovery of *Archaeopteryx*." The same easily can be said for Ostrom himself, whose series of papers (most elegantly, Ostrom, 1976) provided the focus for all later work.

> There is a curious blending of the characters of the various reptilian groups in the Birds; there has been no exclusive adoption of the mode of structure of any one scaly type by these feathered vertebrates; those reptilian qualities and excellencies which are best and highest have become theirs; but how much more! This exaltation of the "Sauropsidan" or oviparous type by the substitution of feathers for scales, wings for paws, warm blood for cold, intelligence for stupidity, and what is lovely instead of loathsomeness,— this sudden glorification of the vertebrate form is one of the great wonders of Nature.
>
> —William Kitchen Parker (1864:56–57)

AVIAN ORIGINS FROM LAMARCK TO 1970

Before evolutionary views were common, the origin of birds (or of anything else for that matter) was not an issue—all things were considered products of divine creation. Furthermore, the morphological gap between birds and other animals was so profound and the fossil record so poor that birds truly seemed to stand apart from the rest of the animal kingdom. The early pre-Darwinian evolutionists were limited to the comparative anatomy of modern forms, a science that was still in its infancy. Lamarck (1809), for instance, derived birds from turtles, presumably on the basis of the rhamphotheca-covered, edentulous jaws; birds themselves (in particular, the penguins) then gave rise to the monotremes.

With the 1859 publication of Darwin's *On the Origin of Species*, evolutionary origins became an important issue, and, in fact, the origin of birds became a principal example of evolution (Evans, 1865; Huxley, 1868b). The discovery of the reptilian bird *Archaeopteryx lithographica* in the Jurassic limestones near Solnhofen, Bavaria, in 1861—less than two years after the publication of the *Origin*—would seem to have come at the best possible time for the Darwinians. But its intermediate characters were not missed by the antievolutionists. Andreas Wagner (1861, 1862:266) quickly wrote the first paper ever published on *Archaeopteryx* (Desmond, 1982) in hopes of "ward[ing] off Darwinian misinterpretations." Wagner considered the fossil (which he named *Griphosaurus*) a feathered reptile of no special relationship to birds. Richard Owen (1862, 1863) described the London specimen and, rather than bemoaning its evolutionary implications, almost rejoiced in its support for his concept of the Archetype, noting that "we discern . . . a retention of a structure embryonal and transitory in the modern representatives of the class, and a closer adhesion to the general vertebrate type" (Owen, 1863:46).

Yet the evolutionary importance of *Archaeopteryx* eventually was seized upon—first by Evans (1865), who saw it as intermediate between birds and reptiles in general, and then Huxley (1868b), who used it in his discussion of the dinosaurian relationships of birds. In fact, Huxley's 1868 paper, his first on bird origins, was framed largely as a proof of evolution. Surprisingly, when Huxley (1868b:73) noted that "but a single specimen, obtained from those Solenhofen slates . . . affords a still nearer approximation to the 'missing link' between reptiles and birds," he was referring not to *Archaeopteryx*, but to the small theropod *Compsognathus* (Fig. 1). He made no mention of any dinosaurian features of *Archaeopteryx* but used it simply as a demonstration of the existence of more reptilian birds. He made comparisons with large dinosaurs such as the ornithischians *Scelidosaurus* and *Iguanodon* and the carnosaur *Megalosaurus*, noting the large number of sacral vertebrae and the general birdlike nature of the pelvis and hind limb. However, the coup de grace of his argument was based on the birdlike characters of *Compsognathus* (viz., the generally gracile skeleton and very birdlike hind limb). Although we often regard Huxley as publishing on "avian origins," Huxley's emphasis was more on placing dinosaurs (and *not* birds) within vertebrate phylogeny. The effect is the same; we read Huxley as stating that birds are related to dinosaurs, but he actually was arguing the converse.

Huxley's (1870a) "Further evidence" paper introduced a few additional

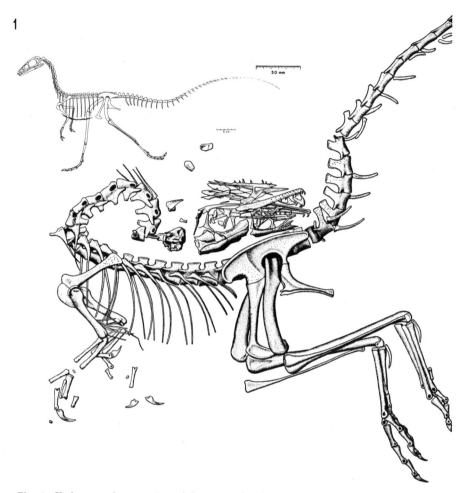

Fig. 1. Skeleton and restoration of *Compsognathus longipes*. (Modified from Ostrom, 1978, courtesy of Bayerischen Staatssammlung für Paläontologie und historische Geologie, Munich, Germany.)

characters supporting the dinosaurian relationships of birds such as tibial torsion and the ascending process of the astragalus. He also noted the similarities in the pelvis of birds and ornithischians such as *Iguanodon* and *Hypsilophodon*—that is, the elongate antacetabular ilium, the obturator process of the ischium, and the retroverted pubis. In fact, although impressed with the avian features of *Compsognathus* in the 1868b paper, he was also struck by the opisthopuby of *Hypsilophodon*, noting (1870a:28) that it "affords unequivocal evidences of a further step towards the bird."

Later that year, Huxley published his classification of dinosaurs

(1870b). Within his Dinosauria were three families: Megalosauridae, Scelidosauridae, and Iguanodontidae. Excluding *Compsognathus* from Dinosauria, he made it "representative of a group equivalent to" Dinosauria. He united the two as Ornithoscelida. He asserted the "affinities" of Ornithoscelida with birds on the basis again of (p. 38) "the peculiarities of the hind limb and pelvis." He did not cite one "ornithoscelidan" that was closer to birds than another. Later, Haeckel (1907) adopted the taxon name and included *Compsognathus* and birds within his Ornithoscelides, a group that was itself subordinated to Dinosauria. Huxley did not consider such a specific position because he viewed the divergence of birds and dinosaurs as taking place in the Paleozoic. In fact, he considered *Archaeopteryx* almost irrelevant to the whole issue (Desmond, 1982), stating (Huxley, 1868a:248) "that, in many respects, *Archaeopteryx* is more remote from the boundary-line between birds and reptiles than some living *Ratitae* are." Huxley (1868b:75) "regarded as certain that we have no knowledge of the animals which linked reptiles and birds together historically and genetically, and that the *Dinosauria*, with *Compsognathus*, *Archaeopteryx*, and the struthious birds, only help us to form a reasonable conception of what these intermediate forms may have been." Later, however, Huxley (1882) vaguely proposed a more direct ancestry of birds (and crocodiles) from very early (perhaps hypothetical) dinosaurs.

Although Huxley deserves credit for his detailed analysis, actually he was preceded by two workers. Gegenbaur (1864) considered *Compsognathus* a phylogenetic intermediate between birds and reptiles on the basis of similarities in the tarsus. Cope (1867), in a brief note that appeared three months before Huxley's first bird-origin paper, considered birds to be related to both dinosaurs and pterosaurs. Penguins, or perhaps ratites, were closer to dinosaurs, and *Archaeopteryx* was closer to pterosaurs. Like Huxley, Cope also was struck by the avian features of *Compsognathus*. The points of resemblance between birds and dinosaurs included the crus, the astragalus, and "a more or less erect position" of the body (Cope, 1867). Whereas Gegenbaur and Cope did little subsequent work on the subject, Huxley's papers stimulated considerable research, and he is rightly credited for the dinosaurian hypothesis.

The controversy resulting from Huxley's papers centered on avian origins, as well as on evolution itself. It is surprising, therefore, that Charles Darwin made little of *Archaeopteryx* and the origin of birds. In later editions of *On the Origin of Species*, Darwin did not showcase *Archaeopteryx* as a long-sought evolutionary intermediate, but merely as a demonstration of "how little we as yet know of the former inhabitants of the world" (Darwin, 1872:315).

Support for Huxley's ideas came largely from outside England. Haeckel, of course, was the most vocal advocate in Germany for both evolution and the reptilian relationships of birds. Whereas earlier, Haeckel (1866) placed birds close to turtles, pterosaurs, and "anomodonts" (a mixed bag composed of dicynodonts, prolacertids, and rhynchosaurs), he later (Haeckel, 1875) included dinosaurs within this nexus (Fig. 2), citing Huxley's comparisons of birds and *Compsognathus*. As mentioned, Gegenbaur (1864, 1878) made comparisons among birds, *Archaeopteryx*, and *Compsognathus* and, partly on the basis of the "tibiotarsus" and long tail, united them in the subclass Saururi within the Sauropsida. Georg Baur (1883, 1884b, 1887) argued persuasively for the

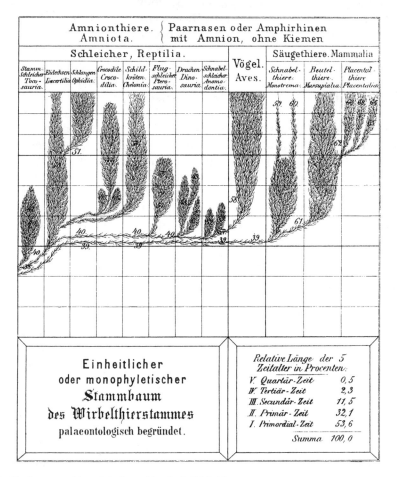

Fig. 2. Phylogeny of Amniota from Haeckel, 1875. The sister-group of birds is Anomodontia ("beaked reptiles"), which includes not only *Compsognathus* (which is not within Dinosauria) but also dicynodont therapsids. Successive out-groups are Dinosauria ("dragons"), Pterosauria ("flying reptiles"), and a group comprising squamates, crocodilians, and turtles.

dinosaurian origin of birds. In 1883, Baur made detailed comparisons of the ankles of birds and dinosaurs (and noted that the ontogeny of the avian tarsus recapitulates dinosaur phylogeny) but offered no specific hypothesis of ancestry within Dinosauria. However, his work on the pelvis (Baur, 1884b, 1887) led him to believe that, because of the presence of a "postpubis" in birds and ornithischians, it is "in the herbivorous Dinosaurs and especially in the ornithopod-like forms [that] we must seek for the ancestry of birds" (Baur, 1884b:1275). W. Dames (1884) took issue with Baur and others, and did not regard any of the resemblances in the pelvis and hind limb as indicative of phylogenetic relationship. Baur (1884a) answered Dames's objections, and Dames (1885) published a bitter reply, criticizing Baur for his "unknown ancestor" of birds and for putting too much weight on too few characters.

In America, O. C. Marsh (1877) originally accepted that birds originated from *within* Dinosauria. However, he later adopted Huxley's less specific hypothesis of common ancestry in the Paleozoic. In "Odontornithes," Marsh (1880:188) reconstructed a hypothetical bird ancestor, "a form [that] would be on the road toward the Birds, rather than on the ancestral line of either Dinosaurs or Pterodactyles." In 1881, Marsh reasserted the similarities of *Archaeopteryx* and *Compsognathus* but noted (p. 340) that "the two forms are in reality widely separated."

In 1879, B. F. Mudge published a note objecting to the dinosaurian ancestry of birds. Mudge (1879:226) complained that only "a few species" of dinosaurs had avian characteristics and that "the dinosaurs vary so much from each other that it is difficult to give a single trait that runs through the whole. But no single genus, or set of genera, have many features in common with the birds, or a single persistent, typical element of structure which is found in both." Mudge's comments are justified if one accepts the then current notion of Huxley and Marsh that birds and dinosaurs share a common ancestry—in current parlance, if Aves and Dinosauria are sister-groups and if certain "avian" features are not also primitive for Dinosauria, then these features are indeed convergently evolved in those dinosaurs possessing them. In that same volume of the *Kansas City Review of Science,* S. W. Williston offered a rejoinder to Mudge in which, without fanfare, he suggested a much more specific view of the dinosaurian origin of birds. Williston (1879:458–459) acknowledged that "scarcely a single trait of structure runs through the whole of Dinosauria; but that fact does not affect the relation existing between the most avian dinosaurs and the most reptilian birds." Williston was among the first to derive birds directly from a specific group of dinosaurs, noting that (p. 458) "true dinosaurs . . . may have given off branches that developed upwards into birds." Thus, Williston answered Mudge by suggesting that avian features arose only once—in the dinosaurian clade of which birds are a member. Although he did not name the dinosaurs close to birds, his description indicates Theropoda. Curiously, Williston (1925) later retreated from the specific position and presented a tree supporting Huxley's hypothesis of the common ancestry of birds and Dinosauria.

Huxley's great adversary Richard Owen (1875) offered a long criticism of Huxley's ideas on dinosaurs and birds. Owen viewed dinosaurs not as erect bipedal forms but as quadrupeds (some of which he regarded as aquatic). He examined some of the similarities between birds and dinosaurs cited in Huxley's papers and compared them with the stegosaur *Omosaurus* (= *Dacentrurus*, the main topic of the paper). It is no surprise that he found Huxley's similarities (as suggested by the more birdlike ornithopods and theropods) to break down when graviportal dinosaurs were considered. Interestingly, Owen (1875) never mentioned *Compsognathus*, Huxley's "missing link" between birds and dinosaurs. Instead, Owen cited resemblances of birds to mammals, in particular, monotremes. For example, according to Owen, some dinosaur-bird characters such as the opisthopuby noted by Huxley are found in only some dinosaurs but all mammals. Although earlier Owen (1870) had ascribed the similarities of birds and pterosaurs to convergence, he (1875:91) later considered pterosaurs to be the reptiles closest to birds but was ignorant as to "how the Rhamphorhynchus became transmuted into the Archeopteryx."

Because Huxley never responded to Owen's paper, Dollo (1883b:87) thought that it was "therefore quite necessary to fall in with the opinion of one or the other naturalist." The basis for most of Dollo's comparisons was the excellent *Iguanodon* material from Bernissart, Belgium. He made extensive comparisons with birds and found them to be remarkably similar (often "identical") in detail to *Iguanodon*. Although Dollo (1882, 1883a,b) often is cited as opposing the relationship of birds to dinosaurs (Osborn, 1900; Ostrom, 1976, 1985; Gauthier, 1986), Dollo's papers (especially 1883b) were directed primarily at proving *la station droite* (in effect, bipedality) in dinosaurs. In fact, he stated (Dollo, 1883b:88) that he "did not care for the present whether the points in common between dinosaurs and birds are coming from heredity or from adaptation." Nevertheless, Dollo (1883b) seems to have favored homology over analogy. He regarded Huxley's work as "classic" and arranged sauropods, ornithopods, and birds in a "phylogenetic series." Thus, rather than being an opponent, Dollo can be considered a supporter of dinosaurian relationships.

Owen was not the only worker to dissent from Huxley. Carl Vogt (1879, 1880) had rather peculiar and contradictory ideas about *Archaeopteryx* and the origin of birds; he was cited by de Beer (1954), Ostrom (1976, 1985), and Gauthier (1986) as supporting the derivation of birds from lizards.

Vogt (1880:452) identified few avian features in *Archaeopteryx* and considered it to be "a Reptile flying by means of feathers and perching with the legs of a Bird." He rejected both Gegenbaur's (1878) union of *Archaeopteryx* and *Compsognathus* into the Sauriuri and Huxley's (1870b) common-ancestry hypothesis. Despite such stated views, Vogt (1880:454) suggested that "the Dinosaurs would lead to the Ratites, the *Archaeopteryx* to the Birds that fly." He did not offer a specific group that led to the *Archaeopteryx*–volant bird line. Vogt (1880:456) "picture[d] . . . the ancestors of the *Archaeopteryx* as terrestrial Reptiles in the form of Lizards." Contrary to the assertions of recent authors, Vogt did not suggest ancestry from Squamata, but only from a generalized reptile. However, Vogt's major objection to the dinosaurian relationships of birds was one that has persisted to today; he considered any resemblances between dinosaurs and birds to be convergent and "only related to the development of the power of keeping an upright position upon the hind-feet" (Vogt, 1880:448).

Vogt's idea of the separate origin of ratites and carinates from different nonavian stocks was not uncommon during this period. George Mivart (1881) noted only in passing his support for such a "double origin"—i.e., ratites from dinosaurs and carinates from pterosaurs. Robert Wiedersheim (1882, 1884, 1885, 1886), on the other hand, published several detailed accounts in which he argued for the diphyly of birds. Although de Beer (1954) and Ostrom (1985) viewed Wiedersheim as sometimes deriving *Archaeopteryx* from lizards (i.e., Squamata) and sometimes from pterosaurs, Wiedersheim's actual position was stable and never involved relationships with Squamata (Fig. 3). According to Wiedersheim, pterosaurs and *Archaeopteryx* shared a common reptilian ancestor; the descendants of *Archaeopteryx* included *Ichthyornis* and all modern carinate birds. Wiedersheim envisioned a common dinosaurian ancestor shared by ratites and *Hesperornis*. These two great clades converge on an early Triassic or late Paleozoic common ancestor. Wiedersheim's argument often is inconsistent internally, and it rested largely on the authority of others (e.g., Marsh's [1880] assertion of ratite affinities for *Hesperornis*). Like Vogt, Wiedersheim was willing to accept large amounts of convergence.

H. G. Seeley was perhaps the first to argue for convergence. He stood up after Huxley read his "Further evidence" paper and "thought it possible that the peculiar structure of the hinder limbs of the Dinosauria was due to the functions they performed rather than to any actual affinity with birds" (in Discussion to Huxley, 1870a:31). Like Cope (1867) and Owen (1875), Seeley (1866, 1881) considered pterosaurs to be "allied" to birds (placing them as a subclass of birds in his 1866 paper) on the basis of elongate coracoids, keeled sterna, pneumatic limb bones, and other features; *Archaeopteryx* was a critical intermediate between pterosaurs and

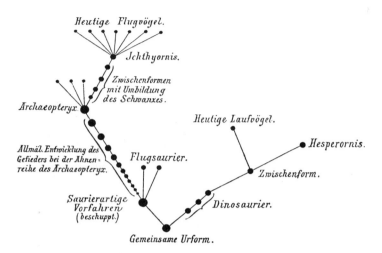

Fig. 3. Phylogeny of birds and their relatives from Wiedersheim, 1885. Living flying birds (*heutige Flugvögel*) are descendents of *Ichthyornis*. The line leading to *Archaeopteryx* shares a common "lizardlike ancestor" with pterosaurs (*Flugsaurier*). Ratites ("living running birds" or *heutige Laufvögel*) and *Hesperornis* represent a lineage descended from dinosaurs.

birds. Huxley (1870b:38) handled Seeley with a convergence argument of his own, stating that the similarities of pterosaurs to birds were the result of "physiological action, and not to affinity."

To summarize thus far, the theory of organic evolution opened the door for the discussion of avian origins. Although priority goes to Gegenbaur (1864) and Cope (1867), Huxley (1868b, 1870a) was the most visible proponent of the relationship of dinosaurs and birds. Huxley was joined by others such as Haeckel (1875), Marsh (1877), Williston (1879), Baur (1883), and Wiedersheim (1885). Many (Vogt, 1880; Seeley, 1881; Mudge, 1879; Dames, 1884; Parker, 1887; Fürbringer, 1888) discounted these resemblances, ascribing them to convergence. Others sought relationships with other groups, most often pterosaurs (Cope, 1867; Owen, 1875; Seeley, 1881; Mivart, 1881; Wiedersheim, 1885). Ostrom (1985:16) credited Fürbringer (1888) with "a compromise explanation—the common ancestor hypothesis." However, as elaborated above, the common-ancestor hypothesis was Huxley's from the beginning. In many respects, Huxley's views seem rather contradictory. He argued cogently for certain characters, such as the pelvis of ornithopods or the tarsus of theropods, as if they indicated monophyly of birds and a particular group of dinosaurs. But at the same time, he considered birds and dinosaurs to share a common ancestry in the Permian. Huxley never explicitly traced all his characters over all Dinosauria. Clearly, to postulate common ancestry of Aves and Dinosauria (i.e., a sister-group relationship) is to postulate convergence in many features of birds and higher dinosaurs.

H. F. Osborn (1900), who supported a "form of the Huxleyan hypothesis," sought to reduce the impact of such convergence by arguing for the origin of birds from an early bipedal dinosaur. Osborn (1900:797), in effect, downgraded the convergence to parallelism by suggesting that "the numerous parallels and resemblances in dinosaur and bird structure, while quite independently evolved, could thus be traced back to a potentially similar inheritance." This quotation articulates a common sentiment among workers in the late nineteenth and early twentieth centuries. Phylogenetic hypotheses often were vague, and people were willing to accept massive convergence. In fact, many workers (e.g., Lowe, 1935) followed Osborn in suggesting that convergence actually *demonstrated* some form of "affinity" as evidence of a shared genetic background.

Robert Broom (1906), on the other hand, suggested that the data should be accepted at face value and presented a phylogenetic tree to which some modern workers would not object: Aves is the sister-group of Theropoda, Pterosauria is the sister-group of these two, and a *"Proterosuchus*-like form" is the ancestor of these three. Broom rejected any notions of convergence in the hind limb, especially the tibia and tarsus—the resemblance is too good. Broom considered the avian-theropod "tibiotarsus" to have evolved in response to weight-bearing (exemplified in the larger theropods). On the other hand, he noted that the ancestor of flying birds must have been very small. Thus, although not explicitly stating it, Broom implied a process of miniaturization in the origin of birds.

The opening decades of this century witnessed a flurry of papers that focused mostly on the origin of flight and almost ignored the phylogenetic question. Pycraft (1906) was one among many to follow Marsh's (1880) scenario for the arboreal origin of avian flight and was the first to reconstruct a hypothetical "Pro-Aves." Pycraft considered the ancestor of birds to be a quadrupedal lizardlike reptile that leaped and parachuted from tree to tree. Similar views were espoused by C. W. Beebe, who in 1915 proposed an arboreal "Tetrapteryx stage in the ancestry of birds" in which feathers projected from both forelimbs and hind limbs, the latter forming a "pelvic wing." As with many authors of this period, Beebe's views on the precise phylogenetic position of birds were unclear, and he regarded "the ancestral stems of alligators, *Dinosaurs*, and birds [as] gradually approaching each other until somewhere, at some time, they were united in a common stock" (Beebe, 1906:8).

Franz Nopsca (1907:234) dissented from these views (see Weishampel and Reif, 1984, for further analysis) and, like Williston (1879), argued instead for a terrestrial origin of flight *"from bipedal long-tailed cursorial reptiles which during running oared along in the air by flapping their free anterior extremities"* (italics in original). Nopsca found no similarities between birds and arboreal mammals such as squirrels and primates but noted great similarity between birds and the obviously ground-dwelling dinosaurs. Nopsca (1923) refuted the evidence of Pycraft, Beebe, and others who argued for an arboreal proavis and reasserted the resemblances of birds and especially theropod dinosaurs. Contrary to Pycraft (1906), Nopsca (1907, 1923) regarded the avian and dinosaurian "cannon bone" clearly as a terrestrial adaptation and cited resemblances with hopping animals like dipodid rodents. However, Nopsca never offered a specific hypothesis of avian relationships; birds were descendants of "Dinosaur-like Reptiles." In his 1929 paper, Nopsca, like Huxley and Osborn, suggested a common ancestry of birds and dinosaurs in the Permian; he agreed with Osborn (1900), without citing him, that the common ancestor presumably resembled something like the prolacertiform *Proterosaurus.*

Nopsca's unwavering support for the cursorial origin of birds and flight was not fully accepted by his contemporaries. Othenio Abel (1911) agreed that birds and theropods possessed important similarities and was among those who argued for a common ancestry. For Abel, however, this common ancestor was an arboreal animal. The avian descendants of this ancestor simply remained in the trees, whereas the theropods became secondarily terrestrial. Some workers sought a compromise and considered the ancestors of birds to be scansorial or partly arboreal. O. P. Hay (1910:22–23), for example, suggested that the proaves were "accustomed to clamber[ing] about over rocks and shrubs and the limbs of trees . . . running or making leaps to catch their prey or to escape capture by their enemies." Hay considered birds to be descended from dinosaurs and even identified the specific group that served as avian ancestors. He regarded ornithischians as being derived from theropods, and sauropods as being an older group. Because, Hay surmised, birds arose earlier than the theropod-ornithischian line, "the sauropods are nearest the stock from which sprang the birds, and it is in their skeletons that we must seek for the primitive common characters" (Hay, 1910:23). W. K. Gregory (1916) also believed that the ancestors of birds were both arboreal and terrestrial but lamented (p. 37) that "the immediate ancestry of the birds is regrettably indecisive." Gregory favored origin of birds from somewhere within the nexus of early dinosaurs and the recently discovered pseudosuchians.

In fact, the description of the Triassic "pseudosuchian thecodont" *Euparkeria* by Robert Broom in 1913 had such an impact that *Euparkeria* deserves to join *Archaeopteryx* and *Deinonychus* as among the most influential fossil evidence in the debate on avian origins. Broom (1913) included within his Pseudosuchia a number of early archosaurs in addition to *Euparkeria,* such as *Ornithosuchus* and *Scleromochlus.* Broom considered the

Pseudosuchia to either "have affinities with" or be ancestral to Theropoda, Pterosauria, and birds. He (with many later workers) regarded these early archosaurs as being simply "primitive enough" to be ancestral to the later forms. Furthermore, the pseudosuchian origin of birds helped answer objections that dinosaurs or pterosaurs were too specialized to serve as avian ancestors. Broom considered this perfectly consonant with his 1906 paper in which he argued "that the bird had come from a group immediately ancestral to the Theropodous Dinosaurs" and the "Pseudosuchia . . . proves to be just such a group" (Broom, 1913:631). Thus, Broom's concept of the pseudosuchian origin of birds is really just a more specific version of Huxley's common ancestry hypothesis. Broom's (1913) characterization of Pseudosuchia marks an important shift of emphasis in the debate away from dinosaurs. The shift often was subtle. For example, if two of Abel's works—one before (1911) and one after (1920) Broom's 1913 paper—are examined closely, one notes that Abel's position had *not* changed substantially; birds and theropod dinosaurs share a common ancestor. But in the later work, the tone of the comparisons with dinosaurs has softened. Whereas earlier, Abel (1911) wrote about the "homology" of bird and theropod hands, he later (1920:389) said that "the hand structure [of theropods] greatly *reminds* one" (italics added) of birds.

Broom's treatment of pseudosuchians had a strong influence on Gerhard Heilmann's discussion of avian ancestry in the 1927 classic *The Origin of Birds*. Apparently it is not well known that Heilmann published a series of long papers (e.g., Heilmann, 1916) in Danish on the same subject. These papers include the major features of the 1927 book and a number of figures and analyses that were omitted from the book (most notably, Cartesian transformations of the skull, limbs, and pelves of pseudosuchians into those of his proavis; Figs. 4–6). Abel (1920) drew heavily from Heilmann's (1916) paper. Nevertheless, Heilmann's book in English (1927) is rightly credited with restructuring the debate. Heilmann was well versed in the anatomy and embryology of modern amniotes, and much of the book is spent with documenting (p. 138) "with absolute certainty . . . that the birds have descended from the reptiles" and have no close phylogenetic relationship with mammals. He also was familiar with the morphology of the important fossil taxa, and Part IV, entitled "The Proavian," is devoted to determining which group is closest to birds. Heilmann assessed pterosaurs, ornithischians, coelurosaurs, and pseudosuchians, in turn, and finally reconstructed in detail his version of the hypothetical proavis. He found that the similarities between birds and pterosaurs generally were superficial. Furthermore, he stated (p. 141) that "the shoulder-girdle [of pterosaurs] has no clavicle, and so the birds cannot possibly descend from these reptiles." This quotation illustrates a guiding principle of Heilmann's book—i.e., Dollo's law of

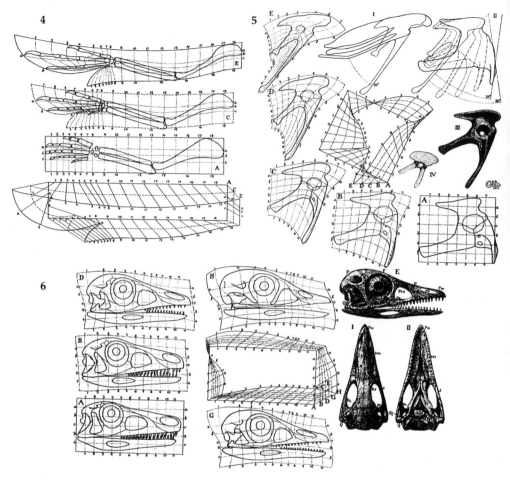

Figs. 4–6. D'Arcy Thompson grids (Cartesian transformations) depicting changes occurring in the evolution of birds from primitive archosaurs (from Heilmann, 1916). (4) Transformations in the forelimb of *Ornithosuchus* (A), Heilmann's hypothetical proavis (C), and *Archaeopteryx* (E). (5) Transformation in the pelvis of *Euparkeria* (A), intermediate forms (B–D), and *Archaeopteryx* (E); I depicts changes occurring from *Archaeopteryx* to *Hesperornis*; II depicts changes occurring from *Euparkeria* to *Archaeopteryx*; III is the hypothetical proavis; and IV is an embryonic gull. (6) Transformation in the skull of a generalized early archosaur (A), intermediates (B–D), Heilmann's proavis (E. I–II), a hypothetical early bird (G), and a schematic modern bird (H).

the irreversibility of evolution: that which is lost cannot be regained. Before beginning his analysis of the fossil groups, Heilmann asserted (p. 140) that "when strictly adhering to this law, we shall find that only a single reptile-group can lay claim to being the bird-ancestor."

Heilmann (1927) was not at all impressed with any of the similarities between birds and ornithischian dinosaurs—certainly none of the sim-

ilarities that are not also found in small theropods. Because the ornithischian pelvis previously had received so much attention in the discussion of bird origins, Heilmann analyzed it in detail. He remarked
(p. 148) that "the mere fact that [the pubis] was directed backward, like
that of the birds, has evidently so hypnotized several scientists that they
have overlooked, or tried to set aside, the many conspicuous differences
between the birds and the Predentates [Ornithischia]." Heilmann considered the ornithischian prepubic process of the pubis to be homologous to
the pubis of other animals and the postpubic process to be a neomorph,
but modern interpretation suggests the reverse.

Numerous workers have noted that Heilmann (1927) cited a large
number of detailed similarities throughout the skeletons of birds and
small theropods such as *Procompsognathus*, *Compsognathus*, *Ornitholestes*,
and ornithomimids. For instance, with respect to the "metatarsals and
toes," he remarked that "the resemblance is so close that we should take
them to be two species within the same genus, and not the representatives of two different classes" (Heilmann, 1927:176). From his analyses he
concluded that (p. 183) "it would seem a rather obvious conclusion that it
is amongst the Coelurosaurs that we are to look for the bird-ancestor."
However, because "the clavicles are wanting" in coelurosaurs, his unflagging adherence to Dollo's law forced him to accept that "these sau-
rians could not possibly be the ancestors of the birds" (p. 183). Nevertheless, Heilmann suggested that the ancestor of birds was "closely akin to
the Coelurosaurs" but "wholly without the shortcomings" (p. 185). Thus,
although many modern workers have interpreted Heilmann (1927) as
rejecting any relationship with theropods, he actually argued for something closer to the common-ancestry hypothesis of earlier workers such
as Broom (1913).

Like Broom (1913), Heilmann (1927) regarded pseudosuchian thecodonts as the group without these "shortcomings." Heilmann reiterated
his adherence to Dollo's law and emphasized the *presence* of clavicles in
these primitive archosaurs. He also discussed the great similarity between the skulls of pseudosuchians such as *Euparkeria*, *Aetosaurus*, and
Archaeopteryx. It should be noted, however, that his reconstruction of the
Berlin *Archaeopteryx* (Fig. 7C) seems to be based largely on imagination.
(The Berlin skull is poorly preserved.) Moreover, the details bear a sus-
picious resemblance to *Euparkeria* and *Aetosaurus*, which would make any
phylogenetic inferences rather circular.

Heilmann (1927) cited other resemblances of pseudosuchians and
birds. For instance, the pubis "excites our interest" because it is twisted
and "looks as if it were about turning backwards, thus exactly fulfilling
our expectations as to the pubis of a bird-ancestor" (p. 189). Reasserting
his belief in the closeness of small theropods to avian ancestry, he envi-

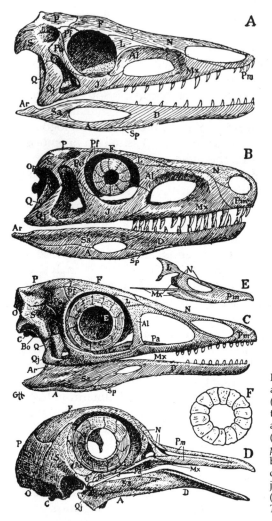

Fig. 7. Perhaps the most reproduced and redrawn figure from Heilmann's (1927) *Origin of Birds* (reprinted courtesy of Dover Publications). (A) Triassic archosaur, *Aetosaurus ferratus.* (B) Triassic archosaur, *Euparkeria capensis.* (C) *Archaeopteryx lithographica,* based on the Berlin specimen. (D) recent pigeon, *Columba livia.* (E) upper jaw of a juvenile duck, *Aythya ferina.* (F) sclerotic ring of a thalattosaur, *Thalattosaurus alexandrae.*

sioned (p. 189) the forelimb of pseudosuchians "passing [through] transitional stages, probably resembling the corresponding forms in the Coelurosaurs." Although Heilmann was forced to reject theropod ancestry because of Dollo's law, he (p. 191) accepted pseudosuchians as avian ancestors simply because "nothing in their structure militates against the view that one of them might have been the ancestor of the birds." He regarded the origin of birds to have proceeded first through a terrestrial, cursorial stage in which the hind limb evolved its avian characteristics (fusion and elongation of metatarsals, and the like) and then an arboreal, climbing stage in which the flight-related forelimb characters appeared.

Interestingly, Heilmann agreed with Abel (1911) that some theropods (e.g., the ancestor of *Ornitholestes* and ornithomimids) also went through an arboreal stage that produced their rather avian tridactyl hands.

Heilmann's (1927) *Origin of Birds* was written authoritatively, well illustrated, and well referenced. Rather than stimulating research on avian origins, it nearly halted further study. The pseudosuchian ancestry of birds (and of virtually all later archosaur groups) became dogma and found its way into the popular literature as well as scientific texts. For example, Heilmann's (1927) figures and restorations (such as Fig. 7) must rank among the most reproduced drawings in the history of paleontological illustration. The pseudosuchian hypothesis was (and remains) a nonspecific hypothesis—birds came from pseudosuchians, but precisely which pseudosuchians is unknown.

Although there was broad acceptance of the pseudosuchian hypothesis, it was not universal. Nopsca (1929), for example, published a response to Heilmann's book reasserting the origin of birds from "dinosaur-like" forms and, more vigorously, the cursorial origin of avian flight. He was strongly opposed to the arboreal proavis and reiterated the cursorial features of birds. There was no "clavicle problem" for Nopsca, because he regarded the avian furcula as a neomorphic ossification; thus, there was no violation of Dollo's law.

J. E. V. Boas (1930) also disagreed with the conclusions of Heilmann (whom he regarded as an "amateur ornithologist") and proposed an interesting alternative. He did not even acknowledge the pseudosuchian hypothesis but reproduced some of Heilmann's figures. Boas's analysis comprised the pelvis of birds and ornithischians and the manus and pes of birds and small theropods. With respect to the similarities in the pelvis, he considered it "unthinkable that such a completely congruous and complex specialization should have taken place twice in the course of phylogenetic history" (Boas, 1930:224). Later, however, after discussing the hands and feet, he remarked (p. 244) that "birds are derived from forms that are quite closely related to compsognathids [in which he included *Compsognathus*, *Ornitholestes*, and ornithomimids], perhaps even were members of this family." Boas resolved this conflict by noting that some ornithischians, such as ceratopsians, lost the postpubic process of the pubis, resulting in a saurischian-like pelvis. He suggested that this was also the case for his "compsognathids" which, by implication, became derived ornithischians. Thus, the ancestor of birds had an ornithischian pelvis and the extremities of a small theropod. Boas's hypothesis seems to have attracted little support from his contemporaries.

Branislav Petronievics published a series of papers (mostly in the 1920's and summarized in Petronievics, 1950) on the morphology of *Archaeopteryx*. He considered the ancestor of birds to be among a "primitive group of lacertilians," but he never made detailed comparisons of birds and squamates or formulated any cogent arguments on avian ancestry. Like many others, Petronievics attributed all similarities between birds, pterosaurs, and dinosaurs to convergence.

Whereas the ideas of Boas and Petronievics seem to be little more than "aberrations" in the post-Heilmann period, those of Percy Roycroft Lowe stimulated an intense debate. Lowe was cited widely as being among the few in the period between Heilmann and Ostrom to argue for the dinosaurian relationships of birds. Lowe's ideas, however, are complex, often contradictory, and were not offered as explicit statements on avian origins. Lowe was interested principally in ratites, and all his comments on avian origins stem from his views on the relationships of ratites to reptiles and other birds. His main thesis, first stated in 1928, was that ratites are *not* secondarily flightless birds that evolved from volant birds via degeneration, but rather that they represent an ancient group that branched off from the avian clade before the evolution of flight. In 1928, Lowe cited Broom and Heilmann and considered (p. 210) ratites to be "direct line relics of an ancient avifauna which marked an early stage in the evolution of the bird from some Proto-Pseudosuchian ancestor." His comparisons of ratites with theropods were not intended to suggest monophyly of the two groups but only analogous patterns of morphological evolution. For example, the reduction of the forelimb in ratites culminating in kiwis is analogous to the same reduction seen in theropods, which culminated in the tyrannosaur condition.

Lowe's 1935 paper is cited most often as indicating his views on the monophyly of birds and dinosaurs. However, he reiterated (p. 400) his belief "in fixing upon the Eosuchia [early diapsids] as, at any rate, forming a provisional ancestral base from which sprang both dinosaurian reptiles and birds" (Fig. 8). Thus, similar features arose independently in birds and dinosaurs, not as a result of similar adaptation, but instead, because there "were resident potential genetic factors" in their eosuchian ancestor which allowed avian or reptilian characters "to crop up almost indiscriminately in any descendent branch" (Lowe, 1935:402). A necessary corollary to Lowe's ratite hypothesis was that *Archaeopteryx* was a flying dinosaur and not a bird at all. He was willing to accept (p. 409) "the diphyletic origin of feathers—a zoological transgression for which I expect no mercy." Lowe (1944) expanded on the dinosaurian nature of *Archaeopteryx* and predicted the discovery of feathers on *Ornitholestes*.

Fig. 8. Phylogeny of birds and most other archosaurs from Lowe, 1935 ("On the relationship of the Struthiones to the dinosaurs and to the rest of the avian class," *Ibis*, courtesy of Blackwell Scientific Publications). Ratites are close to coelurosaurs on the diagram, but their common ancestor lived in the early Triassic and is neither a bird nor a dinosaur.

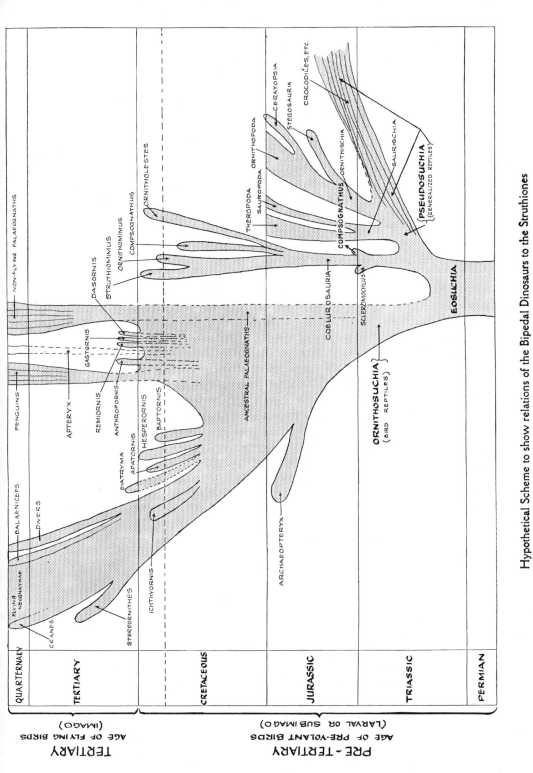

Hypothetical Scheme to show relations of the Bipedal Dinosaurs to the Struthiones

445

The 1944 paper is contradictory with regard to avian origins. At times, Lowe advocated the parallel evolution hypothesis of his earlier papers, but once (p. 522), he suggested a direct ancestor-descendant relationship. When examining Lowe's concept of avian origins, one must remember that his main focus was on ratites. Whereas earlier he had advocated the then-favored pseudosuchian hypothesis, the logical extension of his views on ratites forced him to accept several controversial propositions (e.g., *Archaeopteryx* was not a bird; feathers evolved twice). In summary, it clearly is inappropriate simply to list Lowe with those advocating the origin of birds from theropod dinosaurs—it is both inaccurate and obscures the interesting and idiosyncratic nature of his arguments.

Lowe's conclusions were neither ignored nor accepted. Gregory (1935) and Tucker (1938) offered rebuttals to Lowe, but they were concerned more with ratites than with avian origins. Simpson (1946) also responded to Lowe, asserting views that were to become dogma. According to Simpson, any similarities of birds and theropods are convergent. *Archaeopteryx* was a bird. Flight originated in the trees, and birds evolved from generalized pseudosuchian thecodonts. De Beer (1954, 1956) also refuted the arguments of Lowe, reasoning that *Archaeopteryx* is intermediate between reptiles and higher birds, and that ratites had ancestors that flew.

Lowe's ideas were not universally rejected, however. Nils Holmgren's work on avian phylogeny, published posthumously in 1955, supported some of Lowe's conclusions. For instance, Holmgren (1955) considered the ancestor of ratites to have been flightless. Whereas Lowe (usually) regarded theropods and birds as having taken parallel evolutionary courses, Holmgren posited a direct ancestor-descendant relationship. He answered Heilmann's (1927) complaint of the absence of clavicles in theropods by suggesting that they may have been present and (p. 306) "were either cartilaginous, membranous or merely present as rudiments in embryos." If this were the case, then clavicles were obtained owing to "reappearance or rejuvenation in the birds, after a long period of apparent absence in the Coelurosaurs" (p. 322). Whereas Lowe postulated the diphyletic origin of feathers, Holmgren suggested a monophyletic origin, proposing (p. 307) "that the ancestors of the birds were down-clad reptiles, belonging to a reptile stem issuing from Coelurosaurs." "[In fact] when comparing the hand of *Ornitholestes* with that of *Archaeornis* [= *Archaeopteryx*] it is difficult to avoid the thought that it was the developing wing feathers that caused the lengthening of the fingers in *Ornitholestes* and that in the Coelurosaurs there were the makings of an avian wing" (Holmgren, 1955:309). Regarding flight, Holmgren considered the "proavis" to be divisible into two parts—a cursorial "pro-ratite" and an arboreal "pro-carinate." Both proaves were coelurosaurs, but the procarinate had better developed feathers.

Holmgren's (1955) work generated virtually no controversy; it was almost as though the paper never had appeared. All the reviews that immediately followed (e.g., de Beer, 1956; Swinton, 1958, 1960; Romer, 1966; Bock, 1969; Brodkorb, 1971) do little more regarding avian origins than recite Heilmann's conclusions about *Euparkeria*, *Ornithosuchus*, and the pseudosuchian origin of birds. None of these authors even cites, let alone refutes, Holmgren (1955). The pseudosuchian origin of birds was the unquestioned dogma heading into the 1970's. As Bock (1969:148) noted, "acceptance of [pseudosuchians as avian ancestors] is more by default than by direct demonstration."

AVIAN ORIGINS FROM 1970 TO THE PRESENT

Beginning in the 1970's and continuing to the present, there has been renewed interest in the *specific* position of birds within vertebrate phylogeny. Primary reasons for this recent attention include the discovery of new birdlike archosaurs and a much closer examination of details of archosaur anatomy. Furthermore, the growing popularity of Hennigian systematic methodology (cladistics) has shifted the focus to obtaining more specific hypotheses of relationships and has introduced new organisms and new characters to the debate. As this is relatively recent history and most of these views are well known, I do not examine these efforts in as much detail but focus on the diversity of opinion.

Peter Galton (1970) initiated this period of activity by resurrecting the hypothesis that had been advocated earlier by Baur (1884b, 1887) and Boas (1930)—viz., the ornithischian relationships of birds. Galton's argument rested almost completely on the shared possession of a retroverted pubis and did not cite or discuss the other resemblances noted by Huxley, Baur, or Boas. He regarded ornithischians as being too specialized to serve as the ancestral stock and postulated a mid-Triassic common ancestor that was a cursorial, opisthopubic, bipedal dinosaur with none of the other ornithischian characters. Galton's formulation of the hypothesis was no more successful than Baur's or Boas's in attracting support. In fact, he later abandoned the hypothesis in favor of Ostrom's theropod-bird hypothesis (Bakker and Galton, 1974).

A new view on avian origins appeared when Alick Walker (1972, 1974) suggested that crocodylomorphs and birds constitute a monophyletic group (Fig. 9). Much of his evidence was based on the Triassic "sphenosuchian" *Sphenosuchus*, which he viewed as intermediate in many respects between birds and crocodilians. His evidence included aspects of cranial pneumaticity, cranial kinesis, palatal structure, inner-ear morphology, and the articulations of the quadrate. Walker (1977:320) suggested that the common ancestor of birds and crocodylomorphs was at "a

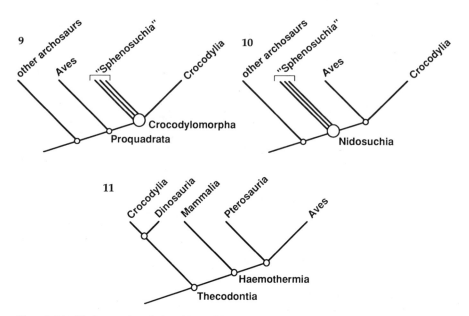

Figs. 9–11. Phylogenetic relationships of birds and various other Amniota based on the analyses of (9) Walker (1972, 1977); (10) Whetstone and Whybrow (1983) and Martin (1983a,b); taxon names are from Whetstone and Whybrow (1983); and (11) Gardiner (1982). In (9) and (10), "Sphenosuchia" is shown as paraphyletic on the basis of the analysis of Benton and Clark (1988).

higher level of organization than that of the Thecodontia" and erected "a new Class, the Proquadrata, to include the Sphenosuchia, Crocodilia and Aves as subclasses."

Support for a sister-group relationship of birds and crocodilians came from K. N. Whetstone and L. D. Martin (1979, 1981), who accepted Walker's (1972) characterization of the braincase articulations of the quadrate. They further proposed that the presence of a fenestra pseudorotundum (a release window associated with the recessus scala tympani) and tympanic pneumaticity represented synapomorphies of birds and crocodilians. Martin et al. (1980) offered another character complex as evidence of crocodilian relationships—i.e., tooth morphology and replacement. The evidence can be summarized as four synapomorphic character complexes: (1) a bipartite articulation of quadrate to braincase, (2) the fenestra pseudorotundum, (3) tympanic pneumaticity, and (4) dental features (Martin, 1983a,b, this volume; Whetstone, 1983; Whetstone and Whybrow, 1983).

Although it is usually referred to as the "crocodilian hypothesis," the position of Walker, Martin, and Whetstone more properly should be called the "crocodylomorph hypothesis," because all crocodylomorphs (i.e., "sphenosuchians" and Crocodylia) are under consideration. These three workers accepted the monophyly of Sphenosuchia, but the recent analysis of Benton and Clark (1988) showed this assemblage to be paraphyletic. Close examination shows that Martin (1983a,b) and Whetstone and Whybrow (1983) proposed a somewhat different hypothesis than did Walker (1972, 1977). Walker considered birds to be the sister-group of crocodylomorphs as a whole (Fig. 9). Martin and Whetstone, on the other hand, regarded birds as the sister-group of crocodilians, with "sphenosuchians" being the sister-group of these two (Fig. 10). Thus, given a cladistic classification, Martin and Whetstone implied that birds are crocodylomorphs.

Walker's (1972) formulation did not initiate an immediate response. Ostrom's (1973) ideas on the coelurosaurian origin of birds, appearing less than a year after Walker's (1972) paper, were not presented as a refutation of Walker but rather as supporting an alternate hypothesis. In fact, most workers (e.g., Bakker and Galton, 1974; Thulborn, 1975) considered Ostrom's analysis so compelling that they dismissed the evidence for crocodylomorph relationships without comment.

It was only after Martin and his students made their contributions that the crocodylomorph hypothesis was evaluated critically. Tarsitano and Hecht (1980) critically examined the crocodylomorph hypothesis and stated objections that have been echoed by many later workers (e.g., Thulborn and Hamley, 1982; Hecht, 1985). They summarized Walker's data and correctly concluded that many of Walker's characters are either primitive or spurious. Tarsitano and Hecht (1980) rejected Whetstone and Martin's (1979) character of the fenestra pseudorotundum but did not discuss the pneumatic characters. McGowan and Baker (1981) questioned the pneumatic synapomorphies because air spaces are found in the skulls of theropods such as *Troodon* (= *Saurornithoides*). Molnar (1985) and Currie (1985) reported cranial pneumaticity in other theropods and also suggested that such evidence diminishes the crocodylomorph hypothesis. Indeed, the presence of cranial air sacs seems to be almost universal in archosaurs (Witmer, 1987). Crocodylomorphs, some theropods, birds, and even pterosaurs possess certain pneumatic features. Clearly, there has been much homoplasy. Early claims (e.g., Whetstone and Martin, 1981; Whetstone and Whybrow, 1983; Witmer, 1984) that possession of tympanic pneumaticity in birds and crocodylomorphs represents strong evidence for monophyly were premature. Archosaurian cranial pneumaticity is only beginning to be understood (Witmer, 1990), and the homology of many sinuses has yet to be demonstrated. Until these homologies are determined, one should remain cautious about any phylogenetic conclusions.

Padian (1982) and Gauthier and Padian (1985) also rejected the crocodylomorph hypothesis, but more on the grounds of parsimony than on the evidence. Put simply, the hypothesis that birds are theropods is better corroborated; thus, any similarities of birds and crocodylomorphs are deduced to be convergences. Gauthier (1986) evaluated the characters offered in support of the crocodylomorph hypothesis and presented the first significant challenge to the dental characters proposed by Martin et al. (1980) and Martin (1983a,b), noting that the absence of these apomorphies in otherwise derived crocodylomorphs suggests that birds and crocodilians acquired the dental features independently.

Perhaps the biggest blow for the crocodylomorph hypothesis came when Walker, the founder of the hypothesis, announced that "the original concept of a particularly close relationship between birds and crocodiles has become so tenuous that it is very difficult to sustain" (Walker, 1985:133). Further study of *Sphenosuchus* and the aetosaur *Stagonolepis* led Walker (1985) to believe that birds and crocodilians had acquired their apomorphic quadrate-braincase articulations independently and in different ways. Because the position of the quadrate head greatly constrains otic morphology, Walker reasoned that the other otic apomorphies also may have arisen convergently. The recognition that the skull of *Sphenosuchus* was not kinetic may have led Walker to recant, because cranial kinesis was such an important part of his earlier argument. Although other advocates of the crocodylomorph hypothesis (e.g., Martin, 1983a,b) always had regarded *Sphenosuchus* as akinetic and agreed with much of the criticism leveled at Walker, the rhetorical value of the abandonment of an idea by its initial proponent is seldom missed in a hot debate, and this is no exception (see, e.g., McGowan, 1985; Gauthier, 1986; Paul, 1988b; Gauthier and Padian, 1989).

Nevertheless, the crocodylomorph hypothesis still is regarded as tenable by Martin (this volume) and, in a modified form, seems to have attracted the support of Tarsitano (1985b, this volume). Although Tarsitano maintained that the immediate ancestor of birds was a thecodont, it seems that he now views that avian ancestor as also being close to crocodylomorphs. Tarsitano (1985b, this volume) did not necessarily exclude theropods from the crocodylomorph-bird clade and suggested that the presence of a vertical basicranium and a basisphenoid sinus in carnosaurs and crocodilians "may indicate their common ancestry in the early Triassic or Upper Permian" (Tarsitano, 1985b:40). Tarsitano (this volume) reiterates his conviction that birds, crocodylomorphs, theropods, and at least some thecodonts (specifically *Postosuchus*) may constitute a monophyletic group, with Aves apparently being the most primitive taxon (because birds plesiomorphically retain a flat basicranium). Unfortunately, Tarsitano has not discussed explicitly the evolution of his characters throughout archosaurs, and it is unclear if he accepts the monophyly of Pseudosuchia, Theropoda, Saurischia, or Dinosauria.

In many respects, the crocodylomorph hypothesis can be considered a specific version of the pseudosuchian thecodont hypothesis. As mentioned, Tarsitano and Hecht (1980) discussed and rejected both theropod and crocodylomorph relationships and offered a "new version" of Heilmann's pseudosuchian hypothesis. They partitioned thecodonts on the basis of the possession of armor and ankle type, and they concluded that birds originated from unarmored mesotarsal thecodonts, a group that also was ancestral to dinosaurs. Their cladogram placed the origin of birds somewhere between *Euparkeria* and *Lagosuchus* (a derived mesotarsal form). They were not explicit as to which taxa are included in their Pseudosuchia and which pseudosuchian was closest to birds. The new pseudosuchian hypothesis is similar to Heilmann's (1927) in that it centers on forms that are deemed "primitive enough," offers no synapomorphies of birds and known thecodonts, and ascribes all resemblances with theropods to homoplasy. It differs in that it predicts that some crocodylomorph apomorphies will be found in the pseudosuchian ancestor of birds.

In support of the pseudosuchian hypothesis, Tarsitano (1985a, this volume) has presented a novel solution to the problem of avian ancestry. Put simply: use aerodynamic principles to determine the most likely model for the origin of flight, create a hypothetical proavis, and then search Archosauria for a match. Hecht and Tarsitano (1982), Martin (1983b), Hecht (1985), Tarsitano (1985a, this volume) viewed the ancestor of birds as small, arboreal, and quadrupedal, as their model for the arboreal origin of flight dictates. Tarsitano (1985a, this volume) therefore has rejected the theropod hypothesis because known theropods, being too large and demonstrably terrestrial, fail the tests predicted by the arboreal theory. Instead, all of these authors point to small Triassic forms such as *Cosesaurus, Longisquama,* and *Megalancosaurus* as being the sort of animals they envision in the arboreal model and that may be close to avian ancestry. Unfortunately, these taxa are poorly known, and some may not even be archosaurs. *Cosesaurus* recently was reclassified as a prolacertiform (Sanz and Lopez-Martinez, 1984). Another problem is the assumption that theoretical analysis of the origin of flight will answer the phylogenetic question. If a cursorial, terrestrial aerodynamic model were accepted, would these authors deny that a "thecodont" could be the terrestrial cursor and then throw their support to the theropod hypothesis? Conversely, what about arboreal theropods? Abel (1911) viewed theropods as originating in the trees. More recently, Paul (1988b) advocated both an arboreal origin of flight and a theropod origin of birds; he even illustrated a feathered *Ornitholestes* clambering about the branches.

Criticism of the pseudosuchian hypothesis came from supporters of theropod relationships. Ostrom (1976) was faced with the Heilmann version of the pseudosuchian hypothesis, which had been dogma for decades. Ostrom found that pseudosuchians indeed were "primitive enough" to be avian ancestors but that no pseudosuchians displayed any unique avian apomorphies. Furthermore, Ostrom noted that the absence of clavicles in theropods (which had forced Heilmann [1927] to accept the pseudosuchian over the theropod hypothesis) was more apparent than real, in that several theropod specimens were preserved with clavicles. Thulborn and Hamley (1982), responding to the Tarsitano and Hecht (1980) version, criticized the latter for their "process-of-elimination" approach and lack of synapomorphies; they discounted the sole synapomorphy of *Archaeopteryx* and *Euparkeria* postulated by Tarsitano and Hecht (1980). Thulborn and Hamley (1982) denied that the new version was more informative than Heilmann's version, noting that *Euparkeria* is among the most primitive and *Lagosuchus* among the most derived thecodonts and that suggesting that birds originated somewhere between the two is not saying much.

Perhaps the sharpest criticism came from Gauthier and Padian (1985, 1989) and Gauthier (1986), who argued that because Thecodontia is a paraphyletic taxon (and thus has the same diagnosis as Archosauria), it is meaningless to state that birds arose from them; such a statement implies only that birds are archosaurs. Although strictly true in a cladistic sense, saying that birds arose from a thecodont states not only that birds are archosaurs but also that they are *not* pterosaurs, crocodylomorphs, or dinosaurs. I agree with Gauthier and Padian, however, that a paraphyletic grade comprising early archosaurs (i.e., Thecodontia) should be abandoned, especially as our knowledge of the diversity and relationships of archosaurs improves. Gauthier and Padian (1985) also disapproved of the traditional composition of Pseudosuchia, which included most of the early archosaurs. Pseudosuchia indeed has become a "taxonomic wastebasket." They provided a new composition, resulting in a monophyletic Pseudosuchia. Unfortunately, this new combination excludes some traditional "pseudosuchians," such as *Euparkeria* and *Ornithosuchus* and, more importantly, includes Crocodylia, which are "true" suchians (= crocodile). It would be better to abandon the name Pseudosuchia altogether, or at least restrict it to meanings closer to its original usage, as advocated by Benton and Clark (1988). Similarly, we should abandon the "pseudosuchian hypothesis" of avian origins. That birds arose from unknown early archosaurs remains a valid hypothesis that could be falsified by demonstration of relationships with a known group; however, it is uninformative and does not suggest specific areas of research.

Perhaps the most surprising development in the debate on the origin of birds came when Brian Gardiner (1982) denied that birds were archosaurs and gave a cladistic justification for Owen's Haemothermia (= Mammalia + Aves) (Fig. 11). Most of Gardiner's data came from soft anatomy, although he did consider a few fossil groups. Pterosaurs were placed as the sister-group of birds. Dinosaurs were united with crocodiles as Archosauria. Archosauria was the sister-group of Haemothermia. Løvtrup (1985) offered his support and contributed a few additional characters. Aside from Janvier (1983), few workers have cited Gardiner, let alone evaluated his data. Gauthier (1986) praised Gardiner for his cladistic methodology but faulted him for his grasp of the morphology and literature. Gauthier et al. (1988) provided a convincing refutation of Gardiner (1982), demonstrating that a consideration of all the evidence (both soft and hard parts) and all the taxa (both living and fossil) embeds birds firmly within a more traditionally construed Archosauria.

Whereas only a handful of recent workers have favored mammals, crocodylomorphs, or "thecodonts" as the sister-group or ancestors of birds, the majority of researchers believe that birds are closely related to theropod dinosaurs. The importance of John Ostrom's efforts in shaping the current debate on avian origins cannot be overstated. In 1973, he laid out the basic plan of his argument in a one-page paper in *Nature*, which he then followed with a series of papers on the origin of birds and of avian flight. Ostrom's 1976 paper in the *Biological Journal of the Linnean Society* rightly can be considered a landmark effort. Ostrom's basic hypothesis is that birds are descended from a specific subset of Theropoda—small coelurosaurian theropod dinosaurs similar to *Ornitholestes* and *Deinonychus*. Ostrom's comparisons almost always made reference to *Archaeopteryx* rather than to recent birds. Ostrom's evidence came principally from the postcranium. (Interestingly, virtually all of the characters of the crocodylomorph hypothesis are cranial.) Although Ostrom found derived similarities throughout the skeleton, the following characters are perhaps the most important to the debate—(1) the phalangeal formula and proportions, and the pattern of digital reduction in the manus; (2) a semilunate carpal; (3) elongation of the forelimb and especially the manus; (4) a rodlike pubis with a distally expanded "foot"; (5) an ascending process of the astragalus; (6) a mesotarsal ankle joint; (7) the reduction of Metatarsal V and the loss of the connection between Metatarsal I and the tarsus; and (8) a reversed hallux. In a related series of papers (e.g., Ostrom, 1974, 1979, 1986) Ostrom championed the "cursorial predator" theory of the origin of avian flight, arguing that the morphology of *Archaeopteryx* is that of a terrestrial cursor and that Solnhofen paleoecology indicates an absence of trees or nearby cliffs from which gliding could take place.

Ostrom's coelurosaurian hypothesis was accepted almost immediately. Bakker and Galton (1974), Bakker (1975), and Thulborn (1975), among others, seized upon Ostrom's findings, often using the hypothesis as the basis for radical new taxonomies or broad physiological inferences.

Although "not trying to create a bandwagon over Ostrom's papers," Cracraft (1977:492) applauded Ostrom for a "masterful job" and "meticulous work." The initial response was not entirely positive, however, and Hecht (1976) claimed that many of Ostrom's characters were of "low weight" or, reminiscent of the nineteenth-century responses to Huxley's ideas, convergences resulting from shared bipedal habits.

Tarsitano, Hecht, and Martin have been the most vocal opponents of the theropod hypothesis and have leveled attacks at several of the important characters employed by Ostrom. Tarsitano and Hecht (1980) and Hecht and Tarsitano (1982) offered a series of arguments in refutation of theropod relationships. They regarded the manual digital homologies of birds and theropods to be different. They ascribed the similarities in the manual phalangeal formula and the presence of an ascending process of the astragalus in *Archaeopteryx* to preservational artifact, but both of these claims have been disproved (Wellnhofer, 1985, and Martin, this volume, respectively). Resemblances in the wrist were viewed as nonhomologous, and much similarity in the shoulder girdle and pubis was denied. Finally, they dismissed the reflexed hallux as a synapomorphy because it is absent in known theropods. Martin et al. (1980) and Martin (1983a) also attacked the character of the ascending process of the astragalus, stating that whereas theropods indeed possess such a process, birds have a separate neomorphic ossification, the pretibial bone. Martin (1983a,b, this volume) has joined Tarsitano and Hecht in questioning the homology of the semilunate carpal.

Although Ostrom (1976) did not present a cladogram, he stated that he tried to use only shared-derived characters when drawing phylogenetic conclusions from his data. Ostrom, however, never named a specific sister-group of birds. Although he noted more similarities between *Archaeopteryx* and dromaeosaurids such as *Deinonychus,* Ostrom (1976:173) concluded that birds descended from "a small unknown *Ornitholestes*-like coelurosaurian dinosaur." Padian (1982:390) provided the first true cladogram of *Archaeopteryx* and other archosaurs in hopes "that the ambiguities of Ostrom's approach can be remedied by more rigorous phylogenetic analysis." In Padian's scheme (Fig. 12), *Archaeopteryx* is the sister-group of deinonychosaurs (Troodontidae + Dromaeosauridae). Although admittedly preliminary and incomplete owing to space restrictions, Padian's (1982) analysis lacks much character analysis and does not include all the relevant taxa. This was rectified in large measure by Gauthier and Padian (1985), who provided cladograms and diagnoses of the archosaurian higher taxa and of saurischian taxa (again with birds as the sister-group of deinonychosaurs). Gauthier (1986) reviewed the data supporting Ostrom's hypothesis and provided detailed character analysis within a phylogenetic framework.

Padian (1982, 1985) and Gauthier and Padian (1985) based their model

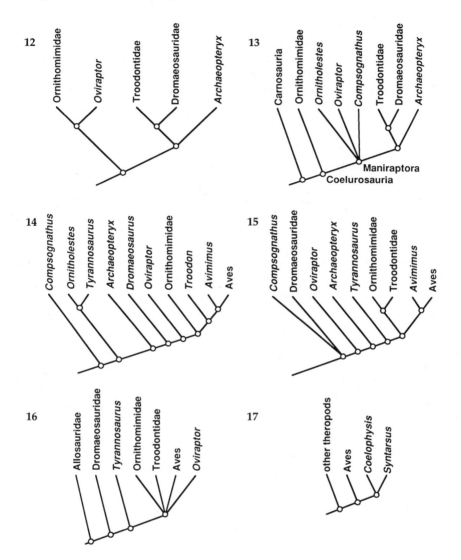

Figs. 12–17. Phylogenetic relationships of birds and various other theropod dinosaurs on the basis of the analyses of (12) Padian (1982), (13) Gauthier (1986), (14) Paul (1988b), (15) Thulborn (1984), (16) Bakker et al. (1988), and (17) Raath (1985). In most cases, these authors included other taxa in their cladograms; I have included only the genera or major higher taxa discussed in the text.

for the origin of avian flight on their phylogenetic conclusions. Like Ostrom, they accepted the possibility of an arboreal origin but favored the cursorial predator hypothesis because the ancestors of birds were terrestrial, obligate bipeds that used their elongate forelimbs in prey manipulation. In fact, they argued that *Archaeopteryx* exhibits no forelimb modifications beyond those seen in deinonychosaurs and that the latter already had the basic down-and-forward forelimb motion that became perfected into the avian flight stroke. They recognized that the arboreal theory may be more intuitively satisfying but argued (as did Ostrom) that their model better accounts for the data and makes fewer appeals to hypothetical intermediates. See Martin (this volume) and Tarsitano (this volume) for views counter to this functional interpretation.

Although Ostrom unquestionably deserves most of the credit for the wide acceptance of the theropod hypothesis, the importance of Gauthier's extensive phylogenetic analysis also must be recognized; it is more than just a formalization of Ostrom's data. As mentioned, Padian (1982) and Gauthier and Padian (1985) published cladograms that embedded birds firmly within Theropoda; Padian's (1982) analysis employed 49 synapomorphies, whereas Gauthier and Padian (1985) reported more than 120. However, Gauthier's 1986 revision of the Saurischia (including birds) is the first paper that provides adequate character analysis (84 characters), the inclusion of many (17) taxa, and a character-taxon matrix. He also justified the broader phylogenetic context of his analysis by including diagnoses and discussions of other archosaurian taxa. Rather than discuss all the characters, I review a few of the more important aspects of Gauthier's (1986) analysis (Fig. 13). Whereas Ostrom built his hypothesis primarily on postcranial characters, Gauthier added 13 cranial synapomorphies (e.g., subsidiary antorbital fenestrae, antorbital tooth row), although a few are not present in birds because they are so specialized. Sixteen characters come from the vertebral column and an additional 14 from the hand and wrist.

As birds are regarded as derived coelurosaurs, Gauthier (1986) focused on relationships within Coelurosauria and not Ceratosauria or Carnosauria (Fig. 13). Unfortunately, he was not able to clarify relationships very much, principally because so many of the taxa are poorly known. Coelurosauria is composed of Ornithomimidae and an unresolved polytomy named Maniraptora. Among maniraptorans, birds and deinonychosaurs are sister-taxa. A problem is that deinonychosaurs may not be monophyletic. Gauthier (1986) united troodontids and dromaeosaurids into the Deinonychosauria on the basis of the shared possession of a raptorial second digit of the foot. Currie (1985), however, presented evidence supporting monophyly of troodontids and ornithomimids on the basis of derived cranial characters also shared with birds. Thulborn (1984) and Bakker et al. (1988) also concluded that troodontids and ornithomimids form a clade independent of dromaeosaurids (Figs. 15–16). Furthermore, Paul (1984, 1988b) and Bakker et al. (1988) argued that troodontids are related more closely to birds than are dromaeosaurids (Fig. 14). Gauthier (1986:47) acknowledged that there is "doubt on the

monophyly of Deinonychosauria." The question arises that if troodontids are indeed related to ornithomimids and birds form a clade with
troodontids, then what becomes of relationships within Coelurosauria?
Another matter complicating Gauthier's (1986) analysis is that he did not
consider the peculiar Mongolian theropod *Avimimus*—a form so birdlike
that Thulborn (1984) and Paul (1988b) regarded it as the sister-taxon of
birds (Figs. 15, 14), and Molnar and Archer (1984) went so far as to
include *Avimimus* within Aves. (Molnar [1985] retreated from this extreme position.)

Despite these problems, Gauthier's (1986) analysis is still the benchmark to which subsequent studies should refer until it is supplanted by a
new, even better corroborated hypothesis. Although polyphyly of the
Deinonychosauria may be possible, it awaits demonstration by phylogenetic analysis of all relevant taxa. Those who disagree with Gauthier's conclusions must (1) discredit Gauthier's entire analysis by demonstrating *numerous* mistakes in character analysis, and (2) propose a
similarly explicit phylogenetic hypothesis incorporating all pertinent taxa
and accounting for Gauthier's characters. It is insufficient to dismiss the
entire analysis by showing problems in a few characters or by proposing
that a single "complex" character indicates different relationships. Likewise, as numerous authors have pointed out, convergence cannot be
invoked in an ad hoc manner but must be deduced from the topology of
the cladogram. At present, supporters of relationships of birds with
either crocodylomorphs, "thecodonts," or mammals have failed to produce a competing cladogram, and in this respect the coelurosaurian
hypothesis is uncontested. Gardiner (1982) came the closest, but Gauthier et al. (1988) successfully refuted mammalian relationships with the
twofold approach advocated above.

Tarsitano (this volume) attempts the first approach by seeking to dis-
count Gauthier's (1986) character analysis of the Maniraptora. Tarsitano
performs the tedious (but necessary) character-by-character evaluation
and objects that many of Gauthier's characters could be described and
illustrated better, a criticism also raised by Martin (1988). Many of Tarsitano's points involve arguments of convergence, and he dismisses
many characters because they are found in pterosaurs or turtles or mammals. Clearly, these comparisons are irrelevant. Even ornithischians are
too distantly related to be useful in character polarization of maniraptorans. Furthermore, he rejects several characters because in these cases
convergence was "easy" or "necessary" for biomechanical reasons. Thus,
in these cases, Tarsitano demonstrates only that convergence is *possible*,
not that it occurred.

Although Ostrom, Padian, and Gauthier argued that birds are related
most closely to deinonychosaurian coelurosaurs, theirs is not the only

hypothesis on the theropod relationships of birds. Ostrom (1976) and Padian (1985) regarded *Archaeopteryx* as being the necessary focus of any discussion of avian origins. This view is not universally held, however. Several workers recently have suggested that *Archaeopteryx* was a feathered coelurosaur that had little to do with the immediate ancestry of birds. As early as 1975, Thulborn (p. 268) referred to *Archaeopteryx* as "little more than a progressive and rather specialized theropod." Thulborn and Hamley (1982) repeatedly questioned the assumption that *Archaeopteryx* was a bird. Finally, Thulborn (1984) presented a cladogram (Fig. 15) in which *Archaeopteryx* is only a moderately derived theropod, and tyrannosaurids, a troodontid-ornithomimid clade, and *Avimimus* are successively closer out-groups of true birds. Similar views were expressed by Paul (1984), whose cladistic analysis concluded that *Archaeopteryx* was an early member of a "protobird" lineage, and that dromaeosaurids, caenagnathids, and troodontids are successively closer to birds (Fig. 14). In fact, Paul (1984:179) went so far as to propose that "dromaeosaurs are probably ground dwelling descendents of the flying archaeopterygids." Paul (1988a) reiterated this claim and expanded his argument in his book on theropods (Paul, 1988b), in which he placed *Avimimus* as the sister-group of true birds.

Avimimus deserves a great deal of attention because it displays a large number of avian features throughout the skeleton. As mentioned, Thulborn (1984) and Paul (1988b) placed *Avimimus* as the sister-taxon of birds (excluding *Archaeopteryx*), and Molnar and Archer (1984) placed it within Aves (including *Archaeopteryx*). Molnar (1985) reevaluated the characters of *Avimimus* and found it more "birdlike" than *Archaeopteryx* but lacking in some of the avian characters (e.g., furcula, opisthopuby) of other theropods. The Soviet paleornithologist Kurochkin (1985) interpreted the significance of *Avimimus* as demonstrating a common "evolutionary potential" among birds and theropods, and supports (in agreement with Paul) the origin of some theropods from *Archaeopteryx* rather than the origin of birds from theropods. Barsbold (1983) also envisioned a great deal of convergence within theropods—a process he termed "ornithization." Barsbold viewed several lineages as becoming birdlike, with *Archaeopteryx* simply representing one of these "ornithized" clades. Kurzanov (1985, 1987), the Soviet paleontologist who first described *Avimimus*, regarded it as feathered (on the basis of an ulnar ridge and elbow kinematics) and perhaps even volant. With respect to avian origins, however, Kurzanov agreed with Kurochkin and Barsbold that convergent evolution of birdlike features within theropods has been widespread, which confounds our efforts to determine the true ancestry of birds. Kurzanov also can be added to the list of those stripping *Archaeopteryx* of its avian status. Kurzanov acknowledged that *Avimimus* may be close to birds but concluded (1985:98) that "the ancestors of the birds are to be found, if not among the theropod dinosaurs, at least within an ancestral group of the theropods."

Kurzanov is not alone among theropod workers in arguing that a Triassic origin of birds is possible. Raath (1985) suggested that *Archaeopteryx* (and all birds) were descended from Triassic "procompsognathid" theropods similar to *Syntarsus* (Fig. 17). Raath (1985) indeed demonstrated a number of similarities between *Syntarsus* and *Archaeopteryx*, but his analysis did not incorporate other taxa, and he was not explicit about character polarities. Chatterjee (1987a) agreed with Raath that these early theropods may be closer than the Cretaceous maniraptorans to avian origins. This hypothesis has the advantage that the theropod sister-group is actually earlier in time than *Archaeopteryx* but must invoke a great deal of convergence. It is yet another hypothesis awaiting a more complete phylogenetic exposition.

CONCLUSIONS

The origin of birds has fascinated scientists since the discovery of *Archaeopteryx* and *Compsognathus*. These fossils helped bridge the gap that confounded nearly all earlier thinkers. The near coincidence of the descriptions of these fossils and the publication of Darwin's *Origin* virtually assured the debate on avian origins a prominent place in the new evolutionary controversy. The list of scientists who published on the topic includes many of the most important paleontologists of all time: Huxley, Owen, Seeley, Haeckel, Gegenbaur, Dollo, Cope, Marsh, Williston, Osborn, Abel, Broom, Nopsca, Gregory, de Beer, and Simpson. Even today, publications on the subject appear regularly. In fact, the number of bird origin papers published since 1970 far outstrips the total number from all previous years.

It is tempting to believe that even without *Archaeopteryx* and *Compsognathus*—armed only with Darwin's theory of common descent—Huxley, Gegenbaur, or Haeckel might have reached similar insights into the origin of birds. But the history of the debate has been characterized by the formulation of new ideas as a result of the discovery of new fossils. The discovery of *Archaeopteryx* and *Compsognathus* resulted in the dinosaurian hypothesis, the discovery of *Euparkeria* produced the pseudosuchian hypothesis, and the discovery of *Deinonychus* was the stimulus for the coelurosaurian hypothesis. Recent finds of small theropods such as *Troodon* and *Avimimus* and primitive archosaurs such as *Longisquama* and *Megalancosaurus* offer prospects of new ideas for the future. Very intriguing are reports of fossil birds from the Triassic of Texas (Chatterjee,

1987b), which, if corroborated, would require a substantial overhaul of our view of archosaurian and avian evolution.

Of course it would be premature to guess whether or not any of these new fossils will be like *Archaeopteryx*, *Euparkeria*, or *Deinonychus* in acting as a catalyst for the development of new views on avian origins. Unless proponents of relationships of birds to crocodylomorphs, early archosaurs, or mammals can produce a detailed and rigorous phylogenetic analysis, it seems likely that theropod relationships will remain a popular notion. There are a number of current versions of this hypothesis, each with a different arrangement of the taxa. There has been a great deal of convergence in theropod phylogeny, as many workers have recognized, and the theropod hypothesis clearly requires considerable refinement. Given the degree of convergence, however, it remains possible that some future discovery in the Triassic or Jurassic may turn up the true ancestor of birds. Perhaps such a find will resurrect the hypotheses of Heilmann, Broom, or Huxley. But appeals to future discoveries are always somewhat empty. Today, the archosaur faunas of the Mesozoic are extremely well known when compared with those known when Huxley first formulated his common-ancestry hypothesis. Furthermore, the broader pattern of archosaur phylogeny is now beginning to be discovered. Although the logistic problems of finding characters, examining all the relevant specimens, analyzing the data, and so forth may seem formidable, it seems possible that we may soon be even closer to a phylogenetic hypothesis representing the truth about avian origins.

Acknowledgments I thank the editors for offering me the opportunity to delve into old, often forgotten literature and to discover the wisdom of our predecessors. Thanks to Drs. Marc Godinot and Hans-Peter Schultze for checking my translations, but any mistakes in the translations are my own. I thank Dr. David B. Weishampel for his careful reading of the manuscript and his interest throughout the project. Over the years I have benefited from conversations and correspondence about bird origins with many workers. Among them are Drs. Sankar Chatterjee, Philip J. Currie, Jacques Gauthier, Larry D. Martin, John H. Ostrom, and Samuel F. Tarsitano. Dr. Larry D. Martin deserves special thanks as he was the person who introduced me to the topic of avian origins and with whom I have spent many pleasant and productive hours discussing birds and other archosaurs. Final thanks go to the librarians at the Welch Medical Library and Eisenhower Library (Johns Hopkins University), who did their best to help me track down numerous important references. Bill Ness and Ann Juneau of the Smithsonian Institutions Libraries were very helpful in obtaining original copies of several illustrations reproduced here.

Literature Cited

Abel, O. 1911. Die Vorfahren der Vögel und ihre Lebensweise. Verh. Zool. Bot. Ges. Wien, 61:144–191.

Abel, O. 1920. *Lehrbuch der Paläozoologie.* Jena: Gustav Fischer Verlag.

Bakker, R. T. 1975. Dinosaur renaissance. Sci. Am., 232(4):58–78.

Bakker, R. T., and P. M. Galton. 1974. Dinosaur monophyly and a new class of vertebrates. Nature, London, 248:168–172.

Bakker, R. T., M. Williams, and P. J. Currie. 1988. *Nanotyrannus,* a new genus of pygmy tyrannosaur, from the latest Cretaceous of Montana. Hunteria, 1(5):1–30.

Barsbold, R. 1983. Carnivorous dinosaurs from the Cretaceous of Mongolia. Trans. Jt. Sov.-Mong. Paleontol. Exped., 19:5–119.

Baur, J. G. 1883. Der Tarsus der Vögel und Dinosaurier. Morph. Jahrb., 8:417–456.

Baur, J. G. 1884a. Dinosaurier und Vögel. Morphol. Jahrb., 10:446–454.

Baur, J. G. 1884b. Note on the pelvis in birds and dinosaurs. Am. Nat., 18:1272–1275.

Baur, J. G. 1887. Ueber die Abstammung der amnioten Wirbeltiere. Biol. Zentralbl., 7(16):481–493.

Beebe, C. W. 1906. *The Bird: Its Form and Function.* New York: Henry Holt.

Beebe, C. W. 1915. A Tetrapteryx stage in the ancestry of birds. Zoologica New York, 2(2):39–52.

Benton, M. J., and J. M. Clark. 1988. Archosaur phylogeny and the relationships of the Crocodylia. Pp. 295–338 *in* Benton, M. J. (ed.), *The Phylogeny and Classification of the Tetrapods,* Vol. 1: *Amphibians, Reptiles, Birds.* Syst. Assoc. Spec. Vol. No. 35 A. Oxford: Clarendon Press.

Boas, J. E. V. 1930. Über das Verhältnis der Dinosaurier zu den Vögeln. Morphol. Jahrb., 64(2):223–247.

Bock, W. J. 1969. The origin and radiation of birds. Ann. New York Acad. Sci., 167(1):147–155.

Brodkorb, P. 1971. Origin and evolution of birds. Pp. 19–55 *in* Farner, D. S., J. R. King, and K. C. Parkes (eds.), *Avian Biology.* Vol. 1. New York: Academic Press.

Broom, R. 1906. On the early development of the appendicular skeleton of the ostrich, with remarks on the origin of birds. Trans. S. Afr. Philos. Soc., 16:355–368.

Broom, R. 1913. On the South-African pseudosuchian *Euparkeria* and allied genera. Proc. Zool. Soc. London, 1913:619–633.

Chatterjee, S. 1987a. A new theropod dinosaur from India with remarks on the Gondwana-Laurasia connection in the Late Triassic. Pp. 183–189 *in* McKenzie, G. D. (ed.), *Gondwana 6: Stratigraphy, Sedimentology, and Paleontology.* Geophys. Monogr. 41.

Chatterjee, S. 1987b. Skull of *Protoavis* and early evolution of birds. J. Vert. Paleontol., 7(Suppl. to No. 3):14A.

Cope, E. D. 1867. An account of the extinct reptiles which approached the birds. Proc. Acad. Nat. Sci. Philadelphia, 1867:234–235.

Cracraft, J. 1977. John Ostrom's studies on *Archaeopteryx,* the origin of birds, and the evolution of avian flight. Wilson Bull., 89(3):488–492.

Currie, P. J. 1985. Cranial anatomy of *Stenonychosaurus inequalis* (Saurischia, Theropoda) and its bearing on the origin of birds. Can. J. Earth Sci., 22:1643–1658.

Dames, W. 1884. Ueber *Archaeopteryx.* Palaeontol. Abh. Bd., 2(3):119–198.

Dames, W. 1885. Entgegnung an Herrn Dr. Baur. Morphol. Jahrb., 10:603–612.

Darwin, C. 1872. *On the Origin of Species.* London: John Murray.

de Beer, G. R. 1954. Archaeopteryx lithographica: *A Study Based upon the British Museum Specimen*. London: British Museum (Natural History).

de Beer, G. R. 1956. The evolution of ratites. Bull. Br. Mus. Nat. Hist. Zool., 4(2):59–70.

Desmond, A. 1982. *Archetypes and Ancestors*. Chicago: Univ. of Chicago Press.

Dollo, L. 1882. Première note sur les dinosauriens de Bernissart. Bull. Mus. R. Hist. Nat. Belgique, 1:161–180.

Dollo, L. 1883a. Note sur la présence chez les oiseaux du "troisième trochanter" des dinosauriens et sur la fonction de celui-ci. Bull. Mus. R. Hist. Nat. Belgique, 2:13–19.

Dollo, L. 1883b. Troisième note sur les dinosauriens de Bernissart. Bull. Mus. R. Hist. Nat. Belgique, 2:85–126.

Evans, J. 1865. On portions of a cranium and of a jaw, in the slab containing the fossil remains of the *Archaeopteryx*. Nat. Hist. Rev. (n.s.), 5:415–421.

Feduccia, A. 1980. *The Age of Birds*. Cambridge: Harvard Univ. Press.

Fürbringer, M. 1888. *Untersuchungen zur Morphologie und Systematik der Vögel*. Amsterdam: Holkema.

Galton, P. M. 1970. Ornithischian dinosaurs and the origin of birds. Evolution, 24:448–462.

Gardiner, B. G. 1982. Tetrapod classification. Zool. J. Linn. Soc., 74:207–232.

Gauthier, J. 1986. Saurischian monophyly and the origin of birds. Pp. 1–55 *in* Padian, K. (ed.), *The Origin of Birds and the Evolution of Flight*. San Francisco: California Academy of Science.

Gauthier, J. A., A. G. Kluge, and T. Rowe. 1988. Amniote phylogeny and the importance of fossils. Cladistics, 4:105–209.

Gauthier, J. A., and K. Padian. 1985. Phylogenetic, functional, and aerodynamic analyses of the origin of birds and their flight. Pp. 185–197 *in* Hecht, M. K., J. H. Ostrom, G. Viohl, and P. Wellnhofer (eds.), *The Beginnings of Birds*. Eichstätt: Freunde des Jura-Museums.

Gauthier, J. A., and Padian, K. 1989. The origin of birds and the evolution of flight. Pp. 121–133 *in* Padian, K., and D. J. Chure (eds.), *The Age of Dinosaurs*. Knoxville: Paleontological Society.

Gegenbaur, C. 1864. *Untersuchungen zur vergleichenden Anatomie der Wirbelthiere: Erstes Heft. Carpus und Tarsus*. Leipzig: Verlag von Wilhem Engelmann.

Gegenbaur, C. 1878. *Grundriss der vergleichenden Anatomie*. Leipzig: Verlag von Wilhelm Engelmann.

Gregory, W. K. 1916. Theories of the origin of birds. Ann. New York Acad. Sci., 27:31–38.

Gregory, W. K. 1935. Remarks on the origins of the ratites and penguins. Proc. Linn. Soc. New York, 46:1–18.

Haeckel, E. 1866. *Generelle Morphologie der Organismen. Zweiter Band: Allgemeine Entwicklungsgeschichte der Organismen*. Berlin: G. Reimer.

Haeckel, E. 1875. *Natürliche Schöpfungsgeschichte*. Berlin: G. Reimer.

Haeckel, E. 1907. *History of Creation*. Vol. II. (Translated by E. R. Lankester.) New York: D. Appleton.

Hay, O. P. 1910. On the manner of locomotion of the dinosaurs especially *Diplodocus*, with remarks on the origin of birds. Proc. Wash. Acad. Sci., 12(1):1–25.

Hecht, M. K. 1976. Phylogenetic inference and methodology as applied to the vertebrate record. Evol. Biol., 9:335–363.

Hecht, M. K. 1985. The biological significance of *Archaeopteryx*. Pp. 149–160 *in* Hecht, M. K., J. H. Ostrom, G. Viohl, and P. Wellnhofer (eds.), *The Beginnings of Birds*. Eichstätt: Freunde des Jura-Museums.

Hecht, M. K., and S. F. Tarsitano. 1982. The paleobiology and phylogenetic position of *Archaeopteryx*. Geobios Mem. Spec., 6:141–149.

Heilmann, G. 1916. Vor nuvaerende Viden om Fuglenes Afstamning. Femte Afsnit: Førfuglen Proavis. Dan. Ornithol. Foren. Tidsskr., 10(2):73–144.

Heilmann, G. 1927. *The Origin of Birds*. New York: D. Appleton.

Holmgren, N. 1955. Studies on the phylogeny of birds. Acta Zool., 36:243–328.

Huxley, T. H. 1868a. Remarks upon *Archaeopteryx lithographica*. Proc. R. Soc. London, 16:243–248.

Huxley, T. H. 1868b. On the animals which are most nearly intermediate between birds and reptiles. Ann. Mag. Nat. Hist., 4(2):66–75.

Huxley, T. H. 1870a. Further evidence of the affinity between the dinosaurian reptiles and birds. Proc. Geol. Soc. London, 26(1):12–31.

Huxley, T. H. 1870b. On the classification of the Dinosauria, with observations on the Dinosauria of the Trias. Proc. Geol. Soc. London, 26(1):32–38.

Huxley, T. H. 1882. On the respiratory organs of *Apteryx*. Proc. Zool. Soc. London, 1882(38):560–569.

Janvier, P. 1983. Le divorce de l'oiseau et du crocodile. Recherche, 14(149):1430–1432.

Kurochkin, E. N. 1985. Lower Cretaceous birds from Mongolia and their evolutionary significance. Pp. 191–199 *in* Ilyichev, V. D., and V. M. Gavrilov (eds.), *Acta XVIII Congressus Internationalis Ornithologici*. Vol. I. Moscow: Nauka.

Kurzanov, S. M. 1985. The skull structure of the dinosaur *Avimimus*. Paleontol. J., 1985(4):92–99.

Kurzanov, S. M. 1987. Avimimidae and the problem of the origin of birds. Trans. Jt. Sov.-Mong. Paleontol. Exped., 31:1–94. [In Russian].

Lamarck, J. B. 1809. *Philosophie zoologique*. (Translated by H. Elliot, 1984.) Chicago: Univ. of Chicago Press.

Løvtrup, S. 1985. On the classification of the taxon Tetrapoda. Syst. Zool., 34(4):463–470.

Lowe, P. R. 1928. Studies and observations bearing on the phylogeny of the ostrich and its allies. Proc. Zool. Soc. London, 1928(1):185–247.

Lowe, P. R. 1935. On the relationship of Struthiones to the dinosaurs and to the rest of the avian class, with special reference to the position of *Archaeopteryx*. Ibis, 5(2):398–432.

Lowe, P. R. 1944. An analysis of the characters of *Archaeopteryx* and *Archaeornis*. Were they reptiles or birds? Ibis, 86:517–543.

Marsh, O. C. 1877. Introduction and succession of vertebrate life in America. Proc. Am. Assoc. Adv. Sci., 1877:211–258.

Marsh, O. C. 1880. *Odontornithes: A Monograph on the Extinct Toothed Birds of North America*. Washington, D.C.: United States Geological Exploration of the 40th Parallel.

Marsh, O. C. 1881. Jurassic birds and their allies. Am. J. Sci., 22:337–340.

Martin, L. D. 1983a. The origin and early radiation of birds. Pp. 291–338 *in* Brush, A. H., and G. A. Clark (eds.), *Perspectives in Ornithology*. Cambridge: Cambridge Univ. Press.

Martin, L. D. 1983b. The origin of birds and of avian flight. Pp. 106–129 *in* Johnston, R. F. (ed.), *Current Ornithology*. Vol. 1. New York: Plenum Press.

Martin, L. D. 1988. Review of "The origin of birds and the evolution of flight." Auk, 105(3):596–597.

Martin, L. D., J. D. Stewart, and K. N. Whetstone. 1980. The origin of birds: structure of the tarsus and teeth. Auk, 97:86–93.

McGowan, C. 1985. Tarsal development in birds: evidence for homology with the theropod condition. J. Zool. London (A), 206:53–67.

McGowan, C., and A. J. Baker. 1981. Common ancestry for birds and crocodiles? Nature, London, 289:97–98.

Mivart, G. 1881. A popular account of chamaeleons. Nature, London, 24:309–312, 335–338.

Molnar, R. E. 1985. Alternatives to *Archaeopteryx:* a survey of proposed early or ancestral birds. Pp. 209–217 *in* Hecht, M. K., J. H. Ostrom, G. Viohl, and P. Wellnhofer (eds.), *The Beginnings of Birds*. Eichstätt: Freunde des Jura-Museums.

Molnar, R. E., and Archer, M. 1984. Feeble and not so feeble flapping fliers: a consideration of early birds and bird-like reptiles. Pp. 407–419 *in* Archer, M., and G. Clayton (eds.), *Vertebrate Zoogeography and Evolution in Australia*. Western Australia: Hesperian Press.

Mudge, B. F. 1879. Are birds derived from dinosaurs? Kansas City Rev. Sci., 3:224–226.

Nopsca, F. 1907. Ideas on the origin of flight. Proc. Zool. Soc. London, 1907(15):223–236.

Nopsca, F. 1923. On the origin of flight in birds. Proc. Zool. Soc. London, 1923(31):463–477.

Nopsca, F. 1929. Noch einmal Proavis. Anat. Anz., 67(12/14):265–300.

Osborn, H. F. 1900. Reconsideration of the evidence for a common dinosaur-avian stem in the Permian. Am. Nat., 34(406):777–799.

Ostrom, J. H. 1969. Osteology of *Deinonychus antirrhopus*, an unusual theropod from the Lower Cretaceous of Montana. Bull. Peabody Mus. Nat. Hist. Yale Univ., 30:1–165.

Ostrom, J. H. 1973. The ancestry of birds. Nature, London, 242:136.

Ostrom, J. H. 1974. *Archaeopteryx* and the origin of flight. Q. Rev. Biol., 49:27–47.

Ostrom, J. H. 1976. *Archaeopteryx* and the origin of birds. Biol. J. Linn. Soc., 8:91–182.

Ostrom, J. H. 1978. The osteology of *Compsognathus longipes* Wagner. Zitteliana, 4:73–118.

Ostrom, J. H. 1979. Bird flight: how did it begin? Am. Sci., 67(1):46–56.

Ostrom, J. H. 1985. Introduction to *Archaeopteryx*. Pp. 9–20 *in* Hecht, M. K., J. H. Ostrom, G. Viohl, and P. Wellnhofer (eds.), *The Beginnings of Birds*. Eichstätt: Freunde des Jura-Museums.

Ostrom, J. H. 1986. The cursorial origin of avian flight. Pp. 73–81 *in* Padian, K. (ed.), *The Origin of Birds and the Evolution of Flight*. San Francisco: California Academy of Sciences.

Owen, R. 1862. On the fossil remains of a long-tailed bird (*Archaeopteryx macrurus* Ow.) from the lithographic slate of Solenhofen. Proc. R. Soc. London, 23:272–273.

Owen, R. 1863. On the *Archaeopteryx* of von Meyer, with a description of the fossil remains of a long-tailed species, from the lithographic stone of Solenhofen. Philos. Trans. R. Soc. London, 153:33–47.

Owen, R. 1870. Monograph of the Fossil Reptilia of the Liassic Formations. Part II. Order Pterosauria. Palaeontogr. Soc. Monogr., 23:41–81.

Owen, R. 1875. Monographs of the British Fossil Reptilia of the Mesozoic Formations. Part II. (Genera *Bothriospondylus, Cetiosaurus, Omosaurus*). Palaeontogr. Soc. Monogr., 29:15–93.

Padian, K. 1982. Macroevolution and the origin of major adaptations: vertebrate flight as a paradigm for the analysis of patterns. Proc. Third N. Am. Paleontol. Conven., 2:387–392.

Padian, K. 1985. The origins and aerodynamics of flight in extinct vertebrates. Palaeontology, 28(3):413–433.

Padian, K. 1986. *The Origin of Birds and the Evolution of Flight*. San Francisco: California Academy of Sciences.

Parker, W. K. 1864. Remarks on the skeleton of the *Archaeopteryx;* and on the relations of the bird to the reptile. Geol. Mag., 1:55–57.

Parker, W. K. 1881. On the morphology of birds. Proc. R. Soc. London, 42:52–58.

Paul, G. S. 1984. The archosaurs: a phylogenetic study. Pp. 175–180 *in* Reif, W.-E., and F. Westphal (eds.), *Third Symposium on Mesozoic Terrestrial Ecosystems, Short Papers.* Tübingen: Attempto Verlag.

Paul, G. S. 1988a. The small predatory dinosaurs of the mid-Mesozoic: the horned theropods of the Morrison and Great Oolite—*Ornithólestes* and *Proceratosaurus*—and the sickle-claw theropods of the Cloverly, Djadokhta and Judith River—*Deinonychus, Velociraptor* and *Saurornitholestes.* Hunteria, 2(4):1–9.

Paul, G. S. 1988b. *Predatory Dinosaurs of the World.* New York: Simon and Schuster.

Petronievics, B. 1950. Les deux oiseaux fossiles les plus anciens (*Archaeopteryx* et *Archaeornis*). Geol. An. Balk. Poluostrva, 18:89–127.

Pycraft, W. P. 1906. The origin of birds. Knowledge and Scientific News, Sept. 1906:531–532.

Raath, M. A. 1985. The theropod *Syntarsus* and its bearing on the origin of birds. Pp. 219–227 *in* Hecht, M. K., J. H. Ostrom, G. Viohl, and P. Wellnhofer (eds.), *The Beginnings of Birds.* Eichstätt: Freunde des Jura-Museums.

Romer, A. S. 1966. *Vertebrate Paleontology.* Chicago: Univ. of Chicago Press.

Sanz, J. L., and N. Lopez-Martinez. 1984. The prolacertid lepidosaurian *Cosesaurus aviceps* Ellenberger & Villalta, a claimed "protoavian" from the middle Triassic of Spain. Geobios, 17(6):741–753.

Seeley, H. G. 1866. An epitome of the evidence that Pterodactyles are not reptiles, but a new subclass of vertebrate animals allied to birds (Saurornia). Ann. Mag. Nat. Hist., Ser. 3, 17(101):321–331.

Seeley, H. G. 1881. Prof. Carl Vogt on the *Archaeopteryx.* Geol. Mag. (ser. 2), 8:300–309.

Simpson, G. G. 1946. Fossil penguins. Bull. Am. Mus. Nat. Hist., 87:1–95.

Swinton, W. E. 1958. Fossil Birds. London: British Museum (Natural History).

Swinton, W. E. 1960. The origin of birds. Pp. 1–14 *in* Marshall, A. J. (ed.), *Biology and Comparative Physiology of Birds.* Vol. 1. New York: Academic Press.

Tarsitano, S. F. 1985a. The morphological and aerodynamic constraints on the origin of avian flight. Pp. 319–332 *in* Hecht, M. K., J. H. Ostrom, G. Viohl, and P. Wellnhofer (eds.), *The Beginnings of Birds.* Eichstätt: Freunde des Jura-Museums.

Tarsitano, S. F. 1985b. Cranial metamorphosis and the origin of the Eusuchia. Neues Jahrb. Geol. Palaeontol. Abh., 170(1):27–44.

Tarsitano, S. F., and M. K. Hecht. 1980. A reconsideration of the reptilian relationships of *Archaeopteryx.* Zool. J. Linn. Soc., 69(2):149–182.

Thulborn, R. A. 1975. Dinosaur polyphyly and the classification of archosaurs and birds. Australian J. Zool., 23:249–270.

Thulborn, R. A. 1984. The avian relationships of *Archaeopteryx,* and the origin of birds. Zool. J. Linn. Soc., 82(1–2):119–158.

Thulborn, R. A., and T. L. Hamley. 1982. The reptilian relationships of *Archaeopteryx.* Australian J. Zool., 30:611–634.

Tucker, B. W. 1938. Some observations on Dr. Lowe's theory of the relationship of the Struthiones to the dinosaurs and to other birds. Pp. 222–224 *in* Jourdain, F. C. R. (ed.), *Proceedings of the Eighth International Ornithological Congress.* Oxford: Oxford Univ. Press.

Vogt, C. 1879. *Archaeopteryx,* ein Zwischenglied zwischen den Vögeln und Reptilien. Naturforscher, 43:401–404.

Vogt, C. 1880. *Archaeopteryx macrura,* an intermediate form between birds and reptiles. Ibis, 4:434–456.

Wagner, A. 1861. Über ein neues, angeblich mit Vogelfedern versehenes Reptil aus dem Solenhofener lithographischen Schiefer. Sitzungsber. Math. Naturwiss. Kl. Bayer. Akad. Wiss. Muenchen, 2:146–154.

Wagner, A. 1862. On a new fossil reptile supposed to be furnished with feathers. Ann. Mag. Nat. Hist., 3rd ser., 9:261–267.

Walker, A. D. 1972. New light on the origin of birds and crocodiles. Nature, London, 237(5353):257–263.

Walker, A. D. 1974. Evolution, organic. Pp. 177–179 *in McGraw-Hill Yearbook of Science and Technology.* New York: McGraw-Hill.

Walker, A. D. 1977. Evolution of the pelvis in birds and dinosaurs. Pp. 319–358 *in* Andrews, S., R. Miles, and A. D. Walker (eds.), *Problems in Vertebrate Evolution.* London: Academic Press.

Walker, A. D. 1985. The braincase of *Archaeopteryx.* Pp. 123–134 *in* Hecht, M. K., J. H. Ostrom, G. Viohl, and P. Wellnhofer (eds.), *The Beginnings of Birds.* Eichstätt: Freunde des Jura-Museums.

Weishampel, D. B., and W.-E. Reif. 1984. The work of Franz Baron Nopsca (1877–1933): dinosaurs, evolution and theoretical tectonics. Jahrb. Geol. B.-A., 127(2):187–203.

Wellnhofer, P. 1985. Remarks on the digit and pubis problems of *Archaeopteryx.* Pp. 113–122 *in* Hecht, M. K., J. H. Ostrom, G. Viohl, and P. Wellnhofer (eds.), *The Beginnings of Birds.* Eichstätt: Freunde des Jura-Museums.

Whetstone, K. N. 1983. Braincase of Mesozoic birds: I. New preparation of the "London" *Archaeopteryx.* J. Vert. Paleontol., 2(4):439–452.

Whetstone, K. N., and L. D. Martin. 1979. New look at the origin of birds and crocodiles. Nature, London, 279:234–236.

Whetstone, K. N., and L. D. Martin. 1981. Reply to McGowan and Baker. Nature, London, 289:98.

Whetstone, K. N., and P. J. Whybrow. 1983. A "cursorial" crocodilian from the Triassic of Lesotho (Basutoland), South Africa. Occas. Pap. Mus. Nat. Hist. Univ. Kansas, 106:1–37.

Wiedersheim, R. 1882. *Lehrbuch der vergleichenden Anatomie der Wirbelthiere.* Jena: Fischer Verlag.

Wiedersheim, R. 1884. Die Stammesentwicklung der Vögel. Biol. Zentralbl., 3:654–668, 688–695.

Wiedersheim, R. 1885. Ueber die Vorfahren der heutigen Vögel. Humboldt, 4:213–224.

Wiedersheim, R. 1886. *Lehrbuch der vergleichenden Anatomie der Wirbelthiere.* 2nd ed. Jena: Fischer Verlag.

Williston, S. W. 1879. "Are birds derived from dinosaurs?" Kansas City Rev. Sci., 3:357–360.

Williston, S. W. 1925. *The Osteology of the Reptiles.* Cambridge: Harvard Univ. Press.

Witmer, L. M. 1984. Bird origins and the siphonial system of birds and crocodilians. Abstract 12, 102nd Stated Meeting of American Ornithologists' Union.

Witmer, L. M. 1987. The nature of the antorbital fossa of archosaurs: shifting the null hypothesis. Pp. 230–235 *in* Currie, P. J., and E. H. Koster (eds.), *Fourth Symposium on Mesozoic Terrestrial Ecosystems, Short Papers.* Tyrrell Mus. Palaeontol. Alberta, Occas. Pap. 3.

Witmer, L. M. 1990. The craniofacial air sac system of Mesozoic birds (Aves). Zool. J. Linn. Soc., 100:327–378.

13 The Question of the Origin of Birds

John H. Ostrom

There is inherent fascination about origins. Some of the most important biological questions deal with the "origin" of life, the "origin" of chordates, human "origins," and the like. The word *origin* itself is such an imposing and seemingly conclusive word in the English language that its use seems to convey authenticity to any statements that follow. That is not my presumption, and for that reason I have framed my title to indicate a question. Why? Because each of the events mentioned here and all other "origin" events are ancient history—i.e., unwitnessed events that are known to have occurred only because we can observe the modern (or fossil) consequences of those events in the form of living cells, chordate organisms, humans, and so forth. Without such consequences there would be no quest. At this remote moment, we are limited to composing historical scenarios based on very incomplete evidence. But, with luck and cautious insight, we may be able to infer an answer that is logical and consistent with the limited evidence at hand. Yet these are only inferences, which are subject to different explanations and new data. *Archaeopteryx* has played a central role in discussions of the origin of birds simply because modern birds are too derived and too remote in time from the event of their origin to provide relevant anatomical evidence about the pre-*Archaeopteryx* condition. Other Mesozoic birds are less remote temporally from the origin event, but they, too, are highly derived from the unknown original avian state. Unlikely as *Archaeopteryx* may or may not be as an ancestral candidate for the Class Aves, it still provides the only available anatomical evidence from rocks of the appropriate age. Were there no specimens of *Archaeopteryx*, the specimens of *Hesperornis* and *Ichthyornis* might be considered as the best evidence pointing toward the ancestral condition for the class, but they would provide a most distorted image of what "proavis" might have been like. Like it or not, the question for which we seek an answer may be beyond resolution, because the fossil evidence is so fragmentary and incomplete, and all living birds are so far removed biologically and temporally from the missing "proavis."

At the outset, I must define certain key words so that there can be no misunderstanding in the discussion that follows. *Origin* means "that from which anything derives its existence," "a source," or "a beginning." For purposes of this paper, the terms *bird* and *avian* refer to particular kinds of amniote vertebrates that are characterized by the following combination of features: (1) an epidermal covering of feathers; (2) obligate bipedal posture and gait; (3) a forelimb and shoulder modified for flight, and not capable of standing support (or forelimbs displaying evidence of former, ancestral flight ability); (4) metabolism endothermic and hypermetabolic; and (5) a pneumatic skeleton (evidenced by the presence of pneumatopores, not just "hollow" skeletal elements) linked with an air sac respiratory system. Of course, not all of these characteristics are individually unique to birds, but collectively they define those animals we call birds and exclude all dinosaurian taxa as presently known.

My final introductory remark is a caution. "Be very careful about what you accept as the truth." It is a common human weakness that we usually see what we want to see—or what we are told to see or hear. Remember the flat earth? To make the point even clearer, let me cite some appropriate examples. Four of the six presently known specimens of *Archaeopteryx* were misidentified—i.e., the first one found (Haarlem specimen, 1855) and every specimen found after the London and Berlin specimens. As originally identified, these were "seen" as either pterosaurs or small dinosaurs. That leads me back to the question of avian origins and where do we start? The obvious first question is: What is the relevant evidence?

THE FOSSIL EVIDENCE

The paleontological evidence is scanty but important nonetheless. Geologically, the oldest fossil remains that most nearly satisfy the criteria

The following pages are deliberately not expressed in the "current" language of cladistics because I am not convinced that it is any less ambiguous or easier for the general audience to comprehend than well-expressed English. I hope that this contribution will be read by (except numbers 4 and 5) of the avian condition are the six specimens of *Archaeopteryx* (Meyer, 1861; Häberlein, 1877; Heller, 1959; Ostrom, 1970; Mayr, 1973; Wellnhofer, 1974, 1988). These specimens were preserved in the Tithonian (Late Jurassic) Solnhofen Limestone of Bavaria, Germany, a sequence firmly dated by associated marine faunas as Malm-Zeta 1–2 (Barthel, 1964, 1978; Janicke, 1969; and others).

All specimens have been studied in detail and reported on by many workers over the past century. Several of the nearly complete skeletons are intimately associated with remarkably distinct impressions of feathers, thereby satisfying criterion 1, above. Geologically younger avian remains such as the classic *Ichthyornis* and *Hesperornis* and *Baptornis* material from the Niobrara Chalk (Coniacian to Campanian) of Late Cretaceous age (Marsh, 1873, 1880) from Kansas and Nebraska also are well known. In addition, there are several other less well known avian taxa, such as *Enaliornis* (Seeley, 1869) from the Upper Greensand (Albion) of England, *Gobipteryx* (Elzanowski, 1976, 1981) from the Campanian (Upper Cretaceous) of central Asia, and *Enantiornis* and *Alexornis* along with other very fragmentary Cretaceous material assigned to Aves. Only a few of these specimens (e.g., *Hesperornis*) also preserve associated impressions of feathers. A taxon recently added to the avian roster is *Palaeopteryx* (Jensen, 1981), which consists of a few limb fragments and a synsacrum from the Morrison Formation (Kimmeridgian-Portlandian) of western Colorado; the age of this formation is approximately equivalent to that of the Solnhofen Limestones. However, the assignment of the *Palaeopteryx* fragments to Aves is judged by many (including myself) to be very doubtful, and even the skeletal identity of the fragments is uncertain. All this material and the aforementioned taxa are younger than the Late Jurassic specimens of *Archaeopteryx*. Also, of all these taxa, only the remains of *Archaeopteryx, Ichthyornis, Hesperornis, Baptornis, Enaliornis,* and perhaps *Gobipteryx* allow for indisputable obligatory bipedal stance— i.e., criterion 2, above. The third criterion is met without doubt in all of the Mesozoic taxa mentioned above except *Palaeopteryx*.

The issue of avian origins has been complicated by the recent announcement by Sankar Chatterjee (pers. comm. in 1986) of supposed

---

many interested parties who are not necessarily familiar with the cladistic jargon. However, I acknowledge that many of my colleagues prefer the cladistic language, and in recognition of that fact, I offer the following without further comment. I leave it to others to explain.

Anatomical states or characters that might be considered as synapomorphies within *Archaeopteryx* and coelurosaurs are as follows: (1) the tridactyl manus configuration, (2) the phalangeal proportions of the manus, (3) the metacarpus configuration, (4) the semilunate carpal, (5) the pes organization, (6) the reflexed hallux, (7) the metatarsus configuration, (8) the mesotarsal ankle joint, (9) the proximal tarsal "ascending process," (10) a fibular crest on the tibia, (11) coracoid morphology, (12) pubic morphology, (13) the pubis orientation?, (14) furcula?, and (15) obligate bipedality.

bird remains (several skeletal parts all disarticulated) from the Late Triassic Dockum Formation (Rhaetic) of west Texas. Because adequate anatomical data have not been published by Chatterjee, it is not possible to
confirm the putative avian identification of the material; hence, no further comments can be made here. If truly avian, the Texas material is
extremely important to the question at hand for obvious reasons; it
predates *Archaeopteryx* by approximately 75 million years! This Texas
discovery certainly raises some giant question marks, but until the specimens are thoroughly studied and documented in publication, the material is irrelevant to the question of avian origins. We are still dependent on
the well-worn specimens of *Archaeopteryx* as the oldest known birds.

THE EVIDENCE OF *ARCHAEOPTERYX*

Most readers are well aware that several recent phylogenetic hypotheses that are based almost entirely on the evidence of *Archaeopteryx* have
generated intense debate about the origin of birds. The most important
hypotheses, together with the first and a few of the most noteworthy
recent references, include the following: (1) Birds were derived from
pseudosuchian reptiles, specifically, thecodontian archosaurs (Broom,
1906, 1913; Heilmann, 1926; Tarsitano and Hecht, 1980; Hecht and Tarsitano, 1982). (2) Birds are related most closely to crocodilians (Walker,
1972, 1974; Whetstone and Martin, 1979; Martin et al., 1980; Martin,
1983a,b). (3) Birds are descended from a theropod dinosaur ancestry
(Huxley, 1868; Williston, 1879; Ostrom, 1973, 1975, 1976). See also Thulborn (1984) for a related opinion.

The remarks that follow are presented in support of the third of these
hypotheses. My views on the question are well known, the evidence
having been presented fully elsewhere. For reviews of the hypothesis
that birds are derived from theropod dinosaurs, see Ostrom (1973, 1975,
1976, 1985) and the more recent papers by Gauthier (1986) and Gauthier
and Padian (1985), who presented essentially the same hypothesis expressed in the currently fashionable cladistic mode. Because no new
evidence has emerged to cause me to change my opinions, I reiterate my
views only as necessary to clarify particular points. I concentrate my
commentary on some of the supposed counterevidence that has been
published recently.

The Teeth of *Archaeopteryx*

Martin et al. (1980) and Martin (1983b, 1985) emphasized the similarity
of the teeth of *Archaeopteryx* to those of the other Mesozoic birds *Hes-*

perornis and *Ichthyornis* in bearing unserrated, *laterally compressed* (my emphasis), tapered crowns that are separated from their expanded roots by a distinct waist. In fact, the teeth of the London, Berlin, and Eichstätt specimens vary within each specimen, as well as between different specimens. In my opinion, the teeth of *Archaeopteryx* do not closely resemble typical teeth of either *Hesperornis* or *Ichthyornis*. Dentitional variation in the Berlin specimen is illustrated in Figure 1. The crowns of the second and fifth upper teeth are rather blunt, broadly straight, and symmetrically tapering. In contrast, those of the third–fourth and the seventh–eighth upper teeth are expanded distally; the crowns taper abruptly, with the anterior profile curving sharply toward a nearly straight posterior edge profile. All the teeth are broad-based and broad in an antero-posterior direction relative to crown length. The premaxillary teeth and the first two maxillary teeth narrow slightly near the crown base, although this constriction is poorly defined and not uniform. The Eichstätt specimen is better preserved but possesses teeth that are narrower, more gracile, and that taper more uniformly than those of the other specimens of *Archaeopteryx*. All teeth are recurved, with both anterior and posterior profile margins converging in smooth backward curves. In both the Berlin and Eichstätt specimens, the teeth seem to be slightly compressed transversely, although all are visible only in lateral aspect. Typical teeth in *Hesperornis* also are only slightly compressed transversely; however, they are recurved more sharply (as are most theropod teeth) than in *Archaeopteryx*. In contrast, those teeth of *Ichthyornis* that I have examined are much more strongly compressed laterally and almost trenchant in form with sharp anterior and posterior edges, but they are not serrated. Although the teeth of neither *Ichthyornis* and *Hesperornis* that I have examined resemble those of *Archaeopteryx*, I recognize that other samples may be more similar.

Martin et al. (1980) also noted that replacement teeth in both *Hesperornis* and *Archaeopteryx* (and presumably *Ichthyornis*) develop in an oval-to-circular resorption crypt in the lingual side of the root of the tooth being replaced. They stressed that "in all of the known toothed birds, the teeth are laterally compressed and unserrated" (Martin et al., 1980:91), concluding with the seemingly logical observation that these dental features seem to be primitive for birds. Ergo, one should expect to see these features in the archosaurian group most closely related to Mesozoic birds. Perhaps, *if* these features are primitive. But, these dental features just as well could be derived in Mesozoic birds. The authors proceeded to argue that similar dental features are found in juvenile stages of some members of the Crocodilia. They noted that this basic tooth is widely distributed within mesosuchians and eusuchians but cited Nash's (1971) description of protosuchian *Orthosuchus* teeth as "conical and unser-

?
Bird/Croc. Teeth Look-Alikes

Ichthyornis dispar

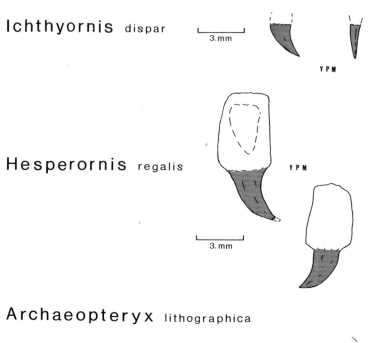

|— 3. mm —|

Y P M

Hesperornis regalis   Y P M

|— 3. mm —|

Archaeopteryx lithographica

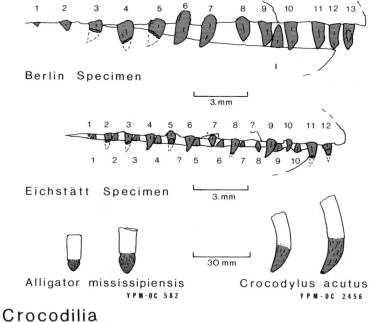

1 2 3 4 5 6 7 8 9 10 11 12 13

Berlin Specimen

|— 3. mm —|

1 2 3 4 5 6 7 8 ? 9 10 11 12

1 2 3 4 ? 5 6 7 8 9 10

Eichstätt Specimen |— 3. mm —|

|— 3O mm —|

Alligator mississipiensis   Crocodylus acutus
YPM-OC 582   YPM-OC 2456

Crocodilia

472

rated." Are the latter laterally compressed? Martin et al. mentioned that only sebecosuchian and pristichampsid crocodilians have compressed and serrated teeth (anomalies?), citing Langston (1973). They contrasted these points with the *usual* dental conditions of theropods—i.e., serrated teeth with straight (= unexpanded?) roots and no constriction between crown and root. The authors did not mention that theropod teeth also are compressed laterally (in varying degrees depending on the taxon, tooth position, and also the ontogenetic age). Martin et al. (1980) noted that tooth replacement in theropods *is* like that of Mesozoic birds and crocodiles, with "elongated" lingual resorption pits, citing Edmund (1960). They then reported that these dental conditions are present in a number of Triassic archosaurs ("thecodontians") and concluded that such dental morphology is primitive for archosaurs and that "the avian/crocodilian condition can be *assumed* [my emphasis] to be derived" (Martin et al., 1980:91). I would suggest another possibility. First, tooth replacement is essentially the same in theropods, crocodilians, and Mesozoic toothed birds—except that theropod resorption pits are "elongated." This seems to be correlated with the high, narrow crowns of most theropod teeth. Furthermore, "oval" versus "elongated" seems to me to be an insignificant (or imprecise) distinction. Second, and more important, some theropods *lack* serrated teeth—specifically, most of the teeth in *Compsognathus* are not serrated, and the teeth of *Ornitholestes* and possibly *Coelurus* are not all serrated. The absence of serrations in laterally compressed teeth of these theropods is anomalous, but then so is the presence of serrated compressed teeth in sebecosuchian crocodiles. The picture that has been painted that assumes the Mesozoic bird-crocodilian dental condition is derived, is more complex than that presented by Martin et al. (1980) and should not be accepted as necessarily true. Tooth morphology seems to be highly plastic, reflecting taxonomic differences, ontogenetic age, diet preferences (Howgate, 1984, 1985), and predation habits of different-sized organisms of different ecological habitats (see Thulborn and Hamley, 1985). Witness the diversity of tooth form in the several species of *Varanus, Iguana,* or *Lacerta,* or that of *Gavialis, Alligator,* and *Sebecus.* There is little similarity of tooth form between *Varanus niloticus* and *Varanus*

Fig. 1. A simple comparison of tooth shape (and root, where visible) of two living crocodilians, two varieties of Odontornithes, and two specimens of *Archaeopteryx*. To this author's eye, there are distinct differences among these three kinds. Also, differences are apparent between the two examples of *Archaeopteryx*. All views are lateral profiles except for the right image of *Ichthyornis*, which is in anterior aspect. *Icthyornis* and *Hesperornis* are taken from Marsh (1880) and confirmed by Yale specimens. *Archaeopteryx* drawings by camera lucida ×3. Crocodilians by camera lucida ×1. Note that scales are not the same for all drawings. YPM = Yale Peabody Museum vertebrate paleonotology collections. YPM OC = Yale Peabody Museum Osteology Collection.

komodoensis. Which is the derived state (or are they equally derived?), and how does one distinguish the derived state in this situation from the primitive condition? Depending upon which out-group is chosen, the polarity may proceed either way. Which is correct?

The Tarsus in *Archaeopteryx*

There is no question that all living birds, *Archaeopteryx*, and theropod dinosaurs (also pterosaurs) are characterized by a mesotarsal ankle joint in which the two proximal tarsals are fused or tightly articulated with the tibia, and also with the fibula in Mesozoic forms. The distal tarsals either articulate loosely or fuse with the elements of the metatarsus in various of these taxa. Martin et al. (1980) challenged my interpretation of the ankle in the specimens of *Archaeopteryx*; I homologized an ascending process (= pre-tibial bone of Martin et al., 1980) of the proximal tarsals of *Archaeopteryx* with the ascending process of the astragalus of the theropod ankle. I still think that homology is correct and refer readers to my response (Ostrom, 1985) to their dissenting argument. Herein, I only repeat my slightly modified concluding remarks from that paper.

> Since both processes [those in theropods and those in *Archaeopteryx*] appear to have formed from ossification centers separate from those of the astragalus *and* the calcaneum, respectively, the processes in fact *may be* the *same* bone which has fused [or articulates] with different [proximal] tarsals in different taxa. But because the tarsal condition is equivocal in the Eichstätt specimen and has not yet been resolved clearly in the Berlin or London specimens, this feature cannot be used either for or against the theropod hypothesis of *Archaeopteryx* ancestry. (Ostrom, 1985:174)

Developmental studies (McGowan, 1984) clearly show that the astragalar ascending process and the pretibial bone are two separate ossifications in birds; the former is typical of ratites, and the latter is unique to carinates. McGowan concluded that presence of the pretibial bone is a synapomorphy for carinates, and that the astragalar process of ratites is a primitive state. These facts do not negate the suggested homology of the ascending astragalar process in theropods and birds. Instead, it merely points to a post-*Archaeopteryx* separation of ratites and carinates, a long-held view. (However, as McGowan concluded, his data reinforce the minority view that ratites are primitive and were not derived from a carinate ancestry.)

It is appropriate to point out here that there appears to be considerable variation in the conditions of the "ascending process" or pretibial bone in theropods. Welles and Long (1977) noted this, but see also Welles, 1983. In *Deinonychus* (YPM 5226) for example, the "ascending process" of the astragalus has an extensive contact with the calcaneum (Ostrom, 1969). This same articulation is even more pronounced in *Albertosaurus arctunguis* (Parks, 1928). In *Gorgosaurus* and *Allosaurus,* on the other hand, the "ascending process" does not have contact with the calcaneum (Lambe, 1917; Madsen, 1977). Tarsitano and Hecht (1980) also attempted to dismiss the tarsal configuration of *Archaeopteryx,* as preserved in the London and Berlin specimens. First, Tarsitano and Hecht (1980:164) suggested that because the dorsal process is on the anterolateral surface of the tibia shaft (as opposed to the anterior surface), "this process is most probably part of the calcaneum" even though the calcaneum had not been recognized with certainty in either specimen. Therefore, they argued (p. 164) that these features "are not homologous but merely functional analogues." Just a few lines down on the same page, Tarsitano and Hecht offered another idea to explain away these embarrassing theropod-like conditions in *Archaeopteryx.* Their explanation was that the entire "ascending process" of *Archaeopteryx* is an artifact! As they expressed it, "Our interpretation of the Berlin and London specimens is that the bony cylinder of the tibio-tarsus has burst open, exposing its calcite-filled interior" (Tarsitano and Hecht, 1980:164). In their words, it is only "calcite"—not bone! This is followed by their *assertion* "that there are only two possible interpretations: either the process is an artifact, or it is non-homologous with the theropod condition" (Tarsitano and Hecht, 1980: 164). They recognized no other possibilities! Contrary to their claim and on the basis of my own observations and those of others, I assure you that the "ascending process" of both specimens of *Archaeopteryx* is genuine bone. Unlike calcitic structures, the processes fluoresce brightly under ultraviolet light like all other bony parts of the preserved skeletons. This condition has been observed by a number of workers and reported most recently by Martin et al. (1980); the condition of the astragalar ascending process is evident in Plate II of de Beer's (1954) monograph. An observant reader also will note that there is a third possible alternative interpretation that was dismissed by Tarsitano and Hecht (1980) in their haste—one that I stated in 1976. My 1976 hypothesis of the pretibial process homology has *not* been falsified, only questioned.

Tarsitano (1985:327) subsequently amended his interpretation by saying, "These ascending processes are not mirror images of each other and are best interpreted as breakage of the tibial walls due to calcite precipitation within the bones." I do not see that this latest (1985) interpretation alters the substance of our two different interpretations—i.e., according to Tarsitano, the object in question is *not* bone and, therefore, there is no astragalar process, whereas according to Ostrom and others, the object under dispute *is* bone and, therefore, most likely represents the astragalar process.

Although not directly part of the issue of tarsal construction, a related feature of importance is the nature of the fibula and its contact (or lack of) with the calcaneum. As I reported in 1976, the fibula in *Archaeopteryx* is unlike that of modern birds in which the bone is abbreviated and does not reach the tarsus. Contrary to statements that the fibula of *Archaeopteryx* is also abbreviated and terminates distally well above the ankle (e.g., see Martin, 1985, this volume), the fibula is preserved intact in the London specimen (both sides; left in the main slab, right in the counterpart slab), the Berlin specimen (right side is preserved in the counterpart slab), and the Eichstätt specimen (in which the right fibula is clear in the counterpart slab). Fragments are missing in the Berlin fibula, but its length indicates that it probably contacted the calcaneum distally. In the London and Eichstätt specimens, there can be no question that the fibula reached the calcaneum; nor is there any question that the "ascending process" (whatever it is) lies *medial* to the calcaneum and the fibula. It also should be pointed out that the Berlin and Eichstätt specimens of *Archaeopteryx* preserve very modest fibular crests on the tibia, a feature that is unique to birds and theropods. To the best of my knowledge, no other archosaurs (or reptiles) possess this feature. It may have something to do with the mesotarsal ankle, or bipedality, or both.

The Hallux of *Archaeopteryx*

The reversed or reflexed hallux is a characteristic of many modern birds. In four of the (now) six known specimens of *Archaeopteryx* in which the hallux is preserved, it is unequivocally fully reflexed—not partially, as Tarsitano and Hecht (1980) originally stated. Subsequently, they corrected themselves, noting that "the hallux of *Archaeopteryx* is at *least* [italics mine] partially reversed and opposable" (Hecht and Tarsitano, 1982:145), and Tarsitano (1985:328) noted that other authors had reported the reflexed hallux condition here, which they associated with an arboreal habit.

Whether full reversal of the hallux occurred in theropods or not is the critical question. In any case, it is obvious that a reflexed hallux did not (and does not) occur in any crocodilian, or "thecodont" yet known. Although I offered (Ostrom, 1978, 1985) an alternative interpretation to that of Tarsitano and Hecht (1980) and Tarsitano (1985) of the well-preserved pes of *Compsognathus*, neither author has responded, although the anatomic conditions that I pointed out are unmistakable and in conflict with their interpretation. These authors argued that the hallux is not reflexed in the theropod *Compsognathus*, whereas I maintain that it was. For the record, let me review the evidence again, as I see it.

The conflict hinges on whether the ungual claw of the hallux is preserved in its natural position or not. I maintain that it is not. This is demonstrated clearly by contrasting the orientations of adjacent collateral ligament fossae on the phalangae of Digit I. *All* of the fossae should be parallel, but as shown in Figures 2–3, they obviously are not. In *Compsognathus*, the collateral ligament fossa in the articulating phalanx proximal to the ungual clearly shows that these two bones are preserved in natural articulation *with each other;* thus, flexion of the ungual was parallel to the long axis of each adjacent phalanx and at right angles to the axis formed by the collateral ligament "hinge" represented by the fossae. *However,* the axis of the collateral ligament fossae of the first metatarsal—the "hinge" against which those two distal phalangae flexed—is unmistakably *perpendicular* to the "hinge" axis of the distal segments. There are two possible explanations. First, the two distal phalangae might have been twisted 90° about their long axis relative to the position of Metatarsal I, or the metatarsal might have been rotated 90° the other way (thereby placing the first metatarsal farther *behind* the second metatarsal)—otherwise the hallux must have possessed a most extraordinary pair of contrasting joints—namely, lateral translation at the proximal joint with the metatarsal, combined with parasagittal flexion between the phalangae at the next joint of that digit. The respective positions of the hallux components, illustrated in Figures 2–3, record beyond any question that the hallux of *Compsognathus* was reflexed. Note that Tarsitano (1985:328) stated, "In conclusion, the position of the hallux phalanges are more reliable than that of Metatarsal I in this specimen of *Compsognathus*," maintaining that the first metatarsal has been displaced, but its phalanges have not. It is obvious that he and I still disagree on the condition preserved in *Compsognathus*.

The condition of the hallux in other theropods is less clear. In most specimens familiar to me, the foot elements have been removed from the enclosing matrix and thus fully disarticulated, so no clue about the original articulations of the foot components is preserved. The hallux in the specimens of *Coelophysis* (e.g., AMNH [American Museum of Natural History] 7223) does not seem to be reflexed. The hallux of *Velociraptor mongoliensis* (AMNH 6518) also is not reflexed. The type specimen of *Saurornithoides mongoliensis* (AMNH 6516), however, shows something peculiar. The left pes is preserved in articulation, revealing the plantar surfaces. Digit I is preserved in articulation with the lateral surface of Metatarsal II and the plane of flexion of the hallux is preserved in the plane of the metatarsus—but *away* from the metatarsus! It would appear that the hallux had been compacted into the bedding plane of the foot, and thus the hallux had been rotated out of its normal position. My best guess is that the hallux of *Saurornithoides* was not reflexed, but that is far from certain. The hallux of *Deinonychus* (MCZ [Museum of Comparative

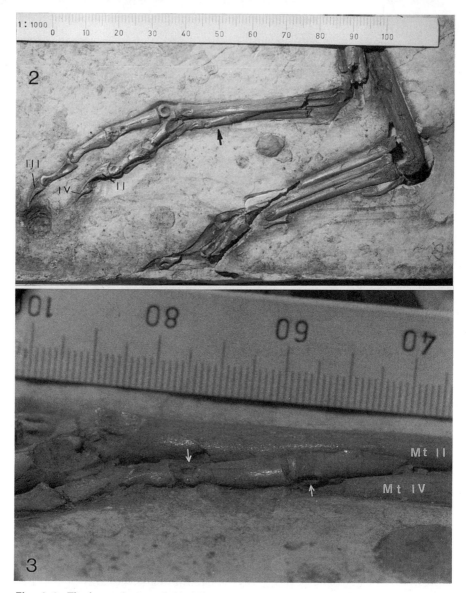

Figs. 2–3. The feet and metatarsals of *Compsognathus longipes* (BSP 1563). (2) Left foot *(above)* and right *(below)* showing the birdlike toe configuration and elongated metatarsals. Arrow points to the left hallux (Digit I). Scale units = 1.0 mm. (3) Microphotograph of the left tarsus showing the hallux in which the proximal phalanx and ungual have been rotated 90° or more about their long axis relative to the first metatarsal, which is firmly and naturally articulated against Metatarsal II—in its normal position on the metatarsal posteromedial surface. The entire metatarsus has been slightly crushed transversely, pressing Metatarsal IV up beneath the hallux metatarsal. It is Metatarsal IV that has been displaced and distorted, *not* Metatarsal I, as interpreted by Tarsitano (1985, this volume). Arrows point to the collateral ligament fossae. The axes of flexion at the two joints of the hallux—which *must* have been parallel in life, but as preserved here are almost perpendicular to each other— provide clear evidence of postmortem deformation. Correcting for that deformation, one can only conclude that the hallux of *Compsognathus* was reflected and originally opposed the three main toes. Scale units = 0.5 mm. Mt = metatarsal. BSP = Bayerische Staatssammlung für Paläontologie, Munich.

Zoology, Harvard] 4371) also seems to have an unreflexed orientation, but the Yale material (YPM [Yale Peabody Museum] 5205) shows distinct articular facets on the medial surface of the shaft of Metatarsal II that suggest a posterolateral contact with Metatarsal I (see Ostrom, 1969:Fig. 75). The case for a reflexed hallux position in theropods seems ambiguous. A few specimens suggest that it was reflexed; many others seem to indicate an unreflected condition; and most material preserves no indication at all. We simply need more data. But the condition in *Compsognathus* is unequivocal; therefore, theropod hallux reflection did exist.

The Carpus of *Archaeopteryx*

Until the true identity of the Eichstätt specimen of *Archaeopteryx* was recognized, details of the construction of the wrist in *Archaeopteryx* were not available. But this remarkable specimen clearly reveals details of the carpus that are extraordinarily similar to those preserved in several theropods—specifically, *Deinonychus, Velociraptor, Stenonychosaurus,* and *Coelurus.* Of special importance is a single carpal bone (= radiale?; the identity of the element is of less importance than its function) that articu-

Fig. 4. Right carpus and metacarpus of the Eichstätt specimen of *Archaeopteryx* viewed in dorsolateral aspect. Arrow points to the distinctive semilunate carpal element articulated closely with the first and second metacarpals, but not the third. So far, this particular element and its articular configuration are known only in certain theropods and *Archaeopteryx.* Scale units = 0.5 mm. I, II, and III identify the metacarpals.

lated closely with Metacarpals I and II, but not with Metacarpal III. The
bone is large and semilunate, and it seems to have articulated directly
with the radius and ulna in all of the taxa mentioned above (see Fig. 4).
This remarkable structural and morphologic resemblance of the *Archaeopteryx* condition to that of several theropods is a classroom example of a
synapomorphy (= homology). To the best of my knowledge, no comparable bone form or character complex occurs in any other group of vertebrates. It occurs *only* in *Archaeopteryx* and certain theropods. There is no
physical evidence available in either the few specimens of *Archaeopteryx*
or the several theropod taxa mentioned above to show whether this
distinctive semilunate bone was ossified from a single element, or the
result of fusion of two or more carpal centers. What is conspicuous,
though, is the unique morphology of the bone (i.e., having a pronounced
asymmetrical proximal ginglymus) and its very tight articulation with the
first two metacarpals but not the third. The strikingly similar configurations of the hand in *Archaeopteryx* and theropods such as *Ornitholestes*,
Chirostenotes, *Velociraptor*, and *Deinonychus*, together with the construction of the wrists, is forceful—indeed compelling—evidence for a *close*
evolutionary relationship between theropods and *Archaeopteryx*. And, *if*
Archaeopteryx is assigned correctly to the Class Aves (as I advocate), that
relationship extends to modern birds—whether by a mainline route, or a
sibling route.

Obligate Bipedality of Birds, Theropods, and *Archaeopteryx*

There are a number of other anatomical details that could be mentioned in support of the "theropod-bird" relationship, but most of these
have been expressed elsewhere (see Ostrom, 1976, for a biased view).
However, there is a final point that I would like to emphasize. That is the
matter of obligate bipedality. Although it may seem to be trivial—and
perhaps (as I warned at the beginning) an instance of "my seeing only
what I want to see"—I consider it extremely important. Birds fly on their
forelimbs alone *only* because somehow those appendages previously had
been "released" from the more usual four-legged stance and locomotion
of typical tetrapods. (Bats, and probably pterosaurs, as well other "flying" tetrapods never were freed fully from the usual quadrupedal posture.) It is quite obvious that (in addition to birds) theropod dinosaurs *also*
acquired the bipedal pose as an *obligate* posture. In my view, that is
not coincidental. The probability that bipedality occurred independently
more than once (and so closely spaced in time) is exceedingly remote, to
say the least (despite the peculiar and unnatural human posture, which
developed much later from a quadrupedal posture). Martin (this volume)
suggests a "slightly sprawled hind limb arrangement" for *Archaeopteryx*

—a suggestion that presents him with some difficulties in deriving birds from an obligate bipedal theropod ancestry. Yet, I think Martin would agree that obligate bipedality preceded even early flight ability in the history of avian evolution. On the other hand, Tarsitano believes that bipedality developed concurrently with the evolution of flight, not before flight. That may be, but it seems less likely to me. First of all, *if* flight developed in a quadrupedal avian precursor, what selective forces would have led to modification of the quadrupedal flight apparatus to a pure forewing flight structure? Or did this early quadrupedal avian flier develop the forelimb wings entirely independently of hind limb adaptation and function? If so, then it was a biped before it was a flier. That would account for the complete functional separation of avian forelimbs and hind limbs. All nonavian fliers, past and present, are or were quadrupedal fliers with greatly diminished terrestrial locomotory mobility and versatility. Birds are the exception.

Yes, there are many varieties of tetrapods that *can* assume a bipedal stance—and even a bipedal gait. Take, for example, the modern lizard *Basiliscus*, which is remarkably fleet as a facultative biped (which might be viewed as a mechanical analogue to the pseudosuchian thecodonts—putative bipeds—that often have been postulated to be the progenitor of proavis). Tarsitano (1983, 1985) noted that the "pseudosuchian thecodonts were not true bipeds," but is this semibipedal mode that he visualizes the most likely source of both avian flight and true bipedality? That poses a problem when we recall that *only* birds (with flight) *and* the theropods (without flight) are (were) *forced* to walk on their hind limbs only. You may describe this as "convergence" or "adaptive similarity," but the common denominator is one of *major structural and behavioral modifications* linked to a most unusual means among tetrapods of moving about. No other group of vertebrate animals comes closer to the avian biomechanical framework than the theropods—no matter how many details someone may distinguish about tooth form, shoulder-socket orientation, tarsal homologies, pelvic form, and the like. Despite the repeated suggestion that certain, unspecified "thecodontians" (whatever they are) tended toward bipedal locomotion, not a single kind has been shown to have been incapable of quadrupedal locomotion. None has been demonstrated to be an *obligate* biped. Nor has a single kind of "crocodilomorph" been shown to be an obligate biped or to have a tendency toward that style of locomotion.

CONCLUDING REMARKS

Although I may be "seeing only what I want to see," I still am convinced (on many grounds) that today's birds are the direct descendant of coelurosaurian theropod dinosaurs. One may call them sibling (= sister) groups, or ancestral-descendant groups. That distinction is beyond our powers of resolution or testing, and it is not of primary importance. What is important is that we recognize that there is a solid body of real evidence that demands careful evaluation of the hypothesis that birds and theropod dinosaurs *are* closely related—more closely related to one another than either is to crocodilians or to "thecodontians."

Acknowledgments I am indebted to many friends and colleagues who provided generous hospitality and assistance over the years. They are too many to name. To those who provided special access to specimens under their charge, my special thanks go to Alan Charig, Angela Milner, Cyril Walker, and Peter Whybrow of the British Museum (Natural History), London; Herman Jaeger, Humboldt Museum für Naturkunde, Berlin; Günter Viohl, Jura Museum, Eichstätt, Germany; Peter Wellnhofer, Bayerische Staatssammlung, Munich; Theo Kress, Solnhofener Aktienvereins, Solnhofen, Germany; Rupert Wild, Staatliches Museum für Naturkunde, Stuttgart, Germany; and C. O. van Regteren Altena, Teyler Stifting, Haarlem, Netherlands. Over the years, I have enjoyed and profited from numerous discussions of issues involving the origin of birds with all of the above, plus many other colleagues and students— e.g., Walter Bock, Paul Bühler, Peter Galton, Jacques Gauthier, Max Hecht, Larry Martin, Storrs Olson, Kevin Padian, Sam Tarsitano, Alick Walker, and others.

Literature Cited

Barthel, K. W. 1964. Zur Entstehung der Solnhofener Plattenkalke (unteres Untertithon). Mitt. Bayer. Staatssamml. Palaeont. hist. Geol., 4:37–69.
Barthel, K. W. 1978. *Solnhofen: Ein Blick in die Erdgeschichte*. Thun, Switzerland: Otto Verlag.
Broom, R. 1906. On the early development of the appendicular skeleton of the ostrich, with remarks on the origin of birds. Trans. So. Afr. Philos. Soc., 16:355–368.
Broom, R. 1913. On the South African pseudosuchian *Euparkeria* and allied genera. Proc. Zool. Soc. London, 1913:619–633.
de Beer, G. 1954. *Archaeopteryx lithographica*. London: British Mus. (Natural History).
Edmund, A. G. 1960. Tooth replacement phenomena in the lower vertebrates. Contrib. Roy. Ontario Mus. Life Sci. Div., 52:1–190.
Elzanowski, A. 1976. Palaeognathous bird from the Cretaceous of Central Asia. Nature, London, 264:51–53.
Elzanowski, A. 1981. Embryonic bird skeletons from the Late Cretaceous of Mongolia. Palaeontol. Polonica, 42:147–179.
Gauthier, J. 1986. Saurischian monophyly and the origin of birds. Pp. 1–55 *in* Padian, K. (ed.), *The Origin of Birds and the Evolution of Flight*. Calif. Acad. Sci. Mem. 8. San Francisco: California Academy of Science.
Gauthier, J., and K. Padian. 1985. Phylogenetic functional and aerodynamic analyses of the origin of birds and their flight. Pp. 185–197 *in* Hecht, M. K., J. H. Ostrom,

G. Viohl, and P. Wellnhofer (eds.), *The Beginnings of Birds*. Eichstätt: Freunde des Jura-Museums.

Häberlein, E. 1877. Neue Funde von *Archaeopteryx*. Leopoldina, 13:80.

Hecht, M. K., and S. Tarsitano. 1982. The paleobiology and phylogenetic position of *Archaeopteryx*. Geobios. Spec. Mem., 6:141–149.

Heilmann, G. 1926. *The Origin of Birds*. London: Witherby.

Heller, F. 1959. Ein dritter *Archaeopteryx* Fund aus den Solnhofener Plattenkalken von Langenaltheim/Mfr. Erlanger Geol. Abh., 31:1–25.

Howgate, M. E. 1984. The teeth of *Archaeopteryx* and a reinterpretation of the Eichstätt specimen. Zool. J. Linn. Soc., 82:159–175.

Howgate, M. E. 1985. Problems of the osteology of *Archaeopteryx*: is the Eichstätt specimen a distinct genus? Pp. 105–112 in Hecht, M. K., J. Ostrom, G. Viohl, and P. Wellnhofer (eds.), *The Beginnings of Birds*. Eichstätt: Freunde des Jura-Museums.

Huxley, T. H. 1868. On the animals which are most nearly intermediate between the birds and reptiles. Ann. Mag. Natur. Hist., 4:2:66–75.

Janicke, V. 1969. Untersuchungen über den Biotop der Solnhofener Plattenkalk. Mitt. Bayer. Staatssamml. Palaeont. hist. Geol., 9:117–181.

Jensen, J. 1981. Another look at *Archaeopteryx* as the world's oldest bird. Encyclia, Utah Acad. Sci., 58:109–128.

Lambe, L. 1917. The Cretaceous theropodous dinosaur *Gorgosaurus*. Canadian Geol. Surv. Mem., 100:1–84.

Langston, W. 1973. Ziphodont crocodiles; *Pristichampsus vorax* (Troxell) new combination, from the Eocene of North America. Fieldiana Geol. Ser., 33:16:291–314.

Madsen, J. 1977. *Allosaurus fragilis*: a revised osteology. Utah Geol. Mineral. Surv. Bull., 109:1–163.

Marsh, O. C. 1873. On a new subclass of fossil birds (Odontornithes). Am. J. Sci., (3)5:161–162.

Marsh, O. C. 1880. Odontornithes: a monograph on the extinct toothed birds of North America. U.S. Dept. Army, Prof. Paper, 18:1–201.

Martin, L. D. 1983a. The origin of birds and of avian flight. Pp. 105–129 in Johnston, R. F. (ed.), *Current Ornithology*. Vol. 1. New York: Plenum Press.

Martin, L. D. 1983b. The origin and early radiation of birds. Pp. 291–353 in Brush, A. H., and G. A. Clark (eds.), *Perspectives in Ornithology*. Cambridge: Cambridge Univ. Press.

Martin, L. D. 1985. The relationship of *Archaeopteryx* toothed birds. Pp. 117–183 in Hecht, M. K., J. Ostrom, G. Viohl, and P. Wellnhofer (eds.), *The Beginnings of Birds*. Eichstätt: Freunde des Jura-Museums.

Martin, L. D., J. D. Stewart, and K. N. Whetstone. 1980. The origin of birds: structure of the tarsus and teeth. The Auk, 97:86–93.

Mayr, F. X. 1973. Ein neuer *Archaeopteryx* Fund. Palaeont. Zeit., 47:17–24.

McGowan, C. 1984. Evolutionary relationships of ratites and carinates: evidence from ontogeny of the tarsus. Nature, London, 307:733–735.

Meyer, H. von. 1861. *Archaeopteryx lithographica* (Vogel-Feder) und *Pterodactylus* von Solnhofen. Neues Jb. Miner. Geol. Palaeontol., 1861:678–679.

Nash, D. 1971. The morphology and relationships of a new genus of crocodilian from the Upper Triassic of Losotho. Thesis. Birkbeck College, London.

Ostrom, J. H. 1969. Osteology of *Deinonychus antirrhopus*, an unusual theropod from the Lower Cretaceous of Montana. Bull. Peabody Mus. Nat. Hist., 30:1–165.

Ostrom, J. H. 1970. *Archaeopteryx*: notice of a "new" specimen. Science, 170:537–538.

Ostrom, J. H. 1973. The ancestry of birds. Nature, London, 242:136.

Ostrom, J. H. 1975. The origin of birds. Pp. 55–77 in Donath, F. A. (ed.), *Annual Review of Earth and Planetary Science*. Vol. 3. Palo Alto: Annual Reviews, Inc.

Ostrom, J. H. 1976. *Archaeopteryx* and the origin of birds. Biol. J. Linnean Soc. London, 8:91–182.

Ostrom, J. H. 1978. The osteology of *Compsognathus longipes* Wagner. Zitteliana Abh. Bayer. Staatssamml. Palaeontol. Hist. Geol., 4:73–118.

Ostrom, J. H. 1985. The meaning of *Archaeopteryx*. Pp. 161–176 *in* Hecht, M. K., J. H. Ostrom, G. Viohl, and P. Wellnhofer (eds.), *The Beginnings of Birds*. Eichstätt: Freunde des Jura-Museums.

Parks, W. A. 1928. *Albertosaurus arctunguis*, a new species of theropodous dinosaur from the Edmonton Formation of Alberta. Univ. Toronto Geol. Ser., 25:1–42.

Seeley, H. G. 1869. Index to the fossil remains of Aves, Ornithosauria, and Reptilia, from the Secondary System of strata arranged in the Woodwardian Museum of the University of Cambridge. Cambridge: Cambridge Univ. Press.

Tarsitano, S. 1983. Stance and gait in theropod dinosaurs. Acta Palaeontol. Polonica, 28:251–264.

Tarsitano, S. 1985. The morphological and aerodynamic constraints on the origin of avian flight. Pp. 319–332 *in* Hecht, M. K., J. H. Ostrom, G. Viohl, and P. Wellnhofer (eds.), *The Beginnings of Birds*. Eichstätt: Freunde des Jura-Museums.

Tarsitano, S., and M. K. Hecht. 1980. A reconsideration of the reptilian relationships of *Archaeopteryx*. J. Linnean Soc. London Zool., 69:149–182.

Thulborn, R. A. 1984. The avian relationships of *Archaeopteryx*, and the origin of birds. J. Linnean Soc. London Zool., 82:119–158.

Thulborn, R. A., and T. L. Hamley. 1985. A new paleoecological role for *Archaeopteryx*. Pp. 81–89 *in* Hecht, M. K., J. H. Ostrom, G. Viohl, and P. Wellnhofer (eds.), *The Beginnings of Birds*. Eichstätt: Freunde des Jura-Museums.

Walker, A. D. 1972. New light on the origin of birds and crocodiles. Nature, London, 237:257–263.

Walker, A. D. 1974. Evolution, organic. Pp. 177–179 *in McGraw-Hill Yearbook of Science and Technology*. New York: McGraw-Hill.

Welles, S. P. 1983. Two centers of ossification in a theropod astragalus. J. Paleontol., 57:401.

Welles, S. P., and R. A. Long. 1977. The tarsus of theropod dinosaurs. *In Studies on Vertebrate Paleontology*. Ann. So. Afr. Mus., 64:191–218.

Wellnhofer, P. 1974. Das fünfte Skelettexemplar von *Archaeopteryx*. Palaeontographica Ser. A, 147:169–216.

Wellnhofer, P. 1988. A new specimen of *Archaeopteryx*. Science, 240:1790–1792.

Whetstone, K. N., and L. D. Martin. 1979. New look at the origin of birds and crocodiles. Nature, London, 279:234–236.

Williston, S. W. 1879. Are birds derived from dinosaurs? Kansas City Rev. Sci., 3:457–460.

14 Mesozoic Birds and the Origin of Birds

Larry D. Martin

The origin of birds has generated more controversy than all other phylogenetic questions except the origin of humans. Modern birds with their wings, feathers, heterocoelous vertebrae, and specialized breathing seem isolated from other vertebrate groups. Thus, it is amazing that a connecting link between birds and their more reptilian predecessors was recognized within two years of the publication of *On the Origin of Species.* This connecting link, *Archaeopteryx lithographica,* has occupied the center stage in the questions of bird origins, and Ostrom (1976b:91) goes so far as to state, "The question of the origin of birds can be equated with the origin of *Archaeopteryx*, the oldest known bird."

The specimens of *Archaeopteryx* usually are identified on the basis of the institution that either originally, or presently, houses them. There are six partial or complete skeletons and one isolated feather. The feather is presently housed in two museums. The main specimen or slab, which may include some mineralized tissues, is at the Munich Museum, whereas the counterslab is in the Museum für Naturkunde in Berlin. The feather was found in the communal quarry at Solnhofen, Bavaria, Germany. The six skeletons are housed in six separate locations: (1) The London specimen was found in 1861 in the Ottmann Quarry near Solnhofen. Currently, it is housed in the British Museum (Natural History). (2) The Berlin specimen was discovered in 1877 in the Durr Quarry at Wegscheidt near Eichstätt. It is in the Museum für Naturkunde in Berlin. (3) The Maxburg specimen was found in 1956 in the Opitsch Quarry near Solnhofen. This specimen is owned privately. (4) The Haarlem specimen was first recognized as an *Archaeopteryx* by Ostrom in 1970. It is housed in the Tyler Museum in the Netherlands. (5) The Eichstätt specimen was found in 1951 in a small quarry near Workersaell, a few kilometers north of Eichstätt. It is held by the Jura Museum in Eichstätt. (6) The Solnhofen specimen was discovered in a private collection in Solnhofen in 1987 and recently has been described by Wellnhofer (1988a). It is housed in the Bürgermeister-Müller Museum in Solnhofen.

Unfortunately, many of the anatomical details of *Archaeopteryx* have not received full description, and the restorations of anatomical structures may be flawed seriously. Some of the best restorations of *Archaeopteryx* are found in Heilmann's (1926) *Origin of Birds*, but even in this classical work, the restoration of the skull and lower jaw is pure fantasy. The neck is too long, and the restoration of the shoulder girdle and the position of the wings are incorrect. There is no factual basis to Kleinschmidt's (1951) restoration of the palate, and almost all the drawings in de Beer (1954) are vague and misleading.

De Beer's (1954) drawings primarily are guides to his excellent photographs that include extensive use of ultraviolet lighting. The London specimen has been cast repeatedly, and the separators used in the casting process seem to act in part as an ultraviolet screen. This was brought to my attention when new ultraviolet photographs were done for me at the British Museum, and newly prepared areas of bone were more fluorescent than were earlier preparations. It seems, then, that the quality of some of de Beer's photographs may be difficult to duplicate. The Haarlem *Archaeopteryx* was photographed with ultraviolet light by Lambrecht (1928), who thought that it was a pterosaur. Lambrecht's was one of the earliest applications of this technique to a fossil vertebrate. The technique is invaluable for the study of *Archaeopteryx*, because bones, calcite, and matrix fluoresce differently, thus permitting the recognition of the margins of bones, which otherwise would be unclear.

Since the publication of de Beer's monograph, four new specimens of *Archaeopteryx* have been recognized. One of these, the Maxburg Specimen, is in private hands and unavailable for scientific study. It was described in some detail by Heller (1959), and I have examined an excellent cast. Unfortunately, the bones are not well exposed, and although Heller's description includes X-ray photographs, these are difficult to interpret. The Haarlem specimen provides no information that is not available from other specimens, but it was the catalyst for Ostrom's extensive studies (Ostrom, 1973, 1974a). Ostrom's photographs of all the specimens are probably the best available and are cited frequently in the present work. The most detailed and accurate description of any single specimen of *Archaeopteryx* is that of the Eichstätt specimen by Wellnhofer

(1974). The fourth specimen recently was studied by Wellnhofer (1988b). I examined it briefly on exhibit, but I have not had it for study.

Almost all of the published accounts of *Archaeopteryx* are by experts in reptilian morphology; undoubtedly, this has influenced their interpretation of the specimens (Ostrom, this volume). When examined from the perspective of an avian biologist, *Archaeopteryx* has proven to be more avian than previously supposed (Martin, 1983a,b). The usefulness of *Archaeopteryx* is in part a factor of its position in avian evolution. I have argued that it is removed from the main line of avian evolution (Martin, 1985, 1987) and maintained it in a separate subclass, the Sauriurae (Appendix I). The monophyly of this subclass, which includes the Cretaceous Enantiornithes, is supported by a suite of derived characters that includes: (1) the reduction of the squamosal; (2) the fusion of the proximal, but not the distal, ends of the metatarsals; (3) the fusion of the proximal metatarsals in a row, rather than having the proximal end of Metatarsal III posterior to the proximal ends of Metatarsals II and IV, as in modern birds; and (4) an unusual association of the scapula with the axial skeleton. These features indicate that *Archaeopteryx* cannot be utilized as the sole model for avian origins, and coupled with the discovery of a Triassic protobird in Texas (Chatterjee, in litt., 1987), serve to separate *Archaeopteryx* further from the question of avian origins.

ADAPTATION IN MESOZOIC BIRDS

Birds are considered archosaurs largely on the basis of their possession of an antorbital fenestra; this is the main reason that we seek their origin among the diapsid reptiles, although no known bird is truly diapsid (contra Gardiner, 1982). Although some workers (Thulborn, 1984) have embedded *Archaeopteryx* within the dinosaurs, it is considered a bird because it has feathers, a furcula, and a pretibial bone. The feathers of *Archaeopteryx* are like those of modern birds in possessing a complex arrangement of barbs and barbules. The flight feathers are asymmetrical (Feduccia and Tordoff, 1979), which indicates their use in powered flight.

The Relevance of Ontogenetic Information

The avian skeleton is highly modified for flight; many of the modifications involve the stabilization of joints through the fusion or loss of individual bones. We can gain some insight into how these changes might have taken place by ontogenetic studies of modern birds. Because of environmental constancy, embryonic development tends to be conservative until late in ontogeny. Anatomical comparisons between the oldest known birds and ontogenetic stages of modern birds supports the idea that primitive characters are retained in early developmental stages, and that embryology is a good indicator of primitive-derived relationships. In fact, the ontogeny of modern birds may preserve something of avian history before *Archaeopteryx* and provide clues to avian origins. Modern birds are so modified that the adults are difficult to compare with reptilian groups. The skull bones are highly fused and their sutures usually cannot be identified, nor can individual elements of the carpometacarpus, tibiotarsus, and tarsometatarsus be determined readily. In order to compare these features, we must rely on a combination of fossil material and ontogenetic stages of Recent taxa, in addition to the rare examples of ontogenetic material of fossil birds. Another related question that should be examined is the relative importance of heterochronic shifts in development (paedomorphosis) in avian evolution and the determination of maturity in fossil birds.

Birds have terminal growth, and it is difficult to understand how the rapid, highly proscribed motions of flight can be maintained in an animal that has joints composed of growing cartilage rather than dense bone. The allometric problems relating wing-surface area to increasing body weight also would seem to limit the possibilities for coupling "continuous" reptilian growth with a flying animal. In modern birds, most growth has ended when the wings become functional and the tarsal bones have fused with the tibia and metatarsals. This also seems to be true for pterosaurs (C. Bennett, pers. comm., 1988). The presence of such fusions in the lower Cretaceous hesperornithiform bird, *Enaliornis*, indicates the terminal growth pattern of this taxon. The growth pattern of the late Jurassic bird, *Archaeopteryx*, is equivocal, because there is considerably less skeletal fusion than in modern birds.

The Nature of the *Archaeopteryx* Specimens

Although the six skeletons of *Archaeopteryx* vary greatly in size, all are allocated to a single species, *A. lithographica,* by Wellnhofer (1988b). Three of these—the London, the Haarlem, and the Maxburg specimens—are about the same size; these specimens generally are considered to be adults and possibly of the same sex. The Solnhofen specimen is the largest of the six examples, and the Eichstätt specimen is the smallest; the latter has been interpreted as a juvenile *Archaeopteryx lithographica* by Wellnhofer (1974). The very complete Berlin specimen lies between the London and Eichstätt specimens in body size. In all *Archaeopteryx* specimens, the cranial and individual jaw bones are unfused and the sutures clearly visible. The carpometacarpus also is unfused. Some of the anterior thoracic vertebrae seem to be fused, but synsacral vertebrae are fused less tightly than in modern birds and number only five. The tail is long and the pygostyle absent. The ribs lack ossified uncinate processes, although it is likely that cartilaginous ones were present. The pubis is opisthopubic, but all the pelvic bones are separated clearly by sutures (Fig. 5); the distal ends of the ilium and ischium are not united. The tibiotarsus lacks a distinct dorsally projecting rotular crest, and the astragalus and calcaneum are incompletely fused to the tibia. The tarsometatarsus is fused proximally but not distally. The tarsal contribution to the tarsometatarsus has not been documented clearly. There may be individual distal tarsals fused to each metatarsal, but it is certain that the large tarsal cap that characterizes modern birds is absent. The foregoing list is essentially a compilation of juvenile skeletal features of modern birds (Figs. 1–2, 4). When these characters occur in adults, they may be regarded as primitive (Figs. 3, 5–6) or as evidence of arrested development. The presence of these features in the adult *Archaeopteryx* leaves few characters by which to determine the maturity of individuals.

The Berlin *Archaeopteryx* has fully developed plumage and a fused tarsometatarsus; therefore, it is reasonable to assume that it is an adult. On the other hand, the Eichstätt specimen is so much smaller than the other specimens that it generally has been considered to be a juvenile. This view is supported by the condition of the tarsometatarsus, which seems less fused than the other specimens; however, the specimen also

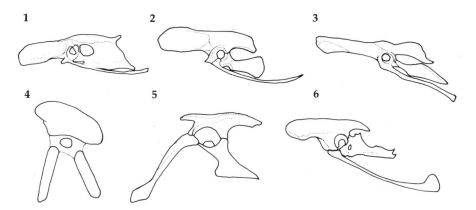

Figs. 1–6. (1) Laterial view of adult *Gallus* pelvis showing elongated anterior ilium, reflected pubis, and fusion of the ischium to the posterior ilium. (2) Juvenile *Gallus* showing sutures within the acetabulum and the lack of fusion of the ischium to the ilium. (3) Upper Cretaceous bird *Apatornis* showing a similar stage to (2) except that sutures are absent in the acetabulum. (4) Embryonic *Larus* with vertical pubis and short anterior ilium. (5) Triassic crocodilian *Protosuchus*, with a pelvis similar to the embryonic *Larus*. (6) *Archaeopteryx* with the pubis reflected. Pubic reflexion, ilium prolongation, and increased fusion appear in the fossil record in the same sequence as they appear in ontogeny. (Modified from Heilmann, 1926; Colbert and Mook, 1951).

seems to have fully developed flight plumage (Wellnhofer, 1974). It is impossible to determine if the teeth were fully socketed because the specimen is crushed. Thus, there seems to be little to support the interpretation of the Eichstätt specimen as a juvenile other than its small size. Walker (1985) listed cranial features that he thought indicated that the *Archaeopteryx* specimens are juvenile. However, adult Mesozoic birds have primitive features that are represented only in the ontogeny of modern birds; loss of these features is correlated approximately with decreasing age of the avifauna (Figs. 1–6). The presence of tooth sockets and adult plumage is adequate to document that the London and Berlin specimens are adult.

The skulls of Mesozoic birds, *Archaeopteryx, Enaliornis,* and *Hesperornis* retain distinct sutures throughout life. The absence of distinct cranial sutures in adults of modern birds may reflect new stresses associated with the use of a horny bill (rather than the toothed jaws of Mesozoic birds), although toothless Paleocene birds also seem to have less cranial fusion than do modern birds (Houde and Olson, 1981).

The sternum, interclavicle, and uncinate processes of the ribs are unossified in *Archaeopteryx,* as they are in juvenile stages of modern birds. The clavicle in modern birds is one of the first bones to ossify and seems to combine both membrane and endochondral elements (Romanoff, 1960). The interclavicle identified in several birds (Romanoff, 1960; Parker, 1891) is incorporated into the keel of the sternum (Fig. 8). Developmentally, the sternum of birds appears as a paired structure, each half of which is posteroventral to, and continuous with, a coracoid. The sternal plates extend ventrally at their midline junction; the ventral extension, along with the interclavicle, forms the keel in carinate birds (Romanoff, 1960). It

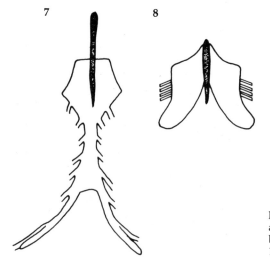

7   8

Figs. 7–8. The interclavicle *(shaded)* and sternum of (7) crocodilian; (8) bird *(Gallus).* (Modified from Stark, 1979.)

seems likely that the ossification of the sternal plates originally permitted the posterior extension of flight muscles originating from the furcula and coracoids. The development of a keel for these muscles at the junction of the sternal plates in advanced fliers is analogous to the development of the high sagittal crest of mammals that provides for the origin of the temporalis muscles. Ossification of the sternum facilitated development of avian breathing using the air sac system. Furthermore, the appearance of ossified uncinate processes at about the same time as ossified sterna may be related to this type of breathing, because these processes are attachment sites for muscles extending to the ribs. Both ossified sterna and uncinate processes occur in *Ichthyornis* and *Hesperornis* from the late Cretaceous.

In adult *Archaeopteryx* the ilium, ischium, and pubis are separated by distinct sutures, but in adult specimens of upper Cretaceous birds such as *Apatornis*, *Ichthyornis*, and *Hesperornis* these sutures are fused. However, in each of the latter three taxa, the distal ends of the ilium and ischium are separated (Fig. 1–6)—a condition that is found in adults of only a few modern birds.

Eggs and Nesting

The only Mesozoic bird that provides any direct information concerning its eggs is *Gobipteryx* from the late Cretaceous (?Campanian) of Mongolia. Seven *Gobipteryx* eggs containing embryos were reported by Elzanowski (1981); other eggs of unknown affinities also are present in the collection (A. Elzanowski, pers. comm., 1979). The eggs were found in sandstone of possible eolian origin; eolian deposits usually develop in regions with few trees. All of the embryos are similar in structure and closely resemble other known material of *Gobipteryx* (Elzanowski, 1981: 169). They indicate that *Gobipteryx* was ground-nesting and that the young were hatched in a relatively advanced state of development. Elzanowski (1981) compared the wing proportions of *Gobipteryx* at hatching with those of the living megapode, *Leipoa ocellata*, which can fly 24 hours after leaving the mound. The scapula and coracoid of the *Gobipteryx* embryos are very similar to those of the Enantiornithes (C. A. Walker, 1981), and I think that *Gobipteryx* is related closely to these birds. The Enantiornithes also resemble *Archaeopteryx*, and I include both in the subclass Sauriurae. We cannot determine if *Archaeopteryx* nested in trees or on the ground, but the structure of its pelvis indicates that *Archaeopteryx* probably laid a very small egg (Heilmann, 1926) and that its young probably were not precocious, as were those of the Enantiornithes.

The evidence suggests that *Hesperornis*, which had restricted mobility on land, must have nested in localized regions such as beaches (Martin and Tate, 1976). Immature hesperornithiform birds are rare, and most of their remains are found in full marine deposits. The young probably did not venture far into the open sea. Some support for this hypothesis can be found in the abundance of adults and juveniles of the hesperornithiform bird *Enaliornis* occurring in Lower Cretaceous, near-shore deposits in England. This material includes stages before the fusion of the astragalus, calcaneum, and pretibial bone to the tibia, and before the development of the high rotular crest on the tibiotarsus characteristic of hesperornithiform birds. On some specimens, the proximal end of the tarsometatarsus is unfused, and the distal tarsals are not fused with the tarsometatarsus. These bones represent approximately the same stage of development as those reported by Martin and Bonner (1977) for a juvenile *Baptornis advenus* from the Upper Cretaceous of Kansas.

Ontogeny and Relationship

In Mesozoic birds, tooth implantation follows a characteristic ontogenetic pattern. In *Ichthyornis*, the teeth of young birds apparently are set in a groove, but with maturation, sockets form in an anterior-posterior sequence along the jaw (Martin and Stewart, 1977). A similar process occurs in crocodilians. In both birds and crocodilians, the formation of the tooth sockets is different than in other archosaurs. This supports a bird-crocodile relationship, as may the embryological relationship of the sternal plates to the interclavicle in birds, which is similar to the condition in adult crocodilians (Figs. 7–8)—both have a simple, rod-shaped interclavicle, and the putative sterna of theropods seems to be single and different in shape. The homology of the foramen pseudorotundum in birds and crocodilians is supported by embryological evidence (Whetstone and Martin, 1979, 1981); although we cannot ascertain the embryology of this feature in extinct reptiles, this does not affect our consideration of it in birds and crocodilians.

Ontogeny also provides an important clue to the evolutionary position of *Archaeopteryx*. In *Archaeopteryx*, the tarsometatarsus is fused only proximally, whereas in modern birds (as well as the Cretaceous Hesperornithiformes) the metatarsals fuse distally before they fuse proximally. The distal tarsals of modern birds fuse together, forming a thick, solid cap over the metatarsals. In *Archaeopteryx*, the distal tarsals seem to be small and individually fused to their respective metatarsals (Wellnhofer, 1974:Fig. 12).

Feathers and Flight

Detailed feather impressions are preserved in the Berlin specimen of *Archaeopteryx*. The best impressions are on the counterslab; these were illustrated by Feduccia (1980) and appear as a beautiful inset in the book *Beginnings of Birds* (Eichstätt: Freunde des Jura-Museums, 1985). These photographs clearly document the asymmetry of the flight feathers used by Feduccia and Tordoff (1979) in their elegant demonstration of powered flight in *Archaeopteryx*. This, coupled with the evidence of the manus claws provided by Yalden (1985) and the analysis of flight mechanics by Norberg (1985) and Rayner (1985), has laid to rest the theory of the cursorial origin of bird flight.

It seems likely that *Archaeopteryx* was a poor flyer by modern standards but was able to make powered flights from tree limb to tree trunk. The slightly sprawled hind limbs and the elongate manus claws would have facilitated trunk climbing. This mode of life would be difficult to derive from obligate bipeds like theropod dinosaurs, because their shortened forelimbs and stiff tails would make climbing vertical tree trunks nearly impossible. It is no accident that the strongest support for the cursorial origin of avian flight comes from researchers committed to a dinosaur origin for birds.

Rietschel (1985) and Stephan (1985) briefly described the flight feathers in *Archaeopteryx*. Feather impressions are found only some distance from the feather insertions; decay of the soft tissue near the insertions probably caused an unfavorable preservational environment. This lack of direct evidence for feather insertion has resulted in controversy over the exact distribution of primaries, although the general distribution is clear. The number of primaries is either 11 or 12 (Rietschel, 1985); Stephan (1985) and Helms (1982) also suggested the presence of a remicle. Stephan (1985:263) stated that "the rather long primaries and secondaries presuppose an efficient manner of closing the wings and of holding them tightly to the body." I pointed out (Martin, 1983b, 1985) that a modern avian osteological system for wing folding is present in *Archaeopteryx*.

Norberg (1985) carries the argument of Feduccia and Tordoff (1979) to its logical conclusion, showing that asymmetrical feathers coupled with feather curvature in *Archaeopteryx* can indicate only "true powered flight." Moreover, microscopic examination of the fabric of the feathers of the Berlin *Archaeopteryx* indicates a basically modern structure. The regular spacing of barbs throughout their length and faint but clear indications of barbules (Fig. 9) demonstrate that the microscopic details of the feathers were among the first avian features to develop.

Postcranial Pneumaticity

Examination of the external surface of the long bones of *Archaeopteryx* indicates that pneumatopores were absent (contra Heller, 1959) and that the long bones were not pneumatic. This conclusion is not surprising

Fig. 9. Scanning electron microscope photograph of a cast of one of the Berlin *Archaeopteryx* feather impressions. A = rachis; B = barb; C = barbule.

3 cm

when we consider that the late Cretaceous flying bird *Ichthyornis* also was nonpneumatic. Many workers have confused hollow bones with the type of pneumaticity found in modern birds. The long bones of virtually all small tetrapods contain large medullary cavities and thus appear hollow. This is not the condition characteristic of modern birds in which the air sac system invades the bones through distinct foramina (pneumatopores). Pleurocoels are weakly developed in the vertebral column of *Archaeopteryx* and are strongly developed in the enantiornithine birds, as well as in hesperornithiforms, *Ichthyornis*, and all modern birds. They are distributed widely among archosaurs, but I have not found a good discussion of their function.

THE SKELETON OF *ARCHAEOPTERYX*

The Skull

The discovery of the Eichstätt specimen of *Archaeopteryx* and the recent preparation of the London cranium (Whybrow, 1982) have increased our knowledge of the skull greatly (Figs. 10–14). Heilmann's restoration based on the Berlin specimen now can be shown to be erroneous except for the general shape of the skull, the size of the orbit, the shape of the antorbital fenestra, and so forth (see Ostrom, 1976b). Heilmann's restoration is remarkably similar to his drawing of *Aetosaurus ferratus* (Heilmann, 1926:Fig. 133), and it seems likely that *Aetosaurus* may have served as a model for restoration of the *Archaeopteryx* skull. Wellnhofer (1974) made the first accurate restoration of an *Archaeopteryx* skull, and his work forms the basis for later restorations. I have combined aspects of his restoration with information from the London cranium and study of the Eichstätt specimen for a new restoration (Figs. 10–11). In this restoration, the

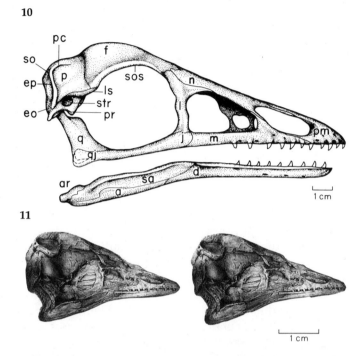

Figs. 10–11. (10) New restoration of Eichstätt *Archaeopteryx* skull and mandible based on study of the specimen and comparison with the London example and Wellnhofer, 1974. (11) Stereophotograph of a cast of the Eichstätt skull. a = angular; ar = articular; d = dentary; eo = exoccipital; ep = epiotic; f = frontal; j = jugal; l = lacrimal; ls = laterosphenoid; m = maxillary; n = nasal; op = epiotic; p = parietal; pc = parietal crest; pm = premaxillary; pp = paraoccipital process; pr = prootic; q = quadrate; qj = quadratojugal; sa = surangular; so = supraoccipital; sos = supraorbital shelf; str = superior tympanic recess.

cranium is more expanded; there is a supraorbital shelf and parietal crests, and the bone that Wellnhofer labeled the squamosal is reinterpreted as an occipital bone. *Archaeopteryx* resembles other birds in having a pointed snout, large orbits (about one-third of the skull length) with ossified sclerotic rings, an expanded braincase, and the complete loss of the upper temporal arch.

The premaxillae of *Archaeopteryx* are small, being only about one-third of the length of the dentigerous portion of the jaws. The dorsal and lateral surfaces are covered by nutrient foramina, and the bone seems to have had four or five unserrated, posteriorly recurved teeth. The premaxilla forms the anterior two-thirds of the dorsal border, as well as about one-third of the ventral border, of the external naris. Dorsally, the posterior third of the nasal opening is formed by the nasal. These paired bones are separated for a short distance anteriorly by the premaxillae. The nasal forms the posterior border of the external naris and meets the dorsal process of the maxilla near this posterior border. The external nares are large (nearly one-third as long as the entire snout), ovoid, and inclined upward at about 30°. The premaxillae overlap the maxillae for a short distance anteriorly, but most of the ventral borders of the external nares are formed by the maxillae. In all other birds including the Hesperornithiformes, the premaxillae contact the nasals and exclude the maxillae from the external nares. This is the condition in pseudosuchians, crocodilians, sphenosuchids, hadrosaurs, ceratopsians, ankylosaurs, hypsilophodontids, stegosaurs, prosauropods, and ornithomimids. I suspect that the condition in *Archaeopteryx* is derived. Some sauropods and many theropods also have maxillary exposure in the external nares, as do pterosaurs. There is a large premaxillary foramen just anterior to the anterodorsal margin of the external nares.

The maxilla is large, extending three-quarters of the length of the snout. It has a tall dorsal process extending up to and bordering the nasal along the dorsal rim of the antorbital fenestra.

One of the most prominent trends in the evolution of the avian skull is the enlargement of the premaxillae. Birds (including *Archaeopteryx*) have large external nares that are bordered dorsally by the premaxillae. This feature is shared by both ornithischian and saurischian dinosaurs but is not present in crocodilians or sphenosuchids. Ellenberger (1977) illustrated a long dorsal process of the premaxilla for *Cosesaurus*, but the naris is small and the feature was not clear when I examined the specimen. Also, I could not determine the length of the dorsal process in *Scleromochlus*. In *Petrolacosaurus*, the naris is small and bordered anteriorly by the premaxilla, which extends posteriorly about half the length of the maxilla. Therefore, dorsal extension is primitive. In *Archaeopteryx*, the premaxillae separate the nasals only in the region above the external narial opening (Fig. 12). In most later birds, the premaxillae completely separate the nasals, which have become small peripheral bones bordering the antorbital fenestrae and the external nares.

Whetstone (1983:Fig. 3) identified as the mesethmoid part of the region considered by Wellnhofer to be the maxilla. As a result, the upper border of the antorbital fenestra is formed by the nasal (as it is in all other birds), rather than by the maxilla, which is the primitive condition in archosaurs. Wellnhofer may have interpreted this region to be part of the maxilla, because that interpretation results in a restoration with two small subfenestra much like those of theropods. Witmer (1987) supported Wellnhofer's interpretation and argued that this configuration resulted from the presence of an enlarged maxillary sinus. However, the absence of any similar structure in the antorbital fenestra of the Berlin specimen argues in favor of the mesethmoid interpretation.

The nasals of *Archaeopteryx* differ from those of all other known birds in being broad and having a long midline contact (Fig. 12). They resemble

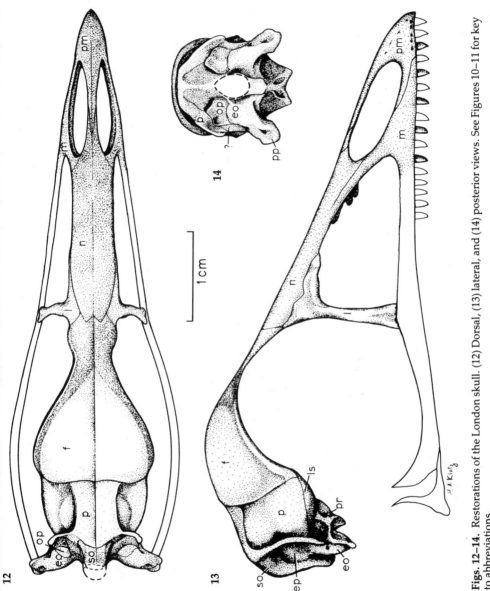

Figs. 12–14. Restorations of the London skull. (12) Dorsal, (13) lateral, and (14) posterior views. See Figures 10–11 for key to abbreviations.

497

the nasals of other birds in two respects. First, they possess long, narrow anterior processes that extend ventrolateral to the dorsal processes of the premaxillae and form about half of the dorsal border of the external nares. Second, the nasal bears a descending process that forms the posterior border and about one-third of the ventral border of the external naris. The two rectangular bones adjacent to the premaxilla on the London center slab appear to be nasals and are slightly arched in a manner consistent with this interpretation. The nasals meet the frontals just above the anterior margin of the orbit.

The antorbital fenestra is large and triangular. Its posterior border is formed by a T-shaped lacrimal (Fig. 13). The lacrimal is similar in shape to that in *Parahesperornis* (Martin, 1984), and in the London specimen, it consists of one bone. (A prefrontal element was suggested by Wellnhofer [1974] for the Eichstätt specimen, but this cannot be confirmed.) Contrary to the suggestion of Tarsitano and Hecht (1980), the lacrimal is not fused to the jugal. It is difficult to see how any streptostylic movement was possible, although it has been suggested by Bühler (1985).

The frontals are flattened anteriorly where they are overlapped by the nasals and abut against the posterior process of the lacrimal. The superior border of the orbit is formed by a thin supraorbital shelf, which is visible on the London specimen and on the mold of the Eichstätt specimen. Wellnhofer (1974) misidentified a portion of the supraorbital shelf as part of the postorbital. The parietal is large for a bird and has distinct, paired parietal crests (Figs. 12–13). The laterosphenoids are large and well ossified (Figs. 10–11, 13).

Whetstone (1983), A. D. Walker (1985), and Witmer (1987) agreed on the identification of a superior pneumatic recess (Fig. 10). The configuration of this recess is unusual in *Archaeopteryx* because it is a shallow, lateral pocket in the prootic. Normally, it would be covered by the squamosal. The whole question of the squamosal is controversial in *Archaeopteryx*, and it now seems impossible that a "normal avian" squamosal was present. Heilmann drew the squamosal much as he did in *Aetosaurus*, but this area is badly damaged in the Berlin specimen and there is no evidence for any bone of the type he described. (See Ostrom's [1979b] excellent photograph of the Berlin skull.) The squamosal proposed by Wellnhofer (1974) lies entirely in the occipital region; this is similar to the position of the squamosal in pterodactyloids (see Wellnhofer, 1978:Fig. 3) but is very different from the situation in either birds or other reptiles. (The bone Wellnhofer identified as the squamosal also includes the paraoccipital process; therefore, it should be identified as the epiotic-exoccipital.) Numerous workers agree that the squamosal of Wellnhofer (1974) and Ostrom (1976b) is an occipital bone or bones (Fig. 14), including the paroccipital process (Martin, 1983a,b; Whetstone, 1983; Tarsitano and Hecht, 1980; A. D. Walker, 1985). Bühler (1985) suggested that the bone labeled laterosphenoid by Wellnhofer might be the squamosal, although Whetstone (1983) and A. D. Walker (1985) considered it to be the prootic. The bone includes the superior pneumatic recess and is clearly the bone identified as the prootic (Fig. 13) in the London specimen; thus, it seems certain that Walker and Whetstone are correct in their evaluation, whereas Bühler is not.

Whetstone (1983) reviewed the newly prepared London cranium and concluded that the squamosal was either absent or reduced. If it was present, it had an unusual configuration and might be the "triradiate bone" that Wellnhofer (1974) described as an impression on the Eichstätt counterslab and identified as a postorbital. Whetstone found no articulations for this bone on the London cranium. If the bone is part of the *Archaeopteryx* skull, it would be more reasonable to accept Wellnhofer's identification than to interpret it as a squamosal that differs widely from that found in any other reptile or bird. Examination of the Eichstätt specimen (Figs. 6, 11) and of the molds that were made for casting leads me to doubt the existence of any "triradiate bone" of the sort restored by Wellnhofer. No missing bone seems to be needed to account for the impression, because fragments of bone (?laterosphenoid and quadrate) of the correct shape and position exist at this position on the main slab.

Whetstone's (1983) speculation that the squamosal was reduced or absent in *Archaeopteryx* is unanticipated and would represent a derived condition, because all modern birds and reptiles that may be closely related to birds have large squamosals. A. D. Walker (1985) suggested that the bone with which the quadrate of the Eichstätt specimen articulates is, in fact, a small squamosal rather than the prootic as interpreted by Whetstone (1983). I find no support for Walker's interpretation. Instead, it seems likely that Walker's putative squamosal is part of either the flat threshold to the posterior tympanic recess or is part of the metotic; in any case, it is a medial structure crushed laterally. Should an additional small bone be present, the resulting unique morphology would not be comparable to any suggested avian outgroup. There is a structure in the occipital crest of the London specimen that might be interpreted as a suture delineating a small squamosal (Fig. 14, indicated by a question mark), but this structure has no parallel in the same region of the Eichstätt specimen; therefore, I think that it is simply a fracture.

In most diapsids, the quadrate articulates dorsally with a squamosal that is displaced lateral to the braincase such that the quadrate has a vertical orientation in posterior aspect. A theoretical problem with this arrangement is that the jaw articulation is located toward the tip of the lever arm (i.e., the squamosal); this necessitates a secure brace with the skull. The upper temporal arch of diapsids seems to provide such a brace.

Thus, the quadrate articulates dorsally with the squamosal; if the quadrate is to be mobile, the only arch that can be sacrificed is the lower one. Loss of the lower arch, in fact, has occurred (possibly many times) in lizards and lizardlike reptiles.

Primitively in tetrapods, the quadrate slants inward proximally and thereby braces the suspensorium against the braincase. In this position, the quadrate may articulate with several bones of the braincase, depending on the size of its head. The complex articulations of the quadrates of birds and crocodilians probably are correlated with the enlarged and complex quadrate heads that characterize these groups. Such an articulation would allow for loss of the middle diapsid arch—a uniquely avian trait that is already present in the Upper Triassic protobird (Chatterjee, 1987) and in *Archaeopteryx*. In *Archaeopteryx*, the head of the quadrate is small (reduced) and appears to have lost completely its connection with the squamosal. In the Eichstätt *Archaeopteryx*, the quadrate is still in articulation, although crushing has faulted the skull so that the cranial roof is tilted laterally. Careful comparison of the Eichstätt and London specimens indicates that the site of the quadrate articulation probably is the small pocket at the anteromedial face of the paroccipital process.

The only clear thing about the temporal region in *Archaeopteryx* is that it must be autapomorphic, and thus it provides practically no information concerning the origin of birds. Efforts to restore a dinosaur-like squamosal-quadrate articulation (A. D. Walker, 1985; Paul, 1984) require the presence of an upper temporal arch for which there is no evidence. On the other hand, the highly derived articulation found in *Archaeopteryx* does not resemble that found in *Sphenosuchus*; A. D. Walker (1985) developed a theoretical argument to show that some shared aspects of the bird-crocodilian quadrate articulation may have been achieved independently. Because these characters in *Sphenosuchus* were central to A. D. Walker's (1972, 1974) views on a special crocodilian-bird relationship, he since has abandoned the idea (A. D. Walker, 1985).

The quadrate is visible in articulation in the Eichstätt specimen, but its medial and anterior margins are obscure or not visible. A. D. Walker (1985) identified the undetermined skull bone on the slab posterior to the London cranium as a quadrate. It is close to the correct size for this interpretation. However, Walker reported that the lower articular surface is missing, and the specimen cannot be easily compared with the Eichstätt quadrate. The bone illustrated by Walker (1985:Figs. 5b–c) is not similar to any other avian quadrate but in some ways does resemble the pterygoid of hesperornithiform birds. However, it may be too large to be a pterygoid.

Witmer (1987) described in detail what is known about pneumaticity in the skull of *Archaeopteryx*. He found evidence for a maxillary sinus and the three tympanic sinuses—i.e., the superior, posterior, and anterior tympanic diverticula. These have the general form found in birds and can be matched in the hesperornithiform birds (Witmer, 1987). Witmer was able to show that the characteristic avian pattern of pneumaticity is found throughout Mesozoic birds. Some of these features have been reported for Triassic crocodilians (Tarsitano, 1985a,b). Cranial pneumaticity has a wide distribution and long history in birds and crocodilians. It also occurs in certain dinosaurs, and at least one late Cretaceous form (*Troodon*) may have a lacrimal sinus and anterior and posterior tympanic recesses (Currie, 1985). This is more similar to the pneumatic system of birds and crocodilians than that found in any other dinosaur, but the homology of these structures to those in birds recently has been questioned (S. F. Tarsitano, pers. comm., 1989). The superior tympanic recess is unknown in dinosaurs (Witmer, 1987). Quadrate-articular pneumaticity is unknown in troodontids and among theropods seems to be found only in the otherwise isolated tyrannosaurids, which lack most of the other pertinent pneumatic features. The occurrence of these cranial features in dinosaurs can be viewed in one of two ways—either (1) as a feature of the common ancestor of birds, crocodilians, and dinosaurs, or (2) as convergent developments within theropods (Tarsitano, 1985a,b). The latter interpretation is consistent with the scattered occurrence of these features.

The superior tympanic recess presently holds as an avian-crocodilian synapomorphy (Witmer, 1987), and the avian configuration of cranial pneumatic sinuses and the course of the carotid within the sinus system and the possession of a fenestra pseudorotundum are all features shared by birds and crocodilians that, as yet, have not been found in combination in any single theropod.

The London skull is separated into a tightly sutured posterior cranium and a lightly sutured snout region. This is also true of hesperornithiform birds. In the London skull, the premaxilla, maxilla, lacrimal, and nasals lie scattered a slight distance from the cranium.

In the lower jaw of the Eichstätt *Archaeopteryx*, the dentary is deflected ventrally (Fig. 10). Ostrom (1976b) pointed out that this deflection is theropod-like. The dentary is slender and bears 11 erupted teeth of which the crowns, but not the roots, are visible. There is a distinct groove below the teeth that widens posteriorly; a row of elongate foramina is located in the groove. A very similar morphology characterizes *Ichthyornis* and *Hesperornis*. The dentary, maxilla, and premaxilla are covered by numerous small, elongate foramina indicating a covering of large scales. A portion of the lateral side pulled off with the counter slab in the middle of the jaw to expose a large splenial that contacted the surangular. As in all other known Mesozoic birds (*Ichthyornis, Hesperornis, Parahesperornis,*

Gobipteryx, and *Enantiornis*), a mandibular foramen is absent. It does not appear that an intramandibular joint of the sort found in *Hesperornis* and *Ichthyornis* was present. Similarly, the prearticular seems to lack the dorsal extension that often occurs on its anterior end in other birds and many other archosaurs. The medial side of the left lower jaw can be seen through the orbit. It consists of three elements—an upper, broad surangular that rests in a groove formed by the prearticular and the angular. The surangular expands anteriorly near its junction with the splenial and dentary. The surangular has a small surangular foramen just anterior to the articular region. There is a relatively long retroarticular process. The angular extends posteriorly to the articular area, and the prearticular is visible ventral to its ventral margin.

Martin et al. (1980) first described the salient features of bird teeth and pointed out their detailed similarity with crocodilian teeth. Edmund (1960) illustrated teeth in the London *Archaeopteryx* specimen—a fact noted both by Martin and Stewart (1977) and by Martin et al. (1980). Only one *Archaeopteryx* tooth could be seen in its entirety, and this specimen (part of the London skeleton) was figured and compared to other bird teeth and to the teeth of a crocodilian (Fig. 16) and a theropod by Martin (1983a, 1985). This drawing does not differ significantly from that published by Howgate (1984) and shows most of the important avian features, including the shape of the crown, lack of serrations, constricted neck between the crown and the root, expanded root, and replacement tooth within a pit that is closed ventrally. Nevertheless, Howgate (1985) denied the avian character of the teeth of *Archaeopteryx*—an opinion with which Ostrom (1985) concurred. The teeth of the London *Archaeopteryx* are somewhat difficult to see with conventional lighting owing to the close similarity of the surrounding matrix, but they do stand out well (as does the rest of the skeleton) under ultraviolet light. As can be seen in the ultraviolet photograph (Fig. 17), the teeth resemble very closely those described for *Parahesperornis* (Fig. 18) by Martin et al. (1980). The triangular flattened crowns, distinct necks, expanded bases, and replacement pits are clear for several examples (Fig. 15) and do not resemble the tooth shape, implantation, or replacement of either *Compsognathus* (which has distinct interdental plates; Ostrom, 1978) or any other dinosaur. The crowns of the teeth in the Eichstätt and Berlin specimens closely resemble one another, and the characters proposed by Howgate to separate them cannot be found in his own drawings of the teeth of these two specimens. Like those of other birds, the teeth of *Archaeopteryx* are set in a groove bounded on both sides by dense bone, and the septa for the sockets are formed similarly to those of the alligator. Wellnhofer (1988) illustrated additional *Archaeopteryx* teeth showing typical avian features.

Currie (1987) claimed that troodontids have birdlike teeth, but judging

15

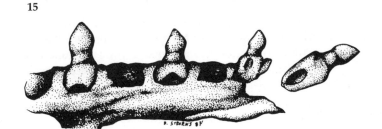

16   17   18

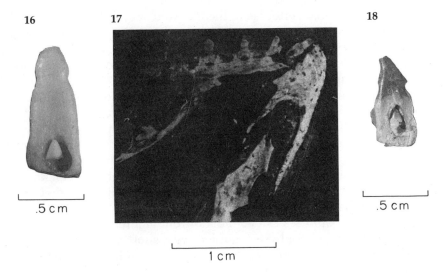

.5 cm   .5 cm

1 cm

Figs. 15–18. (15) Drawing of teeth of London specimen based on ultraviolet photograph (17) of tooth of same specimen. Photographs of (16) alligator and (18) *Parahesperornis* teeth.

from his figures, I do not see that they differ much from other carnosaur teeth. Furthermore, his figure illustrates a typically dinosaurian tooth replacement, rather than the specialized type shared by birds and crocodilians.

Vertebrae and Ribs

Wellnhofer's (1974) identification of nine cervicals and 14 thoracic vertebrae is accurate; Heilmann (1926) erred in his count of 10 or 11 cervicals on the Berlin specimen. The cervicals have unfused, elongate cervical ribs. The cervicals and the neck are comparatively short. Whetstone called to my attention that three vertebrae (probably dorsals 3–5) are fused and have their transverse processes connected. They lie almost in a straight line in the Berlin slab; in part, this accounts for the unusual way that the vertebral column has pulled away from the shoulder girdle. Elzanowski (1981) indicated that Vertebrae 12–14 (numbered from the head) are fused in the skeletons of *Gobipteryx* from the upper Cretaceous of Mongolia. This corresponds to the position of the fused vertebrae in *Archaeopteryx*. I think that *Gobipteryx* had a shorter neck of nine vertebrae (contra Elzanowski, 1981); a neck composed of 13 vertebrae would allow neither adequate room for the scapula nor the fused vertebrae. Hypapophyses do not seem to be present, although they occur in all ornithurine birds and crocodilians. Five vertebrae lie within the pelvis of the Eichstätt specimen (Wellnhofer, 1974:Fig. 7), and 22 make up the tail.

The first five caudal vertebrae of the Eichstätt specimen are short with down-turned transverse processes. Vertebrae 6–18 have flattened intercentra. The prezygapophyses of anterior vertebrae (i.e., to the level of the 11th caudal) are short; elongated prezygapophyses occur on Vertebrae 17–22. Postzygapophyses are elongate on Vertebrae 8–16. The caudals are well illustrated by Wellnhofer (1974). It is noteworthy that Wellnhofer's figure does not portray the pattern of tail-stiffening proposed by Gauthier (1986) for the "Tetanurae," in which elongate prezygapophyses and anterior processes of the intercentra are thought to have resulted in a stiffening effect in certain theropod dinosaurs. Gauthier accounted for the absence of similar structures in *Archaeopteryx* (and thus in all known birds) by invoking vagaries of preservation and the unlikelihood of short prezygapophyses occurring with long postzygapophyses. The completeness of the zygapophyses can be judged from Figures 19–20. *Allosaurus* has elongate prezygapophyses coupled with short postzygapophyses, which demonstrates the possibility of such combinations. In any case, it seems dangerous to consider a feature to be a synapomorphy if (1) it is

19

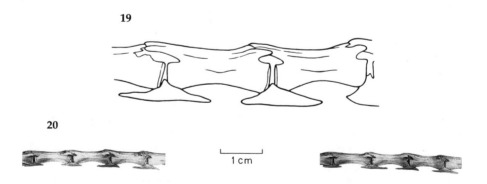

20

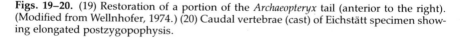

1 cm

Figs. 19–20. (19) Restoration of a portion of the *Archaeopteryx* tail (anterior to the right). (Modified from Wellnhofer, 1974.) (20) Caudal vertebrae (cast) of Eichstätt specimen showing elongated postzygopophysis.

not known to occur in any bird, and (2) the evidence that does exist is contraindicative.

The ribs have widely separated double heads that articulate with the transverse process and the centrum. Gastralia begin just posterior to the coracoids and leave little room for the cartilaginous sternum. The absence of an ossified sternum leaves the supposed presence of that structure in theropod dinosaurs of uncertain phylogenetic importance.

Furcula

"Clavicles" occur in few dinosaurs, but recently furculae have been ascribed to *Oviraptor* (Barsbold, 1983), *Ingenia* (Barsbold, 1983), *Albertosaurus* (Thulborn, 1984), *Allosaurus* (Thulborn, 1984), and *Troodon* [= *Stenonychosaurus*] (Thulborn, 1984). None of these proposed "furculae" could have articulated or functioned in the same way as the furcula of *Archaeopteryx* or any other known bird. Barsbold's restoration of a "theropod shoulder girdle" (Barsbold, 1983:Fig. 18) presents unusual articular relationships among the sternum, coracoids, and putative furcula. Because the scapula is fused to the coracoid in all of the suggested theropod genera, both scapula and coracoid would have to rotate 90° from the position in Barsbold's restoration in order to allow the scapular blade to lie against the ribs. The putative furcula is too narrow to contact the coracoids in this position, and no articulation for such a relationship seems to exist. The rounded anterior edge of the putative sternum would lie at 90° to the rounded posteromedial rim of the coracoid, and no articular surfaces between the sternum and the coracoid are indicated by Barsbold. When the coracoids are rotated to an avian position, the sternum and furcula are left "floating free" in connective tissue unlike the tightly connected coracoids and sternum found in birds. If an avian model with the furculae contacting the coracoids is utilized, then the width of the furcula determines the width of chest. The "furcula" suggested for *Allosaurus* (Thulborn, 1984) is only about 16 cm wide and would indicate an impossibly small chest. The exact nature of all of these bones is uncertain. Even if they should turn out to be shoulder elements, they cannot form an "avian" shoulder girdle.

The furcula in *Archaeopteryx* (Fig. 21) is a large, flat, U-shaped structure with no hypocleidium. Preparation of the proximal end of the right arm of the furcula of the London specimen reveals an elongate, shallow depression of about the right size and shape to receive the biceps tubercle. If the biceps tubercle becomes the acrocoracoid of modern birds (Ostrom, 1976a), this is a possible relationship. The right scapula and coracoid are still in articulation in this specimen—a fact evidenced by the shape of the glenoid fossa at the union of the two bones. This fixes their

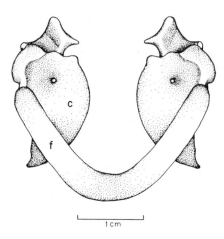

1 cm

Fig. 21. Restoration of anterior view of the shoulder girdle (scapula, coracoid, furcula) of *Archaeopteryx*. c = coracoid; f = furcula.

relative positions in relation to the vertebral column, as the scapula must lie flat against the ribs. In life, the biceps tubercles are located anteriorly and inclined so that the furcula could fit across the chest (Fig. 21). In this position, the furcula extends almost directly ventral from the coracoids and has about the same relationship to the coracoids as the modern avian furcula has to the coracoids and the sternum. In modern birds, the hypocleidium contacts the keel of the sternum, and the furcula provides part of the origin for the muscle mass extending onto the sternum. In *Archaeopteryx*, in which the sternum is unossified, the broad, flat coracoids may have acted more like the sternum of modern birds and the furcula like the keel. This would explain in part the unusual development of the furcula, which is relatively larger than in other birds.

Scapula and Coracoid

Ostrom's (1985:165) contention that the coracoid of *Archaeopteryx* "does not differ in any significant way from the coracoid morphology of typical theropods" is not easily reconcilable with his drawings (Tarsitano and Hecht, 1980:163). In *Archaeopteryx*, the coracoid and scapula are rotated 90° from the normal vertebrate position so that the scapulae lie on the ribs parallel to the vertebral column and the coracoid lies ventrally across the chest. Ostrom (1976a:Figs. 4, 9) correctly figured these relationships for *Archaeopteryx* and (Ostrom, 1974b:Fig. 2) correctly illustrated the scapulo-coracoid of *Deinonychus* in lateral view. We should keep in mind that the coracoid figured by Ostrom (1969a:Fig. 79) is almost entirely restored. In this same paper (Ostrom, 1969a:Fig. 65), there is a good figure of the almost complete coracoid misidentified as the pubis (Ostrom, 1976c:6). In terms of its configuration and attachment to the scapula, this coracoid does not differ greatly from that of other theropods, as I have confirmed from the specimens and as Ostrom (1974b) illustrated (see Martin, 1983b:Fig. 4). Ostrom's (1976b:Fig. 13C) restoration of the scapulocoracoid is distorted. The scapula is shown in lateral aspect, as are his early figures of the whole skeleton, but the coracoid is depicted with an anteroposterior twist and combines lateral and posterior views. In this posture, the joint between the scapula and the coracoid cross at a right angle and no longer articulate, nor can such a highly bent scapulocoracoid be fitted easily to the rib cage.

I agree with Ostrom's association of the "biceps tubercle" on the coracoid of *Archaeopteryx* with the acrocoracoid of modern birds. In modern birds, the furcula contacts a furcular facet on the acrocoracoid and also may contact the scapula. In *Archaeopteryx*, the position and shape of the "biceps tubercle" would allow it to act as the attachment of the furcula, which is grooved posteriorly to fit on the tubercle (Fig. 21). This articulation requires the clavicle and scapula to maintain their relative positions, while in the course of evolution, the coracoid rotated 90° and changed its original relationship with the clavicle. Alternatively, the furcula might have to have articulated with the scapulae. The Solnhofen specimen may support the latter interpretation.

Humerus

The deltoid crest is enlarged and is entirely palmar (Fig. 22). This arrangement can be seen in some crocodylomorphs and dinosaurs but in no other known bird; it seems likely that the condition is primitive.

The hand is elongated along the anconal-palmar plane. The palmar aspect of the distal end of the hand (visible in the Eichstätt specimen; Fig. 24) is typically avian in pattern. It has a rounded internal condyle and a proximodistally elongate external condyle.

Radius and Ulna

The radius and ulna are shorter than the humerus and generally similar to these bones in other birds (Fig. 23). The radius has a typically round cotyle.

Ostrom (1976a:17) concluded that "*Archaeopteryx* was unable to fold the forelimb back against the body." This contrasts with Padian's (1985) argument that the theropod *Deinonychus* could fold its forelimb in the manner of an avian wing. Both of these contentions are unlikely. In the London, Berlin, and Eichstätt specimens, the wings are arranged in the avian folding pattern, and all of the joints are positioned in the proper position for avian folding. The *Deinonychus* specimen cannot be articulated in this

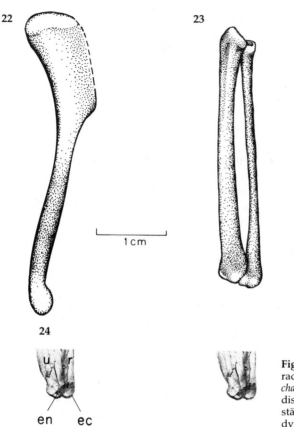

Figs. 22–24. (22–23) The humerus, radius, and ulna of the London *Archaeopteryx*. (24) Stereophotograph of distal end of the humerus of the Eichstätt *Archaeopteryx*. ec = ectepicondyle; en = entepicondyle.

position and, at the same time, retain normal contacts between the joints. (No actual articulation is shown in Padian, 1985:Fig. 3.) The purpose of this folding system is to pack the flight feathers against the body, and unless it is argued that *Deinonychus* had such feathers (a conclusion that could be drawn from Paul's [1984] phylogeny), there seems to be little reason for it to possess a joint system that would render its forelimb useless for most other purposes. Likewise, it is difficult to imagine *Archaeopteryx* with its wings permanently fixed in the outspread flight position or in a "bug-catching" posture (see Ostrom, 1979), as seems to be suggested by Ostrom's (1976a) interpretation.

The muscle attachments on the wing of *Archaeopteryx* are not well demarcated, and the position of the humerus in most of the specimens does not facilitate its description. However, the typically rounded entepicondyle and proximodistally elongated external condyle (Fig. 24) indicate a similar motion between the humerus, and the radius and ulna as is found in modern birds. The ulna has an enlarged, semicircular external condyle distally and a reduced internal condyle, the combination of which provides direct evidence for the avian motion of the manus on the radius-ulna.

If the biceps tubercle is occupied by the furcula rather than by the biceps muscle in *Archaeopteryx*, Ostrom's (1976a) scenario for "raising the sites of origin of the *m. biceps* and *m. coracobrachialis*" might require modification. Although the extensor process is not prominent in *Archaeopteryx*, it is not much lower than the same structure in loons and alcids— taxa that have no difficulty in extending the manus. It seems evident that *Archaeopteryx* had an unusual but fully avian wing that could function in powered flight.

Carpus and Manus

There is no consensus regarding the homologies of the bones constituting the avian wrist and manus (Hinchliffe, 1985), and unfortunately, the issue is central to the theropod theory of avian origins. Ostrom reported what seem to be close similarities between the wrist and manus of *Deinonychus* and of *Archaeopteryx* (Ostrom, 1974a, 1976a). However, there is a significant difference in function between an avian wing and the forelimb of a theropod, which used its manus for food manipulation. Thus, the putative resemblances between avian and dinosaur forelimbs may be more apparent than genuine.

Hinchliffe (1985) recognized four separate bones in the avian wrist. Two of these are the proximal carpals that articulate with the ulna and radius. They have been homologized to the ulnare and radiale of reptiles, but the true ulnare seems to be lost in ontogeny and the bone (avian "cuneiform") contacting the ulna is a pisiform (Hinchliffe, 1985). The radiale probably is the avian "scapholunar." The two carpal bones that fuse with the metacarpals are (1) the "semilunate bone," which is thought to be the Distal Carpal III, and (2) a separate potential distal carpal, "Bone X" (Hinchliffe, 1985).

The Cretaceous bird *Ichthyornis* has a carpus and manus similar to those of Recent birds. The carpus and manus is unknown in Hesperornithiformes, which possess highly modified wings. Only *Archaeopteryx* provides information concerning the early avian wrist and manus, and this is best studied in the Eichstätt skeleton in which all of the elements discussed by Hinchliffe (1985) appear as separate ossifications (Fig. 25). This specimen displays the carpus and manus in both dorsal and ventral aspects. Ostrom (1976b:Fig. 12) illustrated a wrist with two small, rounded bones in the positions of the scapholunar and cuneiform. As figured by Ostrom, both bones appear much smaller and less similar to

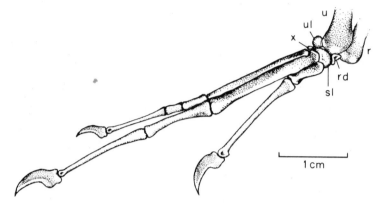

Fig. 25. Restoration of the wrist and manus of the Eichstätt *Archaeopteryx*. r = radius; rd = radiale; sl = semilunate bone; u = ulna; ul = "ulnare"; x = "X-bone" of Hinchcliffe (1985).

those in other avian wrists than they actually are. In the case of both elements, only a small part of the bone is exposed; their actual sizes and shapes can be assessed from Wellnhofer's (1974) drawing of the ventral aspect. Wellnhofer, in this same figure, illustrated a small bone, "?C3," that corresponds in size and position to the "X-bone" of Hinchliffe (1984).

Archaeopteryx has a completely avian wrist with a four-bone carpus (Fig. 25); however, the typical V-shape of the cuneiform is not established. Padian (1985) in his analysis of the forelimb of *Deinonychus* showed a restoration with a small rounded ulnare similar to that figured for *Archaeopteryx* by Ostrom (1976b:Fig. 12), but this figure is the result of an accident in the way the fossil is observed; the ulnare of *Archaeopteryx* is actually much larger and of a different shape. The drawing of this bone in *Deinonychus* by Padian seems to be based on the illustration of the projection of the ulnare past the radiale in Ostrom (1975a:Fig. 6). However, this is also an artifact of the ulnare's being hidden behind the radiale in this view (Ostrom, 1969a:Fig. 61). The ulnare is actually a larger bone with a complex shape (Figs. 26–27).

Ostrom (1969b:9) described the movement of the carpus and manus in *Deinonychus* as follows: "Thus as the two hands flexed at the wrist they also turned toward each other." His original restoration (Ostrom, 1969) of the wrist of *Deinonychus* shows only two bones—a large, semilunate structure that he called the "radiale," and a bone that fits alongside it (Figs. 26–27) so that the metacarpals were articulated into an L-shape. In this arrangement, Metacarpal III is not visible in an anterior (dorsal) view (Fig. 28) unless it is displaced laterally (Ostrom, 1975:Fig. 6). Because the articular surfaces of the "radiale" and "ulnare" lie adjacent to one another, the wrist is rotated (i.e., adducted) about 45°, but not with the

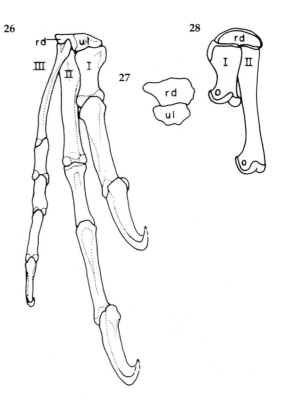

Figs. 26–28. The wrist and manus of *Deinonychus*. (26) Palmar view. (27) Proximal view of radiale and ulnare in articulation. (28) Anconal view. See Figure 25 for key to abbreviations. (Modified from Ostrom, 1969 a,b.)

adduction on the semilunate surface typical of birds. A simple motion makes sense for birds in which the purpose is to tuck the feathers against the body when the wing is folded. One would not expect a similar motion in theropods unless they had large flight feathers. Bipedal dinosaurs that used their manus to manipulate food would need to twist the food toward the mouth. This motion is useless in a bird because the elongated forelimbs prevent the manus from even approaching the mouth. This was always a fundamental flaw in the various models for prey capture by the use of the manus in protobirds. Such rotational motion as is present is between the scapholunar-cuneiform and the radius and ulna. Many dinosaurs have reduced wrists, and according to Ostrom's restorations, the radiale contacted the radius in the same plane as the ulnare. This suggests that the distal carpals may have been lost or fused. In all birds, including *Archaeopteryx*, the semilunate bone forms a unit with the metacarpals, whereas the scapholunar and cuneiform articulate with the radius and ulna. This precludes any homology between Ostrom's "radiale" and the semilunate bone of *Archaeopteryx* or other birds. It is possible that Ostrom's description of the wrist of *Deinonychus* is incorrect, and that a new restoration with the addition of at least one bone could be reordered into a birdlike structure. I doubt that this is so. Ostrom's figures and description of the wrist are convincing, and the needed changes would so restrict the motion of the manus as to make it useless.

Interestingly, Gauthier (1986) seemed to reinterpret Ostrom's wrist restoration in order to bring it in line with avian anatomy. In order to do this, he described Ostrom's radiale as Distal Carpal I and proposed the existence of an avian-like proximal carpal to receive it. As no evidence exists for his suggested homology or of the existence of the additional proximal carpal, I prefer to accept Ostrom's better supported interpretation.

Three metacarpals, with the middle metacarpal being the largest, are present in *Archaeopteryx* as in the modern birds. The homologies of the metacarpals are unclear. Most paleontologists, including me (Martin, 1983a,b), have considered the metacarpals to be I–III, but Hinchliffe (1985) strongly supported their being II–IV on the basis of embryological evidence. Enough is known of the pattern of digit reduction and loss in dinosaurs to support the former interpretation in coelurosaurs, although Thulborn (1984) tried to reconcile this aspect of dinosaurs with avian embryology. The embryological evidence may be more equivocal than has been supposed. If Hinchliffe (1985) is correct in his reasoning, almost all comparisons with dinosaurs disappear in the manus.

As is normal for diapsids in general, Metacarpal "I" is short in *Archaeopteryx*. It also has a flattened shape and a straight contact with Metacarpal "II." There is a rounded bulge in the position of the extensor process. Digit "I" consists of a greatly elongated phalanx and a large claw. The articulation for this digit diverges from Metacarpal "II." Metacarpal "II" is the largest metacarpal, and the "semilunate" bone is centered on it. Digit "II" has two phalanges plus the claw phalanx. The proximal end of Metacarpal "III" is distal to the other metacarpal ends and flattens out against the shaft of Metacarpal "II." Digit "III" has two short proximal phalanges and an elongated third that attaches to the claw phalanx. Recent descriptions of the superficial horny claws by Yalden (1985) seem to establish beyond any reasonable doubt that they were used primarily for climbing.

Pelvis

In many respects, the pelvis of *Archaeopteryx* can be studied best in the London specimen, in which the right side is preserved essentially intact and uncrushed, although the fused pubes are separated as a unit and exposed dorsally (posteriorly) on the main slab. The left ilium and ischium have shifted above and posterior to the right ilium and ischium, and are seen from an internal view, which is difficult to interpret without the aid of ultraviolet photographs (Fig. 29). The ilium of the London speci-

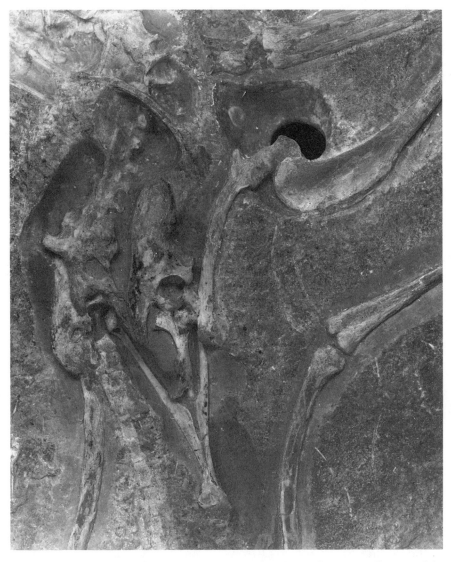

Fig. 29. Ultraviolet photograph of the pelvis of the London *Archaeopteryx*. (Courtey of the British Museum [Natural History].)

men (Figs. 29–31) contains the dorsal and anterior margins of the acetabulum. The posterior and part of the ventral border is composed of the ischium. The pubis would contribute at most about 1 mm to the acetabulum and perhaps did not contribute to it at all. The acetabulum is closed off partially by a wall that extends internally around its rim for about 2 mm. (The anteroposterior diameter of the acetabulum is about 7 mm and

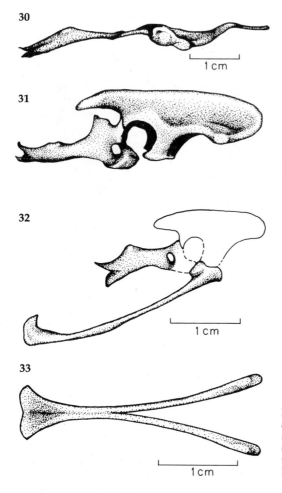

30

1 cm

31

32

1 cm

33

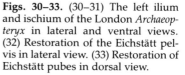

Figs. 30–33. (30–31) The left ilium and ischium of the London *Archaeopteryx* in lateral and ventral views. (32) Restoration of the Eichstätt pelvis in lateral view. (33) Restoration of Eichstätt pubes in dorsal view.

1 cm

the diameter of the internal foramen is about 3.5 mm.) It is not possible to examine the inside of the acetabulum closely in the other specimens of *Archaeopteryx*, although portions of the right pelvis are visible medially on the Berlin specimen (see Ostrom, 1974a:Fig. 7). The posterior ilium does not extend ventral to the upper acetabular edge. The postacetabular ilium extends a short distance posteriorly. It has a flat ventral margin and a curved dorsal one. The ischium bears a square, dorsal process that might have contacted the postacetabular ilium in the undamaged specimen. The postacetabular ilium is about half the length of the preacetabular ilium. The preacetabular ilium expands dorsally and anteriorly. It has a rounded anterior margin that flares ventrally to produce a shelf above a large oval sulcus that extends from the dorsal and anterior margin of the acetabulum nearly to the anterior margin of the ilium. There is a distinct knob on the anteroventral margin of the ilium (Fig. 31), and posterior to it, this margin contains a shallow triangular sulcus that slants inward. The latter may represent the area of origin for m. iliotrochantericus anterior. The ilium of *Archaeopteryx* forms a broad pubic peduncle (Fig. 30) that is slightly inclined posteriorly and nearly meets the ischium.

The ischium has an unusual form. In the London specimen, it is inclined at an angle of about 12° to the ilium (Fig. 31). Its anteroventral margin slopes and contacts the pubis. The anterior margin of the ischium is broad and forms much of the margin of the acetabulum. It contains a large foramen in its lower half. This foramen may be visible as a depression in the Eichstätt specimen (contra Wellnhofer, 1974) similar to the foramen pointed out by Heilmann (1926:14), in the London example. In the Berlin specimen, it lies under the femur. I have not found this foramen in other birds, nor have I found it in any reptile. In the Eichstätt specimen (Fig. 35) it appears not to be completely closed anteriorly. If this is the case, the foramen is comparable to the anterior ischial notch of crocodilians. The ischium become dorsoventrally narrowed posteriorly, and its ventral border turns inward to form a shelf. Toward the distal end of the ischium, the dorsal border forms an elongate thickened knob. The posterior margin contains a V-shaped notch with inwardly turned dorsal and ventral processes.

The pubis of *Archaeopteryx* has generated about as much controversy as all the rest of the skeleton. Most of the debate has involved the angle of the pubis to the ilium. Heilmann (1926:Fig. 10B) restored an angle of about 26°, which is essentially that of the Berlin specimen as it is presently prepared. Ostrom (1976b) rejected this interpretation, arguing that the pubis of the Berlin *Archaeopteryx* had been displaced posteriorly into an unnatural position. Ostrom's maintained that (1) the left ilium, ischium, and pubis are dorsal and posterior to the right; (2) the pubes are fused distally and should move as a unit; and (3) there is a crack between the pubis and the pubic peduncle of the ilium. Thus, as the left side was displaced dorsally and posteriorly, the right pubis was reflected posteriorly and upward.

It is clear and not at all unexpected that there has been some differential movement of the parts of the pelvis owing to compaction; the critical issue is the effect of this differential movement on the angle of the pubis relative to the ilium. First, we should determine how successful this rotation would be in explaining the condition of the specimen if everything worked to maximize the displacement of the pubis in the manner Ostrom described. This can be estimated by comparing the angle between the dorsal borders of the right and left ilium in Ostrom (1976b:Fig. 7A). The maximal angle between these borders is about 30°. Adding this

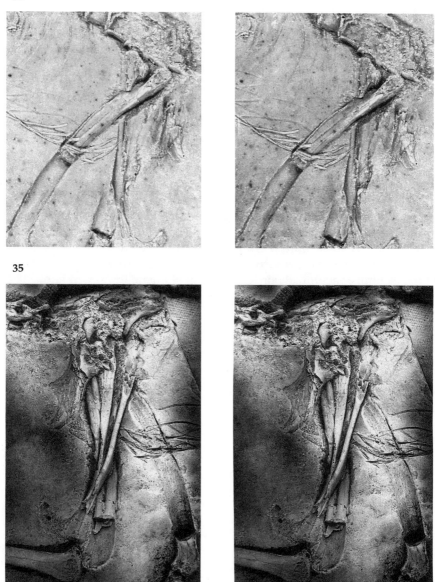

Figs. 34–35. Stereophotograph of casts of the Eichstätt pelvis and femur, counterslab and slab.

516

to the present angle of the pubis results in an angle of 56°, which is 34° posterior to the approximately 90°-position estimated by Ostrom. Such an interpretation also requires the assumption of a wedge-shaped separation of about 3.5 mm between the pubis and pubic peduncle of the ilium. This is wider than the anterior end of the pubis, and it is obvious that rotation of that sort is impossible unless Ostrom considers the anterior end of the pubis to be the pubic-ilial articulation (a view close to Wellnhofer, 1974). However, if one accepts the latter (which would give the nearly 90° illustrated in Ostrom, 1976b:Fig. 8), there is too little displacement of either the anterior end of the right pubis or of the elements of the left side of the pelvis. In other words, it seems physically impossible to derive Ostrom's (1976b:Fig. 8C) restoration of the *Archaeopteryx* pelvis from the Berlin specimen. Judging from the cracks and crushing in the area of the pubic peduncle on the ilium, I would argue that if the pubis is displaced, it is displaced forward (contra Ostrom, 1976b).

Tarsitano and Hecht (1980) are correct in their contention that the best estimation of the pubic angle in *Archaeopteryx* can be acquired from the London specimen. The anterior articulation in its pubes is inclined forward slightly, as it is in the Eichstätt specimen (Figs. 32, 34); the posterior articulation is inclined slightly posteriorly and fits against the ischium. Only the tip of the wedge where these two articulations meet might contribute to the acetabulum. There is a distinct notch just below the posterior articulation that reflects the backward twist of the pubis. Viewed laterally, the pubis is slender and expanded posteriorly on its distal end (Fig. 32). When viewed dorsally, this expansion is cup-shaped in the London specimen and apparently also in the Eichstätt specimen (Fig. 33). The Berlin and the Tyler specimens are strongly crushed laterally; thus, the true dorsal shape of the distal end is not visible, and it is difficult to confirm Ostrom's (1976b:126) statement that the distal expansion was not expanded transversely in those specimens. In dorsal aspect, the London specimen has a distinct depression posterior to the ischial articulation. Heilmann calls this the obturator foramen, homologizing it to a similar feature present in reptiles (pseudosuchians, crocodilians, and the like) but absent in modern birds and theropod dinosaurs. The pubis is flattened or slightly grooved dorsally and rounded ventrally. At approximately the midlength of the pubic shaft, it is expanded inward to form a pubic apron. The apron converges distally and then reexpands to form a spoon-shaped structure. This structure is filled with calcite crystals in the London specimen, but its true outline is evident if the distal end is viewed laterally.

The pubis forms the anteroventral angle of the pelvis just below the pubic peduncle of the ilium. This is the normal position of the pectinal process in birds and the area of origin of the ambiens muscle. However, the pectinal process is formed by the ilium and not by the pubis. The origin of the ambiens muscle was probably on the anterior face of the pubic peduncle above the pubis.

Restorations of the pelvis of *Archaeopteryx* by Ostrom (1976b) and Wellnhofer (1974, 1985) are based on the Eichstätt specimen. This is surprising because, clearly, it is damaged and displaced (Figs. 34–35)—a fact noted by Tarsitano and Hecht (1980). My drawing of the ilium differs from that of Wellnhofer (1974:Fig. 10B) in having the preacetabular margin shorter and more rounded, as is the postacetabular margin (Fig. 32). In Wellnhofer's restoration, the tip of the posterior acetabular border of the ilium lies at about the same level as the tip of its anterior acetabular border. This is dissimilar to the condition in the London and Berlin specimens in which the anterior acetabular border is much lower than the postacetabular border, as it is in modern birds. Examination of the Eichstätt specimen reveals that the ventral margin of the pubic peduncle is obscured by fragments of the right pelvis. The pubes are fused distally as in the London specimen and are displaced from their articulation with the ilium and ischium. The distal ends of the pubes are rotated so that we observe them mostly from a dorsal ("posterior") view. This is apparent on the photographs because the midline of the pubic symphysis is visible and one can easily trace the dorsal and ventral surfaces of the right and left pubis from it (Fig. 35). This means that the view we have of the Eichstätt specimen is much like that of the London specimen. This rotation also can be seen by the relative positions of the proximal ends of the pubes; the right is completely anterior to the pubic peduncle, and the left is displaced posteriorly (Fig. 35; also visible in Wellnhofer, 1974:Fig. 10A, but incorrectly drawn in Ostrom, 1976b:Fig. 20C.2, as exactly in articulation). Ostrom's (1976b:Fig. 20C.2) drawing of the left pubis is too broad anteriorly because it includes matrix and parts of the right femur and ischium. Although this is not especially evident in Ostrom's illustration (1976b:Fig. 20C), it is apparent in Figure 35 and in Wellnhofer, 1974:Fig. 10A and Pls. 23, 7. I compared Wellnhofer's drawing of the proximal end of the left pubis with the specimen, but I could not distinguish it as drawn, nor could I find its outline on the molds of the two slabs. This area is so damaged (as can be seen in Fig. 35) that a secure interpretation of the shape of the left pubis and its articulation with the ilium seems to be impossible without either further preparation or X-ray photographs (Tarsitano and Hecht, 1980). This is not an important problem because a complete right proximal end is preserved on the counterpart slab and can be seen in Figure 34 or Ostrom, 1976b:Fig. 19E. The proximal end has the same shape as that of the London and Berlin specimens (contrary to the restoration in Wellnhofer, 1974). Wellnhofer confused the articulation of the pubis with a fragment of the left ischium; this is apparent if one overlaps his drawings of the pelvis of the main and counterslabs (also see Wellnhofer, 1985:120). The distal ends of the Eichstätt pubes do not form as broad a pubic apron as do those of the London specimen, but they meet distally for a little less than half their length. The distal end is interpreted by Wellnhofer (1974) and Ostrom (1976b) as being unlike the London specimen in having the two pubic expansions tightly fused together to form a bladelike structure. Examination of the Eichstätt specimen shows that when the midline is projected down the symphysis, it is difficult to distinguish whether the pubic expansion is exposed laterally or dorsally. If dorsal, further preparation may reveal the pubic expansion for the right pubis to provide a restoration of the pubes in dorsal aspect— a view similar to that in the London specimen (Fig. 33).

Whereas most of the outline of the ischium of the Eichstätt specimen published by Wellnhofer (1974:Fig. 10) seems to be correct, there seem to be some problems with the proximal end. There is not a good contact between either the right or left ischium and either ilia. There is a deep matrix-filled depression at the position of the ischial foramen in the London specimen. I find no evidence of the orientation of the ischia as restored by Wellnhofer (1974:Fig. 1C, 1985) except their present position, which cannot be natural. None of the acetabular contribution of the ischium nor its anterior ventral border is visible on this specimen.

Ostrom's (1976b) arguments for the position of the pubis as suggested by the Tyler specimen are unconvincing. The only part of the pelvis that is preserved is the distal end of the pubis. The two vertebrae that Ostrom sketched are not in tight articulation; to use one of them to fix the angle of the vertebral column to the distal end of the pubis is doubtful. The absence of the ischium on the slab is to be expected given that it is shorter than the pubis. If we extend the pubis to the length of that of the London specimen, we find that the ischium could have reached the preserved portion of the slab only if the pubis was in articulation with the pelvis. The pubis is clearly in articulation in the Berlin specimen, but in no other; therefore, we should hesitate to accept it as having been in articulation in the Tyler specimen, in which no frame of reference remains. The argument for orientation of the pubis in the Maxburg specimen would be more convincing if the structures were clear in the X-ray photographs. For instance, in my copy of Ostrom (1976b:Fig. 8B), I cannot discern the pubic symphysis, nor can I see the relationships of the proximal end of the pubis. It might be useful to republish that X-ray after logitronic enhancement. If we accept Ostrom's interpretation of the X-ray, both pubes seem to be visible and separated from each other. This is difficult to achieve in an undistorted lateral view, and once again, the evidence for correct orientation is virtually untestable in any rigorous way. As can be seen from cross sections through casts of the shafts of the Berlin specimen, the pubis is not cylindrical (contra Ostrom, 1976b) but is flattened dorsally and curved ventrally. Ostrom's (1976b:Fig. 20A1) photograph of the ilium and ischium of the London specimen is excellent and shows the correct shape of these bones, as can be confirmed by direct observation and studies under ultraviolet light. The sketch by Ostrom (1976b:Fig. 20A2) is not explained in the text. He extended the posterior acetabular contribution of the ilium ventrally in spite of the fact that the junction between the ischium and the ilium is visible *inside* the acetabulum. Furthermore, he removed the entire front end of the ischium despite the fact that the surface of it and the rest of the ischium are continuous. There are structural features that are continuous across parts that Ostrom termed "ischia" to elements that he did not consider to be ischia. Moreover, the whole ischium glows like bone under ultraviolet light, whereas none of the exposed calcite or other matrix does. Other workers who have illustrated this specimen have accepted the anterior margin as correct—a view with which I concur.

The most recent restoration of the pelvis is by Wellnhofer (1985:121), who concluded that it "is not bird-like at all." Wellnhofer's (1985:Fig. 2) sketches of the four *Archaeopteryx* specimens depict bones of fundamentally different shapes and articular relationships. In his sketch of the London specimen, only a narrow wedge of the acetabulum could be formed by the pubis. If the London ischium as he has drawn it were rotated ventrally, as in his restoration (Wellnhofer, 1985:Fig. 4), the acetabulum would lose its round shape; the ischium would not contact the ilium, and the pubis would be excluded from the acetabulum. The sketch of the X-ray of the Maxburg specimen is especially strange, because the pubis seems to be bifurcate. I suppose that the right and left pubes actually are indicated, but if that is the case, the separation of their proximal ends and, hence, the correct shape of the proximal end, is not shown. What appears to be shown is a pelvis with a broad contribution to the acetabulum by the pubis. The drawing of the Berlin specimen shows a small wedge-shaped contribution to the acetabulum by the pubis and a sloping contact between the anterior dorsal portion of the pubis and the ilium. Judging from the shape of the space left for the pubis in the London specimen, the London specimen retains almost exactly the same relationships of the pubis to the rest of the pelvis. This is not so with Wellnhofer's restoration of the *Archaeopteryx* pelvis (Wellnhofer, 1985:Fig. 4) in which the whole shape of the anterior end of the pubis differs from that of the Berlin specimen and also from any of his sketches of it in his Figure 2. In his restoration of the Berlin specimen, the ilium is attached to the anterior end of the pubis, whereas in his sketch of the Berlin specimen (Wellnhofer, 1985:Fig. 2), the anterior end of the pubis is square, and there is a sloping contact for the pubic peduncle. This same morphology is clear on the Eichstätt pelvis (Fig. 34).

It is possible that having the ischium parallel with the postacetabular ilium is primitive for diapsid reptiles. This is the case in *Euparkeria* and many of the Triassic reptiles, including synapsids. The primitive position of the pubes may be perpendicular to the ilium, as is also the case in *Euparkeria*. In that case, most dinosaurs are derived with respect to the positions of their pubes and ischia.

Femur

Ostrom (1985:168) stated that "the right femur is firmly and para-sagitally articulated in the Berlin specimen." However, examination of Figure 36 shows that the head of the femur is not in contact with the dorsal acetabular rim and that this could not be the "life position" of the femur. The medial aspect of the femoral acetabular articulation also shows no articulation with the roof of the acetabulum. (For the condition of the medial side, see Ostrom, 1974a:Fig. 7.) Examination of the ultravio-

36

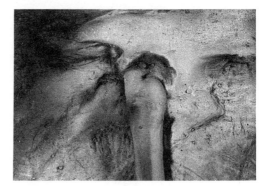

37

Figs. 36–37. (36) The Berlin pelvis with associated femur; note that the head of the femur is not in articulation with the roof of the acetabulum but has been displaced downward. (37) Stereophotographs of cast of the London *Archaeopteryx* femur as fitted into the acetabulum.

let photographs prepared by the British Museum of the London specimen (Fig. 29) clearly shows the internal closure of the acetabulum reported by Martin (1983b); this can be confirmed by observing the medial
view of the left pelvis on the same slab.

This closure is neither characteristic of most modern birds nor of theropod dinosaurs and is coupled with an unusual positioning of the femoral
head in *Archaeopteryx* in which the head is turned forward (not at 90° to
the sagittal plane) and contacts the anterior, rather than the posterior,
wall of the acetabulum (Figs. 37–42). There is no overhanging dorsal rim
of the acetabulum as in most theropod dinosaurs, nor is there a posterior
antitrochanter as in modern birds. This bracing of the femur results in an
outward inclination of the femora (Fig. 37) that may have facilitated
climbing. There seems to be no way to interpret this pelvic structure as
being as cursorially adapted in the manner of modern birds or theropod
dinosaurs. It does, however, support Yalden's (1985) contention that the

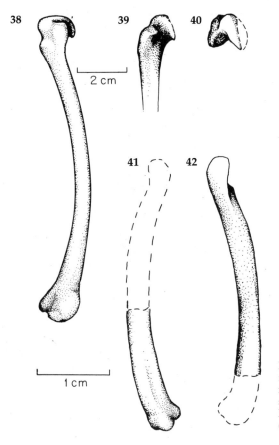

Figs. 38–42. Femora of *Archaeopteryx.* (38–40) The London right femur
in lateral (38), anterior (39), and dorsal (40) views. (41–42) The Eichstätt
left femur in lateral (41) and medial
(42) views.

manal claws of *Archaeopteryx* more likely were an adaptation for climbing than for prey capture.

Tibiotarsus

One of the first lines of evidence supporting a dinosaur relationship for birds was provided by Huxley (1870), who pointed out that the ankles of theropod dinosaurs and birds shared a triangular prominence in front of the distal tibia. Ostrom figured three slightly different restorations of the tarsal region in *Archaeopteryx* (Martin et al., 1980). In all of these, the triangular prominence lies medial to the outer condyle of the tibiotarsus—a position entirely like that of theropod dinosaurs, and dissimilar to that of other Mesozoic birds (*Hesperornis, Parahesperornis, Enaliornis, Ichthyornis,* and *Enantiornis*), in which the triangular prominence is lateral and overlays the calcaneum at least in part. This is also the condition in modern birds. Further study revealed that the restorations in Ostrom (1976b) were the result of a *lapsus,* and that *Archaeopteryx* conforms to the condition in later birds (Martin et al., 1980). It also was possible to demonstrate that this bone was a separate ossification—the pretibial bone—that probably is a synapomorphy of birds.

Such an interpretation of the ascending process removes one of the oldest and most important arguments for a special theropod relationship for birds; however, several attempts have been made to salvage it. One argument, which offers no support to the dinosaurian origin of birds and would render *Archaeopteryx* unlike all other known birds, is that of Tarsitano and Hecht (1980). They recognized that Ostrom's restorations were in error but argued that the ascending process of *Archaeopteryx* was an artifact of preservation. There seems to be no support for this contention. All specimens that show the relevant region of the tibiotarsus (i.e., the London, Berlin, and Eichstätt specimens) have a pretibial bone of the same shape and position as found in other birds. In all cases, the pretibial bone shows up as bone under ultraviolet light and does not have the distinctive fluorescence of calcite found on other parts of the slabs.

Most recently McGowan (1984) presented results from an extensive study of cleared-and-stained birds in support of the conclusions of Martin et al. (1980) that carinate birds do possess an ascending process associated with the calcaneum. McGowan claimed that ratites have a separate ascending process of the astragalus. He argued further that the ratite condition was like that of theropod dinosaurs and, therefore, represents the primitive condition for birds. McGowan's scheme utilized two projections—one from the astragalus (astragalar process), and one from the calcaneum (calcaneal spur). The more primitive of these is the astragalar process, which presently is restricted to ratites. McGowan (1984) claimed that ratites also have an unossified calcaneal spur. He reported that the astragalar process was lost and the calcaneal spur ossified in carinates. Martin et al. (1980) followed other workers (Morse, 1872; Jollie, 1977) in considering the ascending process to be a separate bone (i.e., the pretibial bone) rather than a projection of either the astragalus or the calcaneum. In this case, the process is homologous in ratites and carinates, although it might vary in position. The cartilage precursor of the pretibial bone appears first near the junction of the calcaneum and the astragalus. Initially, it lies over the calcaneum; subsequent growth is largely lateral so that it is confined entirely to the region above the calcaneum with which it eventually fuses. The pretibial bone may vary in the amount of medial growth from one group of birds to another, thereby resulting in variable amounts of overlap of the astragalus. Ratites have a greater astragalar contribution to the tarsal joint than do other birds, and the pretibial bone has more astragalar contact. In virtually all birds (but perhaps not in some ratites), the outer condyle is composed of the calcaneum, whereas the inner condyle and the intercondylar space consists of the astragalus. This arrangement permits us to estimate the relative amounts of overlap of the pretibial bone over the astragalus and the calcaneum even when the two bones are fused. However, the astragalus represents a greater percentage of the functional tarsal joint in some ratites, and this results in the pretibial bone's being confined entirely or mostly to the astragalus. According to some authors (e.g., Olson, 1985), many of the features thought to unite the ratites are actually neotenic and may have developed in parallel. This also may apply to the tarsal joint, because the astragalus begins to ossify before the calcaneum does. If maturity is reached at an early ontogenetic stage, it should be expected that more of the astragalus would be ossified, and thus, that the astragalus would be relatively more important to the tarsal joint than it would be in a bird in which the calcaneum was formed fully.

In early developmental stages, the calcaneum is directly in line with the fibula, but as development progresses, the distal end of the tibia broadens until it overlaps and "captures" the calcaneum. The fibula no longer contacts the ankle and terminates against the lateral side of the tibia. This condition is present in all known birds, including *Archaeopteryx*. The pretibial bone appears at a later stage of development than the astragalus and calcaneum. It usually appears on, or slightly lateral to, the junction. The pretibial often has been termed the ascending process of the astragalus (Romanoff, 1960). In adults of Recent birds, the bone is fused completely with the tibia and tarsal bones, but in Mesozoic birds (even adults), distinct sutures may be visible at its junctions with other bones.

The best review of the tarsal region in theropod dinosaurs is that of

Welles and Long (1974). They recognized five types of theropod tarsi. Each has an ascending process of the astragalus, but the formation of this process is variable; rather than showing progressive change from some primitive type to a more derived and possibly birdlike state, the structure seems to have several independent origins. This is not surprising, because similar structures have evolved in ornithischians (Galton, 1974) and *Lagerpeton* (Welles and Long, 1974). For example, the ceratosauroid type is found in *Ceratosaurus, Coelophysis, Dilophosaurus, Halticosaurus,* and *Syntarsus* (Welles and Long, 1974). This type differs radically from that of birds and other dinosaurs in extending into the distal end of the tibia rather than lying in front of it. It seems possible that the process is not homologous to that of other theropods, let alone the even more dissimilar ascending process (pretibial bone) found in birds. It has now been demonstrated (Martin et al., 1980; McGowan, 1985; Martin and Stewart, 1985) that the pretibial bone of birds ossifies separately from the astragalus and calcaneum, although McGowan (1985) claims that there is some ontogenetic relationship to the astragalus. McGowan (1985) confirmed the observations of many authors (see Martin et al., 1980) that the pretibial bone of carinates fuses primarily with the calcaneum. Martin et al. (1980) and Martin (1983a,b, 1987) showed that this relationship holds in all known Mesozoic birds, including *Archaeopteryx.*

The condition in *Archaeopteryx* can be seen clearly by the way that the calcanae and astragalae have separated from the tibiae in the London specimen. The right calcaneum and pretibial bone are visible in the ultraviolet photograph of de Beer (1954:Pl. IV) where, in part, they are confused with the right fibula. The left astragalus also has separated with the counterslab. No fusion seems to have existed between the astragalus and calcaneum because these two bones are restricted largely to the lateral and medial condyles, respectively (Figs. 43–44). In this case, there can be no doubt that the pretibial bone (Fig. 43) in *Archaeopteryx* is restricted entirely to the calcaneum (Fig. 46). (The "astragalus" of Wellnhofer, 1974:Figs. 11–12, actually must be the calcaneum.) Wellnhofer (1988b) also identified the tarsal with the ascending process as the astragalus, but his Figure 14 shows the astragalus as the bone lateral to the calcaneum on the tarsus. This is a reversal of the normal anatomical position of these two tarsal bones, and it is simpler to assume that the normal avian condition holds for *Archaeopteryx.*

Pes

The feet of modern birds share a unique combination of structures that is unknown in any nonavian archosaur, although individual features may be present. In all modern birds, including the hesperornithiform

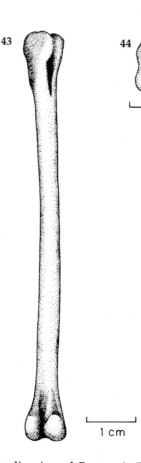

1 cm

Figs. 43–44. The tibiotarsus of the London *Archaeopteryx*. Restoration of anterior (43) and distal (44) views, combining details from both the left and right tibiotarsi.

birds *Enaliornis* and *Baptornis* (Martin and Bonner, 1977), the metatarsals fuse distally before they fuse proximally, and active growth is maintained proximally for a longer time. This arrangement necessitates the development of a tarsal cap to act as an epiphysis on the growing surface of Metatarsals II–IV. These metatarsals are locked together in part by development of flat wedge-shaped contacts with the apex of the wedge, posterior for Metatarsals II and IV and anterior for the middle metatarsal (MT III). This results in minimal anteroproximal exposure for MT III. The avian hypotarsus is formed from the tarsal cap as it droops posteriorly over the metatarsals. A tarsal cap has not been reported in dinosaurs; thus, it is unlikely that they have a hypotarsus, although one was reported in theropods (Thulborn, 1984). Thulborn also listed an intercotylar prominence as an avian feature of the proximal tarsometatarsus. This feature is developed as part of the tarsal cap in ornithurine birds and hence has no homologue in sauriurine birds or for that matter in dinosaurs.

A wedge-shaped locking system appears in the proximal metatarsal contacts of some dinosaurs, including *Avimimus*. It is absent in *Archaeopteryx* and the enantiornithine birds; therefore, its presence in dinosaurs is best explained as homoplasy.

In *Archaeopteryx* and the enantiornithine birds (Sauriurae), the tarsometatarsus is constructed differently than it is in the Ornithurae. In the Sauriurae, the proximal ends of Metatarsals II–IV fuse without fusion of the distal ends. Thus, no tarsal cap is needed and none forms. This eliminates the possibility of a hypotarsus of a normal avian type, and none is known for either *Archaeopteryx* or the Enantiornithes. The distal tarsals may fuse independently to their respective metatarsals, and Wellnhofer (1988b) figured at least two unfused distal tarsals for the Solnhofen *Archaeopteryx*. The metatarsals lie parallel to each other, with MT III fully exposed anteriorly on the proximal end. The unusual nature of this condition in *Archaeopteryx* (Fig. 45) was recognized by Heilmann (1926:20). It seems evident that sauriurine birds developed fused tarsometatarsi after they shared a common ancestor with the Ornithurae and that the common ancestor must predate the late Jurassic.

Proximal or distal foramina are not present, and the papilla for *tibialis anticus* is absent (Fig. 47). The middle metatarsal (III) is the largest and extends distally beyond the others (Fig. 48). Also, it is situated more anteriorly, especially in the Eichstätt skeleton (Fig. 50) so that the distal end is highly arched. This feature, coupled with the slender, elongate

45   46

Figs. 45–46. Stereophotographs of the anterior views of the left foot (45) and left tibiotarsus (46) of a cast of the London *Archaeopteryx*.

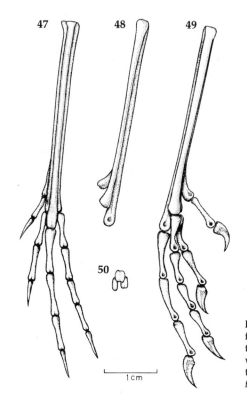

47 48 49

50

1 cm

Figs. 47–50. Restoration of the left foot (47—anterior; 49—lateral) and tarsometatarsus (48—medial view with Metacarpal I attached; 50—distal aspect) of the Eichstätt *Archaeopteryx*.

shape of Metatarsal III (Fig. 47), may distinguish the London and Eichstätt specimens at the species level.

The trochleae for Metatarsals II and IV are at about the same level. Metatarsal I is distally articulated, and the hallux is reflected exactly as in modern birds. This, together with the highly arched distal end (Fig. 50) and the long, curved phalanges, lends support to the hypothesis of an arboreal existence for *Archaeopteryx*.

Ostrom (1976b, 1985) claimed that *Compsognathus* has a distally articulated and reversed hallux as in modern birds. Tarsitano and Hecht (1980) denied that Metatarsal I was reversed, and Ostrom's photographs are decisive on this point. They clearly showed (Ostrom, 1985:Fig. 2) that the articulation of Metatarsals I and II on *Compsognathus* is lateral. In this respect, *Compsognathus* resembles other reptiles and differs from *Archaeopteryx* and other birds in which the metatarsal is twisted so that the articulation with Metatarsal II is posterior to the trochlea (Fig. 49). If Metatarsal I of *Compsognathus* were rotated to a posterior position as in *Archaeopteryx*, the first toe would lie at a right angle to the other digits, rather than opposed to them.

SUMMARY AND CONCLUSIONS

This discussion of Mesozoic birds has been limited largely to the anatomy of *Archaeopteryx*, because both the "dinosaur" and "thecodont" arguments of avian origins have been restricted almost completely to comparisons with *Archaeopteryx*, and many errors of comparison and anatomy have become disseminated widely. I hope that I have clarified at least some of these points such that *Archaeopteryx* (Fig. 51) is viewed as fundamentally birdlike in its anatomy and less like a small dinosaur than some authors have indicated.

I have argued elsewhere (Martin, 1983a,b, 1985, 1987) that I do not consider *Archaeopteryx* on the direct line to modern birds, and that some of its anatomical features are autapomorphous. It is likely that this is true

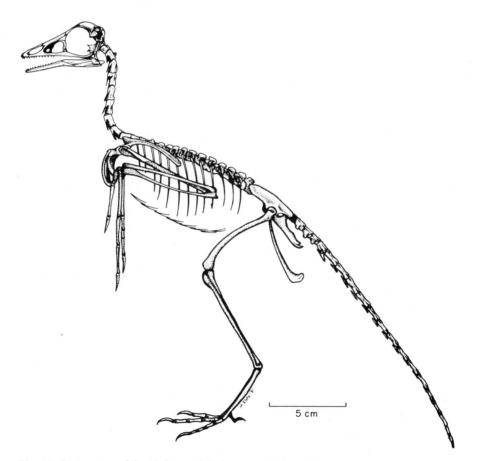

5 cm

Fig. 51. Restoration of the skeleton of *Archaeopteryx lithographica*.

of the quadrate articulation and some details of the basicranium. I also consider aspects of the shoulder girdle and the humerus to be autapomorphic. I suspect that there were contemporary Jurassic birds that, in many respects, were more advanced toward modern birds than was *Archaeopteryx*, and that *Archaeopteryx* might represent a Jurassic "living fossil." Because of this, I urge caution in the use of details of *Archaeopteryx* anatomy to assess relationships or establish character polarities for birds as a whole. I do think that features shared by *Archaeopteryx* and neornithine birds (including the Mesozoic toothed birds in the orders Hesperornithiformes and Ichthyornithiformes) must be considered primitive for birds as a group. In general, *Archaeopteryx* is more "primitive" than are other known birds, although this has been substantially clouded by Chatterjee's (1987) discovery of a protobird in the late Triassic of Texas.

Archaeopteryx has a highly encephalized skull with exceptionally large eyes and a pointed snout. The presence of foramina along the mouth suggest that the mouth may have been bordered by large scales; a horny bill may not have been present. In fact, the loss of teeth may be a result of the development of a birdlike bill because it seems necessary for the edge of the horny bill to extend lingually inside the mouth and over the jaw bones. In the Hesperornithiformes, premaxillary teeth are absent and a horny bill is present. The predentary bone also lacks teeth, but the dentaries and maxillae are toothed. I suspect that the loss of teeth in birds has more to do with a developing pattern of food manipulation than it does with weight reduction or preening. The teeth of *Archaeopteryx* are similar to the teeth of other birds and of crocodilians, and are not particularly similar to the teeth of any known dinosaur.

Archaeopteryx possesses a large antorbital fenestra that is bordered posteriorly by a typically avian lacrimal. Anteriorly, the fenestra seems to include a couple of subdivisions that are similar to those in certain theropod dinosaurs. There seems to be no evidence of a diapsid arch. The quadrate articulation and the basicranial region is dissimilar to those of dinosaurs and crocodilians, and even differs from those of modern birds; therefore, it should be considered autapomorphic. The pattern of cranial foramina and pneumaticity is about the same as in the Hesperornithiformes and modern birds, and attests to the antiquity of these features within the Aves. These features are present in the Triassic protobird and must have evolved at the base of the clade for birds. In general, the skull of *Archaeopteryx* provides little evidence for any of the suggested avian origins that cannot be demonstrated better in other material. It does support the idea that the common ancestor of birds and reptiles must be very ancient.

The lower jaw of *Archaeopteryx* resembles that of other known Mesozoic birds (*Gobipteryx, Hesperornis, Parahesperornis, Ichthyornis*) in lacking a mandibular foramen, which probably also was absent at the base of the avian clade. In fact, if the separation of birds occurred early, the absence of a mandibular foramen may be plesiomorphic. The present evidence showing that birds have a different type of thecodonty from that of dinosaurs may indicate that the common ancestor of birds and dinosaurs was not a thecodont and, thus, would have to be early Triassic (or earlier) in age. The hesperornithiform birds and the Ichthyornithiformes have similar facets on the tips of their dentaries, which suggests the presence of an intersymphyseal bone that connected the distal ends of the dentaries. This predentary in *Hesperornis* is similar to the predentary of ornithischian dinosaurs and has the most limited distribution of all the structures that might indicate relationship. Along with the opisthopubic pelvis, it might be used to resurrect Galton's (1970) ornithischian origin of birds. I do not think that this would be wise, although it does not seem to be a much weaker argument than the one for the widely accepted theropod origin of birds. It is not clear that the lower jaw has an intermandibular joint of the sort found in *Hesperornis* and probably in *Ichthyornis,* and I doubt that such a joint was present. Gingerich (1979) suggested that such a joint in *Hesperornis* might be a synapomorphy uniting birds and certain theropod dinosaurs. However, new and more detailed information on this joint in *Hesperornis* indicates that it is quite different from Gingerich's theropod examples.

The vertebrae of *Archaeopteryx* are amphicoelous, as are those of *Ichthyornis* and *Ambiortus.* Heterocoelous vertebrae occur in the cervical vertebrae of the Lower Cretaceous hesperornithiform, *Enaliornis,* and surprisingly in the protobird from the Triassic of Texas. The neck and the cervical vertebrae of *Archaeopteryx* are short in comparison with those of other birds, and indeed, shorter than those of many coelurosaurs. Pleurocoels are poorly developed and hypophyses seem to have been absent. Gauthier (1986) reported the presence of heterocoelous vertebrae and hypophyses in *Deinonychus,* but examination of Ostrom's (1960) excellent figures of these vertebrae reveals that they are unlike the avian condition.

Heterocoelous vertebrae seem to be restricted entirely to birds, and among modern birds, they are as diagnostic for the class as are the feathers. However, it also is clear that heterocoelous vertebrae have developed within birds more than once.

In some ways, the shoulder girdle and wing of *Archaeopteryx* are organized differently than those of modern birds, and it is not clear that the ancestors of modern birds ever went through an evolutionary stage of flight exactly like that found in *Archaeopteryx.* In *Archaeopteryx,* the flight muscles that largely originate from the sternum in modern birds must have originated from the coracoids and the furcula. The furcula in *Archaeopteryx* is an enlarged structure that not only helped to stabilize the chest but also must have been an important site of origin for flight muscles. It would have reoriented the muscle fibers upward in much the same way the sternum keel does in living birds. The deltoid crest of the humerus is directed forward in a way unlike that in modern birds, and in part, this may have resulted from the fact that in *Archaeopteryx* the wing serves a double function—i.e., it was used in flight and also assisted in climbing. Though this might be expected in other early birds, only *Archaeopteryx* is known to have this peculiar forelimb structure. As in all other birds, the scapula and coracoid are rotated 90° from their usual lateral position, and the glenoid faces outward. This relationship does not exist in any known dinosaur, and there is really no reason that it should. A similar situation does occur in mammalian climbers such as primates and bats, and I think that it is associated closely with the development of an arboreal habitus. The joint structures of the wing and the avian wing-folding apparatus are fully developed. I have pointed out that many of the comparisons between the wing of *Archaeopteryx* and the forelimbs of dinosaurs are erroneous. This is not surprising because most of these features are related to the possession of a feathered wing; thus, I would not expect to find them in dinosaurs unless we also think that these same dinosaurs had flight feathers.

The absence of an ossified sternum and ossified uncinate processes in *Archaeopteryx* indicates that the basal avian stalk lacked these features and that their supposed occurrence in certain dinosaurs has no phylogenetic significance for avian origins. Their absence indicates the absence of avian parabronchi and raises real questions about the efficiency of the respiratory system of *Archaeopteryx*. It is possible that some aspects of modern avian physiology, including homothermy, were not well developed in *Archaeopteryx*, and that the organism was not well suited for sustained flight. The shoulder girdle indicates that *Archaeopteryx* may have had difficulty in taking off from the ground. Its hind limbs seem to be bowed outward, and a truly reflexed hallux is present. Although it is clear that *Archaeopteryx* could progress bipedally, its skills as a runner may have been overestimated. Judging from the pelvis and legs, it is less adapted as a cursor than are most modern birds or bipedal dinosaurs. I can see little remaining evidence to argue for *Archaeopteryx*'s being anything other than arboreal.

Recently, there have been several efforts to position birds in a phylogenetic framework using "cladistic" methodology. This technique provides widely differing results depending on the individual doing the analysis. Gardiner (1982) presented a scheme in which birds and pterosaurs are the sister-group of mammals. This was followed by an interesting representation of a furry, long-armed "protoavis" by Janvier (1983). Tarsitano and Hecht (1980) published a cladogram placing *Archaeopteryx* between

Euparkeria and *Lagosuchus* but gave very little evidence in support of their hypothesis. Tarsitano (1985a) listed five cranial characters shared by birds and crocodilians, of which four are not known in theropods.

Padian (1982) published a cladogram based on about 40 characters showing *Archaeopteryx* as the sister-group of the Saurornithoididae plus the Dromaeosauridae. Gauthier (1986) greatly elaborated Padian's clado-gram and made the "Avialae" [= Aves] the sister-group of the Deinony-chosauria (Troodontidae + Dromaeosauridae). Gauthier increased Pa-dian's character list to 84; however, nearly half of these characters either cannot be demonstrated with certainty to occur in birds as they occur in dinosaurs (40%) or are also characters that occur in crocodilians (10%). Nearly 25% of the characters are considered to have developed con-vergently in pterosaurs and might be expected in other flying verte-brates. (Several of these features can even be found in bats.) The remain-ing 20 or so characters are not impressive, and a phylogeny with birds as the youngest clade is hard to reconcile with Chatterjee's protobird from the early late Triassic of Texas. Such a phylogeny would require the evolution of all other major dinosaur clades before the earliest known occurrence of dinosaurs and the absence of any further radiation of dinosaurs from the early Triassic to their extinction in the late Cretaceous. Thulborn (1984) used 29 characters to obtain a very different cladogram from that of Padian (1982). In Thulborn's scheme, *Archaeopteryx* lies be-tween the carnosaurs *Allosaurus* and *Tyrannosaurus*. The modern birds are closer to the Tyrannosauridae than to *Archaeopteryx*. Paul (1984) provided another cladogram that resembles that of Padian in putting *Archaeopteryx* close to the dromaeosaurs, but his hypothesis is like Thulborn's in plac-ing modern birds closer to a Cretaceous dinosaur *Saurornithoides* than to *Archaeopteryx*.

Paul (1984) reinterpreted some aspects of *Archaeopteryx* anatomy. He listed a "normal theropod squamosal-quadrate articulation, postorbital bar, superior temporal bar, and palate" but provided no support for those contentions, and none seems to exist. Paul also synonymized *Deinony-chus* with *Velociraptor* and *Saurornitholestes*, claiming that the latter are derived from "flying archaeopterygids." There seem to be many mistakes in the distribution of characters in this paper. Of the 18 characters listed as synapomorphies of *Archaeopteryx* and dromaeosaurs, more than half cannot be shown to be shared by the taxa discussed, and the value of the remaining features is questionable.

The ischial shape, the presence of a pubic apron, the partially closed acetabulum, the unossified sternum, the numerous gastralia, and the absence of a mandibular fenestra are primitive features that point to a separation of birds near the beginning of the archosaur radiation. It may be that members of the avian clade were numerous and diverse during the Triassic. I suspect that it included many arboreal forms, possibly including *Longisquama* and *Scleromochlus*. Chatterjee's Triassic protobird is situated stratigraphically at just the right place to support such a scenario.

Unlike most of the other hypotheses of the origin of birds, the avian-crocodilian relationship is not considered one of ancestry, although Whetstone and Whybrow (1983) showed birds within the crocodilian clade. I would expect that birds separated from a common branch with crocodilians before the numerous postcranial autapomorphies of crocodilians developed. Birds would not have been obligate bipeds at this stage, and bipedality would be more a result of the evolution of flight than a precursor to it. I also expect members of this early radiation to share the presence of the following features with crocodilians: (1) a superior tympanic diverticulum (Witmer, 1987), (2) a posterior tympanic diverticulum, (3) an anterior tympanic recess crossed by the carotid, (4) a pneumatic quadrate, (5) a pneumatic articulare, (6) fenestra pseudorotundum, (7) teeth with the crown separated from an expanded root by a distinct waist, (8) the replacement pit in the tooth root closed ventrally until late development, (9) root cementum, and (10) implantation of the teeth in a groove without the development of interdental plates. Characters 2–5 may be found in some dinosaurs but show a scattered distribution, suggesting independent acquisition.

Acknowledgments For allowing me to examine specimens, I thank the late J. P. Lehman, D. Goujet, F. Poplin, and D. E. Russell (Muséum National d'Histoire Naturelle, Paris); A. J. Charig, A. Milner, and C. A. Walker (British Museum [Natural History], London); G. S. Cowles and C. J. O. Harrison (British Museum, Ornithological Department, Tring); A. D. Walker (University of Newcastle upon Tyne); G. Viohl (Jura Museum, Eichstätt); H. Jaeger and H. Fischer (Humboldt Museum für Naturkunde, Berlin); P. Wellnhofer (Bayerische Staatssammlung, Munich); D. S. Peters (Forschungsinstitut Senckenberg, Frankfurt am Main); W. v. Koenigswald (Hessisches Landesmuseum, Darmstadt); Z. Kielan-Jaworowska, A. Elzanowski, and H. Osmolska (Polska Akademia Nauk, Warsaw); P. Ellenberger (Laboratoire de Paléontologie des Vertébrés, Montpellier); J. H. Ostrom and M. Turner (Yale Peabody Museum, New Haven); R. J. Zakrzewski (Sternberg Memorial Museum, Hays); C. B. Schultz, L. G. Tanner, and M. Voorhies (University of Nebraska State Museum); S. L. Olson and C. Ray (United States National Museum, Washington, D.C.); and T. Ferrusquia (Instituto de Geología, Mexico City).

I have benefited from many stimulating conversations with S. Olsen, J. Ostrom, P. Bühler, J. Gauthier, J. Cracraft, A. Feduccia, M. Jenkinson, R. Mengel, P. Brodkorb, H.-P. Schultze, C. Harrison, S. Chatterjee, C. A.

Walker, A. Milner, J. D. Stewart, and K. N. Whetstone. M. A. Klotz and D. Stevens prepared the figures. L. Witmer and R. Chandler critically read the manuscript. Funding was provided by The University of Kansas (sabbatical leave) and University General Research Grant 3251-5038, NSF DEB 7821432 and National Geographic Grant 2228-80. The staff of the British Museum (Natural History) greatly assisted me by making special preparations and photographs of the London specimen.

APPENDIX I. Classification of Mesozoic Birds

Class Aves
  Subclass Sauriurae
   Infraclass Archaeornithes
    Order Archaeopterygiformes
     Family Archaeopterygidae
      Genus *Archaeopteryx* von Meyer
   Infraclass Enantiornithes
    Order Gobipterygiformes
     Family Gobipterygidae
      Genus *Gobipteryx* Elzanowski
      Genus *Horesmavis* Nessov
    Order Alexornithiformes
     Family Alexornithidae
      Genus *Alexornis*
    Order Enantiornithiformes
     Family Enantiornithidae
      Genus *Enantiornis* Walker
      Genus *Avisaurus* Brett-Surman, and Paul
      Genus *Nanantius* Molnar
  Subclass Ornithurae
   Infraclass Odontoholcae
    Order Hesperornithiformes
     Family Enaliornithidae
      Genus *Enaliornis* Seeley
     Family Baptornithidae
      Genus *Baptornis* Marsh
      Genus *Neogaeornis* Lambrecht
     Family Hesperornithidae
      Genus *Hesperornis* Marsh
      Genus *Coniornis* Marsh
      Genus *Parahesperornis* Martin

Infraclass Carinata
  Order Gansuiformes
    Family Gansuidae
      Genus *Gansus* Hou and Liu
  Order Ichthyornithiformes
    Family Ichthyornithidae
      Genus *Ichthyornis* Marsh
    Family Ambiortidae
      Genus *Ambiortus* Kurotchin
  Order Apatornithiformes
    Family Apatornithidae
      Genus *Apatornis* Marsh
  Order Charadriiformes
    Family Graculavidae
      Genus *Graculavus* Marsh
      Genus *Telmatornis* Marsh
      Genus *Anatalavis* Olson and Parris
      Genus *Laornis* Marsh
      Genus *Palaeotringa* Marsh
    Family Cimolopterygidae
      Genus *Cimolopteryx* Marsh
      Genus *Ceramornis* Brodkorb
      Genus *Lonchodytes* Brodkorb
    Family Palintropidae
      Genus *Palintropus* Brodkorb
  Order Procellariiformes?
    Family Tytthostonychidae
      Genus *Tytthostonyx* Olson and Parris

Literature Cited

Barsbold, R. 1923. Carnivorous dinosaurs from the Cretaceous of Mongolia. Joint Soviet-Mongolian Paleontol. Expedition, 19:1–119.

Bühler, P. 1985. On the morphology of the skull of *Archaeopteryx*. Pp. 135–140 *in* Hecht, M. K., J. H. Ostrom, G. Viohl, and P. Wellnhofer (eds.), *The Beginnings of Birds*. Eichstätt: Freunde des Jura-Museums.

Chatterjee, S. 1987. Skull of *Protoavis* and early evolution of birds. J. Vert. Paleontol., 7(3):14A.

Colbert, E. H., and C. C. Mook. 1951. The ancestral crocodilian *Protosuchus*. Bull. Am. Mus. Nat. Hist., 97(3):143–182.

Currie, P. J. 1985. Cranial anatomy of *Stenonychosaurus inequalis* (Saurischia, Theropoda) and its bearing on the origin of birds. Can. J. Earth Sci., 22:1643–1658.

Currie, P. J. 1987. Bird-like characteristics of the jaws and teeth of troodontid theropods (Dinosauria, Saurischia). J. Vert. Paleontol., 7:72–81.

de Beer, G. R. 1954. Archaeopteryx lithographica: *A Study Based on the British Museum Specimen*. London: British Museum (Natural History).

Edmund, A. G. 1960. Tooth replacement phenomena in the lower vertebrates. Contrib. R. Ontario Mus. Life Sci. Div., 52:1–190.

Ellenberger, P. P. 1977. Quelques précisions sur l'anatomie et la place systématique très spéciale de *Cosesaurus aviceps* (Ladinien supérieur de Montral, Catalogue). Caud. Geol. Iberica, 4:169–188.

Elzanowski, A. 1981. Embryonic bird skeletons from the Late Cretaceous of Mongolia. Palaeontol. Pol., 42:147–179.

Feduccia, A. 1980. *The Age of Birds*. Cambridge: Harvard Univ. Press.

Feduccia, A., and H. B. Tordoff. 1979. Feathers of *Archaeopteryx*: asymmetric vanes indicate aerodynamic function. Science, 203:1021.

Galton, P. M. 1970. Ornithischian dinosaurs and the origin of birds. Evolution, 24:448–462.

Galton, P. M. 1974. The Ornithischian dinosaur *Hypsilophodon* from the Wealdon of the Isle of Wight. Bull. Br. Mus. Nat. Hist., 251:1–152.

Gardiner, B. G. 1982. Tetrapod classification. Zool. J. Linn. Soc., 74:207–232.

Gauthier, J. 1986. Saurischian monophyly and the origin of birds. Pp. 1–55 in Padian, K. (ed.), *The Origin of Birds and the Evolution of Flight*. San Francisco: California Academy of Sciences.

Gingerich, P. D. 1979. The stratophenetic approach to phylogeny reconstruction in vertebrate paleontology. Pp. 41–77 in Cracraft, J., and N. Eldridge (eds.), *Phylogenetic Analysis and Paleontology*. New York: Columbia Univ. Press.

Heilmann, G. 1926. *The Origin of Birds*. London: Witherby.

Heller, F. 1959. Ein dritter *Archaeopteryx* Fund aus den Solnhofener Plattenkalken von Langenaltheim/Mfr. Erlanger Geol. Abh., 31:3–25.

Helms, J. 1982. Zur Fossilization der Federn des Urvogels (Berliner Exemplar). Wiss. Z. Humboldt Univ. Math Naturwiss. Reihe, 31:185–199.

Hinchliffe, J. R. 1985. "One, two, three" or "two, three, four": an embryologist's view of the homologies of the digits and carpus of modern birds. Pp. 141–147 in Hecht, M. K., J. H. Ostrom, G. Viohl, and P. Wellnhofer (eds.), *The Beginnings of Birds*. Eichstätt: Freunde des Jura-Museums.

Houde, P., and S. L. Olson. 1981. Paleognathous carinate birds from the early Tertiary of North America. Science, 214:1236–1237.

Howgate, M. E. 1984. The teeth of *Archaeopteryx* and a reinterpretation of the Eichstätt specimen. Zool. J. Linn. Soc., 82:159–175.

Howgate, M. E. 1985. Problems of the osteology of *Archaeopteryx*: is the Eichstätt specimen a distinct genus? Pp. 105–112 in Hecht, M. K., J. H. Ostrom, G. Viohl, and P. Wellnhofer (eds.), *The Beginnings of Birds*. Eichstätt: Freunde des Jura-Museums.

Huxley, T. H. 1870. Further evidence of the affinity between dinosaurian reptiles and birds. Q. J. Geol. Soc. London, 26:12–31.

Janvier, P. 1983. Le divorce de l'oiseau et du crocodile. Recherche, 14(19):1430–1432.

Jollie, M. 1977. A contribution to the morphology and phylogeny of the Falconiformes. Part 4. Evol. Theory, 2:1–141.

Kleinschmidt, A. 1951. Über eine Rekonstruktion des Schädels von *Archaeornis siemensi* Dames 1884 im Naturhist. Museum, Braunschweig. Pp. 631–635 in *Proceedings of Xth International Ornithological Congress*. Uppsala, June 1950.

Lambrecht, K. 1928. Fluorographische Beobachtungen an den "elastischen Fasern" des Pterosaurier-Patagiums. Arch. Mus. Teyler., Ser. III, 6:40–50.

Martin, L. D. 1983a. The origin of birds and of avian flight. Curr. Ornithol., 1:105–129.

Martin, L. D. 1983b. The origin and early radiation of birds. Pp. 291–338 in Brush, A. H., and G. A. Clark, Jr. (eds.), *Perspectives in Ornithology*. Cambridge: Cambridge Univ. Press.

Martin, L. D. 1984. A new hesperornithid and the relationships of the Mesozoic birds. Trans. Kansas Acad. Sci., 87(3–4):141–150.

Martin, L. D. 1985. The relationships of *Archaeopteryx* to other birds. Pp. 177–183 *in* Hecht, M. K., J. H. Ostrom, G. Viohl, and P. Wellnhofer (eds.), *The Beginnings of Birds*. Eichstätt: Freunde des Jura-Museums.

Martin, L. D. 1987. The beginnings of the modern avian radiation. Pp. 9–19 *in* Mourer-Chauvire, C. (ed.), *L'évolution des oiseaux d'après le témoignage des fossiles*. Doc. Lab. Geol. Lyon, Vol. 99. Lyon: Université Claude-Bernard.

Martin, L. D., and O. Bonner. 1977. An immature specimen of *Baptornis advenus* from the Cretaceous of Kansas. Auk, 94:787–789.

Martin, L. D., and J. D. Stewart. 1977. Teeth in *Ichthyornis* (Class: Aves). Science, 195:1331–1332.

Martin, L. D., and J. D. Stewart. 1985. Homologies in the avian tarsus. Nature, London, 315:159.

Martin, L. D., J. D. Stewart, and K. N. Whetstone. 1980. The origin of birds: structure of the tarsus and teeth. Auk, 97:86–93.

Martin, L. D., and J. Tate, Jr. 1976. The skeleton of *Baptornis advenus* from the Cretaceous of Kansas. Smithson. Contrib. Paleobiol., 27:35–66.

McGowan, C. 1984. Evolutionary relationships of ratites and carinates: evidence from ontogeny of the tarsus. Nature, London, 315(6015):159–160.

McGowan, C. 1985. Homologies in the avian tarsus, McGowan replies. Nature, London, 315:159–160.

Morse, E. S. 1872. On the tarsus and carpus of birds. Ann. Lyc. Nat. Hist., 10:1–22.

Norberg, R. A. 1985. Function of vane asymmetry and shaft curvature in bird flight feathers. Pp. 303–318 *in* Hecht, M. K., J. H. Ostrom, G. Viohl, and P. Wellnhofer (eds.), *The Beginnings of Birds*. Eichstätt: Freunde des Jura-Museums.

Olson, S. L. 1985. The fossil record of birds. Pp. 79–238 *in* Farner, D. S., J. R. King, and K. C. Parks (eds.), *Avian Biology*. Vol. 8. New York: Academic Press.

Ostrom, J. H. 1969a. Osteology of *Deinonychus antirrhopus*, an unusual theropod from the Lower Cretaceous of Montana. Bull. Yale Peabody Mus. Nat. Hist., 30:1–165.

Ostrom, J. H. 1969b. Terrible claw. Discovery, 5(1):1–9.

Ostrom, J. H. 1973. The ancestry of birds. Nature, London, 242:136.

Ostrom, J. H. 1974a. *Archaeopteryx* and the origin of flight. Q. Rev. Biol., 49:27–47.

Ostrom, J. H. 1974b. The pectoral girdle and forelimb function of *Deinonychus* (Reptilia: Saurischia): a correction. Postilla, 165:1–11.

Ostrom, J. H. 1975. The origin of birds. Ann. Rev. Earth Plan. Sci., 3:55–57.

Ostrom, J. H. 1976a. Some hypothetical anatomical stages in the evolution of avian flight. Smithson. Contrib. Paleobiol., 27:1–21.

Ostrom, J. H. 1976b. *Archaeopteryx* and the origin of birds. Biol. J. Linn. Soc., 8:91–182.

Ostrom, J. H. 1976c. On a new specimen of the lower Cretaceous theropod dinosaur *Deinonychus antirrhopus*. Breviora, 439:1–21.

Ostrom, J. H. 1978. The osteology of *Compsognathus longipes* Wagner. Zitteliana, 4:73–118.

Ostrom, J. H. 1979. Bird flight: how did it begin? Am. Sci., 67:46–56.

Ostrom, J. H. 1985. The meaning of *Archaeopteryx*. Pp. 161–176 *in* Hecht, M. K., J. H. Ostrom, G. Viohl, and P. Wellnhofer (eds.), *The Beginnings of Birds*. Eichstätt: Freunde des Jura-Museums.

Padian, K. 1982. Macroevolution and the origin of major adaptations: vertebrate flight as a paradigm for the analysis of patterns. Third N. Am. Paleo. Convention Proc., 2:387–392.

Padian, K. 1985. The origins and aerodynamies of flight in extinct vertebrates. Paleontology, 28(3):413–433.

Parker, W. K. 1891. On the morphology of a reptilian bird *Opisthocomus cristatus*. Trans. Zool. Soc. London, 13:43–85.

Paul, G. S. 1984. The archosaurs: a phylogenetic study. Pp. 175–180 *in* Reif, W. E., and F. Westphal (eds.), *Third Symposium on Mesozoic Terrestrial Ecosystems, Short Papers.* Tübingen: Attempto Verlag.

Rayner, J. M. V. 1985. Mechanical and ecological constraints on flight evolution. Pp. 279–288 *in* Hecht, M. K., J. H. Ostrom, G. Viohl, and P. Wellnhofer (eds.). *The Beginnings of Birds.* Eichstätt: Freunde des Jura-Museums.

Rietschel, S. 1985. Feathers and wings of *Archaeopteryx*, and the question of her flight ability. Pp. 251–260 *in* Hecht, M. K., J. H. Ostrom, G. Viohl, and P. Wellnhofer (eds.). *The Beginnings of Birds.* Eichstätt: Freunde des Jura-Museums.

Romanoff, A. L. 1960. *The Avian Embryo.* New York: Macmillan.

Stark, D. 1978. *Vergleichende Anatomie der Wirbeltiere auf evolutionsbiologischer Grundlage.* Vol. 2. *Das Skelettsystem. Allgemeines, Lokomotionstypen.* New York: Springer-Verlag.

Stephen, B. 1985. Remarks on reconstruction of *Archaeopteryx* wing. Pp. 261–265 *in* Hecht, M. K., J. H. Ostrom, G. Viohl, and P. Wellnhofer (eds.), *The Beginnings of Birds.* Eichstätt: Freunde des Jura-Museums.

Tarsitano, S. F. 1985a. Cranial metamorphosis and the origin of the Eusuchia. Neues Jahrb. Geol. Palaeontol. Abh., 170:27–44.

Tarsitano, S. F. 1985b. The morphological and aerodynamic constraints on the origin of avian flight. Pp. 319–332 *in* Hecht, M. K., J. H. Ostrom, G. Viohl, and P. Wellnhofer (eds.), *The Beginnings of Birds.* Eichstätt: Freunde des Jura-Museums.

Tarsitano, S. F., and M. K. Hecht. 1980. A reconsideration of the reptilian relationships of *Archaeopteryx*. Zool. J. Linn. Soc. London, 69(2):149–182.

Thulborn, R. A. 1984. The avian relationships of *Archaeopteryx*, and the origin of birds. Zool. J. Linn. Soc. London, 82:119–158.

Walker, A. D. 1972. New light on the origin of birds and crocodiles. Nature, London, 237:257–263.

Walker, A. D. 1974. Evolution, organic. Pp. 177–179 *in McGraw-Hill Yearbook of Science and Technology.* New York: McGraw-Hill.

Walker, A. D. 1977. Evolution of the pelvis in birds and dinosaurs. Pp. 319–357 *in* Andrews, S. M., R. S. Miles, and A. D. Walker (eds.), *Problems in Vertebrate Evolution.* London: Academic Press.

Walker, A. D. 1985. The braincase of *Archaeopteryx*. Pp. 123–134 *in* Hecht, M. K., J. H. Ostrom, G. Viohl, and P. Wellnhofer (eds.), *The Beginnings of Birds.* Eichstätt: Freunde des Jura-Museums.

Walker, C. A. 1981. New subclass of birds from the Cretaceous of South America. Nature, London, 292:51–53.

Welles, S. P., and R. A. Long. 1974. The tarsus of theropod dinosaurs. Ann. S. Afr. Mus., 64:191–218.

Wellnhofer, P. 1974. Das fünfte Skelettexemplar von *Archaeopteryx*. Palaeontographica (A), 147:169–16.

Wellnhofer, P. 1978. Pterosauria. Pp. 1–82 *in* Wellnhofer, P. (ed.), *Handbuch der Paläoherpetologie.* Stuttgart: Gustav Fischer Verlag.

Wellnhofer, P. 1985. Remarks on the digit and pubis problems of *Archaeopteryx*. Pp. 113–122 *in* Hecht, M. K., J. H. Ostrom, G. Viohl, and P. Wellnhofer (eds.), *The Beginnings of Birds.* Eichstätt: Freunde des Jura-Museums.

Wellnhofer, P. 1988a. A new specimen of *Archaeopteryx*. Science, 240:1790–1792.

Wellnhofer, P. 1988b. Ein neues Exemplar von *Archaeopteryx*. Archaeopteryx, 6:1–30.

Whetstone, K. N. 1983. Braincase of Mesozoic birds: I. New preparation of the "London" *Archaeopteryx*. J. Vert. Paleontol., 2:439–452.

Whetstone, K. N., and L. D. Martin. 1979. New look at the origin of birds and crocodiles. Nature, London, 279:234–236.

Whetstone, K. N., and L. D. Martin. 1981. Common ancestry for birds and croco-diles?—Reply to C. McGowan. Nature, London, 289:89.

Whetstone, K. N., and P. J. Whybrow. 1983. A "cursorial" crocodilian from the Triassic of Lesoth (Basutoland), Southern Africa. Occas. Papers Mus. Nat. Hist. Univ. Kansas, 106:1–37.

Whybrow, P. J. 1982. Preparation of the cranium of the holotype of *Archaeopteryx lithographica* from the collections of the British Museum (Natural History). Neues Jahrb. Geol. Palaeontol. Monatsh., 1982:184–192.

Witmer, L. M. 1987. The cranial air sac system of Mesozoic birds. M.A. thesis. Lawrence: Univ. of Kansas.

Yalden, D. W. 1985. Forelimb function in *Archaeopteryx*. Pp. 91–97 *in* Hecht, M. K., J. H. Ostrom, G. Viohl, and P. Wellnhofer (eds.), *The Beginnings of Birds*. Eichstätt: Freunde des Jura-Museums.

15 *Archaeopteryx:* Quo Vadis?

Samuel Tarsitano

Among the different lines of evidence that have been applied to studies of the origin of birds are anatomical features and aerodynamic data relevant to the origin of flight. Morphological characters have been used in both cladistic and functional analyses to determine relationships among various archosaurian groups, the earliest known bird *Archaeopteryx,* and later birds such as *Hesperornis.* Examples of cladistic approaches to similar phylogenetic problems are the studies by Hennig (1966), Cracraft (1986), and Schoch (1986), and of functional studies, those by Reif (1982) and Reif et al. (1985). Other authors (e.g., Tarsitano, 1985b; Peters and Gutmann, 1985; Rayner, 1985) have sought to identify the ancestor or sister-group of birds by seeking a group whose morphology seems to represent the best aerodynamic model for the origin of flight, given the physiological and aerodynamic constraints that are involved. The results of any analyses now may have to be reevaluated in light of the discovery of an Upper Triassic proavis (S. Chatterjee, pers. comm.)—a birdlike archosaur that seems to differ from *Archaeopteryx* in many ways.

The morphological evidence of avian ancestry is equivocal (Tarsitano, 1985a,b). The digital proportions of *Archaeopteryx* indicate a coelurosaurian ancestry of birds (Heilmann, 1926; Ostrom, 1976a,b, 1985). However, other anatomical features are crocodylomorph-like and suggest a thecodontian ancestry of birds. Among the latter features are the following: (1) the structure of the ear region (Baird, 1970; Wever, 1979; Whetstone and Martin, 1979; Tarsitano and Hecht, 1980; Tarsitano, 1985a,b),

(2) tooth morphology and replacement pattern (Martin et al., 1980; Martin, 1985), (3) braincase construction (Tarsitano, 1985a,b), (4) pneumaticity of the siphonium and braincase (Whetstone and Martin, 1979; Tarsitano and Hecht, 1980), and (5) a coracoid that lies in two planes. Unfortunately, the fossil record lacks early representatives of lineages of the coelurosaurian dinosaurs and arboreal thecodonts which, at present, are the most logical candidates for the ancestors of birds. We are limited to restricted glimpses of the past that suffice only to tell us that birds were derived from archosaurian reptiles.

The avenue of inquiry followed herein is limited to anatomical considerations of the origins of birds. The anatomical analysis germane to the question of the origin of birds has two aspects. The first concerns a cladistic analysis of character-states among the archosaurian reptiles— i.e., a search for the sister-group of birds. The second and more important aspect involves the constraints that each morphological starting point imposes on the possibility of evolving flight. Thecodont and coelurosaur morphologies are sufficiently divergent to have an impact on aerodynamic investigations.

PROPOSED AVIAN-ARCHOSAURIAN AFFINITIES: COELUROSAURS VERSUS THEROPODS

Ostrom (1976b) stated that if *Archaeopteryx* had lacked feathers, it would have been classified as a coelurosaur. Some paleontologists consider the anatomy of *Archaeopteryx* to be virtually identical to that of dromaeosaurian coelurosaurs. In fact, Ostrom (1976a,b) believed that *Archaeopteryx* and coelurosaurs shared so many characters that *Archaeopteryx* and other birds were derived directly from coelurosaurs. Except for the semilunate carpal element, most of the characters used by Ostrom were observed by earlier workers such as Heilmann (1926) and Lowe (1935, 1944). Heilmann (1926) noted similarities in body proportions (but see Yalden, 1984, 1985) and pes construction that led him to suggest that certain coelurosaurs were derived from *Archaeopteryx*—a view that subsequently was adopted as a novel idea by a few paleontologists!

The discovery of an Upper Triassic proavis (S. Chatterjee, pers. comm.) necessitates reexamination, and perhaps abandonment, of many taxa that have been hypothesized to be avian ancestors on the basis of anatomical evidence. The Triassic age of this organism is also of interest, because coelurosaurs do not appear until the Upper Triassic; thus, it would be difficult, if not impossible, for them to have given rise to birds. If the Texas proavis could fly, as Chatterjee (1987) suggested, then the avian lineage must have evolved early in the Triassic prior to the appearance of theropod dinosaurs. Thus, pending new discoveries that demonstrate otherwise, it is temporally impossible for dinosaurs to have given rise to birds.

The Problems Posed by Parallelism

Until a description of the Texas proavis is available, it could be useless to argue about the phylogenetic significance of the anatomy of *Archaeopteryx*. However, morphological study of archosaurs, including *Archaeopteryx*, may elucidate the evolution of flight and offer insight into the prevalence of parallel evolution in archosaurian reptiles (Hecht, 1976). I agree with Wolfgang Gutmann (pers. comm.) and Hecht and Edwards (1977) that parallelism is common in tetrapod evolution, and that by testing models, we will gain a greater understanding of evolutionary patterns. The cladogram of Gauthier and Padian (1985) demonstrates the problem of parallel evolution within the Archosauria. They divided the Archosauria into two groups—the Pseudosuchia and Ornithosuchia. The Pseudosuchia contains the Crocodylomorpha, Rauisuchia, Aetosauria, and Parasuchia in descending order, whereas the Ornithosuchia contains dinosaurs, Pterosauria, *Lagosuchus*, Ornithosuchidae, and *Euparkeria*. The basal group of Pseudosuchia is the Parasuchia, whereas the basal group in the Ornithosuchia is represented by *Euparkeria*. My examination of these basal taxa reveals no morphological similarities of the sinus systems, teeth, pelvis, and ear region shared by crocodylomorphs and birds. I therefore conclude, following Gauthier and Padian (1985), that the presence of similar features in birds and crocodilians must be the result of homoplasy. However, embryological evidence indicates that these characters are homologous (see Retzius, 1884; Parker, 1883, 1891; de Beer, 1937; Baird, 1970; H. Müller, 1963; F. Müller, 1967; Bellairs and Jenkins, 1960; Bellairs and Kamal, 1979) and leads me to question whether the characters used by Gauthier and Padian (1985) to unite coelurosaurs with birds are the result of homoplasy. Moreover, on the basis of our knowledge of archosaurian morphology, the grouping of taxa by Gauthier and Padian (1985) can be shown to be incongruent with the data. Although detailed discussion of each morphological unit would exceed the page limits of this chapter, discussion of selected units will illustrate the problem of separating crocodylomorphs from the avian lineage (cf. Reconsideration of the Anatomical Evidence, below).

Characters change through time, and lineages that depart from basic stocks repeatedly can evolve similar structures at different times, at the same time, or a combination of the two (Simpson, 1961). Considering the resolution of the fossil record and the amount of geologic time involved, it would be difficult to resolve the point of origin of most characters, if in fact there is a single origin. Furthermore, characters may change at different rates in different lineages. For example, the "trend" toward bipedalism discussed by Romer (1966) in the Archosauria is associated with changes in the pelvis and hind limb. If stride length and other characters are advantages in being bipedal (Snyder, 1954; Charig, 1972), then one can understand how different lineages of archosaurs possibly evolved bipedalism at different times. Thus, for Romer, bipedalism could have evolved more than once. The fixing of a specific point of origin for bipedalism may be artificial, if not arbitrary. Likewise, the same may be said for the origins of various other characters. Although this fact may introduce chaos into character analysis, my experience in functional morphology indicates that this unpleasant situation may be commonplace in the evolution of taxa.

There are only certain ways that tetrapod vertebrates or archosaurs could have become bipedal or assumed upright gaits. One repeatedly observes the same adaptations because the same sorts of constraints of muscle attachment, joint strengths and constructions, and material properties apply. Thus, on numerous occasions, tetrapods with upright gaits have evolved trochanters in order to change the direction in which a muscle pulls, or they have moved muscle insertions in order to maintain a low angle of applied force (Tarsitano et al., 1989). The classic tenet that form and function go together sometimes is forgotten. Thus, possession of greater and lesser femoral trochanters by a taxon may reveal something about its stance and gait but perhaps nothing about its phylogenetic relationships. A functional analysis must be superimposed upon a fundamental Hennigian cladistic analysis in order to avoid the labyrinth of morphological reversals and parallelisms. Cladists are more concerned with the existence of patterns of character distribution (Schoch, 1986). Functional morphology seeks to explain how and for what purpose these patterns have emerged. Through typological analysis, recurring patterns can be discovered, but this methodology cannot address the origins of structures or biochemical pathways and thereby deal effectively with homoplasy.

The Status of the Thecodontia

Gauthier (1986) posed two objections to theropods as ancestors of birds. First, he stated that dinosaurs appeared too late and were too specialized to have given rise to birds—an opinion with which I concur on the basis of present evidence. Subsequently, Gauthier dismissed this objection in view of the "abundant" characters shared by theropods and *Archaeopteryx*. He further implied that the opponents of the theropod hypothesis deny a relationship between theropods and birds. However, I (1985a,b) showed that theropods are related to birds but are not their ancestors or sister-group. Gauthier (1986) dismissed a second objection to the theropod ancestry of birds—i.e., that the similarities between theropods and birds are the result of convergence—on the rationale that evidence of convergence and parallelism is lacking. In fact, a number of publications indicate that convergence has occurred in many of the characters used by Ostrom (1976a,b), Gauthier and Padian (1985), and Gauthier (1986) to unite birds and coelurosaurs. Among these studies are those by Welles and Long (1977), Galton (1970), Whetstone and Martin (1979), Tarsitano and Hecht (1980), Hecht and Tarsitano (1982), Martin (1983, 1985), and Tarsitano (1983, 1985a,b). It is interesting that most of the characters used to relate theropods (coelurosaurs) to birds are found in the manus and pes. It can be shown that these characters indeed can be the result of convergence because the patterns present in both taxa represent either levels of organization or systems adapted for the same or similar function.

Gauthier (1986) misunderstood the "cladogram" of Tarsitano and Hecht (1980) in which the authors, on the basis of information available at that time, indicated that there were only three possible placements of *Archaeopteryx.* Their consideration of *Archaeopteryx* as a bird allied to the Thecodontia was a hypothetical arrangement that allowed them to frame the question of avian relationships, rather than a corroborated cladogram of archosaurian relationships. Gauthier criticized Tarsitano and Hecht (1980) for stating that the so-called paraphyletic group Thecodontia contained an as yet undiscovered avian ancestor. He also stated that the thecodont hypothesis should not be considered further because it detracted from the interpretation of relationships among birdlike archosaurs.

The Thecodontia is necessarily paraphyletic because it is, or may be, a grade, and because birds, dinosaurs, crocodilians, and pterosaurs are not relegated to this taxon. The removal of these four groups would make the Thecodontia "temporally" paraphyletic, but this taxon is retained because of its utility as an expression of probable relationships. It would also be necessary for the Thecodontia to exclude the Proterosuchia, the earliest known archosaurs, because then the Thecodontia could be diagnosed with characters to distinguish it from the Archosauria, as Gauthier (1986) properly noted; Gauthier correctly indicated that for this reason the Thecodontia lacks a diagnosis. This is part of a larger problem of typological analysis (Bock, 1965, 1977; Reif, 1982; Reif et al., 1985) and of classification that weights morphological differences and changes in adaptive zones as vehicles for new taxonomic divisions.

Gauthier (1986) claimed to use recency of common ancestry, rather than similarity, to determine classification—a claim with which I disagree. Like all phylogenetic hypotheses, his must be based on specialized similarities (synapomorphies). The claim that his proposed phylogenetic relationships are based on recency of common ancestry is not supported by his methodology, which uses basic typological morphological assemblages. Recency of common ancestry implies a genetic relationship that is inferred only from a morphological analysis.

The exclusion of the Thecodontia from phylogenetic analyses of the origin of birds because the group is paraphyletic eliminates informative material such as *Megalancosaurus* from consideration. The Thecodontia, like other taxa, is composed of lineages. Romer (1966) used the Thecodontia to represent a primitive grade or level of organization within the Archosauria. Gauthier (1986) misunderstood the historical perspective of this taxon and equated it in meaning with smaller monophyletic taxa. He confused attacking a taxonomic name with evaluating the taxa within it. The thecodont hypothesis of the origin of birds was used by Tarsitano and Hecht (1980) and Hecht and Tarsitano (1982) in the Romerian context. According to Hecht and Tarsitano, thecodonts represent a primitive level of organization, and birds are an ancient group that diverged from some unknown or megalancosaurian lineage of Thecodontia. For Ostrom (1976a,b), Gauthier and Padian (1985), and Gauthier (1986), birds diverged much later among coelurosaurs. The term *Thecodontia* implies that the primitive archosaurs are monophyletic. It was not our intention to revise the Archosauria (Tarsitano and Hecht, 1980). Character analysis has begun to clarify the lineages within the level of organization called the Thecodontia (Romer, 1972a,b,c; Walker, 1970, 1972; Bonaparte, 1975, 1978; Tarsitano, 1985a; Gauthier, 1986; Frey, 1988a,b; Parrish, 1987, 1988; Benton and Clark, 1988; Cruickshank and Benton, 1988). These types of problems inevitably arise as one changes from a level of organization classification to one that aspires to be phylogenetic.

The Thecodontia are essentially equivalent to primitive archosaurs, as Gauthier (1986) stated. Thecodonts are characterized by the following: (1) a verticalized basioccipital (except megalancosaurian types) as adults (Tarsitano, 1985a), (2) antorbital fenestrae, (3) a fourth trochanter (Romer, 1956), (4) a calcaneal tuber that is first lateral then posterolateral in advanced quadrupedal forms with more upright stances (Romer, 1956; Charig, 1972; Cruickshank, 1972, 1979; Tarsitano, 1981, 1983; Parrish, 1988), (5) a medially directed tongue process on the calcaneum, (6) a paravertebral shield (except megalancosaurian types) that is biserial in distribution (Frey, 1988a,b), (7) the development of an external mandibular fenestra (Romer, 1956), and (8) the development of a median basicranial sinus. If we exclude the Proterosuchia from the Thecodontia on the basis of the distinct tarsal morphology of the former, then the

Thecodontia is a viable taxon. In some ways, Gauthier's (1986) dilemma concerning the Thecodontia is equivalent to the problem of the taxon Eosuchia, which represents the basal stock of lepidosaurs and perhaps related taxa such as nothosaurs and plesiosaurs (Kuhn-Schnyder, 1963, 1968; Mazin, 1982; Tarsitano, 1982; Tarsitano and Riess, 1982) and represents a level of organization.

Equivocal Characters

The major difficulties found in phylogenetic studies occur in the character analysis. A case in point are the characters used by Gauthier (1986) to relate birds to coelurosaurs, which either (1) occur in carnosaurs and not coelurosaurs, or (2) are so poorly described as to make homoplasy a necessity (e.g., fenestrae pseudorotunda). Other features are absent in the taxa reported to possess them, or simply are symplesiomorphies at various levels within the Archosauria. Below I discuss some of Gauthier's (1986) characters (numbers 69–81) that unite "Maniraptor" coelurosaurs with birds in order to illustrate the difficulties and limitations of typological analysis.

Character 69. Reduction or loss of prefrontal bones occurs many times within the Reptilia and, therefore, has little weight (see Hecht, 1976; Hecht and Edwards, 1977; Schoch, 1986). Owing to the preservation of the material, it is not clear if *Archaeopteryx* has a prefrontal bone.

Character 70. Axial epipophyses are neither described nor labeled by Gauthier (1986) and, therefore, cannot be evaluated. In addition, the fact that they are present primitively within the Saurischia precludes their use as synapomorphies.

Character 71. Prominent hypapophyses in the cervical region occur throughout the Reptilia, including the Crocodilia and other nondinosaurian archosaurs. As yet, they are unknown in *Archaeopteryx;* their absence may be real or an artifact of preparation. As described, the character lacks any information on its functional significance. This is a symplesiomorphic character of little value, or it is a homoplasy.

Character 71a. The occurrence of neural spines and transverse processes only on Caudal Vertebrae 1 through 9 characterizes pterosaurs, turtles, and mammals, but only a few birds. In birds, the character depends upon the number of the caudal vertebrae that are incorporated into the pygostyle. Any reduction of the tail is associated with the reduction of the transverse processes and neural arches and thus can be evolved repeatedly.

Character 71b. Elongated caudal prezygapophyses occur in prosauropods (Galton, 1984); therefore, their presence should be considered symplesiomorphic. Also, the occurrence in *Archaeopteryx* of elongated prezygapophyses in the same region of the caudal series of vertebrae as in coelurosaurs is disputed.

Character 72. A subrectangular coracoid is absent in *Archaeopteryx* and most birds, and occurs only in ratites with degenerate pectoral girdles (Feduccia, 1980). Comparison of the coracoids of *Archaeopteryx* and *Deinonychus* demonstrates that the coracoids of coelurosaurs are reduced, compared with those of *Archaeopteryx* (Fig. 1) (Tarsitano and Hecht, 1980). In *Archaeopteryx,* the coracoid lies in two distinct planes, a configuration not present in every known coelurosaur (Fig. 1) but present in some thecodonts (Tarsitano and Hecht, 1980). Statements to the contrary (e.g., Ostrom, 1985) are not supported by the morphology of these bones.

Character 73. Possession of a forelimb that is nearly 75% of the presacral vertebral length is found in pterosaurs and seems likely to have occurred in the avimorph thecodont, *Megalancosaurus.* This character is poorly coded and could have been evolved repeatedly, depending upon the adaptive zone of the animal. For example, forelimb elongation has been evolved many times in many arboreal animals. It should be noted that the forelimb elongation seen in *Archaeopteryx* is almost twice that of coelurosaurs; therefore, the description of the character-state is misleading.

The resemblance of the manus of *Deinonychus* and *Archaeopteryx* is discussed with this character (see also Gauthier and Padian, 1985). This resemblance concerns the semilunate carpal element, and digital and metacarpal proportions (Heilmann, 1926; Ostrom, 1976a,b). Concerning digital proportions, Tarsitano and Hecht (1980) observed that the topographic first three digits of pterosaurs have similar proportions. In

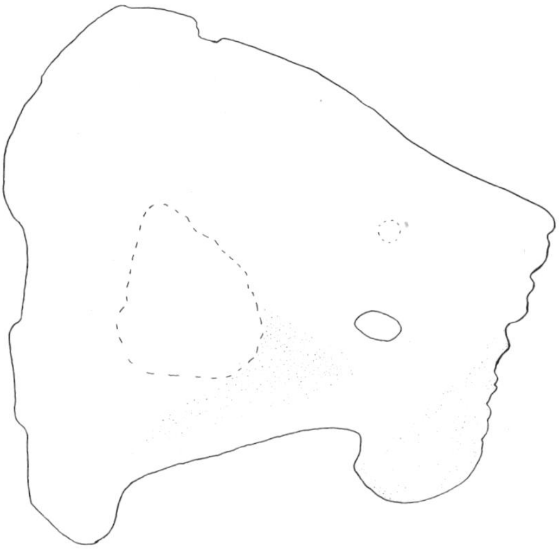

Fig. 1. The coracoid of *Deinonychus.* Note that it is reduced in size and lies in one plane. In thecodonts and *Archaeopteryx,* the coracoid lies in two planes.

some pterosaurs (Wellnhofer, 1980), the first digit has a long proximal phalanx and an ungual much like those of *Archaeopteryx* and *Deinonychus*. The same condition occurs in the avimorph thecodont *Megalancosaurus* (Calzavara et al., 1981; R. Wild, pers. comm.). The second digit of many pterosaurs possesses the same number of phalanges as in coelurosaurs and *Archaeopteryx,* but the digit is smaller than in *Archaeopteryx* and *Deinonychus.* The third topographic digit of pterosaurs is of interest because the phalangeal formula and proportions are the same as those of *Archaeopteryx* (Wellnhofer, 1974) and *Deinonychus,* in which there are two fairly short proximal phalanges, a long penultimate phalanx, and an ungual (Tarsitano and Hecht, 1980). Thus, it seems clear that the formation of a grasping hand proceeds along the same basic Bauplan in archosaurs. This fact is borne out by the manus of *Megalancosaurus,* which although specialized, has a first digit similar to that of *Archaeopteryx,* and as discussed below, possesses a primitive semilunate carpal element (Fig. 2). It should be noted that the digital proportions of *Deinonychus* and *Archaeopteryx* are similar, but not identical. The first phalanx in Digit I in the coelurosaur is massive, whereas in *Archaeopteryx,* it is elongate and slender.

In order to demonstrate that the manus of thecodonts is dissimilar to

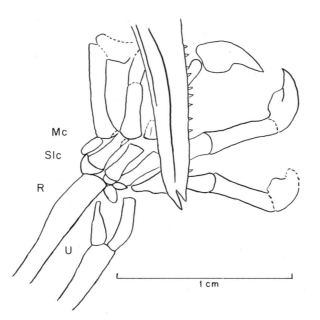

Fig. 2. The carpus of *Megalancosaurus.* Note the presence of a semilunate carpal element or carpal structure that represents the structural antecedent in this thecodont to the semilunate carpal of *Archaeopteryx.* Mc = metacarpal; R = radius; Slc = semilunate carpal; U = ulna.

that of *Archaeopteryx*, Ostrom (1976a, 1985) relied on comparisons of the latter taxon with quadrupedal thecodonts (Tarsitano and Hecht, 1980; Tarsitano, 1985a,b). One would not expect a grasping hand (*Archaeopteryx*) to be similar to a primitive hand used in terrestrial locomotion (quadrupedal, terrestrial thecodonts) (Tarsitano, 1985a). Thus, Ostrom's (1976a,b) comparisons of the manus of *Archaeopteryx* with those of quadrupedal thecodonts seems to have been based on a misunderstanding of these functional differences. The semilunate carpal element occurs in *Archaeopteryx* and *Deinonychus*, related forms such as some carnosaurs, and thecodonts. The character is associated with functional similarity of the manus and is not a good indicator of sister-group relationship.

If the semilunate carpals of coelurosaurs and *Archaeopteryx* are homologous, the common ancestor of coelurosaurs and birds must have had this condition as well; otherwise, the condition could have arisen independently in each group. Although early coelurosaurs lack a semilunate carpal element, their distal carpals may have been antecedent to it. The carpus of *Coelophysis* may represent the ancestral coelurosaur condition; in this taxon, the distal carpal articulated with topographic Digits I and II, and may have been composed of two fused distal carpals (Fig. 3). However, a similar carpal element that articulated with the first two topo-

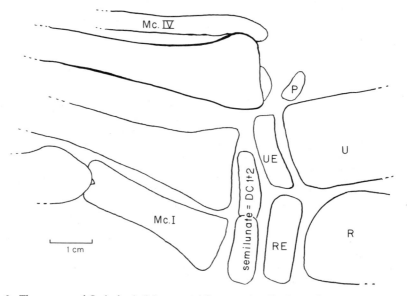

Fig. 3. The manus of *Coelophysis* (Museum of Comparative Zoology 4329). Note the incipient formation of the semilunate carpal, which is composed of two distal carpals (*Coelophysis*) in contrast to recent birds and *Megalancosaurus*, in which the semilunate element is composed of a single, distal carpal. DC I + II = Distal Carpals I + II; McI, IV = metacarpal I, IV; P = pisiform; R = radius; RE = radiale; U = ulna; UE = ulnare.

graphic elements already is present in *Megalancosaurus* (Fig. 2; Calzavara et al., 1980; R. Wild, pers. comm.). It is interesting that neither Ostrom nor Gauthier and Padian discussed the presence of this carpal element in *Megalancosaurus*—a thecodont that, curiously, they did not consider at all.

During the *Archaeopteryx* conference, it became clear that there was disagreement as to the respective identities of the manus and pes of *Megalancosaurus*, a new Triassic thecodont with possible links to bird ancestry (Calzavara et al., 1980; R. Wild, pers. comm.) that was described by Rupert Wild during the conference. The resemblance of *Megalancosaurus* to what many biologists would consider a relative of a proavis seemed to me clear. Some paleontologists at the conference insisted that *Megalancosaurus* was a prolacertid and that the hand was, in fact, a foot. I have examined the prolacertid material in South Africa and found that it lacks the following features of *Megalancosaurus:* an elongated pubes, an ilium having a large preacetabular portion, an antorbital fenestra, and thecodont teeth. Furthermore, the tarsi of prolacertids are quite distinct (Gow, 1975), having a laterally directed calcaneum (Fig. 4). If one compares the tarsus of *Prolacerta* (Fig. 4) with the carpus of *Megalancosaurus* (Fig. 2), it is obvious there is no resemblance whatever. *Megalancosaurus* is not a prolacertid, and the preserved limb is a forelimb.

Thus, it seems that the avian semilunate carpal element that apparently is composed of one distal carpal (Hinchliffe and Hecht, 1984; Hinch-

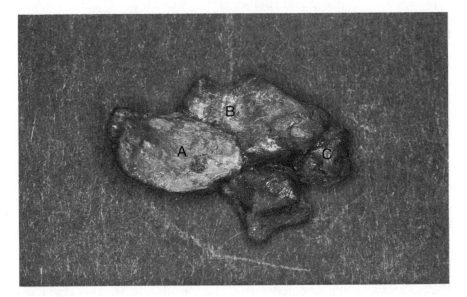

Fig. 4. The tarsus of *Prolacerta.* A = calcaneum; B = intermedium; C = astragalus.

liffe, 1985) is either a symplesiomorphy or the result of homoplasy with regard to its presence in coelurosaurs and *Archaeopteryx*. In this regard, a specimen of *Deinonychus* housed at the Museum of Comparative Zoology is interesting; it is a semilunate carpal of a juvenile that consists of at least two unfused distal carpi. If the avian semilunate element is composed of only one distal carpal, as is shown by autoradiographic studies (Hinchliffe and Hecht, 1984; Hinchliffe, 1985), and the coelurosaurian semilunate element is composed of two, is there not a possible problem in homology? Moreover, according to Ostrom's (1976b) diagrams, the topographic third metacarpal articulates with the semilunate carpal element differently in *Archaeopteryx* and *Deinonychus*. Ostrom (1976b) illustrated the topographic third metacarpal in *Deinonychus* as articulating with the semilunate carpal element ventrally. In *Archaeopteryx*, the articulation is either dorsolateral or most probably lateral with the "ulnare" (Petronievics, 1923). Further preparation of existing material or discovery of another *Archaeopteryx* (Wellnhofer, 1988, 1989) may clarify this issue.

Finally, it is unknown whether birds (including *Archaeopteryx*) and coelurosaurs have the same digits (Tarsitano and Hecht, 1980; Hinchliffe and Hecht, 1984; Hinchliffe, 1985). Traditionally, the digits of coelurosaurs have been identified as I, II, and III (Romer, 1956), but modern embryologists number avian digits as II, III, and IV. This dilemma must be resolved before the homologies of the digits of coelurosaurs and *Archaeopteryx* can be resolved. If birds possess Digits II–IV, then so must *Archaeopteryx*. Thulborn and Hamley (1982) stated that the longest metacarpal must be the third or middle digit in *Archaeopteryx* because of its length; their argument is tautologous and, thus, uninformative.

Character 74. Possession of a bowed ulna occurs throughout the Crocodylomorpha and Lepidosauria, including nothosaurs. This character has evolved so many times that it is best not to utilize it as presently described by Gauthier (1986).

Character 76. A laterally bowed *and* thin (topographic) third metacarpal is absent in *Archaeopteryx* and *Deinonychus*. The character should be revised to exclude lateral bowing, which is not substantiated. Furthermore, these characters are associated with the complex functional character "grasping manus" and should be included as part of Character 73.

Character 77. Ventral curving of the posterodorsal margin of the ilium occurs throughout the Archosauria and mammals. It is either a symplesiomorphy or highly homoplastic and, therefore, is a useless character.

Character 78. The possession of a pubic peduncle that is lower than the ischiadic peduncle is found throughout the Archosauria. If one includes the posteroventrally oriented pubis, the Ornithischia fit this character description quite well (Charig, 1972). However, for some, orientation of the pubis has been difficult to interpret. Study of the London and Berlin material suggests that the pubis in *Archaeopteryx* (Fig. 5) was reflexed (Petronievics and Woodward, 1917; Heilmann, 1926; Tarsitano, 1981, 1983). Ostrom (1976a,b, 1985) and Wellnhofer (1985) mistakenly oriented the pubis ventrally in an attempt to liken it to that of *Deinonychus.* However, the pubis is directed ventrally in early archosaurians such as the Triassic *Euparkeria,* in which the pubis also curves slightly in a posterior direction (Ewer, 1965). Moreover, the fact that early ornithischian dinosaurs have reflexed pubes (Romer, 1956; Charig, 1972) demonstrates that this condition has evolved more than once. Therefore, this character can be considered to be either primitive or homoplastic, and at present, its homology is indeterminate.

The pubis of *Archaeopteryx* is reflexed 140–150°—not at 110°, as stated by Wellnhofer (1988) on the basis of his study of the Solnhofen *Archaeopteryx.* My examination of this material revealed that the pelvis is crushed, the ischia are dislocated and rotated counterclockwise shoving the pubes forward, and the pubes are rotated and deformed. In addition, the left ilium is shoved well above the sacrum. The proximal portions of the ischia lie behind the sacral vertebrae, thereby demonstrating their dislocation. The only way the pubes can be reconstructed in a more vertical position is to use this specimen and the pelvis of the Eichstätt specimen (Fig. 6) as a guide; the pelves of each of these specimens are so distorted that any reconstruction is possible. The same sort of strategy is used by Padian (1983) to reconstruct the pelves of pterosaurs in order to make them bipedal. Walker (1977) and Tarsitano and Hecht (1980) used the Berlin material and the London specimen to make their determination in support of Heilmann's and Petronievics and Woodward's assertion that the pubes were reflexed. The Berlin pelvis is better preserved than either the Solnhofen or Eichstätt material (Figs. 5–6) and shows that the ischia were pointed posteriorly rather than ventrally, and that the pubis was reflexed 150–160°. The shifting of the pubic bones in the Berlin specimen could not be more than a few millimeters. Ostrom (1976a, 1985) did not show that the left ilium, which has been shifted upward, is still connected to the pubis in the Berlin specimen. Thus, the shifting of the left ilium could not distort the position of the right pubis, as Ostrom (1976b) claimed. In this specimen, the left pubis is higher than the right and appears to have taken a more reflexed position than its counterpart. The London material is important because it shows the articulating surface of the right pubis and the conspicuous bend in the proximal portion of the pubic rami (Tarsitano and Hecht, 1980; Hecht and Tarsitano, 1982). In order to articulate the pubes on the ilia, the pubis must be reflexed. Furthermore, the ischia in the London material still are articulated with the ilia and are directed posteriorly.

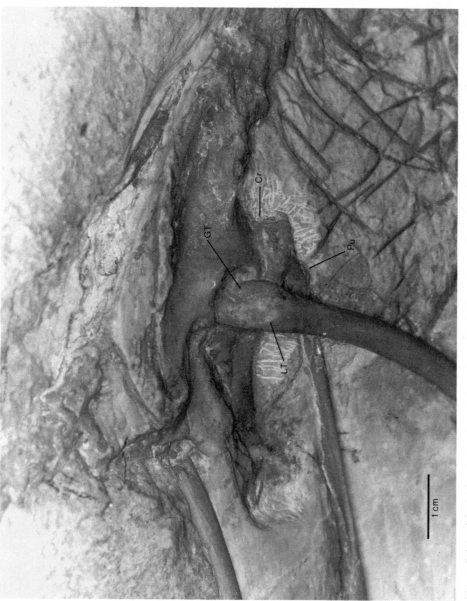

Fig. 5. Pelvis of the Berlin *Archaeopteryx* in lateral view. Note that opistopubis and configuration of the trochanter are unlike these structures in coelurosaurs, suggesting that the stances of the organisms would differ. Cr = crack; GT = greater trochanter; Is = Ischium; LT = lesser trochanter; Pu = pubis.

554

1 cm

Fig. 6. Pelvis of the Eichstätt *Archaeopteryx* in lateral view. Note the altered state (by preservation) of these bones. Pubes and ischia are dislocated, and either turned or rotated.

Character 79. Reduction of the anterior portion of the pubic boot is absent in *Deinonychus;* thus, in this respect, *Deinonychus* is dissimilar to *Archaeopteryx.* Gauthier (1986) seems to have overlooked the significance of this character-state. The pubic boot forms the attachment site for at least the mm. rectus abdominis anterior and posterior (Romer, 1923a,b). Therefore, the orientation of a forward and ventrally directed, or ventrally directed, pubis suggests that it should bear anterior and posterior extensions extending in an anterior-to-posterior direction to accommodate these muscles (Romer, 1923a,b). This is explained by the functional constraints of muscle attachment (Gans and Bock, 1965; Tarsitano and Oelofsen, 1985; Tarsitano et al., 1989). Archosaurs with a posteriorly directed pubis, like those of *Archaeopteryx* and ornithischian dinosaurs, would have a reduced anterior portion of the pubic boot.

Character 80. Possession of an ischium that is two-thirds or less the length of the pubis depends upon the elongation of the pubis in archosaurs and the orientation of the pubis relative to the ischium. As the pubis becomes directed more posteriorly, the ischium tends to become elongated, as it is in ornithischian dinosaurs and birds. It is only partly true that thecodonts do not possess a pubis elongated to the degree found in later archosaurs; the length of their pubes relative to that of the ischia resembles the condition in embryonic birds (Romanoff, 1964). It should be noted that undescribed thecodont material (to be described by Rupert Wild), stegosaurs, and crocodylomorphs fit the character description by Gauthier (1986), as may *Megalancosaurus* (pending Wild's redescription of this taxon). Thus, it seems that this character is a symplesiomorphy or is homoplastic.

Character 81. The presence of an obturator process on the ischium is neither described nor illustrated, and thus it cannot be determined that *Archaeopteryx* and birds have this feature.

Character 82. Partial fusion of the anterior trochanter with the head of the femur makes no sense as stated. It is impossible for a trochanter to fuse with the head that participates in the primary joint. Perhaps what is meant is fusion with the neck of the femur. The greater trochanter of the femur of *Archaeopteryx* lies in nearly the same position as the greater trochanter on the coelurosaur femur because the same muscle would have attached at this point in both taxa. The shape of the trochanters and the angle of the femoral neck (Fig. 5), however, are not the same. Dissection of a crocodilian hind limb is illuminating, because the attachment of the m. puboischio femoralis internus part 2 is in the same position as the anterior (greater) trochanter. As stance becomes more upright, the muscle insertion is moved outward by mineralization of the tendons. This maintains the same angle of attack of the tendons into the bone. The greater (anterior) trochanter of *Archaeopteryx* is shaped differently than that of coelurosaurs, and the posterior (lesser) trochanter does not resemble that of the coelurosaur. Apparently, trochanters can develop in response to new stresses and can change the direction of a muscle or tendon (in mammals and ornithopods) as well as maintain low angles of attack. They can also be considered as outgrowths of the mineralization points of tendons (Tarsitano and Oelofsen, 1985; Tarsitano et al., 1989).

The fact that both ornithischian dinosaurs and mammals evolved similar trochanters because of their upright gaits demonstrates the probability of homoplasy.

Character 83. The loss of the fourth trochanter is involved with a shift in muscle attachment or reduction of the m. caudofemoralis. Thus, when a character is lost, the loss can be approached from different transformational directions. Unless the method of loss is known, the character as coded is homoplastic.

Character 84. The presence of a Pedal Digit IV that is longer than Pedal Digit II also is found in ornithischian dinosaurs that are bipedal and crocodylomorph archosaurs. Thus, this character "may" be associated with an upright gait, leaping, or bipedalism. In any event, the character's distribution does not permit its use as a synapomorphy.

In conclusion, I think that most of the characters used by Gauthier (1986) were not analyzed or coded properly. The character analysis is incomplete, and the functional role of each character has not been considered. Because the functional significance of character-states is not discussed, there is no criterion for using any of the characters as synapomorphies. The actual character description, as well as their distribution within and among the taxa, needs further resolution. In addition, many cranial and postcranial characters that would alter the phylogenetic analysis have been ignored. The same is true for Cracraft's (1986) analysis, in which some characters absent in *Archaeopteryx* are used to unit *Archaeopteryx* with birds. Cracraft (1986) also assumed that deinonychosaurs are the sister-group of *Archaeopteryx* and, thus, chose the incorrect out-group for his phylogenetic analysis. Neither Cracraft (1986) nor Gauthier (1986) considered energetic or aerodynamic constraints on the evolution of flight (possibly because they are difficult to code for computer use), which is critical to the evolution of birds.

RECONSIDERATION OF THE ANATOMICAL EVIDENCE

The Mesotarsal Joint

Gauthier (1986) used the mesotarsal joint as a homologous state or synapomorphy of coelurosaurs and birds. However, advanced mesotarsal joints also are found in sauropodamorphs, ornithischian dinosaurs, and pterosaurs, thereby demonstrating the likelihood of convergence or even the plesiomorphic condition of this character. It now seems clear that because all dinosaurs passed through a level of tarsal organization equivalent to that of crocodilians, there may have been only one way of creating a dinosaurian mesotarsal joint (Walker, 1977; Tarsitano, 1981, 1983; Cruickshank and Benton, 1988). Apparently, this was accomplished by reduction of the calcaneum and the fibular facet of the astragalus to "lock" the fibula back onto the calcaneum (Walker, 1970; Charig, 1972; Tarsitano, 1983). The feasibility of this is illustrated by the tarsus of *Lagosuchus* (Romer, 1972b; Bonaparte, 1975, 1978). This thecodont possesses a crocodilian-like astragalus (Bonaparte, 1975) in which the astragalus with its fibular facet supports the fibula (Fig. 7) and has a reduced (compared with crocodilians) calcaneum. The tarsus of *Lagosuchus* is transitional between the crocodilian and theropod levels of tarsal organization (contra Gauthier [1986], who misinterpreted the tarsal anatomy of *Lagosuchus*); the calcaneum is reduced (an advanced character-state), but the astragalus bears a fibular facet (a primitive state). It is interesting to note that prosauropods also exhibit an anteroposteriorly directed calcaneum with a tuber that is a clear indication that they, too, are derived from a crocodilian (morphological) level of organization (Tarsitano, 1983). Finally, the presence of a mesotarsal joint is primitive for thecodonts (Romer, 1956; Charig, 1972; Cruickshank, 1972; Parrish, 1988) and perhaps for many reptiles (Romer, 1956); this indicates how frequently mesotarsal joints were evolved. Thus, the possession of a mesotarsal joint can be considered to be a synapomorphy, functional homoplasy, or a symplesiomorphy. One must distinguish the type of mesotarsal joint that is under discussion (Parrish, 1987; Cruickshank and Benton, 1988) and realize that the same morphological product may evolve a number of different ways at the same or different times.

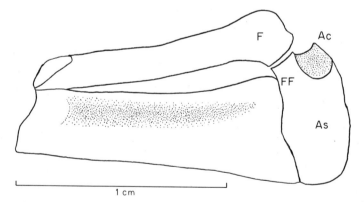

Fig. 7. Posterior aspect of the distal crus and proximal tarsus (calcaneum removed) of *Lagosuchus*, demonstrating the crocodilian nature of the tarsus. Nomenclature after Hecht and Tarsitano (1983). Ac = articulating channel; As = astragalus; F = fibula; FF = fibular facet of astragalus.

The Ascending Process of the Astragalus

It is appropriate at this point to consider the so-called ascending process of the astragalus. The embryology of the avian tarsus indicates that the process develops as a cartilage that attaches during development to the astragalus in terrestrial, bipedal archosaurs (Siegalbaur, 1911; Parker, 1891; Hinchliffe and Johnson, 1980; McGowan, 1984). Ostrom (1976a,b, 1985), Gauthier and Padian (1985), and Gauthier (1986) argued that both coelurosaurs and birds (including *Archaeopteryx*) have an ascending process of the astragalus. Tarsitano and Hecht (1980), Martin (1985), Martin et al. (1980), and Tarsitano (1985b) maintained that birds (including *Archaeopteryx*) possess a pretibial bone—a process associated mainly with the calcaneum. In fact, most carinates studied thus far (with the exception of the problematic Tinamous) have a pretibial bone and not an ascending process of the astragalus. As explained below, the ascending process of the astragalus and pretibial bone may represent alternate character-states of the same character (Martin, this volume).

In the Solnhofen *Archaeopteryx* described by Wellnhofer (1988, 1989), there may be a pretibial bone rather than an ascending process of the astragalus. Wellnhofer (1989:Fig. 14) mislabeled the right and left pes of the Solnhofen *Archaeopteryx*. The right distal tibia, which is better preserved than the left, actually is seen from the posterolateral surface; thus, it is impossible to see an ascending process even if it were present. Because the structure identified by Wellnhofer (if not an artifact) as an ascending process of the astragalus is attached to the calcaneum, it is a pretibial bone. I have examined the Solnhofen specimen and found the tarsus and tibia crushed so badly that it was impossible to identify either an ascending process or pretibial bone. In another specimen of *Archaeopteryx* that purportedly possesses an ascending process, the process is attached to the anterolateral face of the left tibia (Fig. 8).

Recent work by McGowan (1984) on the tarsus of ratites and Tinamous indicates that two processes may form. There is an ascending process (which resembles a descending process in its formation) that fuses with the astragalus in large ground-dwelling birds (Huxley, 1868; McGowan, 1984), and a pretibial bone (labeled "spur") associated with the calcaneum in many carinates. McGowan's study was hampered by the fact that he looked only at ossifications of the tarsus and not their cartilaginous precursors, as did Hinchliffe and Johnson (1980). The attachment points of these tarsal processes must be evaluated during the cartilaginous stages. It is entirely possible that the pretibial bone represents an ascending process that fuses with the astragalus, and that the spur (identified as the pretibial bone) is a new structure in ratites.

Fig. 8. The left tibiotarsus of the Berlin *Archaeopteryx*. The process on the lateral side of the tibia prompted Tarsitano and Hecht (1980) to question the identification of the process. F = fibula; P = ?process or crack.

Further studies may indicate that the ascending process of the astragalus also is attached to the calcaneal cartilaginous precursor in birds. The fact that either the tibiale or fibulare process dominates, depending upon the bird's lifestyle, indicates that the forces generated by the stance and gait of birds influence the development of tarsal locking devices, guides (pseudo–pulley systems), and tendon attachment (Welles and Long, 1977). It should be noted that ornithopods have an ascending process of the astragalus, as well as femoral trochanter formation. This fact demonstrates that this morphology can be, and has been, evolved more than once in the Archosauria in association with an upright, bipedal gait (Galton, 1970; Romer, 1972a,b,c; Charig, 1972; Welles and Long,

1977; Tarsitano, 1983; Martin, 1983, 1985; Parrish, 1988). It seems that bipedal archosaurs develop some type of tarsal locking devices. Those that lack these devices (e.g., pterosaurs) may not have been cursorial bipeds. Early birds (including *"Protoavis,"* fide S. Chatterjee, pers. comm.) lack an ascending process of the astragalus (+ intermedium); therefore, cladistically, the presence of this structure in some later birds and theropod and ornithopod dinosaurs must be ascribed to functional homoplasy. This bears out two points. First, an ascending process associated with the medial proximal tarsus occurs in terrestrial, bipedal archosaurs. Second, the pretibial bone and ascending process are homologous structures that fuse to different tarsal elements; thus, the spur described by McGowan (1984) is a neomorph. I think that all of these character-states were developed independently (i.e., in theropods, ornithopods, and birds) and to solve problems of torque generation about the ankle. They are more significant functionally than phylogenetically.

The Cranial Sinus Systems

The sinus system of birds and crocodilians was studied by Huxley (1869), Owen (1850), Parker (1883, 1891), de Beer (1937), Bremer (1940), Colbert (1946), King (1957), Bellairs and Jenkins (1960), H. Müller (1963), F. Müller (1967), Whetstone and Martin (1979), Walker (1972), Iordanski (1973), Tarsitano and Hecht (1980), Martin (1983), Tarsitano (1985a,b), and Witmer (1987). The sinus system of theropod dinosaurs was examined by Osborn (1906), Russell (1972), Barsbold (1974, 1983), Tarsitano (1985a,b), and Currie (1985). All authors showed that there is a basicranial sinus that minimally extends from the level of the pituitary fossa to the basioccipital. An open sinus characterizes dromaeosaurs and carnosaurs (Fig. 9), and a modified version occurs in adult paracrocodylians, posthatchling crocodilians (Fig. 10), and birds. The sinus seems to have resulted from an invagination of the roof of the braincase floor. It may be a by-product of the formation of Rathke's pocket, or the sinus may represent new pneumatic invaginations; the character is in need of further embryological study. In any case, the sinus does not seem to have been associated primitively with the tympanic cavity, because early coelurosaurs such as *Syntarsus* lack a connection between the basicranial sinus and the tympanic cavity (M. Raath, pers. comm.). For this reason, I questioned the nomenclature of the median eustachian tube system in crocodilians (Tarsitano, 1985a,b). Studies have shown that the sinuses form in similar ways in crocodilians and birds (de Beer, 1937; H. Müller, 1963; F. Müller, 1967). Although we cannot study the embryology of theropod dinosaurs in the same way, it appears that at least tyrannosaurs have remnants of this braincase floor invagination (Fig. 11). Thus, it

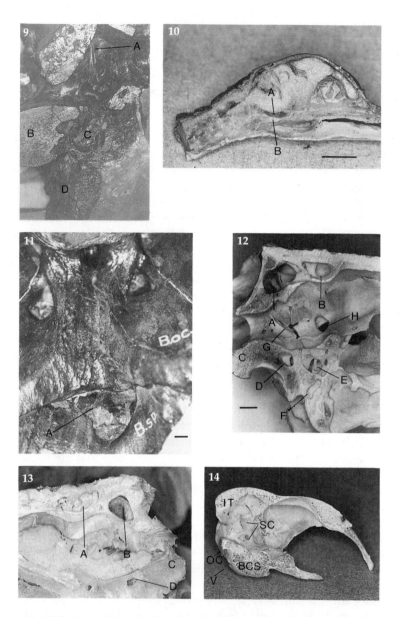

Figs. 9–14. (9) Basicranial sinus in the carnosaur *Tyrannosaurus rex*. A = brain cavity; B = occipital condyle; C = basicranial sinus; D = verticalized braincase floor. (10) Sagittal section through a posthatchling *Caiman*. Note the absence of braincase verticalization and the open nature of the basicranial sinus. A = brain cavity; B = braincase floor. (11) Apparent openings in the basisphenoid connecting the basicranial sinus with the pharynx in *Tyrannosaurus rex*. A = opening into basicranial sinus; B.oc. = basioccipital; B.sp. = basisphenoid. (12) Sagittal section of an alligator braincase, illustrating the transformed neural pocket and intertympanic sinus. A = intertympanic sinus; B = neural pocket; C = occipital condyle; D = basioccipital sinus; E = basisphenoidal sinus; F = median tube leading to basicranial sinus and tube system; G = metotic foramen; H = foramen ovale. (13) Sagittal

562

seems that these sinuses have the same topographic relationship in these lineages and probably have formed in the same way. I believe that the early development of this sinus in all of the aforementioned taxa demonstrates that the taxa are monophyletic. In the coelurosaur *Stenonychosaurus inequalis*, Currie (1985) showed that the basicranial sinus opens laterally into what he considered to have been the tympanic cavity. It is interesting to note that birds lack these openings, whereas crocodilians possess them. This condition may support a crocodilian-theropod connection, but not a theropod-avian one—an idea expressed by Whetstone (1983) that warrants further investigation.

The idea of monophyly among crocodylomorphs, theropods, some pseudosuchian thecodonts (sensu Heilmann, 1926), and birds is supported further by the formation of a quadrate sinus system. Some theropods (e.g., carnosaurs), birds, pseudosuchian thecodonts, paracrocodylians, and crocodilians possess a quadrate sinus (Owen, 1850; Huxley, 1869; Iordanski, 1973; Whetstone and Martin, 1979; Tarsitano and Hecht, 1980; Tarsitano, 1985a,b; Molnar, 1985). A pneumatic quadrate is unknown in coelurosaurs. Possession of a pneumatic quadrate and basicranial sinus is primitive for carnosaurs, poposaurid thecodonts (Chatterjee, 1985), paracrocodylians, crocodilians, and birds. These sinuses are present by at least the Upper Triassic. Carnosaurs also retain this feature and are considered by Chatterjee (1985) to be descendants of the thecodont lineage represented by *Postosuchus*. *Postosuchus* not only has a pneumatic quadrate but also possesses its extension via a siphonium into the articular. The sinus system is the same in birds and paracrocodylians despite Gauthier's (1986) claims that they are not homologous, and the developmental program for the sinus systems is the same in birds and living crocodilians. The morphological differences (i.e., the extent of the system) stem from the fact that the avian quadrate is upright, whereas the crocodilian quadrate slants anteriorly, which is not the case in paracrocodylians.

The morphology and distribution of the siphonial system of paracrocodylians is much like that of birds. Again, all crocodilians, birds, and thecodonts antecedent to carnosaurs have a quadrate-articular sinus; however, coelurosaurs lack this sinus system. The fact that carnosaurs possess this sinus system has no bearing on the condition in coelurosaurs

section of *Gavialus gangeticus* braincase, illustrating the neural pocket and the intertympanic sinus. A = neural pocket; B = intertympanic sinus; C = occipital condyle; D = basioccipital sinus. (14) Parasagittal section of a chicken skull. This is an example of the autapomorphic verticalization of braincase in birds that results from expansion of the air sinus rather than from restructuring to accommodate expanded jaw musculature, as in theropods and crocodilians. BCS = basicranial sinus; IT = intertympanic sinus; OC = occipital condyle; SC = semicircular canals; V = secondarily verticalized braincase.

because Chatterjee (1985) demonstrated the ancestral taxon (or sister-group, with no ancestral connotations) of carnosaurs to be *Postosuchus*-like thecodonts. The questions are: (1) Did coelurosaurs ever have a siphonium? or (2) was it lost early in their history? In a morphological analysis, these questions must be answered in order to support the coelurosaurian ancestor of birds.

To add to the dilemma of unusual character distribution, we have the intertympanic sinus system. (For references, see those for the siphonium.) This sinus extends over the top of the braincase between the tympanic cavities (Fig. 8). It occurs in birds, crocodylomorphs as far back as the Upper Triassic (in their primitive state), and recently was identified in a Cretaceous coelurosaur (P. J. Currie, pers. comm.). However, it must be pointed out that the intertympanic sinus of *Stenonychosaurus* lies above the prootic and slants posteriorly—an autapomorphic condition not found in birds or crocodilians. Thus, the coelurosaurian condition is homoplastic and related to an increase in the volume of the tympanic cavity. Interestingly, carnosaurian sinuses, which are similar to those of crocodilians, lack the intertympanic sinus. We again must answer the question as to whether they ever developed an intertympanic sinus or merely lost it. The character distribution suggests that we do not know the whole morphological story. I would also point out that one must be careful in interpreting hollow bones as pneumatic. The crocodilians are a case in point. The intertympanic sinus in alligatorines (Fig. 12) has a main channel that connects anteriorly to a hollow pocket (Tarsitano, 1985a,b). *Gavialis gangeticus* has such a pocket, which is filled with tissue (Tarsitano, 1987) and, thus, is not hollow or pneumatic, as it is in alligatorines (Fig. 13). Hollow bones in fossils must be interpreted with care.

Before leaving the topic of sinuses, I would like to point out that studies on the development of the true eustachian-tube systems of crocodilians and birds should be examined further. Currie (1985) considered that indentations on either side of the basioccipital in *Stenonychosaurus* represented the pathway of the eustachian tubes. If true, then the course of the eustachian tubes in these coelurosaurs would be autapomorphic because they resemble neither crocodilians nor birds.

Verticalization of the Braincase

The vertical nature of the crocodilian braincase was described by Romer (1956), Iordanski (1973), and Tarsitano (1985a), and that of the theropod, *Tyrannosaurus rex*, by Osborn (1912). I described a verticalization process in the growth of the basicranium in eusuchian crocodilians (Tarsitano, 1985a,b). It is clear that crocodilians and probably all archosaurs that verticalize the basioccipital and basisphenoid do so only *after* hatching. The verticalization process requires up to three years to occur. During the verticalization process, the basioccipital and basisphenoid become vertical relative to the posthatchling condition (Fig. 11). In this regard, the ventrally flat braincase floor in the posthatchling is probably an example of paedogenesis. All theropods except *Compsognathus longipes* also have a verticalized braincase; however, they lack the air tubes characteristic of adult eusuchian crocodilians (Tarsitano, 1985a,b). At this time, I would predict that posthatchling theropods might have had a braincase floor that resembled those of posthatchling eusuchians. The verticalization of at least the basioccipital occurs in some early proterosuchid thecodonts (Charig and Reig, 1970; L. Tataronov, pers. comm.), although *Proterosuchus* appears to have the primitive, flattened basisphenoidal floor. Those coelurosaurs purported to be close to avian ancestry, such as *Dromaeosaurus* and *Syntarsus,* verticalize the basisphenoid and basioccipital. The only coelurosaur that has a flat, ventral, basicranial surface is *Compsognathus longipes,* which supports Ostrom's (1978) claim that it is a juvenile. This fact well may fit quite nicely into the verticalization hypothesis—i.e., that theropods, eusuchian crocodilians, perhaps paracrocodylians, and pseudosuchians verticalized the basicranium and did so within one to a few years after hatching. Verticalization may be a shared-derived character of these taxa, and a feature that argues against separating the Crocodylomorpha from the theropods, as hypothesized by Gauthier and Padian (1985).

Apparently *Archaeopteryx* lacks a verticalized basicranium, although some modern birds verticalize the basioccipital. This condition in birds seems to be secondary, because morphologically, it does not closely resemble the theropod or crocodilian conditions. If verticalization is present in birds, it is associated with pneumaticity of the braincase (Fig. 14) and not the need for jaw-adductor expansion, as in other archosaurs. Thus, verticalization must be defined functionally. The condition of the braincase in the Texas proavis appears to be flattened. However, because the braincase in early birds is flat ventrally, those archosaurs with verticalized braincases can be eliminated from consideration as actual avian ancestors (theropods).

The Middle Ear and the Fenestra Rotundum

Given new finds of dromaeosaurs and their relatives (Barsbold, 1974; Osmolska, 1976; Currie, 1985), more is known about the morphology of the coelurosaurian braincase. However, some of the morphology is still in a confused state. Paleontologists disagree as to the placement of the fenestrae ovalis and rotundum (pseudorotundum of de Beer, 1937). Barsbold (1974) placed both fenestrae posterior in the braincase wall in *Saur-*

ornithoides, whereas Currie (1985) placed them more anteriorly. Whetstone and Martin (1979), Tarsitano and Hecht (1980), and Martin (1983) questioned the identification and placement of the fenestra rotundum in theropods. According to Martin (1983), theropods had a primitive system for pressure release in the scalae tympani. Unlike birds and crocodilians, carnosaurs might have lacked an actual fenestra rotundum with a margin bordered by the metotic process (subcapsular process) (Whetstone and Martin, 1979). Tarsitano (1985b) also stated that the identification of various fenestrae rotunda in theropods might be erroneous because paleontologists had not identified the opening for the lateral head vein; thus, the latter could be confused with the fenestra rotundum. As suggested by new material of *Stenonychosaurus,* their location in coelurosaurs resembles that of crocodiles. In order to determine whether a fenestra pseudorotundum is present in archosaurs, one must verify that the metotic process (= subcapsular process of crocodilians) borders the fenestra (de Beer, 1937; Kuhn, 1971; Walker, 1985). If this strut is preserved in bone, then the fenestra is present; if not, then the structure either was cartilaginous or absent. Obviously, it would be difficult to distinguish the latter. At present, all we can say is that a fenestra rotundum seems to have evolved early in the history of the Archosauria and may be a symplesiomorphy. The fact that similar metotic cartilage formation occurs in birds, crocodilians, and prototherians (Kuhn, 1971) demonstrates that the term *fenestra pseudorotunda* has been misused. The formation of the border of the fenestra is different in lepidosaurs than in archosaurs and early mammals, which demonstrates that the term *fenestra pseudorotundum* might best be applied to lepidosaurs.

The Reflexed Hallux

When found in articulation, the halluces of theropod dinosaurs attach near the midway point of the shaft of Metatarsal II, and the terminal ungual is reduced, compared with those of the other digits (Fig. 15). The condition is different in *Archaeopteryx* (Fig. 15), in which the hallux is attached near the distal end of Metatarsal II, and the ungual is large. The mistaken notion of the reflexed hallux can be traced to Osborn (1899) and his unjustified restoration (by his own admission) of the hallux of tyrannosaurs after *Apteryx.* Since that time, paleontologists have endeavored to restore the hallux in similar fashion despite the evidence provided by the halluces of *Velociraptor, Saurornithoides, Coelophysis, Compsognathus,* and the like. Indeed, it seems strange that a cursorial reptile would possess a reflexed hallux unless it had been retained from some ancestor, as it has been in some birds. Primitive coelurosaurs such as *Syntarsus* and *Coelophysis* lack a reflexed hallux. Instead of discussing the obvious dif-

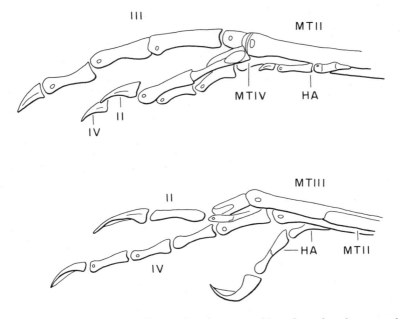

Fig. 15. The pedes of *Compsognathus* and *Archaeopteryx.* Note the reduced nature of the coelurosaur hallux as opposed to the functional hallux of *Archaeopteryx.* Note also that Metatarsal I of theropods attaches to the midpoint of Metatarsal II—a condition not found in *Archaeopteryx.* Recall that early coelurosaurs lack a reflexed hallux. Ha = hallux; Mt = metatarsal.

ferences between the halluces of theropods and *Archaeopteryx,* Ostrom (1985) reported that Tarsitano (1985b) said that the hallux of *Compsognathus* also articulated with Metatarsal IV. Actually, I stated that "Metatarsal I abutted unnaturally against Metatarsal IV," which has a very specific anatomical meaning. Previously, Tarsitano and Hecht (1980), Hecht and Tarsitano (1982), and Tarsitano (1983) ascribed this unnatural position to preservation. Figure 15 demonstrates the differences in character-state between *Archaeopteryx* and coelurosaurs.

The Question of the Broken Phalanx

Some paleontologists (e.g., Howgate, 1984; Thulborn and Hamley, 1982; Ostrom, 1985; Wellnhofer, 1985) have resurrected the "broken phalanx hypothesis" in *Archaeopteryx* and incorrectly ascribed it to Tarsitano and Hecht (1980). These authors did not hypothesize the presence of a broken phalanx in the topographic third digit of *Archaeopteryx;* they merely discussed the *possibility* of such an occurrence for two reasons. First, Ostrom (1976a,b) overlooked Heinroth's (1923) paper on this point. Second, if the phalanx in fact was broken, then the phalangeal formula of

Archaeopteryx would be different from that of advanced coelurosaurs. Wellnhofer (1985) stated that Tarsitano and Hecht (1980) misrepresented Heller (1959) by reporting that Heller wrote that there were two short proximal phalanges in the topographic third digit. Actually, Tarsitano and Hecht referred to Heller's diagram drawn from an X-ray of the Maxburg specimen, which clearly showed a single elongate phalanx in the third topographic digit of one of the hands of the Maxburg specimen. Thus, Heller (1959, 1960) wrote one thing but illustrated something quite different. In an attempt to promote further preparation of the Berlin material, Hecht and Tarsitano (1982) later reported the occurrence of *either* a broken phalanx *or* two short proximal phalanges in the third digit of *Archaeopteryx*. The authors (Hecht and Tarsitano, 1984; Tarsitano, 1985) subsequently responded to Howgate (1984), stating that they did not think the phalanx was broken.

CONCLUSIONS

The relationships of birds have been difficult to assess owing to lack of paleontological material and the need of further anatomical and developmental studies of birds and crocodilians. The following morphological problems involving the theropod ancestry of birds must be addressed by those who favor a coelurosaurian ancestry or sister-group position for birds: (1) the absence of a siphonium and pneumatic quadrate in all coelurosaurs; (2) the lack of clear evidence of a fenestra rotundum in coelurosaurs; (3) nonavian tooth morphology in coelurosaurs; (4) reduced coracoids in all coelurosaurs; (5) the presence of a massive and specialized pelvis and hind limbs that, biomechanically, seem to argue against the development of flight; (6) the presence of the middle temporal arch in coelurosaurs, but its absence in birds; (7) the presence of lateral basicranial sinus openings in coelurosaurs, but their absence in birds; (8) documentation of the presence or absence of a verticalized basicranium in Chatterjee's proavis, and demonstration that any adult coelurosaur possesses a nonverticalized basicranium; (9) proof that the early appearance of Triassic proavis precludes the derivation of birds from coelurosaurs; (10) documentation that the digit homologies of birds and theropods do or do not preclude the derivation of birds from coelurosaurs; and (11) the inclusion of *Megalancosaurus* in analyses of the origin of birds.

Other issues concerning thecodonts also need to be considered because aerodynamic constraints clearly favor an arboreal starting point for the origin of flight (Bock, 1985; Hecht, 1985; Tarsitano, 1985b; Norberg, 1985; Rayner, 1985). Given that no arboreal coelurosaurs are known, the evolution of flight favors *Megalancosaurus*-type thecodonts living in the

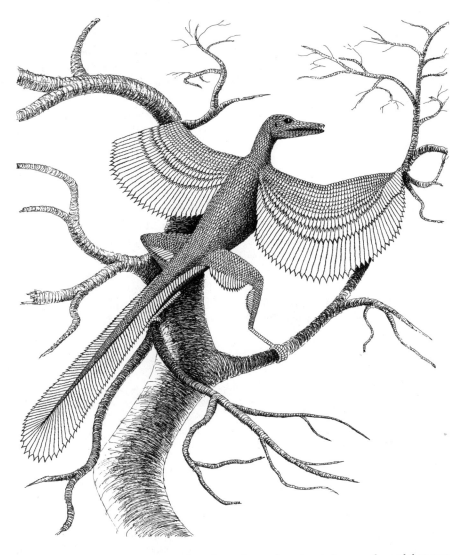

Fig. 16. Diagrammatic representation of an advanced protoavis in a quadrupedal stance.

lower or middle Triassic. Proponents of the coelurosaur ancestry of birds (Ostrom, 1976a,b, 1985) or sister-group relationship between birds and coelurosaurs (Gauthier, 1986) lack a logical coelurosaurian candidate for an ancestor. If they are restricted to coelurosaurs living before the time of *Archaeopteryx* (as some restrict the use of thecodonts), then coelurosaurs lack synapomorphies linking them to *Archaeopteryx* and birds. All similarities between *Archaeopteryx* and coelurosaurs occur after or nearly contemporaneous with the temporal range of *Archaeopteryx*. Because early coelurosaurs share so few synapomorphies with *Archaeopteryx* but mainly symplesiomorphies, it is more parsimonious to consider the few similarities between later coelurosaurs and *Archaeopteryx* to be functional homoplasies. Moreover, should Chatterjee's proavis prove to be an Upper Triassic bird, then I think that we can unequivocally exclude coelurosaurs from the ancestry of birds.

Acknowledgments I would like to thank the persons listed below for their help and support in the preparation of this paper. I am indebted to the following for their kind hospitality and help in providing material for study: Mike Cluver, Jurie Van den Heever (South African Museum); James Kitching, Michael Raath, C. E. Gow (Bernard Price Institute); Angela Milner, Cyril Walker, Alan Charig (British Museum [Natural History]); Wolf Reif and Frank Westphal (Institut und Museum für Geologie und Paläontologie); Rupert Wild (Rosenstein Museum); Bruce Rubige (Bloomfontein); Peter Wellnhofer (Staatssammlungen, München); John Ostrom (Yale University); Hans-Peter Schultze and Linda Trueb (The University of Kansas); Günter Viohl (Willibaldberg Museum, Eichstätt); and F. Müller (Solnhofen Museum). I have benefited from discussions with A. Charig, S. Chatterjee, E. Frey, W. Gutmann, M. Hecht, L. Martin, S. Peters, W. Reif, J. Riess, H.-P. Schultze, L. Trueb, L. Witmer, and R. Wild.

Literature Cited

Baird, I. L. 1970. The anatomy of the reptilian ear. Pp. 193–275 *in* Gans, C., and T. S. Parsons (eds.), *Biology of the Reptilia: Morphology.* Vol. 2. London and New York: Academic Press.

Barsbold, R. 1974. Saurornithoididae, a new family of small theropod dinosaurs from Central Asia and North America. Results of the Polish Expedition, Part V. Palaeontol. Polonica, 30:5–22.

Barsbold, R. 1983. Carnivorous dinosaurs from the Cretaceous from Mongolia. Trans. Jt. Soviet-Mongolian Paleontol. Exped., 19:5–120.

Bellairs, A. d'D., and C. R. Jenkins. 1960. The skeleton of birds. Pp. 241–300 *in* Marshall, A. J. (ed.), *Biology and Comparative Physiology of Birds.* Vol. 1. New York and London: Academic Press.

Bellairs, A. d'D., and A. M. Kamal. 1979. The chondrocranium and the development of the skull in recent reptiles. Pp. 1–263 *in* Gans, C. (ed.), *Biology of the Reptilia.* Vol. 9A. London: Academic Press.

Benton, M. J., and J. M. Clark, 1988. Archosaur phylogeny and the relationships of the Crocodylia. Pp. 289–332 *in* Benton, M. J. (ed.), *The Phylogeny and Classification of Tetrapods.* Vol. 1: *Amphibians, Reptiles, Birds.* Syst. Assoc. Spec. Vol. No. 35 A. London: Clarendon Press.

Bock, W. J. 1965. The role of adaptive mechanisms in the origin of higher levels of organization. Syst. Zool., 14:272–287.

Bock, W. J. 1977. Adaptation and the comparative method. Pp. 57–80 *in* Hecht, M. K., P. C. Goody, and B. M. Hecht (eds.), *Major Patterns in Vertebrate Evolution.* New York and London: Plenum Press.

Bock, W. J. 1985. The arboreal theory for the origin of birds. Pp. 199–206 *in* Hecht,

M. K., J. H. Ostrom, G. Viohl, and P. Wellnhofer (eds.), *The Beginnings of Birds.* Eichstätt: Freunde des Jura-Museums.

Bonaparte, J. F. 1975. Nuevos materiales de *Lagosuchus talampayensis* Romer (Thecodontia–Pseudosuchia) y su significado en el origen de los Saurischia. Chanarense Inferior, Triasico Medio de Argentina. Acta Geol. Lilloana, 13:5–90.

Bonaparte, J. F. 1978. El mesozoica de America de Sur y sus Tetrapodos. Opera Lilloana, 26:1–610.

Bremer, J. L. 1940. The pneumatization of the head of the common fowl. J. Morphol., 1940:143–157.

Calzavara, M., G. Muscio, and R. Wild. 1981. *Megalancosaurus preonensis* n.g., n. sp., a new reptile from the Norian of Friuli, Italy. Atti Mus. Friuli Storia. Nat., 2:49–64.

Charig, A. J. 1972. The evolution of the Archosaur pelvis and hindlimb: an explanation in the functional terms. Pp. 121–155 *in* Joysey, K. A., and T. S. Kemp (eds.), *Studies in Vertebrate Evolution.* Edinburgh: Oliver and Boyd.

Charig, A. J., and O. A. Reig. 1970. The classification of the Proterosuchia. Biol. J. Linn. Soc., 2:125–171.

Chatterjee, S. 1985. *Postosuchus,* a new thecodontian reptile from the Triassic of Texas and the origin of tyrannosaurs. Philos. Trans. R. Soc. London, Ser. B, 309:395–460.

Chatterjee, S. 1987. Skull of *Protoavis* and early evolution of birds. J. Vert. Paleontol. 7, Suppl. to No. 3:14A.

Colbert, E. H. 1946. The Eustachian tubes in the Crocodilia. Copeia, 1946:12–14.

Cracraft, J. 1986. The origin and early classification of birds. Paleobiology, 12(4):383–399.

Cruickshank, A. R. I. 1972. The proterosuchian thecodonts. Pp. 89–120 *in* Joysey, K. A., and T. S. Kemp (eds.), *Studies in Vertebrate Evolution.* Edinburgh: Oliver and Boyd.

Cruickshank, A. R. I. 1979. The ankle joint in some early archosaurs. S. Afr. J. Sci., 77:307–308.

Cruickshank, A. R. I., and M. J. Benton. 1988. Archosaur ankles and the relationships of thecodontian and dinosaurian reptiles. Nature, London, 317:715–717.

Currie, P. J. 1985. Cranial anatomy of *Stenonychosaurus inequalis* (Saurischia, Theropoda) and its bearing on the origin of birds. Can. J. Earth Sci., 22:1643–1658.

de Beer, G. R. 1937. *The Development of the Vertebrate Skull.* Oxford: Oxford Univ. Press.

de Beer, G. R. 1954. Archaeopteryx lithographica: *A Study Based on the British Museum Specimen.* London: British Museum (Natural History).

Ewer, R. F. 1965. The anatomy of the thecodont reptile *Euparkeria capensis* Broom. Philos. Trans. R. Soc. London, Ser. B, 248:379–435.

Feduccia, A. 1980. *The Age of Birds.* Cambridge and London: Harvard Univ. Press.

Frey, E. 1988a. Anatomie des Körperstammes von *Alligator mississippiensis* Daudin. Stuttg. Beitr. Naturkd., Ser. A, 424:1–106.

Frey, E. 1988b. Das Tragsystem der Krokodile—eine biomechanische und phylogenetische Analyse. Stuttg. Beitr. Naturkd., Ser. A, 426:1–60.

Galton, P. M. 1970. Ornithischian dinosaurs and the origin of birds. Evolution, 24:448–462.

Galton, P. M. 1984. An early prosauropod dinosaur from the Upper Triassic of Nordwürttemberg, West Germany. Stuttg. Beitr. Naturkd., Ser. B, 106:1–25.

Gans, C., and W. Bock. 1965. The functional significance of muscle architecture: a theoretical analysis. Ergeb. Anat. Entwicklungsgesch., 38:115–142.

Gauthier, J. 1986. Saurischian monophyly and the origin of birds. Pp. 1–55 *in* Padian, K. (ed.), *The Origin of Birds and the Evolution of Flight.* San Francisco: California Academy of Sciences.

Gauthier, J., and K. Padian. 1985. Phylogenetic, functional and aerodynamic analyses of the origin of birds. Pp. 185–197 *in* Hecht, M. K., J. H. Ostrom, G. Viohl, and P. Wellnhofer (eds.), *The Beginnings of Birds*. Eichstätt: Freunde des Jura-Museums.

Gow, C. E. 1975. The morphology and relationship of *Youngina capensis* Broom and *Prolacerta broomi* Parrington. Palaeontol. Afr., 18:89–131.

Hecht, M. K. 1976. Phylogenetic inference and methodology as applied to the vertebrate record. Evol. Biol., 9:335–363.

Hecht, M. K. 1985. The biological significance of *Archaeopteryx*. Pp. 149–160 *in* Hecht, M. K., J. H. Ostrom, G. Viohl, and P. Wellnhofer (eds.), *The Beginnings of Birds*. Eichstätt: Freunde des Jura-Museums.

Hecht, M. K., and J. L. Edwards. 1977. The methodology of phylogenetic inference above the species level. Pp. 3–51 *in* Hecht, M. K., P. C. Goody, and B. M. Hecht (eds.), *Major Patterns in Vertebrate Evolution*. New York and London: Plenum Press.

Hecht, M. K., and S. F. Tarsitano. 1982. The paleobiology and phylogenetic position of *Archaeopteryx*. Geobios Mem. Sp., 6:141–149.

Hecht, M. K., and S. F. Tarsitano. 1983. The tarsus and metatarsus of *Protosuchus* and its phyletic implications. Pp. 332–349 *in* Rhodin, A. G. J., and K. Miyata (eds.), *Advances in Herpetology and Evolutionary Biology*. Cambridge: Museum of Comparative Zoology.

Hecht, M. K., and S. F. Tarsitano. 1984. Paleontological myopia. Nature, London, 309:588.

Heilmann, G. 1926. *The Origin of Birds*. London: Witherby.

Heinroth, O. 1923. Die Flügel von *Archaeopteryx*. J. Ornithol., 71:277–283.

Heller, F. 1959. Ein dritter *Archaeopteryx*-Fund aus den Solnhofener Plattenkalken von Langenaltheim/Mfr. Erlanger Geol. Abh., 31:3–25.

Heller, F. 1960. Der dritter *Archaeopteryx*-Fund aus den Solnhofener Plattenkalken des oberen Malm Frankens. J. Ornithol., 101:7–28.

Hennig, W. 1966. *Phylogenetic Systematics*. Urbana: Univ. of Illinois Press.

Hinchliffe, J. R. 1985. One, two, three or two, three, four: an embryologist's view of the homologies of the digits and carpus of modern birds. Pp. 141–147 *in* Hecht, M. K., J. H. Ostrom, G. Viohl, and P. Wellnhofer (eds.), *The Beginnings of Birds*. Eichstätt: Freunde des Jura-Museums.

Hinchliffe, J. R., and M. K. Hecht. 1984. Homology of the bird wing skeleton: embryological versus paleontological evidence. Evol. Biol., 18:21–39.

Hinchliffe, J. R., and D. R. Johnson. 1980. The development of the vertebrate limb. Oxford: Oxford Univ. Press.

Howgate, M. 1984. *Archaeopteryx*'s morphology. Nature, London, 5973:104.

Huxley, T. H. 1868. On the animals which are most nearly intermediate between birds and reptiles. Ann. Mag. Nat. Hist., 24:66–75.

Huxley, T. H. 1869. On the representatives of the malleus and the incus of the Mammalia in the other Vertebrata. Proc. Zool. Soc. London, 1869:391–407.

Iordanski, N. N. 1973. The skull of the Crocodilia. Pp. 201–262 *in* Gans, C., and T. S. Parsons (eds.), *Biology of the Reptilia: Morphology*. Vol. 4. London and New York: Academic Press.

King, A. S. 1957. The aerated bones of *Gallus domesticus*. Acta Anat., 31:220–230.

Kuhn, H. J. 1971. Die Entwicklung und Morphologie des Schädels von *Tachyglossus aculeatus*. Abh. Senckenb. Naturforsch. Ges., 528:1–224.

Kuhn-Schnyder, E. 1963. Wege der Reptiliensystematik. Palaeontol. Z., 37:61–87.

Kuhn-Schnyder, E. 1980. Observations on temporal openings of reptilian skulls and the classification of reptiles. Pp. 153–175 *in* Jacobs, L. (ed.), *Aspects of Vertebrate History*. Flagstaff: Museum of Northern Arizona Press.

Lowe, P. R. 1935. On the relationships of the Struthiones to the dinosaurs. Ibis, 13:398–432.

Lowe, P. R. 1944. An analysis of the characters of *Archaeopteryx* and *Archaeornis.* Were they reptiles or birds? Ibis, 86:517–543.

Marsh, O. C. 1880. Odontornithes: a monograph on the extinct toothed birds of North America. Mem. Peabody Mus. Nat. Hist. 1:1–201.

Martin, L. D. 1983. The origin of birds and of avian flight. Pp. 105–129 *in* Johnston, R. F. (ed.), *Current Ornithology.* Vol. 1. New York and London: Plenum Press.

Martin, L. D. 1985. The relationship of *Archaeopteryx* to other birds. Pp. 177–183 *in* Hecht, M. K., J. H. Ostrom, G. Viohl, and P. Wellnhofer (eds.), *The Beginnings of Birds.* Eichstätt: Freunde des Jura-Museums.

Martin, L. D., J. D. Stewart, and K. N. Whetstone. 1980. The origin of birds: structure of the tarsus and teeth. Auk, 97:86–93.

Mazin, J. M. 1982. Affinites et phylogenie des *Ichthyopterygia.* Geobios, Mem. Spec., 6:85–98.

McGowan, C. 1984. Evolutionary relationships of ratites and carinates from the ontogeny of the tarsus. Nature, London, 307:733–735.

Molnar, R. E. 1985. Alternatives to *Archaeopteryx:* a survey of proposed early or ancestral birds. Pp. 209–217 *in* Hecht, M. K., J. H. Ostrom, G. Viohl, and P. Wellnhofer (eds.), *The Beginnings of Birds.* Eichstätt: Freunde des Jura-Museums.

Müller, F. 1967. Zur embryonalen Kopfentwicklung von *Crocodylus cataphratus* CUV. Rev. Suisse Zool., 74:1898–2294.

Müller, H. J. 1963. Die Morphologie und Entwicklung des Craniums von *Rhea americana* Linné, Part II. Visceralskelett, Mittelohr, und Osteocranium. Z. Wiss. Zool., 168:40–118.

Norberg, U. M. 1985. Evolution of vertebrate flight: an aerodynamic model for the transition from gliding to active flight. Am. Nat., 126:303–327.

Osborn, H. F. 1899. Fore and hindlimbs of carnivorous and herbivorous dinosaurs from the Jurassic of Wyoming. Dinosaur contributions, No. 3. Bull. Am. Mus. Nat. Hist., 12:161–172.

Osborn, H. F. 1906. *Tyrannosaurus,* Upper Cretaceous carnivorous dinosaur. (Second communication.) Bull. Am. Mus. Nat. Hist., 22:281–296.

Osborn, H. F. 1912. Crania of *Tyrannosaurus* and *Allosaurus.* Mem. Am. Mus. Nat. Hist., 1:1–30.

Osmolska, H. 1976. New light on the skull anatomy and systematic position of Oviraptor. Nature, London, 262:683–684.

Ostrom, J. H. 1976a. Some hypothetical anatomical stages in the evolution of avian flight. Smithsonian Contrib. Paleobiol., 27:1–21.

Ostrom, J. H. 1976b. *Archaeopteryx* and the origin of birds. Biol. J. Linn. Soc., 8:91–182.

Ostrom, J. H. 1978. The osteology of *Compsognathus longipes* Wagner. Zitteliana, 4:73–118.

Ostrom, J. H. 1985. The meaning of *Archaeopteryx.* Pp. 161–176 *in* Hecht, M. K., J. H. Ostrom, G. Viohl, and P. Wellnhofer (eds.), *The Beginnings of Birds.* Eichstätt: Freunde des Jura-Museums.

Owen, R. 1850. On the communications between the cavity of the tympanan and the palate in the Crocodilia (gavials, alligators and crocodiles). Philos. Trans. R. Soc. London, 27:521–527.

Padian, K. 1982. Macroevolution and origin of major adaptations: vertebrate flight as a paradigm for the analysis of patterns. Proc. Third North Am. Paleontol. Conf., 2:387–392.

Padian, K. 1983. A functional analysis of flying and walking in pterosaurs. Paleobiology, 9:218–239.

Parker, W. K. 1883. On the structure and development of the skull in the Crocodilia. Trans. Zool. Soc. London, 11:263–310.

Parker, W. K. 1891. On the morphology of a reptilian bird, *Opisthocomus cristatus*. Philos. Trans. R. Soc. London, Ser. B, 13:45–83.

Parkes, K. C. 1966. Speculations of the origins of feathers. The Living Bird, 5:77–86.

Parrish, J. M. 1987. The origin of crocodilian locomotion. Paleobiology, 13(4):396–414.

Parrish, J. M. 1988. Joints of crocodile-reversed archosaurs. Nature, London, 331:217–218.

Pennycuick, C. J. 1972. *Animal Flight*. Studies in Biology No. 33. Southampton: Camelot Press.

Peters, D. S., and W. F. Gutmann. 1985. Constructional and functional preconditions for the transition to powered flight in vertebrates. Pp. 233–242 *in* Hecht, M. K., J. H. Ostrom, G. Viohl, and P. Wellnhofer (eds.), *The Beginnings of Birds*. Eichstätt: Freunde des Jura-Museums.

Petronievics, B. 1923. Über das Ulnare im Carpus des Berliner *Archaeornis*. Zentralbl. Mineral. Geol. Palaeontol., Ser. B, 1923:94–95.

Petronievics, B., and A. S. Woodward. 1917. On the pectoral and pelvic arches of the British Museum specimen of *Archaeopteryx*. Proc. Zool. Soc. London, 1917:1–6.

Rayner, J. M. V. 1985. Cursorial Gliding in protobirds an expanded version of a discussion contribution. Pp. 289–302 *in* Hecht, M. K., J. H. Ostrom, G. Viohl, and P. Wellnhofer (eds.), *The Beginnings of Birds*. Eichstätt: Freunde des Jura-Museums.

Reif, W.-E. 1982. Functional morphology on the Procrustean bed of the neutralism-selectionism debate. Notes on the constructional morphological approach. Neues Jahrb. Geol. Palaeontol. Abh., 164:46–59.

Reif, W.-E., R. D. K. Thomas, and M. S. Fischer. 1985. Constructional morphology: the analysis of constraints in evolution. Acta Biotheoretica, 34:233–248.

Retzius, G. 1884. Das Gehörorgan der Wirbeltiere. II. Das Gehörorgan der Reptilien, der Vögel und der Säugethiere. Stockholm: Sampson and Wallin.

Romanoff, A. L. 1964. *The Avian Embryo*. New York: Macmillan.

Romer, A. S. 1923a. Crocodilian pelvic muscles and their avian and reptilian homologues. Bull. Am. Mus. Nat. Hist., 48:533–552.

Romer, A. S. 1923b. The musculature of saurischian dinosaurs. Bull. Am. Mus. Nat. Hist., 48:605–617.

Romer, A. S. 1956. *Osteology of the Reptiles*. Chicago: Univ. of Chicago Press.

Romer, A. S. 1966. *Vertebrate Paleontology*. 3rd ed. Chicago: Univ. of Chicago Press.

Romer, A. S. 1972a. The Chanares (Argentina) Triassic reptile fauna, XIII. An early ornithosuchid pseudosuchian, *Gracilisuchus stipanicorum*, gen. et sp. nov. Breviora, 389:1–24.

Romer, A. S. 1972b. The Chanares (Argentina) Triassic reptile fauna, XV. Further remains of the thecodonts *Lagerpeton* and *Lagosuchus*. Breviora, 394:1–7.

Romer, A. S. 1972c. The Chanares (Argentina) Triassic reptile fauna, XVI. Thecodont classification. Breviora, 395:1–24.

Russell, D. A. 1972. Ostrich dinosaurs from the Late Cretaceous of Western Canada. Can. J. Earth Sci., 9:375–402.

Schoch, R. M. 1986. *Phylogenetic Reconstruction in Paleontology*. New York: Van Nostrand Reinhold.

Sharov, A. G. 1971. Neue fliegende Reptilien aus dem Mesozoikum von Kasachstan und Kirgisien. Tr. Akad. Nauk SSSR Palaeontol. Inst., 130:104–113.

Sieglbauer, F. 1911. Zur Entwicklung der Vogelextremitäten. Z. Wiss. Zool., 97:262–313.

Simpson, G. G. 1961. *Principles of Animal Taxonomy*. New York: Columbia Univ. Press.

Snyder, R. 1954. The anatomy and function of the pelvic girdle and hindlimb in lizard locomotion. Am. J. Anat., 95:1–46.

Tarsitano, S. F. 1981. Pelvic and hindlimb musculature in archosaurian reptiles. Doctoral dissertation. New York: City University of New York.

Tarsitano, S. F. 1982. A case for the diapsid origin of ichthyosaurs. Neues Jahrb. Geol. Palaeontol. Monatsh., 1983:59–64.

Tarsitano, S. F. 1983. Stance and gait in theropod dinosaurs. Acta Palaeontol. Polonica, 28:251–264.

Tarsitano, S. F. 1985a. Cranial metamorphosis and the origin of the Eusuchia. Neues Jahrb. Geol. Palaeontol. Abh., 170(1):27–44.

Tarsitano, S. F. 1985b. The morphological and aerodynamic constraints on the origin of avian flight. Pp. 319–332 *in* Hecht, M. K., J. H. Ostrom, G. Viohl, and P. Wellnhofer (eds.), *The Beginnings of Birds.* Eichstätt: Freunde des Jura-Museums.

Tarsitano, S. F. 1987. The evolution of the Eusuchia. Pp. 214–217 *in* Currie, P. M., and E. H. Koster (eds.), *Fourth Symposium on Mesozoic Ecosystems.* Drumheller: Tyrell Museum.

Tarsitano, S. F., E. Frey, and J. Riess. 1989. On the attachment of tendon to bone and its bearing of the evolution of the skull. Sonderforschungsbereich 230, Natürliche Konstruktionen, Konzepte SFB Heft No. 2. Stuttgart: Kurz & Co. Reprographie GmbH.

Tarsitano, S. F., and M. K. Hecht. 1980. A reconsideration of the reptilian relationships of *Archaeopteryx.* Zool. J. Linn. Soc., 69:149–182.

Tarsitano, S. F., and B. W. Oelofsen. 1985. The invasion of the periosteum. Am. Zool., 25:44.

Tarsitano, S. F., and J. Riess. 1982. Plesiosaur locomotion—underwater flight versus rowing. Neues Jahrb. Geol. Palaeontol. Abh., 164:188–192.

Thulborn, R. A. 1984. The avian relationships of *Archaeopteryx,* and the origin of birds. Zool. J. Linn. Soc., 82:119–158.

Thulborn, R. A., and T. L. Hamley. 1982. The reptilian relationships of *Archaeopteryx.* Aust. J. Zool., 30:611–634.

Walker, A. D. 1970. A revision of the Jurassic reptile *Hallopus victor* (Marsh), with remarks on the classification of crocodiles. Philos. Trans. R. Soc. London, Ser. B, 257:323–372.

Walker, A. D. 1972. New light on the origin of birds and crocodiles. Nature, London, 237:257–263.

Walker, A. D. 1977. Evolution of the pelvis in birds and dinosaurs. Pp. 319–357 *in* Andrews, S. M., R. S. Miles, and A. D. Walker (eds.), *Problems in Vertebrate Evolution.* London: Academic Press.

Walker, A. D. 1985. The braincase of *Archaeopteryx.* Pp. 123–134 *in* Hecht, M. K., J. H. Ostrom, G. Viohl, and P. Wellnhofer (eds.), *The Beginnings of Birds.* Eichstätt: Freunde des Jura-Museums.

Welles, S. P., and R. A. Long. 1977. The tarsus of theropod dinosaurs. Ann. S. Afr. Mus., 64:191–218.

Wellnhofer, P. 1974. Das fünfte Skelettexemplar von *Archaeopteryx.* Palaeontographica, (A), 147:169–216.

Wellnhofer, P. 1980. *Flugsaurier.* Wittenberg Lutherstadt: A. Ziemsen Verlag.

Wellnhofer, P. 1985. Remarks on the digit and pubis problems of *Archaeopteryx.* Pp. 113–122 *in* Hecht, M. K., J. H. Ostrom, G. Viohl, and P. Wellnhofer (eds.), *The Beginnings of Birds.* Eichstätt: Freunde des Jura-Museums.

Wellnhofer, P. 1988. A new specimen of *Archaeopteryx.* Science, 240:1790–1792.

Wellnhofer, P. 1989. Ein neues Exemplar von *Archaeopteryx. Archaeopteryx,* 6:1–30.

Wever, E. G. 1979. *The Reptile Ear.* Princeton: Princeton Univ. Press.

Whetstone, K., 1983. Braincase of Mesozoic birds: new preparation of the "London" *Archaeopteryx*. J. Vert. Paleontol., 2(4):439–452.

Whetstone, K. N., and L. D. Martin. 1979. New look at the origin of birds and crocodiles. Nature, London, 279:234–236.

Witmer, L. M. 1987. The cranial air sac system in Mesozoic birds. Master's thesis. Lawrence: Univ. of Kansas.

Yalden, D. W. 1984. What size was *Archaeopteryx?* Zool. J. Linn. Soc., 82:177–188.

Yalden, D. W. 1985. Forelimb function in *Archaeopteryx*. Pp. 91–97 *in* Hecht, M. K., J. H. Ostrom, G. Viohl, and P. Wellnhofer (eds.), *The Beginnings of Birds*. Eichstätt: Freunde des Jura-Museums.

Section V

MAMMALS

16 On the Origins of Mammals

Miao Desui

More than a century ago William Henry Flower noted (1883:178) that "in the present condition of the world, Mammals have become so broken up into distinct groups by the extinction of intermediate forms, that a systematic classification is perfectly practicable." Similarly, to those who deal mainly with the present diversity of tetrapods, questions such as "What is an amphibian, a reptile, or a mammal" may sound rather naive and somewhat superfluous. "When, however, we pass to the extinct world, all is changed" (Flower, 1883:178). Thus, to those who study evolutionary history of tetrapods, such questions are not only meaningful but also challenging. As all the contributors to this volume are fully aware, the answers to such questions are often painfully uncertain.

Among vertebrate classes, the evolution of mammals from nonmammalian synapsids has been claimed to be the best documented transition in the fossil record (e.g., Kemp, 1982). Yet many problems regarding the interrelationships of various nonmammalian synapsid groups and particularly the closest sister-group of the Mammalia remain unresolved (Hopson, this volume; Kemp, 1988a). Further complications have arisen from recent discoveries of a variety of Mesozoic mammalian remains from several continents. Jenkins (1984) and Clemens (1986) offered excellent reviews of some of those discoveries. In addition, Bonaparte (1986a,b) since has reported on enigmatic Late Cretaceous mammals from Patagonia. This new knowledge of early diversification of the Mammalia upsets the once widely accepted dichotomous separation of early mammals into monophyletic nontherian and therian lineages, and reveals a much more complicated picture of early mammalian evolution. Therefore, the interrelationships among the various groups of early mammals seem to be more obscure now than they seemed a decade ago. More ironically, several authors (e.g., Ax, 1987; Kemp, 1983; Rowe, 1988), even though they embrace the same set of basic principles of phylogenetic analysis, have employed dramatically different criteria in their diagnoses of the Mammalia. Accordingly, their interpretations of the origin and early diversification of mammals also differ.

In this chapter, I discuss (1) the controversial views on the origin and early radiation of mammals from a historical perspective; (2) an osteological diagnosis of the Class Mammalia applicable to fossils; and (3) long-standing problems of monophyletic versus polyphyletic origins of the Mammalia.

HISTORICAL PERSPECTIVE

The scientific recognition of Mesozoic mammals dates to the early nineteenth century (Buckland, 1824). For the first 100 years, scattered reports of Mesozoic mammals were "largely descriptive and taxonomic" (Crompton and Jenkins, 1979:59). It was George Gaylord Simpson who pioneered a systematic exploration into Mesozoic mammals and mammalian origins. Simpson's monumental works of the late 1920's (notably, Simpson, 1928, 1929) led him to believe that mammalian history represents at least four remote and independent derivations "from the ultimate mammalian common ancestry, probably within a group which must be called reptilian by definition" (Simpson, 1929:143). These four independent lineages were Triconodonta, Multituberculata, the ancestral Monotremata, and Theria. Neither of the first two orders left direct modern descendants; the great antiquity of the Monotremata was inferred from the seemingly archaic nature of its living representatives.

Simpson's view was endorsed by Olson's (1944) study of cranial morphology of therapsids. Olson (1944:124) maintained that "there are four lines of therapsids which are developing in parallel lines and each of which contains possible mammalian ancestors." He further suggested that mammals originated from not one, but three to five stocks of therapsids, and that the common ancestry of mammals might be traced to the Late Permian or earlier (see also Hopson, this volume, for details). His hypothesis assumes a polyphyletic origin of the Mammalia, with several independent therapsid lineages crossing the mammalian threshold.

While Simpson's and Olson's views dominated North American paleontology (see also Patterson, 1956), some European paleontologists voiced different opinions. On the basis of the dental configurations of the Jurassic mammals, Bohlin (1945) suggested a diphyletic origin of mammals, with the Pantotheria having evolved from one group of therapsids, and multituberculates, triconodonts, and symmetrodonts from another. Amazingly, Bohlin had not seen a single tooth of a Jurassic mammal when he wrote the paper! Kühne, on the other hand, had firsthand knowledge of a variety of the Triassic and Jurassic mammals (as well as some cynodonts) and voiced various opinions. For example, in 1949, Kühne advocated that *Eozostrodon* (referred to also as *Morganucodon*) and its allies probably gave rise to all mammals except monotremes. Several years later, Kühne decided that the class Mammalia is monophyletic and concluded his monograph on *Oligokyphus* with the following sentence: "In my opinion the Monotremes did not evolve independently from the rest of mammals" (Kühne, 1956:142). Later, he (1958) went further to suggest, though implicitly, that the attempt to draw a demarcation between therapsids and mammals was sterile. One either must classify therapsids as mammals or classify mammals as therapsids. Nevertheless, Kühne (1958) held that pantotheres (including docodonts) were derived from the Liassic triconodonts, and that multituberculates and symmetrodonts each arose from separate therapsid lineages.

Patterson (1956), in his review of the Early Cretaceous mammalian molar evolution, also discussed the origin and relationships of Mesozoic mammals. He essentially followed Simpson and Olson and considered Mesozoic mammals to represent at least five independent crossings of the Reptilia-Mammalia line—viz., Multituberculata, Monotremata, Triconodonta, Docodonta, and Symmetrodonta or Pantotheria.

By 1959, the centennial year of the publication of *Origin of Species*, the polyphyletic origins of mammals seemed to have become entrenched with the appearance of review papers by Simpson (1959) and Olson (1959). In view of the then prevalent neo-Darwinian paradigm, adaptive grade evolution was considered a dominant mode; thus, Simpson's and Olson's conclusions were somewhat inevitable.

Romer (1968:164) once observed that "our advances in knowledge of Mesozoic mammals seem to be on a 'quantum' basis rather than on one of gradual steady progress." I believe that this situation since has changed and now can better be described as a combination of both modes—i.e., a steady and progressive accumulation of new fossil finds throughout the past few decades, punctuated by periodic bursts in the debate on phylogenetic interpretation at approximately 10-year intervals.

The first burst of the debate since the 1960's was triggered by two articles published in *Evolution* (Reed, 1960; Van Valen, 1960). Van Valen (1960:311) thought that "the most important criteria for separating tetrapod classes should be their major adaptive differences" and that "the

Therapsida were probably more mammal-like than reptile-like." Therefore, he proposed to include therapsids in the Mammalia. Reed (1960), on the other hand, considered that any taxon, the Class Mammalia included, should be a clade rather than a grade. To make the Class Mammalia monophyletic, he suggested that all therapsids and certain (but not *all*) pelycosaurs should be classified as mammals. In his reply to these two papers, Simpson (1960) strongly opposed both proposals essentially on the ground of practicability of classification and maintained the criterion of the establishment of a dentary-squamosal jaw joint as the most practical and precise line between the Reptilia and Mammalia. Simpson's seemingly convincing arguments and, more important, his authoritative status put the debate immediately to rest. It should be noted, however, that Simpson did take some of Reed's arguments rather seriously and recast his point of view in a less stringent fashion in a later paper (Simpson, 1961). Simpson nevertheless held to the idea that, by possessing a dentary-squamosal articulation, mammals "are all distinctly more mammalian than reptilian in grade and they are monophyletic at the ordinal level or somewhat below, although polyphyletic at some (not precisely determined) level probably above the family" (Simpson, 1961:91).

Historical controversies on mammalian origins reviewed thus far arose mainly from different interpretations of dental transformations of Mesozoic mammals, based almost exclusively upon isolated dentitions and jaw fragments. However, the purported mammalian ancestors (i.e., nonmammalian therapsids) were known from more abundant, and often complete, skeletons. As a result of this preservational discrepancy, the two groups of the specialists had come to know their respective animals quite differently. Consequently, although the general evolutionary interpretation of therapsid-mammal transition appeared persuasive, the actual and therefore crucial "turning point(s)" remained poorly understood.

Two advancements soon helped narrow the gaps caused by the preservational and specializational biases. First, descriptions of the "triconodontid" cranial materials (Kermack, 1963; Patterson and Olson, 1961) facilitated direct comparative studies on cranial morphology of both advanced cynodonts and the earliest mammals (see also Hopson, 1964). Second, experts on nonmammalian synapsids took great pains in deciphering dental evolution from nonmammalian synapsids (especially the thrinaxodontids) to various groups of Mesozoic mammals (Crompton and Jenkins, 1968; Hopson and Crompton, 1969).

These studies initiated the second burst of the debate. Whereas Patterson and Olson (1961) still were inclined to consider mammals polyphyletic, Kermack (1963, 1967) regarded Mammalia as essentially diphyletic on the basis of the claimed difference in braincase structure between nontherian and therian mammals. However, Hopson (1967), Crompton and Jenkins (1968), and Hopson and Crompton (1969) contended that all mammals probably were derived from a single line of the Late Triassic cynodonts. Hopson and Crompton (1969:17) specifically stated that "our criterion for the monophyletic origin of such a high level taxon as the Class Mammalia would be derivation from an ancestral taxon . . . on the level of family or perhaps lower." Parrington (1967) also reviewed the problem of mammalian origin and suggested probable monophyly of the Class Mammalia. However, he did point out that "the diphyletic evolution of mammals seems probable and polyphyletism is possible" (Parrington, 1967:173). He especially emphasized parallel derivation of a secondary jaw articulation among advanced cynodonts and early mammals, and the fundamental difference in the major jaw-opening muscle (detrahens vs. digastric) between monotremes and modern therians.

In a critique of Kermack's claim that a neat mammalian dichotomy exists on the basis of the position of the foramen ovale, MacIntyre (1967) proposed to restrict the Class Mammalia to therian mammals only. He dubbed so-called nontherian mammals as "quasi-mammals." This proposal never received any serious consideration by the majority of experts on early mammalian evolution. On the contrary, MacIntyre's proposal generated Griffiths's (1978:309) jesting comment that "even if one agreed with his classification the name is not felicitous; if one studies quasi-mammals one might be called a quasi-mammalogist."

Despite the diverse opinions expressed during the 1960's, by the early 1970's a consensus was reached among paleontologists that mammals were divisible into two fundamentally different lineages, nontherians and therians. Nontherian mammals were believed to have a braincase in which the sidewall is formed largely by the anterior lamina of the petrosal; the maxillary and mandibular branches of the trigeminal nerve pass through this lamina. Nontherian mammals include triconodonts, docodonts, haramiyids, multituberculates, and monotremes. Also, it was believed that, despite obvious diversity, all nontherian molars were derived from a morganucodontid-like dentition. In contrast, therian mammals were characterized by the reverse triangulation of the main molar cusps and the presumed expansion of the alisphenoid to form the sidewall of the braincase. It was thought that, invariably, the maxillary and mandibular branches of the trigeminal nerve perforate the alisphenoid in therian mammals. Therian mammals include marsupials, eutherians, eupantotheres, and symmetrodonts as well as their purported antecedent—the kuehneotheriids.

The major disagreement, then, lay in the perceived degree of relatedness between nontherian and therian mammals. As briefly mentioned above, Kermack (1963, 1967) and Kermack and Kielan-Jaworowska (1971) suggested that nontherians and therian mammals evolved independently from different therapsid stocks and, therefore, are related only distantly. However, another group of experts (Crompton and Jenkins, 1968, 1973; Hopson, 1967, 1969, 1970; Hopson and Crompton, 1969; Parrington, 1971) considered that both braincase and molar structures of nontherian and therian mammals could have been derived readily from a common ancestor at or near the therapsid-mammal transition. It should be added that Simpson (1971), though somewhat impressed by the claimed fundamental differences between the two mammalian groups, was uncommitted to either of these two schools of thought.

The third burst of the debate is characterized by what Clemens (1970: 365) called "a bone of contention." Although the hypothesis of a major mammalian dichotomy dominated in the field of vertebrate paleontology throughout the 1970's (e.g., Crompton and Jenkins, 1979), the primary argument for a nontherian versus therian distinction essentially is a purported difference in braincase structure. Several lines of evidence that seem to contradict this hypothesis have emerged gradually during the 1980's. First, Presley and Steel (1976) and Presley (1980, 1981) indicated that the blade of the alisphenoid in therians and the anterior lamina of the petrosal in nontherians have a similar developmental entity and thus are homologous. Second, recent paleontological discoveries revealed that there is no uniform structural pattern of the braincase in either nontherian or therian mammals. For example, the Paleocene multituberculate *Lambdopsalis bulla* possesses a large alisphenoid (Miao, 1988), whereas an Early Cretaceous eupantotherian, *Vincelestes neuquenianus*, retains an extensive anterior lamina of the petrosal (Rougier and Bonaparte, 1988). Furthermore, the great morphological diversity among recently studied Mesozoic mammals (e.g., Archer et al., 1985; Bonaparte, 1986a,b; Chow and Rich, 1982; Crompton and Sun, 1985; Jenkins et al., 1983; Jenkins and Schaff, 1988; Kermack et al., 1987; Rougier and Bonaparte, 1988) simply cannot be categorized by a neat dichotomy. As several authors (e.g., Clemens, 1986; Kemp, 1983, 1988b; Lillegraven et al., 1987; Miao, 1988) have suggested, the hypothesis of an early mammalian dichotomy should be abandoned.

The current debate centers on how pervasive parallel evolution may have been in the origins of mammals. Some authors (e.g., Kemp, 1983, 1988b; Rowe, 1988) prefer to believe that the parallelism was as minimal as possible and, hence, consider the Mammalia to be monophyletic. Others (e.g., Hopson, this volume; Hopson and Barghusen, 1986), though accepting monophyly of the Mammalia, show that a considerable amount of parallel character evolution (even including certain key characters) must be postulated at the therapsid-mammal transition. Still others (e.g., Clemens, 1986; Miao, 1988) remain convinced of the distinct probability of polyphyletic origins of the Class Mammalia. However, recently it has been argued that the repeated controversies on monophyly versus polyphyly of the Mammalia and the persistent obscurity of the therapsid-mammal boundary are largely the result of confusion between the taxon's definition and diagnosis (Rowe, 1987, 1988). Therefore, it seems that the controversy regarding mammalian origins cannot be settled without a general agreement on how to define and diagnose the Class Mammalia.

WHAT ARE MAMMALS?

Ax (1987:217) stated that "there is an oft-repeated question, full of pathos: 'What are mammals?' " Obviously, the question may be answered with a definition, or a diagnosis, or perhaps both. Following Ghiselin (1984), Rowe (1987, 1988) stressed the necessity for decoupling the question into a theoretical aspect (i.e., definition) and a practical aspect (i.e., diagnosis). However, Ax (1987:217) argued that "a single descent community, being ontologically an individual unity in Nature, cannot be defined. It can only be characterized on the basis of the novelties evolved in its stem lineage which were present as autapomorphies in the latest common stem species of the recent representatives." Nevertheless, Rowe (1988:247) proposed to define the Mammalia "as comprising the most recent common ancestor of living Monotremata (Ornithorhynchidae and Tachyglossidae) and Theria (Marsupialia and Placentalia), and all of its descendants."

Although Rowe's definition of mammals seems coherent at first glance, it may not be informative. First, what is the most recent common ancestor of Recent monotremes and therians? Can it be recognized? We have been told by Eldredge and Cracraft (1980:10) that "because ancestors never possess a set of evolutionary novelties unique to themselves, their definition and recognition is, logically, difficult." If the most recent common ancestor of the Recent mammals cannot be defined and recognized, how can one determine its immediate, let alone *all*, descendants? A direct testimony to the failure of this definition is Rowe's (1988) exclusion of some widely accepted early mammals such as morganucodontids and kuehneotheriids from the Class Mammalia. Therefore, two components of Rowe's definition—i.e., common ancestor and all of its descendants— are either unknown or unknowable, given our present state of knowledge.

Second, Rowe (1988:247) further corroborated his definition of mammals by stating: "Thus, if an organism is born to a mammal it is by definition a member of Mammalia, regardless of whether it has hair, mammary glands, or any other character commonly associated with mammals." Admittedly, no one would quibble over Rowe's statement. However, one probably cannot help asking the following questions. What about the "common ancestor"? What group did its parents belong to? Also, it should be stressed that the initial recognition of the platypus as a mammal was based directly upon these sorts of characters—hair and mammary glands (Griffiths, 1978). Moreover, just for the sake of argument, I would like to quote Olson's (1971:188) statement: "Were only parts of the skeleton of *Tachyglossus* preserved in the fossil record and the modern representatives unknown, it would doubtless be considered reptilian without hesitation."

Clearly, Rowe's definition of mammals is based either on the preference to choose the closest sister-groups only among the living forms or on the assumption of monotremes' relatedness to therians. First, the practice of choosing the closest sister-groups only among the living forms obviously would be problematic if monotremes become extinct (an unfortunate but distinct possibility). Should this occur, how should the future zoologists define mammals? Second, mammalian affinity of monotremes originally was determined by several diagnostic characters found in the Recent therians. In this particular context, it is illogical to say that "ancestry is the only criterion that is both necessary and sufficient for taxon membership" (Rowe, 1988:247) while downplaying the relevance of the characterization of evolutionary novelties. Elsewhere, Rowe and his co-authors (Gauthier et al., 1988:182) pointed out: "In our view, Morganucodontidae is not a mammal . . . , because it does not possess the skeletal attributes hypothesized for the most recent common ancestor of the group Monotremata + Theria." Here, they treated some skeletal attributes (rather than direct ancestry) as both necessary and sufficient criterion for excluding morganucodonts from the mammals—an obvious self-contradiction!

From this example, it also becomes evident that Rowe's definition of mammals does not offer a practical solution to our problems. To answer the question "What are mammals?" we still must identify diagnostic features by which the taxon Mammalia can be characterized.

A conventional diagnostic character of Mammalia, i.e., the establishment of a squamosal-dentary jaw joint, was treated thoroughly by Simpson (1960) and widely followed (e.g., Carroll, 1988; Romer, 1966). More recently, several other characters have been proposed to supplement that long-prevailing single-character diagnosis (e.g., Crompton and Sun, 1985; Hopson and Barghusen, 1986; Hopson, this volume). Among the newly added synapomorphies of mammals, two are particularly applicable and practical—i.e., the loss of "alternate" tooth replacement of postcanine teeth, and the possession of postcanine teeth with divided roots.

With an increasing knowledge of cranial material of early mammals, another diagnostic character has emerged: the development of a bony floor to the cavum epiptericum below the ganglia for the trigeminal and facial nerves.

Radical departures from the conventional diagnosis and contents of the Class Mammalia have been advocated recently by Ax (1987) and by Rowe (1988). Ax (1987) reasoned that a particular closed-descent community in nature (i.e., a monophyletic group) should not be characterized arbitrarily or liberally. It must be characterized on the basis of the first evolutionary novelty that occurs in the fossil record and that constitutes an autapomorphy of the recent representatives of the group. In the case of characterizing the taxon Mammalia, Ax (1987) chose the possession of a synapsid-type skull as the sole and sufficient synapomorphy of the closed-descent community that represents the sister-group of Sauropsida. Therefore, Ax's Mammalia comprises what traditionally has been called mammals plus *all* mammal-like reptiles, the Pelycosauria included. Incidentally, this is essentially a restatement of Goodrich's (1916) sauropsid-theropsid division (see also Clemens, 1986).

However, there is the uncertainty concerning the sole synapomorphy (i.e., the possession of a synapsid-type skull) that Ax used to characterize his "Mammalia." A synapsid skull is characterized by a single temporal fenestra bounded dorsally by the postorbital and squamosal bones in its primitive condition (Romer, 1966). As Kemp (1988a:4) ably summarized, "Gaffney (1980) and Reisz (1980) both pointed out that the lower temporal fenestra of diapsids has this construction as well, and therefore the synapsid condition might be seen as plesiomorphic for both diapsids and synapsids and not unique to the latter. Reisz (1986) nevertheless lists it as a characteristic of pelycosaurs, and therefore of synapsids as a whole, with the implication that it is convergent in the diapsids."

Another problem with Ax's proposal is basically semantic. There are two appropriate and readily available terms—*Synapsida* or *Theropsida*—to replace his *Mammalia*. After all, it was Linnaeus who adopted the term *Mammalia* from the Latin word *Mamma* to designate the amniotes that suckle their young (Scott, 1913). According to Gregory (1910:27–28), the "most enduring claim of Linnaeus upon the grateful memory of posterity arises from his recognition of the fundamental importance of the mammae as a class character and from his felicitous coinage of the word 'mammalia' as a class name." We almost can be certain that pelycosaurs and many other so-called "mammal-like reptiles" did not possess mammae and had not yet evolved lactation (Guillette and Hotton, 1986). We had best retain the term *Mammalia* in its original sense and leave the majority of our mammalogist colleagues undisturbed.

Rowe's (1988) analysis included many taxa and numerous characters;

therefore, it is vulnerable to becoming the center of many controversies. It is only fair to say, however, that anyone who would courageously conduct this kind of analysis on such a grand scale is likely to encounter numerous difficulties with the character assessments. Despite admitted ambiguities, Rowe's resultant consistency index of 0.926 is surprisingly high. Careful readings of Rowe's paper indicate that the execution of his methodology might have been flawed. For example, Rowe selected only seven taxa for his PAUP run (see Rowe, 1988:Fig. 3) and, with little justification, interpolated an additional 18 taxa *manually* (see Rowe, 1988:Fig. 4). Among the 18 taxa are included "a number of fossils that have figured prominently in previous discussions of the origin of mammals" (Rowe, 1988:245–246). Rowe (1988:246) claimed that "the addition of these taxa raised the number of equally parsimonious trees to 25, in contrast to the single tree found for the relatively complete taxa. . . . However, it did not change the relationship among complete taxa; differences among the 25 trees were solely in the placement of the deficient taxa." However, it is Rowe's placement of these so-called "deficient taxa" that has departed radically from the mainstream of thought on the early evolution of mammals.

Again, as an example, Rowe (1988) selected seven taxa from the traditionally classified mammalian order Triconodonta and divided them into three clusters. The first cluster, consisting of Morganucodontidae, *Sinoconodon,* and *Dinnetherium,* was not considered part of the Mammalia. The second cluster consists of the triconodont *Amphilestes,* the Monotremata, and three eupantotherian genera. Rowe considered the second cluster to be the first polychotomous radiation of "Mammalia." Rowe's "unnamed" third cluster includes three triconodont genera (viz., *Phascolotherium, Triconodon,* and *Trioracodon*) plus the symmetrodont *Spalacotherium* and the eupantotherian *Peramus.* Rowe also excluded the so-called Theria of Metatherian-Eutherian Grade, an important group in the evolution of therian mammals.

Were Rowe's cladogram to be accepted, much of the important character evolution, both dental and cranial, of early mammals would have to be reinterpreted. For example, the triangulation of main molar cusps, the coiling of the cochlea, and the reduction of the sidewall of the primary braincase (to name only three) would have to be assumed to have evolved in parallel. Alternately, enormous evolutionary reversals would have to be required. Neither seems to be warranted at present.

Rowe (1988:246) further stated that "it is unlikely, moreover, that the deficient taxa are sufficiently informative to overturn a hypothesis based on complete taxa. For example, it is unlikely that an isolated jaw might lead us to believe that placentals are more closely related to monotremes than to marsupials." I would not agree. For instance, the discovery of an isolated mandibular fragment of the monotreme *Steropodon* from the Early Cretaceous of Australia has dramatically changed our view on the "prototherian" affinity of monotremes, as well as the interrelationships of mammals as previously hypothesized on the basis of their living, therefore "complete," representatives (Archer et al., 1985; Kielan-Jaworowska et al., 1987). Unfortunately, Rowe (1988) did not include *Steropodon* in his analysis.

Another example is Sigogneau-Russell's (1989) description of the isolated teeth of haramiyid multituberculates from the Late Triassic of France. Although the multituberculate affinity of haramiyids was suggested first by Marsh (1887) and since has been supported by various authors (e.g., Hahn, 1973; Van Valen, 1976), I (Miao, 1988) was unconvinced until very recently. Sigogneau-Russell's documentation has shown beyond reasonable doubt that haramiyids are indeed multituberculates. This not only disproves Rowe's placement of Haramiyidae outside Mammalia but also falsifies Rowe's chronological calculation of a Bathonian age (the Middle Jurassic) for mammalian origin. Moreover, studies by both Sigogneau-Russell (1989) and Archer et al. (1985) have provided important information contradictory to the phylogenetic relationships of Monotremata and Multituberculata with respect to living Theria that was hypothesized by Rowe (1988:Fig. 3) on the basis of "complete taxa." Sigogneau-Russell's analysis, however, strengthens one of the conclusions that I derived from my study of nonharamiyid multituberculates—"Multituberculates are members of the paraphyletic stem group, and diverged from the main lineage leading to modern mammals prior to emergence of the latest common ancestor of modern mammals" (Miao, 1988:95).

Furthermore, many characters that Rowe (1988) employed to diagnose Mammalia are either equivocal and only partially correct or difficult to apply. Scrutiny of Rowe's diagnosis of Mammalia through character-by-character procedure is not intended here. However, it should be noted that among the 37 characters that Rowe used to diagnose the Mammalia, there is not a single dental character. Given the fact that the major portion of the early mammalian fossil record is represented by isolated teeth, Rowe's diagnosis of Mammalia is of limited application to the fossil record.

In summary, neither Ax's nor Rowe's diagnosis of Mammalia seems particularly helpful. Therefore, I strongly favor Hopson's (this volume) revised list of synapomorphies of Mammalia, which includes several time-honored, highly practical, and diagnostic mammalian osteological characters.

MAMMALIAN POLYPHYLY REVISITED

Much of the historical background for the discussion of mammalian monophyly versus polyphyly was addressed above. However, terms such as *monophyly* and *polyphyly* may have had different meanings to different authors. For example, Mayr (1969:75) once pointed out that "the issue of monophyly has been clouded by various confusions. Some authors have referred to a 'polyphyly' of a taxon when only a polyphyly of the diagnostic character of the taxon was involved, the taxon itself being monophyletic. . . . The class Mammalia is monophyletic because all mammalian lines were derived from the immediately ancestral taxon of therapsid reptiles."

Mayr's viewpoint was shared by Crompton and Jenkins (1979) in their benchmark review of the origin of mammals, wherein they stated (p. 68), "If the reptile-mammal distinction is made on the basis of independently evolved structures (*e.g.*, a three-boned middle ear dissociated from the jaw joint), the result would, by definition, yield polyphyly or at least diphyly. Alternatively, if the boundary were drawn on a character shared by common ancestry (*e.g.*, a squamoso-dentary articulation), the result is monophyly." Incidentally, the evidence from which Simpson (1959) drew his conclusion of polyphyletic origin of mammals was the presumed independent acquisition of squamosal-dentary articulation among early mammalian groups.

Crompton and Jenkins correctly pointed out a true representation of polyphyly in terms of separate ancestry. They stated that "polyphyly here is used to signify that different groups of mammals . . . are derived from *separate* therapsid stocks which also gave rise to other lineages of mammal-like reptiles" (Crompton and Jenkins, 1979:69, italics in the original). However, it should be made clear that these authors did not consider that this was the actual case in mammalian origins and, instead, suggested that various groups of mammals were derived from a single lineage of cynodonts.

Apparently, discussion of monophyly versus polyphyly of the Mammalia depends on how these three key words are defined. First, there are two major alternate definitions of *monophyly*. According to Schoch (1986: 181–182), "A typical classical evolutionary definition reads: A monophyletic group is 'a taxonomic group whose members are descended from a common ancestor included in that group' (Szalay and Delson, 1979, p. 563). By this conception, the members of a monophyletic group must all be descended from a common ancestor included in the group, but all organisms descended from the common ancestor need not be included in the group for it to be considered monophyletic." Incidentally, this is *paraphyly* in cladistic terminology. In contrast, cladists define a *monophyletic group* as a group including a common ancestor and *all* of its descendants (Farris, 1974; Wiley, 1981). Neither definition warrants the recognition of what traditionally has been called the Mammalia as a monophyletic group, because the common ancestor has been assumed to lie in another group, usually referred to as therapsids or cynodonts.

Second, there is little disagreement on the definition of *polyphyly* among different schools of taxonomists. A polyphyletic group is defined as a group "whose common ancestor would not be classified as a member of the same group" (Ridley, 1986:30–31; see also Farris, 1974, and Wiley, 1981). By definition, mammals would be truly polyphyletic as traditionally diagnosed.

Third, as the preceding section reveals, there have been some serious attempts to use a cladistic approach to define or diagnose a monophyletic Class Mammalia which differ dramatically from conventional wisdom. I have indicated that Rowe's definition of Mammalia fails to pinpoint the common ancestor of extant Monotremata and Theria, and therefore cannot in practice delimit a real monophyletic Class Mammalia among the various mammalian and mammal-like groups. Ax's (1987) proposal also seems problematic because of both the semantic problems and the uncertainty about the sole synapomorphy (i.e., the possession of a synapsid-type skull) that Ax used to characterize his "Mammalia."

It should also be noted that Ax's (1987) notion of Mammalia once was proposed by Simpson (1959) as one of the four alternative ways to classify Mammalia. However, Simpson (1959:413) discarded it by stating that "this would evade the issue of polyphyly at its present level, but would probably only transfer it to an earlier level. An arbitrary structural definition would still be necessary, and such definition at the base of the Therapsida or Synapsida might be even more difficult to apply than those now used at the top of those groups."

All arguments presented thus far are largely semantic. Admittedly, at some level there existed a common ancestral population that gave rise to all mammals, no matter how mammals are defined. But in a cladistic analysis, without the prospect of a recognizable common ancestor and the obligatory inclusion of that ancestor into the group in question, is classifying a monophyletic group ever achievable? I think that the contributions to the present volume do provide an interesting and persuasive answer to this question. After nearly two decades of failure to achieve a consensus in mammalian monophyly despite our efforts, it is instructive to read again what Simpson (1971:192) once said: "Is it possible to define a monophyletic group Mammalia including the unknown one ancestral unit and all its descendants but no other species? I submit that this also is obviously impossible and that this concept of monophyly as applied to taxonomy and nomenclature is simply quixotic."

Finally, to emulate Ax (1987), another oft-repeated saying, full of pathos, is "The current controversy cannot be settled until more fossils of crucial importance turn up." I hope that my review demonstrates the converse—i.e., most fossil finds raise more questions than they solve. However, this is *not* a pessimistic view of our painstaking and meritorious endeavors. I firmly believe that, if science progresses at all, it is bound to be marked by the falsification of the established, often more simplistic, hypotheses and the realization of new, ever more challenging problems.

CONCLUSIONS

The origin and early evolution of the Mammalia never has been as exciting and challenging a field of research as it is today. With rapidly increasing knowledge of Mesozoic mammals and broad application of cladistic analysis, a set of synapomorphies characterizing the Class Mammalia was proposed by Crompton and Sun (1985) and Hopson and Barghusen (1986), adopted by Kemp (1988b), and revised by Hopson (this volume). Among these mammalian characters, I recommend the following as particularly practical and useful: (1) the establishment of squamosal-dentary jaw articulation; (2) the loss of "alternate" tooth replacement of postcanine teeth; (3) postcanine teeth with divided roots; and (4) the development of a bony floor to the cavum epiptericum below the ganglia for the trigeminal and facial nerves.

As categorized by these diagnostic characters, the Class Mammalia is in perfect accordance with its traditional concept (Fig. 1), which has enjoyed wide acceptance and has maintained stability in mammalian classification. The more recent proposals of mammalian definition and diagnosis by Ax (1987) and Rowe (1988) are not accepted herein, because they do not appear to have yielded sufficient benefits or any clear progress in our understanding of mammalian phylogeny.

The common ancestor of the various early mammalian groups, probably lacking all of the aforementioned mammalian characters, cannot be recognized and therefore is not included in the Class Mammalia. On the basis of the widely accepted definitions of monophyly and polyphyly, the Class Mammalia is polyphyletic as traditionally defined and categorized.

Acknowledgments I thank H.-P. Schultze and L. Trueb for inviting me to contribute this chapter and especially for their constant encouragement and amazing displays of patience. I am grateful to W. A. Clemens, J. A. Hopson, J. A. Lillegraven, Z. Kielan-Jaworowska, M. C. McKenna, and J. R. Wible for sharing with me their knowledge of, and insight into, early mammalian evolution. I also thank L. D. Martin and H.-P. Schultze for

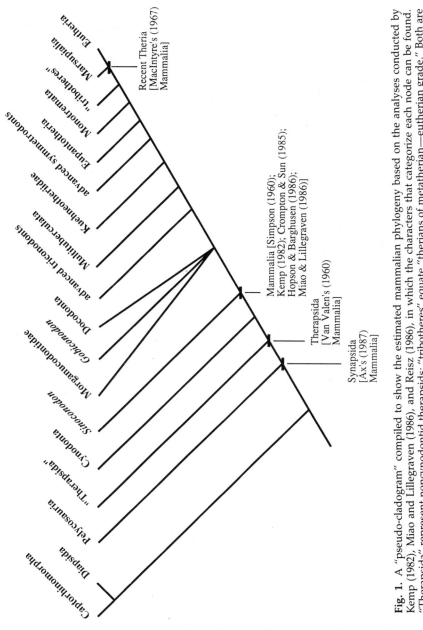

Fig. 1. A "pseudo-cladogram" compiled to show the estimated mammalian phylogeny based on the analyses conducted by Kemp (1982), Miao and Lillegraven (1986), and Reisz (1986), in which the characters that categorize each node can be found. "Therapsida" represent noncynodontid therapsids; "tribotheres" equate "therians of metatherian—eutherian grade." Both are paraphyletic groups. The figure also shows various proposals of the contents of Mammalia. However, because Rowe's proposal is difficult to assimilate into this figure, it should be consulted in Rowe, 1988:Fig. 4.

Capterhinomorpha
Diapsida
Pelycosauria
"Therapsida"
Cynodontia
Sinoconodon
Morganucodontidae
Gobiconodon
Docodonta
advanced triconodonts
Multituberculata
Kuehneotheriidae
advanced symmetrodonts
Eupantotheria
Monotremata
"tribotheres"
Marsupialia
Eutheria

Recent Theria
[MacIntyre's (1967)
Mammalia]

Mammalia [Simpson (1960);
Kemp (1982); Crompton & Sun (1985);
Hopson & Barghusen (1986);
Miao & Lillegraven (1986)]

Therapsida
[Van Valen's (1960)
Mammalia]

Synapsida
[Ax's (1987)
Mammalia]

their help and support. Graduate students in the Division of Vertebrate Paleontology of The University of Kansas have been a constant source of inspiration to me. Particularly, I extend my thanks to J. A. McAllister and M. D. Gottfried, who read the first draft of the manuscript and offered many helpful suggestions. Special thanks are extended to Xiaoming Wang for his assistance in creating the computer-generated cladogram shown in Figure 1, and to Janet Elder for typing the manuscript. I greatly appreciate the kind and constructive reviews by F. A. Jenkins, Jr., J. A. Hopson, J. A. Lillegraven, and Z. Kielan-Jaworowska.

Literature Cited

Archer, M., T. F. Flannery, A. Ritchie, and R. E. Molnar. 1985. First Mesozoic mammal from Australia—an early Cretaceous monotreme. Nature, London, 318:363–366.

Ax, P. 1987. *The Phylogenetic System: The Systematization of Organisms on the Basis of Their Phylogenesis.* New York: John Wiley and Sons.

Bohlin, B. 1945. The Jurassic mammals and the origin of the mammalian molar teeth. Bull. Geol. Instit. Uppsala, 31:363–388.

Bonaparte, J. F. 1986a. A new and unusual Late Cretaceous mammal from Patagonia. J. Vert. Paleontol., 6:264–270.

Bonaparte, J. F. 1986b. History of the terrestrial Cretaceous vertebrates of Gondwana. Act. IV Congres. Argentino Paleontol. Bioestrat., 2:63–95.

Buckland, W. 1824. Notice on Megalosaurus. Trans. Geol. Soc. London, 2:390–396.

Carroll, R. L. 1988. *Vertebrate Paleontology and Evolution.* New York: W. H. Freeman.

Chow, M., and T. H. V. Rich. 1982. *Shuotherium dongi,* n. gen. and n. sp., a therian with pseudo-tribosphenic molars from the Jurassic of Sichuan, China. Australia Mammal., 5:127–142.

Clemens, W. A., Jr. 1970. Mesozoic mammalian evolution. Ann. Rev. Ecol. Syst., 1:357–390.

Clemens, W. A., Jr. 1986. On Triassic and Jurassic mammals. Pp. 237–246 *in* Padian, K. (ed.), *The Beginning of the Age of Dinosaurs.* Cambridge: Cambridge Univ. Press.

Crompton, A. W., and F. A. Jenkins, Jr. 1968. Molar occlusion in Late Triassic mammals. Biol. Rev. Cambridge Philos. Soc., 43:427–458.

Crompton, A. W., and F. A. Jenkins, Jr. 1973. Mammals from reptiles: a review of mammalian origins. Ann. Rev. Earth Planet Sci., 1:131–155.

Crompton, A. W., and F. A. Jenkins, Jr. 1979. Origin of mammals. Pp. 59–73 *in* Lillegraven, J. A., Z. Kielan-Jaworowska, and W. A. Clemens, Jr. (eds.), *Mesozoic Mammals: The First Two-thirds of Mammalian History.* Berkeley: Univ. of California Press.

Crompton, A. W., and A. Sun. 1985. Cranial structure and relationships of the Liassic mammal Sinoconodon. Zool. J. Linn. Soc., 85:99–119.

Eldredge, N. and J. Cracraft. 1980. *Phylogenetic Patterns and the Evolutionary Process: Method and Theory in Comparative Biology.* New York: Columbia Univ. Press.

Farris, J. S. 1974. Formal definitions of paraphyly and polyphyly. Syst. Zool., 23:548–554.

Flower, W. H. 1883. On the arrangement of the orders and families of existing Mammalia. Proc. Zool. Soc. London, 13:178–186.

Gaffney, E. S. 1980. Phylogenetic relationships of the major groups of amniotes. Pp. 593–610 *in* Panchen, A. L. (ed.), *The Terrestrial Environment and the Origin of Land Vertebrates.* London: Academic Press.

Gauthier, J., A. G. Kluge, and T. Rowe. 1988. Amniote Phylogeny and the importance of fossils. Cladistics, 4:105–209.

Ghiselin, M. T. 1984. "Definition," "character," and other equivocal terms. Syst. Zool., 33:104–110.

Goodrich, E. S. 1916. On the classification of the Reptilia. Proc. Roy. Soc. London, Ser. B, 89:261–276.

Gregory, W. K. 1910. The orders of mammals. Bull. Amer. Mus. Nat. Hist., 27:1–524.

Griffiths, M. 1978. *The Biology of Monotremes.* New York: Academic Press.

Guillette, L. J., Jr., and N. Hotton III. 1986. The evolution of mammalian reproductive characteristics in therapsid reptiles. Pp. 239–250 *in* Hotton, N., III, P. D. MacLean, J. J. Roth, and E. C. Roth (eds.), *The Ecology and Biology of Mammal-like Reptiles.* Washington: Smithsonian Inst. Press.

Hahn, G. 1973. Neue Zähne von Haramiyiden aus der deutschen Ober-Trias und ihre Beziehungen zu den Multituberculaten. Palaeontographica, Abt. A, 142:1–15.

Hopson, J. A. 1964. The braincase of the advanced mammal-like reptile *Bienotherium.* Postilla, 87:1–30.

Hopson, J. A. 1967. Mammal-like reptiles and the origin of mammals. Discovery (Magazine of the Peabody Mus. Nat. Hist., Yale Univ.), 2:25–33.

Hopson, J. A. 1969. The origin and adaptive radiation of mammal-like reptiles and nontherian mammals. Ann. New York Acad. Sci., 167:199–216.

Hopson, J. A. 1970. The classification of nontherian mammals. J. Mammal., 51:1–9.

Hopson, J. A., and H. R. Barghusen. 1986. An analysis of therapsid relationships. Pp. 83–106 *in* Hotton, N., III, P. D. MacLean, J. J. Roth, and E. C. Roth (eds.), *The Ecology and Biology of Mammal-like Reptiles.* Washington: Smithsonian Inst. Press.

Hopson, J. A., and A. W. Crompton. 1969. Origin of mammals. Pp. 15–72 *in* Dobzhansky, T., M. K. Hecht, and W. C. Steere (eds.), *Evolutionary Biology.* Vol. 3. New York: Appleton-Century-Crofts.

Jenkins, F. A., Jr. 1984. A survey of mammalian origins. Pp. 32–47 *in* Broadhead, T. W. (ed.), *Mammals: Notes for a Short Course.* Univ. Tennessee, Dept. Geol. Sci., Studies in Geol. 8.

Jenkins, F. A., Jr, A. W. Crompton, and W. R. Downs. 1983. Mesozoic mammals from Arizona: new evidence on mammalian evolution. Science, 222:1233–1235.

Jenkins, F. A., Jr., and C. R. Schaff. 1988. The Early Cretaceous mammal *Gobiconodon* (Mammalia, Triconodonta) from the Cloverly Formation in Montana. J. Vert. Paleontol., 8:1–24.

Kemp, T. S. 1982. *Mammal-like Reptiles and the Origin of Mammals.* New York: Academic Press.

Kemp, T. S. 1983. The relationships of mammals. Zool. J. Linn. Soc., 77:353–384.

Kemp, T. S. 1988a. Interrelationships of the Synapsida. Pp. 1–22 *in* Benton, M. J., (ed.), *The Phylogeny and Classification of the Tetrapods.* Vol. 2: *Mammals.* Oxford: Clarendon Press.

Kemp, T. S. 1988b. A note on the Mesozoic mammals, and the origin of therians. Pp. 23–29 *in* Benton, M. J. (ed.), *The Phylogeny and Classification of the Tetrapods.* Vol. 2: Mammals. Oxford: Clarendon Press.

Kermack, K. A. 1963. The cranial structure of the triconodonts. Philos. Trans. Roy. Soc. London, Ser. B, 246:83–103.

Kermack, K. A. 1967. The interrelations of early mammals. J. Linn. Soc. (Zool.), 47:241–249.

Kermack, K. A., and Z. Kielan-Jaworowska. 1971. Therian and non-therian mammals. Pp. 103–115 *in* Kermack, D. M., and K. A. Kermack (eds.), *Early Mammals.* Zool. J. Linn. Soc., 50 (Suppl. 1).

Kermack, K. A., A. J. Lee, P. M. Lees, and F. Mussett. 1987. A new docodont from the Forest Marble. Zool. J. Linn. Soc., 89:1–39.

Kielan-Jaworowska, Z., A. W. Crompton, and F. A. Jenkins, Jr. 1987. The origin of egg-laying mammals. Nature, London, 326:871–873.

Kühne, W. G. 1949. On a triconodont tooth of a new pattern from a fissure-filling in South Glamorgan. Proc. Zool. Soc. London, 119:345–350.

Kühne, W. G. 1956. *The Liassic Therapsid* Oligokyphus. London: British Museum (Natural History).

Kühne, W. G. 1958. Rhaetische Triconodonten aus Glamorgan, ihre Stellung zwischen den Klassen Reptilia und Mammalia und ihre Bedeutung für die Reichart'sche Theorie. Palaeontol. Z., 32:197–235.

Lillegraven, J. A., S. D. Thompson, B. K. McNab, and J. L. Patton. 1987. The origin of eutherian mammals. Biol. J. Linn. Soc., 32:281–336.

MacIntyre, G. T. 1967. Foramen pseudovale and quasi-mammals. Evolution, 21:834–841.

Marsh, O. C. 1887. American Jurassic mammals. Amer. J. Sci., 33:327–348.

Mayr, E. 1969. *Principles of Systematic Zoology.* New York: McGraw-Hill.

Miao D. 1988. Skull morphology of *Lambdopsalis bulla* (Mammalia, Multituberculata) and its implications to mammalian evolution. Contr. Geol. Univ. Wyoming, Spec. Pap. 4:1–104.

Miao D., and J. A. Lillegraven. 1986. Discovery of three ear ossicles in a multituberculate mammal. Nat. Geogr. Res., 2:500–507.

Olson, E. C. 1944. Origin of mammals based upon cranial morphology of the therapsid suborders. Geol. Soc. America, Spec. Pap., 55:1–136.

Olson, E. C. 1959. The evolution of mammalian characters. Evolution, 13:344–353.

Olson, E. C. 1971. *Vertebrate Paleozoology.* New York: John Wiley & Sons.

Parrington, F. R. 1967. The origins of mammals. Advan. Sci., December, pp. 165–173.

Parrington, F. R. 1971. On the Upper Triassic mammals. Philos. Trans. Roy. Soc. London, Ser. B, 261:231–272.

Patterson, B. 1956. Early Cretaceous mammals and the evolution of mammalian molar teeth. Fieldiana, Geol., 13:1–105.

Patterson, B., and E. C. Olson. 1961. A triconodontid mammal from the Triassic of Yunnan. Internat. Colloq. Evol. Lower Non-Specialized Mammals, Pt. 1, 129–191. Brussels: Paleis der Academiën-Hertogsstraat.

Presley, R. 1980. The braincase in Recent and Mesozoic therapsids. Mem. Soc. Geol. France, New Ser., 139:159–162.

Presley, R. 1981. Alisphenoid equivalents in placentals, marsupials, monotremes and fossils. Nature, London, 294:668–670.

Presley, R., and F. L. D. Steel. 1976. On the homology of the alisphenoid. J. Anat., 121:441–459.

Reed, C. A. 1960. Polyphyletic or monophyletic ancestry of mammals, or: what is a class? Evolution, 14:314–322.

Reisz, R. R. 1980. The Pelycosauria: a review of phylogenetic relationships. Pp. 553–592 *in* Panchen, A. L. (ed.), *The Terrestrial Environment and the Origin of Land Vertebrates.* London: Academic Press.

Reisz, R. R. 1986. Pelycosauria. Pp. 1–102 *in* Wellnhofer, P. (ed.), *Handbuch der Paläoherpetologie*, Teil 17A. Stuttgart and New York: Gustav Fischer.

Ridley, M. 1986. *Evolution and Classification.* London: Longman Group.

Romer, A. S. 1966. *Vertebrate Paleontology.* Chicago: Univ. of Chicago Press.

Romer, A. S. 1968. *Notes and Comments on Vertebrate Paleontology.* Chicago: Univ. of Chicago Press.

Rougier, G. W., and J. F. Bonaparte. 1988. La pared lateral del Craneo de *Vincelestes neuquenianus* (Mammalia, Eupantotheria) y su importancia en el estudio de los Mamiferos Mesozoicos. V. J. Argentinas Paleontol. Vert., pp. 14–15.

Rowe, T. 1987. Definition and diagnosis in phylogenetic system. Syst. Zool., 36:208–211.

Rowe, T. 1988. Definition, diagnosis, and origin of Mammalia. J. Vert. Paleontol., 8:241–264.

Schoch, R. M. 1986. *Phylogeny Reconstruction in Paleontology*. New York: Van Nostrand Reinhold.

Scott, W. B. 1913. *A History of Land Mammals in the Western Hemisphere*. New York: Macmillan.

Sigogneau-Russell, D. 1989. Haramiyidae (Mammalia, Allotheria) en provenance du Trias Supérieur de Lorraine (France). Palaeontographica, Abt. A, 206:137–198.

Simpson, G. G. 1928. *A Catalogue of the Mesozoic Mammalia in the Geological Department of the British Museum*. London: Oxford Univ. Press.

Simpson, G. G. 1929. American Mesozoic Mammalia. Mem. Peabody Mus. Yale Univ., 3(1):1–235.

Simpson, G. G. 1959. Mesozoic mammals and the polyphyletic origin of mammals. Evolution, 13:405–414.

Simpson, G. G. 1960. Diagnosis of the classes Reptilia and Mammalia. Evolution, 14:388–392.

Simpson, G. G. 1961. Evolution of the Mesozoic mammals. Internat. Colloq. Evol. Lower Non-Specialized Mammals, Pt. 1, 57–95.

Simpson, G. G. 1971. Concluding remarks: Mesozoic mammals revisited. Pp. 181–198 *in* Kermack, D. M., and K. A. Kermack (eds.), *Early Mammals*. Zool. J. Linn. Soc., 50 (Suppl. 1).

Szalay, F. S., and E. Delson. 1979. *Evolutionary History of the Primates*. New York: Academic Press.

Van Valen, L. 1960. Therapsids as mammals. Evolution, 14:304–313.

Van Valen, L. 1976. Note on the origin of multituberculates (Mammalia). J. Paleontol., 50:198–199.

Wiley, E. O. 1981. *Phylogenetics: The Theory and Practice of Phylogenetic Systematics*. New York: Wiley.

17 The Nature and Diversity of Synapsids: Prologue to the Origin of Mammals

Nicholas Hotton III

Synapsids are tetrapods of reptilian grade that first appeared in the Middle Carboniferous, gave rise to the first mammals during the Late Triassic, and became extinct in the Early Jurassic. Their mammalian affinity was first recognized by Cope (1878), and synapsids have enjoyed the status of "mammal-like reptiles" since the end of the nineteenth century (Broom, 1932), albeit under a variety of taxonomic names and arrangements. Their current taxonomy, as the Subclass Synapsida with its component orders Pelycosauria and Therapsida, was established by Williston (1925).

Although the history of synapsids has been deciphered largely in terms of its mammalian connection, there is much more to it than the role of synapsids as mammalian ancestors. First, synapsids are noteworthy for their antiquity, which matches that of any amniote lineage, with the more primitive synapsid Order Pelycosauria being identifiable among the earliest reptiles (Reisz, 1972). Second, synapsids were the most abundant and conspicuous terrestrial tetrapods during most of the Permian, when they were counterparts of the dinosaurs of the Mesozoic and the mammals of the Tertiary. Pelycosaurs became conspicuous in terrestrial environments before the end of the Carboniferous and were abundant as top predators until the end of the Early Permian, when they were replaced in greater diversity by the more advanced Order Therapsida. Third, the diversity of therapsids included large numbers of a variety of herbivores, the first tetrapods to contribute significantly as primary consumers to the integration of a terrestrial fauna. Fourth, therapsids were a significant part of the environment of remote mammalian origins, for the fauna dominated by noncynodont therapsids was well established by the time that the mammalian stem-group Cynodontia appeared near the end of the Late Permian.

The present chapter emphasizes these aspects of synapsid evolution by examining synapsids as synapsids, rather than as mammalian ancestors, with special attention being directed to the culmination of synapsid diversity in the therapsids of the Later Permian.

THE NATURE OF SYNAPSIDS

Except for its mammalian connection, the synapsid lineage is isolated. Primitive pelycosaurs are separated by shared derived characters from their contemporaries among the earliest reptiles and in their distinctiveness illustrate the long independence of the synapsid-mammal line from other amniotes. Thus, the evidence of the fossil record supports the redefinition of Synapsida to include mammals, whether such redefinition is based on cladistic (Hopson and Barghusen, 1986) or noncladistic argument (cf. Reed, 1960; Van Valen, 1960). Van Valen suggested that therapsids should be styled "reptile-like mammals," but this characterization carries even more undesirable semantic baggage than the traditional one. Irrespective of their ultimate connection to mammals, synapsids were their own peculiar kind of animal during most of their history, and it is better to dispense with nicknames and refer to them simply as synapsids.

The Synapsid Record

In the course of the transition from pelycosaurs to therapsids, synapsid diversity increased more than threefold at the level of family (Appendix I). This is a conservative estimate based on analyses of pelycosaurs by Reisz (1980, 1986) and Brinkman and Eberth (1983), and of therapsids by Hopson and Barghusen (1986). These analyses are as yet provisional above the level of family, and the authors proposed neither new names of taxa nor new categories, although it is evident that a large number of new suprafamilial categories will be required for Therapsida.

By far the most comprehensive data on the diversity of Permian and Lower Triassic synapsids are to be found in Lower Permian deposits of Texas, Oklahoma, and New Mexico, and Upper Permian and Lower Triassic deposits of South Africa. All of the pelycosaur families recognized by Reisz (1980) are present in Texas, and of the 37 therapsid families recognized by Hopson and Barghusen (1986), 30 are well represented in the Beaufort Group of South Africa (Appendix I). The 16 genera of Gorgonopsia are based on the work of Sigogneau (1970), less the genera transferred in the family Ictidorhinidae to the "Biarmosuchia" by Hopson and Barghusen (1986). African, Asian, and New World deposits younger than the Beaufort Group record the appearance of a variety of very advanced cynodonts and of true mammals, but because such events coincide with the late decline of therapsids, they are beyond the scope of this paper. Following is the traditional biostratigraphic subdivision of the Beaufort, as modified by Kitching (1977):

Lower Triassic
  Cynognathus Zone
  Lystrosaurus Zone
Upper Permian
  Daptocephalus Zone
  Cistecephalus Zone
  Tapinocephalus Zone

Romer's practice of referring the *Tapinocephalus* Zone to the Middle Permian (Romer, 1966) is not widely accepted (Olson, 1962) and so is not followed here.

Reptiles or Mammals?

The nature of synapsids, whether reptilian or mammalian, is defined on the basis of characters that traditionally distinguish living reptiles from mammals. Because synapsids are known only from fossils, such definition depends primarily on osteology, which involves characters related to hearing, feeding, and locomotion. Characters of soft morphology, physiology, and behavior, mostly related to thermoregulation and reproduction, which are given preference in considerations of living forms, must play a secondary role when basic data come from fossils.

Synapsids conform to the traditional definition of reptiles in having several bones behind the dentary in the lower jaw, one of which articulates with the quadrate bone of the skull, and in having a single ossicle, the stapes, in the middle ear. In pelycosaurs, the stapes approaches the quadrate, and during life probably was connected with it by cartilage (Romer and Price, 1940; Brinkman and Eberth, 1983). This is probably the primitive condition, because it is shared by captorhinomorphs and other primitive nonsynapsid reptiles. In turtles and advanced diapsid reptiles, the contact between stapes and quadrate is lost or much reduced, and the slender stapes points toward a concavity at the back of the quadrate, which in most living forms supports a tympanic membrane. Therapsids differ from living reptiles because the condition of pelycosaurs is exaggerated in therapsids; thus, the stapes is in direct contact with the quadrate, bone to bone (Fig. 1). This articulation resembles a diarthrosis, homologous to the joint between the mammalian stapes and incus, the latter being the homologue of the reptilian quadrate. The direct contact between stapes and quadrate is a primary feature by which therapsids are defined as mammal-like.

The configuration of the temporal region of the skull, from which the jaw musculature originates, is an osteological feature that defines all synapsids as mammal-like. In synapsids, the jaw muscles originate from the rim of a temporal opening (Fig. 2) that, as in mammals, is single and is bounded ventrally by the jugal and squamosal (in small pelycosaurs by jugal and quadratojugal, Romer and Price, 1940). In living reptiles, the jaw musculature originates from the rims of two openings in the bony temporal wall (lepidosaurs, archosaurs) or from the inside of the unperforated temporal wall (turtles).

Other osteological features do not become especially mammal-like until, or after, the transition from pelycosaurs to therapsids at the end of the Early Permian. Changes in limb pose, pelvic structure, and bone histology mark the transition, whereas changes in the face and in buccal and nasal passages appear significantly later and for the most part are restricted to cynodonts and advanced therocephalians. Commonly, these features are held to indicate the onset of endothermy and are discussed in

1 cm

Fig. 1. Skull of small dicynodont of Stage I, cf. *Robertia* Boonstra (1948), USNM 22944, palatal aspect. Contact between the stapes (S) and quadrate (Q) is characteristic of therapsids. Note the postcanine teeth (T) and secondary palate (P) formed chiefly by premaxilla.

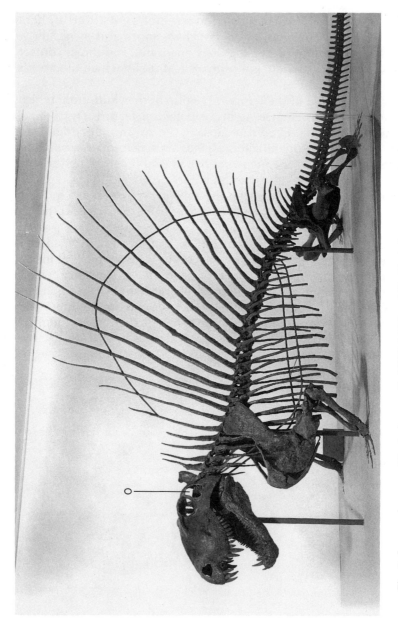

Fig. 2. Large pelycosaur, *Dimetrodon grandis*, USNM 8635, ×0.05. The habitus is lizardlike, and the humerus and femur are scarcely visible from this angle because they project laterally from the body axis. Note the synapsid temporal opening (O).

greater detail in a later section on the contrast between pelycosaurs and therapsids.

Dentitions become more diverse with the origin of therapsids, but therapsid dental specializations are not very mammal-like until the radiation of cynodonts and advanced therocephalians in the Early Triassic. These features, therefore, are treated more appropriately in the context of therapsid diversity. The establishment of multiple roots and the restriction of tooth replacement are delayed until the emergence of true mammals in the Late Triassic, and so are beyond the scope of this paper.

Archaic Characters

The parietal fenestra, an opening that lies between the parietal bones or in the midline between parietals and frontals, is a persistent feature of synapsids. It is universal in pelycosaurs, dinocephalians, and gorgonopsians, present in dicynodonts with a single exception (Hotton, 1974), and present in primitive therocephalians and cynodonts. Its manifestation in synapsids reflects an ancient chordate heritage, for the parietal fenestra is nearly universal in all Paleozoic classes. In tetrapods, by analogy with living lizards and *Sphenodon*, the fenestra housed the parietal apparatus, which presumably was eyelike in structure. Function of the parietal eye remains problematical in many respects, but it is important in thermoregulation and control of reproductive cyclicity in lizards and *Sphenodon*, which, as ectotherms, regulate body temperature primarily by behavior. For a comprehensive review see Quay, 1979. The parietal eye disappears early in tetrapod lineages that become aquatic (turtles, crocodilians, phytosaurs), or endothermic (birds, other archosaurs?, mammals), but not in those that become marine (mosasaurs, plesiosaurs) (Roth et al., 1986). Its persistence in synapsids suggests the persistence of a large behavioral component in thermoregulation, and its decline in cynodonts and advanced baurioids corresponds to the appearance of a number of features that have been related to endothermy. The parietal fenestra disappears earlier, albeit sporadically, in more primitive baurioids, in which its loss is not closely associated with the appearance of features related to endothermy.

Two nonosteological functional complexes can be inferred for synapsids because they are present in mammals but not in nonsynapsid reptiles. They are treated here as archaic because they are also characteristic of living amphibians. These complexes are (1) the mode of excretion of nitrogenous wastes, and (2) the nature of the integument.

The ultimate nitrogen metabolite is ammonia, which, being toxic, is excreted by the kidneys (ammoniotely) and, in agnathans and some gnathostome fishes, by the gills, through which it is diffused into the environment (Schmidt-Nielsen, 1975). It also is excreted by early embryos of a variety of vertebrates and diffused through the shells of amniote eggs (Packard et al., 1977). In adult stages of most vertebrates, however, ammonia is disposed of by conversion to the less toxic urea (ureotely) or uric acid (uricotely). Sharks and most living fish and amphibians are ureotelic, and anurans regularly shift from ammoniotely to ureotely at metamorphosis (Schmidt-Nielsen, 1975), whereas living turtles, lepidosaurs, crocodilians, and birds are uricotelic (Dantzler, 1976). Crocodilians commonly excrete nitrogen as ammonia, and some aquatic turtles can excrete nitrogen either as urea or uric acid, depending on whether they are kept in or out of water, whereas two desert-dwelling anurans have become uricotelic as adults (Schmidt-Nielsen, 1975). This distribution indicates that, in general, the mode of nitrogen excretion is linked to aquatic habit, or at least to abundant environmental water, and may reflect the fact that urea is more soluble than uric acid and therefore requires more water for its disposal. Mammals, alone among living amniotes, are rigidly locked into ureotely and exhibit uricotely only as a pathology (Schmidt-Nielsen, 1975). Marine mammals and many desert-dwelling mammals can concentrate urine, but this process does not conserve water as effectively as uricotely (Schmidt-Nielson, 1975), a strategy that apparently the synapsid-mammal lineage never attained.

The most parsimonious explanation of the distribution of excretory strategies is that ammoniotely and ureotely are primitive, and that uricotely is a synapomorphy of living sauropsid reptiles. This implies that uricotely must have arisen sometime during the later Carboniferous, before the splitting off of diapsids (Peabody, 1952; Reisz, 1977, 1981) from the line that led to turtles, but after the separation of the synapsid lineage from other tetrapods. The argument that uricotely arose before synapsids had split off from other amniotes is rejected as less parsimonious, because it requires a reversal to account for the ureotely of mammals. In this view, mammalian ureotely is a holdover from presynapsid ancestors, in which case synapsids were themselves ureotelic, the conduit by which the condition was transmitted to mammals.

The skin of mammals, like the skin of amphibians and unlike that of living reptiles and birds, is glandular, and the origin of mammalian hair seems to be related to glandular skin. Although hair coexists with scales in a few living mammals, it arises not by modification of scales, but from follicles regularly distributed around the free margins of the scales. Hair differs from scales (and feathers) in the histology and development of its follicles, in its regular association with sebaceous glands, and in the keratin of which it is composed, which is similar to the keratin of living amphibians (Spearman, 1966). There is little evidence to link these data with synapsids, because skin is preserved as a rule only as impression, and keratin only under very unusual circumstances. However, one therapsid specimen is known in which the skin does appear to have been glandular (Chudinov, 1968). In two specimens of *Lystrosaurus* from a single locality in the Beaufort, the muzzles are covered by patches of a thin layer of black material that may be degraded horn, but there is no indication of integument elsewhere. Integumentary structures are unknown in pelycosaurs, although they have been reported in a variety of amphibian contemporaries of pelycosaurs in Texas—e.g., bony scales in *Trimerorhachis* (Olson, 1979) and *Eryops* (Romer and Witter, 1941). Thus, though there is a little evidence to indicate the nature of synapsid integuments, none of it indicates a nonglandular, water-retentive integument like that of living reptiles and birds.

Pelycosaurs and Therapsids Contrasted

Pelycosaurs in general are lizardlike in body form: the trunk and tail are elongate and the limbs are relatively short, projecting laterally from the body axis in the typical reptilian sprawl (Fig. 2). The feet are reminiscent of those of lizards, with the toes splaying out on either side of an elongate fourth digit. Dentitions of pelycosaurs are reptilian; the marginal teeth are of generally similar shape from front to back, single rooted, and replaced continuously throughout the life of the individual. They are implanted on the margins of dentaries, maxillae, and premaxillae in deep grooves, of which the labial walls are higher than the lingual. The dental grooves are divided into sockets, one for each tooth, by partitions of very thin bone. This type of tooth implantation is characterized as subthecodont and is shared by many Paleozoic amphibians (Romer, 1956). There is a movable joint between the basis cranii and the pterygoid bone of the palate, as in living lizards and in primitive reptiles generally. The earliest pelycosaurs are larger than their nonsynapsid contemporaries. They also are distinguished by the temporal opening and a large tabular bone in the temporal region of the skull, features that link them clearly to the succeeding therapsids.

The earliest therapsids, best represented in the earliest Upper Permian of Russia (see Olson, 1962), retain a gracility reminiscent of pelycosaurs, with long body and tail, but Beaufort therapsids are more robust. The tendency toward robustness is most pronounced in advanced dinocephalians (Fig. 3) but occurs very early in dinocephalians discovered in deposits that have been reassigned recently from the Ecca to the oldest Beaufort (Rubidge, 1987, 1988). In Beaufort therapsids generally, the body is thicker and the tail shorter than in pelycosaurs, and the limbs, though proportionately somewhat longer, are stout and the feet are short. The head of the femur is inflected, indicating that the hind limb

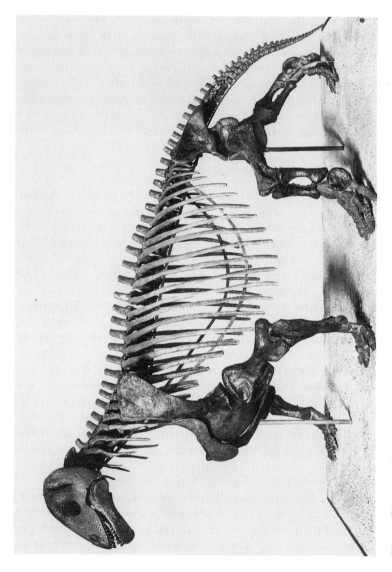

Fig. 3. Dinocephalian, *Moschops capensis*, AMNH 5552, ×0.06. Although *Moschops* is nearly 1 m shorter than *Dimetrodon grandis* (Fig. 2), most of the difference is in the tail. The more robustly built *Moschops* is much the bulkier animal. (From Gregory, 1926, courtesy of the American Museum of Natural History.)

was oriented vertically (Fig. 4) so that the foot fell directly under the body near the midline. Sacral vertebrate have increased from the pelycosaurian number of two or three (Romer and Price, 1940) to four or more, and the pubic symphysis has changed from the primitive compression joint of pelycosaurs to a tension joint (Hotton, 1986). The digits tend to point forward, subparallel to each other, and are subequal in length. Changes in the front limb are comparable but not as marked as in the hind limb. On the basis of this pattern, gorgonopsians and the smallest members of other groups evolved a gracility of quite a different nature from that of

Fig. 4. Dinocephalian, *Moschops capensis*, AMNH 5552, ×0.07. Posterior aspect, to show the vertical pose of limbs. (From Gregory, 1926, courtesy of the American Museum of Natural History.)

pelycosaurs (see below). The therapsid skull has lost all trace of kinetism, the pterygoid being tightly sutured to the basis cranii as in mammals.

The transition from pelycosaurs to therapsids took place across the boundary between the Early and Late Permian, apparently rather quickly, and many of the changes prefigure the mammalian condition, especially with respect to locomotion. Because the vertebral column is an integral part of the locomotor system and in mammals has become highly differentiated, it might be expected to change in concert with changes in hind limb and pelvis. For example, the closer to the midline that the foot falls during locomotion, the less useful the reptilian habit of lateral flexion of the column as a means of increasing stride length. In most therapsids, however, the column is no less flexible than that of pelycosaurs, and suggestions to the contrary (e.g., King, 1981) are refuted by the evidence of articulated skeletons (Figs. 5–6; see also Hotton, 1986). Morphologically, the therapsid column is little modified from the pelycosaurian pattern; centra remain notochordal, zygapophyses lie close to the midline and are obliquely oriented, and most trunk segments bear ribs. There is

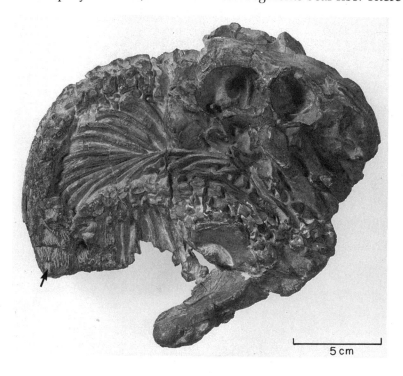

Fig. 5. Stage II dicynodont, *Oudenodon* sp. USNM 335338. Coiled specimen, as found. All of the visible 24 presacral vertebrae are in normal articulation except for a break between presacrals approximately 13 and 14, at the angle indicated by arrow.

little differentiation of thoracic and lumbar regions except in cynodonts such as *Thrinaxodon* and *Diademodon*, in which the posterior dorsal ribs are peculiarly expanded, with overlapping flanges like the uncinate processes of birds. These processes have been postulated to restrict the mobility of the lumbar ribs in relation to the origin of a mammal-like respiratory diaphragm (Brink, 1955, 1958a), but clearly they do not limit the flexibility of the column. A skeleton of *Diademodon* (Fig. 6) looks like a dog asleep on the hearth, and Brink (1958a:Fig. 5a,b) described a skeleton of *Thrinaxodon* coiled up like a sleeping cat.

Remains of pelycosaurs often are found with the dorsal column strongly dorsiflexed, most probably as a consequence of postmortem

Fig. 6. Stage III cynodont, *Diademodon tetragonus*, USNM 23352. Coiled specimen, as found. Fifteen visible presacrals and 14 caudals are in normal articulation. Note platelike dorsal rib (U).

drying and shrinking of dorsal axial musculature. They never have been reported as being coiled like therapsid skeletons, in which the column is gently flexed ventrally and laterally in an attitude it could have assumed during life. From the condition of his specimen of *Thrinaxodon*, Brink (1958a) inferred death and preservation in protected circumstances, perhaps in a burrow.

Brink's inference largely has been vindicated by the discovery of casts of burrows, some of which include skeletons, both coiled and uncoiled, of the dicynodont *Diictodon* (Smith, 1987). The burrows were excavated in a material that now is a uniform mudstone, presumably of floodplain origin, and often are filled with a siltstone that may be continuous with an overlying crevasse-splay deposit. Such complete specimens conform to the form-genus *Daemonhelix* (Smith, 1987), showing a spiral structure of two complete turns in a vertical distance of about 0.75 meters. They end in an ellipsoidal chamber at the bottom, which usually lies just above a layer of calcareous nodules of pedogenic origin. The walls of both passage and chamber are marked by a network of incisions that correspond in size to the digits on the hands of the animal found in the burrows, interspersed with larger gouges that Smith ascribed to the beak of the animal. The specimens of *Diictodon* are about the size of a marmot, and it seems that their burrowing technique was closely comparable.

Simpler burrow casts consisting of a short, straight ramp and ellipsoidal chamber also have been observed by R. M. H. Smith (pers. comm., 1987), but without associated animal remains, nor have burrows yet been found that can be ascribed to other therapsids. However, coiled specimens of *Thrinaxodon* (Brink, 1958a), *Oudenodon* (Fig. 5), and *Diademodon* (Fig. 6) are similar in completeness, attitude, and quality of preservation to the skeletons of *Diictodon* that are found in burrows. Therefore, it seems probable that these animals, like *Diictodon*, died in burrows, although there is no way to tell whether they had dug them or whether they had taken over another animal's burrow. Burrowing may have been quite widespread among the therapsid fauna of the Beaufort, but it need not be expected to have been universal, nor was it. *Lystrosaurus*, for example, is abundant where it occurs at all, and although complete skeletons are common, they are not found coiled. Although the small specimen of *Oudenodon* (Fig. 5) is coiled, complete skeletons of large individuals of that genus are not, nor has coiling been reported in any large therapsid.

As yet, the evidence of burrowing is too new and incomplete to warrant conjecture about its significance in the fauna as a whole, but a few comments are in order. The climate was seasonal during Beaufort deposition, as indicated by well-developed rings in petrified wood, and Bakker (1975) suggested that seasonality was warm-cold because of the high

paleolatitude of the area. For those who would have therapsids endothermic from the beginning of the Upper Permian, warm-cold seasonality
suggests that burrowing was an adaptation for hibernation. However,
direct evidence from sedimentology indicates that the climate was hot
and that seasonality was wet-dry (Keyser, 1966; Stear, 1979). In these
circumstances, if burrowing had anything at all to do with seasonality, it
would permit animals with limited capacity for water conservation to
aestivate in comparative safety. The lack of coiling in *Lystrosaurus* is
consistent with this view, if, as Broom (1932) suggested, *Lystrosaurus* was
more aquatic than other therapsids.

During the transition from pelycosaurs to therapsids at the end of the
Early Permian, changes took place that have been ascribed to the origin of
endothermy. For example, the bones of pelycosaurs show much less evidence of secondary growth than do those of therapsids (de Ricqlés, 1974).
In pelycosaurs, pattern of bone histology is comparable to that of living
ectotherms, in therapsids to that of endotherms (de Ricqlés, 1972, 1974)—
a change that coincides with the acquisition of vertical orientation of the
hind limb. On the grounds that bone histology of therapsids indicates a
rate of turnover of bone tissue that is compatible only with endothermy,
Bakker (1975) asserted that therapsids had become fully endothermic by
the beginning of the Late Permian. Many changes related to endothermy,
however, appeared only much later. In the Early Triassic, the muzzle of
many theriodonts became sculptured so as to suggest rich subepidermal
vascularization and the presence of skin glands, from which are inferred
hair and a rhinarium (Watson, 1931, for baurioids; Brink, 1956, for cynodonts). Fine ridges on the surfaces of the nasal passages of *Thrinaxodon*
(Hopson, 1969) and more advanced cynodonts are supposed to indicate
the former presence of turbinal bones, which in mammals support mucous membrane that warms and moistens inspired air. The parietal foramen persists in primitive therocephalians and the earliest cynodonts of
the Later Permian but disappears more or less gradually and steadily in
later members of both groups. Few of these features have an unequivocal
connection to thermal physiology, and none of them, taken alone, is
definitive of full-blown endothermy. Collectively, however, they suggest
that temperature control became progressively more mammal-like over a
protracted interval of time, for changes in limb pose and bone histology
preceded sculpture of the muzzle, appearance of turbinals, and loss of the
parietal foramen by nearly 20 million years. This pattern conforms to
Olson's proposal of early "incipient homoiothermy" as a common "internal" selective force, to explain the pervasive parallelism in mammal-like
features among the major subdivisions of the Therapsida (Olson, 1944,
1959).

A bony secondary palate is present in *Thrinaxodon* and all later cynodonts and, by analogy with living mammals (Brink, 1956), has been related to endothermy because of its association with putative facial vascularization and turbinal bones. This relationship may be functionally significant in advanced cynodonts and mammals, but it does not explain why dicynodonts, which show few indications of endothermy, evolved a bony secondary palate at least 10 million years before cynodonts and advanced baurioids did so. Nor, for that matter, does endothermy account for the condition of crocodilians, which are ectotherms with an extensive bony palate, or that of birds, which are endotherms without a bony palate. In such disparate animals as mammals and crocodilians, the bony secondary palate provides reinforcement of the muzzle against bending stresses imposed by the action of jaws and dentition (Thomason and Russel, 1986). In Permian cynodonts and baurioids, bony ridges on the medial surfaces of the maxilla also would have served to resist bending stresses and so would prefigure functionally the complete bony palate of later members of their respective lineages. Thus, the advantage of mechanical reinforcement seems to have been the proximal factor in the origin of a bony secondary palate (Thomason and Russel, 1986). Although this argument precludes endothermy as a factor in the origin of a bony secondary palate, it does not foreclose it in later evolution. Once established, the secondary palate would separate air and food passages and so might be expected to allow more scope for endothermy even while it was being modified as endothermy evolved.

Dentitions and Diets

Sphenacodont pelycosaurs provide the most appropriate starting point for a general consideration of therapsid dentitions, because they are the most plausible candidates for therapsid ancestry (cf. Hopson and Barghusen, 1986). In sphenacodonts, marginal teeth are laterally flattened and recurved, reflecting the predatory strategy typical of the group. In primitive forms such as *Haptodus*, two teeth in the anterior third of the maxilla are taller and stouter than the rest, which become gradually smaller both anteriorly and posteriorly. Medial premaxillary teeth are slightly larger than lateral but do not approach the size of the largest maxillary teeth. In more advanced sphenacodonts such as *Sphenacodon* and *Dimetrodon* (Fig. 2), the largest maxillary teeth are abruptly larger than their posterior neighbors, and there are few or no maxillary teeth anterior to them. Medial premaxillary teeth approach the size of the largest maxillary teeth. Few pelycosaurs were as clearly adapted to preying on large animals as the advanced sphenacodonts, but nearly all primitive forms probably required a high-protein diet, feeding on large arthropods or small vertebrates, according to their size.

The only clearly herbivorous pelycosaurs were the caseids, late forms that tended toward large body size but otherwise resembled living herbivorous iguanid lizards in body form, small size of the head, and teeth that were triangular, laterally flattened, and coarsely serrate. Edaphosaurian pelycosaurs are said to be herbivorous because of their bulky bodies and small heads (Romer and Price, 1940), but their dentitions are like those of no living herbivorous lower tetrapods, and this interpretation must be viewed as equivocal. Therapsids were much more successful than pelycosaurs in evolving a variety of herbivorous forms, but because therapsid dentitions are so diversified, they are more effectively dealt with in connection with the groups of which they are characteristic.

The floras that provided food for Permian herbivorous tetrapods are known in a general way, although we do not know what specific food plants were utilized by any animal. Texas deposits that contain fossil pelycosaurs also yield the remains of seed ferns, ferns, conifers, and equisitaleans, and it is probable that less commonly preserved algae, fungi, and mosses also were present in the living flora. The Beaufort rocks of South Africa are notoriously poor in plant fossils (Keyser, 1970), but it is likely that this poverty is more the result of peculiarities of preservation than of biological factors. Assemblages as comprehensive as those of Texas are recorded in deposits below and above the Beaufort Group, which neither begins nor ends with indication of sharp climatic change; it does, however, record a great diversity of herbivorous tetrapods. Therefore, it is probable that therapsids enjoyed the same range of flora that was available to caseids, which would have included plants of tree, shrub, and herbaceous habit (Hotton, 1986).

THERAPSID DIVERSITY

The therapsid fauna of the Beaufort Group of South Africa probably provides a reasonable estimate of global therapsid diversity. It includes more than 80% of known families (Appendix I), and at the present time, it is better documented than any other fauna from an area of comparable extent.

Composition of the Beaufort fauna permits a neat subdivision of time into three stages. Stage I, the oldest, is represented by the *Tapinocephalus* Zone, in which herbivorous and predatory dinocephalians are the most conspicuous elements, small dicynodonts are relatively abundant (Boonstra, 1969), and gorgonopsians and primitive therocephalians are present. In Stage II, coeval with the *Cistecephalus* and *Daptocephalus* Zones, dinocephalians and primitive therocephalians are extinct. Gorgonopsians became the top predators, and dicynodonts and advanced therocephalians evolved a number of specializations that seem to be related to feeding. Primitive cynodonts, which probably arose from undetermined sources within the Therocephalia (Hopson and Barghusen, 1986), appeared before the end of Stage II (Kemp, 1982). Stage III coincides with the *Lystrosaurus* and *Cynognathus* Zones and is marked by extinction of the gorgonopsians, curtailment of dicynodont and therocephalian diversity, and a modest radiation of cynodonts.

Faunal Elements of Stage I

Nowhere is the contrast between robust therapsids and their more gracile predecessors more marked than in dinocephalians (Fig. 3), which also tend to be very large, ranging from the size of a sheep to that of a buffalo. Increase in robustness and size is coordinate with the rotation of the limbs to a parasagittal pose (Fig. 4) and the increase of secondary cortical bone, and so may be related to a change in thermoregulatory strategy. However, the interrelationships among the various factors is complex. Limb pose has at least as much to do with locomotion, and secondary bone with the stress resulting from great weight, as either does with thermoregulation.

Dinocephalians are primitive in their lack of a secondary palate and the relatively small size of the temporal fenestra and dentary but are highly specialized in being large and robust animals. Many also are specialized in the condition of the bones of the skull table and face, which characteristically are thick and dense, almost pachyostotic in appearance. In many cases, reinforcement of the skull table takes the form of large bosses and bumps (Fig. 7), which may have been employed in agonistic behavior such as butting and shoving, and in display (Barghusen, 1975). Dentitions are specialized and quite diverse in Beaufort dinocephalians, which include one family of predators and two of herbivores.

Dentitions of dinocephalians can be traced from the pattern of sphenacodonts through an array of primitive therapsids best known from the Late Permian of Russia, which are also represented in Texas (Olson, 1962) and in South Africa (Hopson and Barghusen, 1986). So far as can be determined, these animals are still rather pelycosaur-like in body form, but they are the oldest synapsids in which the front of the maxilla is the site of a single enlarged tooth, commonly styled "caniniform." In the brithopodids, predatory members of this assemblage, the sphenacodont trend toward enlargement of premaxillary teeth is emphasized, foreshadowing true dinocephalians of South Africa such as *Anteosaurus* (Fig. 8). The anteosaurs are large, stout-bodied, short-tailed predators in which the short face bears long, conical premaxillary teeth that interlocked with their lower counterparts, apparently functioning to tear out large chunks of flesh. The dentition, together with large body size, indicate that *Ante-*

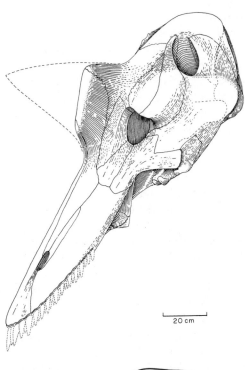

Fig. 7. Skull of dinocephalian, *Struthiocephalus kitchingi*, BPI 284. *Struthiocephalus* is comparable to *Moschops* (Fig. 3) in body bulk and pose, but it carried its much longer head at a steeper angle. (From Brink, 1957b, courtesy of the Bernard Price Institute for Palaeontological Research.)

20 cm

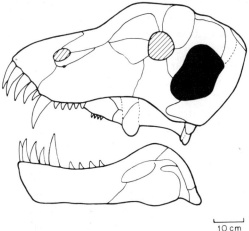

Fig. 8. Skull of predaceous dinocephalian, *Anteosaurus* sp. (From Boonstra, 1969, courtesy of the South African Museum.)

10 cm

osaurus and its kind were adapted to predation on large terrestrial contemporaries.

The prey of anteosaurs included dinocephalians of herbivorous habit such as the tapinocephalians *Moschops* and *Struthiocephalus*. Like other dinocephalians, these animals are large, stout-bodied, and short-tailed, and the anterior teeth are emphasized at the expense of the posterior.

Individual teeth are characteristically spatulate, with a labial ridge of resistant enamel surmounting a lingual shoulder on which dentine is commonly exposed by wear. In *Moschops*, the face is short (Fig. 3), and the adult must have carried its head habitually at least 1 m above the substrate, which suggests that the animal browsed on leafy vegetation of shrubs or small trees. *Moschops*, in fact, is big enough that it could have bent over sizable trees to get at especially enticing tidbits such as the fruit of seed ferns. *Struthiocephalus* (Fig. 7), though comparable to *Moschops* in size, has a long face, with large anterior teeth at the front of an elongate, spatulate rostrum. Brink (1957b), noting that the head habitually was carried with the rostrum pointing downward, argued persuasively that *Struthiocephalus* fed in water, seizing plants growing on the bottom and then raising its head to drain off the water and mud.

Parenthetically, one of the few significant tetrapod elements of the Beaufort fauna that is not therapsid is the anapsid Order Parieasauria. Parieasaurs are large herbivorous reptiles, which must have been potential competition for contemporaneous herbivorous dinocephalians and potential prey for anteosaurs. Like dinocephalians, parieasaurs are large and robust, with stout limbs and a short tail. The teeth are laterally flattened, triangular, and serrate—very similar to the teeth of living iguanid lizards except for size; with their blunt muzzle and flat head, parieasaurs must have borne some resemblance to the marine iguana, *Amblyrhynchus*. They outlived both herbivorous and predatory dinocephalians, surviving to the end of the Permian (*Daptocephalus* Zone).

The dicynodonts, gorgonopsians, and primitive therocephalians of Stage I are all much smaller than contemporaneous dinocephalians; none is larger than a beaver or coyote, and many dicynodonts are smaller than woodchucks. Nor do any have reinforcement of the skull table comparable to that of dinocephalians, although nasal bosses are variously developed in dicynodonts (Fig. 9). Dicynodonts and therocephalians (Fig. 10) exhibit the characteristically robust body form of therapsids, but gorgonopsians (Fig. 11) early evolved a gracility that is more reminiscent of mammals than of lizards. The tail remains relatively short, but the limbs are long and slender, and their habitual vertical pose gives gorgonopsians a rather doglike appearance, implying, perhaps, a vagility approaching that of dogs. More than any other therapsid, gorgonopsians look capable of a mode of life that involved traveling significant distances in the course of foraging.

Dicynodonts: Stages I and II

In the earliest dicynodonts (Fig. 9), which appear early in Stage I, the skull features diagnostic of the group are already manifest. The face is

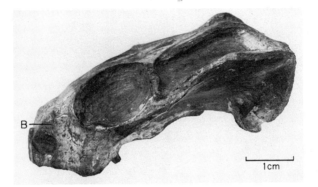

Fig. 9. Skull of small dicynodont of Stage I, cf. *Robertia* Boonstra (1948), USNM 22944, lateral aspect. Nasal boss (B) developed above naris.

short, a bony secondary palate is present, and the marginal dentition is largely replaced by horn-covered cutting and crushing surfaces of bone. A maxillary tusk is present in some genera, absent in others, and in such genera as *Diictodon* and *Kingoria,* present in some individuals but not in others. Many genera are slightly archaic in the shortness of the bony secondary palate and in the retention of a few small teeth on the palatal component of maxilla (Fig. 9) or premaxilla, and on the dentary. The sliding jaw articulation peculiar to dicynodonts (Crompton and Hotton, 1967) is fully developed, and the insertion of temporal musculature indicates great mechanical advantage but slow motion at the front of the jaw. Collectively, these characters indicate that dicynodonts were herbivorous, an indication consistent with their abundance in the fauna.

One adaptive type, represented by the genus *Diictodon* (Fig. 12), is clearly identifiable. In this animal, the muzzle is habitually directed downward, indicating that *Diictodon* tended to forage close to the substrate. When the jaw is fully closed and retracted, the tusks clear the symphysis, and wear facets on the tusks suggest that they were used in grubbing, perhaps for roots or tubers (Hotton, 1986). The terminal phalanges are clawlike, and their effectiveness as digging tools is confirmed by the evidence of burrows (Smith, 1987), which is consistent with the grubbing habit postulated for *Diictodon* and dicynodonts of similar structure. It is likely, of course, that *Diictodon* also fed on low-growing herbaceous plants. Most other dicynodonts of Stage I are not sufficiently well understood to warrant conjecture as to an adaptive type, although some appear to approach *Diictodon* in general structure.

Dicynodonts survived until the end of the Triassic, but their period of greatest abundance and diversity is Stage II. *Diictodon* became very abundant after dinocephalians became extinct, in company with a variety of other small dicynodonts that retain the archaic trait of small, nontusklike

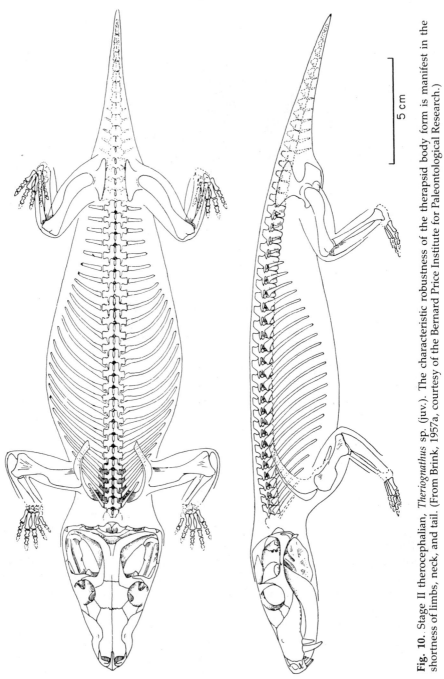

Fig. 10. Stage II therocephalian, *Theriognathus* sp. (juv.). The characteristic robustness of the therapsid body form is manifest in the shortness of limbs, neck, and tail. (From Brink, 1957a, courtesy of the Bernard Price Institute for Paleontological Research.)

5 cm

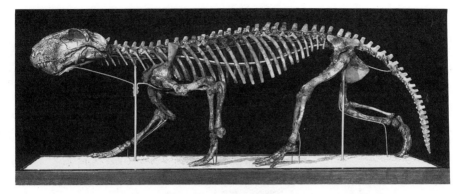

Fig. 11. Stage II gorgonopsian, *Lycaenops ornatus*, AMNH 2240, ×0.07. Shown here are the characteristic dentition, and the length and slenderness of the limbs, which are also manifest in Stage I gorgonopsians. (From Colbert, 1948, courtesy of the American Museum of Natural History.)

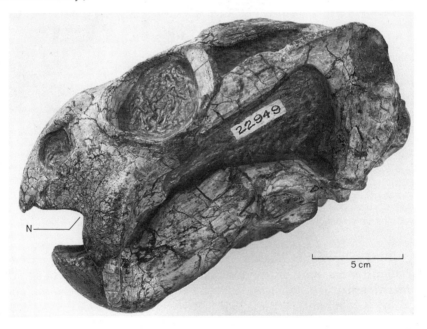

Fig. 12. Skull of stage II dicynodont, *Diictodon* sp., USNM 22949, ×0.57. On the basis of the distinctive notch (N) in the maxilliary margin, this genus is also identifiable in deposits of Stage I, in which it shows a comparable attitude of the head and development of tusks.

teeth. About this time, dicynodonts also started producing genera that rivaled the extinct tapinocephalians in size and probably replaced them as large primary consumers. The genus *Dicynodon* first appeared near the end of *Cistecephalus* Zone time, when it was represented by species of beaver size, but soon produced giants such as the cow-sized *D. leoniceps*

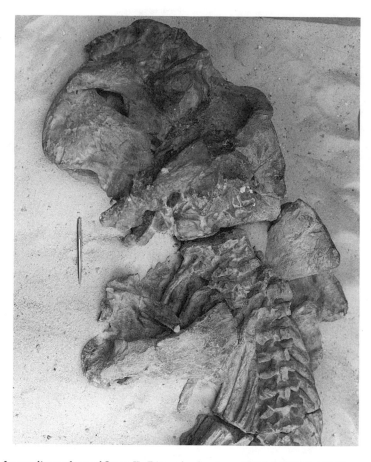

Fig. 13. Large dicynodont of Stage II, *Dicynodon leoniceps*, USNM 299746, ×0.06.

(Fig. 13). The appearance of taxa of large body size cannot be regarded as a secular trend among dicynodonts, however, because forms of squirrel to marmot size (*Diictodon* and others) continued to thrive in all their diversity (Cluver and Hotton, 1981; Cluver and King, 1983).

The diverse morphology of the dicynodonts of Stage II lends itself readily to interpretation in terms of resource partitioning (Hotton, 1986). As noted above, *Diictodon* (Fig. 12) appears to have been specialized for grubbing. In *Oudenodon* (Fig. 14), by contrast, the skull is flat, tusks are absent, and the habitual attitude of the head was horizontal. Head attitude, certain peculiarities of jaw motion occasioned by the lack of tusks, and hooflike, nondigging terminal phalanges suggest that *Oudenodon* browsed on foliage at some distance above the substrate. *Dicynodon*

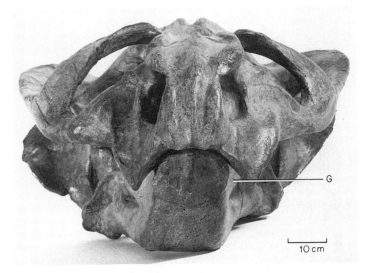

Fig. 14. Skull of large dicynodont of Stage II, *Oudenodon* sp., USNM 22814, anterior aspect. The position of the skull shown here is close to the horizontal attitude at which the flattened head was carried in life. The gap (G) between the side of the jaw and the tuskless caninform process allowed a transverse component of jaw motion, in contrast to the condition of *Diictodon* (Fig. 12) and *Dicynodon* (Fig. 13), in which the large tusks prevented transverse motion.

seems to have been more nearly a generalist. Although its tusks are long, projecting below the dentary symphysis, and show wear patterns consistent with grubbing, the cutting margins of maxilla and premaxilla seem better adapted for shearing foliage than for cutting roots. Probably the most specialized dicynodont of the Lower Permian of South Africa is *Cistecephalus*, ubiquitous at some levels, whose fossorial habit is indicated by its similarity in postcranial detail to the golden mole (Cluver, 1978). The presence of a similar form, *Kawingasaurus*, in a fauna of comparable age in Tanzania (Cox, 1972) reflects the success of a fossorial habit among Late Permian dicynodonts. The genus *Kingoria* is distinctive in its long rostrum, small mouth, and lack of indication of horn on the palate, features that suggest that its forage was softer and came in smaller packages than that of other dicynodonts (Hotton, 1986). It is possible that *Kingoria*, unlike other dicynodonts, took a significant amount of invertebrate prey, perhaps by foraging in forest litter.

Gorgonopsians: Stages I and II

In gorgonopsians, the limbs are advanced in their length and slenderness (Fig. 11) and vertical pose (Fig. 15), and functionally, the feet are as

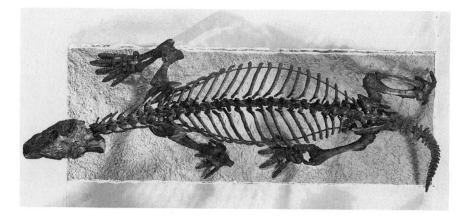

Fig. 15. Stage II gorgonopsian, *Lycaenops ornatus*, AMNH 2240, ×0.07. Dorsal aspect, to show the vertical pose of limbs. (From Colbert, 1948, courtesy of the American Museum of Natural History.)

progressive as the rest of the limb. Much of gorgonopsian structure, however, is conservative. The foot has arrived at its advanced condition not by dropping phalanges to produce the derived phalangeal formula of 2-3-3-3-3 of most therapsids, but by reducing the length of certain phalanges. In consequence, the gorgonopsian phalangeal formula, though variable, remains closer to the pelycosaurian pattern of 2-3-4-5-4. The skull lacks any trace of a secondary palate, and the temporal opening and dentary remain small. The bones of the table are not thickened as in dinocephalians, nor is the skull flattened as in therocephalians. The premaxillary teeth are prominent but a little smaller than the single large caniniform tooth at the front of the maxilla, behind which the maxillary teeth are reduced (Fig. 11). As in sphenacodonts, maxillary teeth are laterally flattened, slightly recurved, and they bear small serrations on the cutting margins.

Gorgonopsians continued to refine their initial specialization as vagile foragers until they died out at the end of Stage II but otherwise did not undergo much diversification. They were small to moderate in size during Stage I and continued to produce taxa of moderate body size during Stage II (Colbert, 1948); however, once dinocephalians had become extinct, some gorgonopsians became large, approaching the size of black bears. In these animals there is an enhanced tendency toward emphasis of the front teeth at the expense of the back, as in anteosaurs. In a few gorgonopsians, the caniniform tooth becomes a serrate saber, with concomitant modification of the dentary for support of the saber tooth and of the jaw joint for increased gape (Parrington, 1955). Ecologically, the large gorgonopsians replaced extinct anteosaurian predators, exploiting the

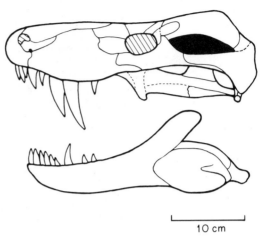

Fig. 16. Skull of stage I therocephalian, *Trochosaurus* sp. (From Boonstra, 1969, courtesy of the South African Museum.)

10 cm

large dicynodonts that had replaced tapinocephalians at the base of the food pyramid.

Therocephalians: Stage I

The therocephalians of Stage I (Fig. 16), like contemporary gorgonopsians, are predators of small to moderate size that are primitive in the small size of temporal fenestra and dentary and in their lack of a secondary palate. They are more primitive than gorgonopsians in their robust body form and in the completeness of their dentitions. Maxillary teeth are laterally flattened and slightly recurved, whereas the premaxillary teeth are conical and smaller than the anterior maxillary teeth. Stage I therocephalians also differ from contemporary gorgonopsians in the flattening of the skull, in a peculiar palatal fenestra at the juncture between pterygoid, ectopterygoid, and palatine, and in many other details of postcranial and skull structure.

Therocephalians: Stage II

Unlike gorgonopsians, the therocephalians of Stage II did not produce very large forms after the dinocephalians became extinct. None is larger than a large dog, and many Stage II therocephalians are about the size of ferrets or skunks. Instead, therocephalians radiated into at least three different morphological types, as represented by primitive baurioids, whaitsiids, and euchambersiids.

The most generalized therocephalians of Stage II are primitive baurioids (scaloposaurs of some authors—e.g., Haughton and Brink, 1954), which lie at the lower end of the size range, lack a secondary palate, and

Fig. 17. Skull of stage II bauriamorph, genus cf. *Ictidosuchoides* sp., USNM 336444.

have a complete but nearly uniform maxillary tooth row (Fig. 17). A single anterior maxillary tooth may or may not be larger than the others. The skull is flat and low, with elongate rostrum and flared temporal region, and with the uniform dentition gives the impression of the skull of a small crocodilian, although the general habitus is not at all crocodilian. The body is elongate and the tail short, and some of these animals look as though they had been quite active on land, the limbs being somewhat elongate and slender. The small size of such forms suggests that their prey consisted primarily of large arthropods and occasional very small tetrapods. A few genera, such as *Tetracynodon darti* (Sigogneau, 1963), in which the face is elongate, the dentition homodont, and the parietal foramen lacking, should be examined carefully for evidence of aquatic habit, for such animals may well have fed on fish or aquatic invertebrates.

The most numerous therocephalians were the whaitsiids, which included species as small as domestic cats as well as the largest of the group. The dentition of whaitsiids is distinctive, with conical premaxillary teeth and a greatly enlarged anterior maxillary (caniniform) tooth, behind which there are no other teeth (Fig. 18). The post-caniniform margins of maxillae (and dentaries) form robust surfaces that may have been covered with horn in life (Fig. 19). The emphasis on premaxillary and anterior maxillary teeth recalls patterns of anteosaurs and gorgonopsians, but with the putatively horn-covered posterior margins of the jaws, the masticatory apparatus is like nothing else on earth, then or now. Whaitsiid habits probably included feeding on sizable prey; Brink (1958b) reported the skull of a dicynodont, cf. *Diictodon*, stuck in the pharyngeal region of a large individual of *Notosollasia* (= *Whaitsia* [Kitching, 1977] = *Theriognathus* [Mendrez, 1974]).

In therocephalians that lack a bony secondary palate, the palatine surface of the rostrum is rather deep and narrow, and in primitive baurioids and whaitsiids, a crista choanalis (Mendrez, 1975) runs the length of the

Fig. 18. Skull of therocephalian, *Theriognathus* sp. (juv.), USNM 23355. Right lateral aspect. Note the edentulous ridge (R) behind the first maxillary tooth. See also Figure 19.

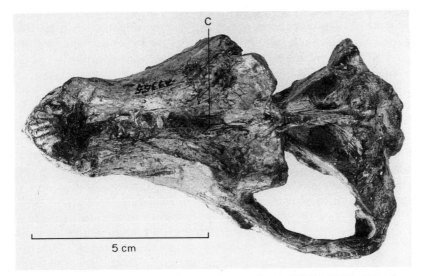

Fig. 19. Skull of therocephalian, *Theriognathus* sp. (juv.), USNM 23355. Palatal aspect. Position (C) of the crista choanalis that suggests attachment of soft secondary palate.

medial surface of the maxilla (Fig. 19, C). The crista would have reinforced the snout against bending stresses, as in the Late Permian cynodont *Procynosuchus* (Thomason and Russel, 1986), and also could have supported a membranous or muscular secondary palate that spanned the narrow space between the maxillae.

The genus *Euchambersia* embodies an extreme specialization, encountered in no other synapsid lineage. The face of this small, rare animal is rather short by comparison with most therocephalians, and in front of the orbit it is occupied by a large, deep pit that communicates anteriorly, via a short channel, with the tooth-bearing margin of the maxilla. The channel ends near the base of the caniniform tooth, which is not only elongate but also bears grooves on its labial and lingual surfaces. Collectively, these features suggest that *Euchambersia* had a venomous bite, the venom being produced by a gland lodged in the facial pit and administered via the grooved caniniform fang. The peculiar structure of *Euchambersia* originally was described by Broom (1932) from a damaged skull and confirmed by Kitching (1977) from a better specimen. So far, these skulls, which lack lower jaws, constitute the only record of the genus, and further details of the venom-delivering mechanism, if that is in fact what it is, have still to be discovered.

Cynodonts: Stages II and III

The first cynodonts (Fig. 20) appear near the end of Stage II as predators not much larger than a ferret but more robustly built; in these respects, they are much like their contemporaries among the primitive baurioid therocephalians. They differ in the greater complexity of posterior maxillary and dentary teeth (Mendrez, 1966), in the lack of a "therocephalian" fenestra in the palate, and in a bony secondary palate longer than in most primitive baurioids. In Stage III cynodonts, these trends continue. The bony secondary palate is yet more extensive, and the posterior cheek teeth become more molariform with the enhancement of crushing as well as cutting and puncturing surfaces on the crowns. In the genus *Thrinaxodon*, of the *Lystrosaurus* Zone, the dentary has become larger at the expense of the postdentary bones, and the lateral walls of the braincase are ossified more extensively.

In the latter half of Stage III (*Cynognathus* Zone), these trends are accompanied by specializations reflecting an adaptive radiation within

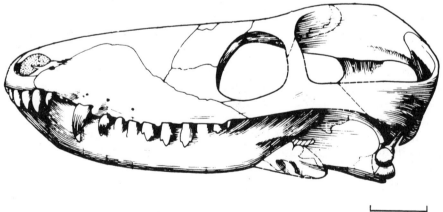

3 cm

Fig. 20. Skull of stage II cynodont, *Procynosuchus* sp. BPI 16. (From Mendrez, 1972, courtesy of the Bernard Price Institute for Paleontological Research.)

the cynodonts. The gomphodont genus *Diademodon*, for example, appears to have been an omnivore and was very abundant at the time. In some respects, its structure (Fig. 21) is reminiscent of that of pigs or bears. The caniniform tooth is large and sharp but ellipsoidal in cross section, rather than laterally flattened and serrate as in gorgonopsians; behind it, the maxillary teeth become progressively broader and more flat-crowned toward the back of the row. A diastema between the caniniform and cheek teeth enlarges with increasing individual size. The temporal opening of *Diademodon* is large, and the flared zygoma bears a prominent masseteric process on its anteroventral surface. The masseteric process resembles that of a pig and indicates a prominent masseter muscle that provided for anteroposterior motion of the jaw during the bite.

At the same time, the presence of a masseter reflects the establishment of a mammal-like muscular sling (masseter vs. internal pterygoid) for the dentary, which helps to reduce the force exerted at the joint (Crompton and Parker, 1978). It is this function of the muscular sling that permits reduction in the size of the postdentary bones before the formation of a mammalian jaw joint between the dentary and squamosal. In *Diademodon*, the cheek teeth are single-rooted, but they erupt at the back of the row and drop out at the front, like the teeth of proboscidean or sirenian mammals. In this respect their replacement is "precociously" mammal-like.

Cynognathus, a contemporary of *Diademodon*, was a long-snouted predator in which the cheek teeth were more clearly adapted for cutting than for crushing. There is no diastema, and each tooth was replaced many times in situ in standard reptilian fashion. The dentary is large in both *Cynognathus* and *Diademodon*, and closely approaches the squamosal posteriorly.

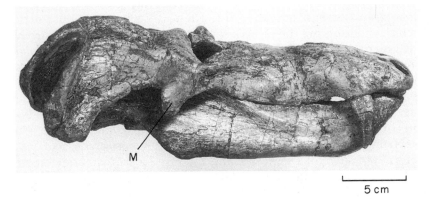

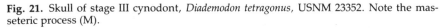
5 cm

Fig. 21. Skull of stage III cynodont, *Diademodon tetragonus*, USNM 23352. Note the masseteric process (M).

5 cm

Fig. 22. Skull of stage III bauriamorph, *Bauria* sp., USNM 23331. Right lateral aspect.

Diademodon and *Cynognathus* are about the size of a large hog and a black bear, respectively; however, small predators about the size of *Thrinaxodon* also are present in fauna of the *Cynognathus* Zone. Some of these are gomphodont cynodonts, allied to *Diademodon*. One is not a cynodont at all but, instead, a surviving baurioid therocephalian. In this animal, *Bauria* (Fig. 22), the premaxillary and first maxillary teeth are sharp and subequal in height, but the flat-crowned cheek teeth are convergent toward the gomphodont pattern of *Diademodon*.

Other Faunal Elements: Stage III

The *Lystrosaurus* Zone differs faunistically from the earlier Beaufort in two respects. First, the dicynodont primary consumers are less diverse taxonomically, having been reduced from at least 10 genera to not more than three; however, they are no less diverse in size, nor less numerous. *Lystrosaurus*, clearly a derivative of *Dicynodon*, is abundant, but the only other dicynodonts are the rare *Myosaurus* (distantly related to *Cistecephalus* [Cluver and King, 1983]) and perhaps *Dicynodon leoniceps*. *Myosaurus* is small and *D. leoniceps* is large; *Lystrosaurus* is represented by species of moderate to large size. Second, gorgonopsians are extinct and large therapsid predators are unknown. The top predator of the *Lystrosaurus* Zone is *Chasmatosaurus*, an aquatic archosaur of rather crocodilian body form. In the succeeding *Cynognathus* Zone, dicynodonts are less abundant, being represented by the fairly common *Kannemeyeria*, like *Lystrosaurus* a derivative of *Dicynodon*, and the rare *Kombuisia*, a derivative of the *Daptocephalus* Zone genus *Kingoria* (Hotton, 1986). *Kannemeyeria* is large and *Kombuisia* is small. The decline of dicynodonts as primary consumers may have resulted, in part, from competition from gomphodont cynodonts, animals that fall between *Kombuisia* and *Kannemeyeria* in body size. *Cynognathus* evidently was a top predator, but not the only one, for it was in competition with a still larger terrestrial archosaur, *Erythrosuchus*, which approached the later rauisuchids in predatory adaptation. With Stage III, marked by the appearance of large thecodont predators, the constriction of dicynodonts as primary consumers, and the extinction of noncynodont therapsid predators, the period of greatest therapsid diversity came to an end.

SUMMARY AND CONCLUSIONS

The history of synapsid reptiles is unique in its continuity from the Early Pennsylvanian to the Early Jurassic and in documenting the transition between two major adaptive grades of vertebrates in detail. It includes the Late Permian history of therapsids, which, in turn, is unique as a detailed record of the first ecologically integrated terrestrial fauna in which tetrapods were primary consumers. Therapsid history begins with two evolutionary developments that are new for tetrapods—one involving locomotion, and the other the effective utilization of plant material as food.

The history of therapsids during the Late Permian is divided into three stages by faunal succession. In Stage I, large primary consumers are represented by titanosuchid and tapinocephalid herbivorous dinocephalians and by the anapsid parieasaurs, and small primary consumers by dicynodonts. The top predators are anteosaurid dinocephalians, which are large enough to be able to feed on anything that moves but are adapted primarily to preying on animals of a size comparable to their own. Smaller predators are gorgonopsians and therocephalians, of a size capable of having exploited the variety of dicynodonts available as small prey, and doubtless the young of dinocephalians and parieasaurs. Stage II is introduced by the extinction of dinocephalians from unknown causes. Apparently in response to this extinction, taxa of large body size appeared among both dicynodonts and gorgonopsians. Dicynodonts replaced tapinocephalids as large primary consumers while continuing to produce taxa that were small primary consumers; their morphological diversity suggests details of resource partitioning. Gorgonopsians replaced anteosaurs as top predators and, at the same time, maintained their role as small predators as well; however, they were less diverse than dicynodonts. Therocephalians reflect the greatest diversity among predators during Stage II; they produced the smallest therapsid predators, some of which doubtless were insectivorous, whereas a few may have been fish eaters. Although therocephalians showed some increase in size after the dinocephalians died out, they probably did not challenge the gorgonopsians as top predators. Stage III marks the beginning of the decline of therapsids as dominant terrestrial tetrapods. Gorgonopsians become extinct, and dicynodonts and therocephalians become much less diverse, although they continued in considerable numbers for a time. Cynodonts underwent a secondary adaptive radiation, during which they may have partially replaced dicynodonts as primary consumers and produced large, as well as small, predators. However, in the midst of their success, their ultimate decline was foreshadowed by the appearance of large thecodonts as top predators in the faunas of Stage III.

Acknowledgments I thank the American Museum of Natural History (AMNH), New York; the Bernard Price Institute for Palaeontological Research (BPI), University of Witwatersrand, Johannesburg; and the South African Museum for permission to use published illustrations, here reproduced as Figures 3, 4, 7, 8, 10, 11, 15, 16, and 20. Thanks are also due to J. W. Kitching and A. S. Brink for their role in acquainting me with the wonders of the Beaufort Group and its fauna. I am indebted to H.-D. Sues for useful suggestions regarding the origin of the secondary palate. I am also indebted, though not especially grateful, to anonymous reviewers for their trenchant criticism of an earlier draft. Support for the collection and preparation of South African fossils in the National Museum of Natural History (USNM) was provided by National Science Foundation grant numbers G14707 and GB1647.

APPENDIX I. Relationships among the Synapsids

Pelycosauria after Reisz (1986); Therapsida after Hopson and Barghusen (1986). All families of pelycosaurs are present in Texas, and therapsids marked by B are well represented in the Beaufort Group of South Africa.
Subclass Synapsida
    Order Pelycosauria
        Unnamed
           Caseidae
           Eothyrididae
        Varanopsidae
        Ophiacodontidae
        Edaphosauridae
        Sphenacodontidae
    Order Therapsida
        "Biarmosuchia" (2 families, 1 B)
        Unnamed

Eotitanosuchia (1 family)
Unnamed
  Dinocephalia
    Anteosauria (1 family B)
    Tapinocephalia (3 families, 2 B)
  Anomodontia
    Venjukovioidea (1 family?)
    Dicynodontia (9 families, Cluver and King, 1983, all B)
  Theriodontia
    Gorgonopsia (1 family: 16 genera, Sigogneau, 1970, B)
    Eutheriodontia
      Therocephalia
        Pristerosauria (1 family, B)
        Eutherocephalia
          Hoffmeyeriidae, B
          Unnamed
            Euchambersiidae, B
            Whaitsiidae, B
            Baurioidea (6 families, all B)
      Cynodontia
        Dviniidae
        Unnamed
          Procynosuchidae, B
          Unnamed
            Galesauridae, B
            Unnamed
              Thrinaxodontidae, B
              Unnamed
                Unnamed
                  Cynognathia (2 families, both B)
                  Chiniquodontoidea (2 families)
                Unnamed
                  Ictidosauria (1 family?, B)
                  Mammalia

Literature Cited

Bakker, R. T. 1975. Dinosaur renaissance. Sci. Am., 232(4):58–78.
Barghusen, H. R. 1975. A review of fighting adaptations in dinocephalians (Reptilia, Therapsida). Paleobiology, 1:295–311.
Boonstra, L. D. 1948. On the anomodont reptiles from the *Tapinocephalus*-Zone of the Karroo System. Pp. 57–64 *in* Dutoit, A. L. (ed.), *Robert Broom Commemorative Volume*. Spec. Publ. R. Soc. S. Afr. Capetown: R. Soc. S. Afr.

Boonstra, L. D. 1969. The fauna of the *Tapinocephalus* Zone (Beaufort Beds of the Karroo). Ann. So. Afr. Mus., 56:1–73.

Brink, A. S. 1955. A study of the skeleton of *Diademodon*. Palaeontol. Afr., 3:3–39.

Brink, A. S. 1956. Speculations on some advanced mammalian characteristics in the higher mammal-like reptiles. Palaeontol. Afr., 4:77–96.

Brink, A. S. 1957a. On the skeleton of *Aneugomphius ictidops* Broom and Robinson. Palaeontol. Afr., 5:29–37.

Brink, A. S. 1957b. *Struthiocephalus* kitchingi sp. nov. Palaeontol. Afr., 5:39–56.

Brink, A. S. 1958a. Notes on a new skeleton of *Thrinaxodon liorhinus*. Palaeontol. Afr., 6:15–22.

Brink, A. S. 1958b. Notes on some whaitsiids and moschorhinids. Palaeontol. Afr., 6:23–49.

Brinkman, D., and D. A. Eberth. 1983. The interrelationships of pelycosaurs. Breviora, 473:1–35.

Broom, R. 1932. *The Mammal-like Reptiles of South Africa and the Origin of Mammals.* London: H. F. and G. Witherby.

Chudinov, P. K. 1968. Structure of the integuments of theromorphs. Doklady of the Academy of Sciences, USSR, Earth Science Section, 179:226–229. [Translation published by American Geological Institute from: O stroyenii kozhnykh pokrovov sveroobraznykh presmykayushchihsya. Dok. Akad. Nauk SSSR, 179(1):202–210.]

Cluver, M. A. 1978. The skeleton of the mammal-like reptile *Cistecephalus* with evidence for a fossorial mode of life. Ann. So. Afr. Mus., 76(5):213–246.

Cluver, M. A., and N. Hotton III. 1981. The genera *Dicynodon* and *Diictodon* and their bearing on the classification of the Dicynodontia (Reptilia, Therapsida). Ann. So. Afr. Mus., 83(6):99–146.

Cluver, M. A., and G. M. King. 1983. A reassessment of the relationships of Permian Dicynodontia (Reptilia, Therapsida) and a new classification of dicynodonts. Ann. So. Afr. Mus., 91(3):195–273.

Colbert, E. H. 1948. The mammal-like reptile *Lycaenops*. Bull. Am. Mus. Nat. Hist., 89:353–404.

Cope, E. D. 1878. The theromorphous Reptilia. Am. Nat., 12:829–830.

Cox, C. B. 1972. A new digging dicynodont from the Upper Permian of Tanzania. Pp. 173–179 in Joysey, K. A., and T. S. Kemp (eds.), *Studies in Vertebrate Evolution.* Edinburgh: Oliver and Boyd.

Crompton, A. W., and N. Hotton III. 1967. Functional morphology of the masticatory apparatus of two dicynodonts (Reptilia, Therapsida). Postilla, 109:1–51.

Crompton, A. W., and P. Parker. 1978. Evolution of the mammalian masticatory apparatus. Am. Sci., 66(2):192–201.

Dantzler, W. H. 1976. Renal function (with special emphasis on nitrogen excretion). Chap. 9, pp. 447–503 in Gans, C., and W. R. Dawson (eds.), *Biology of the Reptilia.* Vol. 5: *Physiology A.* London: Academic Press.

de Ricqlés, A. 1972. Recherches paléohistologiques sur les os longs des tétrapodes. III. Titanosuchiens, dinocephales et dicynodontes. Ann. Paleontol. Verteb., 58(1):17–60.

de Ricqlés, A. 1974. Recherches paléohistologiques sur les os longs des tétrapodes. IV. Eothériodontes et pelycosaures. Ann. Paleontol. Vertebr., 60(1):1–30.

Gregory, W. K. 1926. The skeleton of *Moschops capensis,* a dinocephalian from the Permian of South Africa. Bull. Am. Mus. Nat. Hist., 56(3):179–251.

Haughton, S. H., and A. S. Brink. 1954. A bibliographical list of Reptilia from the Karroo beds of Africa. Palaeontol. Afr., 2:2–187.

Hopson, J. A. 1969. The origin and adaptive radiation of mammal-like reptiles and nontherian mammals. Ann. N. Y. Acad. Sci., 167(1):199–216.

Hopson, J. A., and H. R. Barghusen. 1986. An analysis of therapsid relationships. Pp. 83–106 in Hotton, N., III, P. D. MacLean, J. J. Roth, and E. C. Roth (eds.), *The Ecology and Biology of Mammal-like Reptiles.* Washington, D.C.: Smithsonian Inst. Press.

Hotton, N., III. 1974. A new dicynodont (Reptilia, Therapsida) from the *Cynognathus* Zone deposits of South Africa. Ann. So. Afr. Mus., 64:157–165.

Hotton, N., III. 1986. Dicynodonts and their role as primary consumers. Pp. 71–82 in Hotton, N., III, P. D. MacLean, J. J. Roth, and E. C. Roth (eds.), *The Ecology and Biology of Mammal-like Reptiles.* Washington, D.C.: Smithsonian Inst. Press.

Kemp, T. S. 1982. *Mammal-like Reptiles and the Origin of Mammals.* London: Academic Press.

Keyser, A. W. 1966. Some indications of arid climate during the deposition of the Beaufort Series. Ann. Geol. Surv. S. Afr., 5:77–79.

Keyser, A. W. 1970. Some ecological aspects of the *Cistecephalus* Zone of the Beaufort Series of South Africa. Int. Union Geol. Sci. Comm. Stratigr., Subcomm. Gondwana Stratigr. and Palaeontol. Proc. Pap. Gondwana Symp., 2:687–689.

King, G. M. 1981. The functional anatomy of a Permian dicynodont. Philos. Trans. R. Soc., Ser. B., 291(1050):243–322.

Kitching, J. W. 1977. The distribution of the Karroo vertebrate fauna. Bernard Price Inst. Palaeontol. Res. Mem., 1:1–131.

Mendrez, C. H. 1966. Sur quelques critères de distinction entre thérocéphales et cynodontes. Problèmes actuels de paléont. (évol. des vertébrés). Colloq. Int. C.N.R.S., 163:431–437.

Mendrez, C. H. 1972. Revision du genre *Protocynodon* Broom 1949 et discussion de sa position taxonomique. Palaeontol. Afr., 14:19–50.

Mendrez, C. H. 1974. Étude du crâne d'un jeune spécimen de *Moschorhynus kitchingi* Broom 1920 (?*Tigrisuchus simus* Owen, 1876), Therocephalia Pristerosauria Moschorhinidae d'Afrique australe). (Remarques sur les Moschorhinidae et les Whaitsiidae). Ann. S. Afr. Mus., 64:75–115.

Mendrez, C. H. 1975. Principales variations du palais chez les thérocéphales sud-africains (Pristerosauria et Scaloposauria) au cours du Permien Supérieur et du Trias Inférieur. Problèmes actuels de paléontol. (évol. des vertébrés). Colloq. Int. C.N.R.S., 218:379–408.

Olson, E. C. 1944. Origin of mammals based on the cranial morphology of therapsid suborders. Geol. Soc. Am., Spec. Pap., 55:1–136.

Olson, E. C. 1959. The evolution of mammalian characters. Evolution, 13:344–353.

Olson, E. C. 1962. Late Permian terrestrial vertebrates, U.S.A. and U.S.S.R. Trans. Am. Philos. Soc., 52(2):3–224.

Olson, E. C. 1979. Aspects of the biology of *Trimerorhachis* (Amphibia: Temnospondyli). J. Paleontol., 53(1):1–17.

Packard, G. C., G. R. Tracy, and J. J. Roth. 1977. The physiological ecology of reptilian eggs and embryos, and the evolution of viviparity within the Class Reptilia. Biol. Rev. Cambridge Philos. Soc., 52:71–105.

Parrington, F. R. 1955. On the cranial anatomy of some gorgonopsids and the synapsid middle ear. Proc. Zool. Soc. London, 125(1):1–40.

Peabody, F. E. 1952. *Petrolacosaurus kansensis* Lane, a Pennsylvanian reptile from Kansas. Univ. Kansas Paleontol. Contrib. Vertebr., 1:1–41.

Quay, W. B. 1979. The parietal eye-pineal complex. Chap. 5, pp. 245–406 in Gans, C., R. G. Northcutt, and P. S. Ulinski (eds.), *Biology of the Reptilia.* Vol. 9: *Neurology A.* London: Academic Press.

Reed, C. A. 1960. Polyphyletic or monophyletic ancestry of mammals, or: what is a class? Evolution, 14(3):314–322.

Reisz, R. R. 1972. Pelycosaurian reptiles from the Middle Pennsylvanian of North America. Bull. Mus. Comp. Zool. Harvard, 144:27–62.

Reisz, R. R. 1977. *Petrolacosaurus*, the oldest known diapsid reptile. Science, 196:1091–1093.

Reisz, R. R. 1980. The Pelycosauria: a review of phylogenetic relationships. Pp. 553–592 *in* Panchen, A. L., (ed.), *The Terrestrial Environment and the Origin of Land Vertebrates*. Syst. Assoc. Sp. Vol. 15. London: Academic Press.

Reisz, R. R. 1981. A diapsid reptile from the Pennsylvanian of Kansas. Univ. Kansas Mus. Nat. Hist. Sp. Pap., 7:1–14.

Reisz, R. R. 1986. Pelycosauria. Pp. 1–102 *in* P. Wellnhofer (ed.), *Handbuch der Palaeoherpetologie*, Teil 17A. Stuttgart: Gustav Fischer Verlag.

Romer, A. S. 1956. *Osteology of the Reptiles*. Chicago: Univ. of Chicago Press.

Romer, A. S. 1966. *Vertebrate Paleontology*. 3rd ed. Chicago: Univ. of Chicago Press.

Romer, A. S., and L. I. Price. 1940. Review of the Pelycosauria. Geol. Soc. Amer., Spec. Pap., 28:1–538.

Romer, A. S., and R. V. Witter. 1941. The skin of the rhachitomous amphibian *Eryops*. Am. J. Sci., 239:822–224.

Roth, J. J., E. C. Roth, and N. Hotton, III. 1986. The parietal foramen and eye: their function and fate in therapsids. Pp. 173–184 *in* Hotton, N., III, P. D. MacLean, J. J. Roth, and E. C. Roth (eds.), *The Ecology and Biology of Mammal-like Reptiles*. Washington, D.C.: Smithsonian Inst. Press.

Rubidge, B. S. 1987. South Africa's oldest land-living reptiles from the Ecca-Beaufort transition in the southern Karoo. S. Afr. J. Sci., 83:165–166.

Rubidge, B. S. 1988. A palaeontological and palaeoenvironmental synthesis of the Permian Ecca-Beaufort contact in the southern Karoo between Prince Albert and Rietbron, Cape Province, South Africa. Doctoral dissertation. South Africa: Univ. of Port Elizabeth.

Schmidt-Nielsen, K. 1975. *Animal Physiology: Adaptation and Environment*. London: Cambridge Univ. Press.

Sigogneau, D. 1963. Note sur une nouvelle espèce de Scaloposauridae. Palaeontol. Afr., 8:13–37.

Sigogneau, D. 1970. Révision systématique des gorgonopsiens Sud-Africains. Cahiers de Paléontologie, Éditions du Centre National de la Recherche Scientifique. Paris VII.

Smith, R. M. H. 1987. Helical burrow casts of therapsid origin from the Beaufort Group (Permian) of South Africa. Palaeogeogr., Palaeoclimatol., Palaeoecol., 60:155–170.

Spearman, R. I. C. 1966. The keratinization of scales, feathers, and hair. Biol. Rev. Cambridge Philos. Soc., 41:59–96.

Stear, W. M. 1979. Channel sandstone and bar morphology of the Beaufort Group uranium province in the vicinity of Beaufort West, South Africa. Thesis. South Africa: Univ. of Port Elizabeth.

Thomason, J. J., and A. P. Russell. 1986. Mechanical factors in the evolution of the mammalian secondary palate: a theoretical analysis. J. Morphol., 189:199–213.

Van Valen, L. 1960. Therapsids as mammals. Evolution, 14(3):304–313.

Watson, D. M. S. 1931. On the skeleton of a bauriamorph reptile. Proc. Zool. Soc., Pt. 3:1163–1205.

Williston, S. W. 1925. *The Osteology of the Reptiles*. Cambridge: Harvard Univ. Press.

18 Systematics of the Nonmammalian Synapsida and Implications for Patterns of Evolution in Synapsids

James A. Hopson

The broad outline of the history of the Synapsida, the group traditionally called the "mammal-like reptiles," has emerged gradually from the study of the fossil record over the past 150 years. By the 1930's, the inferred pattern of synapsid evolution was sufficiently detailed to permit an interpretation of the possible evolutionary processes that gave rise to it. From the pattern of large-scale morphologic change in the transition from primitive amniote tetrapods to the earliest true mammals, paleontologists have sought to explain what Broom (1932:331) called "the problem of what has guided the evolution."

Broom was one of the first to read an interpretation of evolutionary cause into the pattern of change in synapsids as he understood it. In his great synthesis of knowledge on the "mammal-like reptiles" of South Africa, Broom (1932) drew a distinction between the evolutionary patterns exemplified by different adaptive types within the synapsids. Most common were the specialized herbivorous and carnivorous "side branches," which he perceived as adapting to the immediate requirements of the environment but succumbing to extinction as the environment changed. Contrasted with these was a persisting central line of "small generalized probably omnivorous types" in which "the evolutionary force is of a different type and seems to have foreseen the future" (Broom, 1932:332). From this stock evolved the mammals, and among the many lineages descended from the basal mammalian stock, one led to human beings. Broom (1932:333) concluded that "amid all the thousands of apparently

635

useless types of animals that have been formed some intelligent controlling power has especially guided one line to result in man."

Working within the framework of the neo-Darwinian synthesis, Olson (1959, 1971) and Simpson (1959) considered the overall pattern of synapsid phylogeny to indicate the gradual acquisition of increasingly mammalian features in *independently evolving lineages* leading to the *multiple* achievement of the mammalian grade of organization. The concept of a polyphyletic origin of mammals was accepted generally by vertebrate paleontologists during the 1950's and 1960's. Such a pattern usually was interpreted in neo-Darwinian terms, with strong selection pressures postulated as acting within separately evolving groups for the perfection of homeothermy. Simpson (1960:171) noted that the parallel changes involved in the development of homeothermy "can hardly have failed to be adaptive" in any environment.

A decidedly non-Darwinian point of view was reflected by Grassé (1977), who accepted essentially the same pattern of parallel evolution of mammalian features as Simpson and Olson, but who interpreted it as indicating a "persevering orientation of evolution" in which Darwinian factors "played no part."

Recently, Kemp (1982) examined the pattern of synapsid evolution and interpreted it in terms of a punctuational theory in which periods of evolutionary stasis alternated with "quantum speciation" events; under this model, the likelihood of similar adaptations appearing independently in separate lineages was deemed "most improbable." But he (1982, 1985) interpreted the long-term pattern of synapsid evolution as exhibiting a gradual, progressive increase in homeostasis, adapting the evolving synapsid lineage "essentially to a highly fluctuating habitat" (1985:182). Kemp's conception of the progressive evolution of increasing homeostasis—i.e., of maintaining an increasingly constant internal environment—is similar to the conception of Simpson (1959, 1960) and Olson (1959) of the gradual perfection of homeothermy during synapsid history.

For the most part, the study of synapsids has involved the working out of adaptive trends in the direction of mammals (Crompton, 1963; Hopson, 1966; Kemp, 1969; Barghusen, 1972; Allin, 1975; Bramble, 1978) without the paying of a great deal of attention to the analysis of detailed phylogenetic relationships. But, in recent years, cladistically oriented evolutionists such as Eldredge and Cracraft (1980), Vrba (1980), and Lauder (1981) have pointed out that the determination of patterns of evolutionary change are highly dependent on the interpretation of phylogeny. As Cracraft (1981:465) noted, "the presence of a trend is phylogeny dependent." Therefore, in order to use the fossil record for interpreting large-scale aspects of evolution, one must have an explicit, detailed, and well-corroborated theory of synapsid relationships. To date, this has not been the case.

Only since 1980 have a number of phylogenetic studies of nonmammalian synapsids appeared that are explicitly cladistic in approach. (For discussions of cladistic theory and methods, see, e.g., Eldredge and Cracraft, 1980, and Wiley, 1981.) Reisz (1980, 1986) and Brinkman and Eberth (1983) analyzed the interrelationships of the primitive synapsids of pelycosaurian grade. Kemp (1982) reviewed the relationships of all synapsids, although he did not present a detailed analysis of characters. More recently, he dealt in a more detailed fashion with the cladistic relations of tritylodontid cynodonts and several groups of primitive mammals (Kemp, 1983). Cladistic studies were published by Battail (1982) on cynodonts, by Cluver and King (1983) on Permian dicynodonts, and by King (1988) on dicynodonts and dinocephalians. Clark and Hopson (1985) reviewed the intrafamilial relationships of tritylodontids, and Sues (1985) reviewed the relationships of tritylodontids to other cynodont groups.

Hopson and Barghusen (1986) published a cladistic analysis of all groups of nonmammalian Therapsida; although this study appeared at the end of 1986, it was essentially completed in mid-1982. Thus, it did not deal comprehensively with systematic studies, such as those of Kemp (1982, 1983), Battail (1982), and Sues (1985), which appeared subsequently. The manuscript was made available to colleagues before publication and has been cited a number of times (e.g., Rowe, 1986 [also see Gauthier et al., 1988]; Rowe and van den Heever, 1986; Clemens and Lillegraven, 1986; Kemp, 1988).

This chapter presents a critical review of recent studies on the phylogenetic relationships of the nonmammalian synapsids, that is, the reptilian subclass Synapsida of traditional, noncladistic, usage (Romer, 1966; Carroll, 1988). As noted by Hopson and Barghusen (1986:84), the only groups considered "natural" in a cladistic classification are those that are strictly monophyletic in the sense that all descendants of a common ancestor are included within the group. They define the Synapsida as a strictly monophyletic taxon by including the Mammalia within it. The remaining amniotes, i.e., the nonsynapsid "reptiles" and the birds, also are united in a monophyletic taxon, the Sauropsida (see Fig. 1). A review of recent work on the interrelationships of synapsids to other amniotes is included, based primarily on the work of Carroll, Heaton, and Reisz. The discussion of the interrelationships of primitive synapsids ("pelycosaurs") is based largely on the recent synthesis of Reisz (1986). The review of therapsid systematics consists of a critical overview of studies, published from 1982 to the present, which appeared after the paper by Hopson and Barghusen (1986) was submitted for

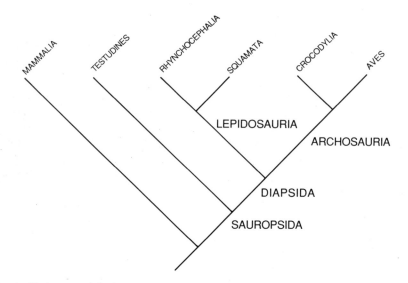

Fig. 1. Cladogram of the living taxa of Amniota.

publication. Among the studies considered are a recently published review of synapsid relationships by Kemp (1988) and an unpublished
doctoral dissertation by Rowe (1986), of which portions have been published recently by Gauthier et al. (1988) and by Rowe (1988).

The present chapter has two principal purposes: (1) to review critically
and in some detail recent hypotheses of relationships among nonmammalian synapsids, pinpointing areas of continuing contention; and (2) to
show how past and current hypotheses of synapsid phylogeny have
influenced interpretations of evolutionary pattern and process within the
group. The discussion of how ideas of synapsid relationships have determined ideas of evolution within the group can be read independently of
the detailed review of synapsid systematics, should some readers wish to
do so.

THE PLACE OF THE SYNAPSIDA WITHIN AMNIOTA AND THE INTERRELATIONSHIPS OF SYNAPSIDA

Synapsida as the Sister-Taxon of Sauropsida: The Extant Groups

For present purposes, I provisionally accept the hypothesis of interrelationships of the living groups of Amniota presented in Figure 1 (also
see Gauthier et al., 1988, for a more detailed discussion). I shall discuss

only two of the features that indicate that the living amniotes are divisible into the Mammalia (which with the extinct members of its clade is called the Synapsida herein) and the Sauropsida. These are the structure of the aortic arches (Goodrich, 1916, 1930) and the structure of the pallium in the forebrain (Ulinski, 1986), both of which provide synapomorphies for a monophyletic taxon containing all living reptiles plus birds.

Goodrich (1916) noted that the systemic aorta of mammals is formed primarily by the *left* fourth aortic arch, whereas in turtles and diapsids (including birds), the principal systemic aorta is formed largely by the *right* fourth aortic arch. Reptiles have two complete systemic aortas arising separately from the heart, but all of the arteries to the head and forelimbs branch from the right arch. Birds possess only the right arch, but this condition is readily derived from the double-arched condition seen in crocodilians (Goodrich, 1916). To the clade consisting of the living reptiles (including turtles and birds), Goodrich applied Huxley's term *Sauropsida*. To mammals, in which the aortic trunk never becomes subdivided, and "synapsidan reptiles" Goodrich (1916:264) gave the name *Theropsida*. He believed that the sauropsidan and theropsidan types of heart and aortic arches "must have evolved from some more symmetrical primitive type, in which the ventricle and the aortic trunk were both single" (Goodrich, 1916:271). From this, Goodrich concluded that the completely divided systemic trunk of sauropsids is much more modified from the primitive condition than is the undivided systemic trunk of mammals and so is unlikely to have given rise to the latter.

Gaffney (1980), following Holmes (1975), disputed the fundamental nature of the differences between the systemic aortas of reptiles and mammals, and argued that the ancestral amniote might have had a partially subdivided aorta, as have modern anurans. However, even if the common ancestor of living reptiles and mammals had, at most, a *partial* subdivision of the aorta (comparable to the condition in anurans), it is still most parsimonious to interpret the complete aortic subdivision of living reptiles as a highly apomorphic condition, one that unites those taxa possessing it as a monophyletic group—the Sauropsida.

An additional feature characterizing living sauropsids (i.e., turtles and diapsids, including birds) is the dorsal ventricular ridge (DVR), a ridge of neural tissue that protrudes into the lateral ventricle of the forebrain (Ulinski, 1986). According to Ulinski (1986:161), it appears that "the central portion of the pallium has been elaborated differently in the diapsid reptiles and turtles, on the one hand, versus the mammals on the other hand." Ulinski prefers the view that this represents a dichotomy in the pattern of elaboration of the pallium in sauropsids and mammals to the view that the mammalian condition is the result of a secondary loss of DVR.

I conclude from this brief discussion, plus the evidence discussed below on the distribution of skeletal features among Paleozoic amniotes, that a synapsid-sauropsid dichotomy represents the most parsimonious interpretation of current evidence of amniote interrelationships. This conclusion would mean that the apparently primitive condition of Jacobson's organ in turtles, cited by Gaffney (1980) as evidence for a sister-group relationship of turtles to a clade consisting of synapsids + diapsids, is better interpreted as a secondary simplification from a more complex general amniote condition. Likewise, Gardiner's (1982) arguments for a sister-group relationship of birds and mammals is not supported by this analysis, though I believe the pattern of distribution of characters in the nonmammalian synapsids provides the strongest argument against it (see Gauthier et al., 1988, for a more detailed criticism of Gardiner's phylogeny of amniotes).

Synapsida as the Sister-Taxon of Other Amniota: The Fossil Evidence

The interrelationships of Paleozoic amniote groups to one another and to their probably nonamniote sister-group, the Diadectomorpha (Heaton, 1980; Brinkman and Eberth, 1983; Heaton and Reisz, 1986), have long been problematic. Carroll (1982, 1986) continually has supported an ancestral position for the Protorothyrididae ("Romeriidae") because of their early appearance in the fossil record (Mid-Pennsylvanian) and his assessment that they lack apomorphic features in comparison with other early amniotes (1982:91). Synapsids are equally ancient, and Carroll (1986:27) considered their skeletal similarity to protorothyridids (see Figs. 5–6) to suggest that they "shared a common ancestry not much earlier than the time of their first appearance in the fossil record." Other Paleozoic amniote groups (Captorhinidae, Diapsida, Mesosauria, Millerosauria, Procolophonidae, and Parieasauria) are considered by Carroll (1982:103) to have been derived independently from protorothyridid ancestors; in his view, no two groups of early amniotes show a special relationship to one another.

A much different set of conclusions with respect to the interrelationships of Paleozoic amniotes was reached by Reisz (1981) and Heaton and Reisz (1986) on the basis of character analyses using explicitly cladistic methods, with primitive character states being determined by out-group comparison. Heaton and Reisz's (1986:404) "determination of the polarity of the different character states is based on morphological comparisons with non-reptilian [nonamniote] out-groups [i.e., Limnoscelidae (Fig. 3), Diadectidae]." Other taxa (e.g., pareiasaurs, procolophonids, millerosaurs, and mesosaurs) were taken into account "in an effort to show that many of the character states . . . determined to be primitive for reptiles [amniotes] have a distribution beyond those groups under discussion [Captorhinidae, Protorothyrididae, Diapsida, and Synapsida]."

The studies of Reisz (1981), Heaton and Reisz (1986), and Kemp (1980a, 1982) emphasized the primitive structure of early synapsids among amniotes (Fig. 4) and the highly derived structure of protorothyridids (Fig. 5), which, according to Heaton and Reisz (1986), share the greatest number of synapomorphies with the Diapsida (see Fig. 2). Captorhinids are the sister-taxon of the protorothyridid-diapsid clade (Heaton and Reisz, 1986), and synapsids are a much more plesiomorphic, distantly related taxon. Although they did not explicitly consider the position of other Paleozoic amniote groups (procolophonids, pareiasaurs, millerosaurs, and mesosaurs), the latter share a number of derived features with the paraphyletic captorhinomorphs and the diapsids, features in which synapsids retain the primitive state. Therefore, it seems at present, at least tentatively, that the basic division of Amniota is into the Synapsida, on the one hand, and the remaining amniote groups on the other, the latter being united as the Sauropsida. This hypothesis of relationships of primitive amniote groups reverses the polarities of many of the character states considered by Carroll (1982, 1986) to be primitive for amniotes.

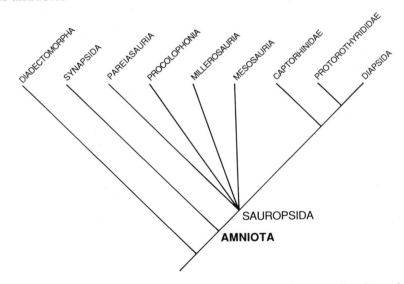

Fig. 2. Cladogram of Amniota, living and extinct, plus Diadectomorpha, the probable sister-group of amniotes. Synapsida includes the Mammalia as a derived subgroup. The branch for Captorhinidae may also include turtles (Testudines).

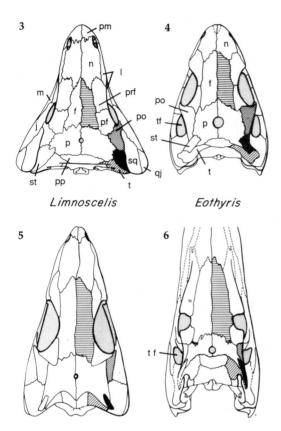

Figs. 3–6. Dorsal views of the skull of a diadectomorph and three primitive amniotes. (3) The diadectomorph *Limnoscelis*. (4) The primitive caseasaurian synapsid *Eothyris*. (5) The protorothyridid sauropsid *Protorothyris*. (6) The eupelycosaurian synapsid *Ophiacodon*. *Limnoscelis* and *Eothyris* possess the following features herein considered primitive for amniotes: frontal *(horizontal hatching)* excluded from orbital rim; postorbital *(stipple)* contacts broad supratemporal *(black)*; and large tabular *(diagonal hatching)* contacts paroccipital process of opisthotic. The modified conditions of these features in *Protorothyris* and *Ophiacodon* are considered to be independently derived within sauropsids and synapsids. f = frontal; j = jugal; l = lacrimal; m = maxilla; n = nasal; p = parietal; pf = postfrontal; pm = premaxilla; po = postorbital; pp = postparietal; prf = prefrontal; qj = quadratojugal; sq = squamosal; st = supratemporal; t = tabular; tf = temporal fenestra. (After Romer, 1956; Reisz, 1986; Heaton and Reisz, 1986; Romer and Price, 1940.)

The Interrelationships of Primitive Synapsida ("Pelycosaurs")

The synapomorphies of primitive synapsids given by Reisz (1980, 1986) are as follows: (1) a lower temporal fenestra present (convergent with Diapsida); (2) a broad, anteriorly tilted occipital plate formed by a large supraoccipital and the paroccipital processes of the opisthotics; (3) the posttemporal fenestrae greatly reduced; (4) a single postparietal; and (5) septomaxilla composed of a broad base and a massive dorsal process.

Reisz (1980), Brinkman and Eberth (1983), and Reisz (1986) reviewed the interrelationships of the principal groups of nontherapsid synapsids (i.e., the grade group "Pelycosauria") using cladistic methods. Reisz (1986:Fig. 41; presented here in slightly modified form in Fig. 7) attempted to resolve the differences between his earlier study and that of

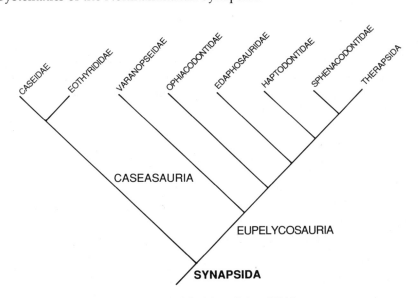

Fig. 7. Cladogram of Synapsida. (Modified from Reisz, 1986.)

Brinkman and Eberth. The following discussion is based primarily on his recent analysis.

According to Reisz, the most primitive known synapsid skull pattern is seen in the Early Permian Eothyrididae (*Eothyris, Oedaleops*), in which the skull is low and broad, the skull roof is gently convex in transverse section, and the antorbital length is about equal to postorbital length (Fig. 4). The frontal either lacks or has a very limited exposure on the orbital margin, and the supratemporal is a large, broad element. In all of these features, eothyridids more closely resemble the primitive amniote morphotype (see Figs. 3–4), as determined by Heaton and Reisz's (1986) cladistic analysis, than they do either protorothyridids or ophiacodontid synapsids (see Figs. 5–6), the two earliest known (Mid-Pennsylvanian) amniote groups (Carroll, 1986; Reisz, 1986). The Eothyrididae is united with the Caseidae as the Caseasauria, which is the sister-taxon of all remaining Synapsida (Reisz, 1986). In caseasaurs, the premaxilla is tilted forward to form a pointed rostrum, which is short in eothyridids but prominent in caseids. In addition, the maxilla participates in the lower border of the orbit, and the external naris is elongated anteroposteriorly. Eothyridids were small and presumably insectivorous or carnivorous, whereas caseids were medium-sized to gigantic herbivores.

The remaining synapsids, both the "pelycosaurian" groups, united as the paraphyletic taxon Eupelycosauria by Kemp (1982) and Reisz (1986), and the Therapsida, are distinguished by the following synapomorphies

(Reisz, 1986): (1) the antorbital region of the skull deeper than wide; (2) the frontal contributing broadly to the orbital margin; (3) the parietal foramen relatively small and located posterior to the transverse midline of the parietal; and (4) the supratemporal reduced to a long, narrow splint held in a groove between the parietal and squamosal. This general morphology, seen in the ophiacodontid *Archaeothyris*, the earliest known synapsid (Reisz, 1972), and in the Early Permian *Ophiacodon* (Fig. 6), was considered primitive for synapsids as a whole by Carroll (1986).

According to Reisz (1986), the Eupelycosauria (Fig. 7; treated here as a monophyletic taxon by the addition of the Therapsida) comprises four "pelycosaurian" taxa—Varanopseidae, Ophiacodontidae, Edaphosauridae, and Sphenacodontidae. The ordering of the first three taxa relative to one another is somewhat uncertain, but a close relationship of Sphenacodontidae to Therapsida long has been recognized (Romer and Price, 1940). As noted below, however, the family Sphenacodontidae, as currently recognized, is probably paraphyletic.

The Varanopseidae, allied by Romer and Price (1940) with the Sphenacodontidae primarily on the basis of similar postcranial specializations, seem to be much more primitive in cranial structure than are the sphenacodontids, and even more primitive than ophiacodontids and edaphosaurids. Reisz (1986) concluded that their postcranial similarities to sphenacodontids are convergent.

The ophiacodontids possess a number of synapomorphies with edaphosaurids and sphenacodontids, including a dorsal process on the stapes that articulates in a socket on the paroccipital process, and an angular with a prominent ventral keel. Edaphosaurids share with sphenacodontids a skull in which the jaw articulation is depressed below the level of the tooth row, the cheek is strongly emarginated ventrally, and the rear of the dentary possesses a well-developed coronoid eminence (see Figs. 8–9 for condition in sphenacodonts). The articular in both has a prominent, ventromedially directed pterygoideus process, below which the prearticular shows a distinctive twisting (Figs. 8–9, 23). The reduction of the quadratojugal in the cheek (Figs. 8–9) is an apparent derived resemblance between edaphosaurids and sphenacodontids. Brinkman and Eberth (1983) restored the quadratojugal in *Edaphosaurus* as being entirely restricted to the region of the quadrate, as it is in sphenacodontids; Reisz (1986) restored it in broken lines to form the lower border of the zygomatic arch but gave as a synapomorphy of Edaphosauridae and Sphenacodontidae, "zygomatic arch formed by jugal and squamosal" (Reisz, 1986:Table 3). Apparently, the precise morphology of the quadratojugal of *Edaphosaurus* is unclear.

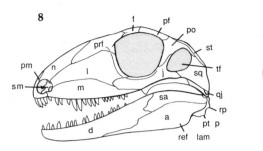

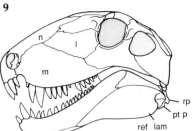

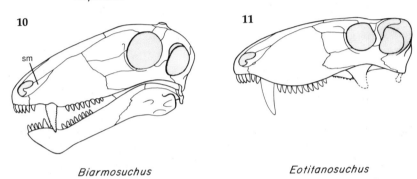

Figs. 8–11. Lateral views of the skull of two eupelycosaurs near the ancestry of Therapsida and two primitive therapsids. (8) The haptodontid eupelycosaur *Haptodus*. (9) The sphenacodontid eupelycosaur, *Dimetrodon*. (10) The biarmosuchian therapsid *Biarmosuchus*. (11) The eotitanosuchian therapsid *Eotitanosuchus*. a = angular; d = dentary; f = frontal; j =jugal; l = lacrimal; m = maxilla; n = nasal; pf = postfrontal; pm = premaxilla; po = postorbital; prf = prefrontal; pt p = pterygoideus process of articular; qj = quadratojugal; ref lam = reflected lamina of angular; rp = retroarticular process of articular; sa = surangular; sm = septomaxilla; sq = squamosal; st = supratemporal; tf = temporal fenestra. (After Currie, 1979; Romer and Price, 1940; Sigogneau and Tchudinov, 1972.)

Sphenacodontidae as the Sister-Taxon of Therapsida

Romer and Price (1940:194) were the earliest to conclude "that the mammal-like reptiles have in all probability descended from the sphenacodontid pelycosaurs." Synapomorphies of the two groups are (Reisz, 1986; Hopson and Barghusen, 1986): (1) the angular bone with a reflected lamina; (2) the zygomatic process of the quadratojugal lost, replaced by a process of the squamosal (possibly shared with edaphosaurids; see above); (3) the retroarticular process of the articular turned downward; and (4) the paroccipital process elongated and directed ventrolaterally.

At present, the family Sphenacodontidae is divided into three sub-families, of which the Haptodontinae is the most primitive because it lacks certain derived features possessed by the other two subfamilies, Sphenacodontinae and Secodontosaurinae. Among the synapomorphies linking the latter two are the withdrawal of the lacrimal from the posterior border of the external naris, with its replacement by a ventral extension of the nasal (Fig. 9), and the elongation of the neural spines (Reisz, 1986). I agree with Reisz that *Secodontosaurus* seems to be related more closely to the sphenacodontines than either is to *Haptodus*. It is uncertain whether *Secodontosaurus* is the sister-taxon of the sphenacodontines, which is suggested by its smaller caniniform teeth and low maxilla, or whether it is closer to *Dimetrodon*, as suggested by its elongated neural spines, than to the sail-less *Sphenacodon*.

Some controversy revolves around which subfamily, Haptodontinae or Sphenacodontinae, is closer to the ancestry of therapsids. Sphenacodontines share with therapsids the possession of enlarged canine teeth in both upper and lower jaws (perhaps secondarily reduced in *Secodontosaurus*) and an enlarged maxilla that contacts the nasal. But because sphenacodontines possess a number of autapomorphies (a descending process of the nasal behind the external naris, enlargement of incisors) unexpected in the ancestry of therapsids, Currie (1979) argued that therapsids more likely evolved from a haptodontine ancestor. However, *Haptodus* (Currie, 1979), the only genus in its subfamily, seems to be distinguished from other genera of sphenacodontids, and from therapsids, on the basis of primitive and perhaps size-related characters only (see diagnosis in Reisz, 1986:74). Therefore, it is more parsimonious to consider the derived features shared by sphenacodontines (including *Secodontosaurus*) and therapsids as synapomorphies indicating a sister-group relationship of the two (as also concluded by Rowe, 1986, and Kemp, 1988). This hypothesis is indicated in Figure 7. Such a pattern of relationships makes the family Sphenacodontidae paraphyletic; it can be made monophyletic by restricting it to the currently recognized Sphenacodontinae plus *Secodontosaurus*. The Haptodontinae, containing *Haptodus* only, then must be raised to family status as the Haptodontidae.

I extend the term *Sphenacodontia* of Romer and Price (1940) to include the Haptodontidae, Sphenacodontidae, and Therapsida (see Fig. 7); as such, it is the sister-taxon of the Edaphosauridae.

THE THERAPSIDA

The Therapsida, including the Mammalia as a derived subgroup, is extremely well characterized. Hopson and Barghusen (1986) listed 11 synapomorphies of the skull and lower jaw, whereas Rowe (1986) listed 21 cranial synapomorphies and an additional 20 from the postcranial skeleton. *Biarmosuchus* (Figs. 10, 13) of the early part of the Late Permian of Russia possesses the primitive therapsid state of most, though not all (see below), of these features.

Most of the synapomorphies of the Therapsida are in the direction expected of a group that is more mammal-like in structure than are the "pelycosaurs." In the skull (Fig. 10), the lateral temporal fenestra is larger than that of sphenacodontids and the lateral surface of the postorbital bone shows evidence of attachment of external adductor jaw muscula-ture (personal observation of *Biarmosuchus*). The upper canine (though not the lower) is greatly enlarged, and the maxilla extends farther dor-sally to contact the prefrontal. The reflected lamina of the angular is deeply notched above, and the anterior coronoid bone (Fig. 23) probably is lost (though its condition is not known in biarmosuchians, and Sigog-neau [1970a:32] suggested its presence in a gorgonopsid). The septomax-illa of therapsids is distinctive, possessing a long posterodorsal process interposed between the nasal and maxilla on the side of the face. In the palate, the pterygoid bones are appressed more closely on the midline than in sphenacodontids, and the interpterygoidal vacuity thereby is greatly reduced.

In addition, primitive therapsids (Figs. 10, 11) possess several derived features that are lost independently in theriodonts (Figs. 27–29) and some advanced dicynodonts. The dorsal process of the premaxilla is greatly elongated, extending far back between the nasals, and the parie-tal foramen is elevated on a chimneylike protuberance.

The postcranial skeleton of primitive therapsids, as characterized by that of *Biarmosuchus* (Sigogneau and Tchudinov, 1972) and *Hipposaurus* (Boonstra, 1965), differs strikingly from that of sphenacodontids in the relative narrowness of the scapula, shortness of the interclavicle, anterior expansion of the ilium, and overall slenderness of the limb bones. The humerus, especially, has a relatively unexpanded distal end in com-parison with that of sphenacodontids, and the head is narrower and more rounded. The femur has a sigmoid shape and a medially inflected head resembling that of modern crocodilians. In addition, though still possessing the primitive amniote phalangeal formula of 2-3-4-5-3 in the manus (Fig. 12) and 2-3-4-5-4 in the pes, the hands and feet of *Biar-mosuchus* (Fig. 13) are functionally mammal-like owing to the reduction to thin discs of the "extra" phalanges in Digits III and IV (and pedal Digit V). The impression one gains from the postcranial skeleton of primitive therapsids is that posture was much less sprawling than is indicated for sphenacodontids, though how much the limbs were drawn in under the body is uncertain.

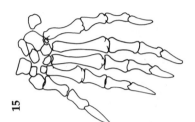

15 *Galechirus*

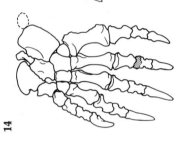

14 *Titanophoneus*

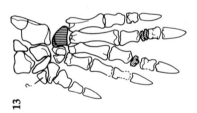

13 *Biarmosuchus*

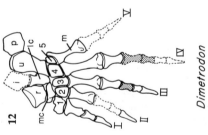

12 *Dimetrodon*

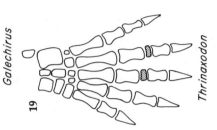

19 *Thrinaxodon*

18 Therocephalian

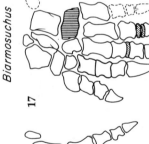

17 Gorgonopsian

16 Dicynodont

648

The Biarmosuchia: A Monophyletic or Paraphyletic Taxon?

The primitive Russian genus *Biarmosuchus* and several later Permian genera, such as *Hipposaurus, Ictidorhinus,* and *Rubidgina* from South Africa, represent the most primitive grade of therapsids. Some of the South African forms are allied as the family Ictidorhinidae on the basis of a number of synapomorphies such as a crista choanalis (rudimentary secondary palate), seen in *Rubidgina* (pers. observ.), *Lycaenodon,* and *Burnetia* (Mendrez-Carroll, 1975), and the possible presence of a preparietal bone (Sigogneau, 1970a,b). Hopson and Barghusen (1986:88) tentatively united the biarmosuchids and ictidorhinids as the "Biarmosuchia," a possibly paraphyletic group "characterized only by possession of primitive therapsid characters."

Further consideration of these primitive therapsids suggests that they possess a few synapomorphies linking them in a monophyletic Biarmosuchia; unfortunately, none of the suggested synapomorphies is clearly unique to the group. The presumed synapomorphies of Biarmosuchia are (1) postcanine teeth with basal swelling and coarsely serrated margins, either posteriorly only (*Biarmosuchus,* pers. observ.; *Rubidgina* sp., Sigogneau, 1970a) or both anteriorly and posteriorly (*Rubidgina angusticeps, Lemurosaurus,* Sigogneau, 1970b); (2) Distal Carpals 4 and 5 fused to form a single large element (*Biarmosuchus,* Sigogneau and Tchudinov, 1972; *Hipposaurus,* Boonstra, 1965); and (3) Distal Tarsals 4 and 5 fused to form a single large element (as in 2).

Our understanding of tooth form in primitive members of other therapsid groups is inadequate, although none seems to duplicate the morphology noted here. Fusion of Distal Carpals and Tarsals 4 and 5 occurs in several other groups of therapsids, but the pattern of distribution of these fusions, as discussed below, indicates that the condition in biarmosuchians occurred independently.

Possession of five distal carpals is primitive for amniotes and occurs in sphenacodontids (Fig. 12). A separate, unreduced fifth distal carpal is

---

Figs. 12–19. Dorsal views of the left manus of a sphenacodontid eupelycosaur and seven therapsids to show the distribution of the fused condition of Distal Carpals 4 + 5 (*vertical hatching*) and the patterns of reduction and loss of phalanges in Digits III and IV (*stipple*). (12) The primitive condition represented by *Dimetrodon.* (13) The biarmosuchian *Biarmosuchus.* (14) The dinocephalian *Titanophoneus.* (15) The primitive anomodont ("dromasaur") *Galechirus.* (16) A primitive dicynodont. (17) A gorgonopsian. (18) A scylacosaurid therocephalian. (19) The cynodont *Thrinaxodon.* i = intermedium; lc = lateral centrale; m = metacarpal; mc = medial centrale; p = pisiform; r = radiale; u = ulnare; 1–5, Distal Carpals 1–5, I–V, Digits I–V; ? = unidentified bone. (After Romer and Price, 1940; Sigogneau and Tchudinov, 1972; Orlov, 1958; Brinkman, 1981; Boonstra, 1966a; Sigogneau, 1970b; Boonstra, 1964; Parrington, 1939.)

retained in all Dinocephalia (Boonstra, 1966b) (Fig. 14) and in primitive anomodonts ("dromasaurs"; Fig. 15). Fusion of Distal Carpals 4 and 5 to form a large element proximal to Metacarpals 4 and 5 occurs in Biarmosuchia (Fig. 13) and in all members of the Gorgonopsia and Therocephalia (Figs. 17–18) in which the carpus is known. In derived anomodonts (dicynodonts), Distal Carpal 5 is absent (Boonstra, 1966a) (Fig. 16). Inasmuch as its absence leaves a gap proximal to the Metacarpal 5, its loss as a separate element within Anomodontia is almost certainly the result of nonossification rather than of fusion with Distal Carpal 4. A similar condition pertains in Cynodontia, in which Distal Carpal 5 is preserved as a reduced element in primitive cynodonts (*Procynosuchus*, Kemp, 1980b; *Thrinaxodon*, Parrington, 1939) (Fig. 19), but is absent in more derived cynodonts such as *Diademodon* (Jenkins, 1971). Distal Carpal 4 in *Diademodon* is not expanded laterally into the space formerly occupied by Distal Carpal 5; therefore, loss of the latter seems to have involved nonossification rather than fusion. If one assumes that reversal from a fused to an unfused condition did not occur, the taxonomic distribution of the fusion of Distal Carpals 4 and 5 occurred independently three times—in Biarmosuchia, Gorgonopsia, and Therocephalia. If reversals from the fused to unfused state are accepted, it is equally parsimonious to assume independent fusions in Biarmosuchia and the common ancestry of Gorgonopsia, Therocephalia, and Cynodontia (Figs. 20, 22), with a reversal to the unfused condition in the ancestry of cynodonts. In either case, the fused condition in Biarmosuchia can be regarded as a synapomorphy of that group. An assumption of fusion of Distal Carpals 4 and 5 in the common ancestor of biarmosuchians and all other therapsids, with separate reversals to the unfused condition in Dinocephalia, Anomodontia, and Cynodontia, involves an additional step; thus, the hypothesis is less parsimonious than that postulating carpal fusion in Biarmosuchia as a separate event from that in Gorgonopsia and Therocephalia.

The pattern of distribution of fused Distal Tarsals 4 and 5 is slightly different; they are fused in biarmosuchians and gorgonopsians and in derived, but not in primitive, therocephalians. Distal Tarsals 4 and 5 are present as distinct elements in all Dinocephalia (Boonstra, 1966b), in primitive dicynodont anomodonts (Boonstra, 1966a), and in primitive therocephalians (Boonstra, 1964). The fifth distal tarsal is absent in at least some primitive anomodonts ("dromasaurs," Brinkman, 1981) and in derived dicynodonts (Romer, 1956). It is also absent in derived cynodonts (Jenkins, 1971), but its condition in primitive cynodonts is unknown. Whether or not one assumes that reversals from the fused to the unfused condition were possible, it is most parsimonious to assume that the derived state was independently evolved three times in therapsids— in Biarmosuchia, Gorgonopsia, and derived Therocephalia.

My conclusion from this discussion is that the Biarmosuchia most

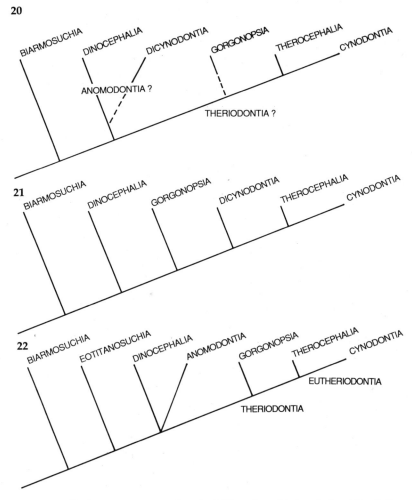

Figs. 20–22. Cladograms representing three theories of interrelationship of the major therapsid groups. (20) Cladogram of Kemp (1988). (21) Cladogram of Rowe (1986, 1988). (22) Cladogram of Hopson and Barghusen (1986).

likely constitutes a monophyletic taxon, diagnosed by nonunique synapomorphies of the carpus and tarsus and by a possibly unique synapomorphy of the postcanine dentition.

The Position of *Eotitanosuchus*

The Russian Permian genus *Eotitanosuchus*, known only from its skull, is considered by Hopson and Barghusen (1986) to be more derived than the Biarmosuchia in possessing a temporal fenestra that is enlarged in comparison with that of biarmosuchians (Figs. 10–11). This enlargement occurred in two ways—(1) by a posterodorsal, and slight lateral, flaring of the squamosal at the rear border of the opening, and (2) by the arching above the level of the orbit of that part of the postorbital bone that forms the anterodorsal rim of the fenestra. The presence of a smooth depression on the lateral surface of the postorbital behind the orbit, bounded anterodorsally and medially by a distinct crest, indicates that some adductor musculature had invaded the lateral margin of the skull roof; this is contrary to the statement of Kemp (1982:103) that there has been no invasion of the intertemporal part of the skull roof by jaw musculature. Attachment of jaw musculature to the dorsolateral surface of the postorbital bone is a feature shared with all more derived groups of therapsids, although an anteromedial invasion of the cranial roof by musculature, which even overrode the postfrontal, occurred only in dinocephalians (Barghusen, 1973).

Provisionally, Kemp (1982) considered *Eotitanosuchus* to lie close to the ancestry of gorgonopsians owing to a similarity of the temporal fenestra in both and because their palatine bones meet on the midline of the palate anterior to the pterygoids. I believe that the temporal fenestra of *Eotitanosuchus* has no special resemblance to that of gorgonopsians and could serve equally well as the precursor to that of dinocephalians such as *Syodon* or of primitive anomodonts such as *Otsheria* and *Venjukovia* (Barghusen, 1973, 1976). Furthermore, my examination of the only known skull of *Eotitanosuchus* suggests that the midline meeting of the palatines is a preservational artifact caused by strong lateral compression of the palate. The toothed bosses on the palatines contact one another and so hide the narrow midline trough in which the vomers and pterygoids meet. A similar midline contact of the palatines in the primitive dinocephalian *Syodon*, which is absent in related forms (see Orlov, 1958:Figs. 12–13), is also clearly the result of lateral compression of the snout (pers. observ.). Thus, I find no difficulty in considering *Eotitanosuchus* to be the primitive sister-taxon of all more derived therapsids, with a special relationship to no particular subgroup among them.

Monophyly of the "Advanced" Therapsida

Hopson and Barghusen (1986) recognized the Dinocephalia, Anomodontia, and Theriodontia as monophyletic taxa that together form an unnamed monophyletic group of advanced therapsids. They considered this group to be more derived than *Eotitanosuchus* in having the temporal roof further reduced in width and the temporal fenestra expanded further laterally. Rowe (1986) listed an additional five synapomorphies of the skull and lower jaw, and 10 more from the postcranial skeleton in support of this taxon. However, my brief review of the distribution of many of Rowe's presumed synapomorphies suggests that only a few are actually diagnostic of the group under consideration. One character that does seem to distinguish the "advanced" therapsids from the Biarmosuchia is the loss of a distinct olecranon process on the ulna (Rowe, 1986); a prominent olecranon is present in the biarmosuchian *Hipposaurus* (Boonstra, 1965). Another synapomorphy of the "advanced" therapsids is the loss of a phalanx in the fifth pedal digit to give a phalangeal formula of 2-3-4-5-3.

One of Rowe's synapomorphies of "advanced" therapsids, the reduction of the phalangeal formula in the manus to 2-3-3-3-3, was discussed in detail by Rowe and van den Heever (1986). They noted that this phalangeal count, which is the ancestral formula in Mammalia, occurs in all dinocephalians, dicynodonts, and therocephalians, as well as in many cynodonts. Only gorgonopsians and some cynodonts appear to retain additional phalanges in Digits III and IV (Figs. 17, 19); these extra phalanges are always compressed discs similar to those seen in *Biarmosuchus* (Sigogneau and Tchudinov, 1972). Rowe and van den Heever (1986:644) argued that the manual phalangeal formula of 2-3-3-3-3 "is most parsimoniously viewed as having arisen once, as an apomorphy of the most recent common ancestor of the unnamed taxon comprising Dinocephalia, Gorgonopsia, Dicynodontia, Therocephalia, and Cynodontia." The presence of additional disclike phalanges in some gorgonopsians and cynodonts "represents either a reversal, in which the ancestral phalanges reappeared but only partially differentiated, or the appearance of *de novo* ossifications" (Rowe and van den Heever, 1986:644). Therefore, the most parsimonious explanation of digital evolution in therapsids requires the assumption of apparent evolutionary reacquisition of a more primitive condition in more than one derived group.

For the present, the conclusions of Rowe and van den Heever cannot be disputed, but several observations suggest to me that the pattern of loss of the supernumerary phalanges was not as clear-cut as they suggested. First, the reduced phalanges of *Biarmosuchus* (Fig. 13) appear to be very much like those of gorgonopsians (Fig. 17), both in location and in morphology; this suggests, though it does not demonstrate, that these elements are homologous vestiges in the process of being reduced. Second, the second phalanx of manual Digit IV in the primitive dinocephalian *Titanophoneus* (Fig. 14) seems to be formed by the fusion of a more proximal vestigial phalanx with a more distal normal phalanx, suggesting that the acquisition of the definitive mammalian phalangeal formula may have occurred independently within the Dinocephalia (see discussion in Rowe and van den Heever, 1986). Third, in the gorgonopsian *Aelurognathus tigriceps*, the fourth digit has a single reduced phalanx that is "deeply constricted" in the middle (Boonstra, 1934:164), giving the appearance of two co-ossified phalanges. Fourth, the phalangeal formula in an undescribed specimen of the primitive cynodont *Procynosuchus* is 2-3-4-4-3, as it is in the primitive *Thrinaxodon*, suggesting that this is the ancestral phalangeal formula for cynodonts as a whole, rather than a divergent specialization within the group.

Kemp (1988) questioned the reality of a monophyletic group of "advanced" therapsids because he questioned the monophyletic origin of the presumed synapomorphy of "a further increase in the size of the temporal fenestra" advanced by Hopson and Barghusen. The probable monophyly of this character-state was discussed in the preceding section. Kemp also questioned the placement of the Gorgonopsia with the Therocephalia and Cynodontia in a monophyletic Theriodontia. This problem will be considered below in conjunction with Rowe's views on gorgonopsian relationships.

With regard to the interrelationships of the component taxa of a presumed monophyletic group of "advanced" therapsids, three recent hypotheses must be considered—i.e., those of Hopson and Barghusen (1986), Rowe (1986), and Kemp (1988). Kemp's hypothesis incorporates the views of King (1988) with regard to the recognition of a monophyletic Anomodontia sensu lato—i.e., a taxon consisting of the Dinocephalia plus the Anomodontia of more restricted usage (Venjukovioidea, Dromasauria, and Dicynodontia). These three hypotheses of therapsid interrelationships (including the more primitive therapsid groups recognized by each author) are presented in Figures 20–22.

The Question of a Dinocephalian-Dicynodont Clade

Hopson and Barghusen found no synapomorphies linking any two of the three groups of "advanced" therapsids that they recognized and so provisionally chose to "express the interrelationships of the Dinocephalia, Anomodontia, and Theriodontia as a trichotomy" (1986:90). Kemp (1988), following King (1988), tentatively recognized the Anomodontia as a monophyletic grouping of the Dinocephalia and Dicynodontia (the latter, in King's usage, including venjukovioids and dromasaurs) on the basis of (1) the loss of the coronoid bone(s); (2) the nonterminal nostrils and the long posterior spur of the premaxilla (modified in higher dicynodonts); (3) the grooved or troughed palatal exposure of the vomers; and (4) the reduction or loss of the internal trochanter of the femur.

The loss of the coronoid bone or bones (it is uncertain at what hierarchical level the anterior coronoid of "pelycosaurs" was lost) from the medial surface of the lower jaw long has been cited as a specialization uniting dinocephalians and dicynodonts (Romer, 1966). Hopson and Barghusen (1986) noted that a coronoid occurs in titanosuchid dinocephalians (Boonstra, 1962), and they suggested that inconsistencies in suture patterns in the lower jaws of the primitive Russian dinocephalians *Syodon* and *Titanophoneus* (as illustrated by Orlov, 1958) could be resolved by supposing the presence of a previously unrecognized coronoid in these taxa. I have since examined the specimens described by Orlov and have identified a single coronoid bone in the anteosaurids (= brithopodids) *Titanophoneus* and *Doliosauriscus*; the coronoid has been lost postmortem on both rami of the *Syodon* mandible, but its position on each side is indicated clearly by a facet on the medial surface of the dentary and prearticular below the tooth row (see Orlov, 1958:Fig. 31). In Figure 24, the coronoid of *Syodon* is restored on the basis of this facet and the preserved element in *Doliosauriscus*.

A further demonstration that the absence of the coronoid is not a synapomorphy of dinocephalians and dicynodonts is that a coronoid is present in several specimens of the lower jaw of the primitive dicynodont (sensu King) *Venjukovia* (pers. observ.). The coronoid of *Venjukovia*,

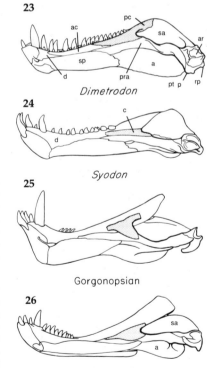

Figs. 23–26. Medial views of the lower jaw of a sphenacodontid eupelycosaur and three therapsids. The (posterior) coronoid bone is indicated by stipple. (23) The sphenacodontid *Dimetrodon*. (24) The dinocephalian *Syodon*. (25) A gorgonopsian. (26) A scylacosaurid therocephalian. a = angular; ac = anterior coronoid; ar = articular; c = coronoid; d = dentary; pc = posterior coronoid; pra = prearticular; pt p = pterygoideus process of articular; rp = retroarticular process of articular; sa = surangular; sp = splenial. (After Romer and Price, 1940; Orlov, 1958; Kemp, 1969; A. W. Crompton, pers. comm.)

though greatly reduced in size, nonetheless demonstrates that the loss of this element is not even a synapomorphy of King's Dicynodontia.

Nonterminal nostrils is a unique resemblance among most dinocephalians (though not *Estemmenosuchus;* Chudinov, 1983) and primitive members of the dicynodont group such as *Otsheria* and *Venjukovia.* But I regard the long posterior spur of the premaxilla, which occurs in biarmosuchians and in *Eotitanosuchus,* to represent the primitive therapsid condition (Hopson and Barghusen, 1986). Likewise, a vomer bearing a ventral groove or trough occurs in ictidorhinid biarmosuchians (Sigogneau, 1970b; Mendrez-Carroll, 1975) and probably also represents the primitive condition for therapsids; unfortunately, the morphology of the vomer in *Biarmosuchus* and *Eotitanosuchus* is inadequately known. The internal trochanter of the femur, although reduced in dicynodonts and more derived dinocephalians, appears to be prominent in the primitive dinocephalians *Titanophoneus* (Orlov, 1958:Fig. 51) and *Anoplosuchus* (Chudinov, 1983:Fig. 43), suggesting that the reduced condition is convergent in the two groups.

Kemp (1988) suggested that because the phylogenetic position of biarmosuchians and *Eotitanosuchus* is uncertain, it is possible that they are also anomodonts (sensu lato), specially allied to dinocephalians and dicynodonts. Although I believe this is unlikely because of the extremely primitive morphology of the Biarmosuchia, it is necessary to test this hypothesis against the alternatives. For the time being, however, I consider a special relationship between dinocephalians and dicynodonts to be only weakly supported.

Rowe (1986) listed seven synapomorphies of an unnamed taxon consisting of Gorgonopsia, Dicynodontia, Therocephalia, and Cynodontia. An analysis of the distribution of all relevant characters, including those of dinocephalians, is necessary before it is possible to determine how well corroborated this grouping is. I would note that the temporal fenestra of primitive anomodonts (i.e., *Otsheria* and *Venjukovia*), gorgonopsians, and eutheriodonts is more expanded both laterally and anteroposteriorly than it is in dinocephalians and more primitive therapsids. Although I do not consider the grouping of the Anomodontia (sensu stricto) with Gorgonopsia and Eutheriodontia to be strongly supported at the present time, I believe it is more probable than a grouping of Anomodontia with Dinocephalia.

The Monophyly of the Theriodontia

As noted above, both Kemp (1982, 1988) and Rowe (1986) questioned the monophyly of the Theriodontia, a taxon accepted by Hopson and Barghusen (1986). All of these authors accept a group consisting of the

Therocephalia and Cynodontia, designated Therosauria by Kemp (1982) and Eutheriodontia by Hopson and Barghusen (1986) and Rowe (1986); however, Kemp (1982, 1988) and Rowe (1986) questioned the traditional association of the Gorgonopsia with them in a monophyletic Theriodontia.

Hopson and Barghusen (1986:Table 1) listed the following five synapomorphies of Theriodontia: (1) the postdentary bones reduced in height and the posterodorsal part of the dentary increased in height so that the latter is left as a free-standing coronoid process; (2) the quadrate and quadratojugal reduced in height; (3) the quadrate and quadratojugal lying in a depression in the anterior face of the squamosal, having no sutural union with the squamosal; (4) the dorsal process of the premaxilla reduced in length (reversal to primitive synapsid condition; convergent on some dicynodonts); (5) the skull reduced in height (convergent on many dicynodonts).

Kemp (1982:162) argued that the coronoid process of the dentary in therocephalians differs in structure from that of gorgonopsians and "quite clearly evolved in parallel." He described it in therocephalians as "flat and broad, rather than triangular in section as in gorgonopsids." More recently, Kemp (1988) reiterated this argument for the independent origin of the coronoid process in these two groups. However, if one compares the lower jaw of a gorgonopsian (Fig. 25) with that of a *primitive* therocephalian (scylacosaurid pristerosaurian; Fig. 26) rather than that of a derived therocephalian (eutherocephalian), there is no fundamental difference between them either in the anteroposterior diameter of the coronoid process or in the way in which the process contacts the surangular. The primitive therocephalian coronoid process is not "broad," nor does the dentary extensively overlap the surangular, as it does in all advanced therocephalians. Even if the difference in cross-sectional shape stressed by Kemp (1982, 1988) is accepted, I am unable to consider this sufficient evidence for nonhomology of the coronoid process in the two groups.

Likewise, the two features of the quadrate-quadratojugal complex cited by Hopson and Barghusen (1986) as synapomorphies of a monophyletic Theriodontia are considered by Kemp (1988:13) "to reflect a similarity that does not really exist." Kemp stated that the quadrate complex of gorgonopsians "is structurally (and functionally) quite distinct from that of the Therosauria [Eutheriodontia]," although he did not specify what the distinctions are. Again, I fail to see why the differences between the quadrate-quadratojugal complex of gorgonopsians, which is certainly specialized for streptostylic movement (Kemp, 1969), and that of therocephalians and cynodonts, which appears not to be so specialized, invalidates the uniquely derived features that they do possess.

Thus, I cannot accept Kemp's rejection of these proposed synapomorphies of the Theriodontia.

Kemp (1988) dismissed the last two synapomorphies of Theriodontia by noting that they also occur in dicynodonts. However, although Character 4—reduction in length of the dorsal process of the premaxilla— occurs in dicynodonts proper, it does not occur in "dromasaurs" (a paraphyletic group of primitive anomodonts), *Otsheria*, or *Venjukovia*, in which the premaxillary process is unreduced from the primitive therapsid condition. Therefore, I see no reason why the reduced condition in theriodonts should not be considered a valid synapomorphy of that group. Character 5 may have been described too broadly by Hopson and Barghusen, but they amplify its description in the text of their paper, pointing out that "the skull takes on a more mammalian form than is seen in other therapsids by becoming lower, with a flatter dorsal profile" (Hopson and Barghusen, 1986:92). In all other groups of therapsids, as in sphenacodontids, the skull is highest in the orbitotemporal region, and the dorsal surface of the snout slopes obliquely downward to the prenarial region. This is also the case in the primitive anomodonts *Otsheria* and *Venjukovia*, although in many dicynodonts the skull is low and relatively flat. Here, too, there can be no question that this resemblance between theriodonts and some dicynodonts is the result of convergence.

Another similarity in skull proportions shared by gorgonopsians and eutheriodonts is that in transverse section, the snout is wider than high and has a broadly convex dorsal profile. In more primitive therapsid groups and in sphenacodontids, the snout always is deeper than wide, with a narrow dorsal surface and steeply sloping sides. Again, this is also the condition in *Otsheria* and *Venjukovia*, though not in some dicynodonts.

In summary, Kemp's criticisms of the characters used to link the Gorgonopsia with the Therocephalia and Cynodontia seem to me to be unsubstantiated.

Rowe (1986) also denied a sister-group relationship between the Gorgonopsia and Eutheriodontia, but in a very different way from that of Kemp. He considered the sister-taxon of Eutheriodontia to be the Dicynodontia, with the Gorgonopsia as the sister-group of this clade. Dicynodontia, as used by Rowe, includes *Otsheria* and *Venjukovia* and corresponds to the Anomodontia of Hopson and Barghusen. Rowe listed 28 synapomorphies linking dicynodonts to therocephalians and cynodonts. However, as Kemp (1988) noted, some of these characters are vague and therefore difficult to evaluate, whereas others are not found in all members of the proposed group. With respect to the latter point, it is especially significant that many of the characters listed by Rowe are not found in the most primitive anomodonts ("dromasaurs," *Otsheria*, and

Venjukovia) or in the most primitive therocephalians (the pristerosaurians of Hopson and Barghusen). For example, an intermandibular fenestra bounded by the surangular, dentary, and prearticular is found in anomodonts, primitive cynodonts, and derived therocephalians (eutherocephalians) but does not occur in pristerosaurians (Fig. 26). Likewise, although a partial secondary palate occurs in *Venjukovia*, dicynodonts, most therocephalians, and primitive cynodonts, no trace of such a structure occurs in *Otsheria* or in lycosuchid therocephalians, two of the most primitive members of their respective clades. Such character distributions strongly suggest that many of the derived similarities of dicynodonts and eutheriodonts arose convergently. Which of the characters cited by Rowe ultimately prove to be unique features of all anomodont and eutheriodont subgroups must await publication and critical evaluation of his full analysis.

To sum up, I believe that at present, the alternatives to accepting an association of Gorgonopsia with Therocephalia and Cynodontia are not sufficiently convincing to require abandonment of the Theriodontia. Kemp (1988) offered negative arguments for removing the gorgonopsians from this association, but he did not provide a definite alternate placement for them. Rowe, on the other hand, offered positive arguments for a sister-group relationship of the Anomodontia and Eutheriodontia, thereby displacing the Gorgonopsia one level down in the phylogenetic hierarchy. Rowe's arguments have not yet been published in full detail, so their merits cannot be evaluated fairly.

The Monophyly of Eutheriodontia and the Interrelationships of Therocephalia

Hopson and Barghusen (1986) listed 11 synapomorphies of a monophyletic taxon composed of the Therocephalia and the Cynodontia. Of these, the following five synapomorphies are unique to these two taxa among synapsids (see Figs. 26, 28–29): (1) the temporal roof completely eliminated so that the temporal fossa is completely open dorsally; (2) the postorbital bone shortened so that it does not contact the squamosal medial to temporal fossa; (3) the parietal expanded posteriorly on the midline behind the parietal foramen, increasing the length of the sagittal crest; (4) the epipterygoid expanded anteroposteriorly; and (5) the posteroventral portion of the dentary forming a thickened lower border that extends below the angular bone and supports the latter in a trough on its medial surface.

In dicynodonts, the temporal roof becomes very narrow, but there is always a remnant of it formed by a lappet of the parietal and the long postorbital bone that partially covers the temporal fossa lateral to the true

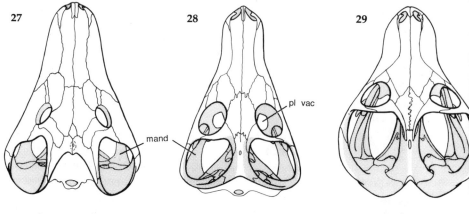

Figs. 27–29. Dorsal views of the skull of three theriodont therapsids: a primitive theriodont and two eutheriodonts. (27) An unidentified gorgonopsian. (28) An unidentified therocephalian. (29) The primitive cynodont *Procynosuchus*. mand = mandible; pl vac = palatal vacuity. (After Kemp, 1969; Mendrez-Carroll, unpublished; Brink, 1963.)

sidewall of the cranial cavity. However, in primitive anomodonts, such as *Venjukovia* (Chudinov, 1983:Pl. 7), the temporal roof is comparable in width to that of gorgonopsians.

The following features are shared with members of nontheriodont groups, notably some or all anomodonts: (6) the postfrontal reduced in size; (7) the loss of teeth on the palatine bone; and (8) lateral to the quadratojugal, the squamosal emarginated from below so that much of the quadratojugal is exposed in lateral view. Character 6 is shared with many dicynodonts, but the postfrontal in the primitive anomodont *Otsheria* does not appear to be much reduced in comparison with that of gorgonopsians. Characters 7 and 8, however, are shared with all anomodonts.

The remaining three characters (Characters 9.5, 9.10, and 9.11 listed by Hopson and Barghusen, 1986:Table 1) may not be present in the Lycosuchidae, the most primitive members of the Therocephalia (J. A. van den Heever, pers. comm.). Therefore, they may characterize only a subgroup within the Therocephalia and be convergently developed in the Cynodontia.

Although Kemp (1972, 1982) previously considered cynodonts to have been derived from a whaitsiid-like ancestor within a paraphyletic Therocephalia, he now accepts a sister-group relationship of Therocephalia and Cynodontia (Kemp, 1988).

Within the Therocephalia, interrelationships of the main subtaxa have proven impossible to resolve thus far. Hopson and Barghusen (1986) recognized two major subgroups of therocephalians—the more primitive Pristerosauria, containing only the family Pristerognathidae, and the more derived Eutherocephalia, containing four monophyletic subgroups: Hofmeyriidae, Euchambersiidae, Whaitsiidae, and the superfamily Baurioidea. Although they considered the Hofmeyriidae to be more primitive than the remaining three groups, new information casts doubt on this conclusion (as noted in an addendum on p. 83 of their paper). Therefore, the interrelationships of the four eutherocephalian groups is unresolved and is best expressed as a tetrachotomy in a cladogram.

The interrelationships of pristerosaurians are equally problematic. They lack the six to eight synapomorphies that characterize the Eutherocephalia, but they possess three autapomorphies that indicate their probable monophyly (see Hopson and Barghusen, 1986:Table 3). Hopson and Barghusen recognized only the family Pristerognathidae in this taxon, but J. A. van den Heever (1986, pers. comm.) recognizes two families of primitive therocephalians—the Scylacosauridae (a senior synonym of Pristerognathidae) and the Lycosuchidae. Lycosuchids are more primitive than scylacosaurids in a number of features in which the latter are more similar to eutherocephalians (van den Heever, 1986). Therefore, the Pristerosauria may be paraphyletic and its supposed autapomorphies either primitive for all therocephalians or independently derived in scylacosaurids and lycosuchids. It is also possible that the resemblances of scylacosaurids to eutherocephalians are convergent, or that the apparently primitive features of lycosuchids are reversals from a more derived state. At present, these problems cannot be resolved.

THE MONOPHYLY AND INTERRELATIONSHIPS OF CYNODONTIA

The Monophyly of Cynodontia

The Cynodontia, including the Mammalia as a derived subgroup (Hopson and Barghusen, 1986), are characterized by an extremely large number of synapomorphies. Hopson and Barghusen (1986:Table 1) listed 28 synapomorphies in the skull and dentition, of which 17 are unique to this group among theriodonts. The monophyly of Cynodontia rarely has been questioned, although K. A. Kermack et al. (1981) and D. M. Kermack and K. A. Kermack (1984) argued that nontherian mammals such as *Morganucodon* cannot have had a cynodont ancestry. Kemp (1983, 1988) and Hopson and Barghusen (1986) pointed out the improbability of this conclusion in view of the large number of cranial synapomorphies (ca. 40) shared by nonmammalian cynodonts and *Morganucodon*.

Many synapomorphies of cynodonts relate broadly to modifications of the food-getting apparatus. The cranial features of basal cynodonts described in the following paragraphs are seen in the Late Permian African genus *Procynosuchus* (Figs. 29, 33), a taxon closely corresponding to the primitive morphotype for all more derived cynodonts (Kemp, 1979; Hopson and Barghusen, 1986). The postcanine teeth possess anterior and posterior accessory cusps and lingual cingula, implying some degree of processing of food in the oral cavity. Probably correlated with this is the presence of a well-developed, though incomplete, secondary palate formed by processes of the palatine and maxilla. (In baurioid therocephalians, the palatine makes only a minor contribution to the secondary palate.) The zygomatic arches are flared laterally and the pterygoid flanges reduced in width; thus, the rear half of the lower jaw lies near the center of the temporal fossa rather than at its lateral margin (Fig. 29). Correlated with the development of a gap between the zygoma and the mandible is the formation of a fossa on the lateral surface of the coronoid process of the dentary (Fig. 33). Barghusen (1968) interpreted this as evidence that a portion of the external adductor muscle mass had extended its insertion ventrally on to the outer surface of the dentary; this is the first indication of the mammalian masseter muscle. Reflecting the increase in mass of the jaw-closing muscles inserting on it, the dentary is elongated posteriorly and overlaps the postdentary bones more broadly than in other theriodonts (though a convergent development occurs in eutherocephalians). In addition, that part of the braincase to which the jaw muscles attached became more resistant to deformation by developing ventrolateral flanges on the roofing bones (frontal and parietal) which meet the dorsal margin of the expanded sidewall of the braincase (formed by the anteroposteriorly expanded epipterygoid and anterior lamina of the prootic). The region around the orbit was strengthened by the secondary redevelopment of a broad contact between the prefrontal and postorbital bones.

The postcranial skeleton of primitive cynodonts is much less well known than the skull. However, the following features, which are seen in *Procynosuchus*, can reasonably be considered primitive for Cynodontia. (1) The dorsal vertebral series is differentiated into thoracic and lumbar regions, with the latter characterized by short ribs suturally attached to the vertebrae. (2) The scapula has out-turned anterior and posterior borders that enclose an elongate vertical trough ("spinatus fossa") between them. (3) The anterior process of the ilium is moderately elongated. (4) The anterior process of the pubis is reduced so that it extends only slightly anterior to the acetabulum. (5) The calcaneum bears a distinct tuber or heel.

Some of these features (2, 5, and perhaps 3) are believed by Kemp

(1980b, 1982) and Rowe (1986) to be absent in primitive cynodonts. However, they are present in a juvenile skeleton of *Procynosuchus* that I am studying. Kemp (1980b) described the postcranial skeleton of a specimen of *Procynosuchus* in which the limb and girdle elements have, in my opinion, suffered diagenetic compression. Therefore, the extreme flatness of the limb bones and unusual breadth and shallowness of the glenoid and acetabulum of the pectoral and pelvic girdles are preservational artifacts. These features are absent in the specimen of *Procynosuchus* (type of *Leavachia duvenhagei*) described by Brink and Kitching (1953) and my juvenile specimen, both of which resemble more typical cynodonts in these respects.

The Interrelationships of the Major Taxa of Cynodontia

The pattern of interrelationships of families within the nonmammalian cynodonts is generally agreed upon by Kemp (1982, 1988) and Hopson and Barghusen (1986), although disagreement persists with respect to the phylogenetic positions of the families Cynognathidae and Tritylodontidae (see Figs. 31–32; also see Sues, 1985, and Rowe, 1986). The relationships of the families Probainognathidae and Chiniquodontidae are also uncertain (Hopson and Barghusen, 1986:83–84, 101–102). These problems are discussed below.

A recent phylogeny of cynodonts by Battail (1982) (see Fig. 30) differs in many details from the phylogenies of Kemp (1982, 1988) and Hopson and Barghusen (1986). Battail's analysis is based primarily on dental characters, and most of his groups are diagnosed by relatively few synapomorphies. The principal unique features of Battail's phylogeny are as follows: (1) the recognition of the reality of a group of small, primitive cynodonts, the Silphedestidae; (2) considering *Dvinia* and *Parathrinaxodon* to be more derived than Procynosuchidae; and (3) the recognition of a monophyletic carnivorous clade, including *Galesaurus*, *Thrinaxodon*, *Cynognathus*, and Chiniquodontidae (including *Probainognathus*), as the sister-taxon of a monophyletic gomphodont clade.

The Family Silphedestidae: An Artificial Taxon

The family Silphedestidae was established by Haughton and Brink (1954) to accommodate several genera of small cynodonts having numerous primitive features in common with small therocephalians ("scaloposaurs"). The specimens assigned to this family are all poorly preserved and therefore difficult to interpret. Hopson and Kitching (1972) concluded that the Silphedestidae is an artificial group, all taxa included in it by Haughton and Brink being based on juvenile specimens of *Pro-*

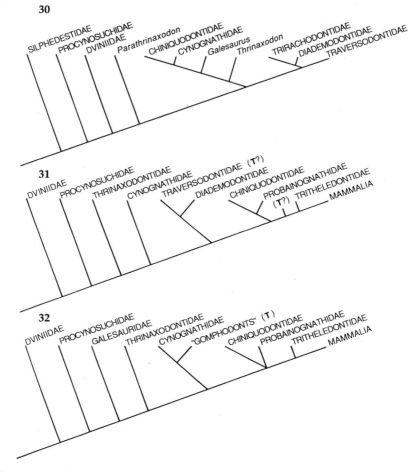

Figs. 30–32. Cladograms representing three theories of interrelationship of the major cynodont groups. (30) Cladogram of Battail (1982). (31) Cladogram of Kemp (1988). (32) Preferred cladogram of the author. T = preferred systematic position of Tritylodontidae; T? = alternate positions of Tritylodontidae.

cynosuchus. However, Battail (1982) argued for the reality of the Silphedestidae as a monophyletic assemblage that he regards as the plesiomorphous sister-group of all other cynodonts.

Nearly all of the distinguishing features of silphedestids can be interpreted as juvenile characteristics. Most of these features also have been used to characterize the Scaloposauridae, a supposed family of small therocephalians composed only of immature individuals (Hopson and Barghusen, 1986:95). Juvenile features of both silphedestids and scaloposaurs include (1) small size, (2) an elongated snout (and short temporal region), (3) an incomplete postorbital bar, (4) weakly differentiated canines, and (5) a broad intertemporal region lacking a distinct sagittal crest. Even the absence of lingual cingula on the postcanines, cited by Battail as a distinguishing character of Silphedestidae, can be considered a juvenile feature, because lingual cingula are absent in babies of the Early Triassic cynodont *Thrinaxodon liorhinus* (Gow, 1985) but are well developed in older individuals.

I am studying an undescribed cynodont specimen from the Late Permian of South Africa which fits the diagnosis of the Silphedestidae in nearly all respects. However, I regard this specimen as an immature individual of *Procynosuchus delaharpeae* (the only species of South African procynosuchid; Hopson and Kitching, 1972; Battail, 1982). Among other features, its braincase is poorly ossified and the three pelvic girdle elements are not fused. This specimen indicates that certain supposed autapomorphies of silphedestids, such as the absence of zygomatic arches and, in some cases, the postorbital bar, are based on postmortem loss; these structures are absent on the right side of the skull but present on the left. The preserved zygoma and postorbital bones are slender and thus readily subject to postmortem destruction.

An additional feature considered autapomorphic for silphedestids by Battail is the absence of the parietal ("pineal") foramen. This is a questionable character because the parietal region is missing or damaged in all specimens attributed to the Silphedestidae (pers. observ.).

The Systematic Positions of *Dvinia* and *Parathrinaxodon* with Respect to the Procynosuchidae

Although *Procynosuchus* seems to lack autapomorphies and therefore approximates the expected primitive morphotype for all later cynodonts, Kemp (1979, 1982, 1988) and Hopson and Barghusen (1986) considered *Dvinia prima* from the Late Permian of Russia to be the most primitive known cynodont. Although uniquely specialized in its dentition, *Dvinia* is more primitive than all other cynodonts, including *Procynosuchus*, in the following features (Hopson and Barghusen, 1986:100). (1) There is limited sutural contact of the prootic with the epipterygoid above the trigeminal foramen. (2) The posttemporal foramen is relatively large. (3) The paroccipital process is anteroposteriorly narrow and lies almost entirely behind the level of the fenestra ovalis. (4) The squamosal lacks a posterior flare in the rear wall of the temporal fossa.

Kemp (1979) cited three features in which *Dvinia* is more primitive (i.e., more therocephalian-like) than other cynodonts—(1) paired vomers; (2) the supraoccipital broader than the postparietal; and (3) the presence of six lower incisors. Battail (1982) and Hopson and Barghusen (1986) questioned the primitiveness of these features. According to Battail (1982), the vomers of *Dvinia* are fused and the supraoccipital is not broader than the postparietal. The high number of six lower incisors is, according to Hopson and Barghusen (1986), an autapomorphy of *Dvinia* inasmuch as the primitive number of lower incisors for theriodonts, including the primitive cynodont *Procynosuchus*, is four.

Battail (1982) considered *Dvinia* to be more derived than *Procynosuchus* in having the "lateral face of the quadratojugal overlapped by the squamosal" and in having the precanine maxillary dentition reduced to a single "vestigial precanine tooth." The significance of the first character is unclear because in *Procynosuchus* the upper end of the quadratojugal lies in a recess in the squamosal (Kemp, 1979:94) so that its lateral face is covered. This feature contradicts Battail's statement that the full height of the quadratojugal of procynosuchids is visible in lateral view. The second character, the presence of a single vestigial precanine tooth, may not distinguish *Dvinia* from procynosuchids because, as discussed below, the number of precanine maxillary teeth in the latter varies from a maximum of three to a possible minimum of zero. In addition, a supposed derived feature of *Dvinia* with respect to procynosuchids is that the secondary palate is partially closed by contact of the maxillary and palatine plates on the midline (Battail, 1982; Hopson and Barghusen, 1986). However, my recent restudy of the two known specimens of *Dvinia*, neither of which is well preserved, indicates that the palatal processes of the maxilla and palatine do not meet on the midline. Therefore, in none of these features is *Dvinia* unequivocally more derived than *Procynosuchus*.

Parathrinaxodon, considered by Battail (1982) to be more derived than procynosuchids and *Dvinia*, was described by Parrington (1936) on the basis of the anterior two-thirds of a skull from the Late Permian of Tanzania. The two halves of the secondary palate appear to have a straight, abutting contact on the midline, and Battail (1982) described the secondary palate as closed along a major part of its length. However, my recent restudy of the sole specimen of *Parathrinaxodon* indicates that the apparent closure of the secondary palate is the result of transverse compression of the snout. In life, the medially curved plates of the secondary palate did not meet on the midline and, thus, were identical to the condition of *Procynosuchus*. The postcanine dentition also is indistinguishable from that of *Procynosuchus*. Battail (1982) allied *Parathrinaxodon* with the more evolved cynodonts on the basis of shared absence of precanine maxillary teeth. I regard the absence of precanines in *Parathrinaxodon* as uncertain, because a suture cannot be distinguished between premaxilla and maxilla; thus, it is possible that one or more of the presumed incisors is a precanine. On the other hand, even if the *Parathrinaxodon* specimen genuinely lacks precanine teeth, I consider this to be insufficient grounds for its familial separation in view of its near identity to *Procynosuchus* in all other respects. Therefore, I consider *Parathrinaxodon* to be a procynosuchid, possibly indistinguishable from *Procynosuchus*.

The Interrelationships of Advanced Cynodontia

As noted by Hopson and Barghusen (1986) and Kemp (1988), the Cynognathidae, Chiniquodontidae, Probainognathidae, the "gomphodont" families (see below), Tritylodontidae, Tritheledontidae, and Mammalia form a clade (see Figs. 31–32) characterized by at least four or five synapomorphies. Kemp (1982) designated a paraphyletic taxon Eucynodontia for the nonmammalian groups of advanced cynodonts but later (1988) made the Eucynodontia monophyletic by including the Mammalia within it.

These authors also agreed that the family Thrinaxodontidae is the sister-taxon of the Eucynodontia. In addition, Hopson and Barghusen (1986) recognized a more primitive taxon, the Galesauridae, intermediate between the Procynosuchidae and Thrinaxodontidae. Galesaurids (*Cynosaurus* and *Galesaurus*) share eight cranial synapomorphies with *Thrinaxodon* and eucynodonts, but they retain an incomplete secondary palate of procynosuchid type. In addition, galesaurids share with thrinaxodontids and some eucynodonts localized platelike expansions on the thoracic and lumbar ribs which Jenkins (1971) termed "costal plates."

Battail (1982) presented a substantially different view of the interrelationships of the post-procynosuchid cynodonts (Fig. 30). He was unconvinced by Kemp's (1979) idea of a close relationship between *Diademodon* and *Cynognathus*—i.e., by the concept of the Eucynodontia. Rather, he regarded the total disappearance of the lingual cingulum on the postcanine teeth as an important character uniting the carnivorous Cynognathidae, Chiniquodontidae, and *Galesaurus*. Although *Thrinaxodon* possesses lingual cingula, its postcanines are more laterally compressed and its cingula narrower than those of *Procynosuchus*. Therefore, Battail considered its cheek dentition to be a divergent specialization relative to that of *Procynosuchus* (and the gomphodont cynodonts), and more similar to that of his carnivorous clade. He regarded the gomphodont cynodonts as the monophyletic sister-group of the carnivores primarily on the basis of the great transverse expansion of their postcanines and the development of occlusion between upper and lower postcanines.

I cannot accept Battail's scheme for two reasons. First, as discussed below, there are at least eight derived features shared by cynognathids, chiniquodontids, and gomphodonts that unite them as the Eucynodontia. Battail's carnivorous clade, in which *Galesaurus* and *Thrinaxodon* are included, is based on only two unique characters—lateral compression of the postcanines and the reduction of lingual cingula. Therefore, Battail's cladogram is much less parsimonious than those of Kemp and of Hopson and Barghusen in which a eucynodont clade is recognized. Second, about half the members of Battail's carnivore clade beyond the level of *Thrinaxodon* possesses lingual cingula on the postcanines. Well-developed cingula occur in the chiniquodontids *Chiniquodon* and *Aleodon* (pers. observ.) and in the probable chiniquodontid *Cromptodon* (Bona-parte, 1972). *Probainognathus* (formerly considered a chiniquodontid) possesses reduced cingulum cusps (pers. observ.). The only chiniquo-dontid known to lack cingula is *Probelesodon*. Thus, Battail's cladogram becomes even less parsimonious by requiring the multiple loss of lingual cingula or their secondary reappearance in chiniquodontids and *Pro-bainognathus*.

Cladograms of the interrelationships of the Eucynodontia were pre-sented by Kemp (1982, 1988; see Fig. 31), Sues (1985), Hopson and Barghusen (1986; see Fig. 32 for a modified version of their cladogram), and Rowe (1986). In the following sections I shall evaluate the alternate interpretations of the interrelationships of eucynodont subgroups on the basis of my restudy of the distribution of characters among these sub-groups. Such reanalysis indicates that the hierarchical levels at which many previously cited characters actually appear differ from those in many or all of the published cladograms. The following topics are dis-cussed: (1) the synapomorphies of Eucynodontia; (2) the position of Cynognathidae; (3) the interrelationships of the major subgroups of gom-phodonts; (4) the positions of Probainognathidae and Chiniquodontidae; (5) the interrelationships of Tritylodontidae, Tritheledontidae (= Ictido-sauria), and Mammalia; and (6) the synapomorphies of Mammalia.

The synapomorphies of Eucynodontia. Hopson and Barghusen (1986:Table 4) listed four synapomorphies of what they call a "post-thrinaxodontid cynodont" clade, which is synonymous with Kemp's (1982) more felici-tous term *Eucynodontia.* The synapomorphies are as follows: (1) the den-tary greatly enlarged so that it closely approaches the jaw articulation and also forms a distinct posteroventral angular region; (2) the vertical por-tion of the surangular and angular reduced in height, and the postden-tary series more rodlike and obliquely oriented; (3) the reflected lamina of the angular further reduced in size from the primitive cynodont condi-tion; (4) the quadrate ramus of the pterygoid greatly reduced or absent.

To this list can be added three additional synapomorphies that Hopson and Barghusen considered to characterize a more restricted group includ-ing Cynognathidae, gomphodonts, Probainognathidae, and Chiniquo-dontidae. Because of the transfer of the last two groups to a clade includ-ing Tritheledontidae and Mammalia (see below), the following become synapomorphies of the more inclusive clade Eucynodontia: (5) secondary jaw articulation formed between the surangular and a flat facet on the descending flange of the squamosal; (6) the dentaries fused at the symphysis; and (7) the pterygoids and basisphenoid forming an elongate ventral basicranial girder. In addition, the following feature cited by Kemp (1988; from Rowe, 1986) also seems to characterize the Eucynodontia—(8) the paroccipital process with a posteroventral ridge behind the jugular foramen (forming a posterior wall to the middle-ear cavity).

Postcranial synapomorphies are difficult to establish because of inadequate knowledge of the postcranial skeleton of most eucynodonts. Nevertheless, a tentative list follows: (9) costal plates absent from the anterior thoracic vertebrae and reduced in size on more posterior thoracic vertebrae; (10) the acromion process on the scapula; (11) the (probable) loss of disclike "additional" phalanges in the manus and pes (known in *Diademodon*); and (12) the (probable) loss of Distal Carpal 5 (known in *Diademodon*).

The position of Cynognathidae. The Early Triassic carnivorous cynodont *Cynognathus* was considered by Kemp (1982, 1988) and Rowe (1986) to be the sister-taxon of all other eucynodonts. In support of the latter group, Kemp (1988) listed four synapomorphies and Rowe (1986) seven. Sues (1985) and Hopson and Barghusen (1986) considered *Cynognathus* to be the sister-taxon of the gomphodont cynodonts only, a group to which the latter authors applied the name *Cynognathia*. Hopson and Barghusen provided only a single synapomorphy of the Cynognathia: a ventrally directed process of the jugal on the anterior root of the zygomatic arch.

Of the characters listed by Kemp and Rowe linking all noncynognathid eucynodonts, only the following one seems to me to be unequivocally unique to all members of this group and distinctly different from the condition in *Cynognathus*—(1) the deep incisure in the dorsal border of the squamosal between the occiput and the posterior root of the zygomatic arch, giving the skull a W-shaped dorsal profile in occipital view. I find several of their other characters insufficiently distinct from the condition in *Cynognathus* to provide a clear-cut differentiation between the two groups. Two features listed by Rowe, the backward flaring of the lambdoidal crest and of the squamosal in the rear of the temporal opening, though not seen in smaller individuals of *Cynognathus* (Broili and Schröder, 1934), are seen in larger individuals (Broili and Schröder, 1935). An additional feature, the posteriorly emarginated coronoid process of the dentary, is doubtfully more derived than the anteroposteriorly expanded, but unemarginated, coronoid process of *Cynognathus*; if anything, the shorter, emarginated coronoid process of other eucynodonts is less modified from the presumed ancestral condition represented by

Thrinaxodon. A second derived feature of noncynognathid eucynodonts, not listed by Kemp or Rowe, is: (2) vomers not exposed between the anterior ends of the maxillae in the secondary palate.

Against these two characters supporting the monophyly of the non-cynognathid eucynodonts can be set the following characters supporting the Cynognathia, a clade consisting of the Cynognathidae and gomphodonts: (1) the jugal with a ventrally directed process on the anterior root of the zygomatic arch; (2) the posteroventral part of the zygomatic arch expanded laterally at, or behind, the level of the quadrate, giving the skull a triangular appearance in dorsal view and creating a broad anteroventral margin to the groove for the external auditory meatus (this feature is most prominently developed in *Cynognathus* and *Diademodon*, but it is also present in the primitive gomphodonts *Trirachodon* and *Pascualgnathus*); (3) the jugal portion of the zygomatic arch dorsoventrally expanded; (4) the internal carotid foramina absent; (5) the dorsal ridge on each lumbar rib is reflected laterally so as to grip the posterior part of the preceding rib; and (6) overlapping costal plates on the lumbar ribs with distinct articular facets.

Kemp (1988) considered the descending process on the jugal to be of uncertain systematic value because it is variably developed in traversodonts and is difficult to distinguish from the thickening of this region in the chiniquodontid *Probelesodon*. Concerning the first point, the jugal process is present in the most primitive traversodont *Pascualgnathus* and in *most* later members of this group; therefore, it should be regarded as an apomorphy of traversodonts. On the second point, a chiniquodontid autapomorphy is the distinct angulation between the ventral border of the maxilla immediately behind the tooth row and that part of the maxilla forming the anteroventral margin of the zygomatic arch; this angulation marked the anterior extent of the masseter muscle on the zygoma (pers. observ.). In cynognathians, the masseter attachment did not extend on to the maxilla but terminated on the descending flange of the jugal. I consider the morphological distinction between the two features to be clearcut and unlike the primitive condition seen in *Thrinaxodon* and retained in *Probainognathus*.

The massive development of the zygoma and the distinctly triangular shape of the skull in dorsal view are developed most fully in the early (Early Triassic) and morphologically primitive cynognathians (*Cynognathus, Diademodon, Trirachodon,* and *Pascualgnathus*). In later traversodonts and in tritylodontids, the zygomatic arches are more parasagittal and may even be constricted in the middle. In contrast, *Probainognathus* and chiniquodontids are more primitive in having less deep and more curved zygomatic arches that lack the distinct posterolateral "cornering" characteristic of cynognathians.

The costal plates on the lumbar ribs of *Cynognathus* and gomphodonts are specialized beyond the condition seen in galesaurids and *Thrinaxodon* in having: (1) the longitudinal dorsal ridge reflected laterally to overlap and grip the rear portion of the preceding rib, and (2) specialized articular facets between adjacent ribs (Jenkins, 1971). In *Probainognathus* and chiniquodontids, the lumbar ribs lack overlapping expansions (Romer and Lewis, 1973); it is more parsimonious to presume that this secondary loss proceeded from the less specialized condition seen in *Thrinaxodon* than from the much more derived condition characterizing *Cynognathus* and gomphodonts.

At present, I consider a clade consisting of *Cynognathus* and the gomphodonts to be better supported than a clade consisting of all eucynodonts minus *Cynognathus*.

The interrelationships of the major groups of gomphodonts. The clade consisting of the gomphodont cynodonts is characterized primarily by the great transverse expansion of the crowns of the majority of upper and lower postcanine teeth and the development of crown-to-crown occlusion (Crompton, 1972; Sues, 1985; Hopson and Barghusen, 1986). Hopson and Kitching (1972) applied the name *Tritylodontoidea* to this group, in which they have been followed by Sues (1985) and Hopson and Barghusen (1986). However, in view of the unresolved controversy over whether the family Tritylodontidae should be included here, I shall use the noncommittal term *gomphodonts* to refer to the undoubted members (i.e., the nontritylodontids) of the clade.

The gomphodonts are well characterized. Hopson and Barghusen (1986:Table 4) listed two synapomorphies for the group, only the first of which (Number 1, below) I now consider valid. The second, "descending flange of jugal greatly enlarged," is equivocal inasmuch as the flange is relatively small in *Pascualgnathus* and other primitive traversodontids. However, at least six synapomorphies of gomphodonts can be cited, as follows: (1) the postcanine tooth crowns greatly expanded transversely, with crown-to-crown occlusion well developed; (2) occluding cheek teeth with true thecodont implantation in which the roots sit in deep sockets and, in life, were held by a periodontal ligament; (3) all transversely expanded postcanines are members of a single sequentially erupting series (*Zahnreihe*), except perhaps in early ontogenetic stages; (4) the postorbital portion of the zygomatic arch deep along its entire length, at its highest point extending above the level of the dorsal rim of the orbit; (5) the zygomatic process of the jugal at least twice as deep as the laterally exposed portion of the squamosal; and (6) separate foramina on the epipterygoid-prootic suture for the maxillary and mandibular rami of the trigeminal nerve.

Three Early Triassic genera—*Diademodon, Trirachodon,* and *Pascualgnathus*—are the most primitive members of the three families of gomphodonts recognized by Romer (1967), Crompton (1972), Battail (1982), and Hopson and Barghusen (1986): Diademodontidae, Trirachodontidae, and Traversodontidae. If the latter family should prove to contain the Tritylodontidae as a highly derived subgroup, as supported by Sues (1985) and Hopson and Barghusen (1986) but not by Kemp (1982, 1983), it becomes paraphyletic. However, for purposes of discussion, the Traversodontidae will be treated provisionally as though it were strictly monophyletic.

The interrelationships of the typical gomphodont cynodonts have been discussed by Battail (1982), Kemp (1982, 1988), and Rowe (1986). Battail (1982) united the Diademodontidae and Trirachodontidae as the sister-group of the Traversodontidae on the basis of the following synapomorphies: (1) the descending flange of the jugal strongly developed, and (2) the crowns of the postcanines with a continuous marginal ring of small cusps and a transversely oriented central ridgelike cusp. Battail argued, correctly I believe, that the Traversodontidae cannot be considered either "advanced" with respect to diademodontids or of diademodontid ancestry, as stated by Hopson and Kitching (1972), because Early Triassic traversodontids are in many respects more primitive than *Diademodon*. Kemp (1982, 1988) considered *Trirachodon* to be a member of the Diademodontidae with the Traversodontidae as the sister-group of the former family; thus, Kemp's concept of gomphodont phylogeny is essentially the same as that of Battail.

Rowe (1986) presented a radically different view of gomphodont relationships. He considered them to be a paraphyletic series of taxa lying between a more primitive chiniquodontid-probainognathid group and a more derived tritylodontid-tritheledontid-"mammaliaform" group (the last of these terms being synonymous with *mammals* of most classifications). Synapomorphies of Rowe's gomphodont-to-mammals clade include some of the dental features noted above, such as thecodont implantation and precise occlusion, plus long, tapering, apically closed roots. He did not attempt to resolve the interrelationships among the more primitive gomphodont genera, leaving them as an unresolved polychotomy. Further discussion of Rowe's views on gomphodont phylogeny are deferred until the problem of tritylodontid relationships is considered.

Hopson and Barghusen (1986) did not attempt to determine the interrelationships among subgroups of gomphodonts. However, on the basis of my unpublished studies of primitive gomphodonts, I consider the Diademodontidae to be the sister-taxon of the Trirachodontidae plus Traversodontidae. Synapomorphies of the latter clade are as follows: (1) the upper postcanines set in from the lateral margin of the snout so that posterior teeth are in line with the lateral face of the transverse flange of the pterygoid and the maxilla forms an overhanging shelf buccal to the tooth row; (2) the ectopterygoid bone small or absent; (3) no anterior wave of "conical" postcanines that sequentially replace "molariform" postcanines; (4) "semimolariform" teeth between "molariforms" and "sectorials" at the posterior end of the postcanine series very few or absent; and (5) the vascular channel on the lateral surface of the prootic, extending from the pterygo-paroccipital foramen to the trigeminal foramen, overlain by a dorsal flange of the prootic to form a closed canal. Characters 1 and 2 undoubtedly are correlated to some degree, with the ectopterygoid having been "squeezed" out of existence by the lingual shift of the tooth row. Nonetheless, *Trirachodon* clearly shares a greater number of derived resemblances with traversodontids than it does with *Diademodon*.

The families Diademodontidae and Trirachodontidae show very little diversity, each possibly being monogeneric. The family Traversodontidae, on the other hand, is extraordinarily diverse (Crompton, 1972; Hopson and Kitching, 1972; Kemp, 1982). I am reviewing the morphology and systematics of the entire group but to date have published only on certain Middle and Late Triassic genera related to *Exaeretodon* (Hopson, 1984, 1985).

The positions of Probainognathidae and Chiniquodontidae. Chiniquodontids are eucynodonts possessing sectorial postcanine teeth but retaining a relatively more primitive skull form than *Cynognathus*. They are characterized most readily by having a secondary palate that is broad posteriorly and extends behind the level of the last postcanine. However, the unique distinguishing feature of the family is the distinct angulation between the ventral edge of the zygomatic process of the maxilla and the anteroventral margin of the zygomatic arch (Hopson and Barghusen, 1986:Table 4). The best known genera are *Probelesodon* and *Chiniquodon* (probably including *Belesodon*) from the Middle and early Late Triassic of South America (Romer, 1969a,b). *Aleodon*, based on a fragmentary skull preserving several heavily worn postcanines from the Middle Triassic of East Africa (Crompton, 1955), was identified as a chiniquodontid on the basis of new material as yet undescribed. In addition, Bonaparte (1972) described the genus *Cromptodon* from the Early Triassic of Argentina on the basis of a pair of very small dentaries with cheek teeth similar to those of *Thrinaxodon* but with broader cingula. Because of its dental resemblance to juvenile individuals of *Aleodon*, I consider *Cromptodon* to be a chiniquodontid.

The genus *Probainognathus* (Fig. 35) was described by Romer (1969c, 1970) as a small chiniquodontid with a glenoid cavity on the squamosal

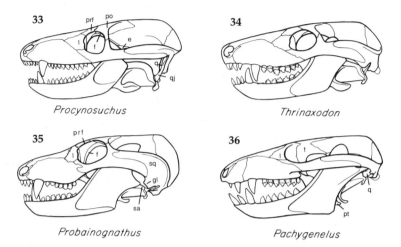

Figs. 33–36. Lateral views of the skulls of four nonmammalian cynodonts. (33) The procynosuchid *Procynosuchus*. (34) The thrinaxodontid *Thrinaxodon*. (35) The probainognathid *Probainognathus*. (36) The tritheledontid *Pachygenelus*. e = epipterygoid; f = frontal; gl = glenoid cavity on squamosal; l = lacrimal; po = postorbital; prf = prefrontal; pt = pterygoid; q = quadrate; qj = quadratojugal; sa = surangular; sq = squamosal. (All original.)

which received the posterior tip of the dentary and part of the surangular. He later referred it to its own family, Probainognathidae, because of its "progressive cynodont features," notably its mammal-like jaw articulation (Romer, 1973). Crompton and Jenkins (1979) have shown since that only the surangular, and not the dentary, enters the squamosal glenoid. Nonetheless, because it combines mammal-like features with an almost total absence of autapomorphies that would bar it from the direct ancestry of mammals, *Probainognathus* is widely regarded as a near-ideal structural antecedent to mammals (Romer, 1969c, 1970; Crompton and Jenkins, 1979; Crompton and Hylander, 1986).

The position of the Chiniquodontidae and Probainognathidae within the Eucynodontia was considered by Kemp (1982, 1988), Hopson and Barghusen (1986), and Rowe (1986). In general, a close relationship between them is accepted by most authors, principally on the basis of the long secondary palate characteristic of both. Kemp (1982, 1988) considered them to be monophyletic and placed them closer to the Mammalia than to the gomphodonts (Fig. 31). As noted above, Rowe (1986) placed them at the base of the unnamed taxon consisting of the noncynognathid eucynodonts, where the Chiniquodontidae, *Probainognathus*, and a gomphodont-to-mammal clade form an unresolved trichotomy.

Hopson and Barghusen (1986) initially considered the Chiniquodontidae and Probainognathidae to be the sister-group of the Cynognathia and, within this larger taxon, linked them as the Chiniquodontoidea on the basis of the elongated secondary palate. However, in an addendum to their paper, they concluded that the evidence linking the Cynognathia and Chiniquodontoidea was weaker than previously supposed. Because the latter might be more closely related to a clade consisting of Tritheledontidae (= Ictidosauria) plus Mammalia, two groups that also possess an elongated secondary palate, this character could no longer serve as a synapomorphy of a monophyletic Chiniquodontoidea. Hopson and Barghusen concluded that the relationships of these two families to each other and to the Cynognathia or Tritheledontidae plus Mammalia must be considered uncertain.

Further study of all of these eucynodont groups has led me to conclude that the Chiniquodontidae and *Probainognathus* do form a monophyletic group with the Tritheledontidae plus Mammalia. The synapomorphies linking them are (1) an osseous secondary palate extending to the posterior end of the tooth row, (2) the descending process of the frontal meeting the palatine in the medial wall of the orbit, (3) the ossified medial wall of the orbit extending further posteriorly (than in primitive cynodonts and cynognathians), (4) the parietal ("pineal") foramen absent, and (5) ribs lacking costal plates. The first character distinguishes this clade from all members of the Cynognathia. The remaining characters, however, distinguish it from *Cynognathus* and only the more primitive gomphodonts. In Middle Triassic traversodontids, the medial wall of the orbit is more fully ossified than in *Probainognathus*, although the frontal does not quite contact the palatine (Kemp, 1980c). In the Late Triassic traversodontid *Exaeretodon* the frontal meets the palatine in the orbit (pers. observ.), the parietal foramen is absent (Bonaparte, 1962), and all ribs lack costal plates.

I have not found any synapomorphies of Probainognathidae plus Chiniquodontidae and so no longer recognize the Chiniquodontoidea as a monophyletic taxon. Neither have I found synapomorphies that would convincingly link one or the other of these families to the tritheledontid-mammalian clade. The secondary palate of chiniquodontids, unlike that of *Probainognathus* but like that of tritheledontids and mammals, extends behind the level of the last postcanine. But *Probainognathus* more closely resembles the latter in having its postcanine tooth rows less divergent posteriorly than in chiniquodontids. The homology of the squamosal glenoid for the surangular in *Probainognathus* and the squamosal glenoid for the dentary in mammals (but not in tritheledontids) is uncertain. For the present, then, I prefer to express the interrelationships of these three taxa as an unresolved trichotomy (Fig. 32).

The interrelationships of Tritylodontidae, Tritheledontidae (= Ictidosauria), and Mammalia. Three groups of Cynodontia, first known in abundance from the Early Jurassic (though represented by rare specimens in the latest Triassic), represent a new, "higher" grade of synapsid evolution than has been discussed thus far. These are (1) the Tritylodontidae, a superficially rodentlike herbivorous group, which survived into the Middle or Late Jurassic; (2) the Tritheledontidae, or Ictidosauria (Fig. 36), a small group of presumed carnivores known principally from the Early Jurassic; and (3) the Mammalia, a group of initially small carnivores that radiated abundantly in the Cenozoic and are the only synapsids to have survived the Jurassic. These groups are more derived than Triassic cynodonts in the following cranial features: (1) the postorbital bar absent, so that the orbit is confluent with the temporal fossa; (2) the prefrontal bone absent; (3) the postorbital bone absent; (4) the quadrate with a concave medial facet; (5) the postdentary bones reduced to a slender rod supported in a deep groove on the medial side of dentary; (6) secondary jaw articulation between the surangular and squamosal absent; (7) the coronoid bone transversely widened; (8) a raised boss on the dentary for support of the coronoid bone; (9) dentaries not fused at the symphysis; and (10) the medial wall of the inner ear ossified, with separate foramina for the cochlear and vestibular branches of the auditory nerve.

In addition, tritylodontids and mammals possess a large number of derived similarities in the postcranial skeleton (Kemp, 1983; Rowe, 1986). Unfortunately, the postcranial skeleton of tritheledontids is poorly known at present. Broom (1932) figured scapulae of two tritheledontid specimens (which he called "Ictidosaurians A and B") and a restoration of the entire skeleton of one of them. Crompton (1958, 1963) described the skull and lower jaw of these specimens as *Diarthrognathus broomi*, but the poorly preserved postcranial skeleton never has been described. Isolated postcranial elements of the tritheledontid *Pachygenelus* currently are being studied by Crompton and me; at present, it is possible to report only that they closely resemble those of morganucodontid mammals (Jenkins and Parrington, 1976) and tritylodontids.

The relationships of these three taxa to one another and to Triassic cynodonts are the subject of a great deal of controversy at present. The long-accepted view has been that (1) Mammalia were derived from a persisting line of small carnivorous cynodonts similar in postcanine morphology to *Thrinaxodon* (Hopson and Crompton, 1969) and *Probainognathus* (Romer, 1969c, 1970); (2) Tritylodontidae were derived from advanced traversodontid gomphodonts (Crompton and Ellenberger, 1957; Crompton, 1972; Hopson, 1984); and (3) Tritheledontidae are structurally very advanced cynodonts, but of uncertain affinities (Barghusen, 1968; Hopson and Crompton, 1969).

Recently, Kemp (1982, 1983) argued for a close relationship between tritylodontids and mammals on the basis of an impressively large number (ca. 20) of derived resemblances between them. The most notable similarities, not seen in Triassic cynodonts, are in the postcranial skeleton. The dental and cranial resemblances of tritylodontids to gomphodonts Kemp regarded as convergences. Rowe (1986) amplified Kemp's analysis and listed no fewer than 47 synapomorphies of tritylodontids and mammals.

With regard to the possible relationships of tritheledontids, Kemp (1982:216) advocated "a relationship with other advanced cynodonts," and in his cladogram he ranked them as the sister-group of mammals. However, he acknowledged that until they become better known, their relationships will remain uncertain. Hopson and Barghusen (1986:Table 4), on the basis of undescribed specimens currently under study by Hopson and Crompton, listed nine synapomorphies of an "Ictidosaurian + Mammalian" clade. Of these nine, Kemp (1988) pointed out that four, and perhaps five, also occur in tritylodontids (Characters 1–4 above, plus prismatic enamel). As noted in the preceding section, a sixth resemblance, the elongate secondary palate, is best regarded as a synapomorphy of these groups with *Probainognathus* and chiniquodontids. This leaves the following three characters uniting tritheledontids and mammals: (1) the basicranium shortened anteroposteriorly, (2) the upper postcanines with an external cingulum, and (3) the zygomatic arch slender along its entire length. Kemp (1988) questioned the systematic significance of these features because all of the known specimens of tritheledontids are small, possibly juveniles. He concluded that until the braincase and postcranial skeleton are described, "the position of the tritheledontids as the sister-group of the mammals must be tentative" (Kemp, 1988:19). A similar conclusion was drawn by Rowe (1986).

On the basis of the total amount of information available (both published and unpublished) on tritheledontids, especially that on the postcranial skeleton, the most parsimonious conclusion as to their relationships is that they are closer to mammals than is any group of "typical" Triassic cynodonts. However, when the tritylodontids are brought into the picture, a tritheledontid-mammal clade indeed must be considered tentative, as Kemp and Rowe pointed out. This is not only because tritylodontids share at least 10 cranial and numerous postcranial characters with both of the other groups, but also because they share many unique resemblances *only* with mammals, and others *only* with tritheledontids. In other words, I believe it is possible that a sister-group relationship between *any* two of these three taxa might be given an equally convincing defense.

The following are tritylodontid-tritheledontid characters not found in mammals: (1) the first lower incisor enlarged; (2) the tip of the enlarged lower incisor fitting into the gap between laterally placed upper incisors;

(3) the premaxillae meeting behind the incisive foramina on the palate; (4) the postcanine tooth rows subparallel; and (5) the postcanine tooth rows extending internally to and behind the anterior border of the subtemporal fossa.

Sues (1985) defended the traditional view that tritylodontids are derived from traversodontid gomphodonts and that their mammal-like features, both cranial and postcranial, are the result of convergence. He reviewed most of the more than 20 features cited by Kemp (1983) as unique resemblances between tritylodontids and primitive mammals, and concluded that fewer than 10 can be accepted as possible synapomorphies. He presented 11 characters of tritylodontids that are shared with various, though not all, genera of Middle and Late Triassic traversodontids.

A number of additional characters uniquely shared by gomphodonts plus tritylodonts can be added to Sues's list, including the last three synapomorphies of the gomphodont clade listed above (see The Interrelationships of the Major Groups of Gomphodonts, above). However, as Kemp (1988) pointed out, some of Sues's supposed synapomorphies occur only in some traversodontid genera, notably *Exaeretodon*, whereas others (primarily dental) occur only in such genera as *Massetognathus*. Therefore, until the interrelationships of traversodontids have been sorted out, arguments for associating tritylodontids with them "will be ambiguous" (Kemp, 1988). I agree with Kemp's conclusion but would point out that as most traversodontids have been described primarily on the basis of dentitions and are as yet very incompletely known in cranial and especially postcranial morphology, a totally satisfactory analysis is not possible at the present time.

A novel attempt at resolving this problem was presented by Rowe (1986), who placed the very derived genus *Exaeretodon* (considered by most authors to be a traversodontid) as the sister-taxon of Tritylodontidae, Tritheledontidae, and Mammalia (the latter group called "Mammaliaformes" by Rowe). Kemp (1988) noted that by placing "the supposed traversodontid *Exaeretodon* on its own" as the sister-group of the three more "advanced" groups, Rowe "actually resolves a number of incongruities." This is because *Exaeretodon* possesses herbivorous cynodont features in its skull and dentition that are similar to features of tritylodontids, in addition to mammal-like features, particularly in its postcranial skeleton, that are similar to features of both tritylodontids and mammals.

As discussed above (The Interrelationships of the Major Groups of Gomphodonts), in Rowe's arrangement the herbivorous cynodonts (i.e., gomphodonts and tritylodontids) form a paraphyletic sequence of branches from the "main" line leading to mammals. This hypothesis of relationships implies one of two things: either the more conservative, carnivorous features of early mammals (and tritheledontids) are secondary reversals to a *Thrinaxodon-Probainognathus* morphology from a specialized gomphodont morphology; or the derived herbivorous features of the multiple gomphodont plus tritylodontid side branches are independently evolved specializations from a persistently carnivorous ancestral morphology. In either case, as with the more traditional view preferred by Sues (1985) and Hopson and Barghusen (1986), Rowe's hypothesis requires an enormous amount of evolutionary convergence.

For the present, a convincing resolution of the problem of interrelationships of gomphodonts, tritylodontids, tritheledontids, and mammals has yet to be demonstrated. Finding a solution would be greatly aided by better knowledge of skull and postcranial diversity in Middle and Late Triassic gomphodonts. *Exaeretodon,* which I would classify as a traversodontid in a monophyletic gomphodont clade (contra Rowe; see Hopson, 1984, 1985), provides the best evidence for a relationship between tritylodontids and more primitive gomphodonts. It represents an intermediate stage in the evolution of the postcranial skeleton from a typical cynodont morphology to an essentially mammalian morphology; therefore, it demonstrates the likelihood that the mammal-like features of tritylodontids evolved independently of those of mammals within the gomphodont clade.

Tritheledontids have the potential to contribute to a solution because, although mammal-like in many features of the skull and postcranial skeleton, they seem in other features to be not only more primitive than tritylodontids and early mammals, but even more primitive than *Exaeretodon* and other gomphodonts. Thus, in any theory of the interrelationships of these groups, a high incidence of convergence, reversal, or both is unavoidable.

Figure 32 presents my preferred cladogram of cynodont interrelationships at the present time. These conclusions are provisional, pending a complete analysis of all available characters using a computerized parsimony program.

I consider tritylodontids to be part of the gomphodont radiation, derived from among Middle Triassic traversodontids with dental features similar to those of *Massetognathus* or *"Scalenodon" hirschoni* (Crompton, 1972), but with postcranial features like those of *Exaeretodon* (Hopson, 1984; Sues, 1985). Therefore, almost by default, I accept a sister-group relationship of tritheledontids with mammals because, although they possess many derived features in common with both mammals and tritylodontids, tritheledontids lack the herbivorous specializations shared by tritylodontids only with gomphodonts.

· *The synapomorphies of Mammalia.* Hopson and Barghusen (1986) listed the following five synapomorphies of Mammalia: (1) the dentary with a distinct, ovoid articular condyle contacting a distinct, concave glenoid cavity of the squamosal; (2) postcanine teeth differentiated into premolars, which undergo a single replacement, and molars, which are not replaced; (3) postcanine teeth with divided roots (convergent with Tritylodontidae); (4) molar teeth with well-developed shear surfaces that form a consistent pattern of wear facets (convergent with some or all gomphodonts); and (5) the quadrate with an elongate stapedial process, the crus longus of the mammalian incus. In light of the recent paper by Crompton and Sun (1985) on the primitive mammal *Sinoconodon*, this list must be modified. These authors demonstrated that *Sinoconodon* is more primitive than the contemporaneous morganucodontids and kuehneotheriids in lacking precise occlusion between upper and lower molars. Therefore, Character 4 above is not a primitive feature of mammals.

Crompton and Sun listed seven synapomorphies of Mammalia, two of which are essentially identical to the first and third characters of Hopson and Barghusen. Another character—"prominent medial ridge and groove on the dentary for the support of the post-dentary bones"—I regard as a synapomorphy with tritheledontids and, thus, not unique to mammals. Also, because all known specimens of *Sinoconodon* have shed their anterior postcanine teeth, leaving only a diastema between the canine and the molars, it is not possible to state that they possessed true premolars at any time during ontogeny. Therefore, Crompton and Sun indicated as a synapomorphy of mammals that they have lost the primitive pattern of "alternate" replacement, which is still seen in tritheledontids. For the present, this is perhaps safer than assuming that *Sinoconodon* possessed true premolars (which in mammals undergo only one replacement) as well as nonreplacing molars. The remaining three synapomorphies of mammals listed by Crompton and Sun are all features of the braincase. An additional uniquely mammalian feature, which they cited in the text but failed to list, is the large promontorium housing the cochlea.

The following is a revised list of synapomorphies of the Mammalia; it is a combination of those characters of Crompton and Sun and of Hopson and Barghusen that I consider valid, to which an additional character is added. (1) The dentary condyle articulating with a squamosal glenoid; (2) the loss of "alternate" replacement of postcanine teeth; cheek dentition possibly divided into premolars, which replace once, and molars, which do not replace; (3) postcanine teeth with divided roots; (4) the quadrate with an elongate, ventrally directed stapedial process, the crus longus of the mammalian incus; (5) the anterior lamina of the periotic bone forming the lateral wall of the cavum epitericum and surrounding the mandibular branch and, possibly, the maxillary branch of the trigeminal nerve; (6) the periotic forming the floor to the cavum epiptericum below the ganglia for the trigeminal and facial nerves; (7) a prootic canal in the periotic; and (8) four lower incisors. The second character is distinguished from the condition characterizing derived traversodontids and tritylodontids, in which the postcanines erupt sequentially but no replacement takes place. The most primitive mammals must have had some replacement inasmuch as it occurs in all nongomphodont (and nontritylodontid) cynodonts, including the majority of mammals.

Tritylodontids also possess divided roots on their cheek teeth, but this is doubtfully homologous to the condition in mammals for two reasons. First, the roots in the upper postcanines of tritylodontids form a posteriorly open U that is distinct from the longitudinally aligned pair of roots in mammals but resembles the single, posteriorly concave, transversely oriented root of such traversodontids as "*Scalenodon*" *charigi* and *Exaeretodon* (pers. observ.). Second, in tritheledontids, the probable sistergroup of mammals, the roots are single rooted (also, see Sues, 1985).

The presence of four lower incisors in *Sinoconodon* and other Early Jurassic mammals represents an apomorphic condition. It differs from the primitive number of three, seen in all Triassic eucynodonts, and the derived number of two seen in tritylodontids and tritheledontids.

DISCUSSION

As noted at the beginning of this chapter, the determination of patterns of evolutionary change depends on the interpretation of phylogeny. Thus, in order to use the fossil record of synapsids to elucidate large-scale evolutionary trends relating to the origin of mammals and the evolutionary processes underlying such trends, one must have an explicit, detailed, and well-corroborated theory of synapsid interrelationships. As is evident from the preceding review of current views on synapsid phylogeny, important areas of disagreement persist among the most recent cladistic analyses of therapsid interrelationships—i.e., those of Kemp (1982, 1983, 1988), Rowe (1986; also see Gauthier et al., 1988; Rowe, 1988), and Hopson and Barghusen (1986). Additional disagreements are evident among students of primitive synapsids (Reisz, 1980, 1986; Brinkman and Eberth, 1983) and early amniotes (Carroll, 1982, 1986; Heaton and Reisz, 1986), but the interpretation of phylogenetic patterns in these groups has not had an impact on evolutionary theory comparable to that of phylogenetic patterns in therapsids and early mammals.

Perhaps the most influential idea derived from phylogenetic patterns in nonmammalian therapsids and Mesozoic mammals is that of large-scale parallel evolution as a pervasive phenomenon in the origin of higher taxa.

This concept was developed in greatest detail by Olson (1944, 1959), who, on the basis of his studies of morphological evolution in nonmammalian therapsids, concluded that "there appear in each of the lines of therapsids, irrespective of their adaptation to one or another facet of the environment, common characters, which are usually more fully expressed sometime after their origin than when they first appear" (1959:348). This interpretation of the therapsid fossil record, taken in conjunction with earlier studies by Simpson (1928, 1929) on Mesozoic mammals, which indicated that several major groups of early mammals showed no convergence at their bases, led Olson (1944) to the idea of a polyphyletic origin of the class Mammalia from several long-independent lineages of nonmammalian therapsids.

The theory of a polyphyletic origin of mammals gained wide support in the 1950's and 1960's and often was cited as supporting evidence for the polyphyletic origin of higher taxa in other vertebrate and many invertebrate groups. Most discussions of the evolutionary processes responsible for such extreme parallelism in therapsid evolution were framed in terms of the neo-Darwinian synthesis, with an emphasis on selection for improved adaptation (but see Grassé, 1977). Olson (1959) and Simpson (1960) postulated the existence of strong selection pressures for the perfection of homeothermy, a "general" adaptation that would confer an advantage in any environment that therapsids occupied. The evolution in all major therapsid lineages of similar features related to improved homeothermy was attributed to constraints of the "epigenotypes"—i.e., "the system of developmental canalizations and feedbacks" (Mayr, 1963:528) that limited the variation expressed in successive therapsid populations, and so led natural selection to yield similar results in separate lineages (see Hopson and Crompton, 1969:16–17).

Olson was well aware that "the validity of the statements concerning independent origin of characters is directly dependent upon the validity of the presumed relationships [of therapsid groups]," and "so far as these are accepted, the statements concerning independent origin of characters should find acceptance" (Olson, 1944:96). Among the most important of Olson's conclusions on nonmammalian therapsid phylogeny as they relate to independent origin of mammal-like characters are the following. (1) Of the four groups studied, the Gorgonopsia are closest to the "ancestral pelycosaurs" and lie close to the ancestry of the Cynodontia. (2) The Therocephalia had a gorgonopsian-like ancestry and are off the line leading to cynodonts. (3) The Anomodontia separated from the other therapsids nearly at the [sphenacodont] pelycosaur level or even from a different ancestral group of pelycosaurs (Olson, 1944:95). Thus, any derived resemblances shared by cynodonts, therocephalians, and anomodonts (or by any two of the three) that are absent in gorgonopsians must have arisen independently within the former groups. In addition, Olson concluded that the Ictidosauria [Tritheledontidae] probably "arose from one of the therocephalian families rather than from the cynodonts" (Olson, 1944:120), indicating that derived features shared by cynodonts and ictidosaurs also arose in parallel (also see Crompton, 1958). The phylogenetic "tree" resulting from this set of "presumed relationships" among therapsids thus resembles a bush with major branches diverging from a sphenacodont stem in the Early Permian and extending upward in time well into the Triassic, with the early mammals forming terminal twigs on several of these branches (Olson, 1944:Fig. 27; 1959:Fig. 1).

Hopson and Crompton (1969) reviewed the problem of parallel evolution in therapsids and the polyphyletic origin of mammals on the basis of new information on the morphology and phylogenetic relationships of nonmammalian therapsids and early mammals acquired since the mid-1950's. By the late 1960's it had become evident that cynodonts were allied to therocephalians rather than to gorgonopsians (Brink, 1960) and that ictidosaurs (tritheledontids) were allied to cynodonts rather than to therocephalians (Barghusen, 1968). This meant that many features formerly considered to be parallelisms or convergences between cynodonts, on the one hand, and therocephalians and ictidosaurs, on the other, now were interpretable as homologies, i.e., synapomorphies. More importantly, knowledge of Late Triassic and Jurassic mammals had increased to the point where it became possible to compare dental patterns and cranial structure between nonmammalian therapsids and mammals and among different groups of early mammals (K. A. Kermack, 1963, 1967; Hopson, 1967; Parrington, 1967; Crompton and Jenkins, 1968). Hopson and Crompton concluded that the evidence supported a monophyletic origin of mammals from a common ancestor that resembled *Morganucodon* in its dentition; in turn, the latter was derived from a lineage of small carnivorous cynodonts dentally similar to *Thrinaxodon*.

Hopson and Crompton recognized an essentially Triassic nonmammalian cynodont radiation interposed between the primarily Permian radiation of primitive therapsids and the Jurassic radiation of primitive mammals, with each radiation being monophyletically derived from the preceding one. Although they acknowledged that parallelism is "a strikingly pervasive phenomenon in therapsid evolution," they believed that the phylogenetic interpretation they presented "indicates that, in general, the greatest amount of parallelism occurs in groups that are phylogenetically most closely related" (Hopson and Crompton, 1969:64). Nonetheless, they accepted a relatively great amount of parallelism within the cynodonts because they considered each group of "advanced" cynodonts, such as gomphodonts (including tritylodontids), ictidosaurs, and the ancestors of mammals, to have been derived independently from a common ancestor at a relatively primitive, essentially thrinaxodontid, structural level. The evolutionary model accepted by Hopson and Crompton (1969) to explain the pattern of synapsid evolution, as they perceived it, was entirely neo-Darwinian—i.e., selectionist, adaptationist, and gradual.

Several major advances in knowledge of therapsids and early mammals occurred subsequent to the 1969 paper of Hopson and Crompton. Of particular significance were descriptions of the small- to medium-sized carnivorous cynodonts *Probainognathus* and *Probelesodon*, intermediate in grade between *Thrinaxodon* and early mammals (Romer, 1969b,c, 1970) and of the cranial and postcranial morphology of morganucodontid mammals (Kermack et al., 1973, 1981; Jenkins and Parrington, 1976).

The first overview of therapsid interrelationships utilizing an explicitly cladistic approach was that of Kemp (1982). Kemp recognized the very advanced nature of the Tritheledontidae within the nonmammalian cynodonts and considered the Tritylodontidae to lie very close to the Mammalia, primarily on the basis of the striking similarity of the tritylodontid postcranial skeleton to that of morganucodontids. Furthermore, he accepted the derived similarities of *Cynognathus*, gomphodonts, and other "advanced" cynodonts as evidence of a shared common ancestry beyond the level of the more primitive thrinaxodontids (Kemp, 1979); in this he disputed the assumption of Hopson and Crompton (1969) that each of these "advanced" cynodont groups evolved independently from a primitive cynodont grade. Kemp also considered cynodonts to have been derived from whaitsiids, advanced therocephalians sharing numerous derived resemblances with cynodonts that are absent in more primitive therocephalians (Kemp, 1972).

Kemp (1982) supported a punctuational theory of synapsid evolution in which periodic mass extinctions were followed by extremely rapid change ("quantum speciation") occurring apparently independently of natural selection and yielding species that were not necessarily well adapted to their environment. Subsequently, these new species gradually became better adapted through natural selection, so that a "correlated progression" of all morphological and physiological systems more or less in concert led to increased homeostasis, "the ability to withstand external fluctuations of the environment by means of regulation of the animal's internal environment" (Kemp, 1982:311). Because of the complexity of a correlated progression, Kemp considered it "extremely improbable that more than one lineage would achieve the same degree of homeostasis at the same time" (Kemp, 1982:332). His conception, he believed, was inconsistent with the older idea of widespread parallel evolution in therapsids. The cases of parallel evolution that he did see in therapsids were "of a very trivial kind," such as secondary palates in dicynodonts, cynodonts, and some therocephalians and the reduction of postcanine teeth in gorgonopsians and pristerognathid and whaitsiid therocephalians.

A consequence of the changing conception of therapsid phylogeny has been to convert many resemblances formerly considered to be convergences (i.e., the result of independent evolution) into homologies (i.e., evidence of relationship). It follows that in each case a different suite of characters, previously interpreted as evidence for an alternative set of relationships, must be considered to be convergences (or else primitive retentions). In general, features that are considered derived in the sense that they are more like those seen in mammals are interpreted by Kemp as evidence of relationship, whereas habitus specializations, such as shared herbivorous adaptations of gomphodonts and tritylodontids, are interpreted as convergences. If such habitus specializations are considered to be relatively superficial with respect to those more obviously related to the achievement of a mammalian level of homeostasis, then Kemp's conclusion that significant parallelism does not occur in nonmammalian therapsids follows from his theory of therapsid phylogeny.

Rowe (1986) carried Kemp's views on therapsid phylogeny even further toward interpreting mammalian features in all groups as synapomorphies (homologies) rather than as probable independent acquisitions. Thus, the mammal-like features not only of tritylodontids but also of gomphodonts (including many postcranial features of *Exaeretodon*) are interpreted by Rowe as synapomorphies with mammals. One effect of this is to remove *Probainognathus* and chiniquodontids much further from mammals than was done by Kemp (1982, 1983). Likewise, mammal-like resemblances (mainly, though not exclusively, of the postcranial skeleton) that were considered to be convergences by other workers are interpreted as synapomorphies by Rowe and used, for example, to link dicynodonts with eutheriodonts (therocephalians plus cynodonts), thereby removing the postcranially more primitive gorgonopsians from close relationship with the latter.

Because the recent cladistic analysis of nonmammalian therapsids published by Hopson and Barghusen (1986) utilized only characters of the head skeleton and teeth, it is of little help in determining whether one should expect a greater amount of convergence to occur in mammal-like features, particularly those of the postcranial skeleton, or in habitus specializations, particularly those of the skull and dentition. However, their study does show that even within the skull and dentition, a great deal of convergent evolution must be hypothesized to have occurred in nonmammalian therapsids.

This chapter has attempted to remedy one deficiency of the Hopson and Barghusen analysis by including postcranial characters. Unfortunately, the postcranial skeleton of many groups is much less well known than is the cranial skeleton, so I have not dealt with it as thoroughly as I would have wished. Because the entire skeleton of many important therapsid taxa has yet to be investigated adequately, significant gaps still exist in the character matrices on which any worker's cladogram of therapsids would be based. These gaps are important because I believe convergent evolution of mammal-like features, including features of the postcranial skeleton, has been a pervasive phenomenon in therapsid evolution; therefore, it is necessary to include in a cladistic analysis information on as many features of as many taxa as possible. Of particular importance is information on the most primitive known members of clades, inasmuch as convergence often cannot be detected if only more derived members are considered. For example, the most primitive members of some therapsid clades—i.e., *Otsheria* and "dromasaurs" among the Anomodontia, lycosuchids and scylacosaurids among the Therocephalia—are much more primitive in morphology than are more derived and better known members of their respective clades. The placement of gorgonopsians within the Theriodontia is disputed (Kemp, 1982; Rowe, 1986) because they seem to be so much more primitive than both eutheriodonts and dicynodonts; yet, gorgonopsians do not appear so primitive when they are compared with the least derived members of the Therocephalia and Anomodontia.

CONCLUSIONS

The purpose of this chapter is to review critically the principal recent hypotheses of synapsid interrelationships in order to point out where a consensus has been achieved and where disagreements remain. In the latter case, I try to identify the points of contention and what new data are required in order to resolve them. A second purpose is to show how theories of phylogeny affect interpretations of evolutionary pattern and, in turn, of evolutionary process. In the phylogeny of nonmammalian synapsids, the predominant question regarding the broad pattern of evolution traditionally has been: How much convergent or parallel evolution has occurred? As noted above, as our understanding of the phylogeny of the group has increased, the amount of required convergence has decreased. However, at the present time there are disagreements on the interrelationships of many synapsid groups that have led to very different interpretations concerning the degree and nature of convergence in synapsid evolution.

Unfortunately, this chapter does not resolve the systematic disagreements it discusses, and consequently, it cannot resolve the problem of how important the independent origin of similar features was in the evolutionary history of synapsids. It is more a progress report than a definitive statement on the pattern of synapsid phylogeny and the significance of convergence in synapsid evolution. In order to arrive at a more definitive statement on these problems, I believe one must observe the procedures discussed below in the analysis of synapsid phylogeny.

First, where possible, *all* relevant taxa must be analyzed. As noted repeatedly in this paper, a special effort must be made to include the most primitive known members of all clades in order to more reliably determine the primitive states of characters for those clades. I believe this is particularly important in the analysis of relationships among synapsids because of the large amount of homoplasy that I perceive to be present among the more derived members of different clades, as documented by Hopson and Barghusen (1986).

Second, as wide a range of characters as possible should be analyzed. If the analysis seeks to test two or more competing cladograms, it must consider *all* of the characters that have been used to support each of the alternatives. In addition, the analysis should consider the distribution of character states *within*, as well as between, well-corroborated monophyletic groups, even if one is not primarily concerned with in-group as much as with between-group relationships. This means including the distribution of character states within clades that are considered to be well off the "main" line to mammals (such as in more derived members of the Dicynodontia or Therocephalia), as well as in the basal members of the clades whose degree of relationship to mammals is being sought. Of equal importance is the inclusion of characters that might be considered autapomorphies of such "side groups" and of no relevance to determining how these groups are ordered on a cladogram with respect to mammals. An example might be features of the specialized postcanine crown patterns of gomphodont cynodonts (including tritylodontids); Crompton (1972) and Hopson and Barghusen (1986) united these forms as a clade largely on the basis of tooth crown features, but Rowe (1986), on the basis of other features, considered them to be a paraphyletic series on the line to mammals. Therefore, unless all available characters are utilized, information about the degree and nature of convergence will be inadequate and conclusions about its significance in synapsid evolution will be unavoidably subjective.

Third, character-state data should be displayed in a character matrix. This permits the amount of incompatibility among character states to be observed at a glance and the effects of missing or disputed data to be assessed.

Fourth, a computerized maximum parsimony analysis should be utilized. Such a procedure will not necessarily provide the "best" answer, but it will bring out a series of equally, or almost equally, parsimonious cladograms that will yield information on which characters are responsible for producing the various alternate cladograms. Such analyses provide a tool for understanding how particular characters affect cladogram topology and therefore their roles, often unexpected, in generating particular alternate cladograms. The principal value of a computerized parsimony analysis, then, is that it forces one to scrutinize and, to the best of one's ability, understand the implications of the characters one uses.

Knowledge of synapsids has grown immensely since 1959 when Olson reviewed patterns of parallel evolution in the group. The theory of a polyphyletic origin of mammals, generally held at that time, is no longer widely accepted, but questions still exist concerning how much independent acquisition of mammalian features occurred in the Synapsida and what its evolutionary significance may have been. As was recognized by Olson (1944), a reliable estimate of convergence must be based on a well-constructed, well-tested hypothesis of phylogenetic relationships. Although this has not yet been achieved, I believe the continuing high degree of activity among students of synapsid phylogeny since 1980 is carrying us much closer to this goal.

Acknowledgments This review of synapsid systematics was motivated by Dr. H.-P. Schultze's invitation to participate in the seminar series on which this volume is based. The project grew much beyond its originally intended limits, leading to many delays in the completion of the manuscript. I gratefully acknowledge the patience and good humor of Dr. Schultze during this period.

Thanks are due Drs. T. S. Kemp, T. Rowe, and G. M. King for providing unpublished manuscripts on synapsid phylogeny to me. The illustrations were prepared by Claire Vanderslice. I have profited from discussion with many colleagues, notably Drs. E. F. Allin, H. R. Barghusen, J. M. Clark, A. W. Crompton, J. Gauthier, T. S. Kemp, J. Wible, and R. Zanon. Drs. Allin, Clark, Wible, and Zanon read the manuscript and provided useful suggestions for improving it. The research on which this review is based was supported by National Science Foundation research grants BSR-8409109 and BSR-8615016.

Literature Cited

Allin, E. F. 1975. Evolution of the mammalian middle ear. J. Morphol., 147:403–438.
Barghusen, H. R. 1968. The lower jaw of cynodonts (Reptilia, Therapsida) and the evolutionary origin of mammal-like adductor jaw musculature. Postilla, Peabody Mus. Nat. Hist. Yale Univ., 116:1–49.
Barghusen, H. R. 1972. The origin of the mammalian jaw apparatus. Pp. 26–36 *in* Schumacher, G. H. (ed.), *Morphology of the Maxillo-mandibular Apparatus.* Leipzig: VEB Georg Thieme.

Barghusen, H. R. 1973. The adductor jaw musculature of *Dimetrodon* (Reptilia, Pelycosauria). J. Paleontol., 47:823–834.

Barghusen, H. R. 1976. Notes on the adductor jaw musculature of *Venjukovia*, a primitive anomodont therapsid from the Permian of the U.S.S.R. Ann. S. Afr. Mus., 69:249–260.

Battail, B. 1982. Essai de phylogénie des cynodontes (Reptilia: Therapsida). Geobios Mem. Sp., 6:157–167.

Bonaparte, J. F. 1962. Descripción del cráneo y mandíbula de *Exaeretodon frenguellii*, Cabrera, y su comparación con Diademodontidae, Tritylodontidae y los cinodontes sudamericanos. Publ. Mus. Munic. Cienc. Nat. Tradic. Mar del Plata, 1:135–202.

Bonaparte, J. F. 1972. *Cromptodon mamiferoides* Gen. et Sp. Nov., Galesauridae de la Formación Rio Mendoza, Mendoza, Argentina (Therapsida—Cynodontia). Ameghiniana, 9:343–353.

Boonstra, L. D. 1934. A contribution to the morphology of the Gorgonopsia. Ann. S. Afr. Mus., 31:137–174.

Boonstra, L. D. 1962. The dentition of the titanosuchian dinocephalians. Ann. S. Afr. Mus., 46:57–112.

Boonstra, L. D. 1964. The girdles and limbs of the pristerognathid Therocephalia. Ann. S. Afr. Mus., 48:121–165.

Boonstra, L. D. 1965. The girdles and limbs of the Gorgonopsia of the *Tapinocephalus* Zone. Ann. S. Afr. Mus., 48:237–249.

Boonstra, L. D. 1966a. The girdles and limbs of the Dicynodontia of the *Tapinocephalus* Zone. Ann. S. Afr. Mus., 50:1–11.

Boonstra, L. D. 1966b. The dinocephalian manus and pes. Ann. S. Afr. Mus., 50:13–26.

Bramble, D. M. 1978. Origin of the mammalian feeding complex: models and mechanisms. Paleobiology, 4:271–301.

Brink, A. S. 1960. A new type of primitive cynodont. Paleontol. Afr., 7:119–154.

Brink, A. S. 1963. A new skull of the procynosuchid cynodont *Leavachia duvenhagei* Broom. Palaeontol. Afr., 8:57–75.

Brink, A. S., and J. W. Kitching. 1953. On *Leavachia duvenhagei* and some other procynosuchids in the Rubidge Collection. S. Afr. J. Sci., 49:313–317.

Brinkman, D. 1981. The structure and relationships of the dromasaurs (Reptilia: Therapsida). Breviora, 465:1–34.

Brinkman, D., and D. A. Eberth. 1983. The interrelationships of pelycosaurs. Breviora, 473:1–35.

Broili, F., and J. Schröder. 1934. Zur Osteologie des Kopfes von *Cynognathus*. Bayer. Akad. Wiss. Math.-Naturwiss. Abt. Sitzungsber., 1934:95–128.

Broili, F., and J. Schröder. 1935. Beobachtungen an Wirbeltieren der Karrooformation: XI. Über den Schädel von *Cynidiognathus*. Bayer. Akad. Wiss. Math.-Naturwiss. Abt. Sitzungsber., 1935:199–222.

Broom, R. 1932. *The Mammal-like Reptiles of South Africa and the Origin of Mammals*. London: H. F. and G. Witherby.

Carroll, R. L. 1982. Early evolution of reptiles. Ann. Rev. Ecol. Syst., 13:87–109.

Carroll, R. L. 1986. The skeletal anatomy and some aspects of the physiology of primitive reptiles. Pp. 25–45 *in* Hotton, N., III, P. D. MacLean, J. J. Roth, and E. C. Roth (eds.), *The Ecology and Biology of Mammal-like Reptiles*. Washington: Smithsonian Inst. Press.

Carroll, R. L. 1988. *Vertebrate Paleontology and Evolution*. New York: W. H. Freeman.

Chudinov, P. K. 1983. Early therapsids. Tr. Paleontol. Inst. Akad. Nauk SSSR, 202:1–227. [In Russian.]

Clark, J. M., and J. A. Hopson. 1985. Distinctive mammal-like reptile from Mexico and its bearing on the phylogeny of the Tritylodontidae. Nature, London, 315:398–400.

Clemens, W. A., and J. A. Lillegraven. 1986. New Late Cretaceous, North American advanced therian mammals that fit neither the marsupial nor eutherian molds. Pp. 55–86 *in* Flanagan, K. M., and J. A. Lillegraven (eds.), *Vertebrates, Phylogeny, and Philosophy.* Contrib. Geol. Univ. Wyoming, Sp. Pap. 3.

Cluver, M. A., and G. M. King. 1983. A reassessment of the relationships of Permian Dicynodontia (Reptilia, Therapsida) and a new classification of dicynodonts. Ann. S. Afr. Mus., 91:195–273.

Cracraft, J. 1981. Pattern and process in paleobiology: the role of cladistic analysis in systematic paleontology. Paleobiology, 7:456–468.

Crompton, A. W. 1955. On some Triassic cynodonts from Tanganyika. Proc. Zool. Soc. London, 125:617–669.

Crompton, A. W. 1958. The cranial morphology of a new genus and species of ictidosaurian. Proc. Zool. Soc. London, 130:183–216.

Crompton, A. W. 1963. On the lower jaw of *Diarthrognathus* and the origin of the mammalian lower jaw. Proc. Zool. Soc. London, 140:697–753.

Crompton, A. W. 1972. Postcanine occlusion in cynodonts and tritylodontids. Br. Mus. Nat. Hist. Bull. Geol., 21:27–71.

Crompton, A. W., and F. Ellenberger. 1957. On a new cynodont from the Molteno Beds and the origin of the tritylodontids. Ann. S. Afr. Mus., 44:1–14.

Crompton, A. W., and W. L. Hylander. 1986. Changes in mandibular function following the acquisition of a dentary-squamosal jaw articulation. Pp. 263–282 *in* Hotton, N., III, P. D. MacLean, J. J. Roth, and E. C. Roth (eds.), *The Ecology and Biology of Mammal-like Reptiles.* Washington: Smithsonian Inst. Press.

Crompton, A. W., and F. A. Jenkins, Jr. 1968. Molar occlusion in Late Triassic mammals. Biol. Rev. Cambridge Philos. Soc., 43:427–458.

Crompton, A. W., and F. A. Jenkins, Jr. 1979. Origin of mammals. Pp. 59–73 *in* Lillegraven, J. A., Z. Kielan-Jaworowska, and W. A. Clemens (eds.), *Mesozoic Mammals: The First Two-thirds of Mammalian History.* Berkeley: Univ. of California Press.

Crompton, A. W., and A.-L. Sun. 1985. Cranial structure and relationships of the Liassic mammal *Sinoconodon.* J. Linn. Soc. London Zool., 85:99–119.

Currie, P. J. 1979. The osteology of haptodontine sphenacodonts (Reptilia: Pelycosauria). Palaeontographica, Ser. A, 163:130–168.

Eldredge, N., and J. Cracraft. 1980. *Phylogenetic Patterns and the Evolutionary Process.* New York: Columbia Univ. Press.

Gaffney, E. S. 1980. Phylogenetic relationships of the major groups of amniotes. Pp. 593–610 *in* Panchen, A. L. (ed.); *The Terrestrial Environment and the Origin of Land Vertebrates.* London: Academic Press.

Gardiner, B. G. 1982. Tetrapod classification. J. Linn. Soc. London Zool., 74:207–232.

Gauthier, J., A. G. Kluge, and T. Rowe. 1988. Amniote phylogeny and the importance of fossils. Cladistics, 4:105–209.

Goodrich, E. S. 1916. On the classification of the Reptilia. Proc. R. Soc. London, Ser. B, 89:261–276.

Goodrich, E. S. 1930. *Studies on the Structure and Development of Vertebrates.* London: Macmillan.

Gow, C. E. 1985. Dentitions of juvenile *Thrinaxodon* (Reptilia: Cynodontia) and the origin of mammalian diphyodonty. Ann. Geol. Surv. South Africa, 19:1–17.

Grassé, P. P. 1977. *Evolution of Living Organisms.* New York: Academic Press.

Haughton, S. H., and A. S. Brink. 1954. A bibliographical list of Reptilia from the Karroo beds of Africa. Palaeont. Afr., 2:1–187.

Heaton, M. J. 1980. The Cotylosauria: a reconstruction of a group of archaic tetrapods. Pp. 497–551 *in* Panchen, A. L. (ed.), *The Terrestrial Environment and the Origin of Land Vertebrates*. London: Academic Press.

Heaton, M. J., and R. R. Reisz. 1986. Phylogenetic relationships of captorhinomorph reptiles. Can. J. Earth Sci., 23:402–418.

Holmes, E. B. 1975. A reconsideration of the phylogeny of the tetrapod heart. J. Morphol., 147:209–228.

Hopson, J. A. 1966. The origin of the mammalian middle ear. Am. Zool., 6:437–450.

Hopson, J. A. 1967. Mammal-like reptiles and the origin of mammals. Discovery, 2(2):25–33.

Hopson, J. A. 1984. Late Triassic traversodont cynodonts from Nova Scotia and southern Africa. Palaeont. Afr., 25:181–201.

Hopson, J. A. 1985. Morphology and relationships of *Gomphodontosuchus brasiliensis* von Huene (Synapsida, Cynodontia, Tritylodontoidea) from the Triassic of Brazil. Neues Jahrb. Geol. Paläontol. Monatsh., 1985:285–299.

Hopson, J. A., and H. R. Barghusen. 1986. An analysis of therapsid relationships. Pp. 83–106 *in* Hotton, N., III, P. D. MacLean, J. J. Roth, and E. C. Roth (eds.), *The Ecology and Biology of Mammal-like Reptiles*. Washington: Smithsonian Inst. Press.

Hopson, J. A., and A. W. Crompton. 1969. Origin of mammals. Pp. 15–72 *in* Dobzhansky, T., M. K. Hecht, and W. C. Steere (eds.), *Evolutionary Biology*. Vol. 3. New York: Appleton-Century-Crofts.

Hopson, J. A., and J. W. Kitching. 1972. A revised classification of cynodonts (Reptilia; Therapsida). Palaeont. Afr., 14:71–85.

Jenkins, F. A., Jr. 1971. The postcranial skeleton of African cynodonts. Bull. Peabody Mus. Nat. Hist. Yale Univ., 36:1–216.

Jenkins, F. A., Jr., and F. R. Parrington. 1976. The postcranial skeleton of the Triassic mammals *Eozostrodon*, *Megazostrodon* and *Erythrotherium*. Philos. Trans. R. Soc. London, Ser. B, 273:387–431.

Kemp, T. S. 1969. On the functional morphology of the gorgonopsid skull. Philos. Trans. R. Soc. London, Ser. B, 256:1–83.

Kemp, T. S. 1972. Whaitsiid Therocephalia and the origin of cynodonts. Philos. Trans. R. Soc. London, Ser. B, 264:1–54.

Kemp, T. S. 1979. The primitive cynodont *Procynosuchus*: functional morphology of the skull and relationships. Philos. Trans. R. Soc. London, Ser. B, 285:73–122.

Kemp, T. S. 1980a. Origin of the mammal-like reptiles. Nature, London, 283:378–380.

Kemp, T. S. 1980b. The primitive cynodont *Procynosuchus*: structure, function and evolution of the postcranial skeleton. Philos. Trans. R. Soc. London, Ser. B, 288:217–258.

Kemp, T. S. 1980c. Aspects of the structure and functional anatomy of the Middle Triassic cynodont *Luangwa*. J. Zool. London, 191:193–239.

Kemp, T. S. 1982. *Mammal-like Reptiles and the Origin of Mammals*. London: Academic Press.

Kemp, T. S. 1983. The relationships of mammals. J. Linn. Soc. London Zool., 77:353–384.

Kemp, T. S. 1985. Synapsid reptiles and the origin of higher taxa. Spec. Pap. Palaeontol., 33:175–184.

Kemp, T. S. 1988. Interrelationships of the Synapsida. Pp. 1–22 *in* Benton, M. J. (ed.), *The Phylogeny and Classification of the Tetrapods*. Vol. 2: *Mammals*. Syst. Assoc. Spec. Vol. No. 35 B. Oxford: Clarendon Press.

Kermack, D. M., and K. A. Kermack. 1984. *The Evolution of Mammalian Characters*. London: Croom Helm.

Kermack, K. A. 1963. The cranial structure of the triconodonts. Philos. Trans. R. Soc. London, Ser. B, 246:83–103.

Kermack, K. A. 1967. The interrelationships of early mammals. J. Linn. Soc. London Zool., 47:241–249.

Kermack, K. A., F. Mussett, and H. W. Rigney. 1973. The lower jaw of *Morganucodon*. J. Linn. Soc. London Zool., 53:87–175.

Kermack, K. A., F. Mussett, and H. W. Rigney. 1981. The skull of *Morganucodon*. J. Linn. Soc. London Zool., 71:1–158.

King, G. M. 1988. Anomodontia. *In* Wellnhofer, P. (ed.), *Encyclopedia of Paleoherpetology*. Part 17C. Stuttgart: Gustav Fischer Verlag.

Lauder, G. V. 1981. Form and function: structural analysis in evolutionary morphology. Paleobiology, 7:430–442.

Mayr, E. 1963. *Animal Species and Evolution*. Cambridge: Harvard Univ. Press.

Mendrez-Carroll, C. 1975. Comparaison du palais chez les thérocéphales primitifs, les gorgonopsiens et les Ictidorhinidae. C. R. Séances Biol. Paris, Ser. D, 280:17–20.

Olson, E. C. 1944. Origin of mammals based upon cranial morphology of the therapsid suborders. Geol. Soc. America Spec. Pap., 55:1–136.

Olson, E. C. 1959. The evolution of mammalian characters. Evolution, 13:344–353.

Olson, E.C. 1971. *Vertebrate Paleozoology*. New York: Wiley-Interscience.

Orlov, Y. A. 1958. The carnivorous dinocephalians of the Isheevo fauna (titanosuchians). Tr. Paleontol. Inst. Akad. Nauk SSSR, 72:1–114. [In Russian].

Parrington, F. R. 1936. On the tooth replacement in theriodont reptiles. Philos. Trans. R. Soc. London, Ser. B, 226:121–142.

Parrington, F. R. 1939. On the digital formulae of theriodont reptiles. Ann. Mag. Nat. Hist., Ser. 11, 3:209–214.

Parrington, F. R. 1967. The origin of mammals. Advmt. Sci. London, Dec. 1967:165–173.

Reisz, R. R. 1972. Pelycosaurian reptiles from the Middle Pennsylvanian of North America. Bull. Mus. Comp. Zool. Harvard, 144:27–62.

Reisz, R. R. 1980. The Pelycosauria: a review of phylogenetic relationships. Pp. 553–591 *in* Panchen, A. L. (ed.), *The Terrestrial Environment and the Origin of Land Vertebrates*. London: Academic Press.

Reisz, R. R. 1981. A diapsid reptile from the Pennsylvanian of Kansas. Spec. Pub. Mus. Nat. Hist. Univ. Kansas, 7:1–74.

Reisz, R. R. 1986. Pelycosauria. *In* Wellnhofer, P. (ed.), *Encyclopedia of Paleoherpetology*. Part 17A. Stuttgart: Gustav Fischer Verlag.

Romer, A. S. 1956. *Osteology of the Reptiles*. Chicago: Univ. of Chicago Press.

Romer, A. S. 1966. *Vertebrate Paleontology*. 3rd ed. Chicago: Univ. of Chicago Press.

Romer, A. S. 1967. The Chañares (Argentina) Triassic reptile fauna. III. Two new gomphodonts, *Massetognathus pascuali* and *M. teruggii*. Breviora, 264:1–25.

Romer, A. S. 1969a. The Brazilian Triassic cynodont reptiles *Belesodon* and *Chiniquodon*. Breviora, 332:1–16.

Romer, A. S. 1969b. The Chañares (Argentina) Triassic reptile fauna. V. A new chiniquodontid cynodont, *Probelesodon lewisi*—Cynodont ancestry. Breviora, 333:1–24.

Romer, A. S. 1969c. Cynodont reptile with incipient mammalian jaw articulation. Science, 166:881–882.

Romer, A. S. 1970. The Chañares (Argentina) Triassic reptile fauna. VI. A chiniquodontid cynodont with an incipient squamosal-dentary jaw articulation. Breviora, 334:1–18.

Romer, A. S. 1973. The Chañares (Argentina) Triassic reptile fauna. XVIII. *Probelesodon minor*, a new species of carnivorous cynodont; Family Probainognathidae Nov. Breviora, 401:1–4.

Romer, A. S., and A. D. Lewis. 1973. The Chañares (Argentina) Triassic reptile fauna. XIX. Postcranial materials of the cynodonts *Probelesodon* and *Probainognathus*. Breviora, 406:1–26.

Romer, A. S., and L. I. Price. 1940. Review of the Pelycosauria. Geol. Soc. America Spec. Pap., 28:1–538.

Rowe, T. 1986. Osteological Diagnosis of Mammalia, L. 1758, and Its Relationship to Extinct Synapsida. Doctoral dissertation. Dept. Paleontology, Univ. of California, Berkeley.

Rowe, T. 1988. Definition, diagnosis, and origin of Mammalia. J. Vert. Paleontol., 8:241–264.

Rowe, T., and J. A. van den Heever. 1986. The hand of *Anteosaurus magnificus* (Dinocephalia: Therapsida) and its bearing on the origin of the mammalian manual phalangeal formula. S. Afr. J. Sci., 82:641–645.

Sigogneau, D. 1970a. Révision systématique des gorgonopsiens Sud-Africains. Pp. 1–416 *in Cahiers de Paléontologie*. Paris: Éditions du Centre National de la Recherche Scientifique.

Sigogneau, D. 1970b. Contribution à la connaissance des ictidorhinides (Gorgonopsia). Palaeontol. Afr., 13:25–38.

Sigogneau, D., and P. K. Tchudinov. 1972. Reflections on some Russian eotheriodonts. Palaeovertebrata, 5:79–109.

Simpson, G. G. 1928. *A Catalogue of the Mesozoic Mammalia in the Geological Department of the British Museum*. London: Brit. Mus. (Nat. Hist.).

Simpson, G. G. 1929. American Mesozoic Mammalia. Mem. Peabody Mus. Yale Univ., 3(1):1–235.

Simpson, G. G. 1959. Mesozoic mammals and the polyphyletic origin of mammals. Evolution, 13:405–414.

Simpson, G. G. 1960. The history of life. Pp. 117–180 *in* Tax, S. (ed.), *The Evolution of Life*. Chicago: Univ. of Chicago Press.

Sues, H.-D. 1985. The relationships of the Tritylodontidae (Synapsida). J. Linn. Soc. London Zool., 85:205–217.

Ulinski, P. S. 1986. Neurobiology of the therapsid-mammal transition. Pp. 149–184 *in* Hotton, N., III, P. D. MacLean, J. J. Roth, and E. C. Roth (eds.), *The Ecology and Biology of Mammal-like Reptiles*. Washington: Smithsonian Inst. Press.

van den Heever, J. A. 1986. A classification of the early Therocephalia. PAL News/ Nuus, 5(1):14.

Vrba, E. S. 1980. Evolution, species and fossils: how does life evolve? S. Afr. J. Sci., 76:61–84.

Wiley, E. O. 1981. *Phylogenetics: The Theory and Practice of Phylogenetic Systematics*. New York: John Wiley and Sons.

Author Index

Subject Index

Page numbers in italic type refer to illustrations.